TECHNOLOGY TRANSFER

SOCIETY FOR THE ADVANCEMENT OF
MATERIAL AND PROCESS ENGINEERING

13TH NATIONAL SAMPE TECHNICAL CONFERENCE

Volume 13

TECHNOLOGY TRANSFER

Mount Airy Lodge
Mount Pocono, Pennsylvania
October 13-15, 1981

Additional copies of this publication may be obtained from
Society for the Advancement of Material and Process Engineering

SAMPE National Business Office
P. O. Box 613
Azusa, California 91702

ISBN 0-938994-19-0

13TH NATIONAL SAMPE TECHNICAL CONFERENCE COMMITTEE

General Chairman	R. F. Wegman, ARRADCOM
Program Chairman	M. J. Michno, Jr., Union Carbide Corporation
Co-Program Chairman	R. A. Golick, Hysol Div/Dexter Corporation
Finance Chairman	F. H. Ancker, Union Carbide Corporation
Arrangements Chairman	M. J. Michno, Jr., Union Carbide Corporation
Publicity Chairman	F. A. Keimel, Adhesives & Sealants Consultant
Luncheon Chairman	W. R. Collins, M & T Chemicals, Inc.
Registration & Membership Chairman	N. Chujko, ARRADCOM
Social Chairman	R. J. Narsavage, Synthatron Corporation
Audio/Visual Chairman	W. J. Russell, ARRADCOM
Transportation Chairman	J. W. Burlage, P.V.O. International
Publication Chairman	A. M. Landrock, ARRADCOM
Co-Publication Chairman	H. E. Pebly, Jr., ARRADCOM

NATIONAL OFFICERS

President	George F. Schmitt, Jr.
Past President	Jack M. Dyer
1st Vice President	L. C. May
2nd Vice President	John T. Hoggatt
Secretary	N. H. Kordsmeier, Jr.
Treasurer	Jerome Bauer

CHAPTER OFFICERS

(New Jersey Chapter)

Chairman	N. Chujko
1st Vice Chairman	R. J. Narsavage
2nd Vice Chairman	R. W. Reiter
Secretary	W. C. Tanner
Treasurer	M. D. Anderson
Sr. National Director	R. A. Golick
Jr. National Director	F. H. Ancker

SESSIONS AND SESSION CHAIRMEN

COMPOSITES I -- APPLIED TECHNOLOGY

L. E. 'Roy' Meade, Lockheed-Georgia Company

Leslie Cohen, McDonnell Douglas Corporation

MATERIALS CHARACTERIZATION I

Jerome L. Bauer, M & T Chemicals/Furane Products

POLYMER INNOVATIONS FOR PROCESS DEVELOPMENTS

Armand F. Lewis, Lord Corporation

James E. O'Connor, Phillips Petroleum Company

ADHESIVES I -- APPLIED TECHNOLOGY

Weldon M. Scardino, Air Force Material Laboratories/MLTN

John Gurtowski, Navair System Command

CHARACTERIZATION OF FABRICATED SYSTEMS

Don Mazenko, Lockheed Missiles & Space Co., Inc.

ENERGY PRODUCTION -- MATERIALS CONSIDERATIONS

Robert J. Martinez, Sandia Laboratories

COMPOSITES II -- APPLIED TECHNOLOGY

Sam C. Aker, Bell Helicopter/Textron

Sam J. Dastin, Grumman Aerospace

MANUFACTURING/DESIGN COST TRADEOFFS KEY TO SUCCESSFUL TECHNOLOGY TRANSFER

Bryan R. Noton, Battelle Columbus Laboratories

ELECTRONICS AND COMMUNICATIONS MATERIAL OPPORTUNITIES

Charles A. Harper, Westinghouse Electric Corporation

Lawrence R. Sparrow, COMSAT Laboratories

ADHESIVES II -- APPLIED TECHNOLOGY

John S. Hutcheon, Hysol Div/Dexter Corporation

MATERIALS CHARACTERIZATION II

Jack R. Vinson, University of Delaware

CORROSION CONTROL

Sara J. Ketcham, Naval Air Development Center

Edward Taylor, Standard Press Steel Company

COMPOSITE STRUCTURES -- DESIGN & FABRICATION

C. W. Schneider, Lockheed-Georgia Company

REINFORCEMENT AND FILLERS

Harry S. Katz, Utility Research Company

SURFACE TREATMENT -- PROCESSES & MATERIALS

Narvel L. Rogers, Bell Helicopter/Textron

PREVIOUS PUBLICATIONS

THE SCIENCE OF ADVANCED MATERIALS AND PROCESS ENGINEERING SERIES

Documents in this series contain papers which have been presented at prior National SAMPE Symposia and cover a wide range of Materials and Process Technology.

Symposium Proceedings		Non-Member Price	Member Price
Volume 1	Filament Winding	$ 3.00	$ 1.00
Volume 8	Insulation-Materials & Processes for Aerospace & Hydrospace Applications	8.00	5.00
Volume 9	Joining of Materials for Aerospace Systems	8.00	5.00
Volume 10	Advanced Fibrous Reinforced Composites	8.00	5.00
Volume 11	Effects of the Space Environment on Materials	10.00	5.00
Volume 12	Advances in Structural Composites	8.00	5.00
Volume 13	Energistic Materials	8.00	5.00
Volume 14	Advanced Techniques for Materials Investigation & Fabrication	10.00	8.00
Volume 16	Materials '71	10.00	8.00
Volume 17	Materials Review for '72	10.00	8.00
Volume 25	The 1980's - Payoff Decade for Advanced Materials	55.00	50.00
Volume 26	Material and Process Applications--Land, Sea, Air, Space	60.00	55.00

NATIONAL SAMPE TECHNICAL CONFERENCE SERIES

Volumes in this series contain papers which have been presented at various technical conferences which deal with specific subject matter related to a narrow segment of the Materials and Process technology.

Technical Conference Series			
NSTC 1	Aircraft Structures & Materials Application	10.00	8.00
NSTC 6	Materials on the Move	30.00	20.00
NSTC 7	Materials Review '75	40.00	30.00
NSTC 8	Bicentennial of Materials	40.00	30.00
NSTC 9	Materials & Processes--In Service Performance	40.00	30.00
NSTC 11	New Horizons--Materials and Processes for the Eighties	55.00	50.00
NSTC 12	Materials 1980	60.00	55.00
NSTC 13	Technology Transfer	60.00	55.00

SAMPE, P.O. Box 613, Azusa, CA 91702

PREFACE

The New Jersey Chapter of the Society for the Advancement of Material and Process Engineering is honored and proud to host the 13th National SAMPE Technical Conference at Mount Airy Lodge, Mt. Pocono, Pennsylvania.

"Technology Transfer", the conference theme, is essential to increased productivity. As highlighted in our keynote address, productivity entails a partnership between universities, industry and government.

Conferences such as this provide a forum for technical exchange between government, university, research institute and industry representatives. Such communication is the basis for technology transfer and increased productivity.

We would like to express our thanks to all those who gave their time and energy to make this conference a success. Particular thanks to Marge Smith and the National Business Office staff.

R. F. Wegman
General Chairman

M. J. Michno, Jr.
Program Chairman

CONTENTS

PAPERS AND AUTHORS

NAMES AND AUTHORS

NAMES AND AUTHORS

NAMES AND AUTHORS

13th National SAMPE Technical Conference
October 13-15, 1981

FLIGHT SERVICE EVALUATION OF ADVANCED STRUCTURES

by

Cecil W. Schneider
Advanced Structures Department
Lockheed-Georgia Company
Marietta, Georgia

Abstract

Advanced composite structures are continually being designed, fabricated and tested in the laboratory. The more promising of these concepts often end up on operational aircraft for extended flight service evaluations as a proof of concept. This paper summarizes several of these flight service programs that have been conducted on Air Force transport aircraft over the past decade. These structures have ranged from simple wing tips to complex center wing sections. Tracking programs initially started out as informal monitoring of flight hours, but have gradually evolved into formal programs with detailed inspection and monitoring activities.

1. INTRODUCTION

Composites are gradually emerging as a viable cost effective structural material for aerospace application. The introduction of composites, or any other new structural material, is paced by new aircraft programs which are fewer and further between than in years past. While advanced structural materials and concepts are continually being developed, fabricated, and laboratory-tested, it is increasingly difficult to obtain in-service experience for these new materials and concepts due to the few new aircraft starts.

This has been especially true for large aircraft in the transport/ bomber size category. Therefore, limited flight service evaluations have been conducted as proof of concept for new structural materials and configurations to gain an in-service performance record prior to design of the next generation aircraft. In addition to generating a maintenance data base, this approach also provides the customer with "hands-on" experience with these advanced materials.

The recent history of flight service evaluations at Lockheed is traced from initial boron "material substitution" wing tips to durable graphite/epoxy wing leading edges. These components have been accumulating a history of exceptional performance in the operational environment.

2. EARLY PROGRAMS WITH BORON

In the mid and late 1960's, the research and development effort was concentrated on boron/epoxy unidirectional material. Exceptionally high modulus and strength were available, but at prices commensurate with the modulus. The material is also difficult to fabricate in forms requiring other than flat or gentle contours. But there are many areas where boron is still cost effective.

C-141 Boron Wing Tip

During this period, Lockheed designed and produced several components under IRAD and contracted funding. The first of these was a wing tip leading edge segment for the C-141 aircraft. This small component, shown in Figures 1 and 2, replaced the sheet metal part on an equal thickness basis. The boron/epoxy part was installed on a Lockheed-owned C-141 aircraft in 1968, but was later removed and reinstalled on a USAF aircraft when the Lockheed aircraft was sold to NASA. This part has been in constant service since that date; however, exact flight time is not known.

C-141 Boron Oleo Door

Another boron/epoxy part fabricated in the late 1960's was a C-141 main landing gear oleo door. This door, shown in Figure 3, is located on the upper surface of the main landing gear pod, and opens when the gear is lowered to permit the main gear trunion to protrude through the upper surface of the pod. The aluminum honeycomb door experienced extensive damage caused by the door banging against the stops on closing. The aluminum skin panels were replaced with an equivalent thickness boron/epoxy laminate to maintain the total panel thickness.

The boron oleo door was installed on a C-141 aircraft by the Air Force in 1969, and has been in continuous use since then. In excess of 23,000 flight hours have been accumulated over this period, with no problems reported. This early flight evaluation component has been monitored only periodically, with no formal tracking or inspection program. Lockheed conducted a detail inspection in 1977, and will conduct another detail inspection in early 1982 when the airplane with the door installed is at Lockheed for fuselage stretch. Periodic reports on accumulated flight time are obtained from field service records to monitor the service history.

C-5 Boron Slat

In the late 1960's, Lockheed designed, fabricated and conducted a flight service evaluation of boron slats for the C-5A aircraft, under contract to the Air Force Materials Laboratory (Reference 1). The goals of this program were to reduce cost and weight while providing flightworthy structure for flight service evaluation. The right-hand inboard slat, shown in Figure 4, was selected for this program. This slat was fabricated of sheet metal, with approximately 1500 parts. The boron slat concept, shown in Figure 5, reduced the part count to 85. The completed slat, shown in Figure 6, had a boron-faced honeycomb "eye-brow" panel with seven ribs. The slat is shown in Figure 7 installed on a C-5A during flight testing. A total of eleven slats were produced and installed, starting in 1970. These slats accumulated over 24,000 flight hours, with a high-time unit of over 4,000 flight hours.

The service performance of the slats was tracked by the Air Force, with periodic inspections performed by Lockheed. Lockheed also prepared a final service evaluation report on the slats to document the in-service maintenance record (Reference 2). Detailed maintenance logs were used to establish a comparison with the metal slat; this showed a projected maintenance action rate of 19 defects per 30,000 hour lifetime for the boron slat, compared to 28 defects for the metallic slat. Unfortunately, the slats were fabricated with then state-of-the-art aluminum honeycomb, which proved to have the same corrosion problems as other aluminum honeycomb structures. The slats were thus removed, concurrent with transfer of the aircraft to another base. The results of this program verified the flight readiness of the boron materials.

C-130 Boron-Reinforced Center Wing

An innovative reinforcement concept for aluminum primary structures was employed to obtain reduced weight and increased fatigue life for the C-130 center wing section. This NASA sponsored program resulted in design, tooling, fabrication, and extended flight evaluation of the reinforcement concept (Reference 3).

As shown in Figure 8, the C-130 center wing is a built-up structure of hat-section stiffened skin. A total of 1200 pounds of aluminum were milled from the skin surface and hat caps of a standard C-130 wing center section. Boron/epoxy unidirectional tape was bonded to the skin and inner surface of the hat-stiffener to reinforce the wing and regain the original wing stiffness. A total of 600 pounds of boron/epoxy were added to each wing, resulting in a 600 pound weight savings for the center wing.

The boron/epoxy laminate was bonded to the aluminum structure using a unique Lockheed-developed "cool-tool" method, whereby the wing was restrained at the ends from thermal growth during the cure cycle. This eliminated any tendency to warp upon cooling due to the differential expansion of the two materials.

Fatigue tests conducted on the center wing section showed a factor of 2 improvement on fatigue life over the original wing, even though the weight was 600 pounds less. Two flight certified center wing sections were fabricated and installed on Air Force C-130 aircraft during production. These aircraft were first flown in 1974, and have since accumulated over 5,000 flight hours each. Periodic detailed inspections conducted by Lockheed have shown no degradation in the wings during this time. These inspections have been conducted by Lockheed at Little Rock AFB concurrent with Air Force phased inspections at approximately six-month intervals. Periodic inspection reports (Reference 4) have been issued to document these inspections.

3. GRAPHITE FLIGHT PROGRAMS

The successful development of graphite/epoxy fiber and fabric materials resulted in a gradual switch from boron to graphite/epoxy structures development. While lower in properties, graphite is much lower in cost and easier to work with during fabrication. Many graphite/epoxy structures have been fabricated over the past several years to develop and verify design, tooling, and fabrication techniques. Only a few of these will be discussed here, to document the flight-service history of those components that have been installed on aircraft.

B-1 Slat

Lockheed designed a graphite/epoxy leading edge slat for the B-1 Bomber under subcontract to Rockwell. This inboard slat was designed to achieve weight and cost reductions. A weight reduction of 15% and an acquisition cost reduction of 34% was demonstrated during the design, tool, and fabrication program. Three full-size articles were produced; one for static test, one for fatigue test, and one for flight evaluation.

The static and fatigue tests demonstrated that all design criteria were exceeded, and the third article was approved for flight. This slat, shown in Figure 9, was installed on the first B-1 aircraft, Air Vehicle Number 1, in early 1977, and accumulated 185 flight hours before the aircraft was deactivated.

C-5 Wing Secondary Structure

High-stiffness composite materials have been used to fabricate integrally-stiffened structures to replace honeycomb panels in many different applications. Much of the light-weight aluminum honeycomb structure fabricated in the late 1960's and early 1970's suffers today from excessive maintenance due to delamination and core corrosion.

Three of these applications that have been accomplished on the C-5 aircraft are depicted in Figure 10. These include wing-to-body fairing panels, fixed trailing edge panels, and a fairing covering the joint between the center and outer wing. All of these honeycomb panels receive extensive abuse during service from frequent removal or from being walked upon.

Two shipsets (four panels each) of wing-to-body fairing panels were designed and fabricated under contract to San Antonio - Air Logistics Center (SA-ALC). These integrally-stiffened graphite/ epoxy panels were installed on two Air Force C-5 aircraft at Dover AFB in 1978. Periodic monitoring of these parts has shown no maintenance activity since that time. These panels, which cover oxygen bottles and are therefore frequently removed, have been approved by the Air force as the preferred spare configuration.

Two shipsets (8 panels each) of trailing edge panels were fabricated in 1979, again under contract to SA-ALC. These integrally-stiffened panels, one of which is shown in Figure 11, were fabricated of graphite/epoxy with fiberglass edge members for attachment to the aluminum substructure. One shipset of panels was installed in 1979, while the other was held as a backup. Periodic inspections are conducted by the Air Force to monitor the progress of the part. The only significant problem encountered to date occured when a mechanic inadvertently stepped on the panel (marked NO-STEP), delaminating some of the stiffeners at the stiffener/skin interface. The wing-to-body fairings and the trailing edge panels have accumulated a total of 6,500 flight hours over the past three years, with the high time unit at approximately 2000 flight hours.

In addition to these experimental components, Lockheed designed, fabricated, and tested a W.S. 120 joint graphite/epoxy fairing on the C-5 "Wing Mod" prototype aircraft. This evaluation resulted in incorporation of this 17-foot-long graphite/epoxy fairing into the C-5 wing modification program, and all C-5 aircraft will have this fairing replace the existing honeycomb panel. This panel has been flying on the prototype Wing-Mod aircraft since 1980 with no reported maintenance activity.

C-141 Composite Leading Edge

Lockheed conducted a tooling and fabrication program for the Manufacturing Technology Division of the Air Force Materials Laboratory which resulted in flightworthy hybrid graphite/fiberglass/epoxy leading edges on the C-141 aircraft (References 6-8). The original inboard leading edge section was fabricated of aluminum honeycomb, as depicted in Figure 12. The 1 3/4-inch thick honeycomb section was replaced by an integrally-stiffened composite part, as shown in Figure 13. Aluminum extrusion edge members, permanently attached to the composite assembly with titanium Hi-Loks, permitted the leading edge to be installed to the wing with the same attachment screws used for the metallic section. A completed leading edge is shown in Figure 14.

The primary design criterion for the leading edge design was increased durability over the aluminum honeycomb part. This was achieved at equal weight and reduced component acquisition cost (Reference 6). Ten flight evaluation articles were produced and installed on Air Force C-141 aircraft at Charleston AFB and McGuire AFB, starting in 1979. To-date, over 15,000 flight hours have been amassed; several units have exceeded 2,000 flight hours. Only two maintenance/repair actions have been reported during this period; both were minor delaminations due to ground-handling damage, and were readily repaired using fiberglass repair techniques with a room-temperature cure adhesive. All repairs were accomplished per a repair guide developed during the contract (Reference 9).

The ten composite leading edges will be tracked during a minimum five-year period by the Air Force Service Evaluation Branch at Pope AFB. Quarterly reports will be issued during this period to provide a detailed record of flight time, inspections performed, and any required maintenance activity. Detail tracking of maintenance and repair activity is also accomplished through the Air Force AFR 61 Maintenance Tracking System, using a unique identification for the composite leading edge to prevent intermixing the data with the honeycomb parts. This will provide a head-to-head comparison of the maintenance history for both the composite and the honeycomb parts.

4. LESSONS LEARNED

The flight evaluation programs discussed here cover a span of over 13 years, with part complexity from simple wing tips to complex primary structure. The in-service performance history of all of these components has been outstanding, confirming earlier maintenance predictions based on laboratory test data. Repairs, when required, have proven to be easier and faster than for honeycomb structure, and generally result in less weight increase.

In addition to providing valuable data on the performance of composite structures under routine operational conditions, these programs have benefitted the Air Force Base and the Air Logistics Center personnel by providing hands-on experience on handling, installing, inspecting, and repairing these advanced structures.

5. REFERENCES

1. W.E. Harvill, et al., "Development Program for Boron Leading Edge Slat for C-5A Aircraft," ASD-TR-72-62, June 1971.

2. R.W. Coleman, "C-5A Boron Slat Service Evaluation Report," Lockheed-Georgia Company Report LG2 US-486.0-1, January, 1973.

3. W.E. Harvill, et al., "Program for Establishing Long-Time Flight Service Performance of Composite Materials in the Center Wing Structure of C-130 Aircraft," NASA CR-112126, Phase I - Advanced Development, Nov 1972

NASA CR-112272, Detailed Design, April 1973
NASA CR-132495, Fabrication, Sept. 1974
NASA CR-145043, Ground/Flight Acceptance Tests, Sept. 1976.

4. J.A. Kizer, "In-service Inspection Report for C-130 Boron-Reinforced Center Wing," NASA CR to be published.

5. J.M. Altman, et al., "Composite, Low Cost Secondary Airframe Structure," AFFDL-TR-76-4, Aug 1976.

6. C.W. Schneider, K.M. Barre, and D.P. Bierce, "Integral Molding Production Verification for C-141 Composite Leading Edge," AFML-TR-79-4213, January 1980.

7. C.W. Schneider, "C-141 Hybrid Composite Leading Edge," 24th SAMPE National Symposium and Exhibition, Vol. 24, May 1979.

8. C.W. Schneider, "Composites - A Solution to Aluminum Honeycomb Maintenance Costs," 12th SAMPE National Technical Conference, Oct 1980.

9. D.P. Bierce, "Repair Techniques for C-141 Composite Leading Edge," Lockheed-Georgia Service Manual Publication No. 1320, Oct 1979.

Figure 1. C-141 Boron Wing Tip Panel

Figure 2. C-141 Boron Panel Installed on Aircraft

Figure 3. C-141 Boron Oleo Door

Figure 4. C-5 Metal Slat

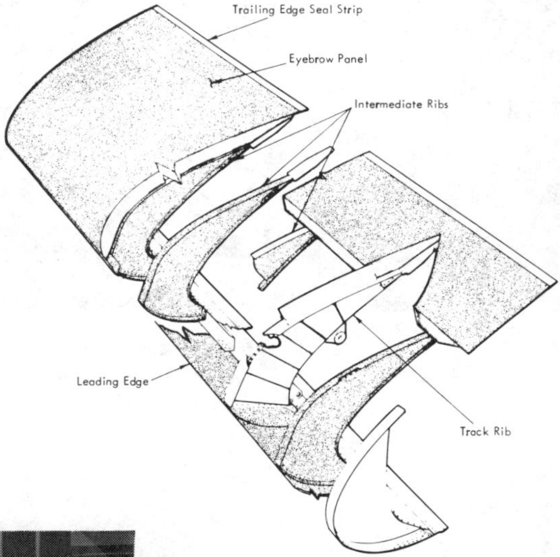

Figure 5. C-5 Boron Slat Concept

Figure 6. C-5 Boron Slat

Figure 7. C-5 Boron Slat Installation

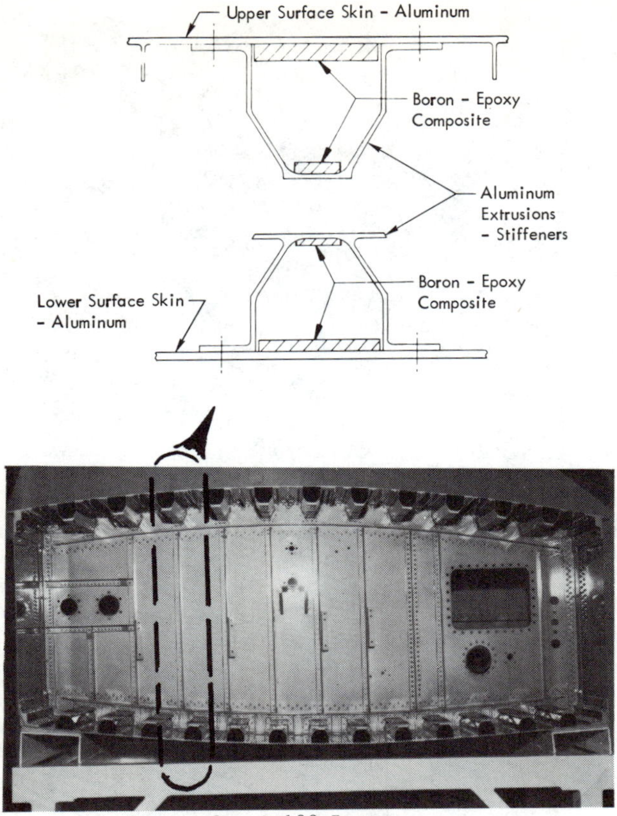

Figure 8. C-130 Boron Reinforced Center Wing Section

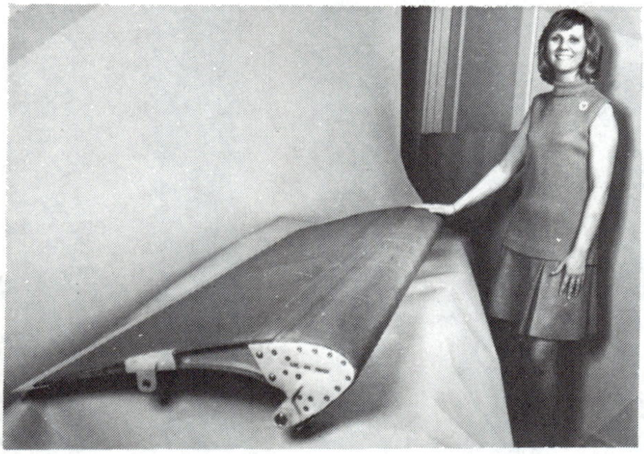

Figure 9. B-1 Graphite Slat

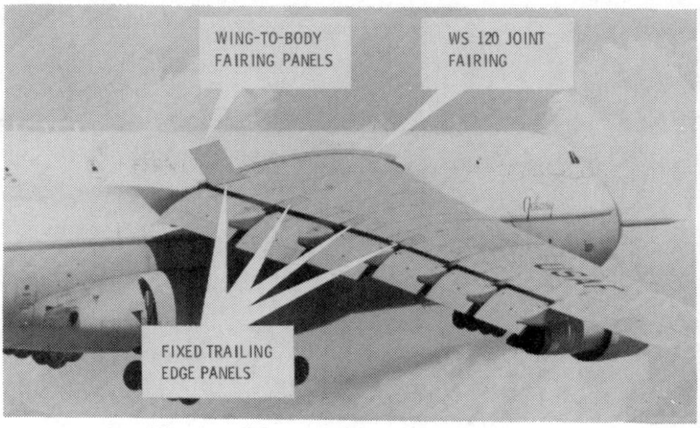

Figure 10. C-5 Composite
Secondary Structure

Figure 11. C-5 Integrally
Stiffened Panel

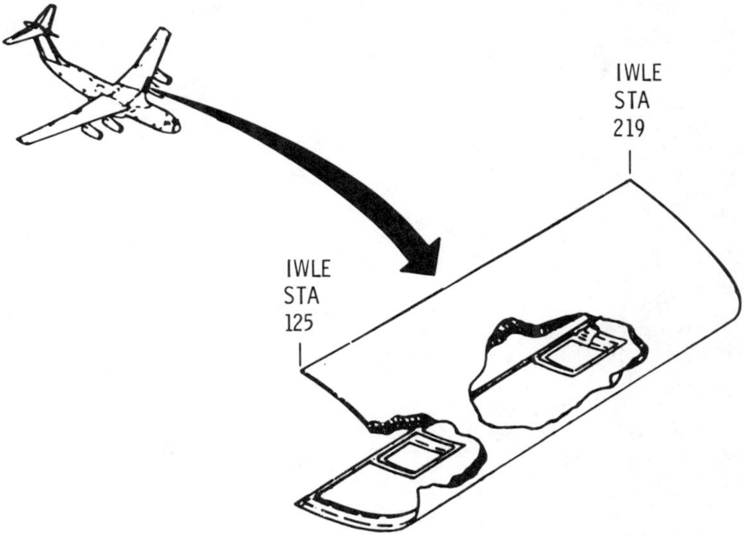

Figure 12. C-141 Honeycomb
Leading Edge

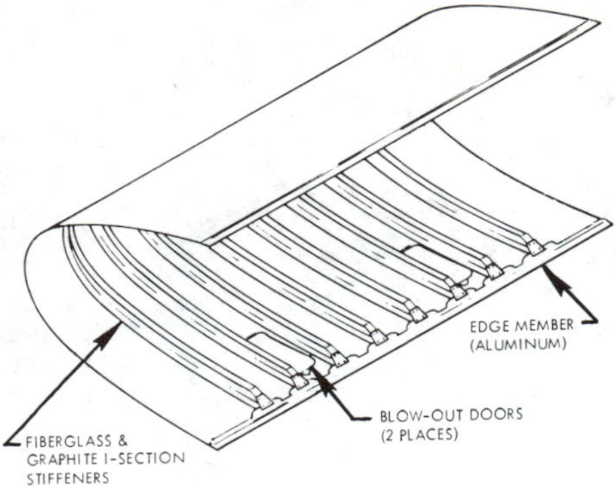

Figure 13. C-141 Composite
Leading Edge Concept

Figure 14. C-141 Composite
Leading Edge

13th National SAMPE Technical Conference
October 13-15, 1981

DAMAGE TOLERANCE AND DURABILITY DESIGN OF COMPOSITE STRUCTURES FOR COMMERCIAL AIRCRAFT

by

Dr. Edmond Chow and Clifford Y. Kam
Douglas Aircraft Company
McDonnell Douglas Corporation
Long Beach, California 90846

ABSTRACT

This paper presents the damage-tolerant and durable design of the composite secondary structures which have been applied to commercial aircraft at Douglas Aircraft Company, and describes the development of composite primary structure based on damage tolerance and durability technology. A description is given of test results as well as the overall developmental methodology that needs to be addressed in order to support the design.

1. INTRODUCTION

The basis of certification of composite structures for commercial aircraft has been established by the Federal Aviation Administration (FAA) in the current regulatory requirements of FAR 25.571, Amendment 25-45, together with Advisory Circulars AC 25.571-1, "Damage Tolerance and Fatigue Evaluation of Structure," and AC 20-107, "Composite Aircraft Structure." For advanced composite to be successfully applied to commercial aircraft, the FAA states in these documents that it must be demonstrated that all applications will achieve a level of safety at least as high as is currently provided by metallic structure, as specified in AC 20-107.

Douglas Aircraft Company has met these criteria in applying high-performance composite materials to secondary structures of commercial aircraft. This has brought a significant weight savings and reduction in fuel consumption.[1] These composites are now being investigated for use as primary safety-critical aircraft components.

2. COMPOSITE STRUCTURES PROGRAMS AT DOUGLAS

Development efforts at Douglas have led to the use of advanced composite materials for the following components of the DC-10: upper aft rudder, vertical stabilizer trailing edge panel, and floor beams.[2-7] In addition, the DC-9 Super 80 has a Kevlar tail cone and Kevlar fan cowl doors, as noted in Figure 1. By December 1981, a Kevlar/polyimide nacelle stang fairing will be flight tested for the DC-9. Other composites to be submitted for flight evaluation for the DC-10 by December 1981 include the nose gear aft door, aileron access panel, and the vertical stabilizer box. These DC-10 components are shown in Figure 2.

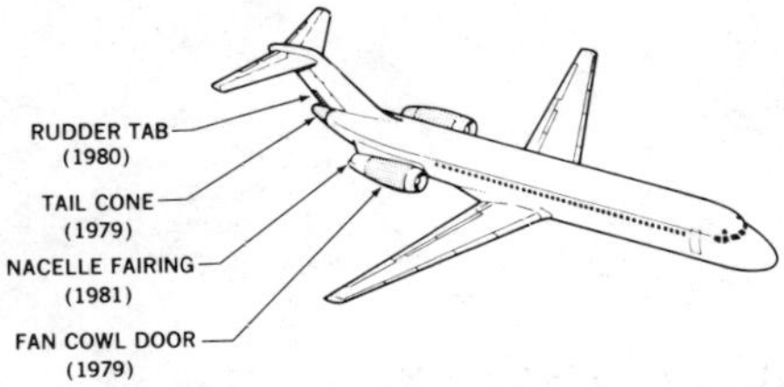

FIGURE 1. DC-9 COMPOSITE APPLICATIONS

The graphite composite rudder was thoroughly ground tested to verify the structural integrity under critical static and dynamic design load conditions. Thermal expansion characteristics were determined experimentally and all critical static load conditions and fail-safe variations were successfully sustained in ground tests.

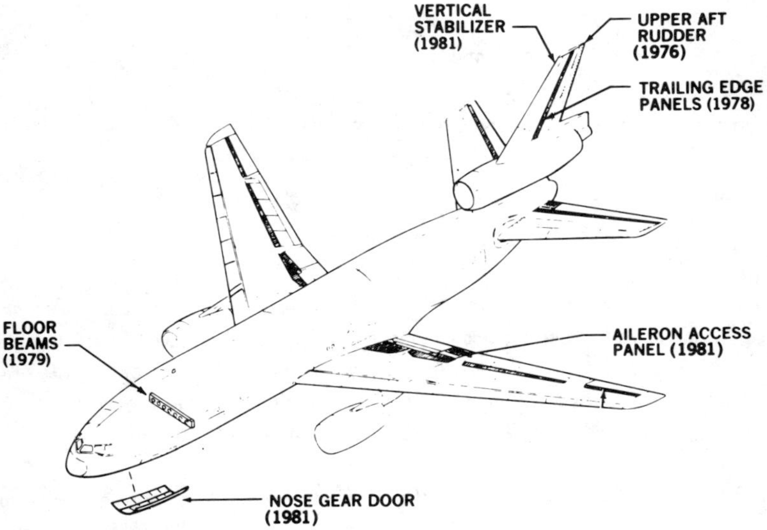

FIGURE 2. DC-10 COMPOSITE STRUCTURE APPLICATIONS

Extensive testing was conducted to satisfy damage tolerance and durability requirement for all these components. The test program was composed of three parts. In the first part, elements were tested to characterize the mechanical properties of composite materials, establish design allowables, and determine reduction factors for the effects of adverse conditions. The second part involved the evaluation of design components to verify the structural reliability of design concepts. The third part was conducted to demonstrate the structural integrity.

A typical test program for DC-10 composite rudders is summarized in Table 1. The test results are presented in Tables 2 through 5 and in Figure 3. The structural arrangement of the rudder system is shown in Figure 4.

TABLE 1
TEST PROGRAM FOR COMPOSITE RUDDER

ELEMENT TESTS STATIC TEST STRENGTH TEST BOLT BEARING TEST FATIGUE TEST FRACTURE TOUGHNESS TEST RESIDUAL STRENGTH TESTS IMPACT TESTS
DESIGN COMPONENT TESTS HINGE FITTING FASTENER TESTS SPAR COMPONENT TESTS BOX COMPONENT STATIC TESTS BOX COMPONENT VIBRATION TESTS BOX COMPONENT FATIGUE TESTS LIGHTNING STRIKE TESTS o SWEPT-STROKE AND RESTRIKE TESTS o DIRECT-STRIKE ATTACHMENT TESTS
DEMONSTRATION TEST RUDDER GROUND TEST THERMAL EXPANSION TEST STATIC TESTS DYNAMIC TESTS FATIGUE TESTS

TABLE 2
SANDWICH-BEAM FATIGUE TEST RESULTS FOR THORNEL 300/5208 MATERIAL (0, 45, 90, –45°)S LAMINATE AT STRESS RATIO, R = –1.0

NASA CR-145068[2]

SPECIMEN CONFIGURATION	SPECIMEN NUMBER	TEST LOAD			NET STRESS		CYCLES TO FAILURE	REMARKS
		N	LB	PERCENT AVERAGE STATIC LOAD	MPa	KSI		
Z3941117-1 (NO HOLE)	1	7900	1776		534.34	77.5	–	STATIC TESTS AVERAGE STATIC LOAD IS 8007N (1800 LB)
	2	7509	1688	--	AVG	AVG	–	
	3	8674	1950				–	
	4	5338	1200	66.7	356.46	51.7	1,000	FAILED IN TEST SECTION
	5	4003	900	50.0	266.83	38.7	5,000	FAILED IN TEST SECTION
	6	3559	800	44.4	237.18	34.4	55,000	FAILED IN TEST SECTION
	7	3203	720	40.0	213.74	31.0	306,000	FAILED IN TEST SECTION
	8	3203	720	40.0	213.74	31.0	113,000	FAILED IN TEST SECTION
	9	3203	720	40.0	213.74	31.0	577,000	FAILED IN TEST SECTION
	10	2802	630	35.0	186.85	27.1	2,471,000	FAILED IN TEST SECTION
	11	2802	630	35.0	186.85	27.1	4,680,000	FAILED IN TEST SECTION
	12	2802	630	35.0	186.85	27.1	1,006,000	FAILED IN TEST SECTION
Z3941117-501 (6.35 mm (1/4-INCH) DIAMETER OPEN HOLE)	1	4946	1112	–	393.0	57.0	–	STATIC TESTS – AVERAGE STATIC LOAD IS 4982N (1120 LB)
	2	4706	1058	–	AVG	AVG	–	
	3	5276	1186				–	
	4	3336	750	67.0	263.38	38.2	11,000	FAILED IN TEST SECTION
	5	3336	750	67.0	263.38	38.2	23,000	FAILED IN TEST SECTION
	6	3336	750	67.0	263.38	38.2	42,000	FAILED IN TEST SECTION
	7	3069	690	61.7	242.70	35.2	10,000	FAILED IN TEST SECTION
	8	3069	690	61.7	242.70	35.2	59,000	FAILED IN TEST SECTION
	9	3069	690	61.7	242.70	35.2	132,000	FAILED IN TEST SECTION
	10	2669	600	53.5	210.29	30.5	1,281,000	FAILED IN TEST SECTION
	11	2669	600	53.5	210.29	30.5	2,392,000	FAILED IN TEST SECTION
	12	2669	600	53.5	210.29	30.5	2,566,000	FAILED IN TEST SECTION

TABLE 3
FRACTURE TOUGHNESS PANEL TEST RESULTS

NASA CR-145068[2]

PANEL NO.	DAMAGE IMPACT ENERGY		ULTIMATE STRESS		REMARKS
	J	IN.-LB	MPa	KSI	
CONTROL	NONE	NONE	365.42	53.00	FAILED AT EDGE OF GRIP REINFORCEMENT
1	1.13	10	520.55	75.50	FAILED THROUGH LAMINATE PANEL 2.54 CM (1.00 IN.) FROM DAMAGE CENTER
2	2.26	20	348.19	50.50	FAILED THROUGH DAMAGE
3	4.52	40	248.21	36.00	FAILED THROUGH DAMAGE
4	(2) AT 4.52	(2) AT 40	417.13	60.50	FAILED THROUGH DAMAGE. THE EDGES OF THE IMPACT AREAS DID NOT HAVE THE SHARP CLEAVAGE FOUND IN PANEL NO. 3

The static test program on the graphite composite rudder was completed without a structural failure of any kind. The critical flight loads were simulated, including loads induced from forward rudder bending and applied airloads but neglecting the minor effects of thermal stresses. The rudder structure was substantiated under critical ultimate stress conditions and fail-safe variations through a series of nine test sequences (see Table 6).

Two critical static load conditions were tested using the setup shown in Figure 5. The test rudder was supported in a horizontal plane by a set of five forward rudder hinge brackets and four dummy control rods. A structural steel test jig duplicated the upper forward rudder interfaces for the hinge brackets and control rods. Deflection of the aft rudder hinge line was achieved using fixed mounts for the lower and upper hinge brackets and movable mounts for the three intermediate brackets. Simulated airloads were generated by a 5-square-inch hydraulic actuator acting through compression whiffling to 71 load points on the surface of the graphite rudder.

TABLE 4

SUMMARY OF BOX COMPONENT STATIC TESTS

NASA CR-145068[2]

TEST SEQUENCE NUMBER	CRITICAL CONDITION			RUDDER CONFIGURATION			APPLIED LOAD (% DESIGN LIMIT LOAD)	CRITICAL STRUCTURE	REMARKS
					FAIL-SAFE				
	RUDDER KICK	ONE ENGINE OUT	DYNAMIC OVER-SWING	ALL WELL	ROD OUT	BOLT OUT			
1	X			X			150	LWR ACTUATOR HINGE FITTING ATTACH TO FRONT SPAR FLANGE; RIB STA 436.20 FLANGE	NO FAILURE
2		X		X			150		
3	X			X			150	SKIN PANEL JUST BELOW THE LWR ACTUATOR HINGE	
4	X				X		100	LWR ACTUATOR HINGE FITTING ATTACH TO FRONT SPAR FLANGE; RIB FLANGE	
5			X			X	100		
6		X				X	100		
7				•			150	RIB SHEAR-WEB	VERIFICATION OF RIB STRENGTH FOR FULL-SCALE STATIC-TEST RUDDER LOADING METHOD
8	X			X			184	RIB STA 436.20 FLANGE	COMPRESSION FLANGE FAILURE MS = +22%

TABLE 5

SUMMARY OF LIGHTNING PANEL TEST RESULTS

NASA CR-145068(2)

TEST SHOT NUMBER	SWEPT-STROKE RESTRIKE POLARITY	RESTRIKE TIME (MILLISECOND)	RESTRIKE RECORD FASTEX FILM FRAME NUMBER	TEST RESULT
1	SWEPT STROKE ONLY		–	SUCCESSFUL
2	NEGATIVE	10	12TH	SUCCESSFUL
3	NEGATIVE	10	12TH	SUCCESSFUL
4	NEGATIVE	12.5	15TH	SUCCESSFUL
5	NEGATIVE	15	18TH	SUCCESSFUL
6	POSITIVE	15	18TH	PARTIAL FAILURE
7	POSITIVE	15	17TH	SUCCESSFUL
8	POSITIVE	15	19TH	SUCCESSFUL

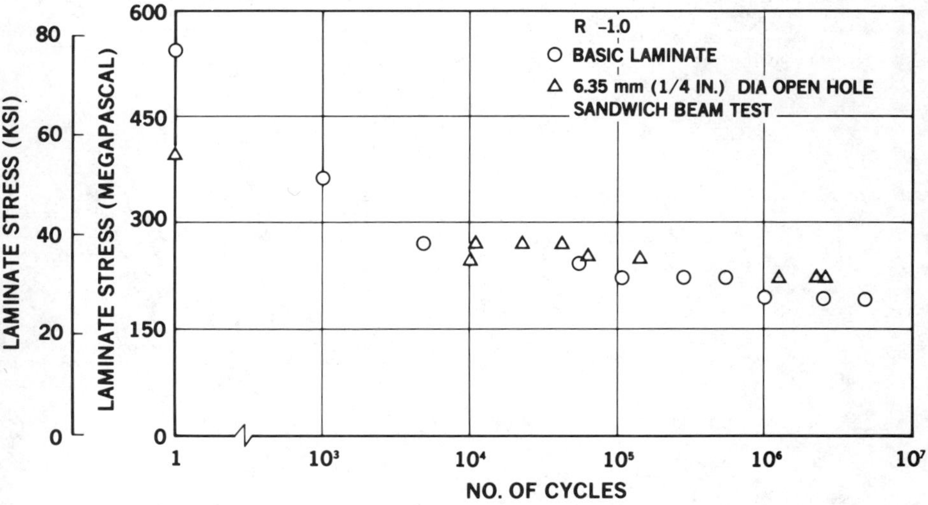

NASA CR-145068(2)

FIGURE 3. FATIGUE CHARACTERISTICS OF THORNEL 300/5208 (0°, 45°, 90°, –45°)S LAMINATE

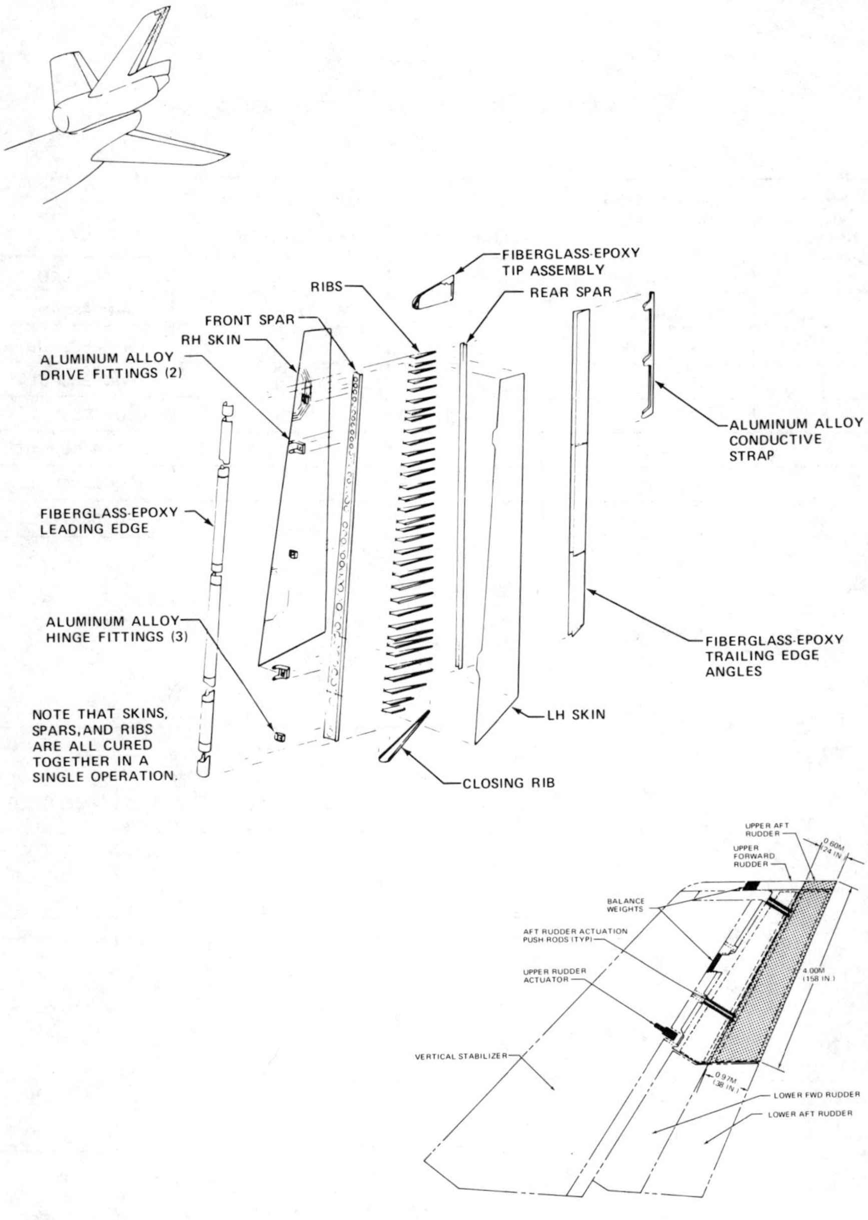

FIGURE 4. STRUCTURAL ARRANGEMENT FOR THE ADVANCED COMPOSITE RUDDER

TABLE 6
SUMMARY OF ADVANCED COMPOSITE RUDDER STATIC TESTS

NASA CR-145068[2]

CRITICAL CONDITION	TEST SEQUENCE	RUDDER CONFIGURATION: ALL WELL	DRIVE ROD OUT	FITTING ATTACH BOLT OUT	RUDDER HINGE BOLT OUT	LOAD LEVEL (% DLL)	REMARKS
RUDDER KICK	1	X				150	NO FAILURE.
	2				X	100	LOWER DRIVE HINGE OUT. NO FAILURE.
	3	X				100	LOWER DRIVE HINGE ROD OUT. NO FAILURE.
	4			X		100	LOWER DRIVE FITTING ATTACH-MENT OUT. NO FAILURE.
ONE ENGINE OUT	5	X				150	NO FAILURE.
	6			X		100	UPPER DRIVE FITTING ATTACH-MENT OUT. NO FAILURE.
	7				X	100	CENTER HINGE OUT. NO FAILURE.
CRITICAL REAR-SPAR STRESS	8					100	LOCAL CONDITION ON REAR-SPAR. NO FAILURE.
ONE ENGINE OUT. UPPER DRIVE HINGE STRESS	9			X		100	UPPER DRIVE FITTING ATTACH-MENT OUT. NO FAILURE.
RUDDER KICK	10	X				415	ATTEMPTED FAILURE TEST. TEST SUSPENDED BECAUSE OF IMPENDING WHIFFLING INSTABILITY.

FIGURE 5. RUDDER STATIC TEST SETUP

Details of the interface between the whiffling and the graphite rudder are shown in Figure 6. Point-loads were applied on alternate ribs of the rudder at mid-rib and at the rib intersections with the front and rear spars. The relatively small size of the aluminum alloy bearing block at the graphite surface afforded excellent visibility of the rudder surface during the tests.

The structural tests were conducted either in a complete configuration or in a failed configuration in which a single structural element such as a drive rod or attach bolt was intentionally omitted from the structure. In the complete configurations, design ultimate loads (150 percent of design limit loads) were applied. In the failed configurations, design limit loads were applied in accord-

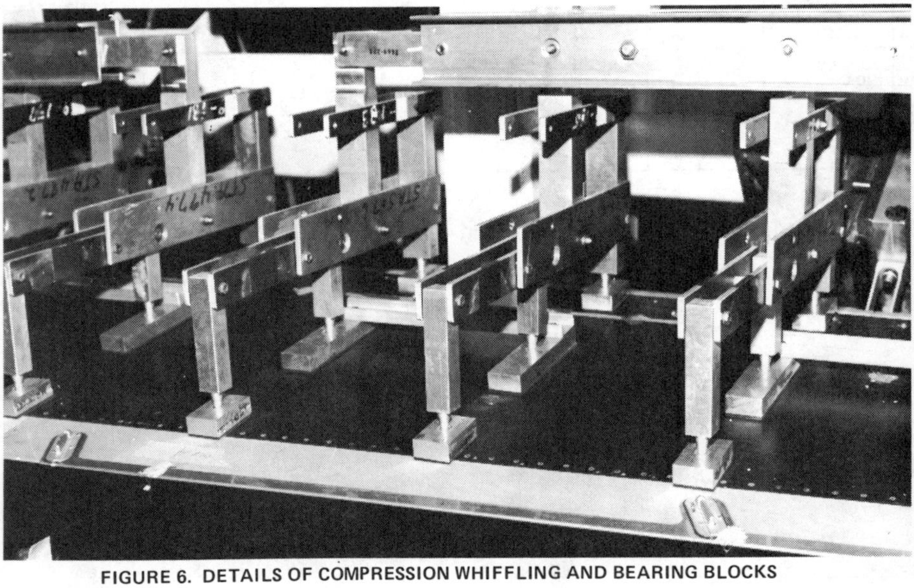
FIGURE 6. DETAILS OF COMPRESSION WHIFFLING AND BEARING BLOCKS

ance with design criteria requirements. Either limit or ultimate hinge line deflections were present at the aft rudder hinge line, depending on the appropriate load level for the test sequence. Governing airloads were applied in increments, and strain and deflection data were recorded after each load application. No structural failures were experienced in the nine test sequences.

The 10th test sequence duplicated one of the ultimate stress conditions, but the applied loads were increased above ultimate in an attempt to determine the failure load and mode for the rudder. The load was increased to 415 percent of design limit load without failure. At this point, the test was suspended because the compression whiffletree system was near critical load conditions. Load-deflection characteristics of the rudder were determined as a basis of comparison for future tests on rudders retrieved from flight service.

The composite rudder was extensively ground tested to demonstrate the structural integrity of all critical details of the design concept. A 33-percent weight reduction of 30 pounds was attained in the composite rudder. Documentation showing compliance with all Federal aviation requirements was submitted to the FAA and certification was received.

3. DEVELOPMENT OF FUTURE PRIMARY COMPOSITE STRUCTURES

Future Douglas airplanes may have large portions of structures made from advanced composites. The potential use of composites is shown in Figure 7. In the future, the composite structural weight could be as much as 60 percent of the total structural weight. Figure 8 shows a projection through the year 2000 of the increasing amount of composite structures which is anticipated on Douglas' new airplanes.

Besides the secondary composite structures, development work on next-generation large composite primary aircraft structure (LCPAS) such as wings and fuselages is in the planning stage of the NASA-ACEE program. The program features a technology development phase followed by a component development phase to design, fabricate, and full-scale ground test large wing sections. The damage tolerance and durability design technology will play an important role in the development.

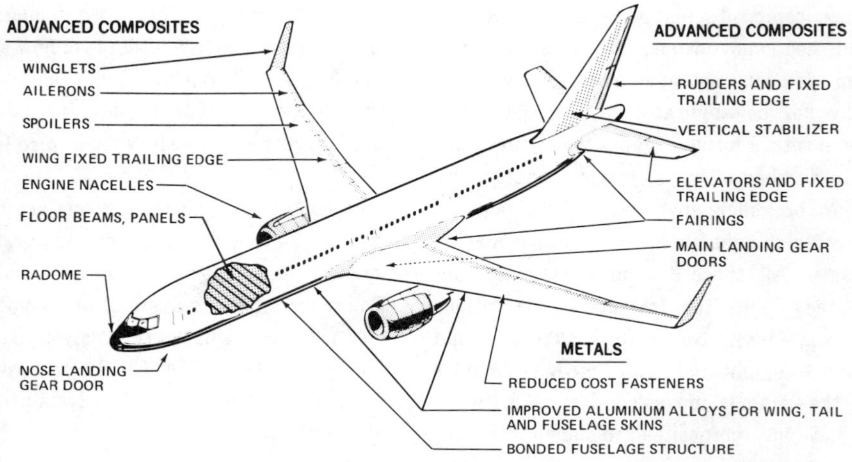

FIGURE 7. AIRCRAFT STRUCTURES TECHNOLOGY WITH 1985 TECHNOLOGY FREEZE

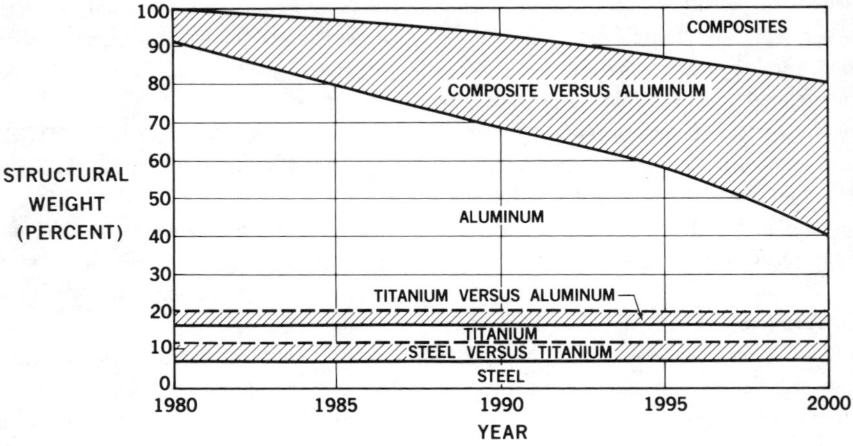

FIGURE 8. STRUCTURAL MATERIAL DISTRIBUTION

The damage tolerance and fatigue evaluation of metallic structure is generally based on linear homogeneous isotropic fracture mechanics. The capability to assure integrity of metallic structures by fulfilling the requirements of FAA and military agencies appears to be well in hand at Douglas. On the other hand, the methodology for damage tolerance and fatigue evaluation of composite structures and the required data base are developing in the industry. Past efforts to demonstrate damage tolerance and durability of composites relied mostly on tests.

Since composite materials are anisotropic and nonhomogeneous, metal-based linear homogeneous isotropic fracture mechanics may not be applicable to advanced fiber composite. Fibrous composites can fracture in a number of ways; by delamination, by fiber breakage, by matrix cracking, by interfacial debonds, and by other mechanisms. All these fractures may be connected. Thus, the fracture configurations associated with composite structures would yield a complicated failure mode. Examples of the common in-service damage between metals and composites are shown in Figure 9. Analytical detection of successive microfailures from microstress analysis and microfailure criteria, and prediction of the coalescence of them to form microfailures are very complicated. However, infinitesimal microfailure from micromechanics analysis could be transformed into a set of equivalent forces and energy which could be accumulated to form initial damage for macromechanics analysis. A recent study of the analysis of fatigue crack nucleation at the free surface of a metal matrix composite medium used the analogy between equivalent body forces and plastic strain gradient.[8]

Figure 10 presents a proposed damage tolerance and durability design task flow chart. The overall methodology for developmental analysis that needs to be addressed in order to support the task is shown in Figure 11.

Two computer analysis systems are being planned for damage tolerance and durability analysis of composite structures, Damage Tolerance Analysis System for Composites (DTASC) and Durability Analysis System for Composites (DUASC). These two systems would automate the analysis process, efficiently handle input and output data, and systematically operate the damage tolerance and durability data base.

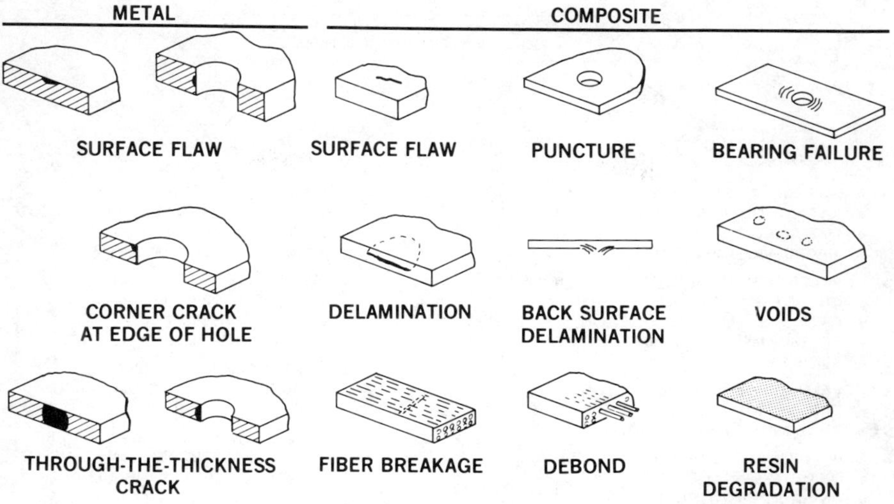

FIGURE 9. EXAMPLES OF IN-SERVICE DAMAGE CONFIGURATIONS

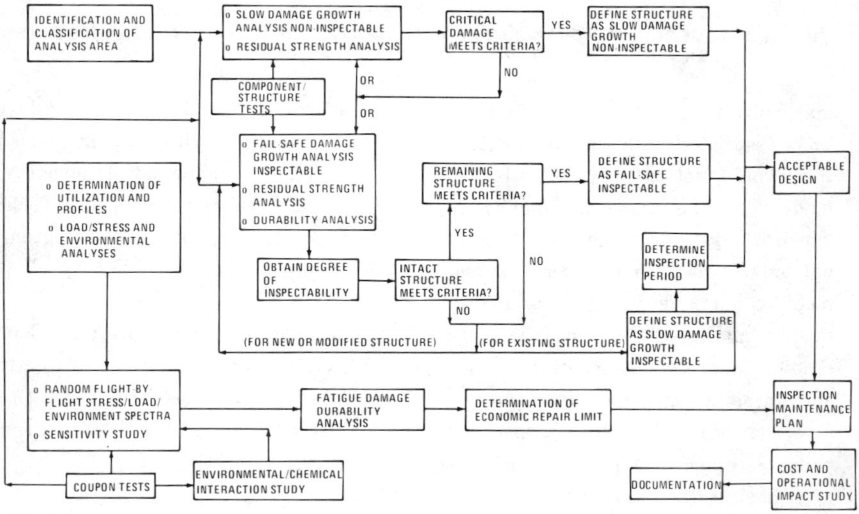

FIGURE 10. A PROPOSED DAMAGE TOLERANCE AND DURABILITY DESIGN TASK FLOW CHART FOR PRIMARY COMPOSITE STRUCTURES

4. CONCLUSIONS

1. Current commercial airframes have a high degree of reliability, and advanced composite structure must meet or exceed this record to be acceptable for general use in primary structure. As an interim approach, advanced composites may be used by conservative application of existing criteria supported by extensive testing and special in-service inspection procedures.[1]

2. Fibrous composites differ from their metal counterparts primarily in anisotropy and nonhomogeneity properties. Fibrous composites can fracture in a number of ways; by delamination, by fiber breakage, by matrix cracking, by fiber debonding, and by other

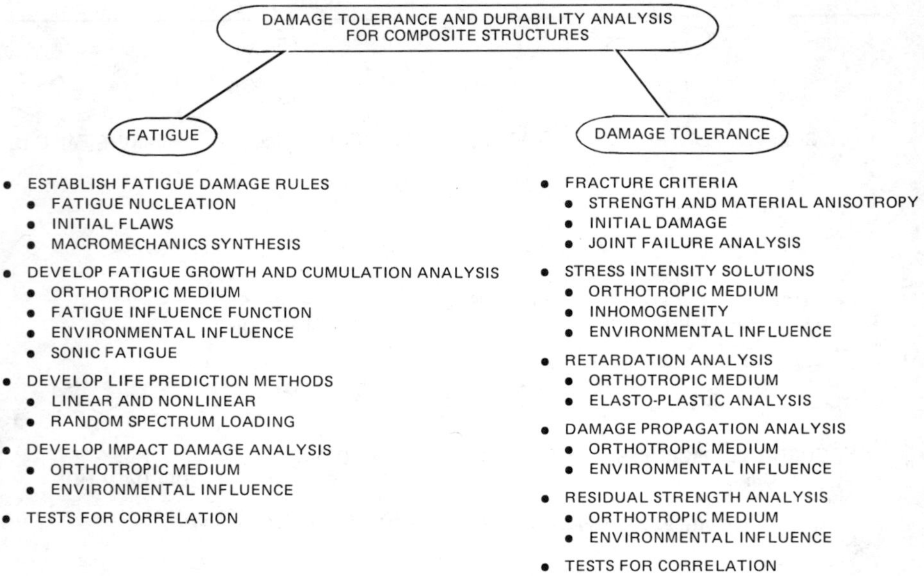

FIGURE 11. OVERALL DEVELOPMENTAL METHODOLOGY OF DAMAGE TOLERANCE AND DURABILITY ANALYSIS

mechanisms. The fracture configurations associated with composite structures could yield a complicated failure mode. The anisotropic and nonhomogeneous nature of the material results not only in requirements for new and modified methods but also for higher quality analysis with more accurate results. Therefore, damage tolerance and durability analysis for composites must be developed to support the primary composite structures for commercial aircraft.

4. REFERENCES

1. Warren, D. "Fail-Safe Design and Reliability in Composite Commercial Aircraft Structure," Third Conference on Fibrous Composites in Flight Vehicle Regime, Virginia, November 1975.

2. Smillie, D. G., "Composite Structure Applications for Commercial Aircraft," Douglas Aircraft Company, Paper 6659, May 1978.

3. Watts, D. J. and Lee, N. R. "Advanced Composite Applications in McDonnell Douglas Commercial Transport Aircraft," 12th National SAMPE Technical Conference, Seattle, Washington, 7-9 October, 1980.

4. Lehman, G. et al, "Advanced Composite Rudders for DC-10 Aircraft – Design, Manufacturing, and Ground Tests," NASA CR-145068, November 1976.

5. Marra, P.J., "Trapped Rubber Processing for a Composite Rudder," Douglas Aircraft Company Paper 6651, September 1977.

6. Dexter, H. B., "Composite Components on Commercial Aircraft, " NASA Technical Memorandum 80231, March 1980.

7. Kam, C., "Effect of Fatigue Loading on Bolt Holes in Graphite-Epoxy Laminates," Douglas Aircraft Company,

Paper 6849, presented to American Society for Testing and Materials, May 1979.

8. Chow, E., "Analysis of Fatigue Crack Nucleation at the Free Surface of an Orthotropic Medium," presented at the 14th National Symposium on Fracture Mechanics, ASTM, Los Angeles, July 1981.

BIOGRAPHIES

Dr. Edmond Chow, a Senior Engineer/Scientist at Douglas Aircraft Company, is responsible for the development of damage tolerance and durability evaluations of composite and advanced metallic structures. Before joining Douglas, he investigated fatigue and fractures in and performed stress analyses of composite structures at Convair/General Dynamics. At Advantech System Corporation, he was in charge of projects involving structural analysis, design, and development of computer programs. He also taught courses in applied mathematics; fracture mechanics; continuum mechanics; numerical methods; and mechanics of solids, plates, and shells. Dr. Chow has a B.S. degree in civil engineering, an M.S. degree in structures, and a Ph.D. degree in mechanics.

Clifford Y. Kam, a Section Chief in Structures Design, directs the structural development and application of advanced composite structures to commercial aircraft at Douglas. He also coordinates the structures development efforts with Materials and Process and Manufacturing R&D. Mr. Kam has been with Douglas since 1952. He holds the M.S. degree in Structural Engineering from the Massachusetts Institute of Technology and the B.S. degree in civil engineering from Marquette University. Mr. Kam is co-holder of a patent for fabricating composite structural components employing integral isogrid stiffening members.

13th National SAMPE Technical Conference
October 13-15, 1981

LIGHT WEIGHT ADHESIVE JOINING OF COMPOSITE STRUCTURES

Kenneth D. Cressy
Supervisor, Technical Services
Furane Products Division
M&T CHEMICALS INC.
Los Angeles, California

Abstract

As light weight composite structures find increased use in transportation and industrial applications, there is increasing need for reduced weight, high strength adhesives to join these composite assemblies. Adhesively bonded structures are demonstrating advantages over riveted or bolted assemblies by providing strength and durability through various environmental exposures. This paper compares properties of rigid, high temperature resistant epoxies with those of flexible, high peel strength urethane adhesives. Emphasis is placed on the compatibility of low density adhesives with low density composite structures.

I. INTRODUCTION

The increasing use of reinforced plastic materials to replace metals especially in transportation vehicles, is creating greater demand for a variety of compatible adhesives. Almost every new flight vehicle includes some structure made with advanced composite materials while certain aircraft are constructed entirely of advanced composites. Joining these metal and plastic components offers new challenge to the adhesive formulator.

Adhesively bonded structures have several advantages over mechanically assembled structures such as reduced number of parts and simpler designs. In addition, adhesives provide increased fatigue resistance and lower maintenance costs. Adhesives yield a more even distribution of stress load over the entire bonded area reducing the concentration of stress which occurs with riveting or bolting. Adhesives may also act as sealants to form barriers against moisture and other environmental contaminants and aid in preventing bimetallic corrosion that often occurs in joining two different types of metal. Another advantage of adhesives is weight reduction, a factor of major importance in energy conservation, especially in transportation vehicles.

Reliable repair of composite structures is being accomplished with structural grade adhesives which have shown improvements in both peel and shear strength. Joint repair of composite structures using fasteners can be accomplished with removal of damaged fasteners and replacement using low density fastener potting compounds. Such field repairs using fast cure adhesive compounds often result in stronger-than-original joints and can be accomplished quickly to return vehicles to use within short time periods. Damaged graphite epoxy structures can be restored to designed structural integrity with graphite/epoxy patch co-cured with epoxy adhesives with medium modulus and good ductility.

2. DISCUSSION

Whether it is repair of damaged structures or assembly of new structures the use of adhesives is becoming more important. Properties of several low density epoxy adhesives are shown in Table I representing typical adhesives used in bonding aluminum and NOMEX honeycomb aircraft parts. Overlap shear strength, as tested on aluminum substrates, range from 1000-2500 psi and each adhesive has a certain uniqueness in handling properties with variations in cure time and temperature.

The increased speeds of aircraft being designed today have pushed material performance requirements to higher limits at higher temperatures. To keep up with these advances it has been necessary for adhesive formulators to utilize chemistries that yield stronger, heat resistant compounds. Table II shows properties of single component, heat cure epoxy adhesives with high temperature properties as demonstrated by high compressive strength at 350°F. In achieving higher temperature performance one other less desirable property, that of modulus of rigidity may cause problems of stiffness or brittleness of bonded joints. Achieving a blend of resilience or toughness and retention of strength above 250°F. has been a challenge.

Much has been published about toughened epoxies and most data shows a fall-off in physical strength as temperatures increase. In addition, even the toughened epoxies become embrittled at temperatures below -55°F. Furane has been developing urethane compounds which overcome both the low and high temperature limitations of epoxies as evidenced by data shown

in Table III. The urethane adhesive (URALANE 5757) has higher strength at -65°F. and R.T. and retains greater strength through 250°F. than a comparable epoxy adhesive. Peel strength of the urethane is much higher as can be expected with elastomeric type compounds, and in applications where high peel strength is most important, a change in mix ratio will develop peel strengths as high as 150 ppi on aluminum substrate. Even though reversion resistance is not usually a problem with epoxies it is shown that this urethane adhesive has superior reversion resistance with no change in durometer hardness after 28 days in an environment of 160°F. and 95% Relative Humidity.
Further review of the data in Table III shows the excellent resistance of the urethane adhesive to both high humidity aging and dry heat aging at 250°F. You will note that dry heat aging produced an increase in both over lap shear strength and T-Peel strength.

Data also shows good adhesion to a variety of substrates such as ABS, polycarbonate, graphite and KEVLAR reinforced epoxy and phenolic. This urethane maintains good strength of adhesion over a wide temperature range, yet is extremely tough in order to withstand the differences in expansion coefficients of the various metal and plastic substrates. The bonding of materials of unequal strength has been successfully accomplished with this elastomeric urethane.

Several factors need to be considered to obtain optimum bond strengths with these urethane adhesives. As with any bonding application, surface preparation is of utmost importance and much has already been published on this subject. The usual aluminum preparation of FPL etch or phosphoric acid anodizing is sufficient for optimum bond strengths with URALANE 5757. The low surface tension of this urethane in the initial mixed phase provides optimum substrate surface wetting followed by gradual polymerization and cross-linking to the final gelled and bonded state. After the initial mixing period, it develops a degree of thixotropy which provides gap filling capability. In bonding substrates of dissimilar materials it is often necessary to fill surface irregularities resulting from mismatched surfaces. Keeping the newly assembled joint under pressure during the initial phase aids in surface wetting and development of maximum adhesion to the substrate surfaces. This surface wetting helps develop maximum sealing permitting the adhesive to also serve as a sealant against environmental contamination.

3. CONCLUSION

The rapidly expanding use of light weight materials in industry today justifies concurrent studies and developments of light weight adhesives. Weight saving is not only important to fly-away vehicles but also to any vehicle requiring energy to move it from place to place. Lighter weight materials are being considered for more and more applications for easier handling and lower shipping costs. The weight saving techniques established for aircraft construction are being utilized in automotive assembly and other industrial applications. Lighter weight metals which are not readily welded along with a wide variety of plastic surfaced composite structures can be easily joined with light weight urethane and epoxy adhesives.

4. ACKNOWLEDGEMENT

The author wishes to acknowledge the work of Steve Marks in preparing data for this paper.

5. BIOGRAPHY

Kenneth D. Cressy is a graduate of the University of New Hampshire and has managed Product Development, Quality Control, Technical Service and Technical Sales functions at Furane Products Division of M&T Chemicals, Inc. for a period of 25 years. Prior to that he was employed by Dow Chemical Company in Freeport, Texas.

TABLE I

LOW DENSITY, TWO COMPONENT
EPOXY ADHESIVES

	87822-A/B	87831-A/B	8623/946	87005A/B-60
Density, g/cc	0.8	0.7	0.7	0.5
Al/Al Tensile Shear, psi	2500	2200	1600	2000
Flammability BMS 5-28	S.E. 15 sec.	S.E. 10 sec.	S.E. 10 sec.	S.E. 10 sec.
Work Life 100 gm. @ 77°F.	90-120 min.	75-90 min.	10-15 min.	60-80 min.
Cure	24 hr. R.T. or 20 min.@ 250°F.	24 hr. R.T. or 1-2 hr. @ 140°F.	6 hr. R.T.	24 hr. R.T.
Consistency	Pourable 100,000 cps	Semi-Paste	Extrudable Non-Sag Paste	Trowellable Paste

TABLE II

LOW DENSITY, ONE COMPONENT
EPOXY ADHESIVES

	1612	1614	1610
Density, g/cc	1.0	0.75	0.5
$\#/ft^3$	62	47	31
Al/Al Tensile Shear, psi	1000	1100	1500
Flammability BMS 5-28	S.E. 15 sec.	S.E. 15 sec.	S.E. 15 sec.
Compressive Strength, psi @ R.T.	18,000	15,000	2,000
@ 350°F.	10,000	7,000	-
Cure:	1 Hour @ 350°F.	1 Hour @ 350°F.	1 Hour @ 350°F.

TABLE III

COMPARISON OF EPOXY AND URETHANE ADHESIVES

	EPOXY[1]	URETHANE[2]
Density, g/cc	1.25	1.18
Lap Shear Strength, psi		
Al/Al @ -65°F.	2365	7450
@ R.T.	2170	2620
@ 150°F.	625	1450
@ 250°F.	200	865
Polysulfone	Not tested	1500
ABS/ABS	1200	915
Phenolic/Graphite	1350	1250
Polycarbonate/Polycarbonate	1750	640
After 30 day Aging @ 160°F., 95% R.H. (Al/Al)	835	1595
After 30 Days Dry Heat Aging @ 250°F. (Al/Al)	3480	3790
T-Peel Strength, pli		
Al/Al	35	89
KEVLAR/KEVLAR	Not tested	21
After 30 day Aging @ 160°F., 95% R.H. (Al/Al)	5	70
After 30 day Dry Heat Aging @ 250°F. (Al/Al)	Not tested	100
Reversion Resistance 28 days @ 160°F., 95% R.H.		
Durometer Hardness D		
Initial	68/55	45/42
14 Days	57/45	45/42
28 Days	50/42	46/38
Ave. Total Loss, Points	-16	-2

(1) 3M-2216B/A

(2) URALANE 5757-A/B

13th National SAMPE Technical Conference
October 13-15, 1981

FILLERS AND REINFORCEMENTS FOR IMPROVED PLASTICS

Harry S. Katz
Utility Research Company
Montclair, New Jersey 07042

Abstract

Fillers and reinforcements have always played an important role in the plastics industry, and have received even greater attention during recent years. Fillers had been considered primarily as a means of lowering the cost of a compound, but can also contribute other benefits such as increased thermal conductivity, reduce coefficient of thermal expansion and mold shrinkage, and improve abrasion resistance. Special fillers can be used to obtain low density, fire retardance, high electrical conductivity, low coefficient of friction, and many other frequently required properties. Recent application of packaging concepts in the mixture of different geometric shapes of fillers has opened up new horizons for improved composite materials and manufacturing procedures.

The use of reinforcements can provide plastics with much higher strength and modulus. Other important improvements are increased creep resistance, thermal conductivity, and heat distortion temperature. Graphite or metal fibers can be used for high electrical conductivity. High performance carbon-graphite filaments can provide specific strength and specific modulus that are higher than structural metals. Hybrid combinations of graphite, Kevlar, and glass filaments can lead to synergistic benefits of performance and cost.

Keywords: Fillers, reinforcements, high strength, high modulus, thermal expansion, fire retardants, conductivity.

1. INTRODUCTION

In 1979, the volume of plastics produced in the United States exceeded the volume of steel, so we have truly entered the age of plastics. Because of this high volume and continued growth, most materials and process engineers will probably be involved in many plastics applications and should therefore be well informed about the effective utilization of these materials. There are many benefits that can be obtained by the judicious use of fillers and reinforcements in plastics, so these composite materials should be considered for every end use.

2. DISCUSSION

Most commercial plastics contain no filler. However, the growth rate of filler and reinforcement usage has been projected to be even faster than that of plastics. Currently, the reported consumption of fillers

TABLE 1

PRICE CHANGES OF RESINS, FILLERS, AND REINFORCEMENTS

Material	Price Level 1973	Price Level 1980	Approximate Rate of Increase
Commodity resins			
Polypropylene	17¢/lb	38	For Resins
Polyethylene	14¢/lb	42	120%
Polyester	20¢/lb	52	
Engineering resins			
Nylon	60¢/lb	134	
Polycarbonate	75¢/lb	122	
Fillers and reinforcements			
Calcium carbonate			
fine	$50/ton	54	For Fillers
intermediate	$32/ton	38	10%
coarse	$18/ton	19	
Talc			
fine	$68/ton	90	
Wollastonite			
F-1	2.4¢/lb	4.0	For Reinforcements
P-1	1.9¢/lb	2.6	80%
Fiberglass roving	35-40¢/lb	72-88¢/lb	

and reinforcements is about 12% by weight that of resins consumed, but is projected to reach about 25% by the year 2000.

A main reason for the increased use of fillers is the fact that their price will rise at a slower rate than that of resins. Table 1 above indicates the pricing trend of resins, fillers, and lower cost reinforcements from 1973 to 1980. Note that the approximate cost increase during that period was well over 100% for typical commodity and engineering resins, but about 10% for fillers and 80% for the listed reinforcements. The price of plastics is closely tied to the price of oil. David Mooberry, general manager of DuPont's Energy and Material Department, has projected that oil could rise to over $100 a barrel by the year 2000.(1)

Some of the reasons for using fillers are shown in Table 2.

TABLE 2

SOME REASONS FOR USING FILLERS

1. Reduce cost.
2. Reduce coefficient of thermal expansion.
   - Reduce mold shrinkage
   - Reduce rejects due to warpage
3. Increase modulus.
4. Increase thermal conductivity.
5. Increase hardness.
6. Improve paint adhesion.
7. Improve abrasion resistance.
8. Use special fillers to obtain low density, fire retardance, high electrical conductivity, low coefficient of friction, etc.
9. Extend the amount of available resin.

In the past, end users were concerned that high levels of filler would result in plastics that would be very brittle and have poor physical properties. However, many current molding compounds have 40 weight % of fillers, and good molding compounds that contain over 75 parts of filler per hundred parts resin may be common in the near future. Calcium carbonate filler has been used at 40 weight % in PVC pipe extrusion compounds, and at much higher levels in polyester formulations for casting synthetic marble. A recent study of flame retardant polypropylene has shown that 40 weight % of fillers can be used with retention of flexibility.(2)

A variety of hollow microsphere fillers are commercially available. These can be used to obtain composites with low density, and low cost per unit volume.

Some of the reasons for using reinforcements are listed in Table 3.

The main reasons for using fiber reinforcements is to increase strength and stiffness of a composite. The ultimate potential for fiber reinforcement can be projected by examination of the properties of whiskers, which are fibers that have been grown under controlled conditions that lead to the formation of an essentially perfect single crystal. These properties are shown in Table 4.

TABLE 3

SOME REASONS FOR USING REINFORCEMENTS

1. Increase strength: tensile, flexural, torsional impact.
2. Increase modulus.
3. Increase creep resistance.
4. Increase thermal conductivity.
5. Increase heat distortion temperature.
6. Decrease coefficient of thermal expansion.
7. Some low-priced fibers can reduce cost.
8. Graphite or metal fibers can provide high electrical conductivity.

TABLE 4

PHYSICAL PROPERTIES OF WHISKERS

Material	Density g/cm^3	Melting point ^oF	Tensile strength psi x 10^{-6}	Young's modulus psi x 10^{-6}	Strength density in. x 10^{-6}
Aluminum oxide	3.9	3780	2-4	70-150	14-28
Aluminum nitride	3.3	3990	2-3	50	13-21
Beryllium oxide	1.8	4620	2.0-2.8	100	31-43
Boron carbide	2.5	4440	1	65	11
Graphite	2.25	6500	3	142	37
Magnesium oxide	3.6	5070	3.5	45	27
Silicon carbide (alpha)	3.15	4200	1-5	70	9-44
Silicon carbide (beta)	3.15	4200	1-5	80-120	9-44
Silicon nitride	3.2	3450	.5-1.5	55	4.2-13

Note that graphite whiskers have a relatively low density, 2.25 g/cm^3, and 140 million psi modulus and 3 million psi tensile strength. If these whiskers were available, and if the technology existed to obtain effective translation of the fiber properties to the composite, molded parts could be obtained that would be magnitudes higher in strength and stiffness than any currently produced. The reality of the current state-of-the-art includes a wide variety of carbon-graphite filaments with moduli that range up to 75-million psi for a high modulus type, and up to about 400,000 psi tensile strength for a high strength type. These excellent properties, and the fact that the cost of these high performance materials have been reduced in recent years from hundreds of dollars per pound to the $30 per pound level, has resulted in the widespread use of this material. Northrop Aviation in California has recently set up a new facility with automated equipment for the fabrication of an "all-graphite-composite-aircraft." Grumman Aircraft and other aerospace companies have used graphite, boron, Kevlar, and glass filament composites for many aerospace structural components. Lockheed is now testing the durability of a L-1011 Tristar aircraft vertical tail fin graphite filament composite.(3) This component weighs 622 pounds, about 25% less than the conventional aluminum tail section it is designed to replace.

Ford Motor Company recently built an experimental car with many carbon-graphite filament composite components.

Hybrid combinations of graphite, Kevlar, and glass filaments can lead to synergistic benefits of performance and cost.(4) The cost of graphite filaments are still relatively high, as compared with Kevlar or glass filaments. Kevlar can contribute improved impact strength. Glass filaments can contribute cost reduction and improved compressive strength.

An important consideration in the use of fillers and reinforcements, is that it is often feasible to obtain synergistic benefits by using combinations of these materials. A study of the mode for maximum packing combinations of different sized particles or different shaped fillers and reinforcements can often lead to improved composites. Dr. John V. Milewski has published a number of papers on packing concepts.(5) Figure 1 is from one of his papers, and shows that a proper ratio in the diameter of spheres and fibers will result in interstitial packing and a minimum void volume. Therefore, there will be a lower resin demand to fill interstices and coat each filler particle. In contrast, if the relative sizes or shapes are wrong, high void volumes will be obtained. About 5 years ago, a small company was started with the plan of selling filler blends that had good packing characteristics. These would be varied combinations of particulate fillers, flakes such as mica, and short fibers. That company has now grown to selling millions of pounds per year. One end use of this material was described in a recent verbal report. This involved a heavy-wall polypropylene liner for a large drum that was injection molded in large quantities. By using 20% of the filler blend, the material cost was reduced 5 to 10%, but because the filler provided higher thermal conductivity and a faster cooling rate, the production cycle was increased 20%; which was even more important than the

FIGURE 1.

ALL CYLINDERS CONTAIN THE SAME VOLUME OF SOLIDS 3 PARTS SPHERES AND I PART FIBERS AT I5/I L/D

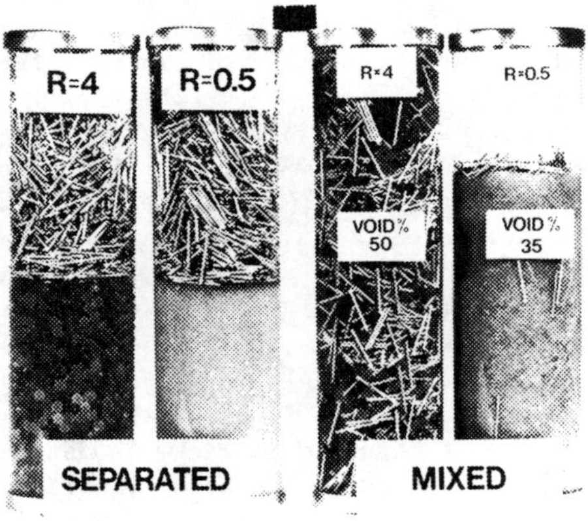

materials cost saving.

An important consideration for the materials engineer is that the filler-matrix or fiber-matrix interface bonding can be a critical factor that will determine to what extent the potential properties of the composite will be achieved and maintained during usage. The use of a coupling agent can provide improved interfacial conditions. The interfacial bond must resist stresses due to factors such as differential thermal expansion coefficients and shrinkage of the resin during cure. Silanes have been the most frequently used type of coupling agent. The silanes provide dual functionality in one molecule, such that one part of the molecule can form a bond to the filler or reinforcement surface, while another part of the silane molecule forms a covalent bond with the resin. A wide selection of silanes are offered by Dow Corning, General Electric, and Union Carbide Corporation. Other types of coupling agents are available. DuPont produces Volan, a methacrylatochromium complex, and Kenrich Petrochemicals Inc. produces a line of organic titanates that have shown remarkable results in some composite systems.

There has been an increasing impetus toward the use of fillers and reinforcement, and this is evident by the large number of recent articles on this subject in technical journals. An article entitled, "Additives 1981 - the Key to Cost-Effectiveness," was published in the May 1981 issue of Plastic Engineering; and a review of "Fillers and Reinforcements 1981" was published in the July 1981 issue of Modern Plastics.

3. REFERENCES

(1) Miller, B., "Bargains in Thermoplastics Custom Compounds," Plastics World, July 1981, 60.

(2) Walter, S., "Filled Polypropylene Can be Flexible - and Flame Retardant, Too," Plastics Engineering, June 1981, 24.

(3) "Graphite Filament Composite Gets an Aircraft Test," Chemical Week, July 15, 1981, 36.

(4) Watts, A. A., ed., "Commercial Opportunities for Advanced Composites," ASTM Special Technical Publication 704, 1980, 66.

(5) Milewski, J. V., "Packing Concepts in the Utilization of Filler and Reinforcement Combinations," Handbook of Fillers and Reinforcements for Plastics, VanNostrant Reinhold Company, 1978, pp. 66-78.

BIOGRAPHY

Harry S. Katz has over 25 years' experience in the plastics industry. His prime involvement and interests have been in reinforced plastics and advanced composites. His industrial experience includes positions as a Materials Engineer and Project Engineer in research and development of plastics components, adhesives, and coatings that have been used in rocket and aircraft engines. Since 1966, Mr. Katz has been a consultant in plastics and composite materials. His firm, Utility Research Company, lists among its clients government agencies and several of the nation's leading industrial firms. Mr. Katz has been granted many U.S. and foreign patents. He is a member of several professional societies.

13th National SAMPE Technical Conference
October 13-15, 1981

NOVEL IMPROVED PMR POLYIMIDES

Ruth H. Pater
National Aeronautics and Space Administration
Lewis Research Center
Cleveland, Ohio 44135

Abstract

N-phenylnadimide modified PMR resins and composites were investigated with the objective of improving the overall flow characteristics of PMR-15 type resins without compromising composite properties. A series of N-phenylnadimide (PN) modified PMR polyimide composites reinforced with Celion 6000 graphite fibers were investigated. The improved flow matrix resins consist of N-phenylnadimide (PN), monomethyl ester of 5-norbornene-2,3-dicarboxylic acid (NE), dimethyl ester of 3,3',4,4'-benzophenonetetracarboxylic acid (BTDE), and 4,4'-methylenedianiline (MDA). Five modified PMR resin systems were formulated by the addition of 4 to 20 mole percent N-phenylnadimide to the standard PMR-15 composition. These formulations and the control PMR resin were evaluated for rheological characteristics. The initial thermal and mechanical properties of the PN modified PMR and the control PMR/Celion 6000 composites were determined. The elevated temperature flexural and interlaminar shear properties and the weight loss characteristics of the PMR composites isothermally exposed at 600°F in air for 1500 hours were evaluated. In addition, the effects of hydrothermal exposure on the glass transition temperatures and the elevated temperature flexural properties of the PMR composites were also investigated. The results of this study clearly showed that the addition of N-phenylnadimide to PMR-15 significantly improved the resin flow characteristics without sacrificing the composite properties. This study further revealed that concentrations of 4 and 9 mole percent PN appear to improve the thermo-oxidative stability of PMR composites.

Keywords: N-phenylnadimide Modified PMR Polyimides, PMR Polyimides, PMR Polyimide Composites, Composite Properties, Flow Properties, Thermo-oxidative Stability, Hydrothermal Exposure

1. INTRODUCTION

The "melt-flow" behavior of the resin matrix exerts considerable influence on the overall processing characteristics and properties of fiber reinforced resin matrix composites. Until the development of the class of high-temperature-resistant resins known as PMR polyimides,[1] the intractible nature of prior art high temperature resins severely restricted their use as matrix materials. Early studies [2,3] demonstrated that PMR matrix resins having a broad range of processing characteristics and properties could easily be synthesized by either varying the chemical nature or the stoichiometry of the monomer reactants. In the study [3] which investigated the resin flow characteristics of fully imidized PMR compositions ranging in formulated molecular weight (FMW) between 1000 and 1500, it was found that the lower FMW compositions exhibited increased resin flow, but with some sacrifice in elevated temperature thermo-oxidative stability. Vannucci [4] reported that the resin flow characteristics of the PMR composition having an FMW of 1500 (PMR-15) could be varied by controlling the degree of resin advancement or imidization.

This paper presents an alternative approach to achieve improved flow PMR polyimides without sacrificing the desirable thermo-oxidative characteristics. The present approach utilizes a chemically compatible flow modifying agent, viz. N-phenylnadimide (PN). The modified PMR resins consist of four monomer reactants: monomethyl ester of 5-norbornene-2,3-dicarboxylic acid (NE), dimethyl ester of 3,3', 4,4'-benzophenonetetracarboxylic acid (BTDE), 4,4'-methylenedianiline (MDA), and the flow modifying agent, N-phenylnadimide (PN). Table 1 shows the chemical structures of these compounds.

The purpose of this investigation was to determine the processing characteristics and properties of PN modified PMR/Celion 6000 composites. The flow properties of the PN modified PMR resins and the physical and mechanical properties of PN modified PMR/Celion 6000 composites exposed to isothermal aging in air at 600°F and humidity conditions are presented.

2. EXPERIMENTAL PROCEDURES

2.1 Materials

Monomethyl ester of 5-norbornene-2,3-dicarboxylic acid (NE, Burdick and Jackson), 4,4'-methylenedianiline (MDA, Eastman), 3,3', 4,4'-benzophenonetetracarboxylic dianhydride (BTDA, Aldrich), 5-norbornene-2,3-dicarboxylic anhydride (NA, Eastman) and aniline (Aldrich) were purchased from the commercial sources as indicated and used as received. Unsized Celion 6000 graphite fiber tows were supplied by Celanese. Dimethyl ester of 3,3', 4,4'-benzophenonetetracarboxylic acid (BTDE) in 50 weight percent methanol solution was prepared by refluxing BTDA in methanol for 2 hours. N-phenylnadimide was prepared according to the following procedure: To a stirred and warm (158°F) solution of 42.0 g (0.26 mole) of NA in 300 ml of acetonitrile was added in one portion a solution of 24 ml (0.26 mole) of aniline in 75 ml of acetonitrile. The mixture was refluxed while stirring for 94 hours. Upon cooling to room temperature, the reaction solution was washed 3 times with 200 ml of saturated potassium carbonate solution. Concentration of the organic solution using a rotary evaporator under reduced pressure afforded 55.4 g (89 percent yield) of endo-N-phenylnadimide as white crystals, m.p. 291-294°F. The spectral data supporting the chemical structure of this compound are given below: ^1H NMR ($CDCl_3$): 1.51 (singlet, 2H, $H_{7,8}$) 2.79 (singlet, 2H, $H_{1,4}$), 3.35 (singlet, 2H $H_{2,3}$), 6.29 (singlet, 2H $H_{5,6}$), 7.28-7.36 (doublet, 5H, aromatic H). IR (KBr): 2880 (C-H), 1765 and 1700 (imide carbonyl), 1592 and 1495 (aromatic), and 740 cm^{-1} (cis-olefin). A DSC scan gave the following: endotherm 292°F (melt) and endotherm 518°F (sublimations).[5]

2.2 Rheological Characterization

Table 2 shows the five PN modified PMR and the control PMR resin compositions used in this study. The molding powders of the six resin systems were prepared by the general procedures given below: A 50 weight percent solution of the monomer mixture in methanol was stirred at room temperature for 1/2 hour to give a clear dark brown solution. The solution was concentrated using a rotary evaporator under reduced pressure first at 122°F for 45 minutes and then at 194°F for 15 minutes. This yielded a brown powder containing traces of solvent. The brown powder was then staged in an air-circulating oven at 400°F for 1 hour to give a brown molding powder which was used for the rheological and DSC studies. The rheological behavior of the molding powders were determined using a Rheometrics Dynamic Spectrometer, Model 7700. Approximately 1 g of each molding powder was placed between parallel plates having 1 inch diameter. The temperature was raised at 7°F/minute to 460°F at which point the sample softened and the gap between the platens was set to 0.055 inch. Immediately, the sample was subjected to torsional shear at a constant frequency of 10 radians/second and 10 percent strain. The sample was heated at 7°F/minute and the shear viscosity was measured and recorded at 2 minute intervals over the temperature range of 460-590°F.

2.3 Thermal Characterization

For the DSC scans, approximately 8 mg of each molding powder was studied at a heating rate of 18°F/minute under a static N_2 atmosphere at 60 psi.

2.4 Resin Flow

The resin flow for each composite system was determined under identical cure parameters as described below. Twelve plies of four-inch square prepreg were stacked unidirectionally in a mold. The stacked plies were staged at 400°F for 1 hour in an air-circulating oven, placed in a cold matched metal die, and then placed in a preheated 600°F press. When the temperature of the mold reached 450°F, 500 psi pressure was applied. The temperature was raised to 600°F at 7°F/minute and the composite system was cured at 600°F under 500 psi pressure for 2 hours. The resin material which flowed as a result of this cure process was determined according to the following equation

$$F = \frac{W_2}{W_1 + W_2} \times 100 \qquad (1)$$

where

F = resin flow, weight percent

W_1 = weight of resin in molded composite sample as determined by acid digestion method

W_2 = weight of staged prepreg - weight of molded composite after removal of resin flash

2.5 Composite Fabrication

The prepreg was prepared by drum winding the Celion 6000 graphite yarn followed by brush application of a quantity of a 50 weight percent methanol solution of resin components calculated to yield finished composites containing 60 volume percent fiber. The tapes were dried on the rotating drum at room temperature for 16 hours, removed from the drum, cut into 3-inch by 8-inch plies, stacked unidirectionally to 12 plies, and then staged at 400°F for 1 hour in an air-circulating oven. The staged lay-up was placed in a cold matched metal die. This was then inserted into a preheated 600°F press. A thermocouple was attached to the matched die to follow the temperature. In order to obtain the molded composites having similar resin contents, the pressurization temperatures for the composites were progressively increased as the concentration of PN increased. For the control sample, 500 psi pressure was applied when the die temperature reached 450°F. For PMR-P1, PMR-P2, PMR-P3, PMR-P4, and PMR-P5, the pressure was applied when the die temperature reached 540°F, 550°F, 560°F, 570°F, and 580°F, respectively. The temperature was then raised to 600°F at 7°F/minute and all of the composites were cured at 600°F under 500 psi pressure for 2 hours and removed from the press when the die temperature reached 350°F. Composites were given a free standing post-cure at 600°F for 16 hours in an air-circulating oven.

2.6 Composite Characterization
The laminates were examined by ultrasonic C-scan for voids and defects prior to testing. The laminates showed essentially void-free scans, and thus were accepted for test specimen preparation. Flexural tests were performed in accordance with ASTM D790 using a three point loading fixture and at a center loading rate of 0.05 inch/minute. Specimen thickness, which ranged between 0.084 and 0.087 inch, resulted in span-to-depth ratios of 23 to 24. Interlaminar shear strengths were measured following ASTM D2344 specifications at a constant span-to-depth ratio of 5. Elevated temperature mechanical tests on dry samples were performed in an environmental heating chamber after the chamber had equilibrated at the test temperature for 15 minutes prior to applying the load. A 15 minute equilibration period was not employed for testing of humidity exposed (wet) specimens. The load was applied to the wet specimens immediately after they had been placed in the loading fixture.

This type of test is referred to as a temperature spike test.[6] For the thermo-oxidative stability study, a series of precut flexural and interlaminar shear specimens were exposed at 600°F in air for various time intervals and then tested. The weight loss data were obtained from the precut flexural specimens. The fiber-resin compositions of the composites were determined by the H_2SO_4/H_2O_2 digestion method, and the densities of the composites were measured using the liquid displacement method.

3. RESULTS AND DISCUSSION

3.1 Processing Characteristics
Figure 1 shows the rheological behavior of the PN modified and unmodified PMR-15 molding powders which had been imidized at 400°F for one hour. It is evident that the PN modified PMR samples exhibited lower shear viscosities over the temperature range 460 to 577°F when compared to PMR-15. The minimum viscosity of the control (PMR-15) was found to be 2940 poise at 550°F while the minimum viscosities of the PN modified PMR samples varied from 1769 for 4 mole percent PN to 884 poise for 20 mole percent PN. This represents a decrease in the minimum shear viscosity ranging from 40 to 70 percent of the minimum viscosity for PMR-15 control. Also, the minimum viscosities of the PN modified PMR resins occurred at lower temperatures than for the PMR-15 control. At 579°F the viscosities of PN modified PMR and PMR-15 were equal.

The minimum viscosity as a function of mole percent N-phenylnadimide is plotted in Fig. 2. This figure shows that the addition of 4 mole percent PN to PMR-15 significantly reduces its viscosity. The figure also reveals that in the PN concentration range of 4 to 20 mole percent a linear relationship holds. A linear relationship between resin flow and mole percent PN was also found in the composite fabrication studies in which the composites were fabricated under identical condition. As indicated in Fig. 3, the resin flow of the composites increased from 3 to 20 percent, when the PN mole percent was increased from 0 to 12 percent. Thus, it is clear that PN is an effective flow modifying agent for PMR-15. This suggests that the melt-flow properties of PMR polyimides can be tailored to meet specific processing requirements by a simple variation of the PN concentration.

In addition to varying the PN concentrations as mentioned above, changing composite pressurization temperatures in the cure cycle can provide an additional method to control the flow properties of the PN modified PMR polyimide composites. As described in Section 2.5, the pressurization temperatures for the control composite and PMR-P1 through PMR-P5 were progressively increased from 450 to 580°F. This was done to allow the PN modified PMR polyimide resins to advance to completely imidized prepolymers. As a result of delaying the application of pressure to higher temperatures in the cure cycle, each of the PN modified PMR polyimide composites exhibited resin flow equal to that of the control system. This is indicated by the resin content of each composite molded (Table 3). This finding suggests that the PN modified PMR polyimides also offer versatility in selecting processing parameters, such as PN concentrations and pressurization temperatures. It is expected that cure pressure can also be varied, but this parameter was not investigated in this study.

The thermal characteristics of the molding powders which had been treated at 400°F for 1 hour are shown in Fig. 4. This figure shows that the thermal behavior of PMR-P1 (curve a) closely resembles that of the control PMR-15 (curve b), suggesting that both resins undergo the same cure characteristics. This is expected since PN and PMR-15 imide prepolymers contain a common crosslinker, viz. the nadic end-group and, thus, the crosslinking reactivity of PN is expected to be the same as that of PMR-15 imide prepolymer. Other PN modified PMR molding powders exhibited similar DSC thermal behavior as PMR-15. The PN modified resins exhibited on-set exotherms near 600°F. This is the temperature at which the crosslinking reaction of the nadic end groups occurs. The endotherm centered around 392°F is assigned to the melt-flow of the imide prepolymers.[7]

The endotherm near 482°F has not been observed in the work published previously.[7] This thermal transition may be ascribed to the melting of polyamic acids, which are precursors of imide prepolymers. The reasons for this suggestion are based on the following observations: (1) this endotherm disappeared from the DSC thermogram of the control sample after additional treatment at 500°F for 1/2 hour; (2) the disappearance of this endotherm occurs with the concomitant enlargement of the endothermic peak near 400°F which is assigned to imide prepolymer. This is evident from a comparison of curve c with curve b shown in Fig. 4. (3) The infrared spectrum of the control sample showed an absorption band around 1650 cm^{-1} assignable to carbonyl stretching vibration of an amide. Because the presence of polyamic acids depends on the extent of staging the molding powders, it appears that the molding powders prepared according to the procedure described in Section 2.4 afforded both polyamic acids and imide prepolymers. This seems to suggest that staging of the PMR monomer mixtures at 400°F for one hour may not result in complete imidization. However, when the mixtures are further treated at 500°F for 1/2 hour, complete conversion of residual polyamic acids to imide prepolymers seems to occur. Polyamic acids in the control PMR-15 polyimide system had not been observed.(8) Although the presence of polyamic acid intermediates in the PN modified PMR and the control (PMR-15) resins investigated in this study appears well founded, additional studies need to be carried out to verify their existence.

3.2 Composite Properties

3.2.1 "As Fabricated" Composites. Table 3 shows the physical properties of the PN modified and unmodified composites. By delaying the application of cure pressure to higher temperatures in the cure cycle, essentially void free PN modified PMR composites were produced as determined by ultrasonic C-scan and by the resin digestion technique. The PN modified and unmodified PMR-15 composites showed only small variations (57 to 59 mole percent) in fiber content. Thus, the fiber contents of the PN modified PMR can be controlled in the same fashion as PMR-15 even though these resins have much higher flow characteristics. The Tg of the modified PMR composites was about 18°F lower than the Tg of the PMR-15 (control) composite. After isothermal aging for 1000 hours, the Tg's of all composites increased by about 36°F. In addition, each of the aged composites invariably exhibited a second glass transition temperature which was found to be approximately 36°F higher than its first glass transition temperature. The presence of two distinctive glass transition temperatures suggests that at least two chemically distinctive polyimides may be formed in the course of the long term thermo-oxidative exposure.

Table 4 summarizes the mechanical properties of the "as fabricated" composites. Although the initial elevated temperature properties of the modified composites were slightly lower than those of the control, the room temperature properties of the modified PMR composites were equal to or superior to the room temperature properties of the control PMR-15 composites. Thus, it may be concluded that quality composites were fabricated from higher flow PN modified PMR prepreg with no sacrifice in the mechanical properties of the composites. The high value of RT flexural strength for the PMR-P5 composite is not readily explainable at this time.

3.2.2 Isothermally Aged Composites - Thermo-Oxidative Behavior. Table 5 summarizes the elevated temperature mechanical properties of the composites exposed and tested at 600°F in air, using precut flexural and interlaminar shear specimens. Table 5 shows that PN modified and PMR-15 composites exhibited excellent strength properties at 600°F after 1000 hours aging at 600°F in air. It is significant that as a result of 360 hours of exposure all of the composites showed a significant improvement in 600°F flexural and interlaminar shear strengths, but no change in flexural modulus. Furthermore, even after 1000 hours, the composites maintained this improvement. Thus, the PN modified PMR composites exhibited a level of thermo-oxidative stability at least equivalent to the unmodified PMR-15. This is significant in that a high flow PMR has been developed with no compromise in the thermo-oxidative stability of the polyimide system.

The 1500 hour isothermal aging data gave some rather unexpected results. Namely, PMR composites containing 4 and 9 mole percent PN (i.e., PMR-P1 and -P2, respectively) exhibited superior thermo-oxidative stability than did PMR-15 and the higher mole percent PN containing composites. This is further illustrated by their superior flexural strength, interlaminar shear strength, and flexural modulus retention (Figs. 5, 6, and 7) and their significantly lower weight loss (Fig. 8) relative to PMR-15 and PMR-P3, -P4, and -P5 composites. Furthermore, the greater thermo-oxidative resistance of the 4 and 9 mole percent PN containing PMR composites is shown in their fiber retention characteristics. As seen in Fig. 9, the PMR-15 control composite revealed many loose surface fibers after the isothermal exposure at 600°F in air for 1500 hours (Fig. 9a). In striking contrast, the PMR-P1 composite contained no loose surface fibers after the same thermo-oxidative exposure as that of the control (Fig. 9b). Thus, it appears that 4 and 9 mole percent PN in PMR-15 have improved the thermo-oxidative stability of PMR-15, and the useful lifetimes of PMR-15/Celion 6000 composites at 600°F in air have been extended by approximately 500 hours.[9]

3.2.3 Hydrothermally Exposed Composites. A series of precut flexural specimens of the PN modified and unmodified PMR-15 composites were exposed to 95 percent relative humidity at 180°F for 360 hours. Table 6 lists the glass transition temperatures and the elevated temperature flexural strengths of the hydrothermally exposed composites. The composites exhibited 1.3 percent equilibrium moisture absorption except for PMR-P1 sample, which absorbed 1.1 percent moisture. Moisture caused a considerable decrease in the glass transition temperature, ranging from 18°F to 54°F. A slower heat-up rate of 18°F/minute compared to 90°F/minute give higher Tg values, indicating that greater drying occurs at the lower heating rate. This is further substantiated by the Tg's of the samples after drying for 15 minutes at 473°F. The Tg's of the dried samples were nearly equal to the Tg's of the composites in the dry as-fabricated condition except for PMR-P1 and -P5. The data demonstrate the reversibility of moisutre effect on the Tg of PMR-15.[6] The moisture resistance of PMR-P1 and -P2 was comparable to that of the control PMR-15 (600°F flexural strengths of about 71.5 and 63 ksi for PMR-P1 and -P2, respectively, compared to about 65 ksi for PMR-15 sample).

These values are approximately 50 percent of the 600°F dry strengths. The other PN modified PMR composites exhibited lower flexural strengths than the above mentioned composite systems. These results are in agreement with the isothermal aging results which showed that PMR composites containing 4 and 9 mole percent PN demonstrated higher composite properties.

It was expected that lower flexural strengths would be obtained for the wet samples because the flexural strengths were measured at a temperature approaching the Tg of wet specimens using a temperature spike test.[6] This method minimizes the drying of the composites. Conventional testing of wet PMR-15 composites usually includes a 15 minute equilibration period at the test temperature. Serafini and Hanson[6] showed that this allows the composite to dry out and does not provide a realistic assessment of effect of moisture. Therefore, the flexural strengths obtained in this study reflect more closely the actual flexural strengths of the wet samples than the values obtained by the conventional testing method.

4. CONCLUSIONS

The processing and properties of N-phenylnadimide modified PMR polyimide/Celion 6000 graphite composites were investigated. On the basis of these results, it is concluded that:

1) higher flow N-phenylnadimide (PN) modified PMR matrix resins have been developed which show no compromise in the elevated temperature properties of Celion 6000/polyimide composites, compared to the control PMR-15.

2) Concentrations of 4 and 9 mole percent PN improve the thermooxidative behavior of the PMR-15 composite systems.

3) Melt-flow properties of PN modified PMR polyimides can be tailored to meet specific applications by either varying the concentrations of PN or the composite pressurization temperatures.

5. REFERENCES

1. Serafini, T. T.; Delvigs, P.; and Lightsey, G. R., J. Appl. Polym. Sci., Vol. 16, 905-915 (1972).
2. Delvigs, P.; Serafini, T. T.; and Lightsey, G. R., NASA TN D-6877 (1972).
3. Serafini, T. T. and Vannucci, R. D., Proceedings of the 30th Annual Reinforced Plastic Composites Conference Sponsored by SPI, Washington, D. C., February 4-7, 1975. Also NASA TMX-71616.
4. Vannucci, R. D., Proceedings of the 9th National Technical Convention sponsored by the SAMPE Atlanta Chapter, Atlanta, Georgia, October 4-6, 1977.
5. Burns, E. A.; Jones, R. J.; Vaughan, R. W.; and Kendrick, W. P., NASA CR-72633 (1970).
6. Serafini, T. T. and Hanson, M. P., ASTM STP in press.
7. Lauver, R. W., J. Poly. Sci., Vol. 17, 2529-2539 (1979).
8. Serafini, T. T., Personal Communication.
9. Delvigs, P.; Alston, W. B.; and Vannucci, R. D., Proceedings of the 24th National SAMPE Symposium, San Francisco, California, May 8-10, 1979. Also NASA TM-79062.

6. BIOGRAPHY

Dr. Ruth H. Pater received a B. S. degree in chemistry from Tamkang College, an M. S. degree in organic chemistry from Southeastern Massachusetts University, and a Ph. D. degree in organic chemistry from Brown University in 1977. Since 1980, she has been associated with the Polymer Matrix Composites Section of the Materials Division at NASA Lewis Research Center. As a research chemist, she is responsible for the synthesis and development of laminating resins and adhesives. Prior to joining NASA, she was employed at the Materials Sciences group of United Technologies Research Center as a research scientist. There, her research activities included epoxy and polyimide resin synthesis and characterization, composite fabrication and evaluation, composite surface analysis, and interface studies. Dr. Pater has published in areas of advanced composite technology encompassing the resin matrix, the interface, and surface analysis.

Table I Monomers used for modified PMR polyimides

STRUCTURE	NAME	ABBREVIATION
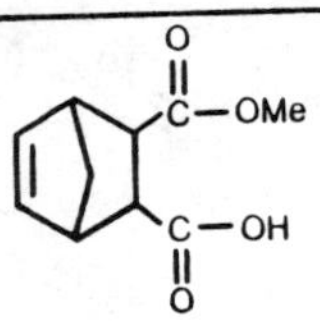	MONOMETHYL ESTER OF 5-NORBORNENE- 2, 3-DICARBOXYLIC ACID	NE
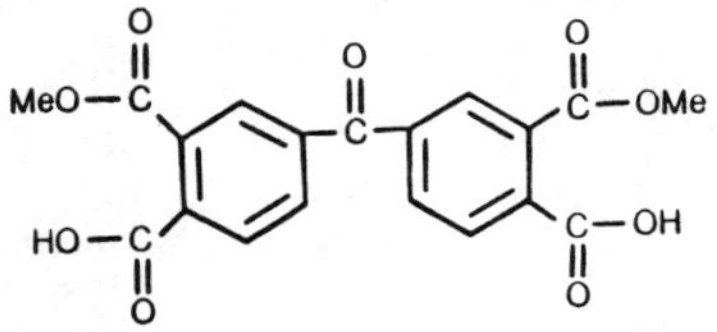	DIMETHYL ESTER OF 3, 3', 4, 4'-BENZOPHENONETETRA-CARBOXYLIC ACID	BTDE
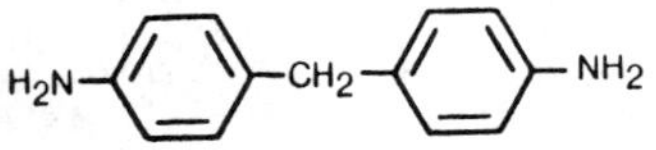	4, 4'-METHYLENEDIANILINE	MDA
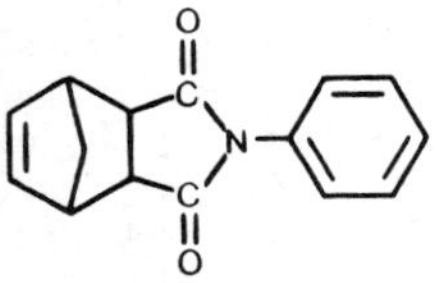	N-PHENYLNADIMIDE	PN

TABLE 2. COMPOSITIONS OF MODIFIED PMR POLYIMIDES

Resin System	PMR-15/PN Mole Percent	PMR-15/PN Weight Percent	Monomer (Mole)			
			NE	MDA	BTDE	PN
PMR-15 (Control)	100/0.0	100/0.0	2	3.087	2.087	0.0
PMR-P1	96/4	99.33/0.67	2	3.087	2.087	0.041
PMR-P2	91/9	98.47/1.53	2	3.087	2.087	0.099
PMR-P3	88/12	97.86/2.14	2	3.087	2.087	0.136
PMR-P4	85/15	97.20/2.80	2	3.087	2.087	0.177
PMR-P5	80/20	96.2 /3.80	2	3.087	2.087	0.251

TABLE 3. PHYSICAL PROPERTIES OF MODIFIED PMR POLYIMIDE/CELION 6000 COMPOSITES

Resin System	Density g/cc[a]	Glass Transition Temperature, °F[b] As Fabricated[c]	1000 hours aged[d] First	Second	Composition Volume Percent[a] Fiber	Resin	Void
PMR-15 (Control)	1.58	666	698	739	59.2	40.2	0.6
PMR-P1	1.56	644	680	707	58.1	41.3	0.6
PMR-P2	1.55	644	680	709	58.8	41.1	0.1
PMR-P3	1.55	644	685	716	57.3	42.2	0.5
PMR-P4	1.55	626	680	707	57.9	41.6	0.5
PMR-P5	1.55	653	680	711	57.1	42.5	0.4

[a] average of three determinations
[b] determined by thermomechanical analysis heating rate of 18°F/minute
[c] post cured for 16 hours at 600°F in air
[d] aged at 600°F in air

TABLE 4. MECHANICAL PROPERTIES OF MODIFIED PMR POLYIMIDE/ CELION 6000 COMPOSITES AS FABRICATED

Composite Property	Resin System PMR-15 (Control)	PMR-P1	PMR-P2	PMR-P3	PMR-P4	PMR-P5
Flexural Strength, ksi[a]						
Room Temperature	231.1	227.1	230.9	235.5	240.9	268.7
600°F	133.9	112.7	118.4	127.5	126.4	129.6
Flexural Modulus, psi x 10^6 [a]						
Room Temperature	16.17	15.66	17.19	15.64	16.62	17.66
600°F	14.58	13.12	13.21	14.26	14.65	13.61
Interlaminar Shear Strength, ksi						
Room Temperature[b]	14.19	17.06	14.04	15.89	16.76	16.09
600°F [c]	6.46	6.13	5.94	6.78	7.53	6.58
Tensile Strength, ksi[b] 10° off axis						
Room Temperature	63.50	64.16	59.36	62.20	64.19	64.22
Tensile Modulus, psi x 10^6 [d]						
Room Temperature[d]	9.14	8.76	8.99	8.64	7.83	10.12
Elongation, Percent[d]	1.07	1.18	1.21	1.45	1.53	1.17

[a] average of three determinations; normalized to 60 volume percent fiber
[b] average of five determinations
[c] average of three determinations
[d] average of four determinations

TABLE 5. MECHANICAL PROPERTIES OF MODIFIED PMR POLYIMIDE/ CELION 6000 COMPOSITES EXPOSED AND TESTED IN AIR AT 600°F

600°F Composite Property		Resin System					
		PMR-15 (Control)	PMR-P1	PMR-P2	PMR-P3	PMR-P4	PMR-P5
Flexural Strength, ksi[a]							
After exposure for	0 Hour	133.9	112.7	118.4	127.5	126.4	129.6
	360 Hours	143.6	143.7	140.9	153.7	161.4	148.0
	600 Hours	145.1	151.2	152.6	151.7	164.6	143.9
	1000 Hours	146.5	153.5	150.4	154.4	163.1	154.8
	1500 Hours	115.6	137.2	137.0	138.2	117.9	121.7
Flexural Modulus, psi x 10^6 [a]							
After exposure for	0 Hour	14.58	13.12	13.21	14.26	14.65	13.61
	360 Hours	15.54	15.96	14.79	15.16	15.2	14.77
	600 Hours	15.71	14.71	14.23	15.14	15.11	14.01
	1000 Hours	12.32	13.00	12.33	12.32	13.48	12.64
	1500 Hours	8.29	11.53	11.04	9.53	8.15	7.09
Interlaminar Shear Strength, ksi[b]							
After exposure for	0 Hour	6.46	6.13	5.94	6.78	7.53	6.58
	360 Hours	6.83	6.67	6.59	7.02	7.90	6.91
	600 Hours	6.72	6.64	6.76	7.10	7.81	7.24
	1000 Hours	6.54	6.59	6.64	7.02	7.98	7.33
	1500 Hours	4.40	5.58	6.19	6.13	5.48	5.55

[a] average of three determinations; normalized to 60 volume percent fiber
[b] average of three determinations

TABLE 6. HYDROTHERMAL EFFECTS ON PROPERTIES OF MODIFIED PMR POLYIMIDE/ CELION 6000 COMPOSITES EXPOSED FOR 360 HOURS AT 180°F AND 95 PERCENT RELATIVE HUMIDITY

Resin System	Moisture Absorbed Wt Percent [a]	Glass Transition Temp., °F				600°F Flexural Strength, ksi[f]
		Dry[b]	Wet A[c]	Wet B[d]	Wet C[e]	
PMR-15 (Control)	1.3	666	617	658	658	65.05
PMR-P1	1.1	644	599	626	613	71.46
PMR-P2	1.3	644	599	621	642	62.94
PMR-P3	1.3	644	621	640	622	56.94
PMR-P4	1.3	626	608	622	622	55.05
PMR-P5	1.3	653	617	630	617	54.40

[a] average of three determinations
[b] dry as fabricated composites tested at a heating rate of 18°F/minute
[c] tested at a heating rate of 90°F/minute
[d] tested at a heating rate of 18°F/minute
[e] after drying for 15 minutes at 473°F and then tested at 18°F/minute
[f] temperature spike tested, average of five determinations, and normalized to 60 volume percent fiber

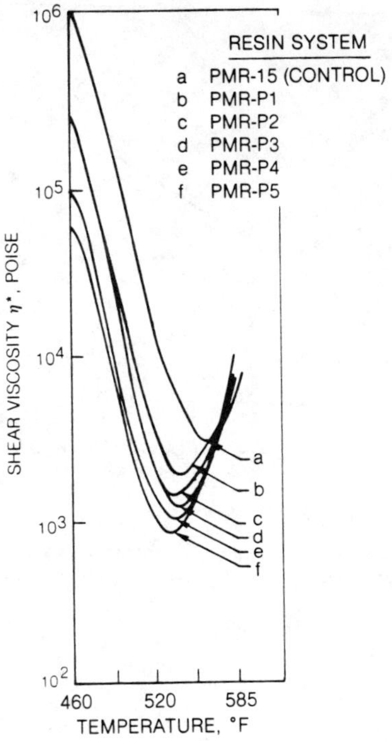

Figure 1 Shear viscosity as a function of temperature for modified PMR polyimide molding powders

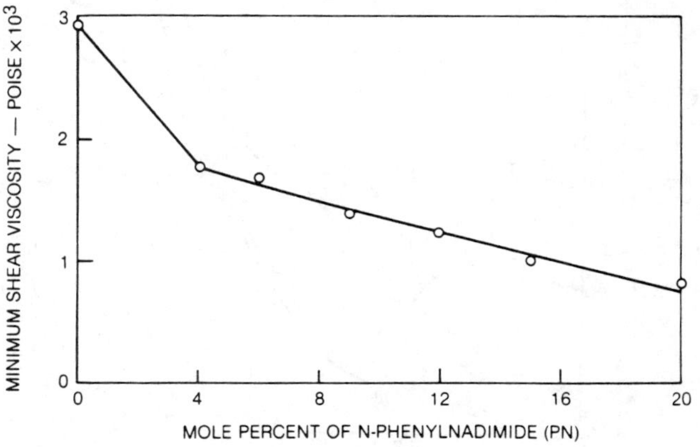

Figure 2 Minimum shear viscosity as a function of concentration of N-phenylnadimide for modified PMR polyimide molding powders

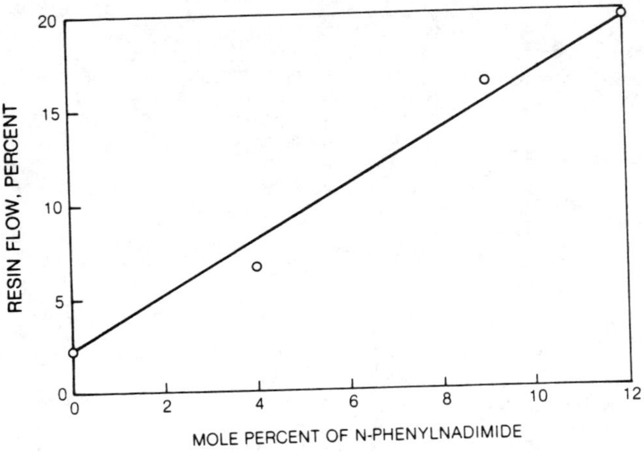

Figure 3 Resin flow as a function of concentration of N-phenylnadimide for modified PMR polyimide/Celion 6000 composites

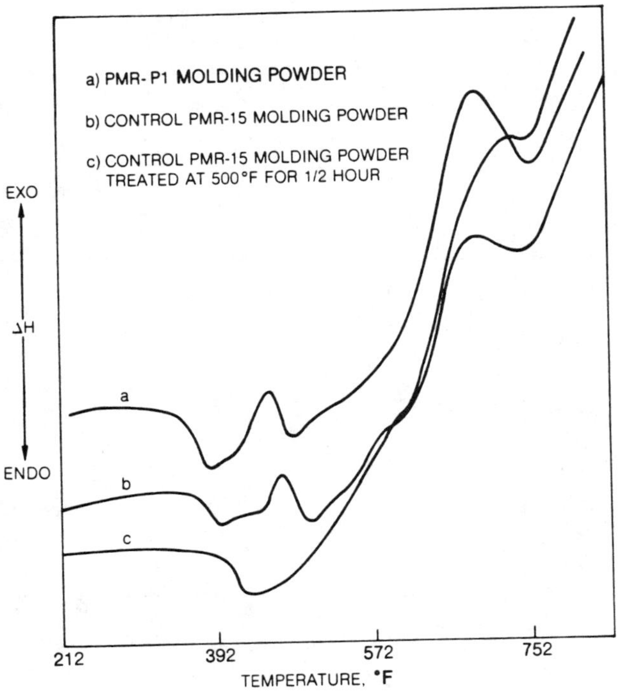

Figure 4 Differential calorimeter scans of PMR-P1 and PMR-15 molding powders

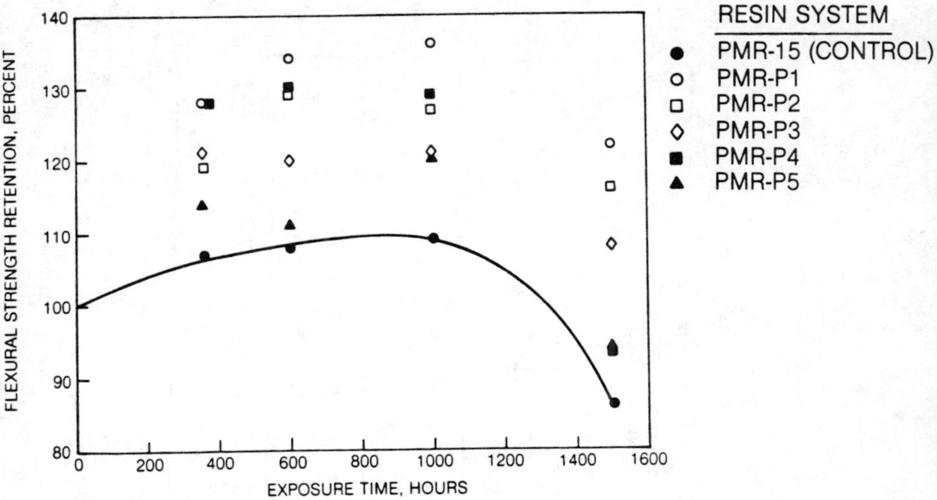

Figure 5 Flexural strength retention of modified PMR polyimide/Celion 6000 composites exposed and tested at 600° F in air

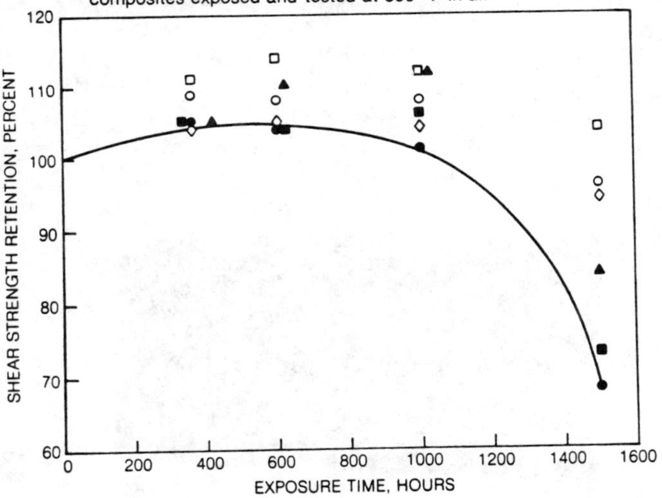

Figure 6 Interlaminar shear strength retention of modified PMR polyimide/Celion 6000 composites exposed and tested at 600° F in air

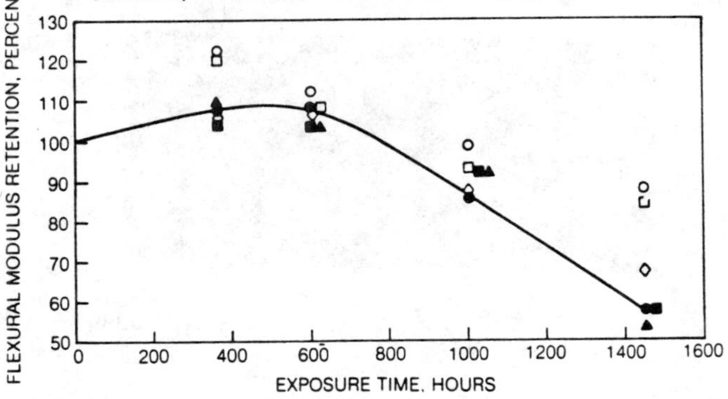

Figure 7 Flexural modulus retention of modified PMR polyimide/Celion 6000 composites exposed and tested at 600° F in air

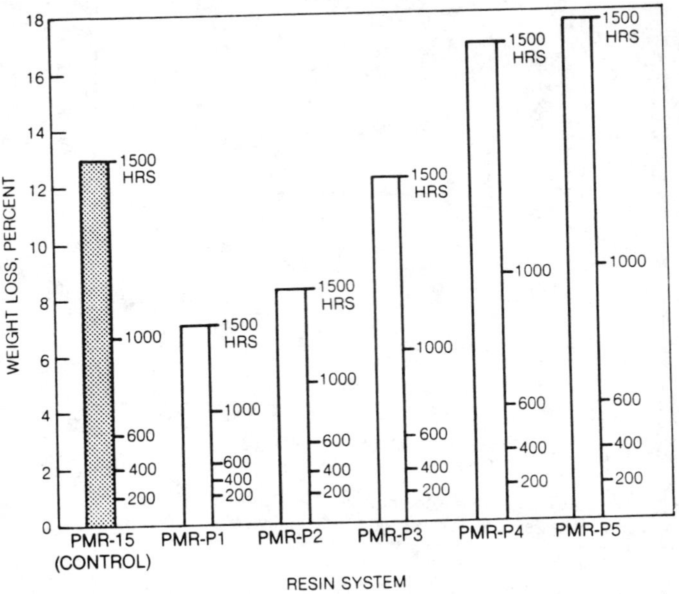

Figure 8 Weight loss characteristics of modified PMR polyimide/Celion 6000 composites exposed in air at 600° F up to 1500 hours

a) CONTROL PMR-15/CELION 6000 COMPOSITE AFTER 1500 HOURS AT 600° F IN AIR

b) PMR-P1/CELION 6000 COMPOSITE AFTER 1500 HOURS AT 600° F IN AIR

Figure 9 Photomicrographs of PMR-15 and PMR-P1/Celion 6000 composites exposed and tested in flexure at 600° F in air

13th National SAMPE Technical Conference
October 13-15, 1981

HPLC EVALUATION OF MY 720

John J. King, Richard N. Castonguay, James P. Zizzi
CIBA-GEIGY Corporation
Ardsley, New York 10502

ABSTRACT

A continuing study of ARALDITE® MY 720 (tetraglycidylated methylenedianiline) by liquid chromatography techniques, including HPLC, is underway at Ciba-Geigy Corporation. This paper reviews the initial liquid chromatography results in this program and presents a new gradient procedure for HPLC. The experimental error involved in the injection of a sample and statistical evaluation of batch to batch variations in the HPLC results are discussed. Correlations between the liquid chromatography data and viscosity and epoxy value are established. Japanese competitive versions of MY 720 are also examined. In addition, discussions concerning the character of MY 720 are presented.

1. INTRODUCTION

High Performance Liquid Chromatography(HPLC) is enjoying widespread use as an analytical tool in the evaluation of polymeric materials. In fact, HPLC is currently being applied as a Quality Assurance(Q/A) or Quality Control(Q/C) instrument for the evaluation of high performance epoxy/graphite prepregs, but before any instrument can be fully utilized for Q/C, what the resulting data means must be understood exactly. Most of today's high performance epoxy/graphite prepregs are based upon Ciba-Geigy's ARALDITE MY 720 (tetraglycidylated methylenedianiline, TGMDA) and HT 976 (diaminodiphenylsulphone). The aim of this paper is to describe the possible make up of MY 720 as well as to present the liquid chromatography work on MY 720 currently underway at Ciba-Geigy.

MY 720 is manufactured by the reaction of methylene dianiline with epichlorohydrin followed by an epoxidation step, Figure 1. This is, of course, an idealized reaction scheme. In any resin, the idealized monomer as well as dimers, oligomers, isomers and unreacted and partially reacted starting materials are usually found. Figure 2 gives some of the possible constituents of MY 720. Other components than those described are, of course, possible. Generally, MY 720 consists of 60% monomer and 35% dimers plus oligomers: the remaining constituents account for 5% of the resin.

Resins by their very definition are expected to contain other components than just the pure monomer. If pure monomer were the only constituent, then the material would be a pure chemical or monomer and not a resin. Most pure epoxy monomer would be very costly to produce and not necessarily beneficial. The presence of constituents other than monomer is what gives epoxy resins their unique processing and reaction characteristics.

Therefore, an understanding of what is in a resin and how it ultimately affects the cure and physical properties is required. Since pure monomer can not be easily manufactured, <u>material consistency</u> becomes of prime importance in order to insure that extreme variations in processing parameters and physical proproperties do not occur.

Consistency is of importance to any resin manufacturer. The ultimate goal of the program currently underway at Ciba-Geigy is to identify and quantify the various components of MY 720 and determine their effects upon resin and resin system physical properties.

The basic analysis at the heart of this program is the HPLC work. The HPLC has the ability to separate out the various components of a resin. The following will review the initial efforts in this area.

2. MATERIALS AND EXPERIMENTAL METHODS

Gel Permeation Chromatographs (GPC) were obtained for twenty-three different batches of MY 720. These were chosen from both Toms River New Jersey and Monthey Switzerland production and are representative. The samples were prepared as 2% solutions in THF with 50 microliters injected onto four microstyrgel columns; 1000,500 and two 100 angstrom columns. THF running at 1.5 ml/min was the mobile phase. A differential refractometer detector was used and the recorder chart speed was set at 16 inches/hour. The output was compared to polystyrene standards for molecular weight determinations.

One batch of MY 720 was run by a HPLC thirty times in order to determine experimental error. Twenty-six different batches were then run in order to determine the batch to batch variations. A 1% solution of MY 720 in Acetonitrile was prepared and 1 microliter was injected onto a Water's Micro Bondapak C_{18} column. The acetonitrile/water gradient conditions utilized are given in Table 1. A variable U.V. detector was used and set at a wavelength of 254 and an AUFS of 1.0.

Viscosity measurements of all of the batches evaluated were done at 50°C by the Cannon-Fenske test method as described by ASTM-D-445.

3. RESULTS AND DISCUSSION

3.1 Gel Permeation Chromatography

A typical GPC curve is given in Figure 3. Four distinct peaks are present and labeled 1, 2, 3 and 4 from left to right. Peak #1 represents the higher molecular weight material and peak #4 is the lower molecular weight constituents. This data was compared to a polystyrene calibration curve and a molecular weight was determined. This was done for twenty-three samples which resulted in the statistical analysis shown in Table 2.

The average molecular weight of MY 720, by this method, is approximately 595 with a reasonably small covariance, indicating a fairly consistent material.

3.2 High Performance Liquid Chromatography

A typical HPLC curve is shown in Figure 4. The eleven peaks of interest are numbered from left to right. Polar materials are the first to come off and non-polar materials are last. Figure #4 also combines the peaks into three families, groups 1, 2 and 3. Group 1 contains peaks #1, 2 and 3, representing polar constituents, probably unreacted or partially reacted starting materials. Peaks #4, 5, 6 and 7 make up group #2 and are associated with TGMDA monomer. Group #3 is the most non-polar of the materials, probably higher molecular weight oligomers and is made up of peaks #8, 9, 10 and 11.

A statistical analysis of twenty-six different batches was done in order to determine how much variations in the peaks or groups of peaks occurred from batch to batch. First, an analysis of experimental error was determined by injecting 30 identical samples and determining the resulting variations. The results of this analysis are given in Table 3. The experimental error involved for each peak, as described by the covariance, runs from 11% to 2% whereas the error involved in the three groups is 3% to 4%. This decrease is likely due to the ease which the groups can be defined versus the difficulty in assessing overlap when defining single peaks. Since the reproducibility of the group data is better than that of a peak by peak analysis and since the groupings have real significance in terms of the properties, much of the discussion in the following sections is carried out in the terms of the groups of peaks:

Group 1 - polar materials, likely unreacted or partially reacted starting materials;

Group 2 - less polar materials, likely monomer and associated materials;

Group 3 - least polar of the materials, likely oligomers.

An analysis of 26 different batches of MY 720 was carried out. Table 4 gives the mean, standard deviation and covariance of the peak areas for the individual peaks as well as for the three groups. The individual peaks have a covariance from 5%, for the monomer peak #7, to 83% for peak #10. The covariance is large for some of these peaks because they tended to overlap in some cases making individual peak areas difficult to calculate. As such, the covariance for the three groupings are more accurate.

Table 5 gives the total average peak area observed for the 26 different batches and, also, the standard deviation and covariance. The covariance is 3.4% which is within the experimental error calculated in Table 3. Therefore, it can be assumed that the 254 wavelength chosen does consistently pickup the same material and that the changes in one peak area can be accounted for by the changes in other peaks.

An examination of Table 4 shows that the monomer group (#2) has the least covariance, amounting to 5%. Considering that the experimental error accounts for 3% (Table 3) then probably only 2% of this covariance accounts for batch to batch variations. Group #3, representing the high molecular weight materials, has a covariance of 16% and Group #1, the impurities, unknowns and partially reacted polar substances has a covariance of 28%. Group #1, #2, and #3 are 2%, 84% and 14% of the total area, respectively. This data is summarized in Table 6.

It is interesting to note that the polar materials in Group #1 are kept at relatively low levels indicative that the material is fairly "clean". The majority of the area is in Group #2 which represents the monomer and related constituents. This is kept closely uniform from batch to batch, an indicating of good monomer batch consistency. The higher molecular weight materials, Group #3, constitute a good percentage of the peak area. Groups #2 and #3 control the resin's viscosity, as will be seen in the next section.

3.3 HPLC Data and Physical Properties

Ciba-Geigy supplies MY 720 in a viscosity range of 7,000 to 20,000 cps (@ 50°C) depending upon the customers preference. There are also other properties that vary from batch to batch including the epoxy equivalent weight and hydrolyzable chlorine content. Table 7 gives the Ciba-Geigy specification values for these properties. The following evaluation attempts to determine if these three properties can be correlated to the LC data obtained.

Table 8 shows the results of a correlation analysis of the HPLC data to viscosity, epoxy equivalent weight and hydrolyzable chlorine content. No apparent correlation was obtained for the hydrolyzable chlorine content but possible correlations did develop for the viscosity and epoxy equivalent weight data.

Peak #10 gave a possible negative correlation to the epoxy value. This peak is in the higher molecular weight group and indicates that as the epoxy equivalent weight goes up the peak decreases in area. The monomer peak (#7) has a positive correlation. These two numbers are consistent, one would expect that as the monomer content decreases and the higher molecular weight material increases then the epoxy value would decrease.

The viscosity data showed stronger correlation. There was a negative relationship between the monomer peak (#7) and viscosity. This demonstrated that as the viscosity increased, the monomer would decrease. Peaks #9 and #10 had a positive relationship. These are the higher molecular weight constituents. As these components increase the viscosity also increases. These observations are in line with what would be expected, that is, as the viscosity increases more higher molecular weight material is present.

This viscosity relationship can be seen more clearly in the grouped data which shows a negative correlation for Group #2, the monomer, and a positive correlation for Group #3, the higher molecular weight material. Group #1 has no correlation with viscosity for Ciba-Geigy material where it is a minor constituent. Based upon the fact that the variation in the monomer content from batch to batch is small (Table 6), then only a small amount of monomer converted to dimers or oligomers controls the viscosity. Group #1, the more polar materials, play only a small part in the viscosity when their concentration is low.

In order to support this viscosity relationship, the four GPC peaks shown in Figure 3 were subjected to the same correlation analysis. Twenty-three batches of MY 720 were run by GPC and the peak areas correlated to the viscosity. The results are given in Table 9.

For the GPC the peaks which can be most readily identified with certainty are #4, the monomer peak, and #1, the oligomer peak. As expected peak #1, the oligomer, shows a strong positive correlation. Peak #4, the monomer, shows a stronger correlation than the HPLC results but it is negative as would be expected. This is in general agreement with the HPLC results. A clearer understanding will emerge as the various peaks are identified.

3.4 TGMDA from Other Manufacturers

Several Japanese companies have been offering offsets to MY 720. Samples of these materials have been run by HPLC and found to be different from Ciba-Geigy production. Table 10 gives the HPLC data for the Japanese and Ciba-Geigy versions of MY 720. The Ciba-Geigy HPLC data, previously developed, gives the area percentages of the three groups and the standard deviations as a percentage of the total HPLC peak areas. These are compared to the area percentages of the various Japanese materials. The materials are different. All of the Japanese materials have a higher percentage of material in Group #1. Figure 5 gives a typical HPLC of the Japanese material. All of the Japanese materials have a higher area in Group #1 due mainly to peak #3.

The monomer family, Group #2 is virtually identical in the four batches examined as compared to the Ciba-Geigy standard. The monomer content of all manufactured types of MY 720 contain approximately the same amount of monomer. If anything, the Ciba-Geigy material may have a higher monomer content. For Group #3, the higher molecular weight constitutes, Epichlon 430 and Ciba-Geigy MY 720 are very similar. The other three versions have much lower amounts of Group #3 which probably relate to a lower viscosity.

Based upon the HPLC evaluation, it can be assumed that the Glyamine 120, the HY 434 and the ELM 434 are all manufactured by a single process. Epichlon 430 is apparently made differently. It can also be assumed that the manufacturing parameters used to produce the Japanese material are quite different from Ciba-Geigy's, thus yielding different materials especially with regard to minor constituents.

4. CONCLUSIONS

An initial review of L.C. data for MY 720 has been completed as well as an initial evaluation of competitive material. The following preliminary conclusions may be drawn:

1. The batch to batch consistency of MY 720 has been found to be better than 95% with respect to monomer content.
2. A reasonable correlation has been established between the epoxy value and viscosity of MY 720 and specific features of the HPLC curves.
3. Small increases in the higher molecular weight constituents appear to cause the viscosity to increase.
4. Competitive Japanese materials exhibit different HPLC curves and may be manufactured by a different process than Ciba-Geigy's MY 720. These Japanese materials have more polar materials and less higher molecular weight components; both of these factors contribute to reduced viscosities. The monomer content of the Ciba-Geigy and Japanese materials are similar.

This paper has been a preliminary study of Ciba-Geigy's program to totally evaluate MY 720. The next step in this program is to identify and quantify each of the major components found in MY 720. These then would be related to the resin's physical properties. It is felt at Ciba-Geigy, that only by first knowing exactly what the HPLC data is saying can this tool be fully utilized as a characterization and Quality Control technique. Ciba-Geigy is committed to do this and will be reporting future results in this area.

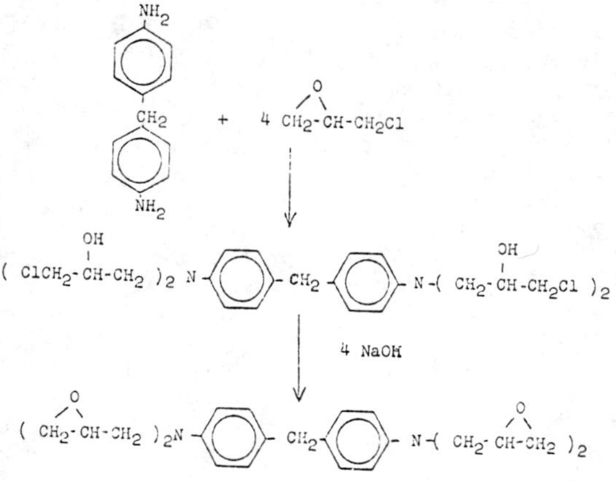

Figure 1: Idealized Reaction Sequence for the Manufacture of MY 720

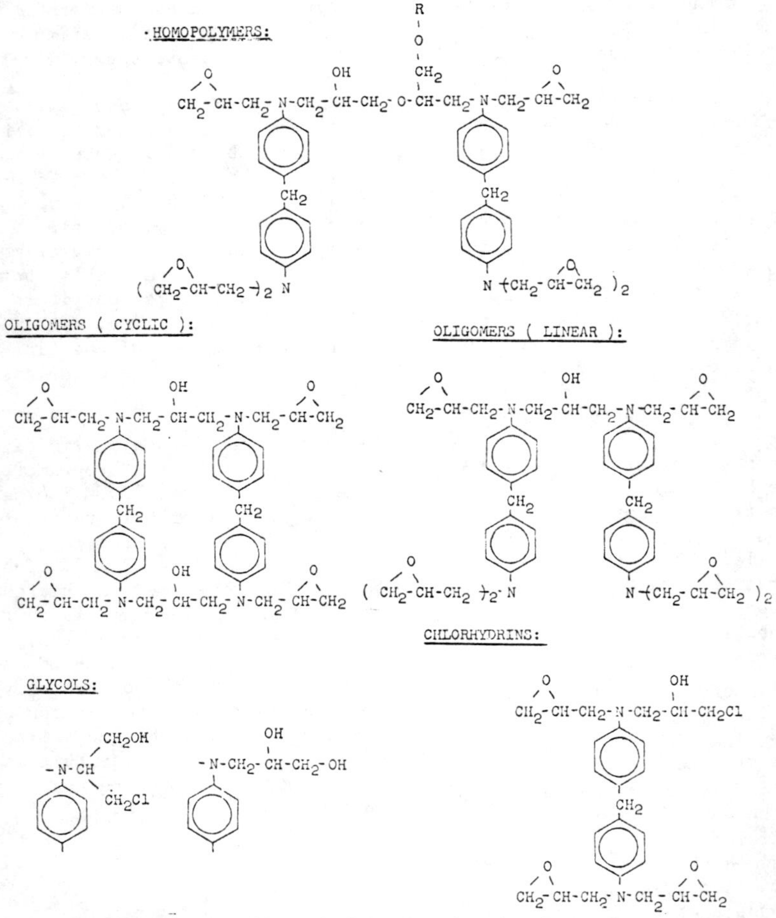

Figure 2: Possible By-Products Formed During the Manufacture of MY 720

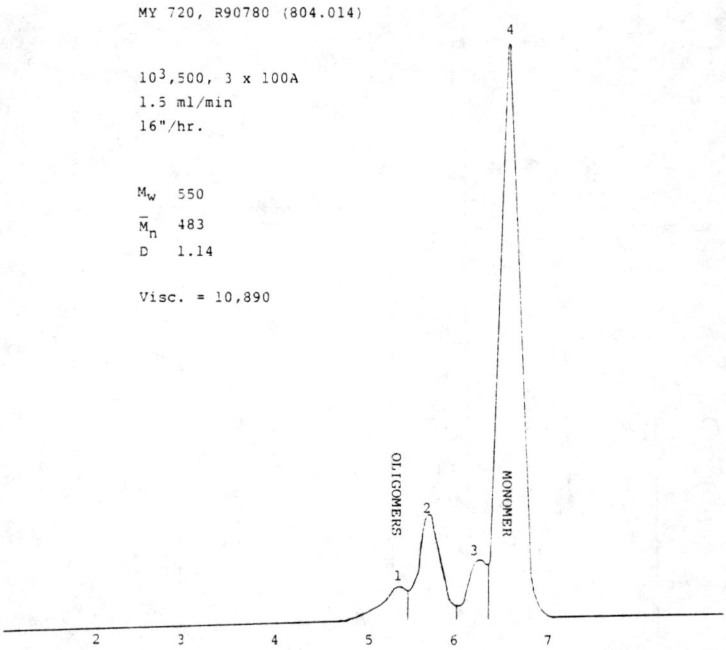

Fig. 3: Typical GPC Curve for MY 720

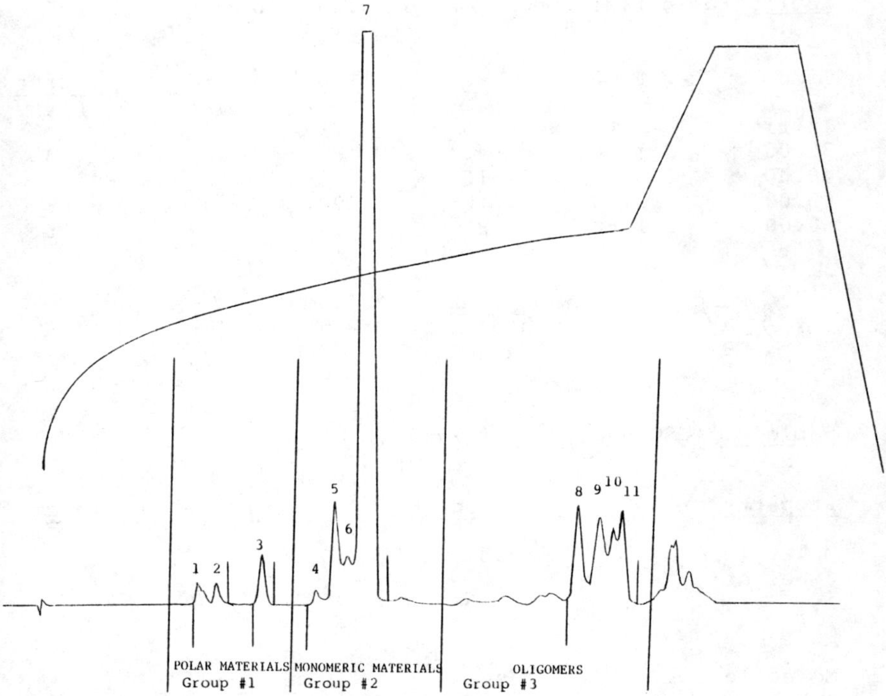

Fig. 4: Typical HPLC Curve of MY 720

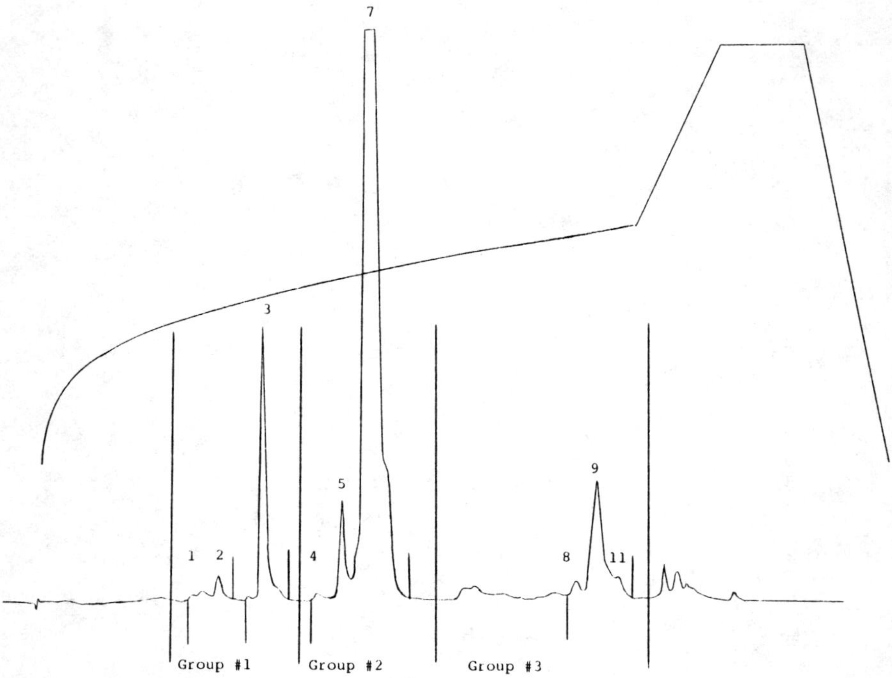

Fig. 5: Typical HPLC of Japanese TGMDA (Glyamine 120)

Table 1: Initial Conditions/Gradient Table

Time	Flow	%A	%B	%C	Curve
Initial	1.60	80	20	0	*
35.00	1.60	40	60	0	04
40.00	3.00	10	90	0	06
45.00	3.00	10	90	0	06
50.00	1.60	80	20	0	06

Pump	Solvent	Comp
A	Water	0
B	CH_3CN	0
C		0

Table 2: GPC Analysis of MY 720

Property	Average	Standard Deviation	Covariance
Molecular Weight	595	57.5	9%
Molecular Number Weight	496	19.5	4%

Table 3: Statistical Analysis of Experimental Error for the HPLC Work

Peak	Average Area	Standard Deviation	Covariance
1	3,500	280	8%
2	3,400	360	10%
3	11,500	290	2%
4	10,600	400	4%
5	30,600	1000	3%
6	14,800	560	4%
7	458,500	14360	3%
8	22,800	760	3%
9	67,900	1690	2%
10	12,000	1300	11%
11	16,400	660	4%
Group			
1	18,300	600	3%
2	515,600	17670	3%
3	120,000	5340	4%

Table 4: Statistical Analysis of Batch to Batch Variations of MY 720 by HPLC

Peak	Average Area	Standard Deviations	Covariance
1	3910	1880	48%
2	4430	1320	30%
3	7440	2370	32%
4	5700	2410	42%
5	30,690	4830	16%
6	12,410	4900	39%
7	576,650	29,790	5%
8	17,930	6130	34%
9	56,390	30,930	55%
10	9260	7720	83%
11	17,030	8270	49%
Group			
1	15,780	4380	28%
2	625,470	31,490	5%
3	100,620	16,390	16%

Table 5: Statistical Analysis of Batch to Batch Variations of Total Peak Areas of MY 720 by HPLC

Average Total Peak Area	-	741,870
Standard Deviation	-	29,350
Covariance	-	3.4%

Table 6: Summary of HPLC Statistical Group Data of MY 720

Group	#1	#2	#3
Experimental Error Covariance	3%	3%	5%
Batch to Batch Covariance	28%	5%	16%
(Batch to Batch) -(experimental error covariances)	25%	2%	12%
% of Total Peak Area	2%	84%	14%

Table 7: Specifications for MY 720

Epoxy Value E/100g	-	0.75 - 0.85
Epoxy Equivalent Weight	-	117 - 133
Viscosity (cps @ 50°C)	-	20,000 maximum
Hydrolyzable Content %	-	.50 maximum

Table 8: Correlation Analysis of MY 720 Viscosity, Epoxy Value and Hydrolyzable Chlorine to HPLC Data

Peak	Viscosity	Epoxy Value	Hydrolyzable Chlorine
1	-.18	-.00	-.20
2	.17	-.02	.26
3	-.30	.36	.20
4	.24	-.12	-.31
5	-.25	.41	.11
6	-.06	.09	.25
7	-.74	.75	.25
8	-.26	.21	-.35
9	.71	-.45	.11
10	.66	-.64	-.16
11	-.02	.51	.01
Group			
1	-.19	-.05	.09
2	-.73	-.03	.18
3	.79	-.15	.17

Table 9: Correlation Analysis of MY 720 Viscosity to GPC Data

Peak	Viscosity Correlation Coefficient
1	.89
2	.62
3	.23
4	-.69

Table 10: HPLC Analysis of Ciba-Geigy Versus Japanese Versions of MY 720

HPLC Group	1	2	3
CIBA-GEIGY MY 720 Grand Average 26 Batches			
Average: Group Peak Areas	2%	84%	14%
Group Standard Deviation	0.5%	3.5%	1.9%
Japanese Materials			
Epiclon 430 Group Peak Areas	4%	81%	15%
Glyamine 120 Group Peak Areas	9%	83%	8%
YH-434 Group Peak Areas	9%	83%	8%
ELM-434 Group Peak Areas	9%	83%	8%

13th National SAMPE Technical Conference
October 13-15, 1981

SERVICE SENSITIVITY OF POLYSULFIDE SEALANTS

Robert J. Stout
General Dynamics Corporation
Fort Worth, TX

Abstract

Tests were conducted to determine the effects of in-service environments on MIL-S-83430 polysulfide sealants. The conditions of interest were sealant chalking and sealant sponging. Sealant chalking is a condition wherein the surface of the sealant in contact with fuel undergoes a chemical and physical change that results in sealant deterioration. This condition is generally progressive through the sealant thickness as a function of time and other parameters.

The original Air Force evaluation of two such sealants to MIL-S-83430 requirements had indicated that chalking would not occur. However, light surface chalking began to occur in military aircraft after short times in service. The parameters creating the chalking were determined, and a chalk-thickness-formation-rate-versus-time was established.

The tendency of MIL-S-83430 sealants to sponge as a result of fuel and temperature exposure was also examined. Sponging is a condition wherein the sealants change from essentially a "solid rubber" state to a "sponge rubber" state if exposed to combinations of elevated temperature, heat-up rate, fuel, water, etc. The effects of various combinations of conditions were determined, and a realistic temperature limit below which sealant sponging does not occur was established.

Keywords: Sealants, Polysulfide, Chalking, Sponging, JP-4, Temperature, Mercaptans

1. INTRODUCTION

MIL-S-83430 establishes the requirements for a high-temperature-resistant filleting and faying surface sealant for integral fuel tanks. The specification requires the sealant to be capable of withstanding long-term exposure from -65° to +250°F, short-term exposure of 4-hour intervals or less to +325°F, and intermittent exposures of 1 hour or less to +360°F. The original Air Force qualification of the sealants (two vendors are qualified) showed that the sealants did not chalk in the jet reference fluid used as a test standard in the specification and that the sealants would meet the temperature parameters. Two incidents encountered in service indicate there are factors that cause the MIL-S-83430 sealants to chalk and also reduce their upper temperature service limit. These two incidents and the corresponding investigations are the subjects of this paper.

2. SEALANT CHALKING

Sealant chalking is a condition wherein the surface of the sealant in contact with fuel undergoes a chemical and physical change that results in sealant deterioration. This condition is generally progressive through the sealant thickness as a function of time and other parameters. The MIL-S-83430 chalking test consists of immersing 1/8 x 1/8 x 5-inch specimens of 14-day-cured sealant in a closed container of jet reference fluid and a container of a layered mixture of 3 percent aqueous sodium chloride and jet reference fluid (all at 140°F) for nine days. Fluids are changed six times. After exposure, the specimens are allowed to dry and are examined for chalking.

As noted earlier, the original Air Force evaluation of sealants to MIL-S-83430 requirements indicated that chalking would not occur with the two products evaluated. However, light surface chalking began to occur in military aircraft after short times in service. This chalking caused concern because of previous experience in which severe chalking was encountered in polysulfide sealants that had been qualified by other companies to lower temperature specifications, such as MIL-S-8802.

The investigation of the chalking began with the collection and examination of samples of sealant and fuel from several aircraft and the initiation of MIL-spec sealant chalk tests by the suppliers and General Dynamics. Examinations of the sealant samples from the airplanes showed that both sealants qualified to the MIL-spec had chalked. The chalk condition was very thin, approximately 0.0002-inch thick, so that it did not pose an immediate service problem. The questions to be answered were how bad would it get over the years and why were the sealants chalking. The test program to provide answers involved fluids (JP-4, JP-8 and JRF), temperature (RT, 140°, 160°F, 200°F), and metallic ion contamination of the JP-4 and JRF. A comparison of the x-ray energy spectrographic analyses of chalked and unchalked surfaces is shown in Figure 1. As the chalk thickness increases, the sulphur content of the surface decreases and calcium carbonate becomes the predominant surface material. Minor quantities of titanium and manganese are also found. A chemical analysis of the chalk showed a similar situation and also traces of iron, nickel, copper, chromium, zinc, and silicon. The trace elements may be due to the exposure of fasteners and other hardware to the JP-4 fuel in the aircraft tanks. The other elements are consistent with the major constituents of polysulfide sealants and thus indicate that the chalk is a concentration of sealant fillers, primarily the calcium carbonate. This material is left as the sealant polymers are gradually extracted by solvent or chemical action of the fuel and contaminates (trace elements) in the fuel.

The erratic results obtained by the investigators involved in the MIL-spec chalk tests indicated that trace elements could play a role in accelerating chalk formation and led to a study of the differences in JRF fluids from various sources. The variations that can be experienced with JRF that may have become contaminated by improper storage are illustrated in Table I. The relative effects of different metals on chalk rate are shown in Table II. One important factor not shown is the role of mercaptans. Mercaptans and the presence of cadmium are known to cause problems with sealants. In fact, the MIL-spec on JP-4 controls the maximum mercaptan content. The importance of mercaptans and metal ions is shown in Table III. Chalking of the MIL-S-83430 sealants will not occur within the nine-day specification test period unless both metal ions and mercaptans are present. Because JRF is normally free of metallic ions, the original tests on the sealants did not reveal their tendency to chalk. The real world of aircraft usage will always involve mercaptans and metal ions; thus, a more realistic and faster chalk-developing test fluid is definitely needed. Although JP-4 would be a desirable base for such a fluid, it chalks too slowly for a reasonably short-time test and the mercaptans in this fluid can vary up to 0.001%. The SAE committee G-9 on Aerospace Sealing has undertaken the task of developing a better standard fluid. One approach under consideration is shown in Table IV. This fluid was used along with unmodified JP-4 and JP-8, in predicting chalk formation rate for the sealants. The rate of chalk thickness versus time and temperature for the fluids is shown in Figures 2, 3 and 4. Microscopic thickness measurements of the chalk formation are illustrated in Figure 5. When the chalk formation rates for several temperatures versus time are known, the chalk thickness for any combination of time and temperature can be predicted by assuming that the chalk rate follows the Arrhenius temperature dependence. In this manner, the expected chalk thickness after 15 years of service was calculated based on the expected service temperature spectrum. In Table V, the calculated thicknesses and the measured thicknesses are shown for two aircraft after 1 and 3 years, and the measured thicknesses are extrapolated to 15 years. The data indicates that the chalk rate model is reasonable. Since the actual sealant thicknesses used in sealing are many times thicker than the predicted chalk thickness, no in-service problem is expected. There is also no need for protective coatings or other measures to protect the surface of the sealant.

3. SEALANT SPONGING

The discovery of sponged sealant in fuel test tanks fabricated by General Dynamics during Air Force testing prompted interest and concern about the parameters that created such a condition. The sponging occurred only in fillet-type sealing or in other areas where the sealant was unrestrained. Faying surface sealant was not affected.

When sponging occurred, the sealants changed from essentially a "solid rubber" state to a "sponge rubber" state. The type of sponging encountered range from a small cell structure to coarse large cells, as shown in Figure 6. Again, both products qualified to MIL-S-83430 were affected, with sponge cell size being the only discernible difference. The thermal history of the sealants prior to the discovery of the condition is shown in Table VI. The temperatures involved were sufficiently below MIL-specification limits that discovery of sponged sealant was unexpected. Various suspect variables that could possibly cause the problem were considered. The search for possible causes included a review of the thermocouple charts, the response of the tanks to the heat cycle was checked for possible evidence of overheating, sealant quality control was reviewed, the sealing process used on the particular test tanks in question was examined, and all related variables were checked. All of these reviews showed that no anomalies had occurred. Additional contradictory information was the fact that no sponging had occurred in the test tanks exposed to a similar heat history at Fort Worth.

A test matrix was then established to determine the role of other variables such as cure time, heat-up rate, sealant thickness, cyclic exposure (in and out of fuel and temperature), sealant mixing methods, etc., in causing the sealant to sponge. The most important variables were found to be temperature and exposure to fuel (or water) just prior to exposure to the higher temperatures. There was some speculation that the texture of the sponge was associated with the method used to mix the sealant base and catalyst. It was also noted that the tendency for sealant to sponge appeared to be batch dependent. The conditions evaluated and the resulting observations are presented in Table VII. Observations from the program are summarized in Table VIII.

It was concluded from the program that fuel-saturated (or water-saturated) sealant has a much lower service temperature limit than implied by the MIL-spec. This discrepancy occurs because test conditions in the MIL-spec do not address the immediate high-temperature exposure of fuel-saturated sealant in fillet-type applications. Actual usage of sealant in supersonic aircraft can involve fuel tanks in which the fuel may have just been depleted before the aircraft accelerates to supersonic speeds. Under these conditions, the temperature of the sealant can increase very rapidly. It appears that a conservative limit for MIL-S-83430 sealants for fillet applications would be in the 240° to 260°F range. At 275°F and higher, sealant sponging may result. As noted previously, this limit would apply only to fillet sealing or sealing in other unrestrained areas. Faying surface sealing is not affected.

4. CONCLUSIONS

Chalking - The study of sealant chalking leads to the following conclusions:

1. Sealant chalk is a concentration of sealant fillers, primarily calcium carbonate.

2. Chalking is caused by fuel extraction of the elastomeric component of the sealant, leaving a layer of inorganic fillers.

3. The phenomenon is accelerated by the presence of mercaptans and

metallic ions in the fuel.

4. Based on the exposure conditions and the service life of a fighter-type airplane, the chalk thickness after 15 years for the sealants investigated is predicted to be between 0.002 and 0.006 inch.

5. This extent of chalking will not significantly affect the sealant's properties or sealability in typical applications.

Sponging - The study of sealant sponging permits the following conclusions:

1. Initiating factors that create sponging in a sealant are saturation of the sealant with fuel or water and subsequent exposure to high temperatures.

2. The tendency of a sealant to sponge is batch dependent. Sufficient tests were not conducted to identify the reason for batch dependency.

3. Sealant cure time greater than seven days prior to exposure to temperature does not reduce sponging.

4. Sponging was less pronounced at 275°F than at 300°F.

5. No sponging occurred at 240°F regardless of prior exposures to fuel or water.

6. Sponging occurred only in fillet sealing applications. Faying surface sealing was unaffected.

5. ACKNOWLEDGEMENT

Acknowledgement is made to H. J. Weltman, M. T. Carroll, Toni Morris, and D. R. Clawson of the Fort Worth Division of General Dynamics for source information used in preparing this paper.

6. BIOGRAPHY

R. J. Stout has been actively involved in the field of non-metallic materials, principally adhesives and adhesively bonded structure, for a number of years. He is currently supervisor of the Materials Engineering Group at the Fort Worth Division of the General Dynamics Corp. where his responsibility encompasses materials and processes activities in metals and non-metallics. He is a graduate of Oklahoma State University with a B. S. degree in Mechanical Engineering.

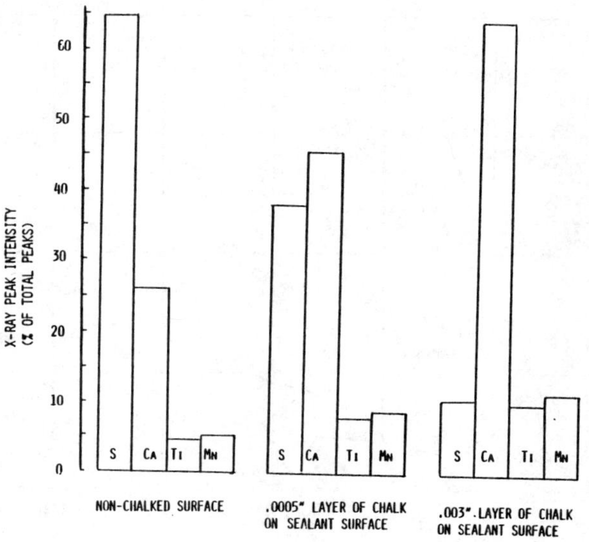

FIGURE 1 TYPICAL X-RAY ENERGY SPECTROGRAPHIC ANALYSIS OF SEALANT SURFACES

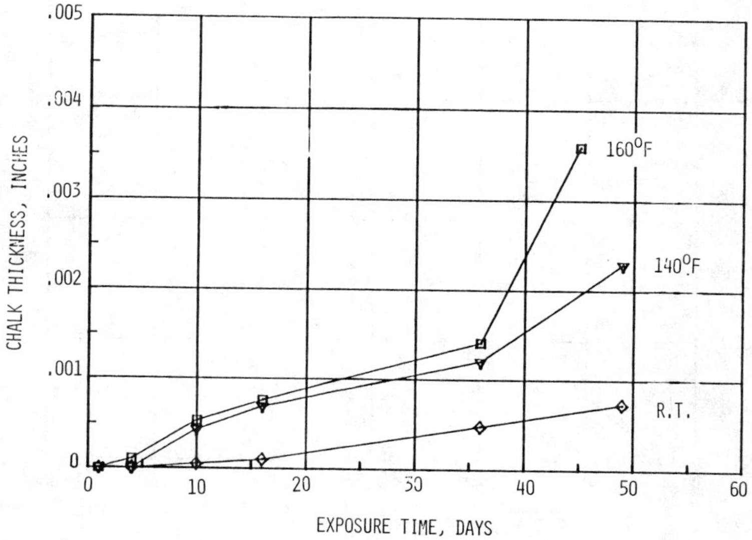

FIGURE 2 CHALK RATE OF SEALANT IN ACCELERATED CHALK FLUID

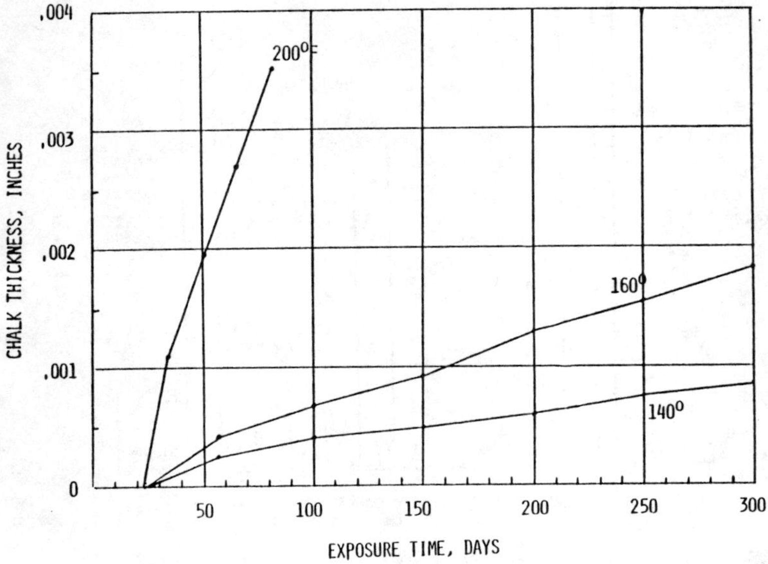

FIGURE 3 CHALK RATE OF SEALANT IN JP-4

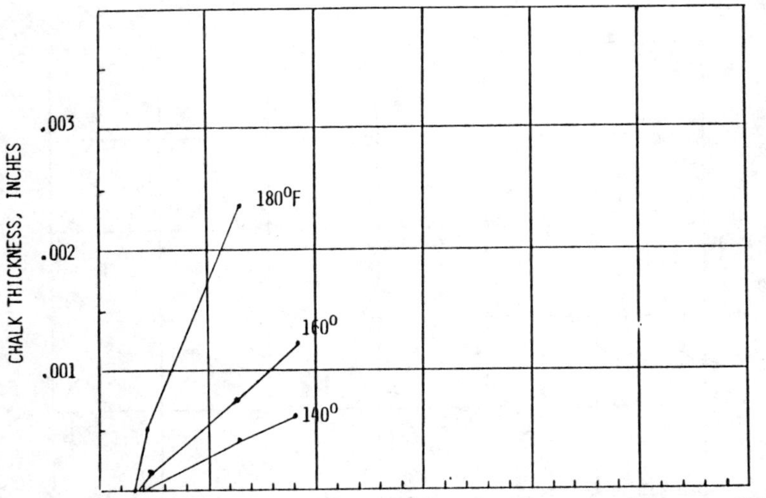

FIGURE 4 CHALK RATE OF SEALANT IN JP-8

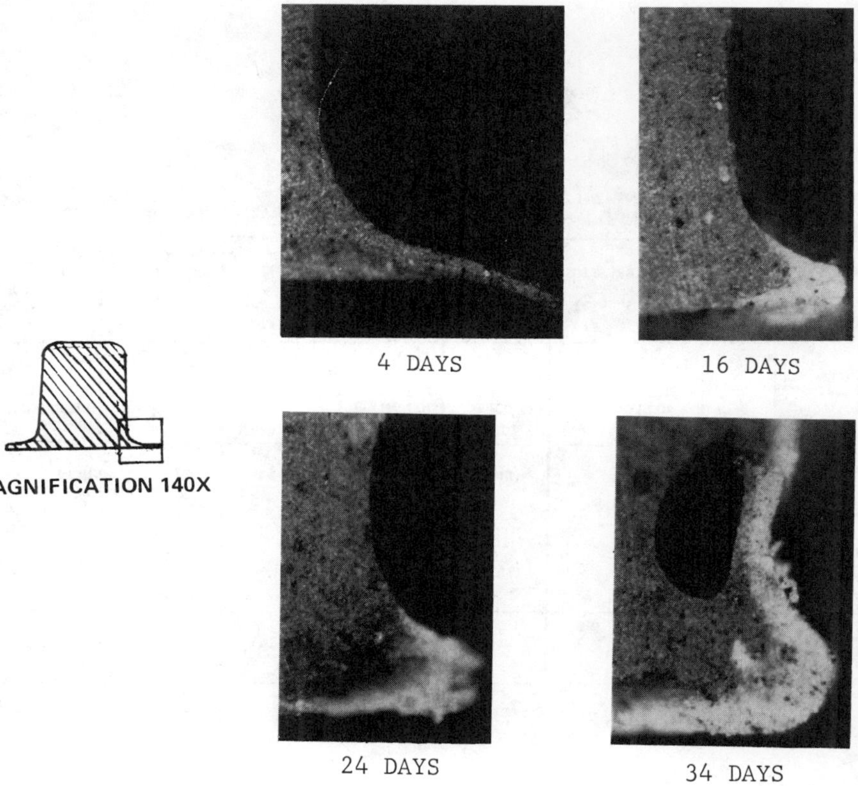

FIGURE 5 CHALK PROGRESSION VS. TIME 200°F JP-4

FIGURE 6 TEXTURE OF SPONGED SEALANT

TABLE I COMPARISON OF BATCHES OF JET REFERENCE FLUID

BATCH NUMBER	CHALK CHARACTERISTICS (TIME TO CHALK ONE 1/8" X 1/8" X 5" STRIP AT 160°F)	MERCAPTAN, WT. %	GAS CHROMATOGRAPH	WATER PPM	CONDUCTIVITY, C.U. (1 C.U. = 10^{-12} OHM^{-1} M^{-1})	ANALYSIS OF NON-VOLATILE RESIDUE
W-786	1 DAY - SLIGHT 3 DAYS - MEDIUM 5 DAYS - HEAVY	.0045	TYPICAL CHROMATOGRAM	150	243	MOD: FE LIGHT: CU, NI, MN, PB
W-880	1 DAY (HEAVY)	.0046	SAME	137	485	HEAVY: FE, CU LIGHT: TI, CR, MN, NI, PB
W-932	1 DAY (HEAVY)	.0042	SAME	146	251	MOD: FE, CU LIGHT: MN, NI, PB
W-942	4 DAYS - NONE	.0048	SAME	141	17	LIGHT: FE, CU, NI, TI

TABLE II EFFECT OF METALS ON CHALK FORMATION

FLUID	METAL	CHALK FORMATION(DAYS REQUIRED)
Std. lab mix JRF	none (control)	trace (13), moderate (15)
Std. lab mix JRF	copper foil	heavy (1)
Lab mix JRF w/o mercaptan	copper foil	none (15)
Std. lab mix JRF	lead foil	moderate (3), heavy (6)
Std. lab mix JRF	cadmium sheet	moderate (5), heavy (10)
Std. lab mix JRF	iron sheet	moderate (8), heavy (15)
Std. lab mix JRF	zinc sheet	trace-mod.(12), heavy (14)
Std. lab mix JRF	nickel foil	trace (10), moderate (15)
Std. lab mix JRF	aluminum foil	trace (13), moderate (15)

TABLE III EFFECT OF METALS AND MERCAPTANS ON CHALKING CHARACTERISTICS OF JP-4 AT 160°F

FLUID	METAL	CHALK FORMATION (DAYS REQUIRED)
Std. JP-4 (0.0005% mercaptan)	none (control)	none (50)
JP-4 (0.001% mercaptan)	none	trace (27)
JP-4 (0.001% mercaptan)	cadmium sheet	trace (28)
JP-4 (0.1% mercaptan)	none	trace (11)
Std. JP-4 (0.0005% mercaptan)	copper foil	moderate (3), heavy (4)
	lead foil	none (50)
	cadmium sheet	none (50)
	iron sheet	none (50)
	zinc sheet	none (50)
	nickel foil	none (50)
Std. JP-4 (0.0005% mercaptan)	aluminum sheet	none (50)

TABLE IV

PROPOSED ACCELERATED CHALKING FLUID

BASE FLUID	JP-4*
ADDITIONAL MERCAPTANS	ADJUSTED TO 0.001%
ADDITIONAL METALS	COPPER TO 1½ PPM CADMIUM 1½ PPM
FLUID CHANGES	TWICE WEEKLY

*JP-4 AS RECEIVED CONTAINED 0.0005% MERCAPTAN, 0.9 PPM COPPER, NO MEASURABLE CADMIUM

TABLE V CHALK THICKNESS MEASUREMENT AND THICKNESS EXTROPLATION

AIRPLANE	SAMPLE DESCRIPTION	CHALK THICKNESS	YEARS SERVICE	CHALK THICKNESS EXTRAPOLATED TO 15 YEARS
A	THICK FILLET	.0002"	1	.003"
B	THICK FILLET OVER CAD-PLATED WING-ATTACH BOLT	.0008"	3	.004"
B	THIN SEALANT LOOSELY ADHERING TO DOME NUTS			
	- FUEL SIDE	.0004"	3	.002"
	- NUT SIDE	.0025"		.012"
B	THICK FILLET	.0005"	3	.003"

TABLE VI

FUEL TANK THERMAL EXPOSURE CYCLE

PRECONDITIONING*

-65°F IN JP-4, 16 HRS
160°F IN JP-4, 80 HRS
200°F, TANK EMPTY, 13 HRS
273°F, TANK EMPTY, 2 HRS

TEST CYCLE*

-65°F IN JP-4, 16 HRS
160°F IN JP-4, 40 HRS
200°F, TANK EMPTY, 6.5 HRS
273°F, TANK EMPTY, 1 HR

* CYCLIC PRESSURE OF 0 TO 5 PSI WAS APPLIED

SEALANT SPONGING WAS FOUND DURING ROUTINE INSPECTION OF THE TANK AT THE COMPLETION OF THE PRECONDITIONING CYCLE AND 1 TEST CYCLE.

TABLE VII
VARIABLES EVALUATION AND DEGREE OF SPONGING ENCOUNTERED

CONDITIONING		PANEL[1] B	PANEL[1] C	PANEL[1] D	PANEL[1] E
CURE TIME AT RT (DAYS) PRIOR TO FUEL OR WATER SOAK		14	14	21	21
JP-4 SOAK @160 ^{o}F (DAYS) PRIOR TO THERMAL EXPOSURE		7	7	7	7
THERMAL EXPOSURE (2 HRS)[2] AT ^{o}F		240	275	275	300
FUEL SOAK					
VENDOR A	BATCH 1	NO CHANGE TO SLIGHT DIFF.	SPONGED	SPONGED	GROSS SPONGING
	BATCH 2	NO CHANGE FROM CONTROL	FINE TEXTURED SPONGING	SPONGED	FINE TEXTURED SPONGING
	BATCH 3	NO CHANGE FROM CONTROL	NO CHANGE FROM CONTROL	NO CHANGE FROM CONTROL	NO CHANGE FROM CONTROL
VENDOR B	BATCH 1	NO CHANGE TO SLIGHT DIFF.	SPONGED	SPONGED	FINE TEXTURED SPONGE
	BATCH 2	NO CHANGE FROM CONTROL	SPONGED	NO CHANGE	SPONGED
	BATCH 3	NO CHANGE FROM CONTROL	NO CHANGE FROM CONTROL	NO CHANGE FROM CONTROL	SPONGED
WATER SOAK					
VENDOR A	BATCH 1	NO CHANGE TO SLIGHT DIFF.	SPONGED	GROSS SPONGING	GROSS SPONGING
	BATCH 2	NO CHANGE	NO CHANGE	SPONGED	GROSS SPONGING
VENDOR B	BATCH 1	NO CHANGE	SPONGED	GROSS SPONGING	GROSS SPONGING
	BATCH 2	NO CHANGE TO SLIGHT DIFF.	SPONGED	SPONGED	GROSS SPONGING
	BATCH 3	NO CHANGE TO SLIGHT DIFF.	SPONGED	GROSS SPONGING	GROSS SPONGING

(1) PANEL A HAD NO EXPOSURE AND WAS USED AS A CONTROL FOR VISUAL COMPARISONS. SEALANT THICKNESS ON ALL PANELS WAS APPROXIMATELY 0.18 INCH. OBSERVATIONS NOTED ABOVE ARE AS COMPARED TO THE CONTROL PANEL.

(2) PANELS WERE PLACED IN THERMAL EXPOSURE IMMEDIATELY AFTER REMOVAL FROM FUEL OR WATER SOAK.

TABLE VII (CONT'D)
VARIABLES EVALUATION AND DEGREE OF SPONGING ENCOUNTERED

HEATING RATE AND THICKNESS EFFECTS

SEALANT THICKNESS INCH	TEMPERATURE °F	HEATING RATE, °F/SEC.					
		VENDOR A			VENDOR B		
		0.05	0.25	1.0	0.05	0.25	1.0
0.1	RT → 275	ALL SPECIMENS SPONGED					
0.2	"	"					
0.1	RT → 300	ALL SPECIMENS SPONGED					
0.8	"	"					

HUMIDITY EFFECTS(1) 25% AND 75% RH

SEALANT THICKNESS, INCH
0.1, 0.2, 0.8

TEMPERATURE, °F RT → 275
RT → 300

HEATING RATE, °F/SEC
0.5, 0.25, 1.0

→ NO SPONGING

(1) SPECIMENS EXPOSED TO ONLY HUMIDITY NOTED (NO FUEL OR WATER) UNTIL WEIGHT STABLIZED BEFORE EXPOSURE TO TEMPERATURE

TABLE VIII

OBSERVATIONS FROM SPONGING TESTS

1. TEXTURE OF SPONGING APPEARS TO BE RELATED TO THE TEXTURE OF AIR BUBBLES EXISTING IN THE CURED SEALANT PRIOR TO FUEL AND TEMPERATURE EXPOSURE.

2. VARIATIONS IN MIXING METHODS (DASHER, ROTARY, HAND MIXING) CHANGE ONLY THE TEXTURE OF THE SPONGE, NOT THE AMOUNT.

3. EXTENSION OF CURE TIME GREATER THAN 7 DAYS PRIOR TO EXPOSURE TO FUEL AND TEMPERATURE DOES NOT RETARD SPONGING.

4. THE SPONGING THAT RESULTS AFTER WATER SOAK IS GENERALLY A MUCH COARSER TEXTURE THAN SPONGING RESULTING FROM FUEL SOAK.

5. SEALANT THICKNESSES BELOW APPROXIMATELY 0.06 INCH DO NOT SPONGE. THE DEGREE OF SPONGING IN THICKNESSES ABOVE 0.06 INCH IS NOT THICKNESS DEPENDENT.

6. HEAT-UP RATES FROM 0.05°F PER SECOND TO 1.0°F PER SECOND DO NOT OFFSET THE DEGREE OF SPONGING.

7. NO SPONGING OCCURRED IN ANY TEST SAMPLE EXPOSED TO 240°F.

8. SPONGING WAS LESS PRONOUNCED AT 275°F THAN AT 300°F. SPONGING WAS INCONSISTENT AT BOTH TEMPERATURES, INDICATING THAT IT MAY BE BATCH DEPENDENT.

9. A "DRY-OUT" CYCLE AFTER FUEL SATURATION AND PRIOR TO HIGH TEMPERATURE EXPOSURE DECREASES SPONGING TENDENCIES.

13th National SAMPE Technical Conference
October 13-15, 1981

POLYMER DESIGN IN RELATION TO ADHESION OF RADIATION-CURABLE COATINGS TO METAL SUBSTRATES

Vincent D. McGinniss
BATTELLE
Columbus Laboratories
505 King Avenue
Columbus, Ohio 43201

Abstract

Radiation curing coating systems are complex formulations consisting of unsaturated monomers, multifunctional unsaturated oligomers, photoinitiators and reactive or nonreactive polymeric components. These components are usually 100% reactive liquid systems that can be applied to a substrate by conventional coating techniques and cured (liquid to solid conversion or formation of three dimensional network structures via free radical initiation-propagation mechanisms) through efficient electrical or light energy interactive processes. This paper will consider certain chemical and structural designs of polymeric systems used in radiation curable coatings technology and the results of polymer design with respect to glass transition temperature (Tg) in promoting adhesive bonding of a radiation curable coating composition to a metal substrate.

1. INTRODUCTION

The development of radiation technology to cure (photopolymerization or interaction with high energy electrons) acrylic/-methacrylic unsaturated monomers, oligomers and polymers into tight network structures introduces several adhesion problems that can be related to or associated with interfacial polymer-substrate failure mechanisms.[1] In the development of final properties for a cured film, we are concerned with total gelation of components into network structures and there are many factors or variables that influence rate of formation and architecture of the network. With the onset of gelation on a non-porous (metal) substrate there is rapid shrinkage and little stress relaxation, all of which affect polymer growth association with the substrate to form adhesive bonding.

It now has been found that radiation curable coating compositions cured by light energy or high energy electron processing equipment, and without the use of additional heat energy, obtain excellent adhesion to metal surfaces providing that the radiation curable coating composition contains certain high Tg (glass transition temperature) polymeric additive materials. These polymeric additives must have glass transition temperatures (Tg) above the ambient application and reaction zone curing temperature in order to affect adhesive bonding capability in radiation curing coatings system.[2]

2. EXPERIMENTAL

The coating compositions used in this study consisted of 10-40% of a high Tg polymeric additive; 60-90% unsaturated acrylic monomer-multifunctional acrylate oligomer mixture; 0-10% of a photoinitiator when the coating system was cured through the use of ultraviolet light energy processing equipment. All adhesion promoting polymeric additives were prepared by conventional polymer synthesis techniques.[3]

Coatings were applied to flat sheet metal substrates and cured with ultraviolet light. After application, the coated metal was placed onto a moving conveyer line and exposed to U.V. radiation from an Ashdee Unit source (2 lamps of 2,000 watts each) placed 5 to 7 inches from the coating. Exposure time was 9 seconds. (Line speeds of 50 to 100 ft. per minute.) The coating compositions were also applied onto a metal sheet substrate and subjected to electron beam from an Energy Science Electrocurtain (150 KV terminal voltage and 3.5×10^{-6} torr vacuum). Each metal sheet was mounted on a moving line of speeds varying from 10 to 40 feet per minute. The dosage was 10 to 3 M rads at 10 M amps current.

Adhesion tests vary considerably depending on the type of coating being tested. By and large, the cellophane tape test is quite common, wherein a piece of adhesive cellophane tape is pressed on an area of the coating which had been scratched, scored or otherwise cut with a sharp edge.

In the present examples the adhesion properties are tested in a procedure believed to be quite severe. Here, the coating in the form of a film was cut into a grid of 1.0 cm x 1.0 cm area (100 mm^2) onto which area a cellophane tape was pressed and pulled quickly. If no film was lifted off or pulled, the rating of 100% adhesion is given; if any part of the film was lifted, then the number of filled squares remaining will determine the extent of adhesion. Thus if 10 squares are pulled off, the adhesion is rated at 90%; if 25 squares are pulled off, then adhesion is 75%.

3. RESULTS AND DISCUSSION

An ethylenically unsaturated binder coating was prepared from the following components: 23 weight parts pentaerythritol triacrylate (PETA), 19 weight parts phenyl cellosolve acrylate (PCA); 1 weight part photosensitizer (2-chlorothioxanthone) and 3 weight parts methyl diethanol amine. (Photosensitize is optional when used with high energy electron processing equipment). To this binder composition was added approximately 16 weight parts of a polymeric additive as described in Tables 1 and 2. This binder system was then further pigmented with 38 parts of rutile TiO_2.

In Table 1 a wide variety of polyester polymer systems were examined for their ability to promote adhesive binding of a radiation curable coating formulation to a metal substrate. In all cases the high Tg values 72 (a); 65 (b); 71 (e); 83 (b); 67 (g) all exhibited 100% adhesion while those polyester polymeric additives having low Tg values or Tg values at or near room temperature 1 (e); 27 (d) did not exhibit adhesive binding capabilities. Similar results were observed for a series of acrylic thermoplastic polymers (Table 2). In all cases where the Tg values were above 20 C excellent adhesion was achieved while these polymeric additives having Tg values below 20 C all exhibited poor adhesion capabilites.

The ability of the Tg value of a polymer additive to control adhesion of a radiation curable coating system is thought to be due to the fact that the high Tg polymer is compatible and soluble within the resinous binder prior to curing but becomes incompatible during the curing process. This incompatibility of the polymeric additive forces it to partition between the metal and cured coating network thus promoting physical adhesion to the metal surface due to the solid high Tg nature of the polymer reverting back to a solid state after curing. In contrast, low Tg polymers which are liquid or mobile at the ambient curing temperatures are encapsulated within the cross-linked binder matrix or are fluid at the metal - cured coating interface and do not provide mechanical adhesion to the metal substrate.

4. REFERENCES

1. V. D. McGinniss and A. Kah, Polm. Eng. and Sci., 17 (7), 498 (1977).
2. V. D. McGinniss, V. W. Ting and A. F. Kah, U.S. Patent No. 4,163,809 (1979).
3. UV Curing Science and Technology, S. Peter Pappas editor, Technology Marketing Corp., Stamford, Connecticut, 1978.

TABLE 1

	Polyester Polymer Additives						
Monomers	a	b	c	d	e	f	g
Isophthalic Anhydride	3 moles						
Hexa-Hydro-Phthalic Anhydride		3 moles	4 moles	6.39 moles	3.6 moles	4 moles	1 mole
Adipic Acid	0.6 moles	0.6 moles	0.8 moles				
Hydrogenated Bisphenol-A	4 moles	4 moles			4 moles	5 moles	2 moles
Cyclohexane Dimethanol			5 moles				
Neo-Pentyl Glycol	0.6 moles	0.6 moles	0.8 moles	5.1 moles	0.6 moles		
Acid Number (final)	24	15	7.5	85	22.4	18	9.6
% OH (final)	0.2	1.6	2.2	0.2	1.0	1.3	4.2
M wt. (calculated)	3727	1654	1401	1220	2075	1852	757
T_gC° low value	58	53	−5	23	64	76	57
high value	72	65	1	27	71	83	67
Adhesion	100%	100%	0%	50%	100%	100%	100%

TABLE 2

SATURATED ACRYLIC RESIN ADDITIVES						
MONOMERS						
Methyl methacrylate	100%					
Styrene		100%				
Isobutyl methacrylate			100%			
Butyl Acrylate				100%		
Hydroxyethyl Methacrylate					100%	
2-Ethylhexyl Acrylate						100%
T_g	105° C.	110° C.	50° C.	−55° C.	55° C.	−112°
Adhesion	100%	100%	100%	0%	100%	0%
Solvent resistance	poor	poor	poor		poor	
Mar resistance	poor	poor	poor		poor	

13th National SAMPE Technical Conference
October 13-15, 1981

DEVELOPMENTS IN REACTIVE PROCESSING OF NEW POLYMERS BASED ON BOTANICAL OILS

J. A. Manson, L. H. Sperling, Shahid Qureshi, and A. M. Fernandez
Materials Research Center, Lehigh University, Bethlehem, PA 18015

Abstract

In the processing of interpenetrating polymer networks (IPNs), a crosslinked polymer I (e.g., an elastomer) is first swollen with a monomer II (e.g., one yielding a plastic) mixed with a crosslinking agent, and the monomer mixture is then polymerized. To eliminate one of the two steps required, a reactive processing technique has been developed to yield simultaneous interpenetrating networks (SINs) in which the precursors of both polymers I and II are mixed together and the two networks formed more nearly simultaneously. This technique has been successfully applied to the preparation of a wide range of SINs including several based on elastomers derived from botanical oils such as castor, linseed, and vernonia, and glassy polymers such as polystyrene. In this paper, the technique is described, along with the kinetics and the effects of process variations on phase inversion, morphology, and properties of the final products. It is shown that products ranging from toughened plastics to reinforced elastomers can be readily obtained.

Keywords: reactive processing; simultaneous interpenetrating polymer networks; renewable resources

1. INTRODUCTION

Interpenetrating polymer networks (IPNs) are combinations of two polymers in which each network is synthesized and crosslinked in the presence of the other.[1,2] They thus constitute a means for achieving a type of polymer blend structure in which the phase continuity and dimensions can be controlled by varying the crosslink density and the polymerization conditions. In a sequential IPN, monomer I is polymerized (together with a crosslinking agent), and monomer II is added (together with a crosslinking agent) and polymerized in the presence of polymer I. While a voluminous technical and patent literature has developed,[1] and while a wide range of interesting properties have been reported for IPNs, simplification of the 2-stage process mentioned would be advantageous in lowering costs.

One such approach has been to attempt the more nearly concurrent synthesis of the two networks to form simultaneous interpenetrating networks (SINs). In this case, two monomer systems (or prepolymers) (along with respective crosslinkers and initiators) are mixed in the same vessel, prereacted (though not usually precisely at the same time), and poured into a mold in which reaction is completed. By varying the extent of prereaction, the morphology and hence the mechanical properties can be varied widely; in any case the polymerization process is considerably simplified by use of this reaction molding technique.

Indeed some of the SIN systems should be adaptable to reactive injection molding (RIM) techniques, which are already finding many applications in the low-cost and rapid processing of a wide variety of reactive systems, especially urethanes.[3,4] Before investigating means of effecting especially rapid polymerizations of SINs, we have studied several examples of SINs. Recent emphasis has been given to systems based on elastomers derived from polyfunctional botanical oils, as examples of raw materials based on renewable resources, combined with a glassy polymer such as polystyrene.[5-12]

In this paper we review past and present research on this topic at Lehigh, and indicate the effects of polymerization conditions on morphology and properties.

1.1 SIN Techniques

One of the first SINs was prepared by polymerizing an epoxy resin with phthalic anhydride as hardener in the presence of a polymerizing n-butyl acrylate/diethylene glycol dimethacrylate system.[3,4] An experimental design was selected to investigate the optimum combination of epoxy prereaction time, concentration of initiator for the acrylic monomer, and concentration of crosslinker. It was found that optimum strengths and toughness were achieved not when the two polymerizations were proceeding at about the same rate, but when one gel point was reached before the other. This finding was attributed to the fact that essentially simultaneous gelation led to phase sizes that were too small to permit the toughening of the brittle epoxy by the elastomeric polyacrylate.

Other SINs and related materials have also been based on combinations such as the following, in which at least one of the components is prepolymerized prior to SIN formation polyurethane/polymethacrylate[14], polyurethane/polymethacrylate[15], polyurethane/polyester, polyester-urethane/polystyrene[16], and polyurethane/epoxy.[17] Products have ranged from toughened plastics to reinforced elastomers. Not only has increased toughness been reported, but also improved thermal stability [polyurethane/polymethacrylate[18]], tensile strength, and adhesive bond strength [polyurethane/epoxy[17]].

2. MATERIALS, SYNTHESES, AND TESTS

The following oils have been used successfully: castor (CO), vernonia (VO), and epoxidized linseed (ELO), crambe (ECr), lesquerella (ELq), and lunaria (ELu). Castor and lesquerella oils contain free OH groups, and vernonia epoxy groups; all the oils contain C=C bonds. Castor oil was prepolymerized through its hydroxy groups[5,7] in situ using three different crosslinking agents: (a) 2,4-tolyene diisocyanate (TDI) to yield a polyurethane network (COPUN); (b) sebacic acid to yield a polyester network (COPEN); and (c) a mixture of sebacic acid and TDI to yield a mixed ester/urethane network (COPEUN). By changing the stoichiometry it was possible to prepare prepolymers having isocyanate, carboxy, or hydroxy end groups, which could then be crosslinked through further reaction with themselves or a crosslinking agent. A typical reaction sequence is shown for a polyester/polystyrene system in Fig. 1. Thus the desired prepolymer (based on equivalent amounts of castor oil and sebacic acid) was first dissolved in a mixture of styrene containing benzoyl peroxide and 55% divinyl benzene in the approx. wt. ratio 80:0.4:0.8. After stirring under nitrogen for 5-10 min, samples were poured into a mold before gelation, heated for 24 hr at 80°C to react the styrene, and then for several more hours at 180-200°C to complete polymerization of the elastomer.

As shown in Fig. 2, the epoxidized oils were prereacted with sebacic acid (SA) or dimer acid (DA) until close to the gel point, mixed with the same styrene mixture used above, and polymerized first at 80°C and then at 140°C.

3. DISCUSSION

It was important to determine the homopolymer gel times, for pouring must take place before gelation. As shown in Table 1, gel times varied widely, depending on the oil and the crosslinking agent.

Table 1. Gel times (at 140°C) for typical homopolymer elastomeric networks.

Oil	%crosslinker		gel time,min	
	SA	DA	SA	DA
linseed	26	73	45	60
crambe	25	72	60	120
lesquerella	24	68	60	130
lunaria	23	67	140	180
vernonia	29	--	420	---

3.1 Process Conditions and Phase Relationships

As might be expected, phase separation, and sometimes phase inversion, occurs during the polymerization process. The sequence of morphological changes depends on the compositions, solubility, reaction rates, and stirring conditions. In fact the final properties (see below) depend very much on the phase relationships developed as a result of the interplay of these factors.

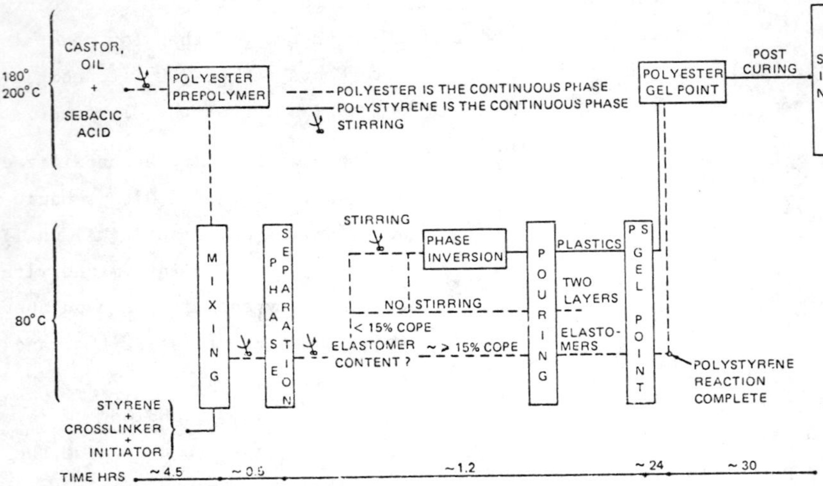

Fig. 1. Process flow sheet for synthesis of castor-oil-polyester-based SINs.[7]

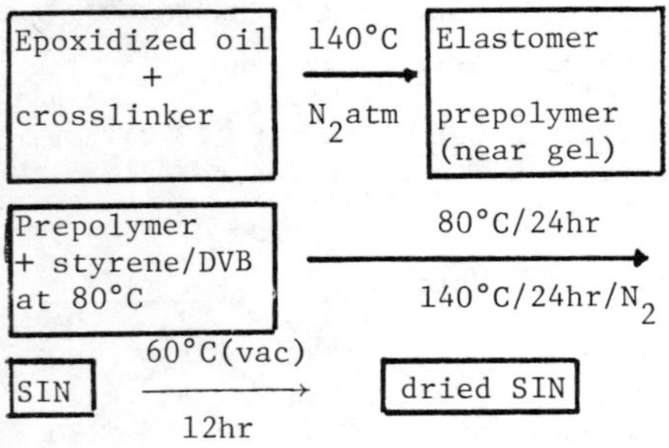

Fig. 2. Process flow sheet for synthesis of SINs based on epoxidized oils.

Morphologies were characterized by thin-section electron microscopy as described earlier.[6-9] Dynamic mechanical spectra were obtained at 110 Hz using an Autovibron DDV-IIIc. Stress-strain behavior was examined using an Instron tester at a rate of 0.085 mm/s; notched impact strengths were determined using an Izod tester (ASTM D256-73). Fatigue crack propagation (FCP) tests were made using compact-tension specimens following procedures described elsewhere[21], at a frequency of 10 Hz and a load ratio (min/max) of 0.1. FCP rates were plotted as the crack growth rate/cycle, da/dN, as a function of ΔK, the range in the stress intensity factor; ΔK is given by $Y\Delta\sigma\sqrt{a}$, where Y is a geometrical term, $\Delta\sigma$ the range in applied stress, and a the crack length.

With castor-oil systems, the polymerizing styrene first yields a dispersion of the polystyrene/styrene solution in a continuous matrix of elastomer.[5] As the polymerization of styrene proceeds, a phase inversion may or may not take place, depending on the composition and stirring conditions. Thus for systems containing more than about 15% of elastomer, no phase inversion will occur, regardless of stirring, and the elastomer remains the continuous phase.

In contrast, with systems containing about 10-15% of elastomer, what happens depends very much on the existence and duration of stirring. With vigorous stirring, a phase inversion soon occurs, as in the synthesis of high-impact polystyrene,[20] so that the polystyrene phase becomes continuous. However, if stirring is not provided, the system separates into two layers, the top being rich in a continuous elastomer phase and the bottom being rich in a continuous polystyrene phase. If stirring is provided, but stopped at various times, the sequence depicted in Fig. 3 occurs. Layering takes place, but the nature of the

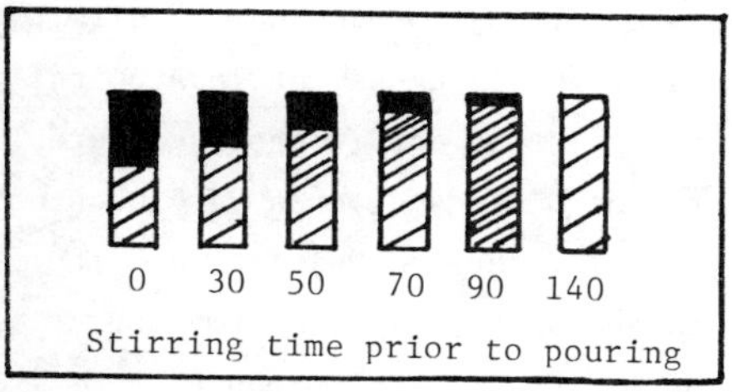

Fig. 3. Layering effect during synthesis of a 10/90 COPEN/PSN SIN.[7] Shadowing proportional to hardness.

layers depends on the duration of stirring prior to observation. For stirring times up to 90 min, the volume of the upper layer (elastomer continuous) gradually decreases until at a little more than 90 sec. (∿110 sec) the upper layer disappears and a phase inversion takes place so that the polystyrene phase becomes continuous. The consequences of these events are clearly evident in Fig. 4, which shows

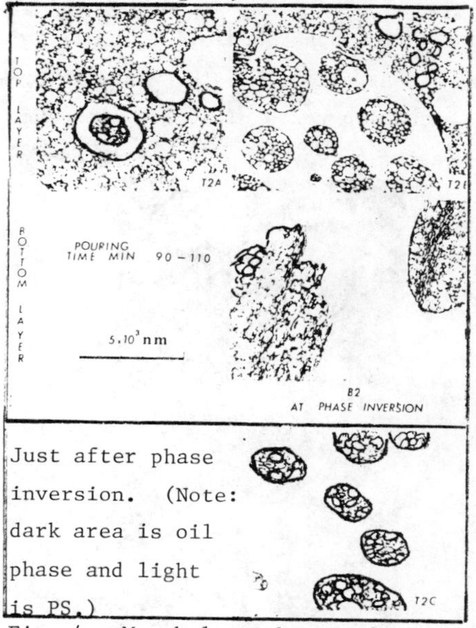

Fig. 4. Morphology changes in a 10/90 COPE/PSN SIN. Sample poured at or near phase inversion stage.[7]

micrographs of upper and lower layers just before phase inversion, approximately at the phase inversion point, and just afterwards.

Thus in the period of time between phase inversion and gelation of the PS continuous phase, the ultimate morphology becomes fixed. Interestingly, the width of this "window" depends on the elastomer content. The higher the prepolymer content, the shorter the time interval between phase inversion and gelation, probably because phase inversion then requires a greater conversion of styrene to polystyrene. In any case, the phase morphology is rather coarse, with discontinuous phases having sizes in the range of 0.1-0.2 µm — a range greater than that noted for IPNs having a similar composition.

With the other botanical oils at concentrations of ∿10% phase inversion also takes place, as with the castor oil systems (Fig. 5). As

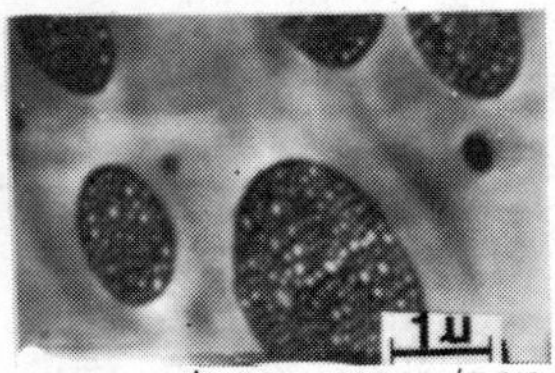

Lunaria: 10/90 ELuODAN/PSN

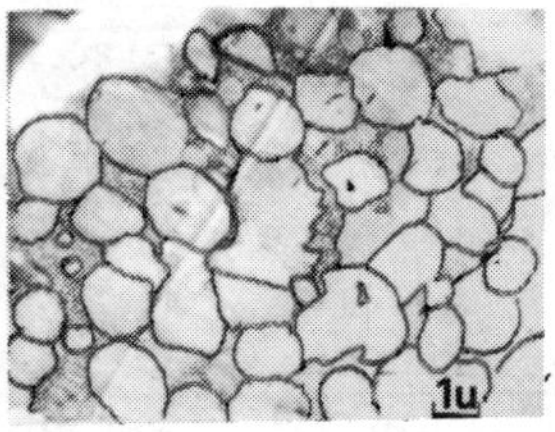

Lesquerella: 15/85 ElqODAN/PSN

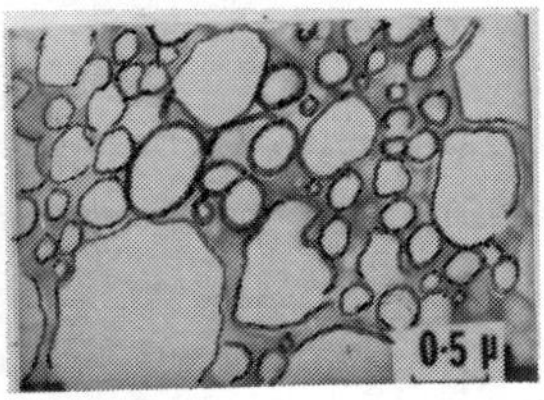

Linseed: 40/60 ELODAN/PSN

Fig. 5. Morphology of typical oil-based SINs.[11]

long as the oil is prepolymerized prior to dissolution in the styrene PS domains are found within the elastomer phase, as with the castor-oil systems and, for example, high-impact polystyrene[2,20] On the other hand, at least with linseed oil, omission of the prepolymerization step leads to elastomer domains that contain no PS inclusions. Again, phase inversion is encouraged by vigorous stirring.[11]

As the concentration of elastomer is increased, the phase continuity becomes dependent on the nature of the oil. At an elastomer content of 15%, epoxidized crambe and lunaria oils give discontinuous elastomer phases, as did castor oil. In contrast, epoxidized lesquerella and linseed oils yield a continuous elastomer phase and an incomplete phase inversion (not shown), respectively. It is thought that the higher viscosities of these oils may inhibit phase inversion; a higher stirring rate does effect phase inversion with epoxidized-linseed-oil systems.[11] However, at an elastomer concentration of 40%, all systems exhibited discontinuous polystyrene phases, which contained inclusions of elastomer whose sizes depended on the oil. Hence in all casés, at least with the stirring conditions used here, the morphology depends very much on the oil, the composition, and the extent of prepolymer formation.

3.2 <u>Viscoelastic Response</u>

All SINs examined so far exhibit well-defined peaks in the loss modulus, E'', corresponding to both the polymerized oil and the polystyrene phases (Fig. 6). Typical values are given in Table 2 (for additional data, see refs. 5 and 10-12). Gehman studies confirm the results for the ELO systems.[11]

With all the castor-oil systems[5], the two T_gs are shifted inward from those for the homopolymers. This behavior is common in polyblends that exhibit a significant degree of

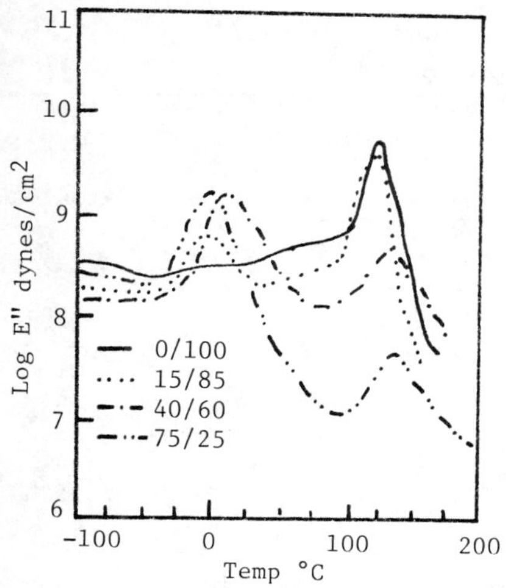

Fig. 6. Dynamic mechanical spectra for some oil-based SINs.(11)

Table 2. T_g Behavior of Botanical-oil-based SINs.(5,11)

Composition	Lower T_g,°C	Upper T_g,°C
PSN(Control)	---	116-120
COPEN/PSN (40/60)	-58	114
COPEN	-66	---
COPEUN/PSN (40/60)	-30	114
COPEUN	-50	---
COPUN/PSN (40/60)	- 6	115
COPUN	- 4	---
VOSA (50/50)	-45	120
VOSA	-45	---
ELOSAN/PSN (15/85)	+20	116
ELOSAN	---	---
ELODAN/PSN (15/85)	- 6	130
ELODAN	- 3	---

phase mixing on the molecular scale; calculations(5) suggest up to 20% mixing, especially of the PS in the elastomer phase. With the vernonia-oil system(12) on the other hand, little or no such mixing appears to take place; similar behavior is evident in the case of epoxidized-linseed-oil systems crosslinked with sebacic acid. However when dimer acid is used to crosslink epoxidized linseed oil, the T_g of the PS is increased by the presence of the elastomer phase - an anomalous response noted previously with acrylate/epoxy SINs.(34) In this case, the anomaly disappears when the elastomer is not prepolymerized prior to SIN formation, suggesting that high constraints on the PS network must be exerted by the ELO network.

3.3 Stress-strain, Impact, and Fatigue Behavior

As is the case with other elastomer/plastic IPNs or SINs,(1,5) stress-strain behavior ranges widely, from that of a toughened plastic to that of a reinforced elastomer (Fig. 7 and Table 3), depending on the concentration of the oil-based phase. Stress-whitening is characteristic, suggesting some combination of crazing and shear response. The detailed response may also be seen to correspond to the morphologies discussed above. Thus, with 10% of an oil-based phase, all systems behave like toughened plastics, the PS network being continuous, while at 40% of the oil-based phase, the behavior is characteristic of reinforced elastomers. With an intermediate content of oil-based phase (15%), the behavior depends on the oil used, epoxidized lesquerella oil yielding a reinforced elastomer (elastomer phase continuous) and the other oils (PS

Table 3. Stress-strain, Relaxation, and Impact Characteristics of SINs(5,11)

Base Oil	Specimen	TS,MPa	ε_B,%	E,MPa	IS,J/m
PSN control	for non-castor systems	44.8	4	830	16
Linseed	10/90 ELODAN/PSN	41.8	6	689	34
Linseed	10/90 (not prep.)	39.9	7	746	27
Crambe	10/90 ECrODAN/PSN	37.2	10	758	59
Lunaria	10/90 ELuODAN/PSN	34.5	5	726	43
Lesquerella	10/90 ELqODAN/PSN	35.4	6	778	45
Lunaria	15/85 LEuODAN/PSN	----	--	---	65
Lesquerella	15/85 ELqODAN/PSN	33.5	45	203	81
Linseed	40/60 ELODAN/PSN	17.8	37	51	--
Crambe	40/60 ECrODAN/PSN	14.9	63	76	--
Lesquerella	40/60 ELqODAN/PSN	13.2	56	21	--
Castor	10/90 COPEN/PSN	37.0	16	1520	68
Castor	10/90 COPEUN/PSN	25.5	19	1090	44
Castor	10/90 COPUN/PSN	36.7	6	1680	25
PS Control	for castor systems	46	2	2360	13

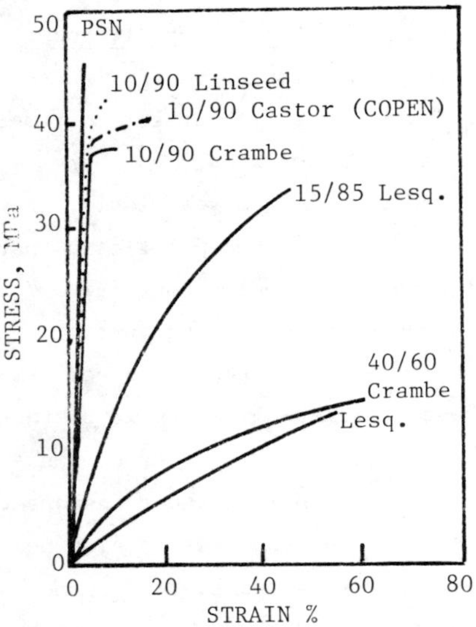

Fig. 7. Stress-strain curves for typical oil-based SINs.(11)

phase continuous) tough and leathery materials.

Impact strength tends to follow the toughness reflected in the area under the stress-strain curves. There is some evidence that impact strength is higher the lower the T_g of the oil-based phase. This relationship is not surprising for the generation of crazes around the elastomer particles is more effective the more rubbery the particles are at the high strain rates involved in the test. It also seems likely that, as with high-impact PS, maximum toughening at a given elastomer content is obtained when PS is occluded in the elastomer phase; thus the impact strength in a 10/90 ELO system is higher when a prepolymer is used and a secondary PS phase is seen within the ELO phase.

Fatigue characteristics also reflect a generalized toughening due to the presence of an oil-based phase; preliminary results are shown in Fig. 8, and compared with the response of typical high-impact-PS (HIPS) and ABS resins. Even at low elastomer contents, the FCP rates are all lower at a given value of ΔK (and hence $\Delta\sigma$) than for the control PS network, and are at least as low as those for the HIPS and ABS. Ultimate failure was essentially by softening and tearing, indicating a high degree of ductility, even though the PS phase was crosslinked.

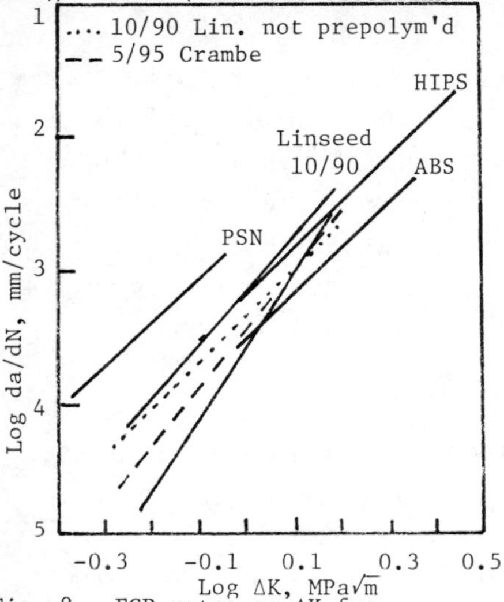

Fig. 8. FCP rates vs ΔK for typical oil-based SINs.[11]

4. CONCLUSIONS

Several conclusions may be drawn:

1. Use of the SIN technique, a reactive molding process, can greatly simplify the process of synthesizing interpenetrating polymer networks to obtain toughened plastics, leathery materials, or reinforced elastomers.

2. The SIN technique can be applied to the case of botanical oils that contain functional groups suitable for crosslinking or that can be suitably functionalized.

3. While the rates of polymerization of these oil-based systems are probably inherently slower than those of materials currently used in RIM applications, it should be possible to enhance the rates considerably by judicious catalysis.

4. In any case, more reactive SIN systems should be candidates for processing by RIM techniques.

5. ACKNOWLEDGMENTS

The authors wish to acknowledge financial support from the National Science Foundation, Grant No. PFR 7827336, and the many contributions by former members of the team: Drs. N. Devia-Manjarres, G. Yenwo, and A. Conde, and Mr. J. Pulido. We greatly appreciate the kindness of Drs. L. H. Princen and H. Kleiman (USDA, Peoria, IL) in supplying samples of the oils used.

6. REFERENCES

1. L. H. Sperling, "Interpenetrating Polymer Networks," Plenum, New York, 1981.
2. J. A. Manson and L. H. Sperling "Polymer Blends and Composites," Plenum, New York, 1976.
3. R. E. Camargo, C. W. Macosko, and M. Tirrell, Org. Coatings Plast., 44(1), 265 (1981).
4. W. E. Becker, "Reaction Injection Molding," Van Nostrand-Reinhold, 1979.
5. L. H. Sperling, N. Devia, J. A. Manson, and A. Conde, ACS Symp. Ser., no. 121, 407 (1980).
6. N. Devia, J. A. Manson, L. H. Sperling, and A. Conde, Polym. Eng. Sci., 1978, 18(3), 200.
7. N. Devia, L. H. Sperling, J. A. Manson, and A. Conde, Macromolecules, 1979, 12(3), 360.
8. N. Devia, L. H. Sperling, J. A. Manson, and A. Conde,

Polym. Eng. Sci., 1979, 19(12), 870, 878.

9. N. Devia, L. H. Sperling, J. A. Manson, and A. Conde, J. Appl. Polym. Sci., 1979, 24, 569.

10. L. H. Sperling, J. A. Manson, Shahid Qureshi, and A. M. Fernandez, I&EC Prod. Res. Dev., 20 163 (1981).

11. Shahid Qureshi, J. A. Manson, and L. H. Sperling, Org. Coatings Plast. Chem., 43(2), 7 (1980) in press, 1981.

12. A. M. Fernandez, L. H. Sperling and J. A. Manson, ibid, in press, 1981.

13. R. E. Touhsaent, D. A. Thomas, and L. H. Sperling, J. Polym. Sci. 46C, 175 (1974).

14. S. C. Kim, D. Klempner, K. C. Frisch, and H. L. Frisch, Macromolecules 10, 1187 (1977).

15. G. Allen, M. J. Bowden, D. J. Blundell, F. G. Hutchinson, G. M. Jeffs, and J. Vyvoda, Polymer 14, 597 (1973).

16. H. L. Frisch, K. C. Frisch, and D. Klempner, Polym. Eng. Sci. 14, 646 (1974).

17. H. L. Frisch, J. Cifaratti, R. Palma, R. Schwartz, R. Forman, H. Yoon, D. Klempner and K. C. Frisch in Polymer Alloys, D. Klempner and K. C. Frisch, eds., Plenum, NY 1977.

18. S. C. Kim, D. Klempner, K. C. Frisch, and H. L. Frisch, J. Appl. Polym. Sci. 21, 1289 (1977).

19. G. M. Yenwo, Ph.D. Thesis, Lehigh University, 1977.

20. C. B. Bucknall, Toughened Plastics, Applied Science, London (1977)

21. S. L. Kim, et al., Polym. Eng. Sci., 17 194 (1977).

7. BIOGRAPHICAL SKETCHES

John A. Manson received his B.Sc. in honors Chemistry and Physics (1949) and his M.Sc. and Ph.D. in Chemistry (1951 and 1945) from McMaster University. After post-doctoral research in Chemical Engineering at the University of Michigan he joined the Air Reduction Co., Inc. where he became supervisor, polymer research. He joined Lehigh University in 1966; he is now professor of chemistry and director, polymer materials laboratory. His interests include structure and engineering behavior of polymers and composites.

Leslie H. Sperling received his B.S. degree in Chemistry from the University of Florida (1954) and his M.A. (1957) and Ph.D. (1959) from Duke University. Following employment as a research chemist for the Buckeye Cellulose Corporation, and postdoctoral research at Princeton University, in 1967 he joined Lehigh University where he is now professor of chemical engineering and staff member, polymer materials laboratory. His research interests include the structure, properties, morphology, and behavior of polymers with emphasis on IPNs and graft polymers.

Shahid Qureshi received his B.Sc. (1970), and M.Sc. (1972) degrees in chemistry from Punjah University, Pakistan, and a postgraduate university diploma (1977) from the Tokyo Institute of Technology. He is currently a Ph.D. student in Polymer Science and Engineering at Lehigh, with research interests in the synthesis and morphological and mechanical behavior of polymers.

A. M. Fernández received her B.S. degree in chemistry from Universidad Nac. de Córdoba, Argentina (1973). In 1974-1976 she joined the Universidad Nac. de Misiones, Argentina as an instructor; in 1976-1978 she joined the Universidad Industrial de Santander, Colombia as a graduate student in Polymer Science. In 1980 she joined Lehigh University as a graduate student studying polymers made from botanical oils.

13th National SAMPE Technical Conference
October 13-15, 1981

BONDING METALS WITH ACRYLIC STRUCTURAL ADHESIVES

Denis J. Zalucha
Hughson Chemicals
Lord Corporation
Erie, Pennsylvania

Abstract

Acrylic structural adhesives are two-part, room temperature curing compositions consisting of (1) addition polymerizable monomers, (2) polymers dissolved or dispersed therein, and (3) initiator systems capable of generating free radicals at the ambient temperature. Working times can range from about 30 seconds to several hours. Much progress has been made in incorporating or scavenging oil, so as to reduce or eliminate the need for surface cleansing of metals. Accelerated test data point to good durability for many acrylic-bonded assemblies.

Key Words

Structural Adhesive
Acrylic Adhesive
Metal Bonding

1. INTRODUCTION

Acrylic structural adhesives are relative newcomers to the adhesive bonding industry. They first appeared in the U.S. in the late 1960's where they were initially promoted for bonding engineering plastics. Original commercial products were somewhat brittle and adhered to only a limited range of materials. These novel materials had enough attractive attributes such as the rapid cure at room temperature, easy processing through automatic dispensers and the option of the "no-mix" method of use that much time and effort has been invested over the past decade in making this technology into a reliable, widely applicable family of adhesives. This paper describes some of the basic acrylic adhesive chemistry, the three major stages in their evolution and points out some of the things they can or cannot do.

2. DISCUSSION

The whole idea of acrylic adhesives begins with the fact that certain ethylenically unsaturated small molecules will undergo a free radical chain growth polymerization at room temperature. The "workhorses" of acrylic adhesives have been methacrylic acid (I) and methylmethacrylate (II) although many other similar materials are used for a variety of reasons.

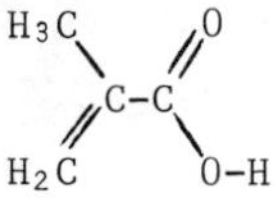

I

methacrylic acid

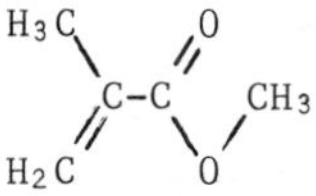

II

methylmethacrylate

When polymerization is initiated by free radicals, long polymers form but remain dissolved in a sea of unreacted monomer. The polymerization continues as free radicals are introduced but the mass remains fluid until enough polymer is formed (about 20% of the monomer) that the remaining monomer is insufficient to dissolve it. At this point the mixture has "gelled" and the polymerization rules change so that the remainder of the polymerization occurs very rapidly, taking the material to a rigid solid.

This is an attractive starting point for an adhesive. The material can be cured quickly at room temperature, is initially an easily handleable fluid and remains fluid for some time after the initiation of cure before it gels and rapidly solidifies.

The currently available acrylic structural adhesives based on this free radical polymerization mechanism have many features in common. All consist of two parts which must be joined, but not necessarily mixed, at the time of use. Each has at least one component consisting of a solution of polymers in the curable methacrylate monomers. Also dissolved in this component is one reactant of a pair which react at room temperature to produce free radicals. The second free radical reactant is carried in the other portion of the adhesive which may also be a polymer in monomer solution or might be a simple solution of the second reactant. These are the required constituents, but many other materials such as pigments, dyes, plasticizers, fillers, stabilizers or viscosity control agent are commonly added. The adhesives can be made in viscosities ranging from sprayable to nearly solid pastes.

Many reactant pairs producing free radicals are known. Typical pairs might be an organic peroxide and a transition metal ion or a tertiary amine. The requirement is that they be reasonably stable separately but react rapidly when combined.

When the two portions of adhesive are joined, the methacrylate monomers polymerize leading to a cure profile shown in Fig. I. The abscissa is time and the ordinate could be average molecular weight, temperature, viscosity electrical resistance or any other property

Fig. I

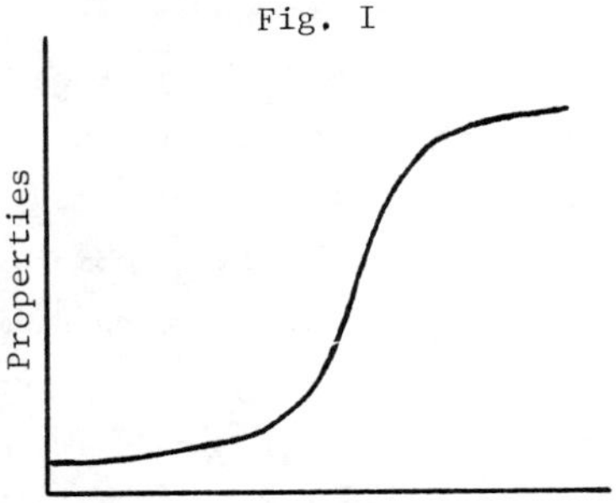

varying with increasing molecular weight. Depending on the formulation, the time to the rapid rise can be in the range of about 30 seconds to several hours. The beginning of the rise corresponds roughly to the gel time or working life. After this the viscosity becomes too high for adequate wetting of the adherend surfaces.

Three methods of accomplishing the "joining" of the two parts of the adhesive are available. The first is the common "mix-in" method in which parts A and B are mixed prior to application. This is done either by hand or with automated dispensing equipment. Mix ratios vary from 1:1 to about 20:1 depending on the adhesive.

The second common method involves a "surface activator". One part of the adhesive is a monomer-polymer solution. The second part is an activator solution of one of the free radical reactants, the other being in the first part. The activator solution is applied to one adherend surface and the solvent, if present, is allowed to evaporate. Some such activators remain liquid while some form stable dry films which can be stored for weeks or months before bonding. The adhesive is applied to the second surface. Polymerization begins on contact of the adhesive with the activator and proceeds as in Fig. I even though the two adhesive components are not mixed. This method is attractive in that it offers both long working time for making large assemblies and short cure time once the bond surfaces are joined. Capital expenditures for metering and mixing equipment are avoided and the simplicity can improve reliability in the hands of unskilled workers.

The third, much less common method of use is called the "A/B" method. It is something of a hybrid of the two other methods. Both portions of the adhesive are polymer-monomer solutions. One portion is applied to each adherend and, as in the surface activator, polymerization begins when the bond is closed and the two adhesive components meet.

Nearly all acrylic adhesives currently being marketed fall into one of three general classes: (1) General Purpose, (2) Second Generation or (3) High Performance. The classes differ in subtleties of chemistry, many of which are proprietary. Activators for the various types are not necessarily interchangeable. There are obvious differences in performance among the three groups and some are more appropriate than others in certain applications. Table I summarizes some shear strength data for typical "middle of the road" representatives of each group. The numbers are not intended to be performance specifications since by special formulating it is possible to change performance within any group to handle special problems.

The original acrylic adhesives were known as "modified acrylic structural adhesives." A modern representative of this type is listed as a "General Purpose" acrylic in Table I. These adhesives are modified and reinforced with elastomers and reactive polymers to achieve adhesion and durability. These adhesives perform very well on engineering plastics and can provide adequate adhesion to many metals if one is willing to do some surface preparation. Polymerization of these adhesives appears to be inhibited by zinc and copper so that General Purpose acrylics are not normally used to bond copper or zinc or alloys containing large percentages of them.

Second Generation acrylics (sometimes known as Reactive Fluid adhesives) appeared in the mid-1970's. These came about as a response to the need in general manufacturing for adhesives to bond mill finish metals without the need of surface preparation. The initiation chemistry was changed to include an oil tolerant surface activator and oil tolerant polymers are used in the adhesive. Representative shear strength performance is shown under "Second Generation" in Table I. Adhesion to oily ferrous metals and aluminum alloys is significantly improved over the earlier General Purpose adhesives. Adhesion to copper and zinc is still less than ideal although this problem can be overcome to some degree with special primers. These adhesives are normally used in the "no-mix" mode since it appears that some of the advantage of the oil tolerant activator is lost when it is mixed into the adhesive rather than applied to the adherend surface. Specially formulated equal mix systems with good metal adhesion are available.

The third class, High Performance Acrylics appeared in the late 1970's. In this case specially synthesized monomers have been used to produce a system capable of

TABLE I

Acrylic Adhesive Shear Strength
ASTM D 1002-72

Adherend	Preparation	General Purpose	Second Generation	High Performance
6061-T6 Al	None	5.5 MPa	15.9 MPa	30.4 MPa
	Shot Blasted	29.7	11.7	34.5
2024-T3 Clad Al	None	4.8	9.0	32.4
	Shot Blasted	26.2	23.5	31.0
SAE-1010 Steel	None	2.1	19.3	38.0
	Shot Blasted	29.7	24.1	41.4
Copper	None	9.0	5.5	21.3
SAE-72 Brass	None	3.5	6.2	16.6
SMC (Polyester)	Abraded	3.7	0.1	4.1
ABS	None	3.5	4.8	4.1
Polycarbonate	None	8.3	6.9	4.8
poly MMA	None	4.1	4.1	4.1

bonding unclean surfaces. The initiator chemistry was again reworked to facilitate cure against copper and its alloys. Zinc surfaces are sometimes a problem although primers and surface treatments to enhance adhesion are available. These adhesives retain the excellent adhesion to engineering plastics and provide a generally higher level of adhesion to metals than the earlier technologies.

Some type of environmental resistance test or stress endurance test is routinely run on nearly all acrylic adhesives. With the advent of the High Performance acrylics, more sophisticated testing has started. These adhesives have reached adhesion levels where they might be suitable for such application as bonding automotive body sheet or aircraft structures. Such testing is time consuming, but some initial encouraging results have been realized.

The High Performance adhesive listed in Table I has been evaluated for resistance to sinusoidal loading at 30 Hz according to ASTM D3166-73. Adherends were mill-finish 6061-T6 aluminum. What appears to be an endurance limit has been found at about 30%-35% of the ultimate failing load (Fig. II).

Results of testing resistance to aggressive environments are also encouraging. For example, bonds to unprepared 6061-T6 aluminum survive 1000 hours exposure to 55°C condensing humidity retaining 80-90% of the initial strength. The loss appears in part the result of the plasticizing effect of the moisture. Stressed to 25% of their ultimate failing strength, similar structures survive several days in 55°C condensing

Fig. II

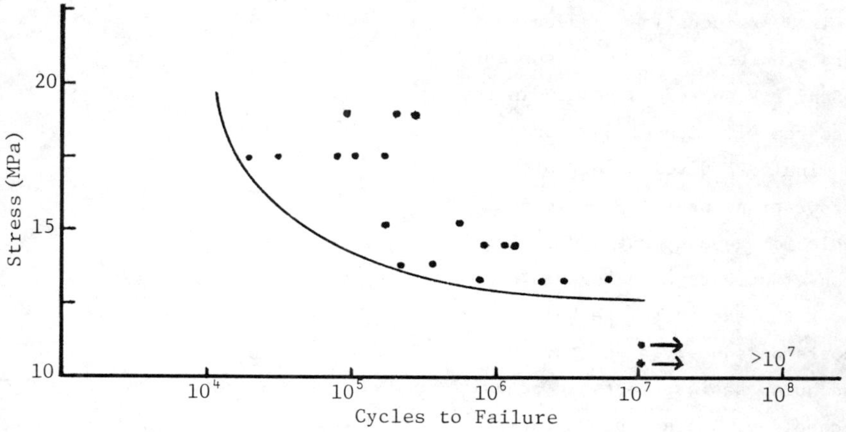

humidity. Here we have determined, through ISS/SIMS analysis using ^{18}O labeled water, that the failure occurs cohesively in the metal oxide surface.

Such performance on unprepared metals is entirely adequate for many industrial and manufacturing uses. The failures occurring in the oxide layers imply that further benefits can be had by providing a more controlled, stable metal surface for the adhesive. In unstressed tests on phosphoric acid anodized 2024-T3 Alclad aluminum, the High Performance adhesive retained approximately 95% of its original shear strength (ASTM D 1002-72) after 1000 hours exposure to 60°C condensing humidity. Peel strength was slightly higher than initially in the same test. All failures were cohesive. Phosphoric acid anodized 2024-T3 Alclad lap shear samples with an average failing strength of 23.5 MPa have been held at 6 MPa stress in 55°C condensing humidity for up to 500 hours. While about 25% of samples failed between 400 and 500 hours, those surviving retained from 70% to 98% of the original failing strength.

3. SUMMARY

In their short history acrylic structural adhesives have undergone three major steps in their evolution. In a little over a decade they have progressed from laboratory curiosities to sophisticated bonding systems. Their simplicity in use, fast, room temperature cure and adhesion to unprepared metals make them appealing for high speed assembly line work. Early test results show that this technology holds promise in the fabrication of large, durable metal structures where they might replace current heat cured adhesives saving time, energy and avoiding heat degradation of adherends.

Biography

Denis J. Zalucha is a graduate of Purdue University. He is currently a Senior Research Associate in the Research Division of Hughson Chemicals, a division of Lord Corporation, where he is responsible for development and evaluation of structural adhesives. He has authored several technical papers on adhesives and testing and holds basic patents in adhesive technology. He has broad experience both in the chemistry of adhesives and in their use in engineering and manufacturing. Dr. Zalucha is a member of the Society of Manufacturing Engineers, The Adhesion Society and the American Chemical Society.

13th National SAMPE Technical Conference
October 13-15, 1981

ANAEROBIC ADHESIVES: PROPERTY MODIFICATION FOR IMPROVED PROCESSING

Bruce D. Murray
Manager, Technical Service Center
Loctite Corporation, Newington, Connecticut

Abstract

Anaerobic adhesives have existed as commercial products for nearly thirty years. Beginning with a single formulation, anaerobic adhesives and sealants have been modified so that hundreds of formulations are available. Modifications have been achieved primarily to fit anaerobic technology into new application areas. Products have been controlled in areas of viscosity, cure speed, polymer hardness and adhesive strength. While many of the modifications have been achieved by simple formulation, others have required synthetic methods to combine anaerobic chemistry with other technologies. Methods for modifying anaerobic technology and the resulting products are discussed and related to properties and applications.

1. INTRODUCTION TO ANAEROBIC ADHESIVES

Anaerobic adhesives were first formulated nearly thirty years ago as liquid materials which would self-cure between the threads of nuts and bolts or other rigid metal assemblies and provide a strong locking or holding force. This technology has been improved to provide a wide range of formulations which now are used in many areas, not only on metal but also on ceramics, glass and many thermoset plastics. Present application areas cover a multitude of industries and functions and anaerobic technology has become a major element in the assembly of many parts.

Anaerobic adhesives are stabilized by air contact. A combination of metal surface contact and the exclusion of air initiate a free radical polymerization to a thermoset acrylic polyester polymer.[1] Thus, on a practical basis, anaerobic products can be stored at room temperatures for extended periods of time and cure as a single component adhesive or sealant without requirements for mixing or heat. In a liquid state, anaerobic formulations are generally non-flammable, have low odor levels, and are classified as mild skin sensitizers. Cured polymers generally are acceptable for operation in contact with acids, bases, and any organic solvents. The polymer does not degrade readily at temperatures up to 300°F. The result is a series of anaerobic formulations which can be used in many environments and improves the durability of a multitude of assemblies.

Locking threaded assemblies remains an important application area for these adhesives. Nuts, bolts, studs and screws can be effectively locked against vibration loosening by the "jamming" action of the hardened acrylic polyester polymers. Locking strength can be controlled through modification of the cured polymer so that assemblies locked against vibration loosening can be disassembled with ordinary tools. Assemblies requiring "tamper-proofing" can be permanently locked with high strength anaerobic threadlocking compounds which "jam" and bond threads together. Generally, this is not necessary as several studies[2,3] have indicated that even low strength anaerobic threadlocking compounds are among the most effective methods of preventing thread loosening due to parts shifting, vibration or side sliding. The most recent advance in threadlocking technology has been microencapsulated anaerobic products which provide a reliable inspectable locking force without the need for online liquid application.

Anaeroibc formulations have also shown capability for sealing a multitude of assemblies. These assembly areas are: threads, flanges, which are normally sealed with gaskets; and porosity in metal castings, welded assemblies or powdered metal parts. Liquid anaerobic products wet metal surfaces well and fill small gaps between the metal parts before curing. They cure without significant shrinkage, resulting in a positive seal. In addition, anaerobic sealing materials assist in holding parts together.

Requirements for sealing applications have led to the development of a wider range of viscosities of anaerobic products and cure systems which can be adjusted to accommodate both very fast systems and slow systems that allow adjustment of articles to be sealed. Cured anaerobic sealants have been formulated to release easily from metal surfaces and, thus, facilitate repair of sealed flanges.

The term "retaining" has been developed to explain the function of assembling smooth cylindrical parts. Examples of these parts are gears, wheels and pulleys on shafts, electric motor components and bearings. To achieve acceptable strength on "retaining" applications, anaerobic polymers have been strengthened by formulating adhesive agents into the product. These more adhesive products have resulted in a technology capable of "bonding" cylindrical articles together and achieving higher strength levels than traditional methods. Anaerobic products are also used to supplement the properties of traditional methods, such as press fits, splines, and keyways.

The most recently developed application area for anaerobic products is the area of structural adhesive bonding. Retaining technology fulfills requirements of the "ideal bond line", i.e., smooth, cylindrical parts which are subjected to only shear stress. Flat structural bonds are subjected to shear, impact, cleavage and peel stress. Anaerobic structural adhesives required modifications to improve elongation, surface adhesion, and crack resistance, or toughness. Modifications of curing properties which lead to improved cure through bondline gaps have allowed anaerobic structural adhesives to be used on larger parts and to serve as a "cushion" between surfaces which are subject to shock stresses. Recent combinations of urethane and anaerobic acrylic chemistry have resulted in resin systems which not only cure by the anaerobic cure mechanism, but also have outstanding properties in areas of surface adhesion, impact resistance, peel resistance, and environmental resistance.

Compared to other types of structural adhesives, such as epoxies or urethanes, anaerobic structural adhesives offer little new in areas of strength and durability. They are equivalent to the best of these adhesives, but the feature which makes them most desirable is the cure capability which makes use of anaerobic surface activators rather than mixing or heat. Anaerobic structural adhesives can achieve strong durable bonds to metals, glass, ceramics and thermoset plastics without extensive processing, mixing or heat cures. It is these cure characteristics which often lead to the use of an anaerobic structural adhesive over other types.

In addition to resin modifications, some anaerobic structural adhesives make use of a unique ultraviolet light sensitive cure system to provide improved cures and superior adhesion to glass. This results in a simplified process for glass assembly systems and also an excellent method for coating and potting electronic components. Ultraviolet adhesive formulations also overcome a limitation of a "wet" surface, which is the result of air contact. Ultraviolet formulations can provide simple processing and bond durability in areas of glass and metal bonding and high intensity light can cure adhesives exposed to air.

2. ANAEROBIC TECHNOLOGY AND MODIFICATIONS

Anaerobic products can be modified by formulation or by synthetic methods. Modifications of "machinery" adhesives (formulations for threadlocking, sealing and retaining) are generally the result of formulating to achieve optimum properties in areas of cure speed, viscosity, adhesion, or hardness (threadlocking strength). Each formulation change can require a different technique or method and specific property changes in one area may affect others. Generally, when properties are modified, the formulation changes can follow in a perceived direction.

Some areas have required more than simple formulation and newer products have departed from the acrylic diester first used for anaerobic products. The effective development of structural adhesives required cured polymers which were highly adhesive, crack resistant (tough), and environmentally resistant. New synthetic methods toward urethane modified anaerobic monomer systems were required to achieve this toughness.

The simpliest anaerobic formulation consists of a monomer system (acrylic diester) and a peroxide polymerization initiator. For practical purposes, this formulation also has accelerators to give a practical cure speed on production lines. This anaerobic formulation would have a viscosity of approximately 15 cps, achieve fixture cure on steel surfaces in approximately 60 minutes, and would reach high shear strength (2000 psi). This formulation serves as the basis for other anaerobic formulations. It is well suited for locking small threaded parts and penetrating into the porosity of welded metal or powdered metal assemblies. It is not suited for large threaded parts, as the thin liquid would flow from the gaps readily. Thus, it would be unacceptable for gasketing, threaded parts over 1/2" diameter, or sealing tapered threads as might be found on a pipe fitting. The strength of this basic anaerobic is such that it would be unsuitable for allowing removal of many fasteners.

Thus, to achieve successful use of an anaerobic in other applications, modifications would be necessary. These modifications occur in several areas:

2.1 Viscosity

Viscosity modifications are usually achieved easily with anaerobic formulations. Thermoplastic polymers are generally soluble in acrylic diesters used for anerobic formulations and, thus, raising viscosity is often as simple as dissolving the proper polymer in the basic monomer system. Figure 1 illustrates the effectiveness of this method. As new soluble polymers were discovered, the effective range in viscosity increased until thick pastes with viscosities over 1 million cps could be made successfully.

Increasing viscosity of anaerobic products is not without its problems, however. Often polymers have limited solubility and, thus, have limited capability to achieve viscosity control. Also, many thermoplastic polymers are cured in such a way that they contain residual curing agents which can cause instability in the anaerobic formulation, limiting shelf life. Polymeric modifiers can also cause problems with strength changes. On the other hand, in combination with a plasticizer, the proper polymeric modifiers can assist in achieving a proper balance of threadlocking torque and viscosity.

Fig. 1 Viscosity Range, Centipoise, Anaerobic Adhesives 1960-1980

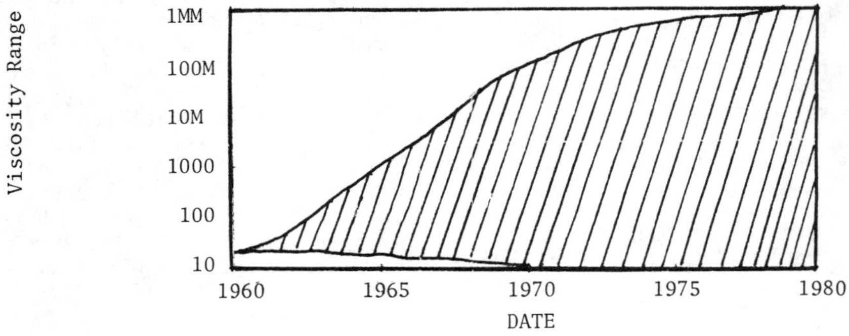

In addition to simple viscosity regulation, the development of thixotropy can allow significant control of the anaerobic formulation on an assembly line. Generally, fumed silica is used successfully to develop thixotropic ratios of up to 10:1. Resulting products on assembly lines are easily applied and can stay in place on the threads of a fastener or even on vertical surfaces. Thus, the proper control of viscosity and thixotropy can eliminate the "mess" of working with liquids on production lines and create formulation tailored to specific uses.

2.2 Strength

As stated above, the basic anaerobic threadlocking formulation is moderately strong. Control of product strength by formulation is achieved by use of plasticizers or "adhesive agents". Figure 2 illustrates the effect of plasticizers and adhesive agents. In areas of threadlocking, these differences may be inconsequential as ultimate strength is not always related to locking capability and, in locking and sealing applications, high strengths are usually not required and could be detrimental if fasteners are destroyed on disassembly.

In locking threaded assemblies, plasticizers affect both breakaway (initial moving) torque and prevailing (continuing) torque. Table 1 illustrates the effect of plasticizers on each area. The use of plasticizers can have major effects on the relationship of each, but can either gain or lose in effectiveness when used in combination with other modifiers. Generally, plasticizers lose effectiveness as the percentage increases. Thus, with anaerobic compounds, a minimum strength level as well as a maximum can be reached.

The effect of surface adhesion for anaerobic machinery adhesives is generally minimal. Products achieve locking capability by "keying into" imperfections in the surfaces and providing an effective 100% surface-to-surface contact between components. The polymer prevents movement of the parts and thus prevents loosening. If parts are shifted and moved beyond the compressive strength of the polymer, the polymer is crushed and will allow parts to move under a controlled force. Improvements of adhesion through addition of organic acids can improve resistance to initial movement in high strength retaining compounds. Upper strength limits to anaerobic formulations with adhesive agents generally occur at approximately 3500 psi on steel. Synthetic methods for new polymers are used to achieve higher strengths.

Fig. 2. Shear Strength Range Anaerobic Adhesives with Formulated Plasticizers/Adhesive Agents

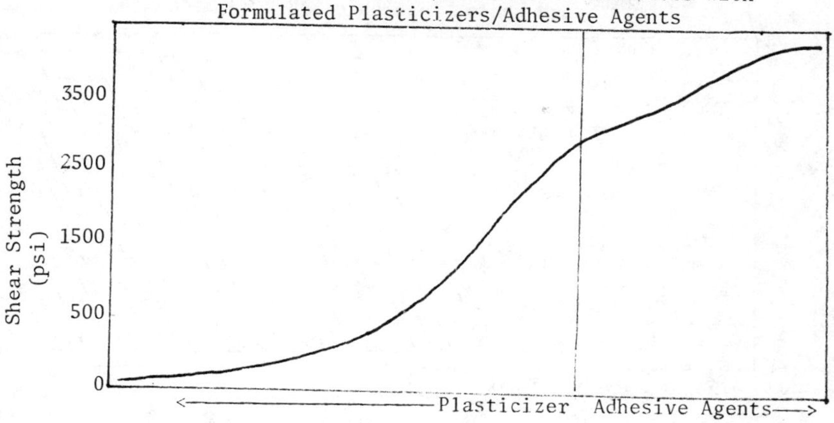

Other use factors, such as the metal surface, surface finish, cure technique and part design can also have significant effects on adhesive strength and many of these will negate or supplement the effect of adhesive agents.

2.3 Cure Rate

Many adhesive users decide that they need the fastest possible cure system to achieve their production line objectives. However, some users of anaerobic adhesives find it necessary to achieve slow cure to allow adjustment of their machinery parts, hours or even days after assembly. Thus, the concept of formulating anaerobic adhesives to specific use calls for a range of cure systems to fit particular functions. Initially, anaerobic formulations had a "fixture" or "finger tight" time of approximately three hours. Initial attempts at modifying cure speeds were aimed at reducing cure speed. Success came easily as modified cure systems were developed to reduce fixture time to 15 minutes (Figure 3). The use of "primers" or surface activators further reduced fixture time to the area of five minutes before modified cure systems could achieve this with no "primer". The most recent developments in "primer" technology have reduced cure times to a practical limit of 30 seconds. Cure rates faster than this could result in assembly problems as the product between the threads of a fastener could cure before the fasteners are completely assembled. Capabilities have been reduced to approximately ten seconds.

Anaerobic products can also be formulated to achieve fixture times of 48 hours or more. The objective in these extended cures is to allow part adjustability during assembly. Slow curing formulations have been developed primarily for pipe thread assemblies which may require adjustment over a period of a full day after assembly. Not only the cure initiators but also the viscosity

Table 1. Effect of Plasticizers on Breakaway/Prevail Torques for Anaerobic Threadlockers. Removal Torque, In. Lbs., 3/8-16 Steel Fasteners.

Strength	Low Viscosity*	Moderate Viscosity+
Weak	60/60	75/35
↓	60/150	150/45
	60/260	200/100
Strong	---	250/250

*15 cps +1100 cps

Fig. 3. Cure Speed Range, Anaerobic Adhesives 1960-1980

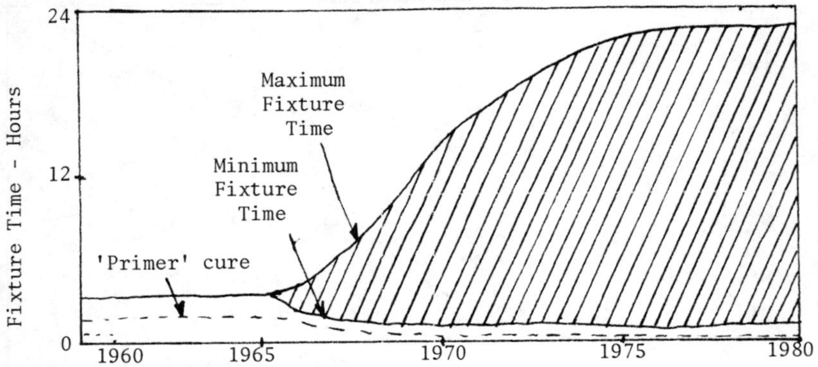

regulators, plasticizers and other fillers can enter into the control of the cure rate.

2.4 Structural Adhesives and Adhesion

Properties of an adhesive or sealant are often interdependent on many aspects of the formulation. This is especially the case with structural adhesives. Structural adhesives require a combination of surface adhesion, flexibility, "toughness", and crack resistance not existing in the "machinery" adhesive area. Early anaerobic machinery adhesives could not meet these requirements. Approximately fifteen years ago, urethane modified acrylic technology showed promise as structural adhesives which could achieve high strength on metals. However, these initial attempts at structural adhesives were still brittle and were excluded from high stress use areas.

In 1975, further advances were made and an improved line of structural adhesives were introduced.[4] Capabilities were extended with improved elongation and crack resistance, better impact and peel resistance, and improved surface adhesion.

In improving the performance of structural adhesives, the major factor was not improvement in adhesion, but improvement in the properties of the polymer. Table 2 compares properties of adhesive based on the original urethane modified acrylics, and the "second generation" anaerobic structural adhesives.

Improvements in properties of the second generation structurals are a result of improved "toughness" in the polymer. The adhesives are not necessarily superior in bond strength, but can cure better through large gaps and absorb stresses without shattering or cracking. Anaerobic structural adhesives achieve strengths and durability equal to "the best rubber-modified epoxies" and also offer production line users the benefit of an anaerobic product which can be used easily and dispensed automatically.

It is these new proprietary resin systems which offer improvements in bonding performance and can modify the capabilities of machinery adhesives to achieve significantly wider capabilities of anaerobic adhesives as assembly tools.

2.5 Applications of Anaerobic Adhesives and Sealants

Applications of anaerobic adhesives and sealants have extended use to nearly every area of metal assembly. The property modification capability allows anaerobic manufacturers to formulate products which simplify assembly procedures and improve performance of the assembled parts. Modifications have been made to the original thin anaerobic liquid to allow many new applications:

- faster cure allows liquids to penetrate porosity in welded oil-filled transformers to prevent leakage and pollution.

Table 2 Cured Property Comparison, Anaerobic Structural Adhesives

Property	First Generation	Second Generation
Tensile Shear	1800-4500 psi	2000-4500 psi
Tensile	2000-8000 psi	3000-8000 psi
Impact Strength	4-5 ft. lbs.	4-45 ft. lbs.
Fatigue Strength	800-12000 psi	800-1500 psi
T-Peel	3-5 PIW	5-60 PIW
Temperature Range	-65°F-400°F	-65°F-450°F
Gap Curing (Maximum)	.020"	.060"
Elongation	.2 - 15%	50%
Thermoplastic Bonding	Poor	Fair
Oil Sensitivity	Poor	Fair

- a very high viscosity anaerobic threadlocker can lock seating rings into valves used in chemical plants, thus preventing leakage.

- a very high strength, high viscosity formulation is used to lock hull integrity studs in submarines.

- thixotropic anaerobic formulations are used as gasketing materials for gearboxes and transmissions in automotive, truck and tractor assemblies.

- a slow curing, highly filled anaerobic compound is used to seal tapered threaded fittings in hydraulic control systems for materials handling equipment.

- anaerobic retaining compounds with adhesion promoters are used in five places as the only adhesive method for assembling electric motors for double-insulated hand tools.

- high strength anaerobic structural adhesives are used to assemble ceramic circuits into high speed projectiles for the military. The anaerobic adhesive does not crack or fail even as the projectile is fired from a cannon.

- rapic curing anaerobic adhesives, which will cure under ultraviolet light, bond metal trim to automotive windows completely and reliably within minutes after exposure to ultraviolet light.

Anaerobic adhesives have touched the lives of most people in the industrialized world. The capability to formulate and modify properties has furthered the creation of products and new concepts to fit the necessary parameters of many application areas. As anaerobic technology continues to improve, we can expect further modifications and more available properties for improved performance in all areas.

REFERENCES

1. Elliott, J. R.; Hauser, M.; and Murray, B. D.; "Anaerobic Adhesives" Chapter 33, pp 560-568. Skeist I. (ed.) Handbook of Adhesives Second Edition 1977. Van Nostrand Reinhold Company.

2. Batson, R. E.; and Tokarski, J. G.; The Technology of Threaded Fasteners, Loctite Corporation, 1974.

3. Haviland, G. S.; "A Logical Approach to Secure Bolting and Locking" SME International Show and Conference, Cleveland, OH April 29, 1980. Society of Manufacturing Engineers, Dearborn, MI.

4. Baccei, L. M. and Murray, B. D.; "Anaerobic Adhesives, New Properties" SME Technical Paper, Ad. 75-592 Society of Manufacturing Engineers, Dearborn, Michigan.

Bruce Murray is Manager of the Eastern Technical Center for Loctite Corporation. He has been with Loctite for more than fifteen years as a chemist, product manager, and Technical Service manager.

Areas of responsibility have included cyanoacrylate adhesive product development, vacuum impregnation process development, product and commercial development of anaerobic structural adhesives, and Technical Service activities for various areas of the United States. Present responsibilities cover technical service for anaerobics, cyanoacrylate and other adhesive technologies in the Eastern United States.

He is a member of the American Chemical Society and Society of Manufacturing Engineers.

13th National SAMPE Technical Conference
October 13-15, 1981

Linear Tensile Members as Composites and Kevlar for Ropes, Lines, Guys and Cables

Dr. E. Scala
Cortland Cable Co., Cortland, N.Y.

Abstract

The properties and applications of Kevlar to linear tension members such as ropes, guys and cables are reviewed. The important tensile property advantages are compared to today's standard materials. The dynamic properties of the Kevlar are compared for energy absorbtion, creep and cyclic stressing. The effects of wear, abrasion, corrosion, temperature and chemical effects are discussed.

1. INTRODUCTION

Lines, ropes, guys and cables are some of the terms commonly used for linear tension members that must be flexible. A plethera of vocabulary has accumulated over the centuries, arising from the large range of industries, applications, and markets served. The textile and cordage fields have their own terminology, as do the wire rope, electrical wire, and cable industries. In recent years the composites technology has added their own. The synthetic fibers (nylon, polyester and polypropylene) have long since displaced the natural fibers as the basic yarns in cordage. The galvanized steel dominates the wire rope and electromechanical cable tonnage. Of the modern high strength fibers, namely, glass, graphite and aramid, it is the latter that cuts across all the fields. This paper deals primarily with one patented DuPont aramid fiber, KEVLAR, and the last ten years of experience in the manufacture and testing. The comparative properties of the Kevlar are summarized in the data taken from

DuPont Company literature (Table 1, and Figures 1, 2 and 3. The Kevlar is available only as a yarn, made up of many parallel filaments, each about 12 microns in diameter. The Table 2 lists the forms that the DuPont Company has settled on after ten years of prototype production and marketing. This table is included to emphasize the advantages and limitations which establish the modular build up of strength and elongation that can be designed into the tension members. There is considerable data also available on the chemical resistance, tension-fatigue, and good high temperature properties which make it far superior to other synthetic fibers, and virtually chemically inert relative to steel. The incredible variations of applications of Kevlar include bullet-proof vests, boat hulls, sails, hausers, and fiber-optics strength members. This paper will concentrate on the lines, ropes, guys and cables to illustrate the Kevlar as a key member in composites applied to linear tension members.

For both military and industrial operations, the rope and cable are vital technologies, though not as dramatic, nor funded as are other structural members. They have been in the background, advancing slowly and conservatively. The U.S. Navy has probably been most responsible, encouraging industry and doing its own R & D, though predominently through commercial laboratories. An examination of the literature for technical information, brochures and journal papers finds a great gap between the practical sales type information and the newer papers that treat finite element analysis and statistical behavior or Weibull plots. In most instances, the worlds are too far apart and do not understand each others language. The properties listed for synthetic ropes have not changed greatly in the past twenty years. In wire rope literature one finds the acronym IPS specifying "improved plow steels". Only an old metallurgist would know what that is, and no modern student of materials is likely to be impressed with its tolerances, chemically or mechanically. For that matter, the basis for buying nylon, polyester, or Kevlar for ropes would appear ludicrous compared to the care taken for carbon fibers, or S-glass used in composites. In fact, you would find the formula for aramid fibers only by research laboratory attempts to analyze them. These points will not be elaborated on here, but the quality control or thorough materials engineer will discover soon enough the limited control over the quality. This is not new, for those who have had to deal with the ceramics industry and qualifying glass fibers. The significance of this point, is the difficulty in assuring reproductive high quality over long periods and large volumes. One can also expect changes in properties from different suppliers and different purchase orders. Yet the rope industry considers synthetic fibers as scientifically controlled, compared to bales of imported natural fibers. While the textile and cordage industries do extenxive QC testing, the recourse is minimal.

Because of the low elongation of the higher strength fibers such as Kevlar, the processing, properties, and methods of application become more critical. The sublities of design and processing affect the range of properties that will be reviewed here:

- Tensile properties of break strength, elongation and elastic modulus.
- Dynamic properties such as impact strength, energy absorption, creep, spring constants, cyclic stress properties.
- Wear, abrasion, and corrosion properties.
- Temperature, moisture, and chemical effects.

2. TENSILE PROPERTIES

The early cordage industries had extensive demand for high quality ropes and hausers from the ship building industry, and the continual replacement of rigging on whalers, cargo ships, and in the grand finale of the clipper ships. The Russian hemp from the Baltic was the best, supplanted in the late 19th century by manila from the Phillipines. All other hemps and sisals were inferior but used in tonnage amounts for industry and agriculture. The high strength flax and silk had special uses, but too expensive except for critical low volume applications, as in threads and bow strings. When archery was prime as artillery, the flax or linen was vital to victory. But both the hemp and the flax had to use a matrix or coating as combined lubricant and abrasion protection. The North Carolina tar or pine resin was the key to the long life of rigging that could survive years at sea, and beeswax kept the bow string. There is an interesting corollary here to the reinforced plastics and the best of Kevlar today, namely, that the coating had to be applied to the fiber and yarn to get filament wetting. Soaking or impregnating the line or rope after fabrication never works as well for uniform wear resistance and optimum strength. For every line made, whether a thread or large rope, the coating is key to better properties, and is invariably the proprietary item that the industry guards.

For Kevlar this has been a hard wax, applied to the filament material as it is made, similar to the lubricants applied to the newly formed glass fiber. Conventional yarn lubricants are used for protection in textile forming operation. Urethane coatings have been used as well for abrasion protection and matrix, with respective penalties of short life for lubes, and high weight and cost for urethanes. The Kevlar 29 with a wax finish, Type 960 is the best for the rope yarns. (Table 1).

The higher modulus Kevlar 49 is used for the greater stiffness in reinforced plastic structures, such as boat hulls and antenna tower guys. As will be discussed later, the 49 also has a lower creep rate. Although break strengths for the two Kevlars are listed as the same, the 49 within an optimized epoxy can show substantially higher average break strengths, with recent reports of over 600,000 psi in the filaments. In ropes using Kevlar 29, average values of only 250,000 to 300,000 psi are obtained, which reflect a fiber strength with only shear interaction by friction, compared to the shear interaction by matrix bonding.

As a point of reference, the Kevlar tensil strengths are three times that of nylon or polyester, and can match the smaller diameter steel wire ropes.

The method of tensile testing is quite critical, and there are established procedures (Fed. Std. 191) for yarns, textiles, cords, ropes, etc. The Kevlar, however, is particularly sensitive to the grips or method of termination.

The combination of stress concentrations, compression, and abrasion cause most Kevlar tensile failures to occur at or near the termination, reducing the measured break strength by 10% to 20%. The most practical method for testing large numbers in quality control is the use of the capstan or drum, but the most efficient or high strength values are obtained by eye splices, using a version of the compression braid-type grip, variously referred to as the Kellum's grip, "Chinese finger", or helical wrap (Wall/Hood type). The latter involves forming an eye at each end, and a long counterhelical wrapping back over the standing line or center of the test section. For this present discussion the standard capstan type grip is assumed, but it can produce average strength values of up to 10% less than the eye splice. Note that tuck-type eye splices typical for nylon or polyester can result in lower average values for Kevlar and greater scatter in data. Industrial data on ropes rate strengths according to their preferred terminations, but there are MIL specified minimums.

One area of application for the Kevlar that has required extensive testing has been the sonobuoy cable. This is a composite cable that consists typically of a stranded tinned copper conductor with an extruded polyethylene, propylene or copolymer insulation, and an outer Kevlar 29 strength member, with or without an extruded, braided or coated outer jacket. The fiber optic version of such a cable would use a parallel Kevlar 49 strength member, for minimum elongation. While test samples are from 10" to 10 ft., the cables are used in continuous lengths of from 100 ft. to over 20,000 ft.

The best way to summarize the Kevlar property is to examine the data using a Weibull Plot [1]. Figure 4 is an example that compares a number of features. A Weibull plot relates the cumulative failure probability ($\ln(1-p)^{-1}$) to the log of the break strength ($\ln s$). A very steep slope and straight line would mean a very consistent behavior and one dominant failure mechanism. This is seen in the solid line the which is a braided Kevlar 29, using 10" gage lengths and $2\frac{1}{2}$"d Instron capstan grips. For comparison the data to the right represents Kevlar 29 of 15,000 denier, a heavy yarn used for larger cables. This is quality control data using the same test procedures, but from yarn right off the tube, as received from DuPont, without any significant twist. (Note: This yarn is made of ten ends of 1,500 denier, and is best used with an optimum twist of 0.3 to 0.5 twists per inch for higher average break strength.) For various reasons, the yarns are frequently not used as twisted single strands, and for quality control is tested as received. In this instance, it also illustrates the larger scatter that is obtained with untwisted or unbraided Kevlar. There appears to be two breaks, or three slopes, though the raw data could be interpreted as a curve sloping to the right.

Nevertheless, a varying mechanism of failure is suggested, with lower strength bundles less erractic, or consistent, lower strength. The higher strength bundles are more erratic (lower slope). This slope is the parameter "m" and the lower slope would have a higher numerical value.

The Weibull Equation is:

$$P(s) = 1 - e^{-(s/s_o)^m (L/L_s)}$$

where s is break strength, s_o the normalizing parameter such as an average B.S. or 2/3 the B.S., L is the cable length and L_s the sample test length, and m is the slope parameter. Since there is great interest in the effect of cable length, this equation suggests that a cable 1,000 ft. long or three orders of magnitude longer than the samples, would have a significant effect. The Weibull equation can be used to compare strengths and lengths, $s/s_o = (L_s/L)^{1/m}$ so that the greater the slope m, or greater the scatter of data, the greater the decrease in strength of long lengths. Roger A. Holler of the Naval Air Development Center, at Warminster, Pa. (2) had calculated that 5,000 ft. lengths could have strengths of 60% to 75% of the strengths determined by 5 ft. sample testing. This data in Figure 4 suggests that the lower slope or m value for the braided line would signify less of a strength loss for long lengths compared to the parallel strands. There is a good deal more testing needed of long lengths, which is now being accumulated. The effect of construction is well known on ropes and cables, relating to both the wire-rope-type or cabled-construction, the braid, and the parallel designs. Statistical predictions become increasingly useful for Kevlar cable evaluation, and for the understanding of modes of failure.

3. DYNAMIC PROPERTIES OF SYNTHETIC FIBER ROPES

The dynamic properties of nylon and polyester are quite complex due to the inherent time dependant and rate dependant characteristics [3]. K. R. Bitting [4] examined in great detail the behavior of lines used for mooring buoys; nylon, polyester, and polypropylene lines of over ½" diameter that are cycled at working loads 10% to 30% of break strength and at high loads of 30% to 60%. The effects of water are significant on nylon, but the effects of abrasion dominate. These are well known fiber failure modes identifiable by scanning electron micsoscopy, in both ropes and textiles. The various sizes and constructions of rope change the amount of abrasion, such as plaited vs. double braided.

A key parameter examined was the Dynamic Stiffness or Spring Constant (lb/in/in x 10^4). This measures the energy absorbed as pounds per inch of line length for a given total length. The object was to check the effect of long time exposure (5 years) as mooring lines in sea water. There were several significant conclusions, such as loss of 50% of original tensile strength for both nylon and polyester lines after 5 years, with a 50% increase in strain-at-failure. The spring constant for a nylon double-braid increases 100% after the 5 year exposure, while polyester shows only a 10% to 25% increase. While an extensive degradation is indicated, these are really rather phenomenally good considering the conditions.

Kevlar, when properly pre-conditioned, does not show any significant change in Spring Constant unless worked at loads over 30% of B.S.

Snap loading at under 30% of B.S. does not result in fatigue failures for synthetic ropes, but higher loads show extensive fatigue damage. Snap loading is defined as a cyclic loading through 0 and for electro-mechanical cables invariably causes conductor failure.

There has been data on Kevlar lines that indicates up to 28% loss of B.S., after one year as a mooring line [5]. This represents the most pessimistic data which may have construction and deployment effects rather than Kevlar degradation. There is also data that shows no change in filament strengths, due to the water and load cycling.

There is abundant data to show that at the lower load ranges of 10 to 20% of B.S., there is not likely to be much degradation from tension-tension, or sea water, but the facts are not all in. Certainly there is not the stretch or spring rate change as in the other synthetic fiber ropes. Degradation from self-abrasion is always involved and proper design will keep this below 20%, depending on loads and cycles.

The comparative creep rates are shown in Fig. 2. These illustrate the excellent behavior of Kevlar 49, recognizing that this is for data on loads of 50% of B.S., far higher than a properly designed structure.

The high loads were used in preliminary tests because they were the lowest loads to get meaningful numbers for short time tests (under 1 year). With factors of safety of 5, or less than 20% B.S., the creep would be well under 0.05% per year. While not as low as steel or carbon fiber, it would be superior to steel under corrosion conditions, and far better than any other of the other synthetic fibers.

There have been many studies on the energy absorption of Kevlar compared to steel and nylon ropes. While it can not be said that the Kevlar has no snap-back, its characteristics are much safer than steel or other synthetic ropes. Work is currently being done to accumulate more information.

4. WEAR AND ABRASION

The Kevlar cables require both an internal lubricant, and an external jacket. The cables are most often jacketed with braided nylon or polyester to keep weight and cost down. Although extruded jackets are used when low dielectric has to be maintained (as in radio antenna guys), the typical poly-urethane or ethylene jackets are inferior to braided jacket abrasion resistance, and subject to longitudinal splitting, and non-uniform thicknesses. The combined use of bonded braided jackets with impregnation of TPR or uethane is excellent for severe abrasion conditions on sheaves or pulleys or against concrete. The use of the type 960 finish supplied on the Kevlar is excellent, making the cables of Kevlar superior to steel in working over sheaves, as well as tension-tension fatigue. They will far exceed nylon ropes and when properly designed stand up better than galvanized improved plow steel. This is particularly significant in electro-mechanical cables and conditions that combine bending fatigue and corrosion problems as in marine use.

When very small diameter lines are needed with high wear, such instruments, or carriage drives on typewriters, the treated nylon or polyester braids are best. But they creep and stretch.

Kevlar has been designed and used for these applications where over 3 million cycles of life is a common requirement.

Typically, however, one can not expect to retrofit Kevlar into systems made for steel or synthetic ropes.

Proper design changes have to be made for optimum use.

High strength fibers such as Kevlar also fail under conditions referred to as static fatigue or stress rupture. The classical example is glass, which if loaded to more than 70-80% of B.S., the rods or fibers will fail on standing in less than one year. Figure 3 compares the synthetic fibers, although the nylon and other polymer fibers fail with a creep of the material, while the Kevlar appears to rupture, similar to glass and carbon fiber. In effect, one has to be careful that a design with Kevlar does not have a cumulative loading that allows more than 60% of B.S. for times of 1 year and over. The typical designs with safety factors of 3 to 5 will not experience any rupture problem, allowing for the creep mentioned above.

5. TEMPERATURE AND CHEMICAL RESISTANCE

The list of chemicals and conditions fill many tables, and the reader is referred to DuPont Kevlar literature, such as the information Bulletin K-506A. In brief, the Kevlar has long useful life under load at temperatures under 300°F. Above 350°F the loss of break strength begins to be greater than 10%. The Kevlar does not melt, but tends to char at 500 to 800°F. Except for higher concentrations of strong acids, it will not show much deterioration, and no loss of strength with most solvents, water, weak acids. But the tests have to be performed to be certain. High pressures with steam tend to degrade the polymer, so that use of Kevlar cables for deep well drilling will result in problems of strength loss. Generally, it is far better than any of the other synthetic fibers, and is similar thermally to its related aramid fiber Nomex. Kevlar has a resistance to corrosion, sea water, etc. which is far superior to the steels and stainless steels and can definitely compete in price with the latter. While limited on the high temperature side compared to steel, there is no embrittlement at low temperatures. It can be used in liquid nitrogen, with a small increase in modulus. The arctic application of Kevlar cables has been successful, as well as storage at temperatures below -60°C.

References:

1. Weibull, W., A Statistical Function of Wide Applicability. (Jnl of Appl.Mech. Vl.18,1951)
2. Holler, Roger A., Naval Air Development Center, Warminster, Pa. (Private Communication).
3. DuPont Tech. Inf. Bull. No. X92 and X99.
4. Bitting, Kenneth R., The Dynamic Behavior of Nylon and Polyester Line, U.S.Coast Guard R & D Ctr., Avery Pt., Groton, Ct.(Report No. Cg-D-39-80)
5. Foster, George. Bedford Inst. of Oceanography, Dartmouth, N.S. Canada. (Private Communication)

BIOGRAPHY

E. Scala is a consulting engineer in Ithaca, New York and was formerly Professor of Metallurgy and Materials at Cornell University. Dr. Scala (D. Eng. Yale University) has served as a consultant to many corporations and government agencies, advisory boards, etc. Prior to joining the Cornell Faculty in 1961, he had been a metallurgist at Chase Brass and Copper Co., Manager of Materials at Avco Corp., and Director of the U. S. Army Mechanics and Materials Research Center at Watertown, Mass. For the past 25 years he has been working with high strength fibers and composite materials. He started the Kevlar line and cable work at the Cortland Line Company ten years ago, as Director of its Advanced Products Division which has now become the Cortland Cable Company, Cortland, New York.

Figure 1
"Free" Length Comparison
(Length at which Strength Member Breaks of Its Own Weight = Tensile Strength / Density)

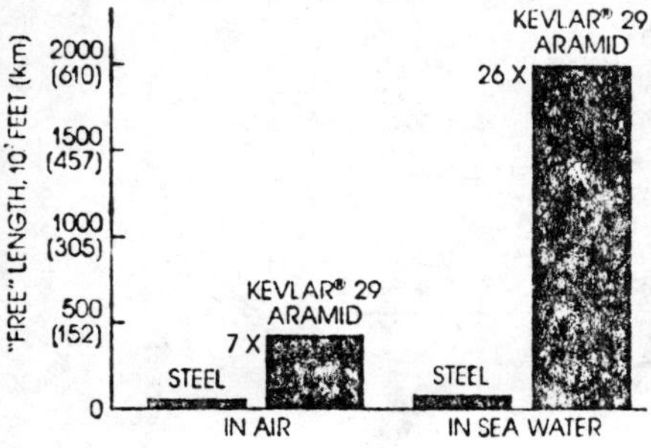

Figure 2
Creep at 50% Ultimate Break Strength

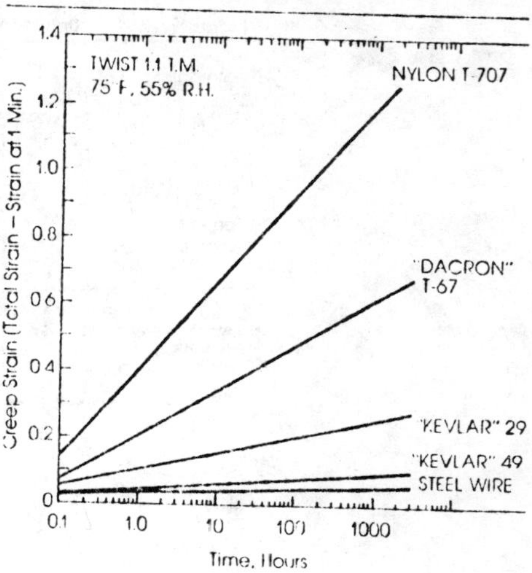

Figure 3
Tensile Strength
vs.
Time Under Fixed Load

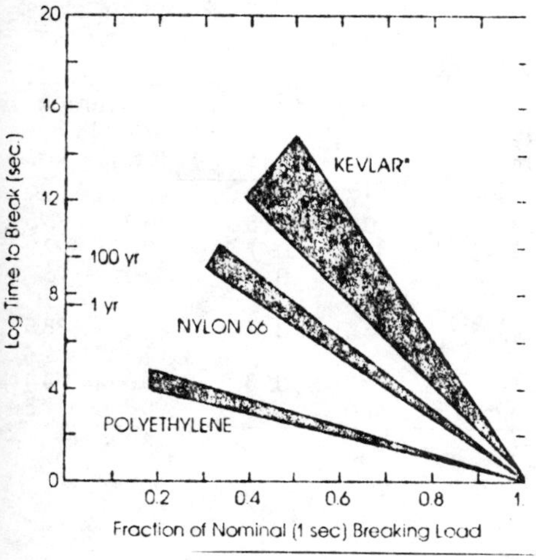

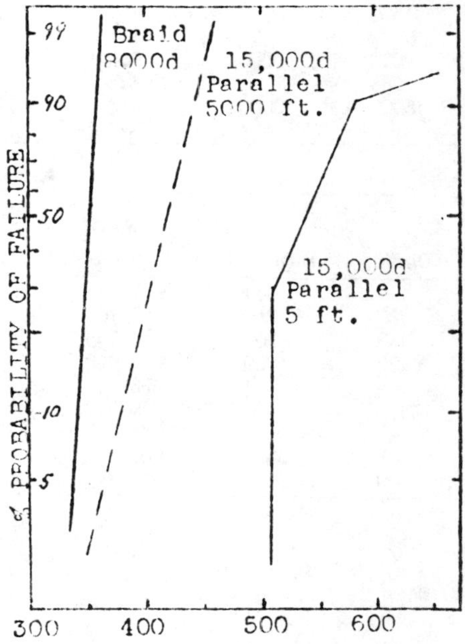

FIG. 4 WEIBULL PLOT OF KEVLAR CABLE TENSILE STRENGTH. (SCHEMATIC ADAPTED FROM R.A. HOLLER REF 2)

TABLE I

Property	Unit	KEVLAR 29	KEVLAR 49	GIPS WIRE
Tensile Strength**	lb/in2	400,000	400,000	285,000
	(MPa)*	(2,760)	(2,760)	(1,960)
Modulus	lb/in2	9×10^6	18×10^6	29×10^6
	(MPa)*	(62,000)	(124,110)	(200,000)
Density	g/cm3	1.44	1.44	7.86
Elongation at break**	%	3.6	2.4	1.1
Filament diameter inch		0.000478	0.000474***	0.020
	(mm)	(0.01214)	(0.01204)	(0.508)

*MPa = MN/m^2 = lb./in.^{2} x 6.89 x 10^{-3}
**Twisted yarn – ASTM D2256 (KEVLAR), single wire (GIPS).
***Except 2130 denier – filament diameter = 0.0005096 inch
GIPS – galvanized improved plow steel
NOTE: These properties are nominal values as tested by a specific method and are not for use as product specifications.

TABLE II

KEVLAR 29

Denier	Filaments	Available Types	Yield yd/lb.	Yield m/kg	Nominal Yarn Dia. 10^{-3} in	Nominal Yarn Dia. mm	Min. Break Strength lbs.	Min. Break Strength kg.
200	134	964	22320	45000	7.8	0.20	8.0	3.6
400	267	964	11160	22500	11.0	0.28	19.0	8.6
1000	666	961	4464	9000	17.3	0.44	43.0	19.5
1500	1000	960 961	2976	6000	21.2	0.54	62.0	28.0
9000	6000	960 961	497	1000	52.0	1.32	360.0	163.3
15000	10000	960 961	298	600	67.1	1.70	615.0	278.9

KEVLAR 49

Denier	Filaments	Available Types	Yield yd/lb.	Yield m/kg	Nominal Yarn Dia. 10^{-3} in	Nominal Yarn Dia. mm	Min. Break Strength lbs.	Min. Break Strength kg.
195	134	965, 968	22895	46155	7.7	0.20	7.7	3.49
380	267	965, 968	11749	23684	10.9	0.28	16.0	7.25
1140	768	965, 968	3916	7895	18.4	0.47	49.0	22.2
1420	1000	965, 968, 978	3144	6338	21.0	0.53	59.0	26.7
2130	1000	965, 968, 978	2097	4228	25.4	0.65	85.0	38.5
4560	3072	968	980	1973	36.8	0.93	250.0	113.4
7100	5000	968	630	1268	47.0	1.19	390.0	177.3

13th National SAMPE Technical Conference
October 13-15, 1981

AN UPDATE ON SYNTACTIC CORE AS A NEW COMPOSITE MATERIAL

Dr. D. Landman
Hysol Division/The Dexter Corporation
P. O. Box 312, Pittsburg, CA 94565

Abstract

Further work on a new, convenient form of syntactic core tape material for use in composite construction is described. Initial work comparing two thicknesses of core with sandwich construction using E-glass, graphite and Kevlar R skins showed that sandwich properties vary considerably with skin reinforcement type, and that the level of adhesion by the resin to the fiber is a controlling factor. In addition, sandwich shear properties were, in most cases, inversely proportional to core thickness.

Later work comparing interply versus sandwich core and no core construction with E-glass, graphite and Kevlar R skins suggests that sandwich and interply structures have comparable strength and stiffness properties. Syntactic core does not appear to markedly degrade composite properties; it considerably enhances compressive properties. Interply structure is better than sandwich construction because it leads to non-catastrophic failure in flexure. Syntactic core offers an alternative approach to composite construction since less fiber layers are needed to build up stiffness. A 350°F resin system was used as the composite and core material.

1. INTRODUCTION

Earlier this year preliminary data was presented on a newly available low density syntactic core film.[1] This material is an epoxy resin system incorporating high strength hollow glass microballoons. It can be made in roll form with widths ranging from one to three feet and thicknesses from ten to fifty mils.

It has been recent practice to build up composite stiffness by using more layers of fiber or by using thick lightweight cores to build thicker structures. Syntactic core offers an alternative approach: basically, strength requirements are first determined, followed by stiffness requirements. Instead of using more fiber layers, the stiffness is built up with syntactic core to give a thin structure which is strong and lightweight.

In the course of our investigations into the use of syntactic core in combination with other composites, we decided to study the effects of different structural configurations on overall composite properties. This report details the results of our findings.

2. GENERAL PROPERTIES OF SYNTACTIC CORES

Syntactic core is a lightweight controlled thickness tape and a space-filler. It is a good adhesive system and can be co-cured with conventional prepregs. Tables 1 and 2 show tensile shear and flatwise tensile strength values for various sandwich structures incorporating a 20 mil layer of a 350°F curing syntactic core, ADX 819. The adhesive strength is very good over the temperature range -65°F to 180°F.

Syntactic core is easy to handle so that assembly costs for parts are reduced. Since syntactic core increases stiffness to weight ratio, parts can be machined more easily. For example, when an "all" Kevlar [R] composite is cut with a band saw, the edges are badly frayed; when syntactic core is used with Kevlar [R], the cut edges are much neater. Other benefits to using syntactic core include: less warpage of composites, lighter parts and improved impact resistance on a weight basis.

Syntactics have greater densities compared to conventional core materials. However, their strength to weight ratio exceeds other core materials (Table 3). As a result, syntactics offer greatest advantage where core thicknesses less than 100 mils are required. This is the thickness where it is difficult to cut most conventional core materials. To date, applications where syntactic core is used benefit from this feature.

Syntactics tend to be notch sensitive because they are brittle. This presents no problems when syntactics are used in combination with face sheet materials which are usally compliant. In addition, syntactic cores are compatible with a wide variety of materials. For example, aluminum cores in combination with graphite face sheets are subject to galvanic corrosion; syntactic cores are not.

3. APPLICATIONS

Syntactic core is used in the advanced composite aileron[2] and vertical fin[3] in Lockheed's L-1011. In the aileron, the syntactic is used for form a sandwich-structured face sheet with graphite fabric. This gives a structure with almost 50% less substructure and over 20% weight reduction compared to the all metal aileron.

A second application, also by Lockheed, is in the construction of a remote piloted vehicle called the Aquila.[4] The construction in this case is also a sandwich one with Kevlar[R] face sheets. The entire vehicle including the rib substructure is made in this fashion, giving a total weight of about 50 lbs. for a payload of over 200 lbs.

Georgia Tech is presently investigating the use of syntactic core in a lightweight aircraft. Their intention is to fly around the world to set a gas mileage record. The anticipated construction for their "Project Hawk" is Kevlar/syntactic core sandwich.

4. TEST PROGRAMS

The objective of the first part of the test program was to determine the effect of different core thicknesses on overall composite properties. The second part was concerned with the effect on composite properties of fabric/syntactic core interply* construction versus a fabric/syntactic core sandwich construction.

The materials used as skin reinforcements in this test program included: Sytle 1581 Woven E-glass (resin content 48-53%); Style 285 Kevlar[R] ** fabric (resin content 57-58%); unidirectional Thornel [R] *** T300/3000 graphite tape (resin content 34-35%); Style A193P (Hercules) woven graphite (resin content 36-37%). These fabrics were impregnated with the same 350°F resin system as used in ADX 819. The core material used in all cases was ADX 819. Core thicknesses were built up to 20 or 50 mils by using 10 and/or 20 mil thick ADX 819.

* Interply construction refers to alternating layers of fabric and syntactic core.

** Kevlar[R] is a registered trademark of E. I. du Pont de Nemours & Co.

*** Thornel[R] is a registered trademark of Union Carbide Corporation.

Figure 1 shows the sandwich construction used for the first part of this program. Woven fabric face sheets were layed up in alternating 0°/90° layers; unidirectional graphite tape face sheets were layed up 45°/0°/135°/45°/0°/135°. The syntactic core thickness was 20 or 50 mils.

Figure 2 shows the various constructions used for the second part of this program. The woven fabric face sheets were layed up 0°/90°; unidirectional graphite tape face sheets were layed up all 0°.

Figure 3 shows a preferred lay-up configuration for making controlled thickness laminates. Coupons were cut for testing from 11" x 13" laminates. The laminates were cured in a vented autoclave at 350°F and 85 psi pressure for one hour.

Kevlar was cut using a band saw with the teeth in reverse direction. All other coupons were cut using a diamond slab saw. Fine sandpaper was used to remove frayed edges. When necessary, tabs or metal blocks were bonded to the coupons using a room temperature curing secondary adhesive. As pointed out previously[1], difficulties were experienced in acquiring data where secondary bonding was necessary, because the composite structures were too strong.

Tests in this program were: longitudinal tensile (ASTM D-3039), compressive (Fed. Std. 406-1021) and flexural (ASTM D-790) properties, as well as, short beam shear (ASTM D-2344) and flatwise compressive strengths (Mil·Std. 401A). The tests were run at 75°F and 180°F. The results of some of these tests will be highlighted in this report. Further details are available in a preliminary data package.[5]

In order to calculate the various strength properties, it was assumed that these are dominated by the strength of the face sheets. Consequently, for comparative purposes, normalized stress and normalized modulus were calculated as indicated by equations (1) - (3).

$$\text{Normalized Stress} = \frac{P}{W \times T} \qquad (1)$$

$$\text{Normalized Modulus} = \frac{\text{Normalized Stress}}{\text{Strain}} \qquad (2)$$

$$T = N \times t \qquad (3)$$

Where, P = Ultimate Load

W = Width of Specimen

T = Normalized Thickness

N = Number of plies of face sheets

t = Nominal thickness per ply

5. RESULTS

5.1 The effect of different core thicknesses on composite properties. In the previous report [1] of this work it was shown that syntactic cores in combination with composite skins provided structures with substantial strength properties. The strength depends on the type of fiber, adhesion to the fiber and thickness of the core.

The short beam shear (Table 4), normalized longitudinal tensile (Table 5) and normalized longitudinal compressive (Table 6) properties of 20 versus 50 mil syntactic core sandwich structures further illustrate these points. For example, from the short beam shear and longitudinal tensile properties (Tables 4 and 5) it can be seen that the composites with 50 mils syntactic core are marginally lower in strength than those with 20 mil core. In the case of Kevlar, syntactic core significantly enhances the longitudinal compressive properties (Table 6). Compared to unidirectional graphite/syntactic core sandwich composites, the non-core containing composites are considerably weaker (compare last two entries in Tables 4,5 and 6).

5.2 Interply versus sandwich construction. When we began this work, our major concern was with building structures of approximately equal thickness. As can be seen from Figure 2, this meant that in some cases it was necessary to increase the number of fiber layers (syntactic core layers were always of the same number). In the case of the composites with no syntactic core, a considerable number of additional fiber layers were needed to build up to the same thickness as the core containing composites. Consequently, comparisons based on the "normalized" properties [equations (1) - (3)] are especially important. Details of the thicknesses and specific gravities of samples made for this study are shown in Tables 7 and 8. For purposes of easier identification, the symbols I, S and N are used in the text and tables to denote interply, sandwich and no core ("all"composite) constructions.

Tables 9 and 10 show the normalized longitudinal tensile and compressive properties for the various I, S and N constructions. As expected, the strength and modulus properties are strongly dominated by the nature of the fiber used. Thus the order of greatest to weakest tensile strengths is unidirectional graphite, woven graphite, Kevlar and woven glass, as expected (Table 9). Similar observations are applicable for compressive strengths. In the case of tensile properties, the syntactic core does not appear to markedly degrade overall properties; for compressive properties, syntactic core considerably enhances properties.

Turning to a comparison of interply (I) versus sandwich (S) constructions, it can be seen that there are no marked differences. The interply structures are stronger in tension, while the sandwich structures are stronger in compression. Short beam shear and flexural properties on these materials have also been run. These results have been included in the preliminary data package.[5] We have also done 180°F tests which will be the subject of another report.

Table 11 shows a comparison of the flexural properties for interply versus sandwich constructions. Basically, the same trends as for the tensile experiments are observed. However, examination of the Instron curves and failure modes for interply versus sandwich construction produced some interesting results. Figure 4 shows the load/extension curves for flexure of interply (I) and sandwich (S) beams. The sandwich structure behaves in a fairly Hookean fashion until ultimate load is reached. At this point catastrophic failure of the beam occurs. The interply structure also behaves in a Hookean fashion to ultimate load (the lower load value for interply is exaggerated in Figure 4). However, instead of failing catastrophically, the ultimate load is recovered. This process continues until essentially all the fiber layers are broken. In other words, the syntactic core shields one fiber layer from the next during loading. Construction of this kind should be extremely useful.

6. CONCLUSIONS

The results presented in this paper regarding the effects on composite properties from using different thickness of syntactic core further correlate with those presented earlier.[1] At least in the case of unidirectional graphite laminates, syntactic core construction enhances physical properties on an equivalent number of face sheet basis. The syntactic core acts as a very thin spacer which effectively separates fiber face sheets and improves stiffness properties, as well as maintaining strenth properties of the composite. The flexural results for interply versus sandwich construction, strongly point out the desirability of the interply construction for non-catastrophic failures.

7. REFERENCES

1. "Syntactic Core - A New Composite Material", D. K. Klapprott and D. Landman, 26th National SAMPE Symposium, 795 (1981).

2. Lockheed-California Co./NASA Contract NASI-15069.

3. Lockheed-California Co./NASA Contract NASI-14000.

4. "Mini-RPV Being Developed for Army", Aviation Week and Space Technology, 54, Jan. 7, 1980.

5. "Preliminary Data Package for the Use of Syntactic Core in Composite Construction", D. Landman, Hysol Division/The Dexter Corporation (1981).

8. ACKNOWLEDGEMENTS

It is a pleasure to acknowledge the work of my associate Mr. Paul R. Schreiner who has done much of the laboratory work described in this paper. Many thanks to my colleague Mr. David K. Klapprott who helped with suggestions and proofing of the paper. Finally this work was done under the auspices of Hysol Division/The Dexter Corporation at Pittsburg, California.

Table 1: Average Tensile Shear Strength of Composites with 20 mil Syntactic Core Sandwich Construction

Facing Material	Core Tensile Shear Strength (Ksi) -65°F	75°F	180°F
Woven Glass	1.8	2.0	1.8
Kevlar	1.5	1.4	0.9
Woven Graphite	1.7	1.8	1.7
Unidirectional Graphite	1.5	1.7	1.2
Aluminum		2.4	

Table 2: Average Flatwise Tensile Strength of Composites with 20 mil Syntactic Core Sandwich Construction

Facing Material	Flatwise Tensile Strength (Ksi) -65°F	75°F	180°F
Woven Glass	3.8	3.7	3.4
Kevlar	1.5	1.9	1.5
Woven Graphite	1.6	2.8	2.5
Unidirectional Graphite	1.8	3.0	2.2
Aluminum		3.5	3.4

Table 3: Properties of Various Core Materials

Material	Density (lb/ft^3)	Shear Strength to Weight Ratio (ft)	Compressive Strength to Weight Ratio (ft)
Honeycomb	1.5-8	480-1080	360-2280
Balsa	5-10	360-420	170-240
Urethane Foam	2-30	180-360	120-780
Syntactic Core	30-50	840-1440	1560-2520

Table 4: Average Short Beam Shear Strengths of Composites at 75°F made with two thicknesses of Syntactic Core in Sandwich Configuration.

Facing Material	Short Beam Shear Strength (Ksi) 20 Mil Core	50 Mil Core
Woven Glass	6.4	6.5
Kevlar	3.4	3.1
Woven Graphite	6.5	5.3
Unidirectional Graphite	4.3	4.3
Unidirectional Graphite with no syntactic core	3.0	

Table 5: Average Longitudinal Tensile Properties of Composites at 75°F Made with Two Thicknesses of Syntactic Core in Sandwich Configuration

Facing Material	Normalized Tensile Strength (Ksi)		Normalized Tensile Modulus (psi x 10^6)		Elongation %	
	20 Mil	50 Mil	20 Mil	50 Mil	20 Mil	50 Mil
Woven Glass	46.6	45.9	3.2	3.1	1.8	1.9
Kevlar	53.0	45.5	3.9	3.6	1.4	0.9
Woven Graphite	113.6	79.2	10.0	9.6	1.1	0.8
Unidirectional Graphite	87.6	85.0	6.6	7.7	1.3	1.7
Unidirectional Graphite with no syntactic core	78.0		6.8		1.2	

Table 6: Average Longitudinal Compressive Properties of Composites at 75°F Made with Two Thicknesses of Syntactic Core in Sandwich Configuration

Facing Material	Normalized Compressive Strength (Ksi)		Normalized Compressive Modulus (psi x 10^6)	
	20 Mil	50 Mil	20 Mil	50 Mil
Woven Glass	62.1	63.7	3.4	4.2
Kevlar	15.1	21.6	2.8	4.8
Woven Graphite	57.5	56.4	8.3	10.0
Unidirectional Graphite	82.3	56.4	7.1	8.3
Unidirectional Graphite with no syntactic core	44.8		9.8	

Table 7: Average Thickness of Cured Composite Laminates Containing Syntactic Core

Facing Material	Average Cured Thickness of Laminates (in.) Interply	Sandwich	No Syntactic Core
Woven Glass	0.095	0.110	0.087
Kevlar	0.098	0.085	0.093
Woven Graphite	0.092	0.106	0.090
Unidirectional Graphite	0.110	0.114	0.085

Table 8: Average Specific Gravities at 75°F of Cured Composite Laminates Containing Syntactic Core

Facing Material	Average Specific Gravity of Laminates (lbs/ft^3) Interply	Sandwich	No Syntactic Core
Woven Glass	0.0481	0.0458	0.0686
Kevlar	0.0356	0.0396	0.0489
Woven Graphite	0.0432	0.0382	0.0575
Unidirectional Graphite	0.0332	0.0340	0.0568

Table 9: Average Normalized Longitudinal Tensile Properties at 75°F for Composites with Interply, Sandwich and No Syntactic Core Construction

Facing Material	Normalized Tensile Strength (Ksi)			Modulus (psi x 10^6)			Elongation(%)		
	I	S	N	I	S	N	I	S	N
Woven Glass	39.2	49.0	58.0	9.4	3.3	3.5	1.2	1.8	2.1
Kevlar	51.6	44.4	62.4	4.7	5.3	4.2	1.3	1.2	1.4
Woven Graphite	90.1	101.1	112.1	10.8	9.7	11.0	0.9	1.1	1.0
Unidirectional Graphite	174.8	183.6	185.2	18.8	19.1	19.0	0.9	1.0	0.8

I: Interply Construction (See Figure 2).

S: Sandwich Construction (See Figure 2).

N: No Syntactic Construction (See Figure 2).

Table 10: Average Normalized Longitudinal Compressive Properties at 75°F for Composites with Interply, Sandwich and No Syntactic Core Construction

Facing Material	Normalized Compressive Strength (Ksi)			Modulus (psi x 10^6)		
	I	S	N	I	S	N
Woven Glass	89.7	82.5	52.8	4.3	4.2	4.2
Kevlar	26.4	23.0	20.5	4.1	5.1	4.3
Woven Graphite	83.3	70.4	51.7	9.4	8.9	9.1
Unidirectional Graphite	127.6	103.8	66.0	12.2	22.3	18.3

I: Interply Construction (See Figure 2).

S: Sandwich Construction (See Figure 2).

N: No Syntactic Construction (See Figure 2).

Table 11: Average Flexural* Properties at 75°F for Composites with Interply and Sandwich Construction

Facing Material	Flexural Strength(Ksi) I	Flexural Strength(Ksi) S	Modulus (psi x 10^6) I	Modulus (psi x 10^6) S
Woven Glass	59.5	50.0	2.5	3.1
Kevlar	40.9	51.1	2.4	2.8
Woven Graphite	71.5	68.3	5.0	6.1
Unidirectional Graphite	111.4	102.0	6.6	9.1

I: Interply Construction (See Figure 2).

S: Sandwich Construction (See Figure 2).

* Three point bending experiment on 2.5 in. x 1.0 in. x 0.125 in. specimen

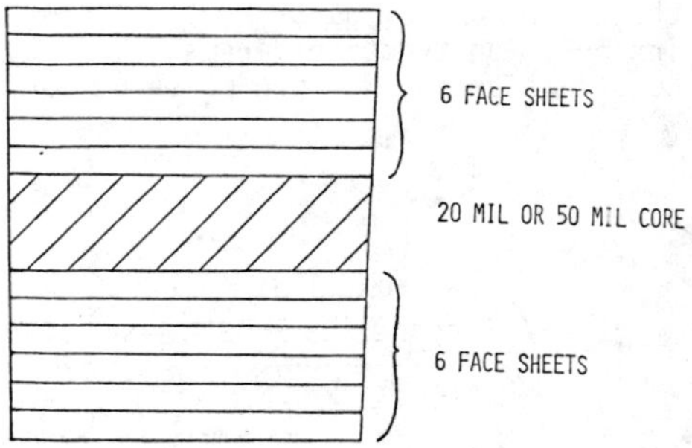

Figure 1: Sandwich Construction for Part 1 of Program to Determine the Effect of Different Core Thicknesses on Overall Composite Properties

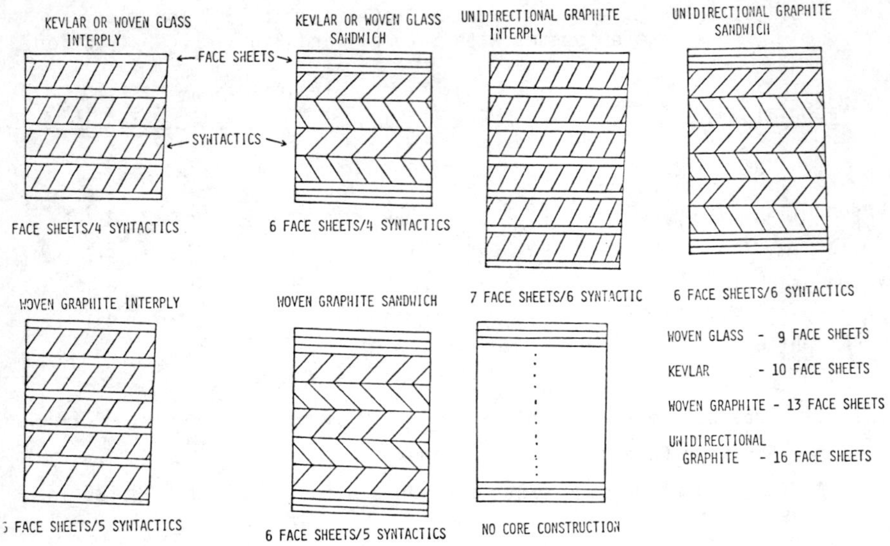

Figure 2: Various Constructions Used for Part 2 of Program to Determine the Effect on Composite Properties of Fabric/Syntactic Core Interply Construction Versus Fabric/Syntactic Core Sandwich Construction and No Core Construction.

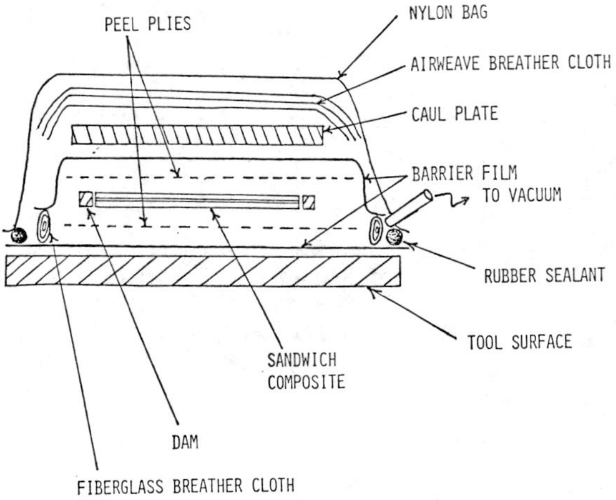

Figure 3: Suggested Lay-up Configuration for Making Controlled Thickness Laminates with Syntactic Core.

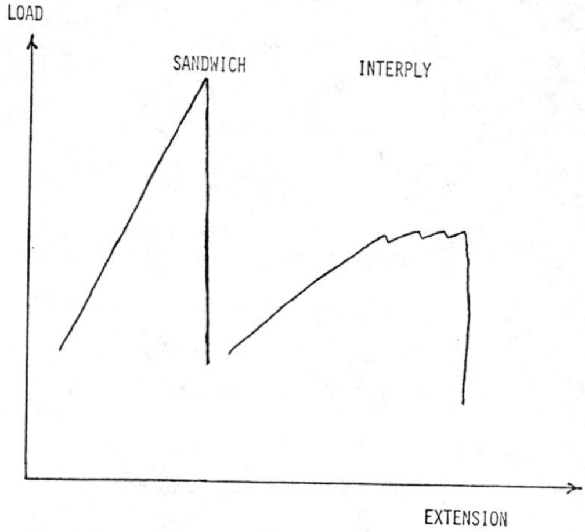

Figure 4: Load/Extension Curves Run on an Instron for Flexure of Sandwich Versus Interply Constructions

BIOGRAPHY

David Landman is an Associate Scientist at Hysol Division, The Dexter Corporation located at Pittsburg, California. He is responsible for development of syntactic foam materials. Prior to joining Hysol, he was employed by Milliken Chemical where he was responsible for technical development of epoxy curing agents. Dr. Landman received his B. S. and Ph.D. in chemistry at the University of Sydney, Australia.

13th National SAMPE Technical Conference
October 13-15, 1981

A NEW RESIN FOR FIELD REPAIR

C.B. Delano and A.H. McLeod
Aerotherm Division of Acurex Corporation
Mountain View, California 94042

F.J. Riel
Rohr Industries Incorporated
Chula Vista, California
and
R.H. Greer
Rohr Industries, Incorporated
Riverside, California

Abstract

Several low energy cure resin systems for field repair of aircraft have been investigated. The selected epoxy system provides good room temperature adhesive strength after overnight cure at room temperature and good adhesive strengths at 300°F after moderate post cure. A new method of fabricating graphite composites at room temperature has been developed for field repair of graphite composite structures. Metal-to-metal and composite-to-metal bond strengths are reported.

1. INTRODUCTION

Repair techniques in normal use in field and depot applications for failed or damaged aircraft components have been relatively primitive compared to the primary bonding procedures used for manufacturing new hardware. Over the years, various techniques have been developed to upgrade field repair technology. For example, special repair carts have been designed to provide vacuum bag pressure, plus heat lamp or heating blanket cures. Two documents address the problem of structural repair of various types of damage(1,2). The first report covers all degrees of damage from minor scratches to major rebuilding of large areas of damage. The documents list a variety of acceptable adhesive systems for room temperature-to-elevated temperature cures. Room temperature curing systems however, are discouraged except for minor dent or scratch repairs. The second report includes direct evaluation of a room temperature curing adhesive and two heat curing systems. The report concludes that the room temperature curing system evaluated is not acceptable, and that unacceptability applies to ambient curing adhesives in general. The major reason for rejecting the room temperature curing systems was poor performance in humid environments under stressed conditions. Thus, a heat cure under vacuum bag pressure was defined as the preferred repair procedure. Both reports describe a phosphoric anodized procedure suitable for local surface treatment.

Acurex and Rohr undertook the development of resin systems which could be cured by low energy processes under sponsorship of the Materials Laboratory of the Air Force Wright-Aeronautical Laboratories. Processing requirements included simple cure procedures by unskilled personnel with no cumbersome equipment such as autoclaves. Performance properties equivalent to heat-cured state-of-the-art systems were sought. The developed resin(s) is required for use both as a composite matrix and as an adhesive for repair of both composite and metal structure of military aircraft. Two performance temperatures were specified: 180°F-to-200°F for a Class I system and 300°F for a Class II system.

The efforts described in this paper summarize the excellent progress to date toward the development of a resin system for field repair. Although much work remains to be done, we are very encouraged by our initial results. The data currently available indicate the feasibility of a single repair kit for use at room temperature and both Class I and Class II performance temperatures.

2. SCREENING OF COMMERCIALLY AVAILABLE AND NEWLY DEVELOPED RESIN SYSTEMS

We found no low energy curing system available commercially which met the requirements for a Class II system. Some commercially available room temperature curing systems had marginally acceptable properties for Class I performance, however none had acceptable environmental durability. Screening efforts on commercially available and Acurex-developed systems eventually led to selection of a low moisture absorption two-part epoxy system. The selected system has the familiar individual parts with long storage lives, commercially available ingredients, good environmental durability, contact and/or vacuum bag pressure processing and holds potential for spike cures and very rapid repair.

The results of the final screening efforts on the best resin candidates are given in Table 1. Neat resin specimens (1/4 inch squares) were subjected to high temperature post cure (for cure completion) and then used to determine the resin's moisture weight gain and softening point after 24 hour water boil. The measurement of the distortion temperature of a specimen under a 450 psi compressive load (DTUL) was found to be very sensitive to moisture plasticization (softening) effects. Neat resin samples upon redry at 300°F after 24 hour water boil frequently cracked. Although resin cracking upon redry is not completely understood. It appears that resins which do not crack upon redry are superior to those that do crack. Barcol Hardness measurements of the dry resin castings suggests useful resin modulus for graphite fiber composites. In combination with the "wet" DTUL measurements, an upper use temperature (wet) for such composites is suggested. Lap shear strengths of the best systems varied from 1300 to nearly 2900 psi with heat cure. Formulation of the heat-cured systems for room temperature cure increased the room temperature lap shear strengths with some sacrifice in high temperature performance and moisture resistance. In particular, it should be noted in Table I that all systems formulated for room temperature cure show an increase in moisture weight gain and DTUL loss due to 24 hour water boil from their respective unformulated compositions.

Potential cure methods for the best candidate systems include surface activation for acrylic systems wherein catalyst is applied to one surface and resin to the second surface. Radiation cure, heat cure, as well as several other cure methods hold potential for field repair. Blind-side repair for radiation cure systems may not be practical.

3. THE SELECTED RESIN SYSTEM AND ALUMINUM SINGLE LAP BOND STRENGTHS

The epoxy novolac, DEN 438 cured with isophorone diamine (next to last entry in Table 1) was selected for further development efforts due to its excellent moisture resistance and promising DTUL (1.7 weight percent gain and DTUL of 311°F after 24 hour water boil). The final compositions of the formulations of the isophorone diamine cured epoxy novolac for room temperature cure for a composite matrix resin, AF-R-40 and an adhesive resin AF-A-42 are given in Figure 1. (AF-R-40 is Air Force Resin-40 and AF-A-42 is Air Force Adhesive-42.)

The single lap shear properties of these resin systems with 2024-T3 aluminum are given in Table 2. Of note was the increase in elevated temperature strength after 160°F humid aging for the room temperature cured systems. This is due to exposure to the heat of the humid aging conditioning which caused a further cure of the systems. Initial shear testing showed that severe strength reduction occurred with the AF-A-42 system at about 140°F if the system were only room temperature cured. Although slow heat cure schedules would provide elevated temperature performance without significant softening, we elected to achieve our

Class I and Class II performance temperatures by heat cure for rapid turn-around times. Various cure schedules versus the developed DTUL's are given in Table 3. Table 3 results show that overnight cure at 140°F leads to a DTUL of 201°F and probably provides a Class I performance system. One hour at 180°F was selected for the Class I system and two hours at 300°F for the Class II system.

Composite properties are not discussed in detail in this paper. However, it is appropriate to mention the unique process which was used to fabricate composites from the AF-R-40 resin. A thick resin layer was applied to Mylar and a picture frame wrapped five turns with graphite broad goods placed on the resin layer. With careful attention to vacuum bagging procedures, a 10-ply graphite-AF-R-40 composite was produced. These unidirectional Celion composites (small amount of Dacron fill fibers) were used for the straps in the double strap specimens described below.

4. ALUMINUM-COMPOSITE DOUBLE STRAP BOND STRENGTHS

Several methods have been investigated for transitioning the load from the structural member to be repaired into the straps or "patch." Tapered edges(3) and carefully cut prepregs with specified layup sequences(4) have been used. We investigated use of surface plies of several materials such as S-glass between the tangs and straps to achieve load transitioning. We found that although very high loads per pounds per inch of width could be achieved with certain geometries, calculation of the load at failure in unit stress was not straight forward since overlap distances could not be accurately determined because of the surface reinforcement. With heavy, strong fabrics failure usually occurred between the fabric and adherend in the "zero" thickness glueline. To simplify interpretation of the wet stress durability test results, bondline thickness of 5-to-10 mils, as practiced in metal-metal bonding, were selected.

After several scouting experiments we finally selected the specimen geometry shown in Figure 2 which provided a minimum 5 mil bondline, failing loads in unit stress (psi) due to finite overlaps and failures in the bondline.

Table 4 gives the strengths of the aluminum-composite double strap bonds resulting from Class I and Class II cure cycles. The construction was aluminum tangs with graphite straps as indicated in the table. Class I strengths are adequate and air aging up to 500 hours showed little degradation of the bond strength. Class II testing indicated that the adhesive system continued to cure for the first few hours of exposure to 300°F. Bond strengths may be questionable for use of the adhesive at this temperature. Four specimens were tested at each data point.

The observed failure modes are described in Figure 2 and failure modes of the specimens at each data point are given in Table 4. Failure to the graphite strap (failure mode 5) is not a test of the AF-A-42 since the resin at that point in the specimen is the AF-R-40 matrix. Failure between the 1659 E-glass used for bondline thickness control and the 116 E-glass used for galvanic isolation was usually not observed.

Stress durability test results of the aluminum-composite double strap bonds is given in Table 5. The 875 psi sustained stress under ambient conditions had little if any influence on the residual bond strengths up to 1000 hours. Bond strength reductions were observed with moisture exposure. All failures occurred in the AF-R-40 matrix at the strap's surface (failure mode 5) and the 116 E-glass may have had an influence on the moisture sensitivity of the bonds. It is noted that extensive investigation of surface treatments did not occur for addressing this failure mode and application of the galvanic isolation fabric may have included wetting problem since the composite surface was only MEK wiped and not sanded prior to bonding.

5. COMPOSITE-COMPOSITE DOUBLE STRAP BOND STRENGTHS

The construction of these bonds was identical to that used for the aluminum-composite bonds except the 116 E-glass used for the galvanic isolation described above was not employed. 3501-6/AS fiber composites were used for the tangs and the AF-R-40/Celion fiber composites for the straps. Their construction is indicated in Table 6. Table 6 gives the bond strengths for the Class I and Class II cure cycles of these specimens. Although strengths are somewhat less than the previously described aluminum-composite double strap specimens, Class I strengths appear adequate whereas Class II strengths at 300°F may be low. Bond strengths were not effected by thermal aging at either 180°F or 300°F.

The AF-A-42 did not adhere well to the 3501-6/AS fiber composite as evidenced by the adhesive failures to that substrate (failure mode 1). Only eight specimens of the 64 represented in the table showed any evidence of cohesive failure. Occasional failure in the strap was observed.

Table 7 gives the stress durability testing under ambient and moisture exposures of the composite-composite double strap bonds. As discussed above with the initial testing, little information about the strength of the AF-A-42 adhesive can be inferred from the data since failure modes were principally to the 3501-6/AS fiber tang surface and in the strap. Occasional cohesive failures were observed however, air entrained in the bondline may be the cause of these failures. Sustained stress under ambient conditions had little effect on the residual bond strengths whereas moisture conditions appear to result in reduced bond strengths. This may be a problem with the 3501-6/AS matrix. Further testing is needed to define the origin of the wet strength loss.

6. CONCLUSIONS

Bond strengths of the AF-A-42 are adequate for Class I field repair applications and possibly Class II field repair applications. 500 Hour thermal aging at 300°F did not significantly reduce bond strengths.

Ambient stress durability tests indicated that the AF-A-42 is not significantly effected by the 875 psi sustained stress up to 1000 hours.

Stress durability under humid conditions for 1000 hours (140°F, 95+ percent RH) appears good with aluminum/AF-R-40 composite double strap bonds. With 3501-6/AS-composite/AF-R-40 composite double strap bonds, further efforts are required to understand why residual bond strengths were reduced under humid aging conditions.

A room temperature curing primer system needs to be developed for use with the AF-A-42.

The efforts described in this paper were sponsored by the Materials Laboratory of the Air Force Wright-Aeronautical Laboratories under contract F33615-79-C-5038.

7. REFERENCES

(1) "Adhesive Bonded Aerospace Structures Standardized Repair Handbook," AFML-TR-77-206, December 1977.

(2) "Repair of Bonded Primary Structure," AFFDL-TR-78-79, June 1978.

(3) "Repair Guide for Large Area Composite Structure Repair," AFFDL-TR-79-3039.

(4) Sage, G.N., "Aspects of Bonded Joint Design in Carbon-fibre Reinforced Plastic," Adhesive 3, Applied Science Publishers LTD, London, 1978.

8. BIOGRAPHIES

Chadwick B. Delano is the Assistant Manager of the Materials Department of the Aerotherm Division of Acurex Corporation. He has been responsible for a number of programs which cover a wide span of composite and adhesive research and technology, including environmental resistance, vulnerability protection and other requirements. He received his B.S. degree in Chemistry from San Diego State College in 1959.

Alfred H. McLeod is a Staff Chemist for Aerotherm Division of Acurex Corporation. He is responsible for a number of resin formulating efforts for a number of applications which include adhesives, potting compounds and composite matrices. He received his B.S. degree in Chemistry from San Diego State College in 1957.

Dr. Frank J. Riel is Program Manager of a number of DoD and Rohr sponsored programs involving application of structural adhesives to high performance aerospace structures using aluminum, titanium, steel and advanced composite materials. Prior to joining Rohr, he was Technical Director of Whittaker R&D Division in San Diego, California. He received his Ph.D. degree in Organic Chemistry from Purdue University in 1951.

Robert H. Greer is Materials and Process Engineer for Rohr Industries, Incorporated. He is responsible for organizing and supervising activities on materials, engineering and development tasks involving structural adhesives and advanced aerospace structures. He received his B.A. degree in Chemistry from San Diego State College in 1953.

TABLE 1. SUMMARY OF BEST RESIN CANDIDATES

		Resin's 24 Hour Water Boil Properties[b]		Condition of Redry Specimens	Barcol Hardness	Room Temperature Al-Al Single Lap Shear Strength, psi	
Resin	Potential Cure Methods[a]	DTUL, °F	Weight Gain, (%)	Post Cured	Post Cured	RT Cured	Post Cured
VINYL ESTERS							
DEN 438-methacrylate	2	302	2.3	Cracked	41	--	1390
Epon 828-methacrylate	2	291-311	3.4-3.7	Cracked	39-45	--	1354-1699
Epon 1031-methacrylate	2	401	2.9	Cracked	45	--	1345
URETHANE METHACRYLATES							
PAPI	2	293	---	OK	46	--	1950
Isophorone diisocyanate	2	280	2.6	Cracked	46	--	1467
Above formulated for RT cure	1+2	235	5.6	Cracked[c]	52(122°F)	2140	1360
Trimethylhexane diiso-cyanate	2	178	3.1	OK	43	--	2037
Above formulated for RT cure	1	144	6.3	Cracked[c]	37(122°F)	3030	1590
URETHANE ACRYLATES							
PAPI	1+2, 3[d]	216	2.9	Cracked	40(122°F)	1950	1380
Isophorone diisocyanate	1+2, 3[d]	183	4.2	Cracked	50(122°F)	2900	1510
Trimethylhexane diiso-cyanate	1+2, 3[d]	163	4.8	Cracked	47(122°F)	3060	2010

TABLE 1. Concluded

Resin	Potential Cure Methods[a]	Resin's 24 Hour Water Boil Properties[b]		Condition of Redry Specimens	Barcol Hardness	Room Temperature Al-Al Single Lap Shear Strength, psi	
		DTUL, °C	Weight Gain, (%)	Post Cured	Post Cured	RT Cured	Post Cured
EPOXIES							
Hysol 9340	1	127	5.5	OK	--	2800	2870
Masterbond EP33UN	1+2	---	---	--	--	4210	2290
0510/Isophorone diamine	2	329	3.3	OK	--	871	2880
DEN 438/isophorone diamine	2	311	1.7	OK	--	580	1714
Above formulated for RT cure (AF-R-40)	1+2	248	2.8	OK	43	1530	2740

[a] Cure methods, 1 = ambient conditions, 2 = heat required, 3 = possibly radiation
[b] Post cured systems, two hours at 300°F
[c] Other formulations did not crack upon redry
[d] Radiation cures probably also require heat to complete polymerization

TABLE 2. ALUMINUM-ALUMINUM SINGLE LAP SHEAR STRENGTHS, psi[a]

Specimen Conditioning	Room Temperature Cure Cycle Test Temperature				Class II Cure Cycle Test Temperature			
	-65°F	RT	180°F	300°F	-65°F	RT	180°F	300°F
				AF-R-40				
Initial, dry	1310	1530	635	250	1890	2740	2310	2330
Humid aged[b]	--	3290	2040	1030	--	2410	2810	840
				AF-A-42				
Initial dry	696	3810	590	290	2280	2050	2241	1598
Humid aged[b]	--	2850	--	350	--	2500	--	520

[a]Surface treatment, 10V phosanodized, Z 6040 silane
[b]Humid aging, 10 days at 160°F and 95+% RH

TABLE 3. DTUL VERSUS CURE CONDITION OF AF-R-40

Cure Condition	DTUL
Overnight at room temperature	140°F
50 Minutes at 140°F	180°F
2 Hours at 140°F	187°F
6 Hours at 140°F	189°F
16 Hours at 140°F	201°F
30 Minutes at 180°F	235°F
2 Hours at 300°F	320°F

TABLE 4. ALUMINUM-COMPOSITE DOUBLE STRAP BOND STRENGTHS[a]

(2024-T3 Bare Aluminum Tangs/AF-A-42-1659 E-glass/AF-R-40-116 E-glass-Celion Straps)[b]

Specimen Conditioning[c]	Class I Cure Cycle Test Temperature			Class II Cure Cycle Test Temperature		
	-65°F	RT	180°F	-65°F	RT	300°F
0.5 Hour	4397(14)/ 2,2,2,5	3392(12)/ 2,2,2,5	2292(5)/ 2,5,5	4419(13)/ 5,5,5,5	3568(15)/ 1,1,2,5	1390(22)/ 5,5,5,5
50 Hours			2565(8)/ 1,1,2,5			1106(25)/ 2,2,5,5
100 Hours			2587(8)/ 1,2,2,2			2125(23)/ 2,2,2,2
200 Hours			2291(12)/ 2,2,2,5			1682(19)/ 2,2,2,5
300 Hours			2234(11)/ 2,2,2,5			1801(7)/ 2,2,2,2
500 Hours			2259(10)/ 2,5,5,5			2229(5)/ 2,2,2,2

[a] Strengths in psi, (coefficient of variation, percent)/see Figure 2 for failure modes e.g., 1, 2
[b] Surface treatments: aluminum: 10V phos anodized, Z 6040 silane; composite: grit blast, Z 6040 silane
[c] Time in air at test temperature

TABLE 5. ROOM TEMPERATURE STRESS DURABILITY TESTING OF ALUMINUM-COMPOSITE DOUBLE STRAP BONDS[a]

(2024 T3 Bare Aluminum Tangs/AF-A-42-1659 E-glass/AF-R-40-116 E-glass-Celion Straps)[b]

Specimen Conditioning[c]	Class I Cure Cycle	Class II Cure Cycle
100 Hours, ambient	3155(8)/5,5,5,5	3487(9)/5,5,5,5
100 Hours, humidity	2125(32)/5,5,5	3304(21)/5,5,5,5
250 Hours, ambient	3314(5)/5,5,5,5	3257(15)/5,5,5,5
250 Hours, humidity	2696(4)/5,5,5,5	2750(29)/5,5,1+5,1+5
500 Hours, ambient	2157(18)/5,5,5,5	---
500 Hours, humidity	---	---
750 Hours, ambient	2497(28)/5,5,1+5,1+5	3268(30)/5,5,1+5,1+5
750 Hours, humidity	2079(31)/5,5,5,5	2653(24)/5,5,5,5
1000 Hours, ambient	2981(11)/5,5,1+5,1+5	2743(46)/5,5,5,1+3
1000 Hours, humidity	2436(9)/5,5,5,5	2436(19)/5,5,5,5

[a]Residual strengths in psi after hours indicated at 875 psi sustained stress/failure modes principally adhesive at strap surface (see Figure 2).
[b]Surface treatment: aluminum: 10V phos anodized, Z 6040 silane; composite: grit blast, Z 6040 silane
[c]Humidity conditions: 10 days, 95+% RH, 140°F

TABLE 6. COMPOSITE-COMPOSITE DOUBLE STRAP BOND STRENGTHS[a]

(3501-6/AS Tangs/AF-A-42-1659 E-glass/AF-R-40-Celion Straps)[b]

Specimen Conditioning[c]	Class I Cure Cycle Test Temperature			Class II Cure Cycle Test Temperature		
	-65°F	RT	180°F	-65°F	RT	300°F
0.5 Hour	3016(12)/ 1,3 ea 1+6	2322(4)/ 2,2,2,2	1700(18)/ 1,1,1,1	3713(8)/ 1,1,1+6,1+6	2182(21)/ 1,1,1,1	1370(29)/ 1,1,1,1
50 Hours			2258(10)/ 1,1,1,1			1346(18)/ 1,1,1,1
100 Hours			2063(23)/ 1,1,1,1			1466(28)/ 1,1,1,1
200 Hours			2350(9)/ 1,1,1,1+6			1633(18)/ 1,1,1,1
300 Hours			2214(2)/ 1,1,1,1			1489(7)/ 1,1,1,1
500 Hours			2298(32)/ 1,1,1,1			1630(12)/ 2,2,2,2

[a]Strengths in psi, (coefficient of variation, percent)/see Figure 2 for failure modes e.g., 1, 2
[b]Surface treatment: grit blast, Z 6040 silane
[c]Time in air at test temperature

TABLE 7. ROOM TEMPERATURE STRESS DURABILITY TESTING OF COMPOSITE-COMPOSITE DOUBLE STRAP BONDS[a]

(3501-AS Tangs/AF-A-42-1659 E-glass/AF-R-40-Celion Straps)[b]

Specimen Conditioning[c]	Class I Cure Cycle	Class II Cure Cycle
100 Hours, ambient	2383(32)/1,1,1,1	2653(35)/1,1,1,1
100 Hours, humidity	2311(7)/1,1,1,1+6	2144(23)/1,1,1+6
250 Hours, ambient	2128(18)/1,1,1,1	3373(23)/4 ea 1+6
250 Hours, humidity	1670(18)/1,1,1,1+6	3168(21)/4 ea 1+6
500 Hours, ambient	2000(17)/1,1,2,2	2465(18)/1, 3 ea 1+6
500 Hours, humidity	1350(16)/1,1,1,2	
750 Hours, ambient	2650(23)/1,1,1,2	2954(6)/1,1,1,1
750 Hours, humidity	1551(29)/4 ea 1+6	1862(45)/4 ea 1+6
1000 Hours, ambient	2525(30)/1,1,1,1	2278(39)/1,2,1+6,2+6
1000 Hours, humidity	1393(10)/1,1+6	1333(6)/1,1

[a]Residual strengths in psi after hours indicated at 875 psi sustained stress/failure modes as indicated in Figure 2

[b]Surface treatment: grit blasted, Z 6040 silane

[c]Humidity conditions: 10 days, 95+% RH, 140°F

Constituent	Percent by Weight	
	AF-R-40	AF-A-42
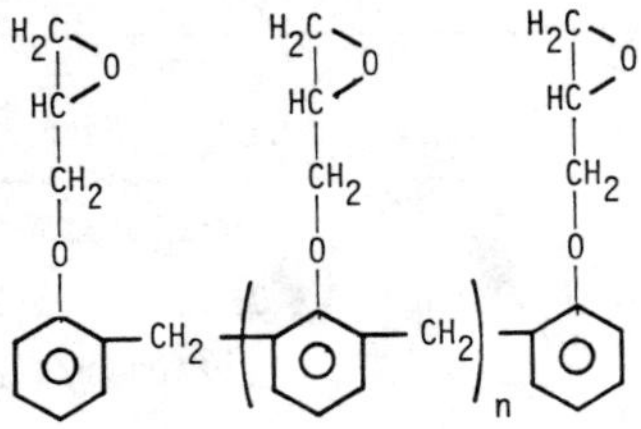 DEN 438	64.6	57.3
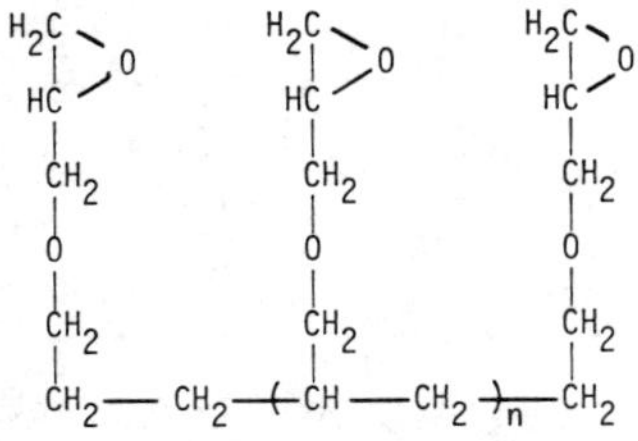 Poly(allyl glycidyl ether) (PAGE)	16.2	14.3
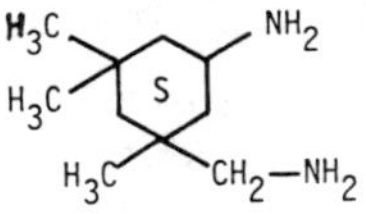 Isophorone diamine	14.9	12.6
$H_2N-(CH_2)_6-NH_2$ Hexane diamine	4.3	3.8
ATBN 1300 X 21 Diamine terminated polybutadiene-acrylonitrile	0	4.0
AA 351 Diamine terminated polyoxytetramethylene	0	8.0

Figure 1. Constituents of AF-R-40 and AF-A-42

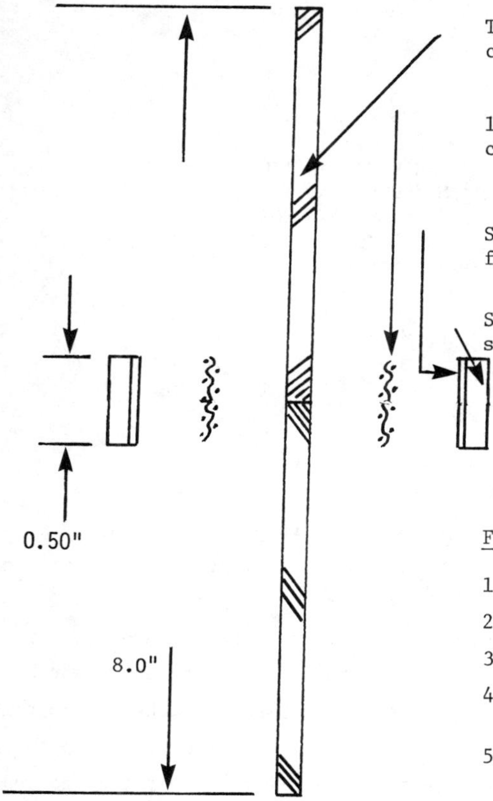

Failure Modes (see tables)

1. Adhesive failure at tang surface
2. Combined 1 and 3 failures
3. Cohesive failure
4. Failure at 116 E-glass surface (rarely observed)
5. Adhesive failure at strap surface (116 E-glass removed from strap)
6. Failure in strap

Note: 1) All stock was 65 ± 2 mils in thickness.
2) With combined failure modes e.g. (1+6), it is not possible to distinguish between primary and secondary failures.

Figure 2. Double strap lap shear configuration

13th National SAMPE Technical Conference
October 13-15, 1981

INVESTIGATION OF SURFACE PREPARATION METHODS AND 250°CURING ADHESIVE SYSTEMS AS RELATED TO PROVIDING DURABLE ADHESIVE BONDED JOINTS FOR THE 300 SERIES STAINLESS STEELS

Lance G. Look
The L.O.M. Corporation
Traverse City, Michigan

Abstract

Service related experience, with the 300 series stainless steels, has shown less than desirable durability of adhesive bonded joints subjected to a high humidity environment under load. In pursuit of a solution to this problem, various surface preparation methods, adhesive films, and primer combinations were investigated. The total effort examined various combinations which included eight 250°F curing adhesive films and one 350°F curing film, with and without nine different adhesive primers, ten combinations of surface preparation methods, and the effects of final rinse temperature. Wedge specimens subjected to 130°F and condensing humidity were employed to evaluate the candidate products and surface preparation methods. This initial effort led to a matrix of four surface preparation methods and four adhesive systems for further investigation. Wedge test specimens were subjected to the same 130°F condensing humidity and a similar group to water emersion. Two distinct results were obtained. One surface preparation method consistently demonstrated resistance to both environments, for all adhesives, and one adhesive system demonstrated similar resistance for all surface preparation methods.

1. INTRODUCTION

The singularly unique application of structural adhesive bonding, which is subjected to combinations of extremes in range of climatic conditions, corrosive chemical and abrasive operating environments, and the most rapid accumulation of high load cycles, exists in the helicopter rotor blade. It is this same application, forced by the necessity to improve structural fatigue life, which served to pioneer much of the development of material, process, and tooling techniques for the adhesive bonding of primary structure. It is also an application for which extensive service experience and historical data is available.

Service related delaminations of adhesive bonded stainless steel components, in particular the 300 series, have been recognized for a number of rotor blade structures. The most frequent cases have existed in areas of high differential strain of components such as perimeters of abrasion strips and doubler tips. These areas of peak load transfer would be expected to be more susceptible to earlier failure but only under conditions of load magnitudes far greater than the known operating conditions. Recognition of cases involving abrasion strip edge delamination preceeded other incidences of component delamination, because these were among the earliest applications of stainless steel bonding in quantity production of rotor blades. Stainless steel, 301 in particular, received wider utilization over cobalt and nickel because of its cost to performance advantage. It was quite common, during the Vietnam conflict, to witness perimeter edge voids which exhibited an adhesive failure mode, at the stainless steel interface, on blades returned from service for overhaul and repair. The rate of progression of these delaminations was low, due to the fact, that the applications involved relatively thin members which provided a low incremental stiffness change to the basic underlying structure and were generally provided full surface bondline interface with the basic blade structure. The resulting strains in the adhesive bondline, due to the combined centrifugal loading and bending moments, were of a low magnitude. These facts, combined with the frequency of component replacement due to service related erosion, provided circumstances which did not generally allow edge delaminations to progress to significant proportion prior to repair station replacement of the component.

In more recent years, stainless steel components have become a more integral part of the total primary structure in a number of blade designs. With this occurrence, the operational bondline stresses have been increased. As with the cases of the abrasion strip application, premature delaminations occurred in the field at highly stressed areas of doubler tips and the most outboard end of the structures. The mode of failure, in all cases, was adhesive to the stainless steel surface, indicative of a surface treatment problem. However, all supportive quality control coupon tests, for the structures and the surface preparation lots of material, had shown high shear and peel values exhibiting cohesive mode of failure. In addition, laboratory tests, subjecting the same structure to repeated load cycles of much higher amplitude, failed to duplicate the field problem.

Little general correlation could be found to exist with service time, in terms of flight hours, and commencement of delamination. Correlation did exist, however, with geographic area of operation, maintenance, and cumulative time in an area of operation. Specifically, delaminations occurred after prolonged exposure to geographic areas with a high humidity environment and/or in situations of frequent washings of the structure to remove chemical spray or other surface residue. Further to this, structures which were painted with more brittle lacquer systems demonstrated far greater susceptibility to the problem than did the same or similar structures which were treated with heavy coatings of more elastic epoxy and/or polyurethane paint systems. This correlation, of delamination under load, with high humidity and moisture, was indicative of deficiencies in bond joint resistance to hostile environmental conditions which was, in all ways, similar in nature to that which has been recognized, and in more recent years, resolved for the structural adhesive bonding of aluminum alloys. The use of the newer generation adhesives, adhesive primers, and the implementation of improved controlled surface preparation processes, as exemplified by the Boeing phosphoric anodizing process, has served to correct the greater portion of this problem for the aluminum alloys. The phenomenon's relation to protective surface coatings was further evidence of the problem. At areas exhibiting a high strain differential between components, such as a doubler tip, the thin more brittle surface coatings would fracture around the tip perimeter of the doubler and thereby expose this highly loaded area to the environment.

The application of heavy coatings of more elastic epoxy and polyurethane systems, to these structural areas, is recognized to significantly delay, if not preclude, the onset of delaminations as well as arrest the further propagation of failure in repaired areas. However, because of the vulnerability of these coatings to removal by severe erosion or other mechanical means, as well as the possible reduction in effectiveness with aging, the use of these coatings should be regarded as the provision of protection to existing structures, and added insurance to new designs employing improved surface preparation methods and adhesive systems. It was, and still is, concluded that this form of protection should not be regarded as the primary solution to the problem.

The task then became that of determining which surface preparation processes, adhesives, and adhesive primers would display resistance to failure under the recognized conditions of high load at high humidity. The use of the wedge test was decided upon to be the principle means of evaluation. Lap shear and peel tests were decided to be the secondary methods of evaluation.

2. TEST PROGRAM

2.1 Test Program Goals

The test program goals were established to be the determination of which combinations of surface preparation, adhesive, and adhesive primers would provide a high degree of resistance to moisture or humidity attack under load. The ultimate goal was to refine the better candidate systems to provide minimal crack growth of a predominantly cohesive failure mode. It was also desirable to develop, in this program, a process which would provide consistent results between material lots. It was decided that particular attention was to be focussed upon those adhesive and adhesive primer systems which had provided the most favorable results in the USAF PABST Program. Because these adhesives would receive broader future utilization in adhesive bonded aluminum structures, the compatibility of these adhesive systems with other materials was considered to be of importance.

2.2 Data Search

A literature search was performed of recommended surface preparation methods for the 300 series stainless steels. These data included procedures and processes of several airframe manufacturers and structural adhesive suppliers. In addition to this, pre-plating processes were examined because of the commonality of the recommended prebond chemical etch methods with some plating preparation and passivation processes. It was noted that many acid baths and sequences of operation were identical except for variations between specifications on time, temperature and solution concentration. Representative chemical etch and mechanical abrasive processes were selected from the available data.

An examination was made of existing film adhesive and adhesive primer systems which have either extensive past or present production utilization. A number of these systems as well as two more recently developed adhesives were chosen to be included in the test program.

2.3 Screening Tests

2.3.1 Controls Controls were applied to test coupon and test method, standardizing of degreasing method, water for solutions and rinse operations, and adhesive film type and thickness.

The primary test method, employed for process evaluation, was the environmental wedge test. The test environment was 130°F and condensing humidity for an exposure time of one hour. Because of the wealth of comparable data for aluminum alloy bonding, it was decided to match the same bending stiffness of that test specimen. Therefore, the stainless specimen thickness was .093 inch for an E of $26x10^6$ psi as compared to the 1/8 inch thick aluminum standard specimen. The remaining specimen dimensions were 1 inch width and 6 inches in length. A 1/8 inch thick wedge was pressed between the two bonded strips to a depth of one inch from the end to generate the initial crack. The alloy chosen for test was 301 1/2H stainless steel. Degreasing operations, which preceeded all surface preparation methods, included a first step solvent cleaning. This was followed by a cold alkaline soak and rinse. A final hot alkaline solution immersion was then employed with a final rinse and inspection for water break. All rinse operations and make up for all aqueous solutions employed deionized water. The resistivity was 250,000 ohms mininmum, and the system employed sodium polishers. A final rinse followed by an oven drying cycle at 150°F followed at the conclusion of all surface preparation processes.

All adhesives, subjected to testing, were .06 $\#/ft^2$ weight. All 250°F curing adhesives were supported films.

2.3.2 Test Combinations The adhesive films and adhesive primers which were the subject of initial screening tests are presented in Table 1.

TABLE 1
ADHESIVES & PRIMERS

ADHESIVES	& CODE	PRIMERS	& CODE
FM 123-5	1	BR 123	A
FM 53	2	BR 53	B
FM 73	3	BR 127	C
FM 76	4	BR 1009-50	D
FM 1000	5	EC 3924	E
AF 126	6	EC 3909	F
AF 163	7	EC 2320	G
M-1113	8	M-2271A	H
EA9628	9	M-6726	I
		None	N

One 350°F curing adhesive was included for comparative purposes. Table 2 presents a summary of test processing procedures for surface preparation. To be remembered is that the previously described degreasing operations preceeded implementation of these surface preparation procedures, and final rinse and 150°F oven drying concluded the process for all procedures.

TABLE 2
SURFACE PREPARATION PROCESSES

CODE	1st PROCESS	2nd PROCESS
I	10% H_2SO_4 @140-160°F	10% HNO_3 & 2%@HF@CR.T.
II	" " " "	10% HNO_3 @ R.T.
III	PreBond 700	--------------------
IV	10% H_2SO_4 @140-160°F	22.5% H_2SO_4 & 2.5% $Na_2Cr_2O_7$
V	" " " "	.2-1.0% H_2CrO_4
VI	22.5% H_2SO_4 & 2.5% $Na_2Cr_2O_7$	PreBond 700
VII	25% HNO_3 & 2% $Na_2Cr_2O_7$	.2-1.0% H_2CrO_4
VIII	220-240 Grit Al Oxide @40-60 psi	--------------------
IX	" " " " " "	Phosphoric Acid Anodize
X	" " " " " "	Chromic Acid Anodize

The initial screening tests were not inclusive of all 900 possible combinations of adhesives, adhesive primers, and surface preparation processes. Certain combinations were known to be incompatible from previous experience, and some combinations were of limited scope of interest. As a result, 92 combinations were explored. The initial acceptance criteria was a crack growth of .20 inch or less with the mode of failure, or crack growth, to be predominantly cohesive. Candidate combinations which exhibited a crack growth beyond the dimensional limit and/or an extensive adhesive failure mode were considered to have failed to qualify from further examination.

2.3.3 Results The test coupon combinations and results, in terms of failed and passed, are presented in Table 3. It is of interest to note that, with but a few exceptions, those specimens which passed exhibited cohesive crack growths well below the .20 inch limit. In a similar vane, those specimens which failed exhibited extensive crack growth of predominantly adhesive failure mode. Only a few cases existed in which initial segregation of the two catagories was difficult to assess. In these cases additional specimen tests proved to be conclusive.

TABLE 3
SCREENING TESTS - CANDIDATE SYSTEMS

SURFACE PREPARATION PROCESS	ADHESIVE & PRIMER COMBINATIONS	
	FAILED	PASSED
I	1N,1A,1C,2N,2A,2B, 2C,2D,2E,2F,2G,3C, 5N,5D,7E,7G,9C	8H
II	1C,2B,3C	8I
III	2N,2B,2C,5N,5D	-----
IV	1N,1C,2N,2C,2H,3H, 6E,9C	2B*,3C*,5D*,8H
V	3N	2B,2C,3C,3I,4C,8H, 8I, 9C
VI	2N,2B,2C,5N,5D	-----
VII	1C,2B,3C	-----
VIII	2B,3N,3C,8N,3I, 2I,9C	8H,8I
IX**	-----	3C,8H,8I
X**	-----	3C,8H,8I

Notes: *Successful to 130°F Condensing Humidity Wedge Test, but not to 24 hour water emmersion
**Limited tests to determine improvement over Process VIII

2.4 Candidate System Tests

2.4.1 Candidate Selection Based upon the results of the initial screening tests, as presented in Table 3, four adhesives and four primers, in particular combination, appeared worthy of further investigation. Three surface preparation methods appeared to provide promising results. These surface preparation processes were IV, V and modifications of VIII.

It has been noted, during the initial test program, that an occassional lot of stainless steel specimen material would provide better results than others. This occassional variation was most dramatically displayed in specimens which were subjected to processes which most consistently provided extensive crack growth, of an adhesive mode, such as Processes I and II. It was decided that any practical process should provide consistent results which were independent of normal variations in the manufacture of the stainless steel. Therefore it was decided to include Process I, in the final evaluation, for the purpose of providing a control to identify any such relationship with the material which would otherwise artifically enhance the test results. The combinations of process, adhesive, and adhesive primer which were selected for further testing are presented in Table 4.

TABLE 4
FINAL TEST

PROCESS	ADHESIVE	&	PRIMER
I	FM 53	&	BR 53
IV	FM 73	&	BR 127
V	M1113	&	M2271A & M6726
VIII	EA9628	&	BR 127

2.4.2 Test Results At the completion of this phase of the test program, in excess of 450 environmental wedge specimens had been tested. The test results are presented in Table 5. These data related average crack growth, in inches, and the principle mode of failure. The initial crack length, before elevated temperature and humidity exposure, was reasonably consistent for all specimens tested in this later phase of the program. Two very consistent sets of results were evidenced by the test data. One result, which was significant in satisfying the program goals,

TABLE 5
AVERAGE CRACK GROWTH - INCHES
& PREDOMINANT FAILURE MODE*

ADHESIVE/PRIMER	I	IV	V	VIII
FM 53/BR 53	.235 A**	.188 C/A	.173 C	.183 A
FM 73/BR 127	1.550 A	.135 C/A	.109 C	.307 A**
M1113/M2271A	.105 C	.103 C	.088 C	.119 C
M1113/M6726	.103 C	---	.098 C	.100 C
EA9628/BR 127	1.180 A	.170 A	.070 C	.490 A

Notes: *A = Adhesive failure, C = Cohesive failure
**Extensive Variation and inconsistency in results

was that Process V provided low specimen crack growth, of a cohesive failure mode, for all adhesive and adhesive primer combinations. The second trend was that Metalbond 1113 adhesive, with either M2271A or M6726 primer, provided similarly consistent environmental resistance for all surface preparation methods examined in this phase of testing. Representative specimens employing the surface preparation of process VIII, dry abrasive blast, are presented in Figures 1 & 2. The first two specimens of Figure 1 and the last two of Figure 2 are representative of very poor environmental resistance for this surface preparation process.

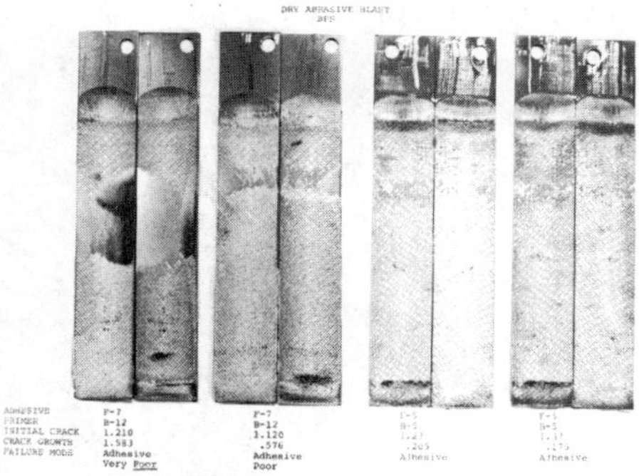

FIG. 1 DRY ABRASIVE BLAST-PROC. VIII
L-R, Specimen 1&2-FM 73/BR 127
Specimen 3&4-FM 53/BR 53

The characteristically good results provided by Metalbond 1113 is illustrated by comparing the specimens presented in Figures 3 and 4. These Figures present typical results for the chemical etch sequence of Process I. This process had shown poor environmental resistance for all adhesive and primer

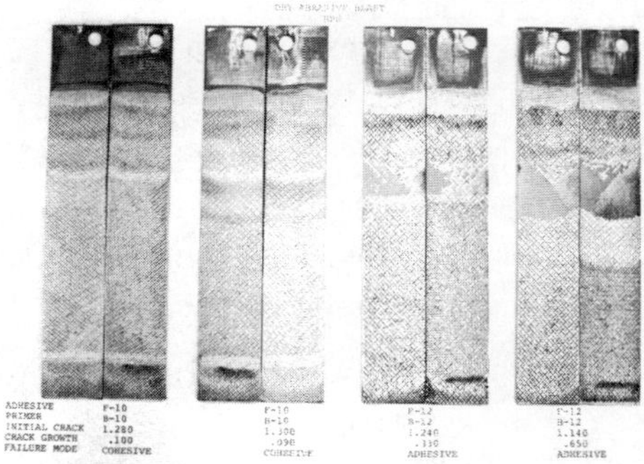

FIG. 2 DRY ABRASIVE BLAST-PROC. VIII
L-R, Specimen 1&2-M1113/M2271A
Specimen 3&4-EA9628/BR 127

combinations tested with the exception of that one adhesive system with either M2271A or M6726 primers.

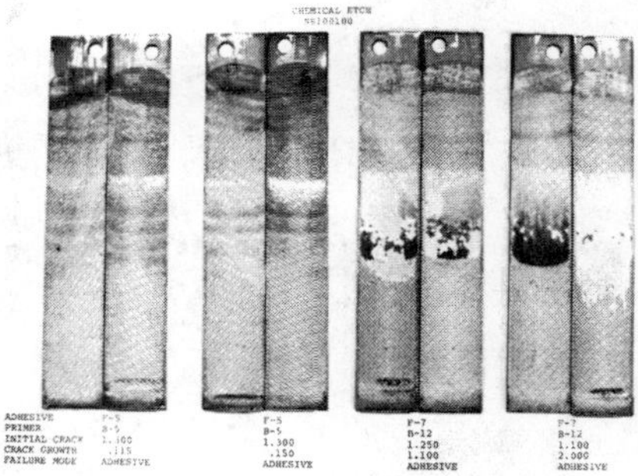

FIGURE 3 CHEMICAL ETCH-PROCESS I
L-R, Specimen 1&2 FM 53/BR 53
Specimen 3&4 FM 73/BR 73

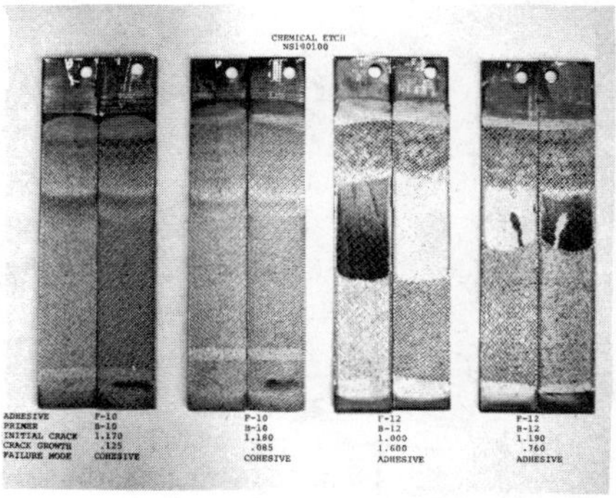

FIGURE 4 CHEMICAL ETCH-PROCESS I
L-R, Specimen 1&2 M1113/M2271A
Specimen 3&4 EA9628/BR127

Representative wedge specimens for Process V are presented in Figures 5 and 6. It is to be noted that all four adhesive and adhesive primer combinations demonstrated a high resistance to the elevated temperature and condensing humidity test conditions.

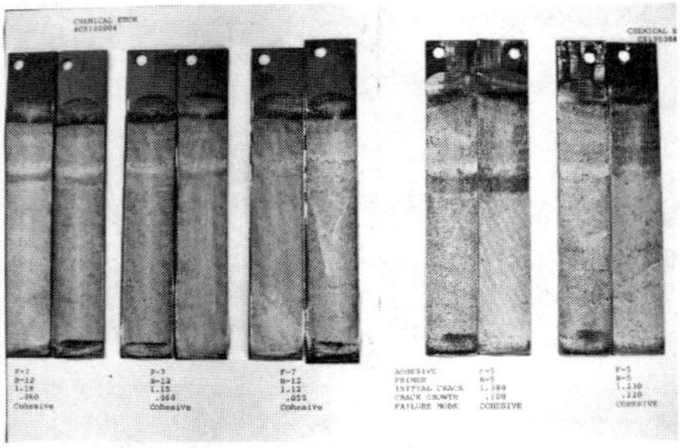

FIGURE 5 CHEMICAL ETCH-PROCESS V

L-R, Specimen 1,2,&3 FM73/BR127
Specimen 4&5 FM 53/BR 53

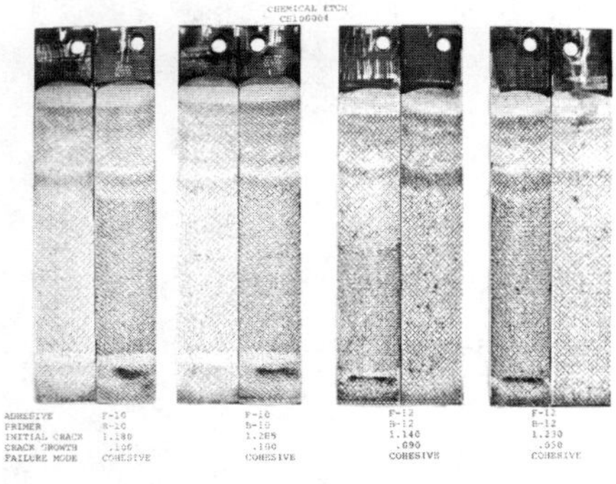

FIGURE 6 CHEMICAL ETCH-PROCESS V

L-R, Specimen 1&2 M1113/M2271A
Specimen 3&4 EA9628/BR127

2.4.3 <u>Process Refinement</u> The history of problems, related to the prebond surface preparation processing of the 300 series stainless steels, had shown that it was characteristically more difficult to provide reliable, consistent, results with the thinner gage materials. Peel tests of 0.004 to 0.012 stainless specimens, representative of a number of structures, had a wide scatter of results in terms of load magnitude and mode of failure.

It was decided to employ a modified environmental wedge specimen, which would test the effectiveness of a surface preparation processes on the thinner gage stainless. To do this, two laminates of 0.007 to 0.009 inch 301-3/4H stainless steel were bonded in the center parting plane of the standard environmental wedge specimen as shown in Figure 7. Refinements of the more successful Process V was of particular interest. The one adhesive system chosen as a control for these tests was FM 73 with BR 127 primer. After experimentation with the concentration of the second solution of that process, and rinsing techniques, a series of specimens were subjected to test.

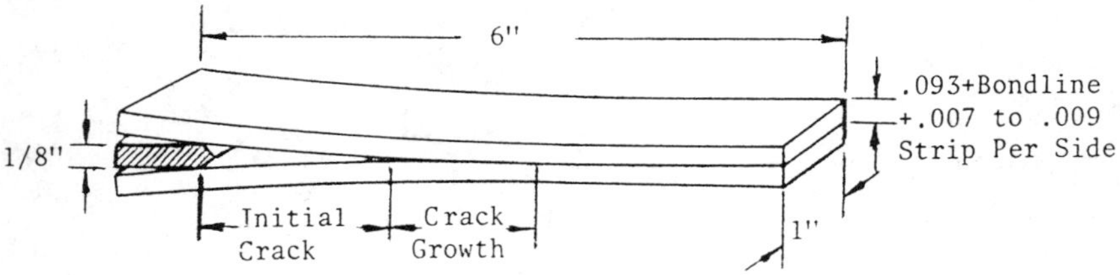

FIGURE 7 MODIFIED WEDGE SPECIMEN

A total of 350 specimens of the modified configuration presented in Figure 7 were tested.

The crack growth data was grouped in terms of the percentage of specimens within an incremental variation of crack growth of a magnitude of 0.01 inches, and the average crack growth within that 0.01 inch increment was recorded. As an example, 17.8% of the specimens exhibited a crack growth between 0.090 and 0.099 inches with an average growth in that group of .0932 inches. These grouped data are presented in Figure 8.

2.5 Process Description-Process V

The first steps of the process involve thorough degreasing operations. The details, of this, are discretionary to the user. The only requirement is that the cleaned parts exhibit no water break after final rinsing.

The first etching solution is comprised of 10% H_2SO_4, per Federal O-S-809B, and 90% deionized water, both by volume. The solution temperature is maintained at 150° to 170°F. A solution conditioning, prior to utilization, requires dissolving 0.7 to 0.9 ounces, of the stainless alloy, per gallon of solution.

The part is immersed into the solution with care not to contact any tank surface. When the surface of the part attains a uniform dull black appearance it is removed, and immediately immersed into the second solution.

The second solution is maintained at room temperature. It is comprised of .2 to .6%, by weight, Chromic

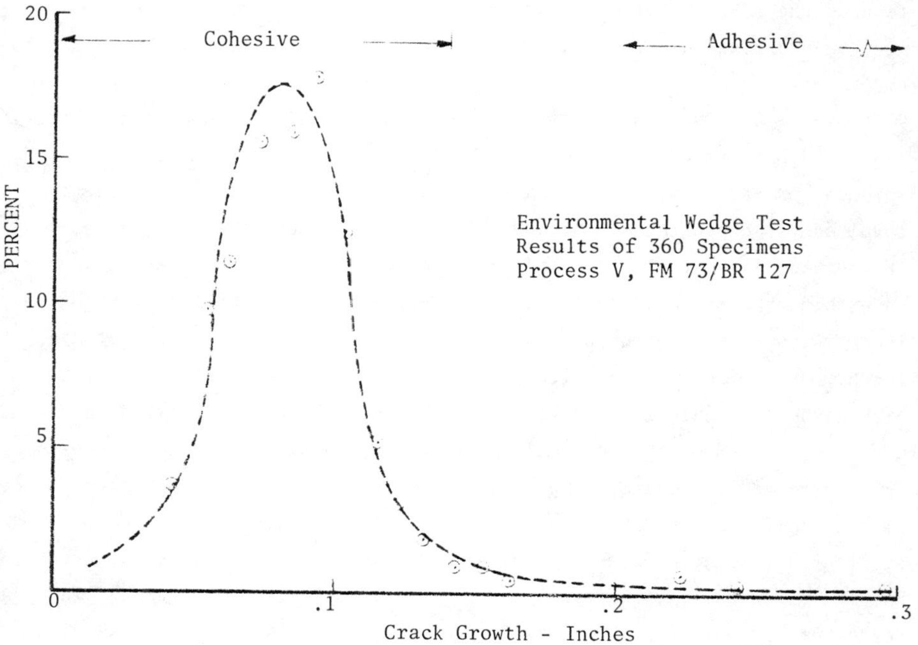

FIGURE 8 CRACK GROWTH - PERCENT OCCURRENCE

Acid per Federal O-S-595B and the balance deionized water.

The part is repeatedly dip rinsed in the dilute chromic acid solution until the dull black surface is shed, leaving a uniform flat dark grey appearance. When less than three dips is required to remove the products of the first etch from the surface, the chromic acid solution must be replaced.

The part is then immersed into a cold rinse tank of deionized water to fix the surface. This is followed by a cold deionized water rinse and check for water break. The parts are finally dryed in a ventilated oven at 150°F. Parts, processed by these means, must be primed within a 24 hour period after final operations.

In all steps, prior to the application of the adhesive primer, care must be exercised not to abrade the prepared surface.

3.0 CONCLUSIONS

A variety of surface treatment methods, preparatory to the structural adhesive bonding of the 300 series stainless steels, are employed within the airframe industry. These surface preparation processes can be catagorized into chemical etching and mechanical-abrasive processes. Most of the chemical systems relate to acid etching and, although differing slightly in concentrations, solution temperature, and immersion timing, share a great degree of commonality. Most of these same preparation methods are common to accepted processes prior to plating operations.

Close adherence to most of these specifications will provide coupon test specimens which will, in the laboratory, provide acceptable if not high value lap shear and peel results. Cohesive mode failures are generally realized with the exception of cases in which such high shear values, of the adhesive joint, have resulted in yield of the adherend. The introduction of the environmental wedge test, however, as a control for process acceptance, can provide quite different results which are of a practical concern.

The developmental test program, discussed in this text, has provided one chemical etch surface preparation process which has demonstrated consistent results in resistance to an elevated temperature high humidity environment. This particular process also provides the practical advantage of a high correlation of test results with surface color and texture characteristics for each processing sequence. Sample specimens were subsequently supplied to the US Air Force Materials Laboratory. These specimens were subjected to standard test procedures including extended time enviromental wedge specimen tests of an exposure period of 14 days. The results of these tests, and in particular the wedge tests, were consistent with our findings in that the crack extension was of a completely cohesive failure mode.

Production application of Process V has been monitored over its use in over two hundred structures. The environmental wedge test specimen results, performed as part of the quality control function, have been consistent with the results previously presented in Figure 8. Lap shear tests of 100 production structrues have shown results of 4800 to 6920 psi with an average of 5940 psi. An interesting aspect of the lap shear tests is that, although there is an apparrently large range in the shear results, the corresponding tensile stress of the adherend has been consistently within a narrow range at or above the material lot yield strength.

The second, and unanticipated, result of the test program is certainly noteworthy. The Metalbond 1113, with at least the two primers tested, provided very favorable environmental wedge tests results for a variety of surface preparation methods. We have subsequently extended tests of this product to examine very simple surface preparation procedures for effecting structural repairs, and have achieved very favorable results.

The Process V, presented in this paper, should not be considered as the only successful means for stainless steel surface preparation. It is one for which sufficient data now exists to provide a high degree of confidence. Our company is currently examining a second step procedure for the dry abrasive blast method which appears to open this method to a wider group of adhesives and primers. Also the choice of the four adhesives and primers, employed in our second phase testing, should not be construed as exclusionary of other film adhesive products.

Sample tests of other adhesive systems, in conjunction with surface preparation in accordance with Process V, have demonstrated successful results.

4. ACKNOWLEDGEMENTS

The author wishes to acknowledge the courtesy and contribution of Mr. Ted Reinhart of the US Air Force Material Laboratory for the test work performed at that facility.

5. BIOGRAPHY

Mr. Look is a mechanical and structures design engineer. He has a B.S. degree in Mechanical Engineering and holds a DER status with the FAA in structures, which is inclusive of analytical substantiation - static and fatigue, detail design approval, the witnessing of official tests, and approval of MRB actions. Mr. Look's background includes positions with the Advanced Concepts Division of Lockheed, Advanced Design at Hughes Helicopter, and Chief Engineer-Design at Parsons Corporation prior to the formation of the L.O.M. Corporation, of which he is President. He holds numerous patents inclusive of helicopter rotor systems, high speed deployable structures, and munition devices. His background in the specific area of helicopter rotor and control system design and development spans 20 years.

13th National SAMPE Technical Conference
October 13-15, 1981

INTERFACIAL CHARACTERISTICS OF 350°F ADHESIVE BONDING

Daniel B. Arnold
Erick E. Peterson
Boeing Commercial Airplane Company
Seattle, Washington

ABSTRACT

Previous investigations reported reduced lap-shear strength or stress durability of some 350°F curing adhesive systems on phosphoric acid anodize (PAA) modified primers were utilized, the reported reductions in mechanical properties were more severe.

This investigation determined the structure and chemistry of the PAA and FPL oxides; and investigated the mechanical properties and component interactions of bonded joints containing 350°F curing adhesive systems combined with PAA and FPL. On .0625 in. thick adherend machined lap-shear specimens, FPL had greater strength. However the significant reduction in strength,. reported earlier, were not substantiated. Because of the significant normal stress gradients present in this specimen other investigators have recommended that it not be employed to measure shear strength. On .125 in. thick adherend machined lap-shear specimens, PAA had greater shear strength. Stress durability on PAA was superior to or equal to that of FPL.

Analysis determined that the failure modes were independent of the surface preparation, and depended on the intrinsic properties of the adhesive or primer.

1. HISTORICAL BACKGROUND OF INVESTIGATION

The state of the art of adhesively bonded aluminum has been significantly improved in recent years. Adhesives and primers have been developed to optimize strength, toughness, and environmental durability. Prebond surface preparations have been developed to yield stable oxide substrates. Specifically, Forest Products Laboratory (FPL) etch has been replaced by phosphoric acid anodize (PAA), noncuring epoxy primers have been replaced with corrosion-inhibiting adhesive primers (CIAP), and moisture-sensitive adhesives have been replaced with moisture-resistant adhesives. These improvements have provided the technology base for expanded usage of adhesive bonding and commitment of bonding to primary aircraft structure (fig. 1). The Primary Adhesively Bonded Structure Technology (PABST) program at McDonnell-Douglas and the bonded honeycomb structure on the Boeing YC-14 AMST are examples of the aircraft industry's commitment to adhesive bonding.

The greatest improvement has been obtained with 250°F-curing, 180°F-service, modified epoxy adhesives, CIAP primers, and PAA substrates. An apparent "incompatibility" was reported for one 350°F-curing adhesive/primer system—BR-227A/FM-61 on PAA oxide. The characteristics of the incompatibility were reduced lap-shear strength and a slick-appearing, potentially interfacial failure mode. The failure mode appeared to be interfacial between the primer and oxide surfaces when examined visually and with optical microscopes. A subsequent investigation demonstrated that this incompatibility occurred with anodically produced oxides, but not with FPL-etched surfaces (table 1). Also, an analysis of the failure modes using the scanning electron microscope (SEM) showed the failure modes on all the anodically produced oxides to be cohesive within the primer, with a very thin layer of primer over the oxide surfaces (ref 1, 2). Figure 2, an SEM photomicrograph, shows a typical failure that has a very thin layer of primer over the oxide surface and bridging across islands of oxide. Use of BR-227A, containing an elastomeric additive, resulted in some reduction in lap-shear

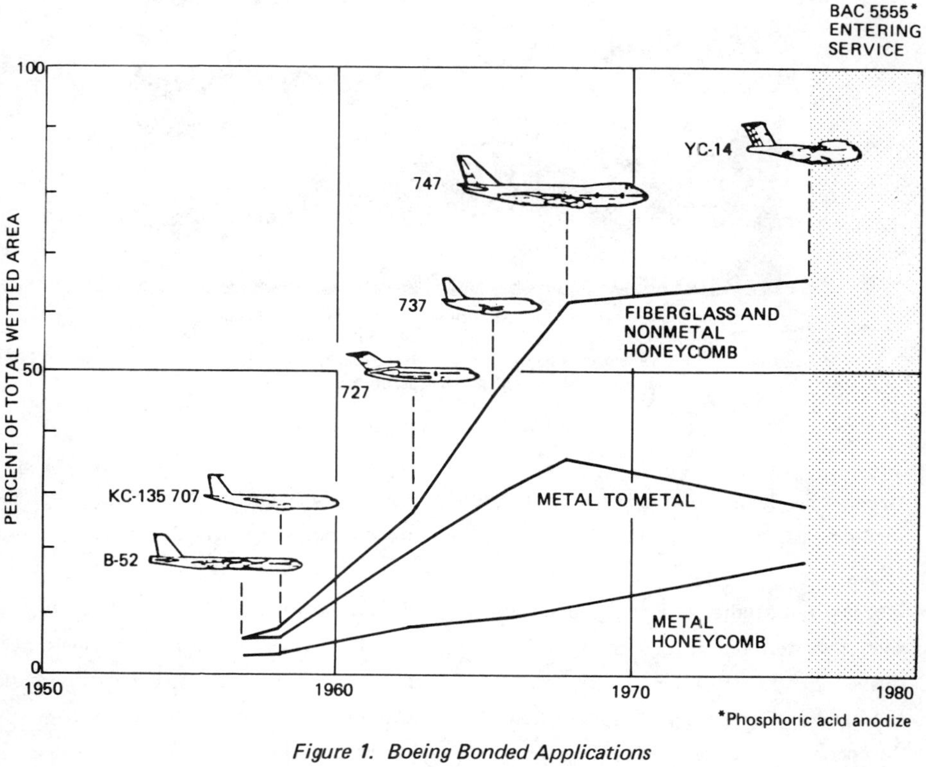

Figure 1. Boeing Bonded Applications

Table 1. Lap-Shear Strengths for Different Aluminum Surface Treatments vs BR-227A and BR-227 Primers

Primer	(Solids (%)	Average (5) lap shear strengths (lb/in^2)			
		Surface treatments			
		FPL	PO_4 20V	PO_4 5V	CrO_3 22V
BR-227A	15* 30**	3,293 2,664	2,317 2,535	2,263 2,198	2,565 2,348
BR-227	15 30	3,577 3,312	3,290 3,402	3,306 3,450	3,279 3,202

*Primers at 15% solids were sprayed on aluminum
**Primers at 30% solids were dipped on aluminum
FM-61 Adhesive was employed

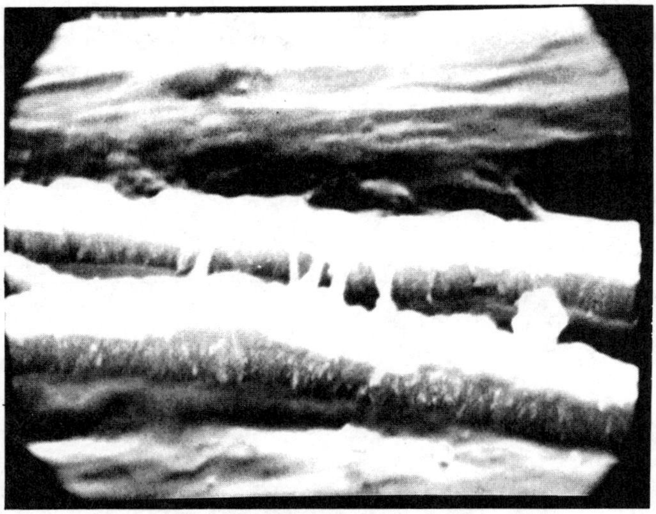

Oxide cross section of surface of failed lap shear specimen that has been anodized in phosphoric acid at 20V and primed with BR-227A. An elastomeric polymer is evident on surface of oxide and stretched across the oxide islands

Figure 2. Oxide Cross Section

strength in comparison to BR-227 on all surface preparations tested.

More recent studies at Boeing have shown that the method of primer application influences the behavior of BR-227A on anodized surfaces. Spray application of the primer produces reduced lap-shear values, whereas "curtain coating" application produces greater lap-shear values.

BR-227A primer is obtained by modifying BR-227 with a nitrile rubber elastomer. Historically, the modification was made to obtain a primer with higher viscosity during the final cure.

The reduction in lap-shear strength of the non-rubber-modified primer (BR-227) on PAA in comparison to FPL was not as pronounced as for BR-227A.

During the course of Air Force Contract F33615-74-C-5056 (ref 3), a similar "incompatibility" was reported for another 350°F-curing adhesive/primer system, PL-728/ PL-729-3. This incompatibility was a reduction in the measured lap-shear strength of the adhesive system on PAA in comparsion to FPL when clad adherends were used (2024-T3 and 7075-T6 clad). For the same alloys bare, the measured lap-shear strengths of this adhesive system on both surface preparations are equal. The mean lap-shear strength of this adhesive system on 7075-T6 bare/PAA exceeds that of 7075- T6 bare/FPL. However, the difference is not statistically significant. The statistical significance of the data was determined using the statistical test for the difference between two means employing Student's t-distribution and a 1% critical value (ref 4, 5). This test was used throughout this investigation to analyze the statistical significance of differences in mechanical test data.

Neither of the two investigations cited report reduced stress durability of the two adhesive systems on PAA in comparison to FPL. Quite to the contrary, the durability on PAA is as good as or better than that on FPL.

An independent investigation conducted by Schwartz (ref 6) showed that lap-shear specimens bonded with PL-728/PL-729-3 on PAA surfaces failed prematurely when exposed to 95°F salt spray while under a sustained stress at 50% of their room-temperature ultimate strengths. The same adhesive/primer system bonded to FPL-etched surfaces showed much better stressed-durability performance. When the PL-728 primer was omitted from the bondline of the PAA specimens, the stressed-durability performance was equal to or better than that of PL-728 on FPL-etched surfaces. The differences in lap-shear strength of PL-728/PL-729-3 on PAA in comparison to FPL using two alloys are not statistically significant. The alloys used were 2024-T3 and 7075-T6 bare.

Failure mode analyses were not conducted during either of the studies that investigated PL-728/PL-729-3.

The results of the three investigations are summarized as follows:

1. Marceau (ref 1) reported significantly reduced room-temperature lap-shear strengths on anodically produced oxides when BR-227A was used.
2. Marceau (ref 1) reported good stress durability for systems using BR-227A and BR-227.
3. Marceau et al. (ref 3) reported reduced room-temperature ultimate lap-shear strength on 2024-T3 clad and 7075-T6 clad PAA in comparison to 2024-T3 clad and 7075-T6 clad FPL when PL-728/PL-729-3 was used.
4. Marceau et al. (ref 3) reported good stress durability on PAA and opti-

mized FPL for PL-728/PL-729-3. The residual strength of PL-728/PL-729-3 on PAA was greater than that on FPL for 2024-T3 (clad or bare) and 7075-T6 (clad or bare). Some of the differences may be statistically significant.

5. Marceau (ref 1) reported that the failure mode of test specimens containing BR-227A was cohesive within the primer.
6. Schwartz (ref 6) reported inferior stress durability for the PL-728/PL-729-3 adhesive system on PAA in comparison to FPL.
7. Schwartz (ref 6) reported room-temperature lap-shear strengths for PL-728/ PL-729-3 on PAA and FPL that did not differ statistically. On 2024-T3 bare adherends, he reported greater mean shear strength for PAA than for FPL.

These differences in reported results are summarized in Table 2. The stress-durability test results reported for PL-728/PL-729-3 are not consistent.

2. HYPOTHESIZED MECHANISMS OF BONDED JOINT FAILURE

The differences in test results discussed above raise the following questions:

1. Do high-temperature adhesive systems (i.e., those that cure at 350°F) have reduced room-temperature ultimate lap-shear strength on PAA in comparison to FPL?

Table 2. Comparison of Investigation Results for PL-728/PL-729-3 System

	Marceau investigation	Schwartz investigation
Lap-shear test	Equivalent strength on bare aluminum FPL superior on Clad aluminum	Equivalent strength on bare aluminum No tests on Clad aluminum
Stress durability test	PAA superior	FPL superior

2. Do these adhesive systems have reduced stress durability on PAA in comparison to FPL?
3. Why does BR-227A have a greater reduction in shear strength on anodically produced oxides than does BR-227?
4. Are rubber-modified primers incompatible with PAA?

Marceau and Schwartz proposed some mechanisms they believed might explain the phenomena they observed.

Schwartz proposed that some primers-particularly those that contain elastomers -do not totally penetrate the thick, porous columnar PAA oxide, resulting in a zone of unimpregnated oxide that is weak in shear. When the joint is loaded in shear, the unsupported oxide fails (ref 6) (see fig. 3).

Marceau proposed that the high-molecular-weight rubbers in rubber-modified adhesive primers could not penetrate the porous PAA oxide, whereas the lower molecular weight components could. This would result in an elastomer-rich zone above the thick PAA oxide. The elastomer-rich zone can function as a weak boundary layer (ref 1) (see fig. 4).

Marceau proposed a second mechanism, illustrated schematically in Figure 5, that would probably operate in conjunction with his first mechanism. This mechanism postulates that premature failures do not occur until the oxide fractures, thus creating stress risers in the primer at these points. Thick oxides produced by anodizing break into larger pieces than do thin oxides, such as produced by FPL etch, thereby creating zones of higher stresses at these points of oxide fracture. These stresses, coupled with a thin, weak boundary layer of polymer just at the oxide surface, can cause premature failure (ref 1).

3. REQUEST FOR INVESTIGATION RECEIVED BY BOEING

To investigate the failure mechanisms and further determine the mechanical properties of higher temperature adhesive primers on porous anodic oxides, Boeing was awarded this contract.

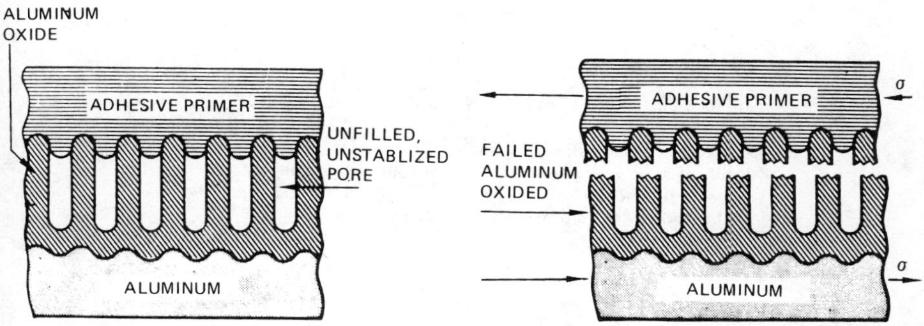

(A) BONDED SYSTEM UNSTRESSED

(B) BONDED SYSTEM UNDER SHEAR STRESS

Hypothesis of oxide failure in a bonded system with a thick duplex oxide when exposed to shear stresses

Figure 3. Proposed Oxide Failure Model (First Mechanism)

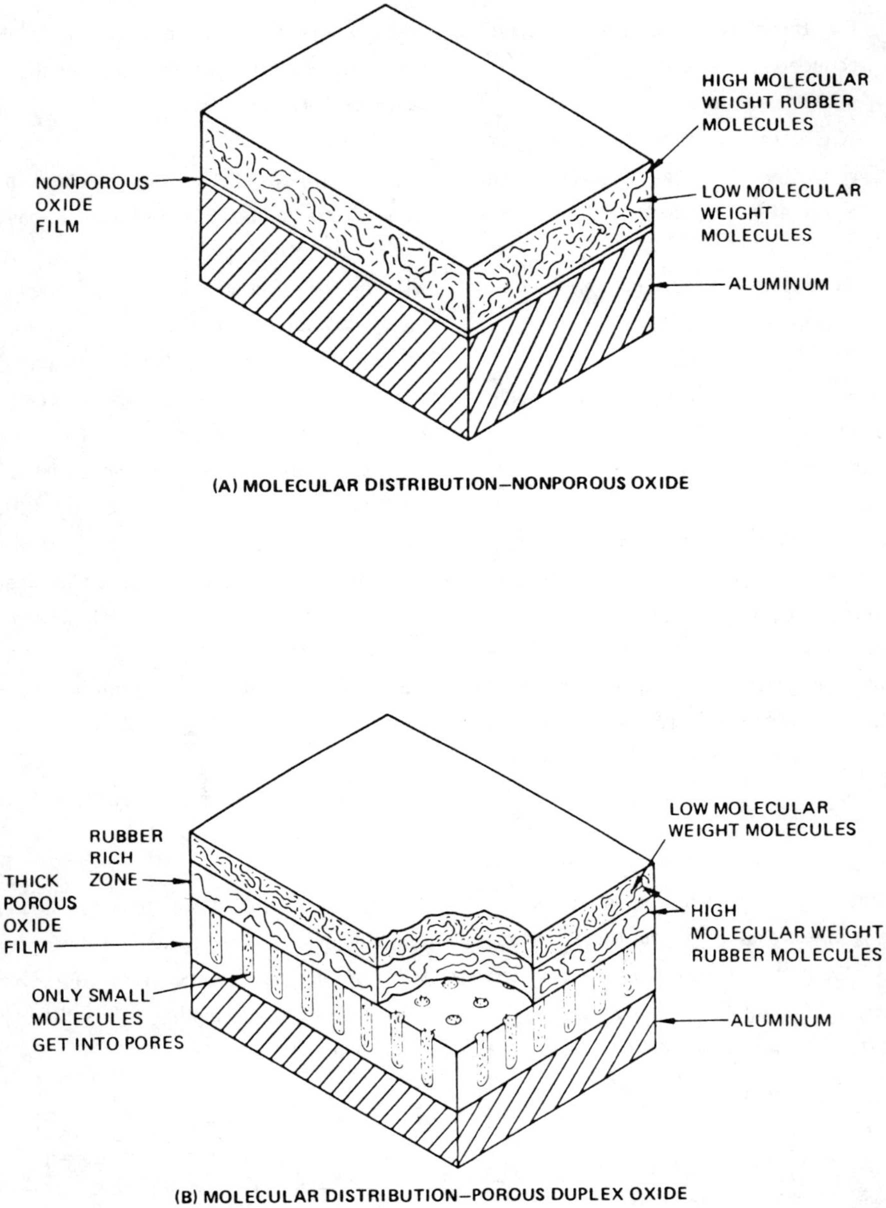

Influence of oxide surface characteristics on molecular distribution at the oxide/polymer interface: (A) nonporous surface with low tendency to influence molecular separation; (B) porous oxide with strong tendency to influence separation of small molecules from large rubber molecules.

Figure 4. Molecular Distribution on Porous and Nonporous Oxide Surfaces (Second Mechanism)

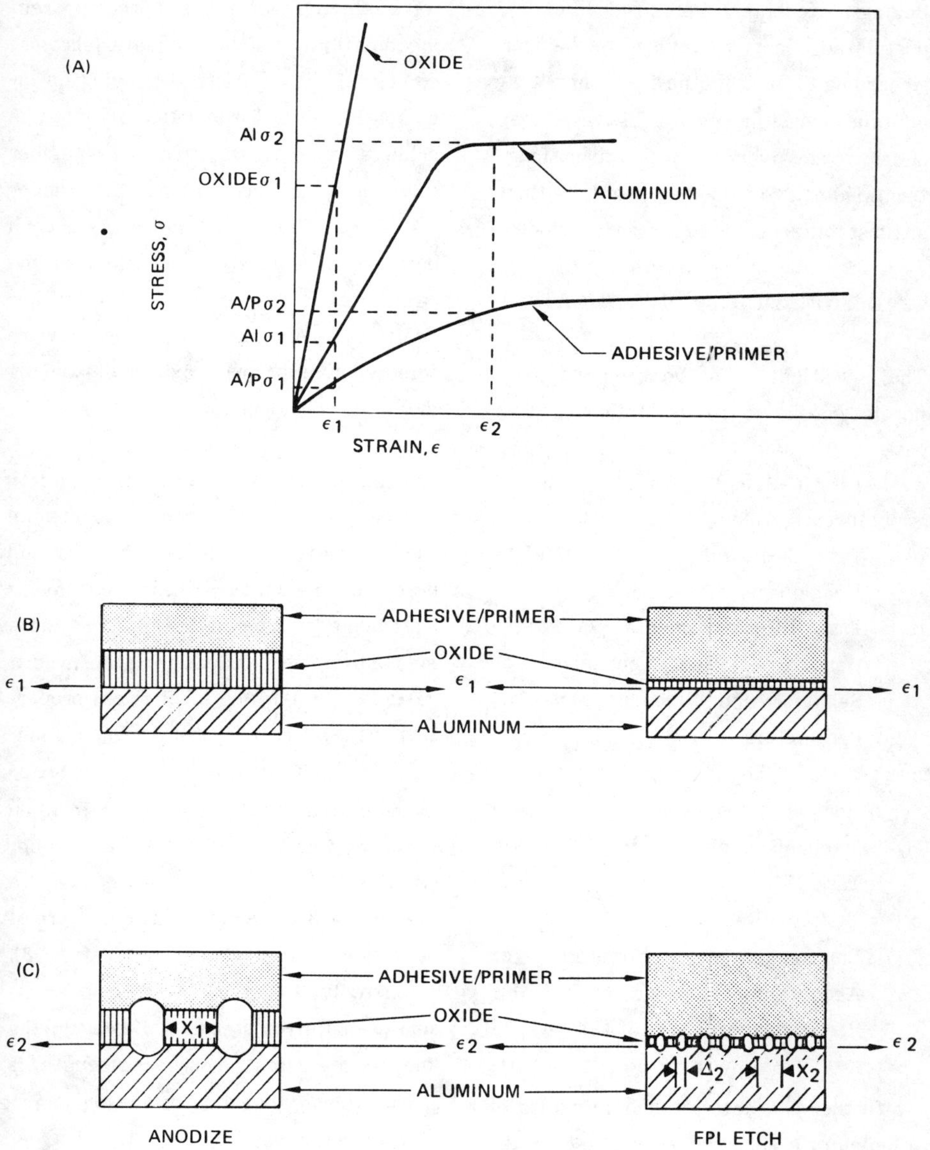

Illustration of how the stress-strain characteristics of the different components in a bonded system influence failure: (A) relative stress-strain characteristics of the three components; (B) system at strain, ϵ ; and (C) system at strain ϵ_2 where oxide has failed.

Figure 5. Stress/Strain Characteristics of Bond System Components (Third Mechanism)

Boeing's technical proposal (ref 7) postulated one more failure mechanism. According to this hypothesis, primers that contain elastomeric additives may undergoexcessivedeformationduringlong-termloading. This loading might fracture an unstabilized, anodic, porous oxide.

4. DESCRIPTION OF INVESTIGATION

The investigation to be reported later herein consisted of the following five tasks:

- Task I—Selection of Materials for Investigation
- Task II—Postulation of Candidate Mechanisms of Behavior
- Task III—Experimental Investigations to Define Mechanisms of Behavior and Relevant Materials Parameters
- Task IV—Engineering Characterization and Evaluation to Validate Mechanisms of Behavior Defined in Task III
- Task V—Range of Materials and Environmental Parameters for Acceptable Bonded Joint Performance

To further characterize anodic oxides of aluminum, a subcontract was let to the Physics Department of Washington State University. For a discussion of this investigation, see Appendix D.

Each task will be described briefly (see fig. 6).

Task I defines the materials and processes used in the investigation. Test materials were selected that exhibited the apparent incompatibility, and these materials were compared with systems displaying no incompatibility. Three pairs of adhesive primers were investigated. One member of each pair did not contain rubber additives. The second member of each pair had a composition identical to the first, except that it did contain rubber additives. Anodization processes were selected that produce oxides of varying morphology and chemistry.

Task II defines the four original candidate failure mechanisms proposed to explain the interaction of various primers and porous anodic oxides. These mechanisms have been described in detail. Two more mechanisms were added during the investigation to explain the mechanical properties of the adhesive systems. One mechanism is an explanation of the reduction in shear strength of an epoxy resin when it is modified with a nonreactive rubber additive (the rubber-modified epoxy is a true composite, because it is a mixture of heterogeneous phases) (ref 8). Admittedly, the theory is semiquantitative at best. However, the knowledge that the shear strength is reduced and that this reduction can be estimated is very important.

The second mechanism hypothesized is that the overall shear strength of an adhesive-bonded joint also is dependent on the primer/adhesiveinteraction. Thishypothesis is important because the different primers were evaluated initially with different adhesives on the same surface preparations. Consequently, experimental

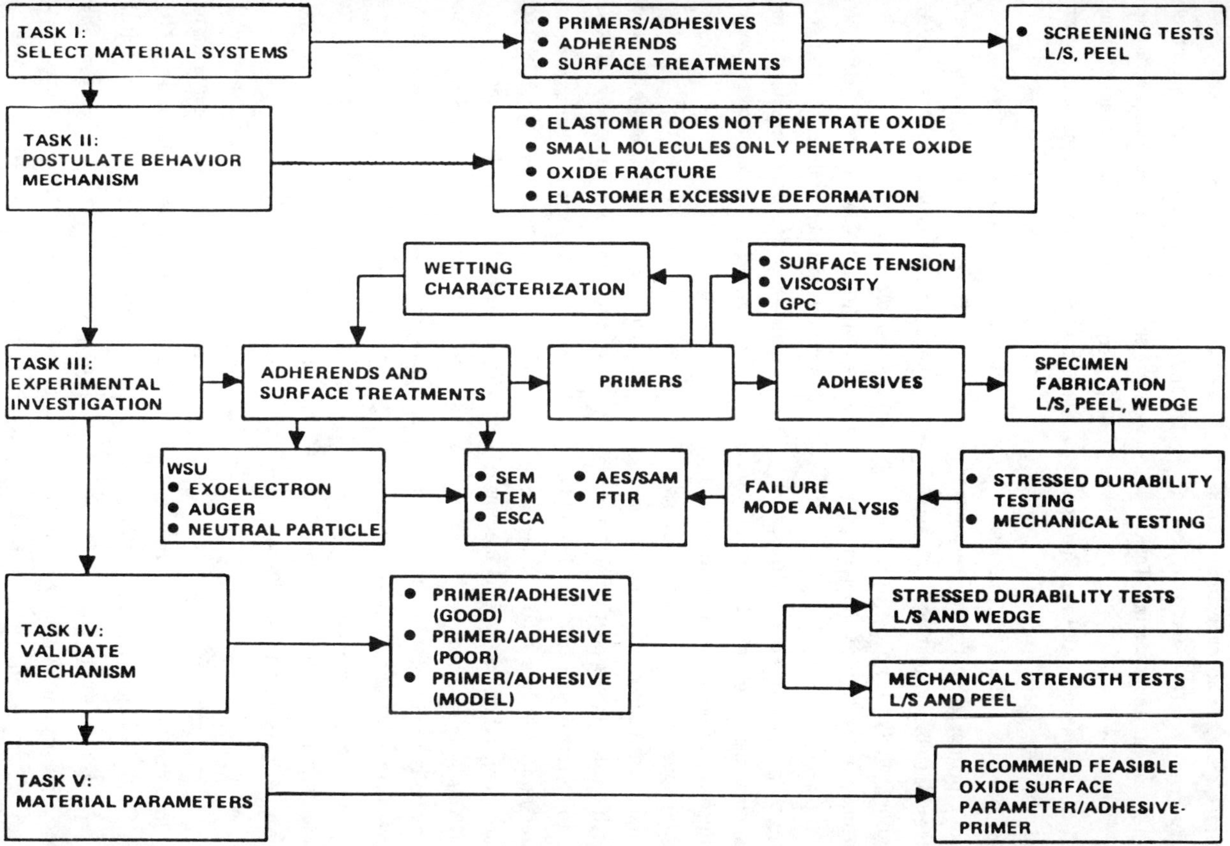

Figure 6. Program Task Flow

results that could be interpreted as resulting from interactions between the oxide and primer actually might result from primer/adhesive interactions. This possibility greatly influenced the design of the mechanical tests during Task IV.

Most of the experimental investigation occurred during Task III. The main activities of Task III are as follows:

1. Oxide/primer/adhesive interactions
2. Stress-durability testing
3. Failure analysis
4. Discussion of failure mechanisms

The main purpose of these activities was to determine the mechanical properties and failure mode of specific adhesive systems on specific surface preparations for bonding. Various types of mechanical tests were used. It was particularly important to determine if the mechanical properties and/or failure mode of a specific system depend on the type of loading mode. If this is true, then the important loading mode is the one for which bonded joints are designed. Many bonded assemblies are designed to transfer the load in shear. Ideally, one strives to minimize, if not eliminate, peeling stresses in bonded assemblies.

Equally important are the relative mechanical test values for those primer/adhesive systems that can be compared. For example, the mechanical properties of two different adhesive primers on the same surface preparation cannot be compared if they were bonded with different adhesives. Relevant relative comparisons of mechanical values are necessary because of the requirement to establish the materials parameters for acceptable bonded joint performance.

The purpose of Task IV was to further verify the mechanisms of behavior identified in Task III. This was accomplished by performing the test plan shown in Figure 7. The significant aspects of this plan are as follows:

1. Two different alloys were used because they create different oxide morphologies.
2. Specimens were bonded with an adhesive primer and its elastomer-modifiedcounterpart. Intermediate- and high-modulus adhesives were used.
3. The Bell peel test and the unsymmetrical wedge test also were used to test the integrity of the bonded joints.
4. The shear strength of both adhesive systems on PAA and FPL was tested with thick- and thin-adherend machined lap-shear specimens (fig. 8) to measure the effect of different ratios of normal stresses to shear stresses on the measured shear strength.
5. Sustained stress and unsymmetrical wedge specimens were exposed to 160°F, 100% relative humidity (RH) to determine the environmental durability of the adhesive systems under investigation. These two tests were performed on PAA only for the following reasons:
   a. Tests were performed to determine if primer and/or adhesive can cause interfacial incom-

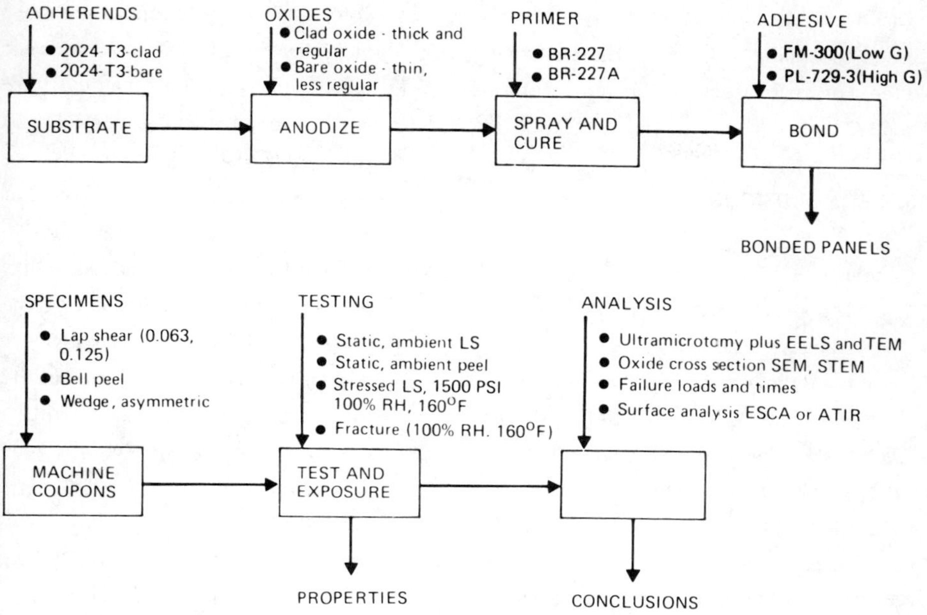

Figure 7. Task IV Test Plan

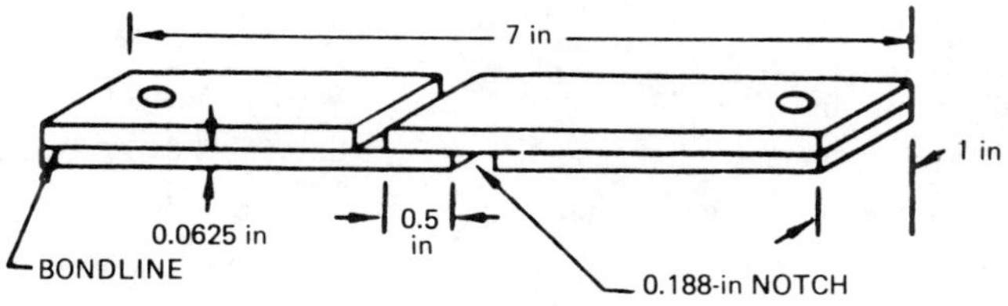

Figure 8. Machined Lap-Shear Specimen

patibility during environmental testing. No potential interfacial compatibility has been identified during environmental testing on FPL-processed adherends. Consequently, it was not tested.

b. Results of Task III indicated that failure mode was independent of surface preparation. Consequently, it was not necessary to test these primer/adhesive combinations on both surface preparations.

c. Bonded specimens using FPL-processed adherends were not fabricated, to reduce costs toward the end of the contract.

The wedge test evaluates the fracture toughness of the adhesive joint. The sustained-stress test evaluates the load-carrying capacity of the joint during environmental exposure.

Task V established the materials and processes necessary to improve the performance of elevated-service-temperature bonded joints.

5. CONCLUSIONS

Conclusions of the program are:

1. The shear strength of 350°F-curing adhesive systems on PAA is greater than that on FPL. This finding is significant because initial lap-shear testing during this program with thin-adherend lap-shear specimens implied that the converse was true. Because of the high normal stress gradients in thin-adherend lap-shear specimens, the adhesive bond fails in peel and not in shear. Failure analysis of the tested specimens determined that loading of the fracture surface was mode 1. Because of the high normal stress gradients, recommendations have been presented that thin-adherend lap-shear specimens not be used to measure shear strength. Our experimental data support this recommendation.
2. It was determined that high-temperature adhesive primers must have good toughness and strength if environmental durability is desired. Rubber additives can improve the toughness of epoxy primers. Unfortunately, this addition often results in a loss in shear strength. An example of this is BR-227A. Analysis determined that the rubber additive in BR-227A is incoherent (i.e., the rubber particle functions like a void). On the other hand, the rubber particles in PL-728 are coherent (i.e., there is good mechanical linkage between the rubber particle and the matrix resin). Furthermore, we determined that the matrix resin of PL-728 is ductile. The adhesive primer PL-728 has good durability and shear strength because of the matrix ductility and coherent rubber additive. Conse quently, a ductile matrix and coherent rubber additive are two good ways to obtain good toughness and high shear strength.
3. The PL-728/PL-729-3 system has greater strength and durability than either the BR-227/FM-61 or BR-227A/FM-61 systems.
4. The failure mechanism of BR-227/FM-61 is an interfacial failure between the primer and the phenolic side of FM-61 or a cohesive failure in the phenolic side of FM-61. This is a duplex adhesive, one side of which is an epoxy and the other a nitrile/phenolic. Teamed with other adhesives, BR-227 primed, bonded specimens demonstrated markedly improved shear strength and environmental durability.
5. The model primers, formulated for this investigation, and FM-300 demonstrated excellent shear strength and environmental durability.

6. RECOMMENDATIONS

Recommendations of the program are:

1. There must be industry-wide agreement as to what constitutes the compatibility criterion between

350°F-curing adhesive primers and surface preparations for bonding. Currently, this agreement does not exist. The position taken in this report is that the shear strength, as measured by the thick-adherend lap-shear specimen, is the appropriate criterion. Many bonded assemblies are designed to transfer their load in shear. Other people contend that the shear strength, as measured by the thin-adherend lap-shear specimen, is appropriate. This specimen has a significant normal stress gradient.

2. The controversy concerning testing bonded joints with thin-adherend lap-shear specimens must be resolved. The position taken in this report is that the thin-adherend lap-shear specimen is not a suitable measure of bonded joint shear strength. It is a mixed-mode test that is difficult to interpret.

3. More information is needed to describe how the surface preparation affects the mechanical strength of bonded joints. The shear strength of the 350°F-curing adhesive systems that were tested is greater on PAA than on FPL. When the normal stress gradients present are increased, 350°F-curing adhesive systems have greater strength on FPL than on PAA.

4. This investigation has determined that if 350°F-curing adhesive primers are to possess good environmental durability, they must be tough. However, toughening the system should not result in significant loss of shear strength. Using coherent rubber additives (i.e., those that provide good mechanical linkage between the phase-separated rubber particles and the matrix resin) and/or using ductile matrix resins appear to be two good approaches to obtaining good strength and toughness. It is recommended that these approaches be further developed for 350°F-curing adhesive primers.

5. The PL-728/PL-729-3 system is recommended over the BR-227A/FM-61 or BR-227/FM-61 systems. Primer PL-728 has good matrix ductility.

6. Teaming BR-227 with a different adhesive would result in higher strength bonds. This may be productive for some applications. The toughness of the bonded joint would be limited by the low ductility of this primer.

7. The two model primers should be evaluated further as potential new "production" primers. These primers have good strength and durability.

7. ACKNOWLEDGEMENTS

The work performed under this program was sponsored by Air Force Wright Aeronautics Laboratories, Nonmetallic Materials Division, contract F33615-78-C-5133. Mr. Herb Schwartz was the program monitor. The final report, AFWAL-TR-80-4176, has been issued.

8. BIOGRAPHIES

Daniel B. Arnold - Mr. Arnold is currently the manager of the Boeing Model 757 Chemical Technology. Previous assignments have included manager of Boeing Military Airplane Company Materials and Process Technology and program manager of the subject contract.

E. E. Peterson - Mr. Peterson received his B.S. in Chemistry from the University of Washington. His work includes studies of materials and processes for structural bonding. He is Technical Leader of the AFWAL funded contract, "Relationship of Interfacial Compatibility to Durability of Adhesive Bonded Joints (F33615-78-C-5133).

REFERENCES

1. J. A. Marceau, "An SEM Analysis of Adhesive Primer Oriented Failures on Anodized Aluminum," 23rd National SAMPE Symposium and Exhibition, Los Angeles, CA, May 2-4, 1978.
2. Boeing Letter B-8800-429L dated January 7, 1977, from J. A. Marceau to H. S. Schwartz, AFWAL/MLBC.
3. J. A. Marceau and J. C. McMillan, "Exploratory Development on Durability of Adhesive Bonded Joints," Technical Report AFML-TR-76-173, October 1976.
4. P. G. Hoel, Introduction to Mathematical Statistics, John Wiley &Sons, New York, NY, 1954, pp. 211-228.
5. C. R. C. Standard Mathematical Tables, ed. C. D. Hodgman, Chemical Rubber Publishing Company, Cleveland, OH, 1959.
6. H. S. Schwartz, "Effect of Adherend Surface Treatment on Stressed Durability of Adhesive Bonded Aluminum Alloys," presented at Symposium on Durability of Adhesive Bonded Structures, U.S. Army Armament Research and Development Command, Dover, NJ, October 27-29, 1976; also published in SAMPE Journal, vol. 13, no. 2, March/April 1977.
7. J. A. Marceau and D. B. Arnold, "Technical Proposal--Relationship of Interfacial Compatibility to Durability of Adhesive-Bonded Joints," Boeing Document No. D180-24255-1, May 15, 1978.
8. R. M. Christensen, Mechanics of Composite Materials, John Wiley & Sons, New York, NY, 1979, pp. 292-294.

13th National SAMPE Technical Conference
October 13-15, 1981

COMPATIBILITY OF 350° CURING HONEYCOMB ADHESIVES WITH PHOSPHORIC ACID ANODIZING

Erick E. Peterson, Daniel B. Arnold
and Melvin C. Locke
Boeing Commercial Airplane Company
Seattle, Washington

Abstract

The adhesive bonding industry and aircraft repair facilities are adopting phosphoric acid anodizing as the baseline surface preparation for aluminum honeycomb structure. For existing design parts, it is desirable to continue usage of existing adhesive and primer systems. A program was initiated at Boeing to test the compatibility of these existing adhesives with phosphoric acid anodized surfaces. Testing consisted of conventional lap-shear and honeycomb peel testing and included stress durability of metal/honeycomb panels. The work on this program was sponsored by AFWAL/FD, Mr. William Shelton, in conjunction with McDonnell Douglas Corporation.

Keywords: Bonding, Honeycomb Panels, Environmental Durability, Peel Strength, Optimization.

1. INTRODUCTION

Boeing conducted an extensive review of the causes of inservice delamination of bonded structure about 10 years ago. This review revealed that the majority of the inservice metal-to-metal disbonds occurred at the primer/adherend (oxide) interface. Subsequent investigation determined that an environmentally weakened oxide, whose properties were not consistent, was a significant cause of the inservice failures of bonded structure.

A variety of activities were undertaken to improve the environmental durability of bonded structure. (There is an excellent review of these activities by McMillan[1].) The main activity was the development of more consistent and environmentally resistent oxides to be employed as surface preparations for bonding.

Initially the standard Forest Products Laboratory (FPL) etch was optimized to improve the consistency of the oxide in the bonded joint. Later this process was replaced with the phosphoric acid anodize (PAA) process which is even more environmentally stable. Thousands of wedge specimens were tested during the scale-up and qualification of this process. During the implementation of this process, 250°F curing adhesive systems were employed almost exclusively to bond the aircraft assemblies.

Initial testing of adhesive systems which cure at 350°F on PAA indicated that there was a reduction in shear strength of these systems on PAA compared to FPL. A reduction in sustained stress durability was also reported for one adhesive system on PAA compared to FPL. Very limited failure analysis indicated that the failures of 350°F adhesive systems on PAA might be between the primer and PAA oxide or in the PAA oxide itself. Consequently, limited mechanical testing and cursory failure analysis gave rise to the common belief that PAA is incompatible with some 350°F primers.

Boeing received another contract to investigate the mechanical properties and failure mode of these 350°F curing adhesive systems on PAA and FPL. This other contract investigated metal-to-metal bonded structure only. The results of this other contract are:

A. The adhesive systems which were tested demonstrated as good or better sustained stress durability on PAA compared to FPL. Usually the performance on PAA was superior.

B. When properly measured, these adhesive systems have greater shear strength on PAA compared to FPL.

C. The failure mode of the adhesive systems is independent of the surface preparation.

The previous investigation concludes that 350°F curing adhesive systems shear strength and durability on PAA is generally superior to those same properties on FPL. This other investigation and its conclusions are discussed in detail in its final report[2].

The current investigation was awarded to Boeing to study the relative mechanical properties and environmental durability of square edge metal/metal honeycomb bonded panels whose skins were processed with PAA and FPL. The effect of varying the PAA process conditions on metal-to-metal shear strength and environmental durability was also to be determined. Consequently this investigation also addressed the relative performance of 350°F curing adhesive systems on PAA and FPL.

This investigation consisted of the

following tasks.

Task 2A - Cyclic environmental durability of metal/metal honeycomb square edge bonded panels. The specimen configuration and cyclic environment are discussed in section two of this report.

Task 2B - Honeycomb peel testing of metal/metal honeycomb bonded panels at room temperature and at -67°F.

Task III - This is a failure analysis of relevant specimens from the other tasks. This task is incorporated into the discussion of the other tasks in section two.

Task IV - This is an investigation of the effect of varying the PAA process conditions on the metal-to-metal shear strength and fracture toughness.

2. EXPERIMENTAL INVESTIGATION

2.1 Peel Strength Determination

2.1.1 Alloys, Surface Preparation and Adhesive Systems The panel skins were fabricated from clad and bare 2024-T81 and clad and bare 7075-T76 aluminum. The bonded panels contained 5052 aluminum alloy honeycomb core whose nominal density is 4.5 pounds per cubic foot[3]. The surface treatments for bonding were the optimized Forest Products Laboratory (FPL) etch process[4] and the Phosphoric Acid Anodize (PAA) process[5]. The adhesive systems that were evaluated are listed in Table 1. The source of these adhesive systems is also given.

2.1.2 Specimen Configuration and Experimental Conditions The configuration of the metal/honeycomb peel specimens is given in Figure 1. The testing was conducted at room temperature and at -67°F. The climbing drum honeycomb peel test was performed on a random sample of three specimens.

2.1.3 Experimental Results The results of the peel tests will be discussed but the massive amount of data which was generated will not be presented. This approach was employed to keep the report length within the prescribed page limit. Except for slight variations in ranking caused by statistical fluctuations in the peel strength of adhesive systems which have similar peel strength, the ranking of the adhesive systems is the same on the four alloys. Furthermore, the adhesive systems have the same ranking on both surface preparations. The adhesive systems, listed in order of decreasing peel strength, follow:

- BR-227A/FM-61
- MB-6725-1/MB-329
- EA-9205/EA-9649
- RB-500/RB-398
- BR-400/FM-400
- HT-424F/HT-424
- -/HT-424
- EC-3917/AF-130
- -/AF-31
- EC-2174/AF-31

Two adhesives - AF-31 and HT-424 - were employed with and without an adhesive primer to bond the metal/honeycomb panels.

Except for RB-500/RB-398, the peel strength of the panels processed with the two surface preparations is statistically equivalent. The t test at 95% confidence was employed to test the data[6]. Although panels processed with FPL and bonded with RB-500/RB-398 had statistically better peel strength than panels processed with PAA and bonded with RB-500/RB-398, the difference in peel strength is not significant in reality.

The peel strength of the adhesive systems generally increases when the test is conducted at -67°F. The four exceptions to this trend are the three 3M[R] adhesive systems (one of which is unprimed AF-31) and BR-227A/FM-61.

2.1.4 Failure Analysis No specimens failed interfacially between the primer and the oxide. Consequently the failure mode is independent of the surface preparation. The failure mode of the adhesive systems is given in Table 2. The reason why the 3M[R] adhesives foamed is not known at present. The speciments with better peel strength possessed moderate to good filleting to the core. A photograph of a representative specimen possessing good filleting is shown in Figure 2. A photomacrograph of good adhesive to core filleting is shown in Figure 3.

2.2 Environmental Durability of Bonded Panels

2.2.1 Alloys, Surface Preparations and Adhesive Systems See Section 2.1.1 for a discussion of these materials and processes.

2.2.2 Specimen Configuration and Experimental Conditions The environmental test panel configuration is shown in Figure 4. One panel was flexed in a three point bending test until one bonded surface delaminated. Two other panels were then flexed in the fixture shown in Figure 5 at one-third of the ultimate strain of each panel. The panels were exposed to the environment described in Figure 6 for a total of 2041 cycles.

2.2.3 Experimental Results By the end of the test, ten panels had failed. The number of cycles at which a panel failed are listed in Tables 3 and 4. The residual strength in pounds for the non-failed panels is also listed in Tables 3 and 4. More panels processed with PAA failed for the first two adhesives than did panels processed with FPL. However, more adhesive systems failed on panels whose adherends were processed with FPL than with PAA. The mean residual strength of the non-failed panels (for both surface preparations) is listed in Table 5. The differences in mean residual strength in Table 5 are not statistically significant. The t test at 95% confidence was employed again to test the data.

2.2.4 Failure Analysis The non-failed panels were X-rayed for water absorption, thru-transmission ultrasonically (TTU) tested for disbonding. Some of the panels bonded with EC-2174/AF-31 and EC-3917/AF-130 have significant amounts of water absorption and disbonding. None of the other adhesive systems have significant amounts of water absorption or disbonding. Water in five percent of the core or more is considered significant. Delamination to five percent of the core or more is considered significant. There was no significant difference between panels processed with PAA or FPL for either water absorption or delamination. The failed panels were not X-rayed for water absorption because it is not possible to determine if absorption or delamination caused the waters presence.

None of the failed panels disbonded in the oxide layer or between the oxide and the primer. All failures were cohesive in the primer or between the adhesive and the core. A failed panel processed with PAA is shown in Figure 7. A failed panel processed with FPL is shown in Figure 8. The failure mode of the failed panels is independent of the surface preparation Consequently the two surface preparations have equivalent environmental durability on metal/honeycomb bonded panels.

2.3 PAA Process Variation Analysis

2.3.1 Alloys, Adhesive System and PAA Process Conditions The alloys chosen for this task were bare 2024-T3 and bare 7075-T6 aluminum. Clad aluminum was not used because of its sacrificial tendency in bonded joints. Based on the peel strength of the adhesives, their failure mode in peel, and the environmental durability through 1700 cycles, MB-6725-1/MB-329 was chosen to investigate the effect of variations in PAA process on the mechanical properties of metal-to-metal bonded joints. Later one of the environmental panels processed with PAA and bonded with MB-6725-1/MB-329 failed. As the panel failed between the adhesive and the core (Figure 7), testing for this task was continued with this adhesive system. A full factorial design, utilizing the PAA process parameters listed below, was employed to determine the effects of variations in the PAA process on the mechanical properties of bonded metal-to-metal joints:

Temperature	65°F, 75°F and 90°F
Time	10 and 20 Minutes
Voltage	6 and 15 Volts

2.3.2 Specimen Configurations and Experimental Conditions The specimens shown in Figure 9 were fabricated with MB-6725-1/MB-329 and the 12 combinations of PAA process parameters. The lap-shear testing was conducted at room temperature and at 350°F. The wedge specimens were exposed for thirty days to 160°F, 95 to 100 RH. The crack growth was measured initially and then at 1 hour, 1 day, and 7 days. Five replicates of each specimen at each test condition were evaluated.

2.3.3 Experimental Results

The results of the analysis of the mechanical tests, but not the specific data, will be presented to keep the report within the page limit. The wedge test data will be discussed first.

The crack growth data at 7 days were analyzed employing analysis of variance[7]. The data was analyzed using the interactive software package STATPK[8] on a Control Data Corporation Cyber 175 mainframe.

The two PAA process conditions which affect the fracture toughness of the bonded joint are the second order interaction between time and temperature, and voltage. Both conditions **are significant at 95% confidence**. Voltage is not significant at 99% confidence. Figure 10 illustrates the second order effect of time and temperature. The minimum crack growth point occurs at specification anodization time and voltage. The test anodization temperature closest to the specification value (77°F) has the lowest crack growth. By comparison, the 10 minute anodization time produces the maximum crack growth. The crack growth curves for 2024-T3 bare aluminum have the same general shape. The maximum crack growth (.35 inch in one week) on either alloy is quite acceptable.

The lap-shear specimen configuration is given in Figure 9. An adherend thickness of 0.125 inches was used because employing thinner adherends result in poor estimates of the bonded joint shear strength. This phenomenon is discussed in detail in Air Force report AFWAL-TR-80-4176[2].

An analysis of variance was employed again to analyze the data. The interaction between time and temperature, and voltage are important factors. However, the time-temperature interaction does not affect the shear strength to the extent it affects the fracture toughness. Voltage has a greater effect on shear strength than on fracture toughness. Two other factors affect the shear strength but not the fracture toughness of the adhesive joint-the alloy and the interaction between the alloy and temperature. A tentative explana tion of why the aluminum alloy type affects the shear strength follows.

The aluminum adherends plastically de form (i.e., they bend) because of the eccentric load on the specimen during testing. Bending of the adherends increases the normal stresses and decreases the shearing stresses on the bonded joint. Increasing the normal stress decreases the measured shear strength. Increasing the shear stres increases the measured shear strength

The greater yield strength and work hardening of 7075-T6 bare aluminum compared to 2024-T3 bare aluminum result in less bending of 7075-T3 bare adherends compared to 2024-T3 bare adherends. Less bending of the 7075-T6 adherend results in greater measured shear strength.

The second new factor is a synergistic interaction between alloy type and temperature. This interaction can be discerned by examining the mean shear strength which is plotted in Figures 11 and 12. At 65°F and 75°F reducing voltage and time has an opposite effect on the shear strength of the two alloys.

A t test was performed between the "15 volt-20 minute" and the "6 volt-10 minute" shear strength data at each temperature. The latter anodize condition represents the maximum variation in the PAA process. These differences are statistically significant at 65°F and 75°F for 2024-T3 bare aluminum (at 95% confidence). None of these differences are statistically significant for 7075-T6 bare aluminum. Although some statistically significant differences in shear strength exist, these differences are not large enough to indicate an excessive reduction in bonded joint strength.

The maximum reduction in shear strength is about 5 percent and the average standard deviation is about 2 percent. The maximum standard deviation is 5 percent.

The 350°F shear strength on 7075-T6 is plotted in Figure 13. The specification voltage and time produce the greatest lap-shear strength at a particular temperature. The difference in shear strength at the three temperatures for specification time and voltage is not significant. There is considerable variation in the 350°F lap-shear strengths of 2024-T3 bare aluminum at different PAA process variations. Currently, this is not understood. By comparison, the variation in shear strength on 7075-T6 bare is much less and is not significant.

The 350°F shear strength of this adhesive system on PAA-treated adherends processed per specification is about 85% of the room-temperature value. These are quite high values.

2.3.4 Failure Analysis

The wedge specimen failed cohesively in the adhesive with stable crack growth during the test. The specimen was split open to photograph it. Along the length of the specimen, where the specimen was split open, there is some failure close to one metal adherend. This is an artifact caused by opening the specimen. Two wedge specimens are shown in Figure 14. The left hand pair is on 7075-T6 bare aluminum. The right hand pair is on 2024-T3 bare aluminum.

The lap-shear specimens failed in the

mixed mode fashion which is indicative of a good bonded joint[2]. The joint failed cohesively in both the adhesive and primer. Two lap-shear specimens are shown in Figure 15. The specimen on 7075-T6 bare is on the left and 2024-T3 bare on the right.

3.0 CONCLUSION

The conclusions of the investigation are:

1. The adhesive systems have equivalent peel strength and durability on PAA and FPL.

2. The adhesive systems failure mode in peel is independent of the surface preparation.

3. The environmental panels which disbonded in test, failed between the adhesive and the aluminum core.

4. No adhesive system failed interfacially between the adhesive primer and the oxide for either surface preparation.

5. Variation of the PAA process parameters does not significantly affect the fracture toughness of metal-to-metal bonded joints.

6. Variation of the PAA process parameters does not significantly affect the lap-shear strength of metal-to-metal bonded joints. Although some differences are real, they are not significant.

7. Variation of the PAA process parameters has a greater effect on the lap-shear strength of bonded joints having bare 2024-T3 adherends than it does on bonded joints having bare 7075-T6 adherends. This is not understood at present.

ACKNOWLEDGMENT

This investigation was funded by the Air Force Wright Aeronautical Laboratories as part of the Primary Adhesively Bonded Structure Technology (PABST) program. The Douglas Aircraft Company is the prime contractor. The Boeing Company performed the work under subcontract, PABST 800 Test Program, to the Douglas Aircraft Co. Mr. William Sheldon (AFWAL, FD), Air Force Flight Dynamics Laboratory of the Air Force Wright Aeronautical Laboratories, is the technical monitor.

BIBLIOGRAPHY

1. J. C. McMillan, "Surface Preparation-The Key to Bondment Durability" in AGARD Lecture Series No. 102, Bonded Joints and Preparation for Bonding, Technical Report AGARD-LS-102.
2. E. E. Peterson and D.B. Arnold, "Relationship of Interfacial Compatibility to Durability of Adhesive-Bonded Joints," Technical Report AFWAL-TR-80-4176, March 1981.
3. Boeing Material Specification BMS 4-4, Rev. J, "Honeycomb Core 5052 Aluminum Alloy."
4. Boeing Process Specification

BAC5514, Rev. P, "Common Bonding Requirements for Structural Adhesives."
5. Boeing Process Specification BAC 5555, Rev. E, "Phosphoric Acid Anodizing of Aluminum for Structural Bonding."
6. P. G. Hoel, Introduction to Mathematical Statistics, John Wiley & Sons, New York, NY, 1954, pp. 211-228.
7. D. Durand, Stable Chaos, General Learning Press, Morristown, N.J., 1971, pp. 360-382.
8. Boeing Computer Services Document No. 10201-019, Revision 1, Mainstream[R]-CTS Interactive Statistics Package (STATPK), March, 1978, pp. 51-55.

BIOGRAPHIES

E. E. Peterson received his B.S. in Chemistry from the University of Washington. His work includes studies of materials and processes for structural bonding. He is the Technical Leader of the PABST 800 Test Program (Contract S&S-19770-C).

D. B. Arnold received his B.S. in Chemical Engineering from the University of Washington. Currently he is supervisor of the 757 Chemical Unit. Previously he was supervisor of Materials and Processes for the Boeing Military Airplane Company. He is the Manager of the PABST 800 Test Program.

M. C. Locke received his B.S. in Chemistry from the University of Washington. His work includes studies of bond durability and surface-polymer interactions. He wrote the proposal for the PABST 800 Test Program.

Table 1 Adhesive Systems Tested

Primer	Adhesive	Vendor
BR-227A	FM-61	American Cyanamid Company
BR-400	FM-400	Same as above
HT-424F	HT-424	Same as above
	HT-424	Same as above
RB-500	RB-398	Ciba Geigy
EA-9205	EA-9649	Hysol Division, The Dexter Corporation
EC-2174	AF-31	Minnesota Minning and Manufacturing Company
	AF-31	Same as above
EC-3917	AF-130	Same as above
MB-6725-1	MB-329	Narmco Manufacting Incorporated

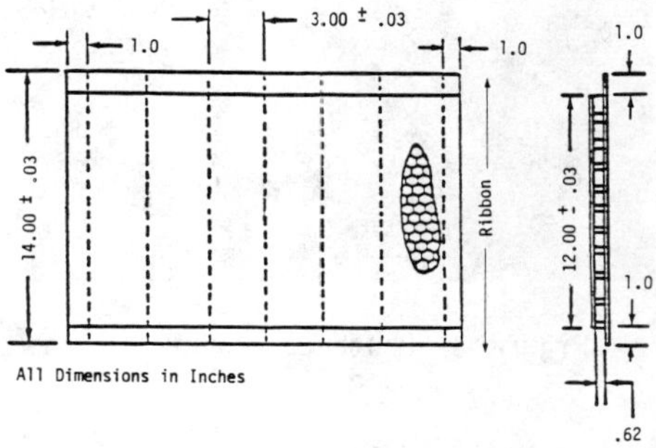

Figure 1 Honeycomb Peel Specimen Configuration

Table 2 Honeycomb Peel Specimen Failure Mode

PRIMER	ADHESIVE	PAA	FPL
EC-2174	AF-31	ADHESIVE FOAMING	ADHESIVE FOAMING
EC-3917	AF-130	ADHESIVE FOAMING	ADHESIVE FOAMING
EA-9205	EA-9649	COHESIVE PRIMER	COHESIVE PRIMER
BR-227A	FM-61	MIXED MODE CORE	MIXED MODE CORE
BR-400	FM-400	COHESIVE PRIMER	COHESIVE PRIMER
HT-424F	HT-424	INTERFACIAL CORE	INTERFACIAL CORE
MB-6725-1	MB-329	COHESIVE PRIMER	COHESIVE PRIMER
RB-500	RB-398	COHESIVE PRIMER	COHESIVE PRIMER
------	AF-31	ADHESIVE FOAMED	ADHESIVE FOAMED
------	HT-424	INTERFACIAL CORE	INTERFACIAL CORE

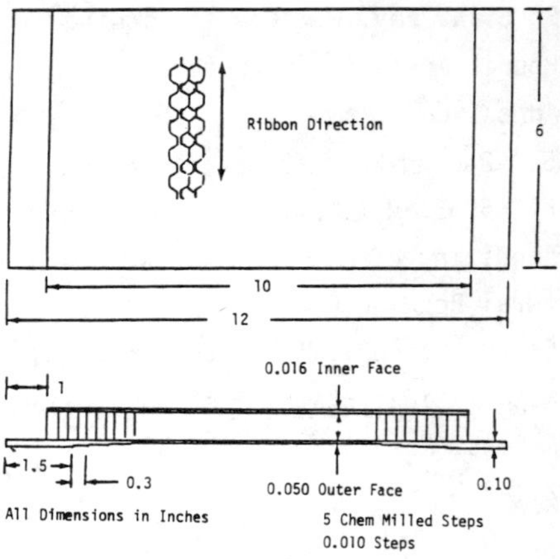

Figure 4 Environmental Test Panel Configurations

Figure 2 Photograph of Peeled Metal/Honeycomb Panel

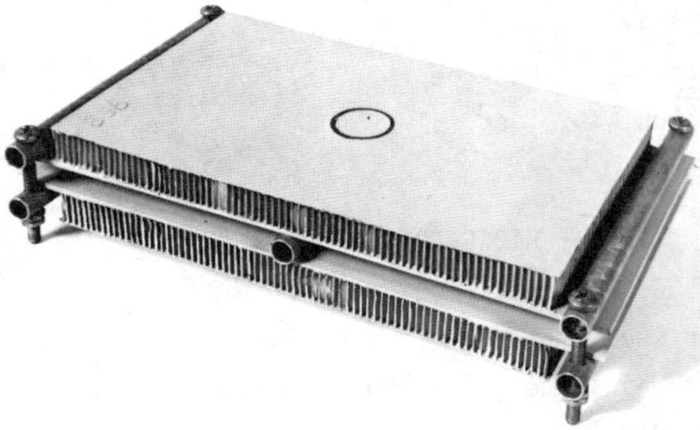

Figure 5 Flexed Environmental Panels

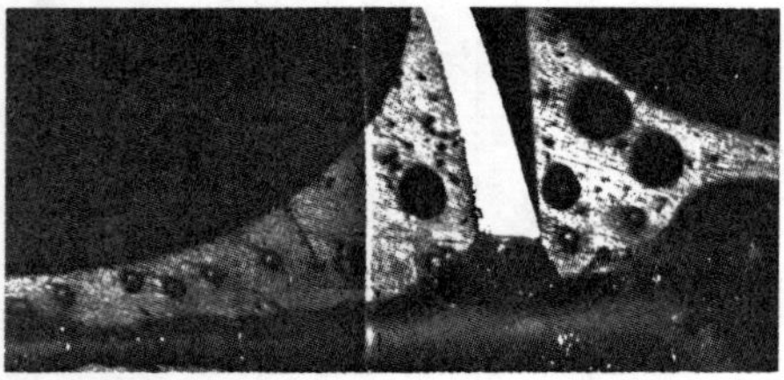

Figure 3 Photomacrograph of Good Filleting

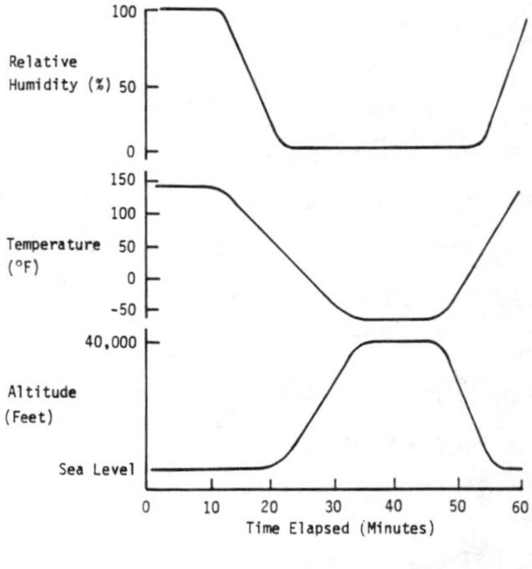

Figure 6 Simulated Flight Cycle

Table 3 Environmental Panel Residual Strength and Failures PAA Surface Preparation

PAA SURFACE TREATMENT

PRIMER ADHESIVE	2024-T81 CLAD	2024-T81 BARE	7075-T76 CLAD	7075-T76 BARE
EC-2174 AF-31	815 CYC 386	400 CYC 240	247 297	242 156
EC-3917 AF-130	176 97	400 CYC 165	1343 CYC 238	580 CYC 205
EA-9205 EA-9649	886 968	976 985	849 785	828 908
BR-227A FM-61	1026 1046	1011 1060	915 925	1042 1080
BR-400 FM-400	623 649	595 489	530 563	680 445
HT-424F HT-424	485 495	493 420	333 361	348 346
MB-6725-1 MB-329	649 750	1836 CYC 640	586 564	469 479
RB-500 RB-398	431 431	340 388	503 473	453 535
AF-31	317 328		81 94	
HT-424	360 381		348 397	

NUMBER OF CYCLES STATE WHEN A SPECIMEN FAILED
OTHER NUMBERS INDICATE RESIDUAL STRENGTH IN POUNDS

Table 4 Environmental Panel Residual Strength and Failures FPL Surface Treatment

FPL SURFACE TREATMENT

PRIMER ADHESIVE	2024-T81 CLAD	2024-T81 BARE	7075-T76 CLAD	7075-T76 BARE
EC-2174 AF-31	183 268	580 CYC 342	254 291	214 136
EC-3917 AF-130	164 142	580 CYC 135	43 128	222 205
EA-9205 EA-9649	865 882	1076 998	818 847	940 939
BR-227A FM-61	1053 1032	1020 971	1032 1000	1343 CYC 722
BR-400 FM-400	623 579	400 CYC 715	652 453	610 413
HT-424F HT-424	427 477	100 566	419 416	530 469
MB-6725-1 MB-329	689 680	369 407	621 686	497 491
RB-500 RB-398	514 505	393 473	483 480	638 586
AF-31	210 270		302 282	
HT-424	365 333		359 327	

NUMBER OF CYCLES STATE WHEN A SPECIMEN FAILED
OTHER NUMBERS INDICATE RESIDUAL STRENGTH IN POUNDS

Table 5 Mean Residual Strength of Non-Failing Environmental Panel

PRIMER	ADHESIVE	PAA	FPL
EC-2174	AF-31	261.7 ± 76.0	241.1 ± 69.2
EC-3917	AF-130	176.2 ± 52.6	148.4 ± 58.6
EA-9205	EA-9649	898.1 ± 74.5	920.6 ± 85.6
BR-227A	FM-61	1013.1 ± 61.1	975.7 ± 114.9
BR-400	FM-400	571.8 ± 80.7	577.9 ± 108.1
HT-424F	HT-424	410.1 ± 71.8	425.5 ± 142.1
MB-6725-1	MB-329	591.0 ± 99.3	555.0 ± 130.4
RB-500	RB-398	444.3 ± 62.1	509.0 ± 74.5
------	AF-31	205.0 ± 135.9	266.0 ± 39.6
-------	HT-424	371.5 ± 21.8	346.0 ± 18.8

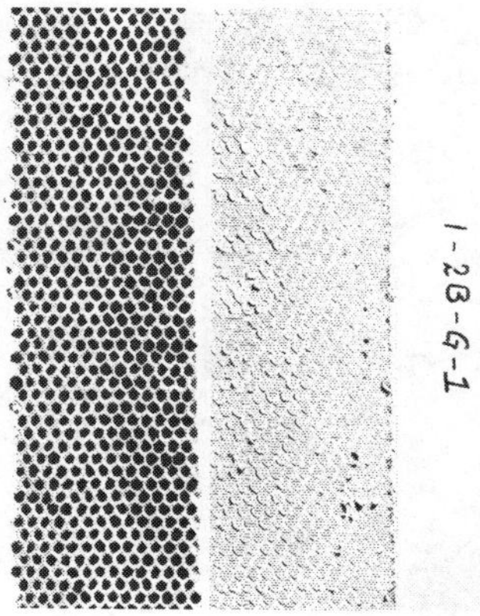

Figure 7 Photograph of Failed Environmental Panel Processed with PAA

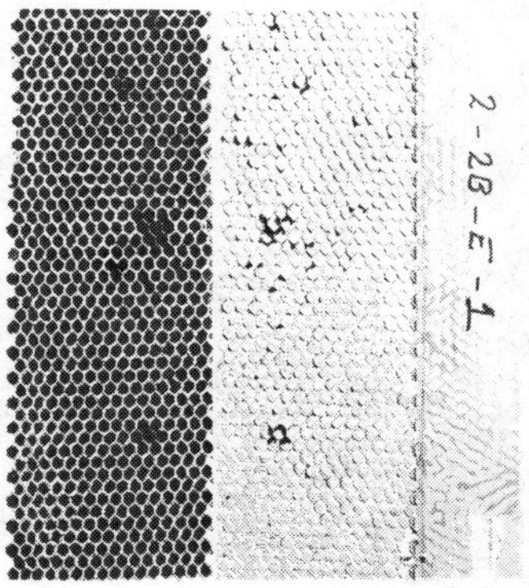

Figure 8 Photograph of Failed Environmental Panel Processed with FPL

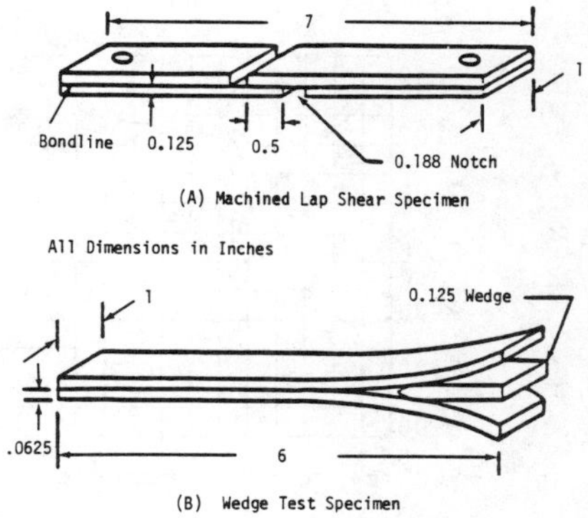

Figure 9 Test Specimen Configuration

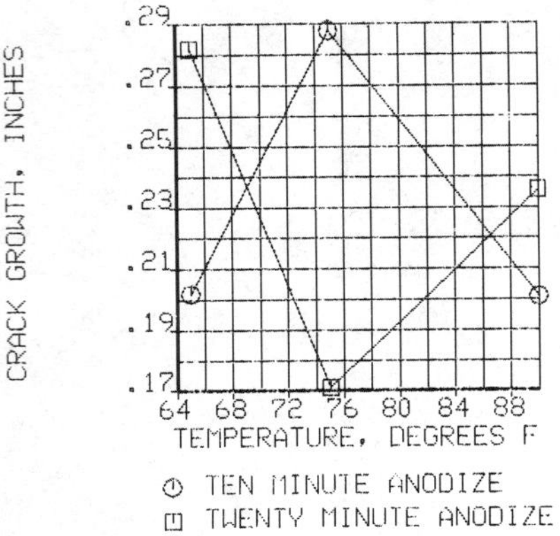

Figure 10 Mean Crack Growth vs. Time and Temperature for 7075-T6 Bare Aluminum

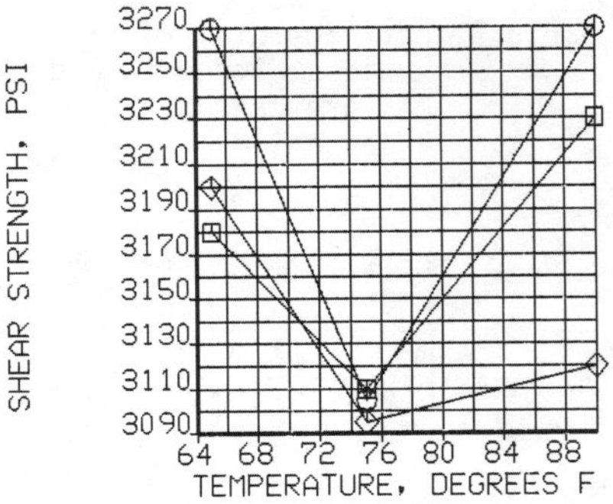

Figure 13 Shear Strength at 350°F vs. PAA Process Variations on 7075-T6 Aluminum

THESE SYMBOLS APPLY TO GRAPHS 11, 12 AND 13.

- 15 VOLT, 20 MINUTE ANODIZE
- 6 VOLT, 20 MINUTE ANODIZE
- 15 VOLT, 10 MINUTE ANODIZE
- 6 VOLT, 10 MINUTE ANODIZE

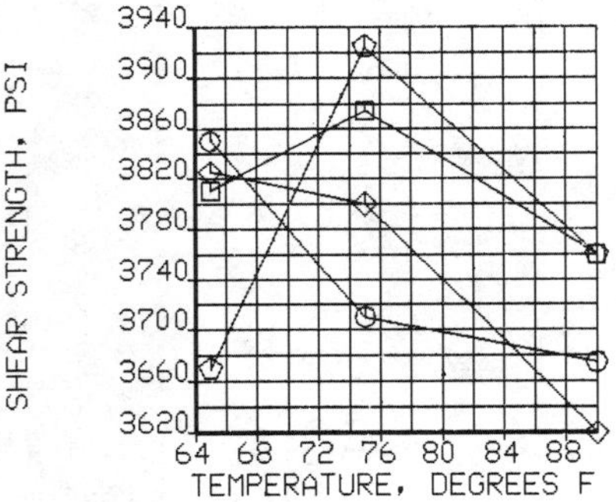

Figure 11 Shear Strength vs. PAA Process Variations on 2024-T3 Bare

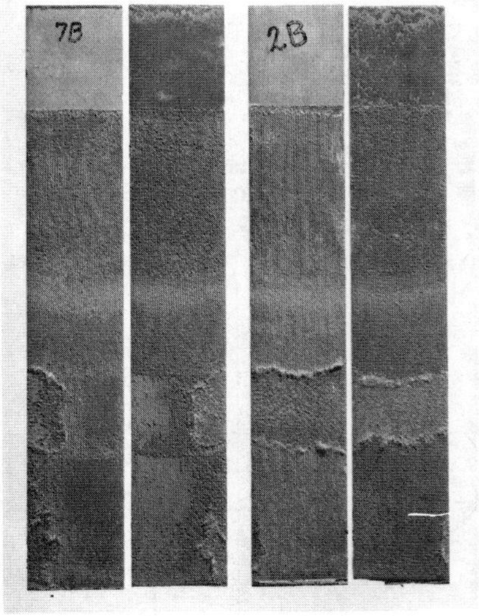

Figure 14 Photograph of Task IV Wedge Specimens

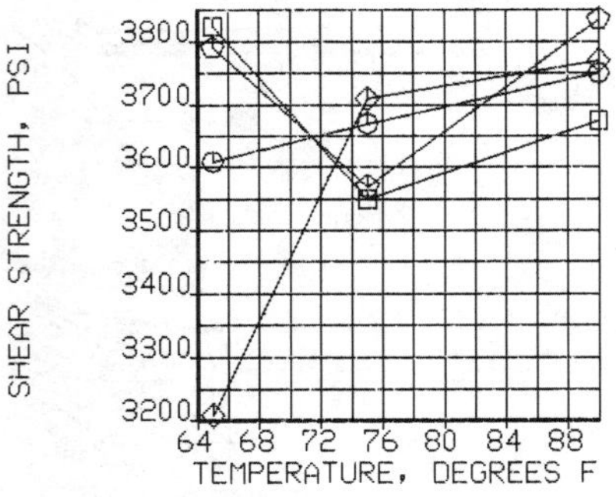

Figure 12 Shear Strength vs. PAA Process Variations on 7075-T6

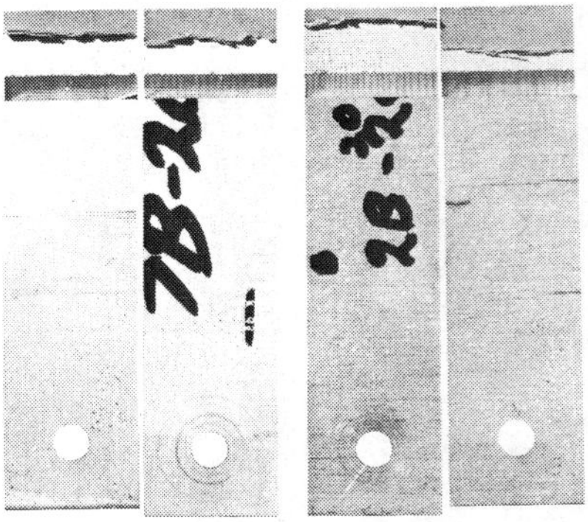

Figure 15 Photograph of Task IV Lap-Shear Specimen

13th National SAMPE Technical Conference
October 13-15, 1981

CHARACTERIZATION OF COMPOSITION VARIATIONS IN A STRUCTURAL ADHESIVE

Kenneth G. Kibler
Materials Research Laboratory
General Dynamics
Fort Worth, Texas 76101

Abstract

Controlled composition and processing variations of RB398 adhesive have been evaluated with typical mechanical-type receiving inspection tests, current physiochemical methods, and additional performance tests. The results of these investigations are used to assess the sensitivity of processing and performance behavior to material variations, and to guide the implementation of physiochemical quality control procedures.

Keywords: Chemical screening, adhesive testing, quality control, structural adhesives.

1. INTRODUCTION

Traditional control of polymeric materials such as adhesives and graphite-epoxy composites has relied on physical/mechanical testing. Conventional mechanical testing results for such materials, via stress-strain, peel, flexure, etc., are often quite incomplete indicators of material quality. As pointedly emphasized by various workers, e.g., Hinrichs and Thuen[1], mechanical results may be strongly influenced by specimen fabrication and testing techniques.

We are presently engaged in a project for Process Control of Adhesives and Sealants, which is one of the Technology Development efforts of the F-16 Technology Modernization Program. This particular project seeks to realize more cost-effective and reliable control of aircraft structural adhesives and fuel tank sealants by implementing a new physiochemical control and inspection system.

Our approach is to augment current physical/mechanical tests, which check conformance of materials to specification in terms of test specimen performance, with physiochemical tests which provide data on the composition and processability variations of materials within specified limits. The former test procedures are labor-intensive and provide little insight into the essential characteristics of the materials being tested. The proposed methods may be effectively automated and will provide information of a more fundamental nature useful in diagnosing the causes of performance anomalies.

Figure 1 depicts our planned receiving inspection system for adhesives and adhesive-primer materials. In this paper we will summarize our investigations to date to establish correlations between compositional changes, processability, and structural-environmental performance of one of our qualified 350°F-cure structural adhesives, RB398, manufactured by the Composite Materials Department of Ciba-Geigy. The project includes similar investigations for another of our 350°F-cure systems, Narmco's MB329-7. That work will be summarized in a future report.

As the tasks summarized here are completed, we expect to define material acceptance specifications in terms of composition, formulation, and cure limits, and use those specifications in the system shown in Figure 1.

2. EXPERIMENTAL RESULTS

Various programs have previously investigated aspects of physiochemical detectability for constituents in nominal and varied matrix resins and adhesives.(2-6) More recent SAMPE volumes, for example, continue the documentation of our attempts to understand material structure-property relationships through state-of-the-art instrumentation and application.

We here deal with Reliabond RB398 scrimmed adhesive, a multi-component epoxy system cured with dicyandiamide (DICY) hardener, and with structural and processing characteristics strongly influenced by aluminum and other fillers.

Table 1 lists the variations of RB398 formulated for us by Ciba-Geigy. The variations are not exhaustive, but were constructed on the basis of supplier and user experience with material criticalities. The levels of variation were selected to make both physiochemical and mechanical detectability likely; they exceed any expected supplier variation. The listed variations include only one variable in each case. Batches 16 and 17 are nominal batches which are "cooked" longer than usual by the supplier.

TABLE 1 ADHESIVE VARIATIONS

Batch	Variant
0	Control - Fine Hardener
1	-50% " "
2	-25% " "
3	+50% " "
4	Control - Coarse Hardener
5	-50% " "
6	-25% " "
7	+50% " "
8	+25% Solid Epoxy
9	-25% " "
10	+40% Liquid Epoxy
11	-40% " "
12	+25% Aluminum
13	-25% "
14	+40% Flow Control Filler
15	-50% " " "
16	+30 Min. Cook at 190°F
17	+60 Min. " " "

To determine the effect of these variations, we employed the spectrum of tests listed in Table 2. The mechanical tests will, for the most part, be recognizable as typical adhesive receiving inspection tests in many cases. To these are added chemical, rheological, environmental, and durability tests.

2.1 Physiochemical Tests

Figure 2 shows typical reverse phase liquid chromatographic (LC) results obtained for the material using a Spectra Physics 8000B HPLC with a water/acetonitrile (AN) gradient as follows:

Time:	% AN	% H_2O
0	40	60
40	99	1

46	99	1
47	40	60
57	40	60

An ODS-2 column (with pre-column) was used with a 1 mg/ml sample dissolved in THF and a 2 ml/min flow rate. Detection was at 230 nm.

TABLE 2 TEST METHODS

- o High Pressure Liquid Chromatography
- o Fourier Transform Infrared
- o Karl Fischer Moisture Analysis
- o Weight Range
- o Flow at 300°F
- o Thermogravimetric Analysis
- o Thermal Expansion, Dry and Saturated at 75% RH
- o Gel Time at 295°F
- o Double Overlap Shear at 270°F
- o Single Overlap Shear at -65, RT, +270°F
- o Flatwise Tension at RT
- o Stressed Durability at 140°F, 98% RH
- o Dynamic Viscosity (Rheometrics)
- o Moisture Absorption Characteristics
- o Flexural Sandwich Beam at 270°F
- o Fluid Tightness at RT

Figure 2 shows the breakout of RB398 and three of its "resin" constituents. The typical HPLC detectability of varied constituents is shown in Figure 3 for the adhesive hardener. As indicated there, a calibration curve was necessary due to the nonlinearity of UV absorbance with sample concentration. Coefficients of variation (C.V.) are considerably higher for lower concentration.

The right half of Figure 3 shows hardener detectability for all batches of the special materials. The 10% detection variability indicated for a nominal amount of hardener (0% variation) is typical for work of this type.

The detectability of organic variations in the adhesive appears to be reliably performed by HPLC, at least at levels of variations which can be shown to affect performance or processing. Different supplier staging of the materials (e.g., Batches 16 and 17) was not detectable by HPLC, but we have noted no serious anomalies in performance or processability for the "over-cooked" batches.

Figure 4 shows similar procedures and results for hardener changes measured with a Nicolet MX-1 Fourier-Transform IR Spectrometer. The samples were dissolved in acetone, filtered, and scanned in a liquid cell.

As-received moisture contents of each roll of material were measured by Karl Fischer titration. The measured weight percents of water are given in Table 3. Typical standard deviation is 0.01-.02%.

TABLE 3 AS-RECEIVED MOISTURE CONTENT OF VARIED ADHESIVES

Batch	$\%H_2O$
0	0.139
1	.159
2	.146
3	.131
4	.158
5	.172
6	.213
7	.147
8	.156
9	.146
10	.182
11	.146
12	.152
13	.146
14	.254
15	.125
16	.132
17	.131
Mean	.158 ± .032

All batches fall within one standard deviation of the mean except for

Batches 6 and 14. All batches were lab-mixed under carefully-controlled conditions, so the variation in Batch 6 is probably not significant. The high value for Batch 14 suggests that wet filler may be the main contributor to adhesive moisture content.

In combination with other factors, it is possible for a high moisture content to result in unacceptable flow/fillet characteristics during bonding, so it is probably wise to screen incoming materials for moisture content. In the present investigation, however, no definite correlation was evident between moisture and other measured properties.

Water content of resin matrix materials has often been measured by various workers with a DuPont 902 Moisture Analyzer. We found that instrument to be quite susceptible to cell plugging when used with these adhesives, probably due to light end polymers released during heating.(7) Gas chromatographic or Karl Fischer procedures should thus be used.

The results of other physical-type measurements were similar for all batches. As-received weight range, thermogravimetric behavior to decomposition, and thermal expansion characteristics were typical of nominal production materials, and presented no surprises.

2.2 Performance Tests

Double-overlap shear strengths measured at 270°F for each of the varied adhesive materials are shown in Figure 5. Also shown, for comparison, are results on batches of nominal production material. The lower dashed line is the minimum strength requirement, while the upper dashed is the typical average value for materials purchased for production.

One notes that, except for two batches, all the "off-formulated" materials are quite acceptable in terms of this test. The unacceptable batches, 1 and 5, represent 50% reductions in hardener content. Other data on these batches also show unacceptable fillet characteristics in skin-to-core bonding.

Figure 6 shows the strength trend with hardener content quite clearly. If 4600 psi (the dashed line) is the acceptable lap shear minimum, then the hardener content may vary by -25% to +50%, a far wider range than nominal batch-to-batch variability. There also appears to be no dramatic effect due to hardener particle size for this material, based on this test.

Also shown in Figure 6 are flow properties of these same batches. This property is determined as the distance (inches) the material flows down an inclined aluminum plate in a 300°F oven. We have found it to be a quick and simple indicator of expected processing and filleting properties of the material.

The Figure 6 results indicate all batches with varied hardener are quite similar in flow except those with 50% deficiency, and those would be rejectable on this basis.

If we examine a different strength parameter such as single-overlap shear, the results are not nearly as pleasing as the smoothly-varying double-lap shear data in Figure 6.

Figure 7 shows single-lap shear strengths for the batches with varied hardener. There are no real trends apparent, but this is not surprising for this type of material when tested in a single-lap joint, for a significant component of peel is probably present. RB398 has very desirable elevated temperature properties, but one does not expect good peel strength.

If we look at the single-lap shear strength for all the varied batches (Figure 8), few trends are evident except for the very poor performance of Batches 11 and 14. The variations in each of these batches produced extremely viscous materials, and the supplier had particular trouble coating them. We will see from the flow tests reported in the following section that these materials could be readily screened from the others without these strength tests.

We should note that Batches 11 and 14 were received much later than the other materials due to the difficulties in coating them. Those two materials were thus not tested in some cases, then, due to scheduling problems.

The highest single-lap shear strengths were obtained for Batch 10 (+40% liquid epoxy), but when the materials were tested with a smaller single overlap area (0.25 in^2 vs. 0.5 in^2) (see Figure 9), Batch 10 showed the lowest single lap shear strength.

Figure 8 indicates a low RT strength for Batch 5, but Figure 9 shows Batch 5 to have a quite respectable single lap shear strength at 140°F. One should probably conclude that the dominant factors determining these single lap shear strengths may not be material characteristics, once a minimum strength is exceeded. We are likely measuring variabilities in specimen/surface preparation.

In order to explore performance properties beyond "strength-type" tests, stressed durability investigations were conducted at 140°F/ condensing humidity conditions on 0.125 in. adherend single overlap specimens. The fixture holds 4 coupons and one solid aluminum blank which is strain gaged. The coupons were loaded to 75% of the Batch 0 strength at 140°F until two of the four coupons fail or the test is stopped. Survivors were then statically failed, as shown in Figure 9.

The times-to-failure from this test are shown in Figure 10. The first interesting feature of the results is the poor performance of the control batch (0), although the nominal batch (4) with coarse hardener performed rather well. Batch 12 with high Al content was the best performer and it also had good lap-shear strengths (Figures 5 and 8).

Batch 9 (-25% Liquid Epoxy) had poor durability, but rather good shear strengths. Consider also the group of coupons from Batches 4-7, where the coarse-ground hardener is varied from nominal to ±50% of nominal. The times-to-fail are quite similar for each case (except an early failure for Batch 5), yet we have seen the considerable variation in lap shear strengths in Figures 5 and 8.

One must view "mechanical" data with great caution. Although tests reported here show certain <u>coupon</u> property differences, it is not at all clear that <u>material</u> differences are being consistently measured. This adhesive appears very forgiving in many cases of rather substantial constituent variations. Variabilities in coupon preparation and testing may overshadow typical material variations.

2.3 Processability Tests

Figure 11 summarizes gel time results for these materials measured at 295°F. No significant differences may be noted between hardener-varied batches, but the changes between Batches 8-9, and 10-11, respectively, are interesting. Changes in those two epoxy contents affect when the material gels (and other processing features we will examine). We have seen in Figure 8 what a poor performer is Batch 11.

Figure 12 shows the measured flow properties (inclined plate at 300°F, as previously described). In most cases more than one measurement was conducted, so the range of values is indicated. On the basis of these results, Batches 1, 5, 11, and 14 would certainly be suspect, if not immediately thrown out.

Since moisture content can affect flow properties in many cases, the Karl Fischer results for moisture content are also shown in the Figure. High flow (Batches 1, 5) accompanies "average" moisture content in some cases, while low flow (Batches 11,14) materials have both "average" and unusually-high moisture values. One suspects the measured flow variations are primarily due to the intentional chemical changes in these cases, but some moisture interaction cannot be ruled out.

To explore more fully the rheological behavior of these materials, a dynamic mechanical analysis was performed on most batches using a Rheometrics RDS instrument with 25mm parallel plates. The methods used were similar to those reported by Thuen and Hinrichs.(7)

Samples of adhesive from which the carrier was removed were placed between the RDS plates at 50°C. The temperature was ramped to 160-170°C at 2°C/min, and 15% strain at a frequency of 5 rad/sec was applied. Instrumental problems in some cases prevented recording the measured storage modulus and thus the loss tangent. Figures 13-15 compare the measured dynamic viscosities between certain sets of variations.

In Figure 13 are plotted the viscosity variations in Batches 4, 5 and 7 during the thermal ramp. Many of the previously-noted features of these batches are demonstrated here. All three batches finally gel at the same time/temperature region around 160°C, but the low-hardener batch reaches a much lower viscosity for a longer time and is slower to react than the other batches. This substantiates and elucidates the gel time results (Figure 11) and flow test data (Figure 12). Note that the +50% hardener appears to have little influence on the rheology as measured here.

Figure 14 compares viscosities (measured as above) for the ±25% solid epoxy variations. As expected, an increase in the solid (higher molecular weight) epoxy increases the starting and flow viscosity, but seems to then retard the cure, requiring longer to gel.

Similarly, Figure 15 shows the effects of variation in flow control filler. Though it is only present at a level of typically a few percent by weight of the total filled adhesive, a +40% variation (say 2% by total weight) causes an almost complete loss of processing properties (previously screened by the flow test shown in Figure 12).

If we tabulate the minimum viscosities for the batches, as measured by the Rheometrics experiment, and compare with typical flow test results, the correlation shown in Figure 16 results. (Viscosity data were not obtained for the low flow Batch 11). These results suggest that a flow test requirement should provide excellent screening of incoming batches via a very simple and inexpensive test. It is interesting to note that flow results for production material typically run between 2.5 and 4.0 inches, i.e. within a standard deviation of the average we have measured here.

Rheological measurements such as those obtained by Rheometrics or comparable instruments are highly important in establishing the response of adhesive materials to chemical, processing, and environmental variations. The results are clearly important for process optimization, e.g., and for structure-property diagnostics.

For routine receiving inspection tests of established materials, however, simpler (and cheaper) flow tests may provide adequate screening.

3. CONCLUSIONS

A series of physiochemical, mechanical, and rheological tests on controlled variations of a typical 350°F-cure aerospace structural adhesive suggest several important features relating to the inspection and control of such materials.

Mechanical tests may often show coupon property differences which are not caused by material property differences; i.e., coupon preparation and testing variables may overshadow material variations. These materials are, in fact, rather forgiving of modest upsets in their formulation.

Variations in organic constituents are generally detectable by high performance liquid chromatography, at least at levels which can be shown to affect performance or processing.

Material changes which dramatically affect flow/processing properties also will affect one or more performance properties. Screening may thus be accomplished prior to coupon fabrication.

Cost- and quality-effective inspection tests for these adhesives appears likely by insuring (1) nominal compositional integrity, and (2) consistent flow/processing characteristics.

ACKNOWLEDGEMENTS

The work reported herein was sponsored by the F-16 Technology Modernization Program, Contract F33615-80-G-0007/0002 and 0003. Many scientists of Materials and Structures Technology, General Dynamics, have contributed substantially to this work, including C. L. Amerson, T. S. Creasy, J. M. Cartwright, F. H. Wuellner, B. O. McCauley and D. B. Sampley. Mr. Weldon Scardino is the AF Project Engineer for this work.

REFERENCES

1. Hinrichs, R., and Thuen, J., National SAMPE Symposium, Vol. 24 (1979), p. 404.

2. Carpenter, J.F., ibid., p. 446.

3. Wickham, A.A., Rice, D.D., and Dubois, R.J., ibid., p. 506.

4. Hagnauer, G.L., Polymer Composites 1, 81 (1980).

5. Britton, R.A., "Acceptance Criteria for Structural Adhesives," AFML-TR-78-53 (May 1978).

6. Sewell, T.A., "Chemical Characterization and Quality Control for an Adhesive," N00019-79-C-0064 (Feb. 1980).

7. Thuen, J., and Hinrichs, R., SAMPE Journal, p. 6 (Sept./Oct. 1980).

BIOGRAPHY

Dr. Kibler is a Senior Engineering Specialist in the Materials Research Laboratory of General Dynamics. He serves as project leader and principal investigator for in-house investigations dealing with physiochemical characterization of polymeric systems and for contracted research in the process control of adhesives and sealants.

Dr. Kibler has also managed AF-sponsored programs and IRAD projects dealing with time-dependent environmental behavior of composites, the response of graphite-epoxy composites to laser radiation, viscoelastic characterization of epoxy materials, and the effects of environment on the physical properties of composites. He received his Ph.D. in Physics from the University of Iowa in 1966.

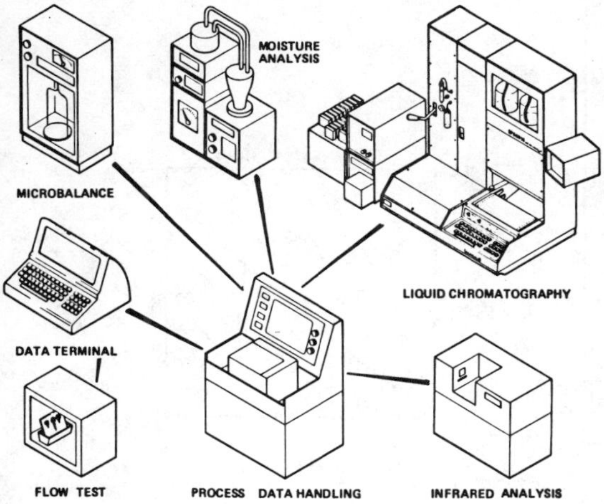

Figure 1 Primer/Adhesive Receiving Inspection System (PARIS)

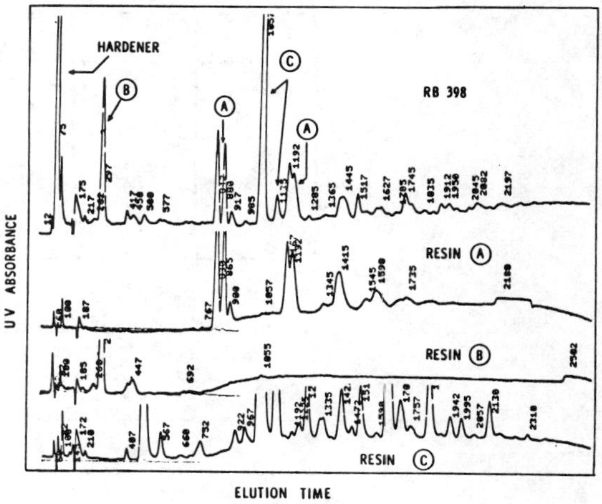

Figure 2 HPLC Separation of RB398 Adhesive

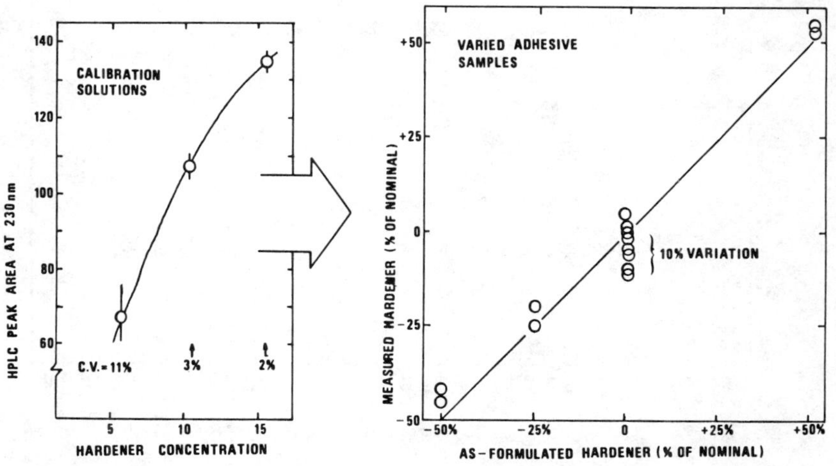

Figure 3 Liquid Chromatographic Measurement of Hardener Variations

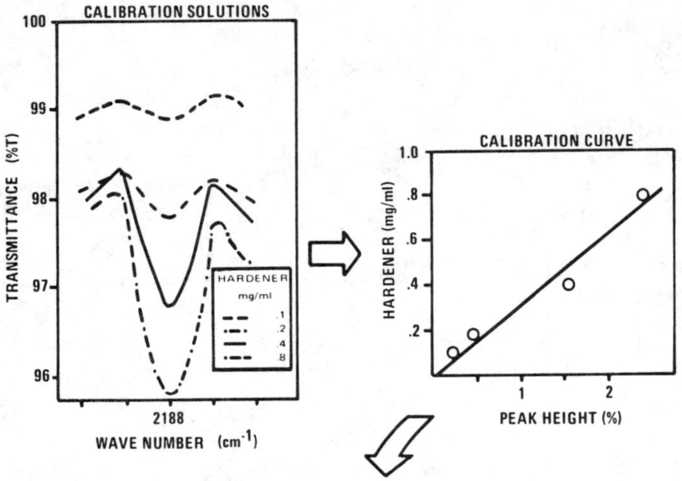

HARDENER VARIATIONS

BATCH NO.	RELIABLE FORMULATION	GD MEASUREMENT
4265-0	STD	STD
–7	+50%	+50%
–6	-25%	-26%
–5	-50%	-45%

Figure 4 Infrared Measurement of Hardener Variations

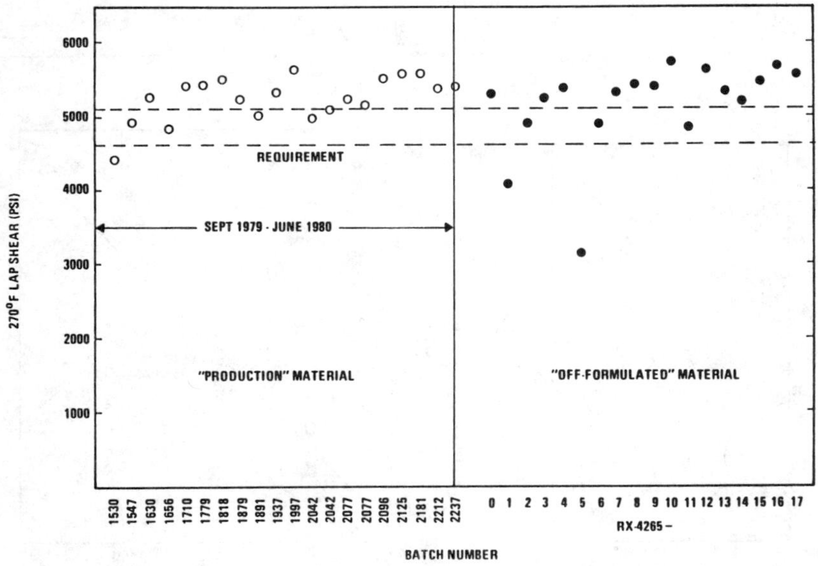

Figure 5 Double-Overlap Shear Strengths at 270°F

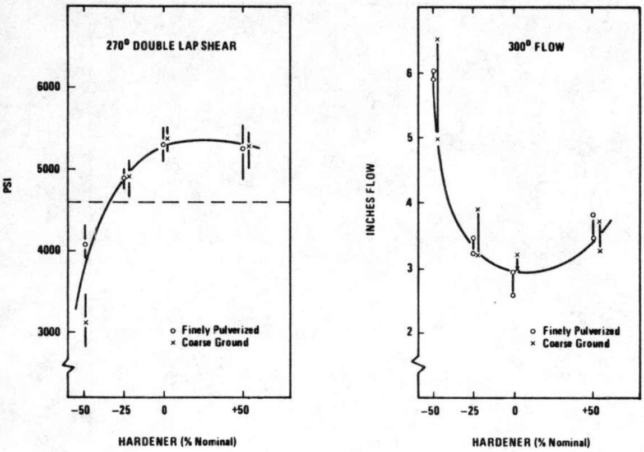

Figure 6 Double-Lap Shear Strength and Flow vs. Hardener Content

Figure 7 Single-Lap Shear Results vs. Hardener Content

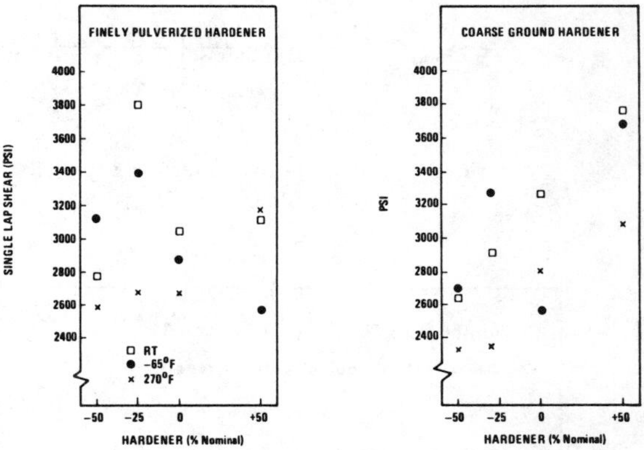

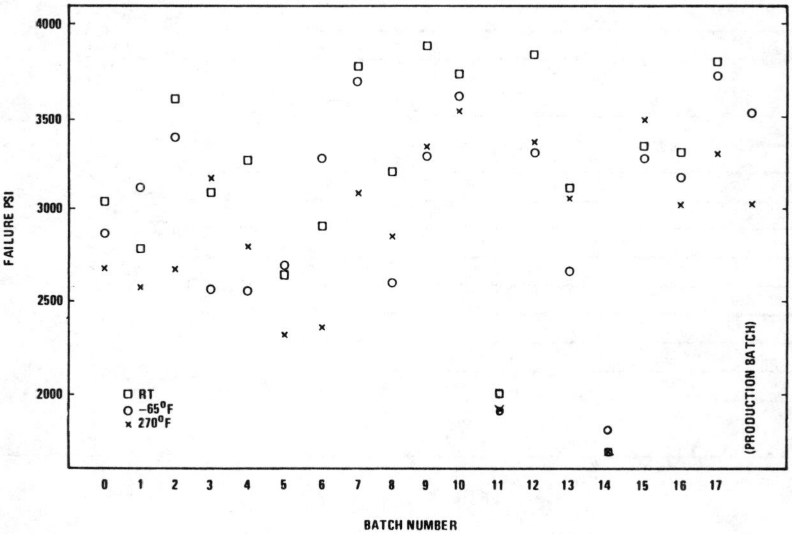

Figure 8 Single-Lap Shear Strengths for All Varied Adhesives

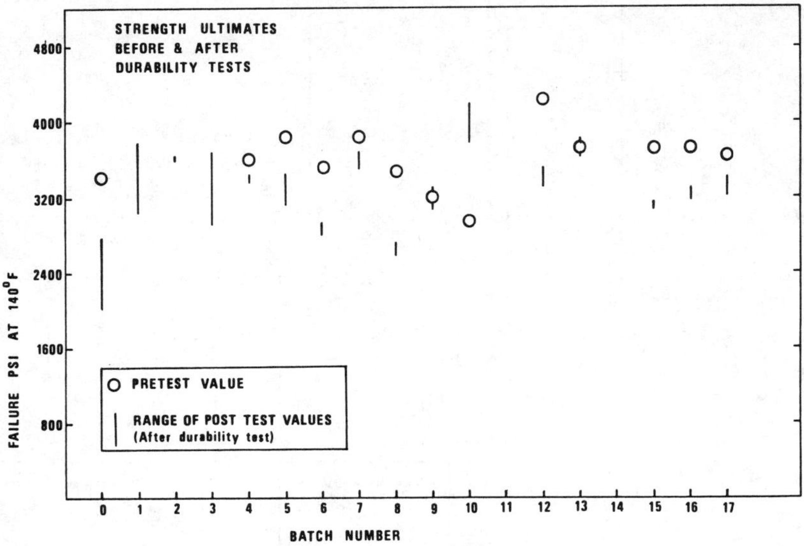

Figure 9 Single-Lap Shear Strengths Before and After Durability Testing

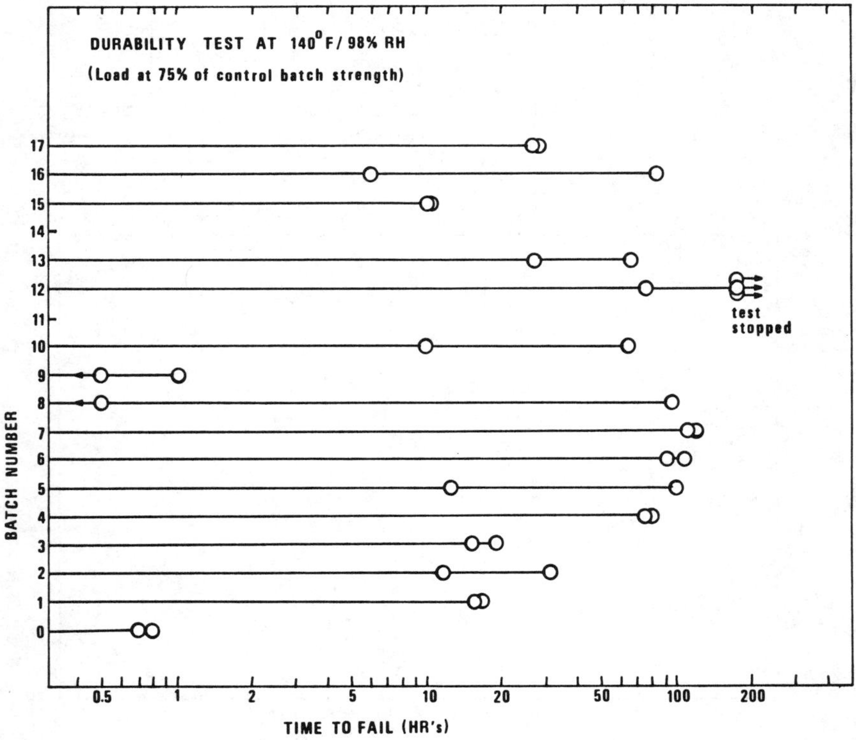

Figure 10 Durability Test Results (Constant Load)

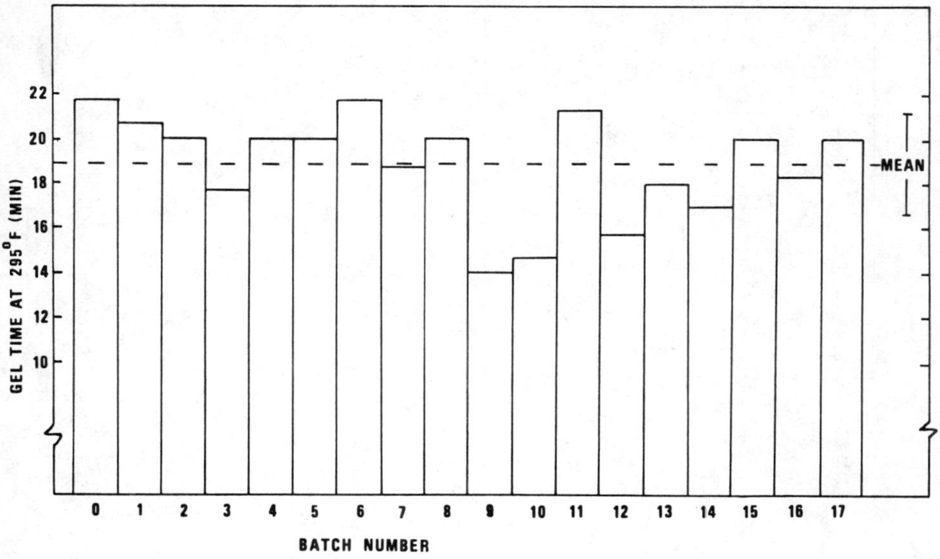

Figure 11 Gel Times at 295°F

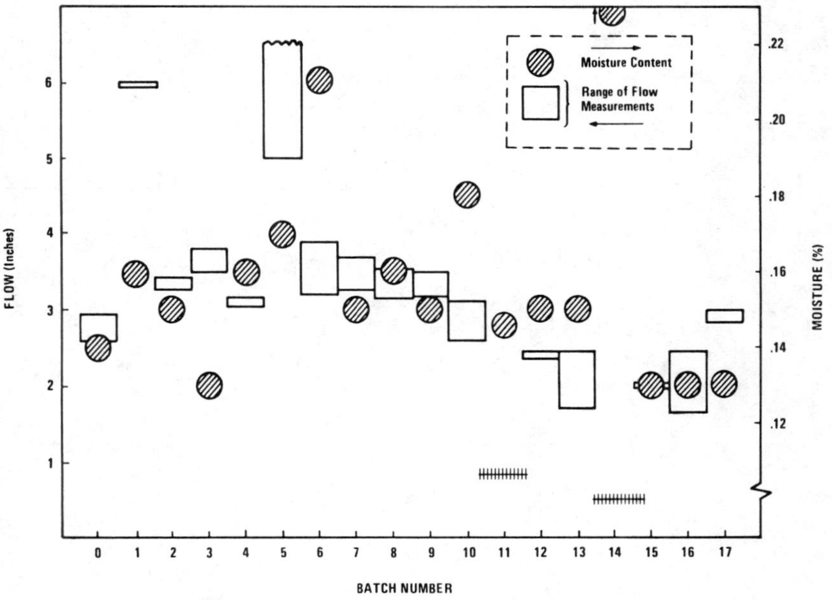

Figure 12 Flow Properties and Moisture Content

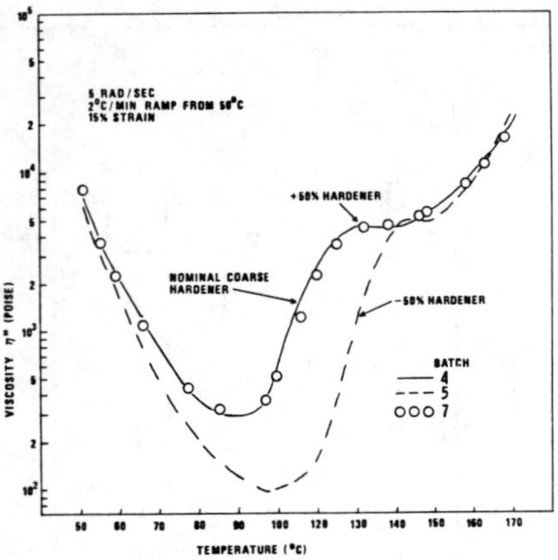

Figure 13 Viscosity Varies with Hardener Changes

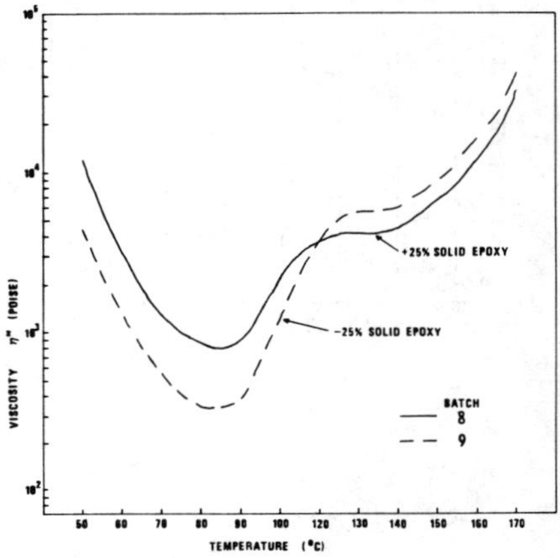

Figure 14 Viscosity Varies with Solid Epoxy Changes

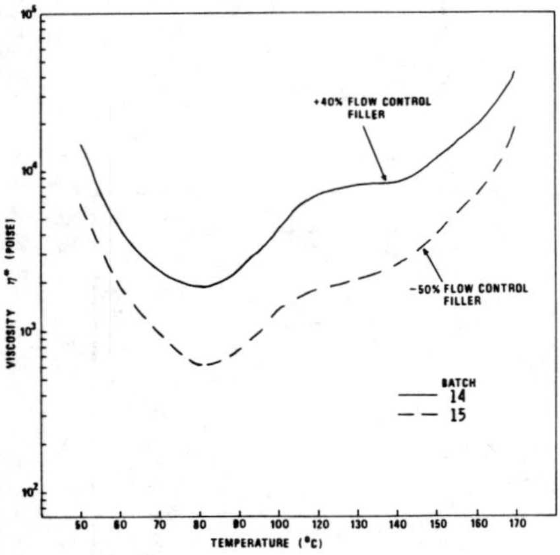

Figure 15 Viscosity Varies with Flow Control Filler Changes

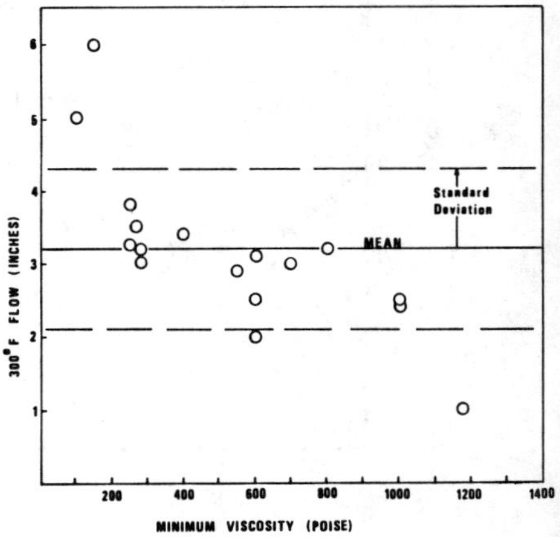

Figure 16 Flow/Viscosity Correlations

13th National SAMPE Technical Conference
October 13-15, 1981

Development of Advanced Composite Spacecraft Antenna Reflector Structures

Roger A. Stonier
FORD AEROSPACE & COMMUNICATIONS CORPORATION
Western Development Laboratories Division
Palo Alto, California

Abstract

An overview is given of the development of advanced composite spacecraft antennas. The use of graphite epoxy and Kevlar epoxy materials is now fairly standard for all spacecraft antenna structures. These advanced composite materials are ideal because antennas must be strong, stiff, and lightweight, and be dimensionally stable when exposed to the extreme thermal environment of space. A number of applications are described, including: the high gain antenna reflector for the Viking Mars Orbiter Spacecraft, the first flight graphite composite antenna; the 3.7 meter (12 ft) diameter antenna reflector for the Voyager (Mariner Jupiter Saturn) Spacecraft, the largest solid graphite flight antenna ever developed; high frequency dimensionally accurate antenna reflectors developed for a number of commercial satellite programs; and the large deployable antennas for the INTELSAT V satellite.

These applications include simple sandwich shells, and large diameter ring and rib stiffened shell structures. Characterization testing of the materials and the process development required to fabricate these advanced composite antenna structures is discussed.

1.0 INTRODUCTION

Graphite and Kevlar composite materials are now being utilized extensively in spacecraft structures. These materials were first used in antenna reflector structures in the early 1970s and are standard on almost all recent spacecraft antenna designs. The outstanding physical and mechanical properties of graphite and Kevlar materials offer significant advantages for both conventional and advanced antenna designs. Antenna structures are always weight-critical and require high stiffness and dimensional stability over the thermal environment the antenna must operate under while in space. Graphite and Kevlar composites are unique in providing a near-zero thermal coefficient of expansion and high specific strength and specific modulus properties.

The following discussion examines the use of graphite and Kevlar composites in antenna reflector structures. Materials and processing developments and advancements are described, as well as design aspects of advanced composite antennas. Specific advanced composite flight antenna applications at the Western Development Laboratories Division of Ford Aerospace & Communications Corporation (FACC) are illustrated.

2.0 BACKGROUND

The U.S. Air Force Materials Laboratory (AFML) sponsored some of the first programs for development of graphite composite antenna reflectors (Ref.1). A number of aerospace companies then began independent programs resulting in fabricated hardware (Ref.2). Most were approximately 1.0-1.2 meter (3-4 ft) diameter and were fairly heavy designs (Figure 1). FACC developed the first flight graphite composite reflector structure for the Viking Mars Orbiter Spacecraft during 1972-73. Since that time reflector technology has advanced to current applications requiring improved surface accuracy and better dimensional stability at minimum weight. Current antenna structures are usually offset-fed parabolic segments, and most are deployable. Advanced composite flight spacecraft antennas have been fabricated by a number of companies in the United States (Refs. 3 thru 8) and companies in England, France, Germany (Refs. 9,10), Italy, and Japan.

Figure 1. Early Development 1.2 Meter Diameter Satellite Antenna

3.0 MATERIALS CHARACTERIZATION

The materials used in spacecraft antennas are selected in order to obtain a structurally efficient design with minimum weight and maximum dimensional stability. All flight materials must meet stringent vacuum outgassing requirements.

3.1 Advanced Composite Materials

Reflector structural designs are usually stiffness critical to achieve the required high natural frequency. A near zero thermal coefficient of expansion is also desirable. Therefore high modulus and ultra-high modulus graphite fiber materials, which also have the lowest expansion coefficients, are usually used. Kevlar composites are utilized when a low thermal conductivity is required or in reflector designs requiring microwave transparency, such as frequency selective or polarization sensitive surfaces. The graphite and Kevlar composite materials that are used in antenna structures are obtained in prepreg form with formulated epoxy or modified epoxy matrix resin systems. Unidirectional tape and sheet materials, in addition to thin woven fabrics, are used extensively. Very thin prepreg materials are used in reflector dish structures to achieve minimum weight.

Early graphite antenna reflectors used 0.13 mm (0.005 in) unidirectional tape materials with high modulus PAN or Rayon precursor graphite fibers. Skins were usually designed with a balanced quasi-isotropic layup, producing very heavy structures. Later the use of thin 0.06 mm (0.0025 in) unidirectional prepreg sheet materials and skin designs based on a 0°, 90°, 90°, 0° layup resulted in very lightweight antenna structures. Honeycomb sandwich designs with 2 ply 0°, 90° unsymmetrical skins and a total skin thickness of 0.13 mm (0.005 in) have been developed. Table 1 shows mechanical property data for these thin ply HMS graphite epoxy skins.

Ultra-high modulus GY-70 graphite fiber has been used in more recent reflector designs. Thin ply 0.06 mm (0.0025 in) GY-70 graphite epoxy prepreg sheet materials have recently been used on a number of programs. Typical mechanical properties of this material are shown in Table 2. Recent programs have also utilized a thin 0.13 mm (0.005 in) GY-70 woven fabric. Sandwich reflectors have been developed which

Table 1. Tensile Properties of Fiberite 934/HMS Graphite Composite Laminates (0.06 mm/ply or 2.5 mil/ply Graphite)

Specimen Fiber Orientation	Test Orientation	Ultimate Stress		Modulus	
		MN/m^2	lbf/in.^{2} x 10^{-3}	GN/m^2	lbf/in.^{2} x 10^{-6}
0°, 90°, 90°, 0°	0°	558	81	117	17
	90°	614	89	117	17
0° ± 60°	0°	434	63	83	12
0°, 90°	0°	600	87	124	18
	90°	538	78	110	16

Table 2. Mechanical Properties of Fothergill & Harvey Code 69 Epoxy Resin/GY-70 Graphite Unidirectional Composite Laminates (0.06 mm/ply or 2.5 mil/ply Graphite)

Test Temperature, °C (°F)	Typical Tensile Properties				Typical Short Beam Shear Strength	
	Ultimate Stress		Modulus			
	MN/m^2	lbf/in.^{2} x 10^{-3}	GN/m^2	lbf/in.^{2} x 10^{-6}	MN/m^2	lbf/in.^{2} x 10^{-3}
23 (73)	518	75	250	36	41	6
177 (350)	414	60	230	33	35	5

utilize this graphite fabric in combination with a thin Kevlar Style 120 fabric. These thin skins, when cocured to Kevlar honeycomb core, produce a structure with a near-zero thermal coefficient of expansion. Mechanical properties are shown in Table 3.

Many future designs will utilize the new high modulus and ultra-high modulus pitch graphite fibers.

3.2 Honeycomb Core Materials

Aluminum honeycomb core is generally utilized as the core material in reflector sandwich structures. The core has excellent shear properties, is available in a variety of densities, provides excellent thermal conduction between sandwich skins, is easy to form and machine, and facilitates electrical grounding of the structure. Standard aluminum honeycomb core in densities between 32 kg/m^3 (2.0 lbm/ft^3) and 50 kg/m^3(3.1 lbm/ft^3) have been used in most designs, and even lower core densities are available. Hexcel's Flexcore, a material that can be easily formed to the compound contour of a reflector, has also been used. The only disadvantage of aluminum honeycomb core is that it is a major contributor to the high thermal coefficient of expansion of the overall sandwich composite. Use of a new Kevlar honeycomb core material produces a low expansion sandwich; however, only one type of Kevlar core now is available in 34 kg/m^3 (2.1 lbm/ft^3) density and it does not have optimum shear properties.

Future high performance reflector structures will require improved Kevlar core and new graphite honeycomb core materials.

3.3 Adhesive Materials

Structural epoxy film adhesives have been used for bonding precured composite skins to honeycomb core. Most systems cure at 120°C (250°F). Figure 2 shows the

Table 3. Tensile Properties of Narmco 5208/GY-70 Twill/Kevlar 120 Woven Fabric Hybrid Composite Laminate

Test Orientation	Ultimate Stress		Modulus	
	MN/m^2	$lbf/in.^2 \times 10^{-3}$	GN/m^2	$lbf/in.^2 \times 10^{-6}$
0°	152	22	54	7.8
90°	138	20	48	6.9

Note: GY-70 twill fabric nominal thickness of 0.13 mm (0.005 in.)
Kevlar style 120 fabric nominal thickness of 0.10 mm (0.004 in.)

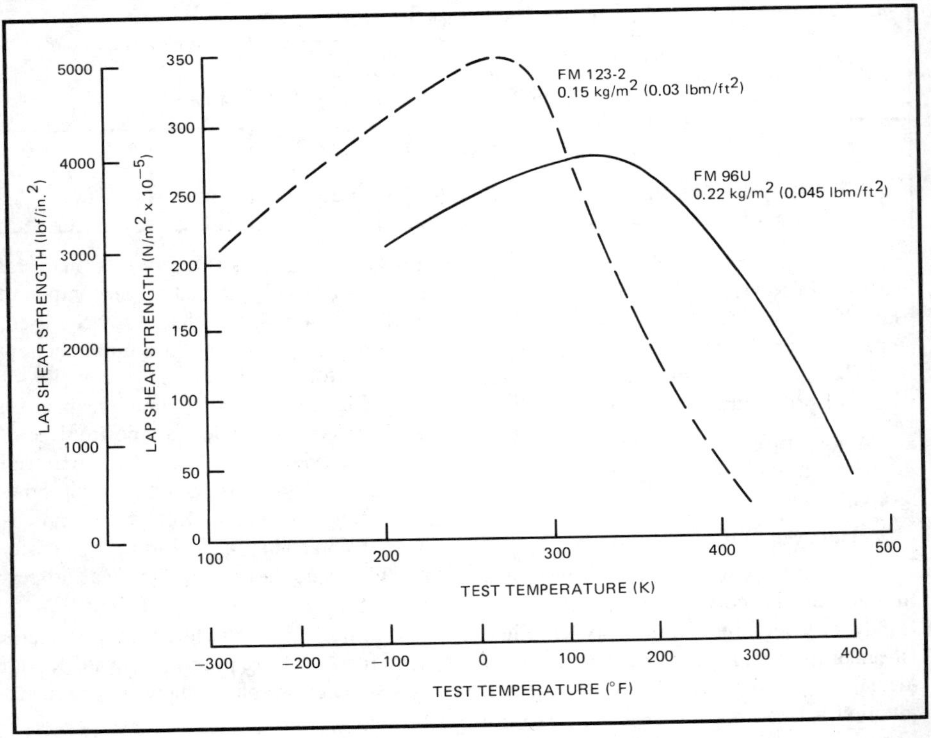

Figure 2. Film Adhesive Shear Strength

effect of test temperature on the shear strength of a moderate temperature and an elevated temperature adhesive film. Film adhesive weights as low as 0.10 kg/m^2 (0.02 lbm/ft^2) are now being used for reflector sandwich structures.

A structural epoxy paste adhesive is required for bonding stiffener ribs to the reflector shell. Figure 3 shows the effect of test temperature on the shear strength of several of these adhesive systems.

4.0 PROCESS CHARACTERIZATION

Advanced composite reflector skins are processed using conventional vacuum bag/autoclave molding techniques. Manufacturing and processing techniques and tooling designs generally depend on the reflector size and the surface accuracy requirement. The surface accuracy requirement is established based on the electrical design, particularly the operational microwave frequency. The composite skins are generally cured on low expansion ceramic, graphite, or Invar tooling, so that the machined parabolic contour of the tool does not change at the elevated cure temperature. The use of higher expansion metal tooling may be acceptable for some applications, depend-

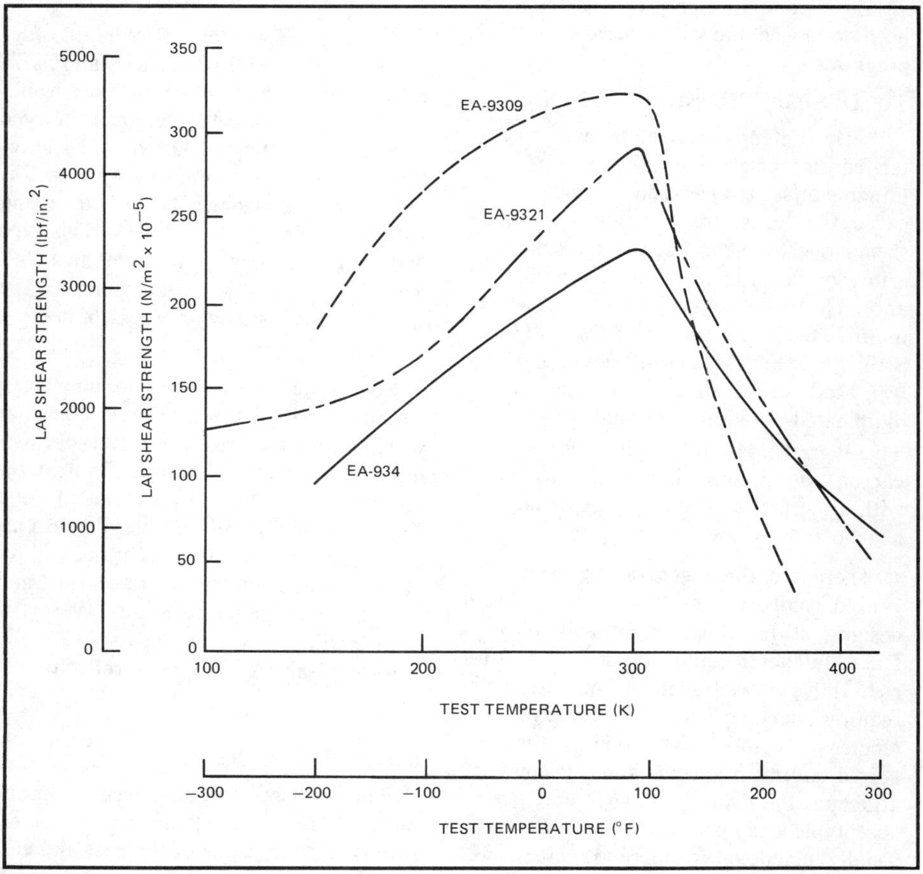

Figure 3. Paste Adhesive Shear Strength

ing on how large a deviation from the ideal surface contour can be tolerated on the as-manufactured part.

Most composite prepreg materials have been cured at 177°C (350°F). The first reflector structures utilized precured graphite composite skins that were then adhesive bonded to honeycomb core to produce the completed sandwich structure. Recent antenna structures have been cocured using conventional laminating prepreg systems in combination with a bond ply, usually a film adhesive. The bond ply provides for bonding to the honeycomb core and also acts as a barrier to restrain the resin flow of the epoxy prepreg during its low viscosity phase. New controlled flow advanced composite prepreg systems have been developed and will be used on future programs.

5.0 DESIGN CONSIDERATIONS

Early antenna reflectors were center-fed parabolas. Newer programs are utilizing offset-fed parabolid segments, because the feeds, transmission lines and their support structure do not block the aperture or degrade the electrical performance. This is particularly significant when multiple feeds are used, as on the INTELSAT V satellite. Reflectors tend to partially shadow themselves when sun-illuminated, creating thermal gradients. These gradients cause distortions which can be minimized by designing a structure with a thermal coefficient of expansion as near to zero as possible.

There are three general types of advanced composite solid antenna reflector designs: stiffened laminate shell; unstiffened "full depth" sandwich shell; and stiffened thin sandwich shell. A laminate shell requires added stiffeners for even small diameter reflectors. This provides the required natural frequency and a means for attaching the dish to the spacecraft with acceptable load distribution paths. Designs usually include one or more graphite rings (depending on diameter) with intermediate graphite stiffening ribs. The stiffening structure parts are fabricated as separate details and are secondarily bonded to the reflector shell using a room temperature or a moderate temperature curing structural epoxy paste adhesive.

Weight efficient honeycomb sandwich dishes can be designed up to approximately 1.2-1.5 meter (4-5 ft) diameter with "full depth" core and no additional stiffening structure required. Overall sandwich thickness is usually 1.3-1.5 cm (0.5-0.6 in) thick. Attachment to the spacecraft is through inserts bonded into the honeycomb sandwich. Higher density honeycomb core is usually required around the inserts for adequate shear strength, and graphite doubler plies may be required on the skins at these points due to higher skin stresses.

For larger diameter reflectors, in order to achieve the required stiffness and natural frequency, the "full depth" core would have to be so thick that a design of this type is not weight efficient. For these large reflector structures, a thin sandwich shell is utilized with secondarily bonded rib or ring stiffened members on the back. This type of design is significantly lighter than a "full depth" core design. Attachment to the spacecraft is usually to the rib or ring structure.

Solid reflector structures are now being manufactured to the maximum size that will fit within the space constraints of existing launch vehicles. Larger reflector sizes for spacecraft like ATS, Galileo, and TDRSS have utilized furlable rib mesh reflector structures. Future programs will require improved dimensional accuracy and very large diameter reflectors. Many of these programs will utilize foldable solid segment graphite composite reflector designs.

6.0 APPLICATIONS

The application of advanced composite materials to spacecraft antenna reflectors is illustrated by some of the past and current flight programs at Ford Aerospace & Communications Corporation (FACC).

6.1 Viking Mars Orbiter

The first flight graphite composite spacecraft antenna reflector was developed for the Jet Propulsion Laboratory and orbited Mars in July 1976 on NASA's Viking Orbiter Spacecraft. The 1.47 meter (58 in) diameter dual frequency X/S-band parabolic antenna reflector is shown in Figure 4. This reflector structure is an unstiffened sandwich shell with 0.25 mm (0.010 in) graphite epoxy faceskins. The skins are made up of four 0.064 mm (0.0025 in) thick plies of unidirectional high modulus graphite (HMS) with multiple gores oriented 0°, 90°, 90°, 0°. Construction details are shown in Figure 5. Skins are precured on the same ceramic layup tool and the honeycomb core is bonded to the front skin and then machined to the required contour prior to back skin adhesive bonding. The antenna dish is hard-mounted to the spacecraft at three equally spaced support points with titanium attachment inserts potted into the sandwich. Local graphite doubler buildups around each insert are integrally molded into the skin. The antenna weight is 4.63 kg (10.2 lbm) including reflector feed and feed support truss. Several flight quality antennas were fabricated and used in the qualification test program and for flight and spare units. Fabrication details of this program have been reported (Ref.11).

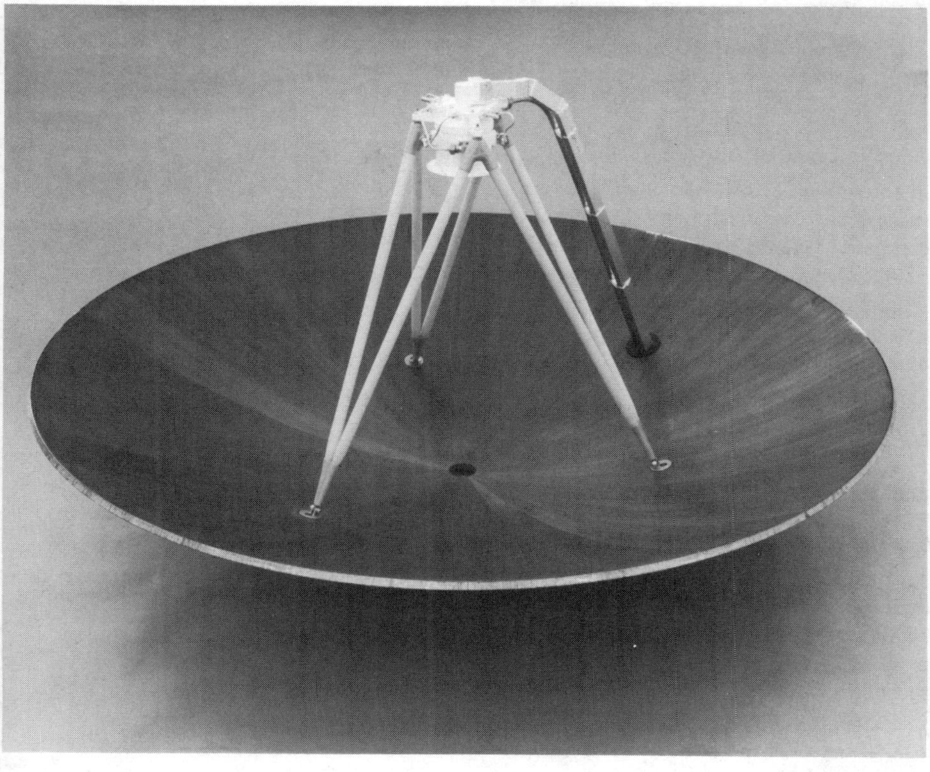

Figure 4. Graphite Composite Antenna for the Viking Mars Orbiter Spacecraft

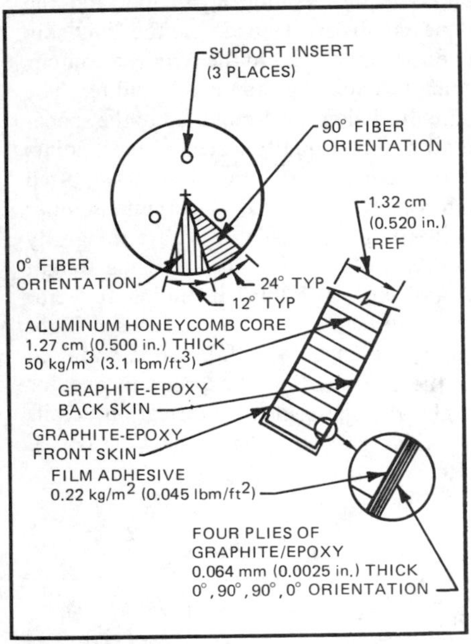

Figure 5. Reflector Construction of Viking Mars Orbiter Antenna

This program was a major advancement in antenna reflector technology, because it successfully demonstrated that graphite composite antenna structures could be fabricated that were lighter in weight and offered improved performance and were still cost competetive with conventional metallic and bonded metallic designs.

6.2 Voyager (Mariner Jupiter Saturn)

The largest graphite composite antenna reflector ever manufactured was developed under contract to NASA-Jet Propulsion Laboratory for the Voyager spacecraft. Each of the two flight spacecraft contain a 3.7 meter (12 ft) diameter graphite epoxy dual frequency X/S-band high gain antenna. The antenna shown in Figure 6 consists of a high efficiency shaped primary and secondary reflector system. The primary reflector shell is constructed of 0.25 mm (0.010 in) thich graphite epoxy faceskins and tapered 32 kg/m^3 (2.0 lb/ft^3) density aluminum honeycomb core. High modulus graphite (HMS) prepreg with a 0°, 90°, 90°, 0° stacking sequence is used in the reflector shell skins. The fabrication process consists of precuring the graphite epoxy front skin and cocuring the back skin on the machined honeycomb core.

The reflector is stiffened and supported on the back side by a 1.7 meter (5.6 ft) diameter sandwich ring and sandwich diaphragm panel. The skins for both of these structures consist of a 0° ± 60° stacking sequence using high modulus graphite (HMS) prepreg material. Attachment of this stiffening structure to the reflector shell is accomplished by both mechanical fasteners through potted inserts and by adhesive bonded clips. Construction details are shown in Figure 7.

The secondary subreflector shown in Figure 8 contains a frequency selective surface (FSS). It consists of Kevlar epoxy faceskins with RF transparent Nomex honeycomb core. This was the first use of a FSS subreflector on a spacecraft antenna.

Figure 6. Graphite Composite 3.7 Meter Diameter Antenna for the Voyager (Mariner Jupiter Saturn) Spacecraft

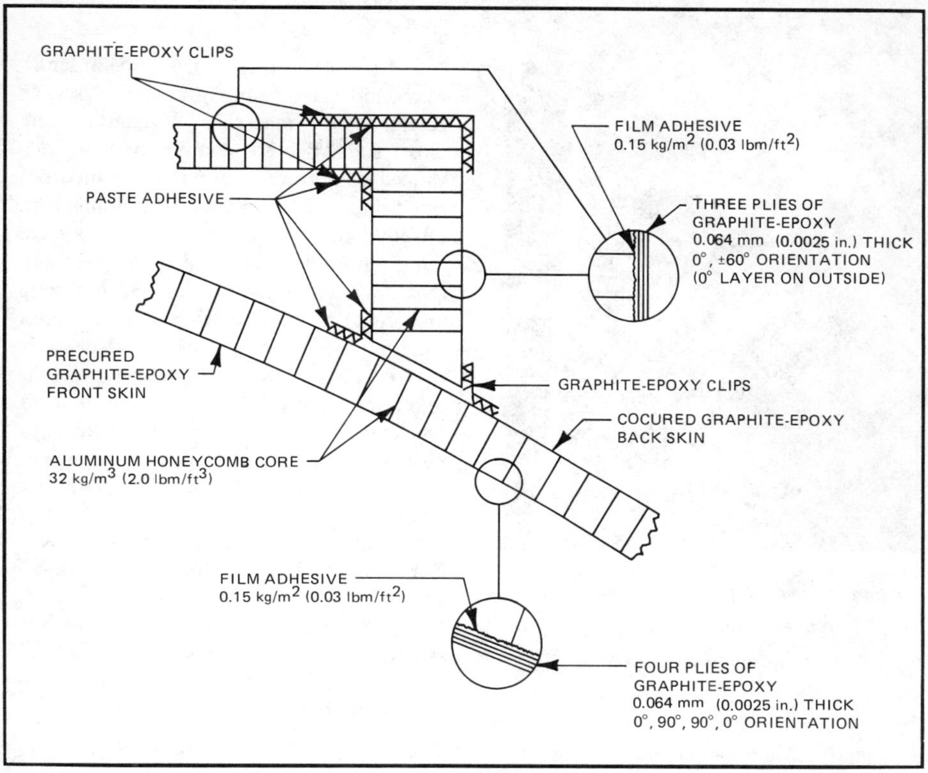

Figure 7. Reflector Construction of Voyager (Mariner Jupiter Saturn) Antenna

It was designed to reflect X-band signals with the antenna operating as a Cassegrain system, and also to transmit S-band signals with the feed at the reflector prime focus. Frequency selectivity of the subreflector is accomplished with aluminum dipoles on the Kevlar faceskins. Development of this subreflector has been previously reported (Ref.12).

The tubular truss structure supporting the subreflector, S-band feed, and a low gain antenna are also fabricated of high modulus graphite epoxy material. Fiber orientation is predominately longitudinal to achieve maximum stiffness. The cone shell which supports the X-band feed at the vertex of the reflector is also graphite composite.

The weight of the Voyager antenna subsystem is 43 kg (95 lbm) including: reflector (with stiffening ring structure), feed horns, feed support structure, subreflector with tubular support truss, and a low gain antenna.

The antenna experiences the low temperatures of deep space (-200°C) and thermal gradients (80°C) due to solar illumination and shadowing. The antenna also is exposed to aerodynamic heating (120°C) during launch due to early release of the protective fairing.

The antenna is also exposed to the high radiation levels of the Jupiter environment. Radiation testing of the graphite and Kevlar materials showed no significant degradation of properties.

Figure 8. Voyager (Mariner Jupiter Saturn) Frequency Selective Surface Composite Antenna Subreflector

A qualification model antenna was subjected to environmental tests including thermal cycling in vacuum, sine vibration, and acoustic vibration.

6.3 Japan Communication Satellite (CS)

The Japanese medium-capacity communication satellite (CS) contains a single despun folded-horn antenna. This advanced antenna system was developed under contract to Mitsubishi Electric Corporation (MELCO) for the National Space Development Agency of Japan (NASDA). The antenna is designed to operate at four different frequency bands. The antenna shown in Figure 9 consists of: a 1.5 meter (6 ft) graphite epoxy sandwich reflector; a graphite monocoque shell which supports the reflector; a graphite ring-stiffened Kevlar conical horn; and a Kevlar cylinder. The reflector surface was a computer generated surface defined by 18,000 points. This surface produced a K-band antenna pattern that was shaped to cover the Japanese islands.

All the structural parts of this antenna were fabricated using graphite epoxy or Kevlar epoxy materials. It contains the most sophisticated combination of advanced composite materials ever incorporated into a single spacecraft antenna. Six different advanced composite materials are integrally molded or bonded together into the completed assembly. The Kevlar cone and cylinder contain 0° and ±45° plies with a tapered wall thickness. Kevlar is used because of its low thermal conductivity characteristics which minimizes heat loss from the satellite. The reflector support shell is constructed with a quasi-isotropic $(0° \pm 60°)_s$ graphite composite layup. The reflector structure consists of a sandwich with varying thickness honeycomb core. The reflector faceskin layup consists of four plies of graphite with a 0°, 90°, 90°, 0° fiber orientation. Reflector structures were manufactured with a 0.05 mm (0.002 in) rms deviation from the theoretical contour.

Figure 9. Japan Communication Satellite (CS) Antenna

6.4 Japan Engineering Test Satellite (ETS II) and Japan Experimental Communications Satellite (ECS)

Both of these NASDA satellites were developed for MELCO and have a multi-frequency antenna system using graphite epoxy antenna reflectors and advanced composite antenna support structures. The ETS II K-band reflector (Figure 10) was fabricated using a graphite epoxy sheet moldling compound (SMC) containing random chopped GY-70 graphite fibers. This produced a structure with isotropic properties. The reflector is 28 cm (11 in) diameter and weighs 0.5 kg (1.1 lbm). The antenna support structure is Kevlar.

Figure 10. Japan Engineering Test Satellite (ETS-II) K-Band Antenna

The ECS satellite has two graphite composite antenna reflectors (Figure 11). Both reflectors are designed with two ply 0°, 90° graphite skins. Thin ply HMS graphite prepreg is used. The support and positioning structure for both antennas (Figure 12) consists of a graphite and Kevlar molded hybrid structure.

6.5 INTELSAT V

The INTELSAT V satellite contains 11 different antennas mounted on an antenna tower module structure (Figure 13). Four offset-fed paraboloidal antenna reflectors are used for communication coverage. The

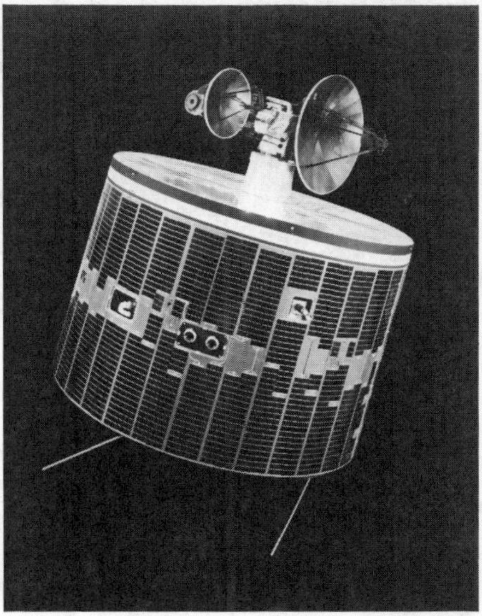

Figure 11.

Japan Experimental Communications Satellite (ECS)

Figure 12.

Graphite/Kevlar Composite Support Structure for Japan ECS Communications Satellite

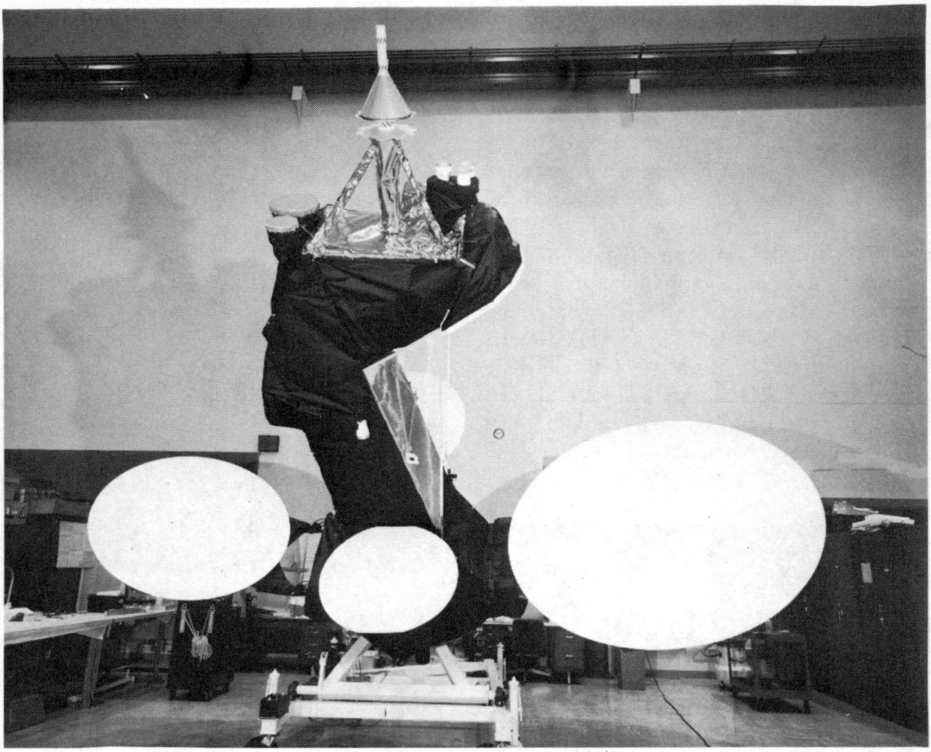

Figure 13. INTELSAT V Satellite Antenna Module

two large Hemi/Zone antenna reflectors are stowed during launch and deployed on orbit. These reflector structures were developed and fabricated for the FACC satellite by TRW. The larger antenna reflector is 2.4 meter (96 in) diameter and weighs 10.7 kg (23.5 lbm). The small dish is 1.55 meter (61 in) diameter and weighs 4.4 kg (9.7 lbm). Both reflector structures consist of a thin honeycomb sandwich shell with a backside stiffening rib structure. The skins of both the shell and ribs consist of a thin GY-70 twill graphite fabric and a layer of Kevlar Style 120 fabric. Kevlar honeycomb core is used to produce a lightweight low expansion structure.

The two medium sized spot beam antennas were supplied by Selenia (Rome, Italy) and the graphite composite reflectors were manufactured by British Aerospace Dynamics Group (Stevenage, England). The design of these reflectors is similar to the Viking antenna reflector design. Two precured GY-70 graphite composite skins are adhesive bonded to constant thickness honeycomb core. One of these reflectors is deployable and both are steerable in order to move the spot coverage on the earth to the desired locations.

6.6 INSAT-I

Graphite composite antenna reflectors were developed for the Insat-I domestic satellite system of India. The reflectors were developed and manufactured for the FACC-built satellite by Selenia. Both antenna structures are offset-fed parabolas, and both are stowed against the satellite and deployed into position after launch. The larger C/S-band reflector is rectangu-

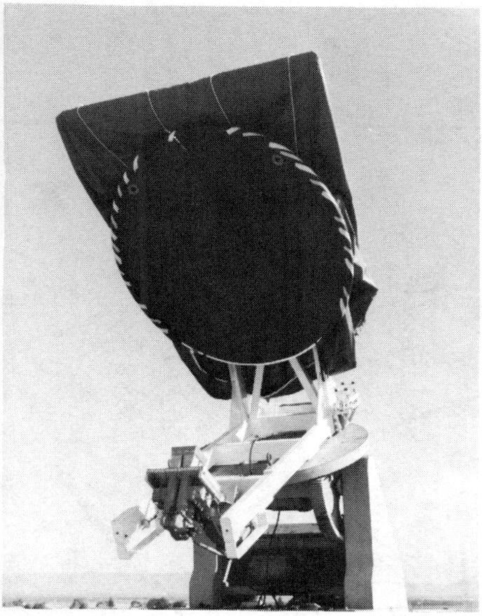

Figure 14. C-Band Antenna Reflector for the Insat-I Satellite

lar in shape with dimensions of 1.6 meter x 1.5 meter (63 in x 60 in). The C-band reflector shown in Figure 14 has a circular shape and is 1.4 meter (54 in) diameter. Both of these structures consist of a 0.64 cm (0.25 in) thick honeycomb sandwich shell with a stiffening and deployment rib structure, also of sandwich construction, bonded to the back of the reflector (Figure 15). An ultra high modulus (GY-70) thin ply graphite prepreg is used with a 0°, 90°, 90°, 0° ply orientation in the skins of the sandwich shell and a $(0° \pm 60°)_s$ orientation in the rib skins. Processing includes precuring the front skin, followed by bonding constant thickness aluminum honeycomb core with a 120°C (250°F) film adhesive. The back skin is then cocured directly on the honeycomb core. The rib structure is secondarily bonded with a room temperature curing epoxy adhesive system. The larger C/S-band antenna reflector weighs 6.2 kg (13.6 lbm) and the smaller C-band dish weighs 4.4 kg (9.7 lbm). All of the flight antennas have been completed for two Insat-I satellites. The first satellite will be launched in early 1982.

7.0 CONCLUSIONS

Advanced composite spacecraft antenna reflector structures were first developed in the early 1970s. Since that time, it has been demonstrated on program after program that graphite and Kevlar composites are the best available materials for high performance spacecraft antenna reflectors. The technology has evolved with each new spacecraft program. The designs, materials, and manufacturing processes which are now available can produce the ultralightweight near-zero expansion antenna reflector structures which are required for today's programs. However, future programs will have even more demanding requirements of: more accurate reflector surfaces; better dimensional stability; larger reflector diameters; and longer life. Further advancements in the state-of-the-art of graphite and Kevlar composite materials and processes, and antenna design concepts will be essential to meet the challenge of the future.

8.0 REFERENCES

1. *Advanced Composite Material Study for Millimeter Wavelength Antennas,* Technical Report AFML-TR-71-205, Goodyear Aerospace Corporation.

2. Harold L. Hillesland, *Development of an Advanced Composite Antenna Reflector,* Fourth National SAMPE Technical Conference and Exhibition, Palo Alto, California, October 1972, Volume 4, pp. 235-248.

3. J. W. Young, Jr., and T. A. Dougherty, *Application of Composite Materials for Fabrication of Spacecraft Communication Antennas,* XXVII Congress, International Astronautical Federation, Anaheim, California, October 1976.

4. J. A. Fager, *Application of Graphite Composites to Future Spacecraft Antennas,* AIAA Paper No. 76-238, AIAA/CASI 6th Communications Satellite Systems Conference, Montreal, April 1976.

5. J. S. Archer and L. O. Berman, *Advanced Composites in Spacecraft Design,* AIAA Paper No. 76-237, AIAA/CASI 6th Communications Satellite Systems Conference, Montreal, April 1976.

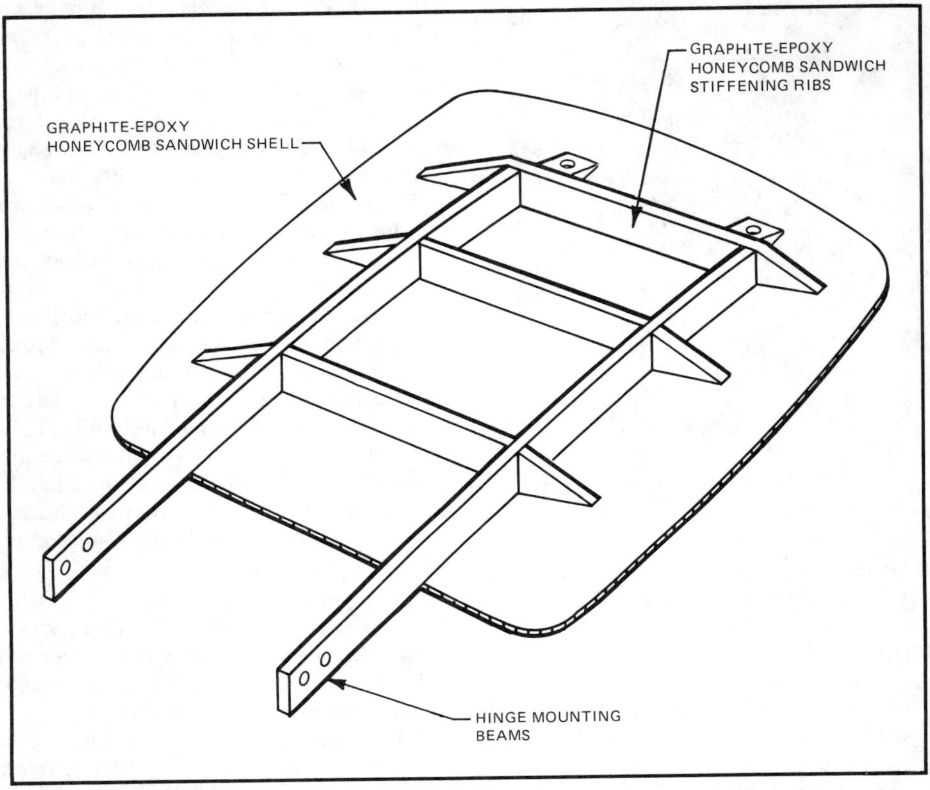

Figure 15. Rib Stiffened Insat-I Graphite Composite Antenna Reflector

6. A. Knoell and G. Krumweide, *Structural Development of the NIMBUS-G/ SEASAT A SMMR Graphite Epoxy Antenna Reflector,* Second International Conference on Composite Materials, Toronto, Ontario (Canada), April 1978; Published by AIME, Warrendale, Pa., 1978, pp. 1377-1391.

7. Gounder N. Raj, *Advanced Composites Applications in RCA Satellites,* Twenty-Sixth National SAMPE Symposium and Exhibition, Los Angeles, California, April 1981, Volume 26, pp. 216-231.

8. H. L. Hillesland, *Advanced Composite Hardware Utilized on the INTELSAT V Spacecraft,* Twenty-Fifth National SAMPE Symposium and Exhibition, San Diego, California, May 1980, Volume 25, pp. 202-211.
9. Bernd Abt and Hermann Rieger, *High Precision Graphite/Epoxy Antennas for Communications Satellites,* Twenty-Sixth National SAMPE Symposium and Exhibition, Los Angeles, California, April 1981, Volume 26, pp. 848-860.
10. K. Keen, P. Molette, B. Pieper, C. M. Herkert, and W. Schaefer, *Development and Testing of a New CFRP Antenna Reflector for Communication Satellites,* Raumfahrtforschung, Band 20-Heft 4, July/August 1976, pp. 173-181.
11. Roger A. Stonier and Harold L. Hillesland, *Fabrication of a Graphite Epoxy Antenna for the Viking Orbiter Spacecraft,* Nineteenth National SAMPE Symposium and Exhibition, Buena Park, California, April 1974, Volume 19, pp. 596-608.
12. Roger A. Stonier, *Development of a Low Expansion, CompositeAntenna Subreflector with a Frequency Selective Surface,* Ninth National SAMPE Technical Conference, Atlanta, Georgia, October 1977, Volume 9. pp. 504-518.

9.0 BIOGRAPHY

Roger A. Stonier is Senior Staff Scientist at Ford Aerospace & Communications Corporation's Western Development Laboratories Division. Recently he was Manager of the European Graphite Composite Program for the Insat-I Satellite. He was formerly Supervisor of the Antenna Materials Engineering Section where he was responsible for materials, processing and manufacturing of advanced composite antenna structures and microwave components. While with the California Institute of Technology - Jet Propulsion Laboratory he worked on the first advanced composite spacecraft application, and was responsible for the materials engineering on the first graphite composite spacecraft antenna structure. He was associated with the Boeing Company as a Senior Research Engineer and Lead Engineer in the Commercial Airplane Group, and also worked for Aerojet General Corporation as a Development Chemist. He received his Bachelor of Science in Chemistry from the University of California in 1962 and is a Licensed Professional Engineer. He is a member of SAMPE and is currently on the Editorial Review Board of the SAMPE Journal.

13th National SAMPE Technical Conference
October 13-15, 1981

RTV SILICONE FOAM FOR SPACECRAFT ELECTRONIC ASSEMBLIES

George R. Dallimore and
Robert J. Milligan
Lockheed Missiles and Space Co.
Space Systems Division
Sunnyvale, California

Abstract

Discoloration and decomposition of a silicone foam material used in a spacecraft electronic wire-wrap assembly occurred during exposure to a thermal-vacuum environment. Subsequent investigation disclosed that the silicone foam material needed extensive post curing or extended ambient aging to fully inactivate the reactive constituents prior to vacuum exposure. This report describes the work performed in chemically characterizing the silicone foam raw materials, the identification of the reactive species involved in the degradation process, and the processing conditions required to prevent the decomposition reaction.

"Keywords": Silicone foam, vacuum degradation, silane, curing.

1. INTRODUCTION

An open cell, ten pound per cubic foot density, silicone foam, Dow Corning S-5370 material, has been used since 1962 at Lockheed's Space Systems Division for the protection of wire-wrap matrices in electronic equipment. A typical wire-wrap matrix in an electronic black box is shown before foaming in Figure 1. The silicone foam provides both vibration damping to the wire-wrap pins and interconnecting wires, and electrical isolation of the pins and wires from each other and from any conductive debris.

The DC S5370 material is a two-part, RTV (room temperature vulcanizing) silicone material that begins foaming within one minute after mixing. Dow Corning's recommended mix ratio is six parts stannous octoate catalyst per hundred parts of S-5370 base resin and their recommended cure is 24 hours at room temperature. Lockheed processed the DC S-5370 material per Dow Corning's recommended mixing and curing conditions from 1962 until 1969, when a problem occurred with outgassing and migration of low molecular weight silicone species and excess stannous octoate catalyst during thermal vacuum exposure. At that time, the stannous octoate catalyst ratio was decreased to 2 phr (parts per hundred resin), and the curing conditions were changed to one hour at room temperature plus 24 hours at $65^{o}C$ to complete the polymerization reactions and to oxidize the stannous octoate to stannic oxide (SnO_2).

In early 1980, a foamed wire-wrap assembly was subjected within four days of curing to a special 48 hour thermal-vacuum conditioning treatment at $65^{o}C$ and 10^{-4} torr to further reduce outgassing of the DC S-5370 foam material. Upon removal from the vacuum chamber, the normally tan colored, resilient foam was observed to be partly grayish and somewhat brittle on the exposed surface. Further examination of the foamed matrix disclosed that the foam became progressively more black and friable below the surface. Pin-to-pin electrical insulation resistance measurements at 500 vdc showed the black/friable foam was still a good insulator. However, its friable nature made it unlikely that the foam structure would remain intact under exposure to vibration and shock conditions.

The black/friable foam condition had not been previously encountered during the eighteen years successful usage of this foam material. However, foamed assemblies were not normally subjected to thermal vacuum conditions within four days of curing. Usually two weeks would have elapsed between foaming and thermal vacuum testing of the completed assembly. A failure analysis investigation was initiated to determine the cause(s) of the black/friable condition. The first part of the investigation was a study of the silicone foam polymerization reactions and possible failure mechanisms.

2.0 SILICONE FOAM CHEMISTRY

2.1 Silicone Foam Polymerization Reactions

The DC S-5370 silicone foam material is formed by the reaction of the base resin, a polysiloxane polymer, with stannous octoate catalyst. Infrared (IR) and mass spectroscopy (MS) analyses revealed that the base silicone resin is a complex mixture of silanol (SiOH) terminated polysiloxanes, silane (SiH) containing oligomers, and tetrapropyl silicate $Si(O\ C_3H_7)_4$ cross-linker. The main chemical reactions occurring during the foaming process are as follows:

$$\equiv SiOH + \equiv SiH \xrightarrow[H_2O]{Sn} H_2\uparrow + \equiv Si{-}O{-}Si\equiv \qquad (1)$$

$$4\equiv SiOH + Pr{-}O{-}\overset{\displaystyle O{-}Pr}{\underset{\displaystyle O{-}Pr}{Si}}{-}O{-}Pr \xrightarrow[H_2O]{Sn}$$

$$4PrOH + \equiv Si{-}O{-}\overset{\displaystyle O{-}Si\equiv}{\underset{\displaystyle O{-}Si\equiv}{Si}}{-}O{-}Si\equiv \qquad (2)$$

$$\equiv SiOH \quad + \quad \equiv SiOH \xrightarrow{Sn}$$

$$\equiv Si\text{-}O\text{-}Si\equiv \quad + \quad H_2O \qquad (3)$$

Reaction (1) between silanol terminated silicone polymers with silane containing oligomers liberates hydrogen which functions as the blowing agent to form the cellular structure. Reaction (2) between the silanol terminated polymer and the tetrapropylsilicate serves to cross-link the polymer into a three dimensional network. Reaction (3) between residual unreacted silanol groups results in additional chain extension or further cross-linking. All of the main reactions shown above plus numerous side reactions are occurring during the foaming process.

2.4 Possible Failure Mechanism

The black color of the degraded foam could be caused by: (1) free tin metal, (which because of its dispersion throughout the foam would be formed in a sub-colloidal state and would appear black); (2) black stannous oxide, SnO; (3) a colored tin-silicone species; (4) or a colored octanoic acid degradation product. It was initially suspected that the black color was most probably caused by either free Sn or by black SnO. The friability can be explained either by: (1) reversion, i.e., chain scission into lower molecular weight segments that are no longer rubbery; or (2) further cross-linking which would result in a more glassy-brittle structure. It appeared from the initial observations of the absence of the black/friable condition on surfaces directly exposed to the air, that an active ingredient still present in the deeper sections was not reacting with oxygen or water from the air; or that the tin catalyst had been oxidized to catalytically inactive white stannic oxide, SnO_2 near the surface but not in the deeper sections. Upon exposure to heat in a non-oxidizing vacuum atmosphere, either the non-oxidized/hydrolyzed active ingredient or the tin in the deeper sections (or both) caused the black/friable foam condition.

2.4.1 Reversion Reactions. The incomplete oxidation or decomposition of the excess tin catalyst in the deeper sections to a catalytically inactive form was a leading suspect. The excess stannous octoate would normally at first hydrolyze to SnO and octanoic acid followed by oxidation to SnO_2. Heating in vacuum might form the black SnO and the heated octanoic acid might decompose the ≡Si-O-Si≡ backbone of the silicone polymer chain resulting in friability.

Another reversion mechanism would be the attack on the silicone polymer backbone by the propyl alcohol released by direct hydrolysis of excess or partially reacted propyl silicate.

The hydrolysis proceeds slowly at ambient conditions when moisture from the atmosphere diffuses into the foam and reacts with residual propyl silicate liberating propanol. The odor of propanol is readily detected for days even after foam specimens have been cured for 24 hours at 65°C. The propanol trapped in the foam matrix might be causing reversion to occur during thermal-vacuum conditioning when the removal of the water would favor the following reaction:

$$\equiv Si\text{-}O\text{-}Si\equiv \quad + \quad 2PrOH \xrightarrow{Sn}$$

$$\equiv SiOPr \quad + \quad \equiv SiOPr \quad + \quad H_2O\uparrow \qquad (4)$$

However, the vapor pressure/temperature curve of n-propanol (BP 97°C) and water are essentially the same and the propanol should come off at the same rate as water in thermal-vacuum unless the propanol was more strongly chemsorbed or hydrogen bonded than the water.

2.4.2 Cross-Linking Reactions. A second degradation mechanism would be further crosslinking between residual silane or silanol materials and a tin-silicone intermediate during the dry, nonoxidizing thermal vacuum conditions as follows:

$$\equiv SiH \quad + \quad \equiv Si\text{-}O\text{-}Sn\text{-}O\text{-}Oct \xrightarrow{VAC}$$

$$OctOH \; + \; Sn \; + \; \equiv Si\text{-}O\text{-}Si\equiv \qquad (5)$$

$$\equiv SiOH \quad + \quad \equiv Si\text{-}O\text{-}Sn\text{-}O\text{-}Oct \xrightarrow{VAC}$$

$$OctOH \; + \; SnO \; + \; \equiv Si\text{-}O\text{-}Si\equiv \qquad (6)$$

3. SILICONE FOAM CHARACTERIZATION

The possibility that the black/friable condition could be caused by either a reversion process or further cross-linking occurring during thermal-vacuum conditioning led to the obvious need for further chemical characterizations of the silicone foam curing processes, and of the normally tan/resilient versus the black/friable foam materials.

3.1 Silicone Foam Curing Study

3.1.1 Processing Conditions. It was initally suspected that the particular lot of material used in foaming the matrix assembly might be the cause of the black/friable condition. All available (1975 through 1979), lots of foam material were tested by preparing free-foamed specimens in 400 ml (4 inch deep) glass and polypropylene beakers, curing in the normal manner and within 4 hours exposing them to the 48 hour, 65°C, and 10^{-4} torr vacuum conditioning procedure. These foamed specimens, including some lots prepared at 2, 4 and 6 phr catalyst levels, all exhibited the same phenomenon of being tan/resilient on the surface and progressively becoming black and/or friable starting at approximately ½ inch below the surface. To test the hypothesis of an incomplete cure, foam specimens were prepared, cured for various time periods at various temperatures and relative humidities, and ambient aged for various periods prior to exposure to the 48 hour, 65°C, 10^{-4} torr vacuum environment. The various curing conditions, ambient aging, and test results are shown in Table I. As can be seen from these data, ambient aging for 7 days after normal cure, curing in a humidity chamber at 65°C and 50% RH, curing for 48 to 72 hours at 65°C, or curing at 85° to 100°C for 24 hours all prevented the black/friable condition from occurring. These processing condition studies confirmed that some unreacted constituents in the normally cured foam were responsible for the black/friable condition. To understand the mechanism of formation of black/friable foam material further studies were conducted on the reactions occurring during cure and ambient aging.

3.1.2 Products Offgassed During Cure and Ambient Aging. A series of experiments were performed to determine the chemical changes occurring during oven cure and ambient aging that allowed the longer (48 to 72 hours) 65°C cured, higher temperature (85 to 100°C) cured, or longer ambient (7 days) aged foams to survive the thermal-vacuum conditioning without becoming black and friable. The initial experiments sought to identify chemical reactions occurring in the foam by identifying what reaction products were offgassing during curing and ambient aging. The offgassed products analyzed by IR and Gas Chromatography (GC) during the 1 hour ambient cure, during the 24 hour cure at 65°C, and during the first 24 hour ambient period after the oven cure, were primarily n-propanol, with traces of ethanol and silane. Only propanol and no further ethanol or silane were detected during subsequent ambient aging. The ethanol was confirmed by Dow Corning as coming from traces of this solvent that are present in the raw materials. The silane is one of the active ingredients used in the formulation and the propanol is from the hydrolysis of the propyl silicate cross-linker. The propanol offgas data is plotted in Figure 2 and shows that the excess propyl silicate continues to hydrolyze and form reactive silanol groups for a considerable time period. Whether the newly formed silanol groups continue to condense and/or react with residual silane forming further crosslinks was not determined. The propanol offgas data gave some indications of what was happening during cure and ambient aging but was not conclusive in itself.

3.1.3 Effects of Thermal-Vacuum Environments. A study was made of the effects of thermal-vacuum environments on DC S-5370 foam specimens prepared at 2 phr catalyst ratio. Tests were run at temperatures of 38 to 65°C and at vacuums of 0.1 to 10^{-6} torr in various glass apparatus so that the onset of the blackening condition could be visually observed. Heating of the normally cured foam material at $65^{o}C$ and 10^{-6} torr in a Mass Spectrophotometer for 14 hours did not produce the black/friable condition. Heating of the cured foam material at $65^{o}C$ and 0.1 to 1 torr in a laboratory vacuum oven for periods up to 5 days did not produce the black/friable condition. Heating of variously cured specimens at temperatures of 38 to $65^{o}C$ and vacuums of 10^{-1} to 10^{-4} torr in an all glass Cahn Microbalance for various time periods revealed that the black/friable condition started after approximately 4 hours exposure. It was also noted that approximately 60 to 80% of the total weight lost during the thermal-vacuum exposure occurred prior to the 4 hours. The various curing, ambient aging, thermal-vacuum exposures, and test results obtained on the Cahn Microbalance are tabulated. in Table II. As can be seen from these data, ambient only cured foam does not become black/friable at $65^{o}C$ if the pressure is 10^{-1} torr or higher. Normally cured foam does not become black/friable at $65^{o}C$ if the pressure is 10^{-3} or higher, nor if the temperature is $52^{o}C$ or less at 10^{-4} torr pressure.

Heating of normally cured foam material in a Quadrapole Mass Analyzer (QMA) at 10^{-3} to 10^{-4} torr and analyzing the offgassed products disclosed that the initial major offgassing constituents were at mass/charge ratios (m/e) of 18 (water) and at 73 (trimethylsilane radical, $(CH_3)_3$ Si -). The 73 peak is typical of all silicones and there were additional minor silicone peaks at m/e's of 192 - 195. After 30 minutes only the water peak was significantly reduced. Since the Cahn microbalance work showed that the blackening was occurring after 4 hours, then neither the water nor the n-propanol (which has the same vapor pressure as water), should still be present. This suggested that a relatively non-volatile species, such as the partially reacted silane (≡SiH), could be causing the black/friable condition.

3.1.4 <u>After Cure, Ambient Aging Study</u>. The second set of cure study experiments looked at what chemical reactions were still occurring in the foam material itself after various cure cycles and during subsequent ambient aging. These analyses were accomplished by taking core samples from the foam specimens and subjecting them to MS analyses directly. Preliminary analyses showed that the propanol, water, octanoic acid (from the catalyst), and the 73 m/e ratio silicone species could be monitored during ambient aging using MS analyses. It was also noted that hydrogen continued to evolve from an ambient only cured specimen (at 2 phr catalyst ratio) for 2 days. Analysis of the data generated by the MS analyses showed that this analysis technique was not the best method for determining in-situ aging reactions because of the complex nature of the silicone polymer. In general, the water content was initially low because of the heat cure, but the foam rapidly equilibrated with moisture from the air. The propanol content decreased gradually with time similarily to the offgassed propanol analysis (see Figure 2). The octanoic acid content gradually increased with time due to the hydrolysis of the remaining stannous octoate catalyst. Specimens prepared at 6 phr catalyst had markedly higher (as expected), octanoic acid content than did the 2 phr specimens.

It was apparent from the vacuum microbalance data that a relatively non-volatile reactive constituent in the foam was responsible for the black/friable condition. Such a reactive constituent could be the silane (SiH) which was initially present in excess and should be relatively non-volatile since it would probably be partially reacted into the polymer matrix. None of the MS analyses showed any peaks that could be directly related to the SiH content. A wet chemical method was found wherein the unreacted SiH materials could be reacted using a solution of potassium hydroxide (KOH) in methanol and the liberated hydrogen determined quantitatively on the GC. A series of foam specimens were made up and treated with the alcoholic KOH solution and the residual unreacted SiH determined. The results of these analyses are shown in Figure 3. As can be seen from these data the residual silane decreases with the longer cure times at $65^{o}C$, with the $65^{o}C$/50% RH cure, and with the $100^{o}C$ cure. It appears from the data in Figure 3 and in Tables I and II, that foam cured to a residual silane content below 35 micromoles/gram should remain tan and resilient after thermal-vacuum exposure. The data shown in Figure 3 indicates that the best cure (of those studied), is the 24 hour cure at $100^{o}C$. This cure produced significantly less residual propanol, silane (indicative of a more complete cure), and octanoic acid (indicative of a more complete decomposition of the tin octoate catalyst).

3.2 Chemical Analyses - Black/Friable versus Tan/Resilient. A number of chemical analyses techniques were brought to bear on the questions of what is the black material and what caused the friability, (i.e., reversion or further cross-linking). IR, MS, MS/GC and ESCA (Electron Spectroscopy for Chemical Analysis), were repeatedly tried on various samples of black/friable and tan/resilient foams without success. Lengthy x-ray diffraction analyses showed the presence of elemental beta tin and non-stoichiometric tin oxides, Sn_2O_3 and Sn_3O_4 in the black/friable and not in the tan/resilient foams. These analyses showed conclusively that the black/friable foam contains finely divided widely dispersed elemental beta tin that is partially oxidized. Additional experiments showed that the black tin could be made to disappear by baking (oxidizing) in air for a few hours at 200°C or for a few days at 125°C during which the black tin was converted to white stannic oxide (SnO_2).

Several thermal analysis methods were employed on the black/friable and tan/resilient materials that showed conclusively that the friability was due to further crosslinking and not due to reversion. Thermogravimetric analysis (TGA) (see Figure 4) of the black/friable and tan/resilient materials showed that the black/friable material is more highly crosslinked than the tan/resilient material since it has a lower initial (up to 350°C) weight loss.

High temperature Differential Scanning Calorimetry (DSC) analyses were run on black/friable and tan/resilient foam in nitrogen. These showed a small exotherm at 300°C for the tan/resilient material followed by a broad endotherm at 380°C for both the tan/resilient and black/friable materials. The exotherm at 300°C indicates that the tan/resilient material still has unreacted components that are crosslinking at the 300°C temperature whereas the black/friable material does not. The broad endotherms at 380°C are indicative of thermal decomposition of the basic foam material.

4. CONCLUSIONS

It was concluded from the extensive characterization studies that the presence of an incompletely reacted silane (SiH) component in the base resin was causing the foam to become black (from free tin metal being split out of the polymer matrix) and friable (due to further crosslinking). The residual silane in the standard cured (24 hours at 65°C) foam continues to react predominately with water vapor at ambient conditions. Thus standard cured foamed that has sufficient ambient exposure (one week, minimum), does not become black/friable during thermal-vacuum conditioning, nor does foam that has the excess silane content reduced through further curing at 65°C, or curing at higher temperatures (85 to 100°C), or curing under higher humidity conditions (65°C/50% RH).

5. BIOGRAPHIES

Mr. Dallimore is a Staff Engineer in the Materials & Processes Engineering Department, Space Systems Division, Lockheed Missiles and Space Company. Mr. Dallimore has 27 years experience with non-metallic materials and processes associated with electronic equipment. For the past 20 years, he has specialized in the field of dielectric and insulating materials as applied to various types of electrical/electronic equipment for aerospace applications at Lockheed.

He has conducted extensive evaluation, testing, and failure analysis investigations into all types of coating, foaming and potting/embedding materials and processes. Mr. Dallimore has a B.A. degree in General Science (Chemistry major) from the University of Rochester and is a member of SAMPE.

Mr. Milligan is a Specialist with the Materials and Processes organization of the Space Systems Division of Lockheed Missiles and Space Company. Prior to his current position, he was employed at Acurex Corporation, Mountain View, CA, and at Whittaker Corporation Research and Development Division, San Diego, CA, where he was involved in numerous activities covering the fields of material and environmental engineering and organic and polymer chemistry. Mr. Milligan received a B.A. degree in Chemistry from the University of California, Riverside, and a M.S. degree in Organic Chemistry from San Diego State University. He is a member of the American Chemical Society and SAMPE.

TABLE I

RESULTS(1) OF VARIOUS MIX RATIOS, CURING AND AMBIENT AGING CONDITIONS

Catalyst Ratio (phr)	Ambient Cure (hrs)	Curing Conditions (hrs/^{o}C)	Ambient Aging (hrs)	After 48 hrs/65°C/10^{-4}torr	
				Black	Friable (2)
2	-	-	96	yes, very	yes
2	-	24/65	48(3)	grayish	yes
2	1	24/65	-	yes	yes
2	1	24/65/50% RH	-	no	no
1	26	24/65	-	yes	yes
2	24	24/65	-	yes	yes
4	1	24/65	-	gray	yes
6	1	24/65	-	no	yes
2	1	24/65	48	gray	yes
2	1	24/65	168	no	no
1	0 & 1	48/65	-	no	no
2	0 & 1	48/65	-	no	no
2	1	72/65	-	no	no
2	1	24/85	-	no	no
2	1	24/100	-	no	no

NOTES: (1) DC S-5370 silicone foam in glass or polypropylene beakers

(2) Determined by removing specimen and handling

(3) Removed from beaker

TABLE II

RESULTS [(1)] OF CAHN MICROBALANCE EXPERIMENTS

Ambient Cure (hrs)	Curing Conditions (hrs/°C)	Ambient Aging (hrs)	Vacuum Conditions (hrs/°C/torr)	Black	Time (hrs)	Friable
24	-	-	$20/38/10^{-4}$	mottled	-	-
24	-	-	$20/52/10^{-4}$	yes	-	yes
24	-	-	$10/65/10^{-4}$	yes	4	yes
24	-	-	$20/65/10^{-2}$	yes	-	yes
24	-	-	$20/65/10^{-1}$	no	-	no
24	-	-	$20/80/10^{-4}$	yes	-	yes
19	6/65	2	$9/52/10^{-4}$	no	-	no
24	20/65	0	$10/65/10^{-4}$	yes	4	yes
1	24/65	2	$20/52/10^{-4}$	no	-	no
1	24/65	2	$20/65/10^{-3}$	no	-	no
1	24/65	2	$20/65/10^{-2}$	no	-	no
1	24/65	24	$20/65/10^{-3}$	no	-	no
1	24/65	24	$20/65/10^{-2}$	no	-	no
1	24/65	24	$20/65/10^{-4}$	yes	-	yes
1	24/65	48	$20/65/10^{-4}$	yes	4	yes
1	24/65	72	$10/65/10^{-4}$	yes	4	yes
1	72/65	2	$15/65/10^{-4}$	no	-	no
1	72/65	48	$10/65/10^{-4}$	no	-	no
1	72/65	72	$12/65/10^{-4}$	no	-	no
1	24/105	2	$10/65/10^{-4}$	no	-	no

NOTES: (1) DC S-5370 silicone foam, 2 phr catalyst

(2) Time when black condition becomes visible

(3) Determined by probing specimen after vacuum exposure

FIGURE 1 TYPICAL WIRE-WRAP MATRIX IN AN ELECTRONIC ASSEMBLY

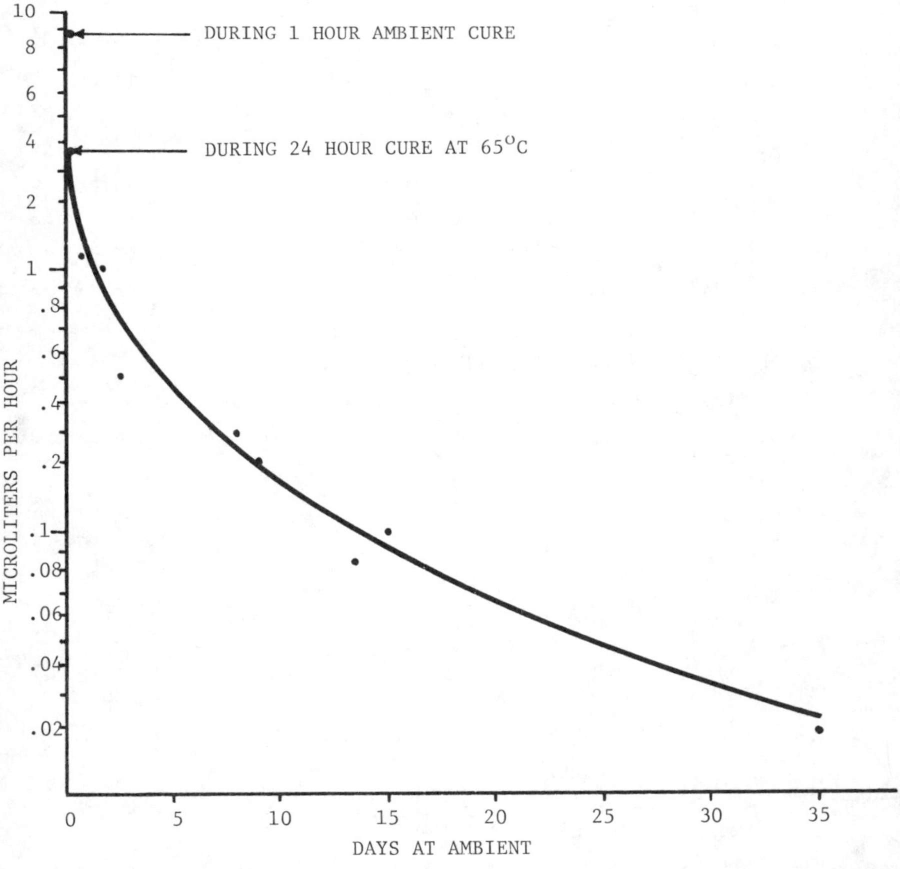

FIGURE 2 N-PROPANOL OFFGASSED DURING CURE AND AMBIENT AGING

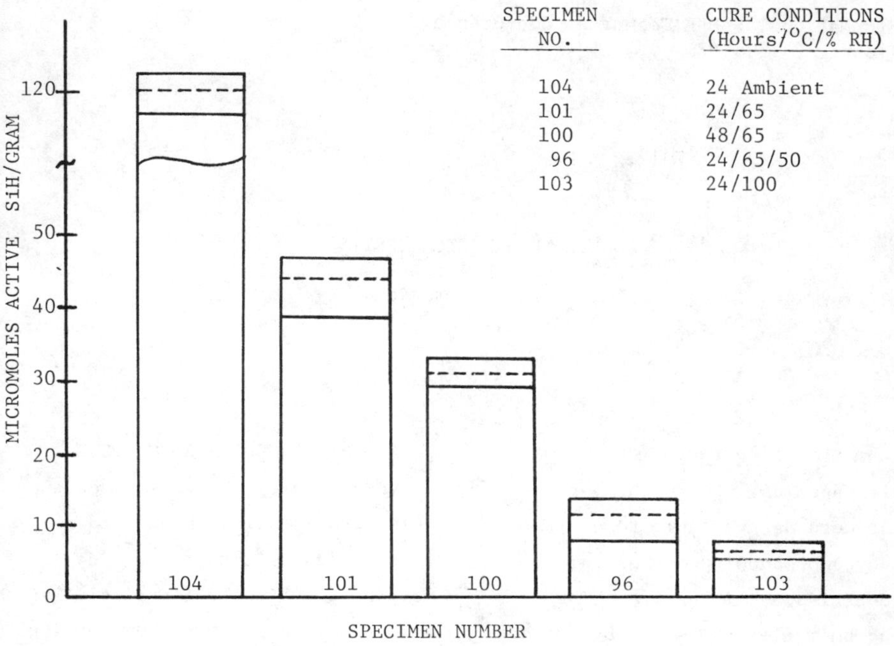

FIGURE 3 ACTIVE SILANE (≡SiH) CONTENT OF CURED FOAMS

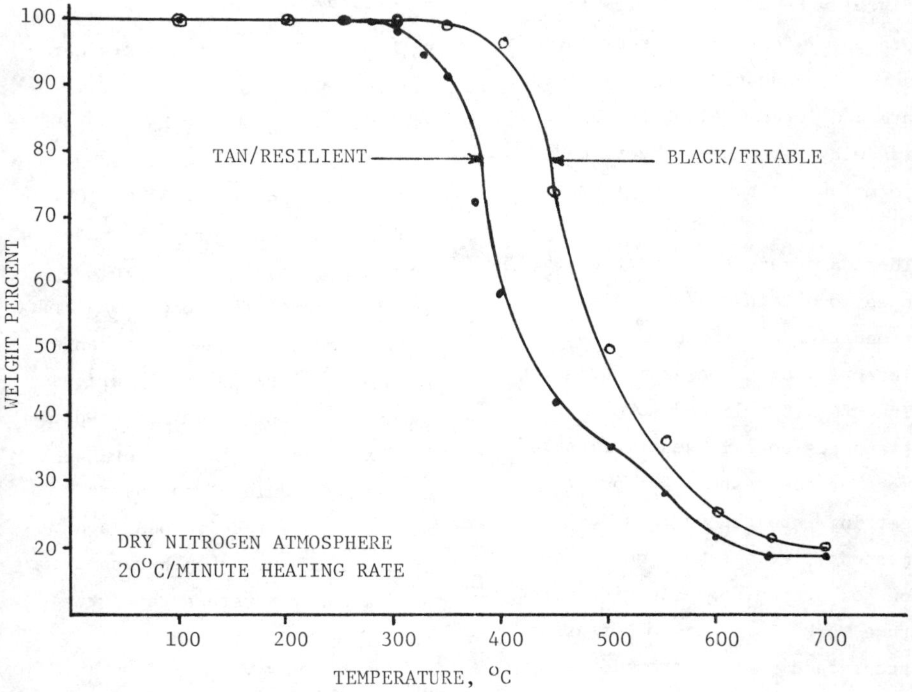

FIGURE 4 THERMOGRAVIMETRIC ANALYSES, BLACK/FRIABLE AND TAN/RESILIENT

13th National SAMPE Technical Conference
October 13-15, 1981

HONEYCOMB SANDWICH TESTING

Thomas N. Bitzer
Hexcel Corporation

Abstract

This paper deals with the various test methods available to evaluate honeycomb sandwich mechanical properties and integrity. The long beam flexure, short beam flexure, plate shear, flatwise compression, edgewise compression, flatwise tension, and climbing drum peel tests are discussed, and several nondestructive test methods are mentioned. Also, the basic sandwich structural theory and formulas are briefly explained.

1. INTRODUCTION

There are many tests that can be used to obtain the mechanical properties and the structural integrity of a honeycomb sandwich. The MIL-STD-401B and ASTM standards contain descriptions of most of the tests. But before getting into the various test methods, let's take a quick look on how a honeycomb sandwich structure works in order to better understand what information the tests provide.

2. SANDWICH CONCEPTS

A typical sandwich consists of two thin, dense facings bonded to a thick, lightweight core. Each component by itself is relatively weak and flexible but when working together they provide an extremely stiff, strong, and lightweight structure.

Some of the facing materials normally used are aluminum, steel, fiberglass, graphite, kevlar, and plywood, and their thicknesses range from approximately 0.010" to 0.5". Honeycomb cores are made from aluminum alloys, stainless steel, fiberglass, nomex, and kraft paper, and their densities range from about 1 to 40 pcf. Various types of adhesives may be used to bond the facings to the core such as films, liquids, and pastes. Most of these require heat and pressure to cure them but some two part adhesive systems can cure at room temperature.

Now a little structural theory.

When a sandwich panel is loaded, such as a beam, the facings take the bending load (one skin in compression and the other in tension) while the core takes the shear load as shown below.

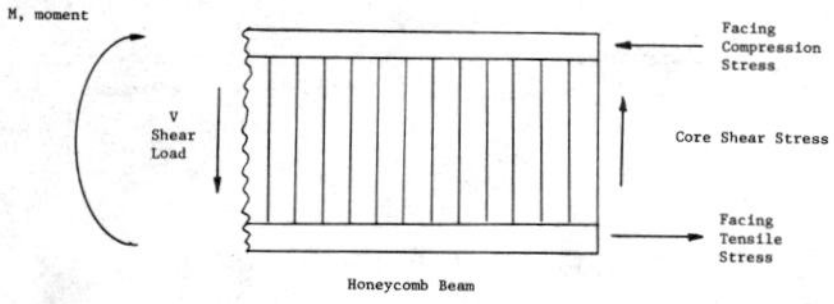

FIGURE 1

The facing stresses (compression or tension) are assumed to be uniformly distributed over the thin facings, and the core stress is uniform over the honeycomb thickness. It should be noted that since the facings do take a small part of the shear load that the shear load is divided by "h" and not "t_c".

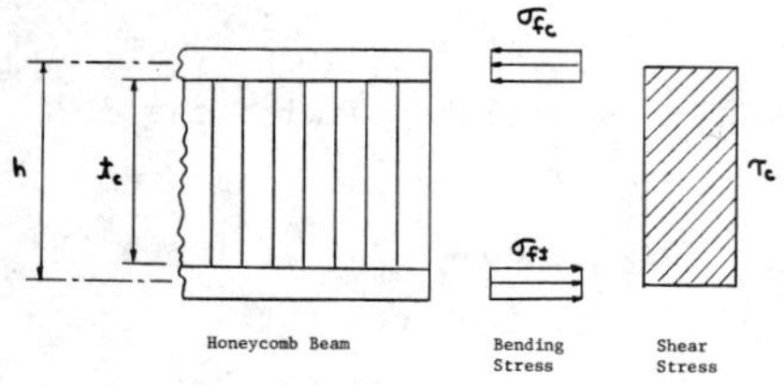

FIGURE 2

3. SANDWICH FORMULA

The basic equations for a honeycomb sandwich with similar facings are as follows:

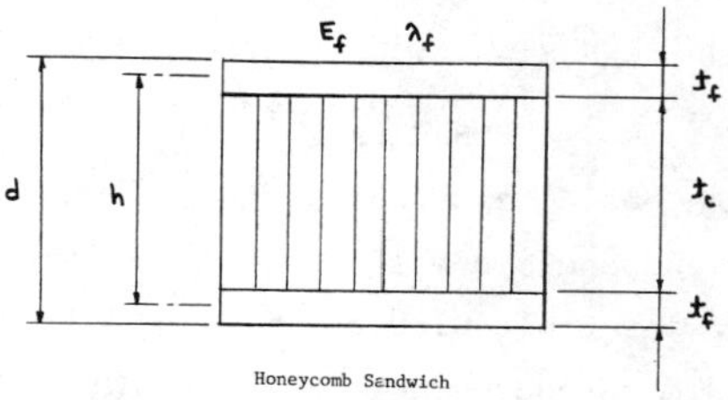

FIGURE 3

Facing Stress —— $\sigma_f = \dfrac{M}{t_f\, h}$

Core Stress —— $\tau_c = \dfrac{V}{h}$

Deflection —— $\Delta = \underbrace{\dfrac{2 K_b P L^3 \lambda_f}{E_f\, t_f\, h^2}}_{\text{Bending}} + \underbrace{\dfrac{K_s P L}{h\, G_c}}_{\text{Shear}}$

Where:

M = moment, in-lbs/in

V = shear load, lbs/in

P = total load, lbs/in

L = span length, inches

K_b = bending deflection coefficient

K_s = shear deflection coefficient

t_f = facing thickness, inches

t_c = core thickness, inches

d = sandwich thickness, inches

h = distance between facing centers, in.

E_f = facing modulus, psi

λ = 1-Poisson's ratio squared $(1-\mu^2)$

G_c = core shear modulus, psi

You may have noticed that the sandwich panel deflection consists of two parts; the bending deflection and the shear deflection. Normally the latter is quite small, being around one or two percent of the total; however, in some cases (thick panels with short spans) it can be a significant percentage.

4. MECHANICAL TESTING

The basic tests for honeycomb sandwich panels are listed in the following table with the pertinent ASTM method numbers and MIL-STD-401B paragraphs.

HONEYCOMB SANDWICH MECHANICAL PROPERTY TESTS

Test	ASTM Method	Mil-Std-401B
Long Beam Flexure	C393	5.2.4
Short Beam Flexure	C393	5.2.4
Plate Shear	C273	5.2.2
Flatwise Compression	C365	5.2.1
Edgewise Compression	C364	5.2.1.1
Flatwise Tensile	C297	5.2.3
Climbing Drum Peel	D1781	5.2.6

TABLE I

Let's first discuss the long beam flexure test. This test is used to determine the facing compression or tensile stress and panel stiffness. It can be a "three point" or "four point" test as shown below. The basic difference is seen in the beam's shear and moment diagrams - the first only having the maximum moment at a point under the load while the latter has the maximum moment constant between the two load points with no shear stresses. If the facings are of different thicknesses or materials the beam can be positioned to cause the weaker skin to fail in compression or tension - compression being the top skin and tension the bottom.

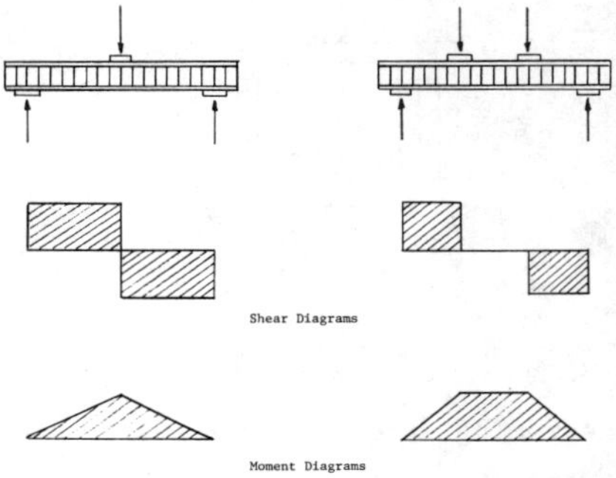

FIGURE 4

One test set up we use at Hexcel is an 18" span with 2" between loads as shown in Fig. 5. A strain gage can be mounted between the two load pads to determine the facing modulus, and on sandwiches with thin skins, rubber pads should be used under the load pads to prevent local failures of the facings or honeycomb. Also, the honeycomb should be orientated with the "L" direction in the span length to help assure obtaining facing failures. A dial gauge or transducer can be mounted under the beam to obtain load-deflection graphs.

Figure 5

It is best, if possible, to use a span length so the maximum beam deflection is not much more than the sandwich thickness. If the deflection becomes large, the normal equations for determining facing stresses will not be correct; they will give stress levels that are lower than actual due to neglecting the moment caused by the horizontal component at the supports.

The short beam flexural test is used to obtain the shear strength of the honeycomb. The core shear modulus can also be determined if the facing properties are known; however, this method will not provide as accurate a value as the plate shear test and is not the recommended way. Basically there are two common loading arrangements: quarter point or third point loading. The spans and load pad widths must be calculated so that facing failures or local core compressive failures do not occur before the honeycomb fails in shear. The usual width of these specimens are three inches, and a loading rate is used that produces failures in 3 to 6 minutes. The core can be orientated with the "L" or "W" direction in the span length, depending on which core shear property is being determined - remember in normal hexagonal honeycomb the "L" shear properties are approximately twice the "W" properties. Again, panels with thin skins should have rubber pads or thin metal shims under the load pads to prevent high local core compressive stresses.

Plate shear tests can be performed on the honeycomb panel by cutting out specimens 2" x 7.5", 2" x 6", or 3" x 12" and bonding them to 0.5" thick steel plates. The 3" x 12" specimens should be used on cores with 1" or larger cell sizes in order to test as many complete cells as possible. This test provides the honeycomb's shear strength and shear modulus and can be run in compression (the 2" x 7.5" sample) or tension (the 2" x 6" sample) as shown in Figures 6 and 7. Usually these tests are done on honeycomb without facings, and these are the values reported in Hexcel's honeycomb property literature. We have not found any difference in the results in testing in tension or compression and prefer the latter test because it is simpler and quicker to set up on the test machine.

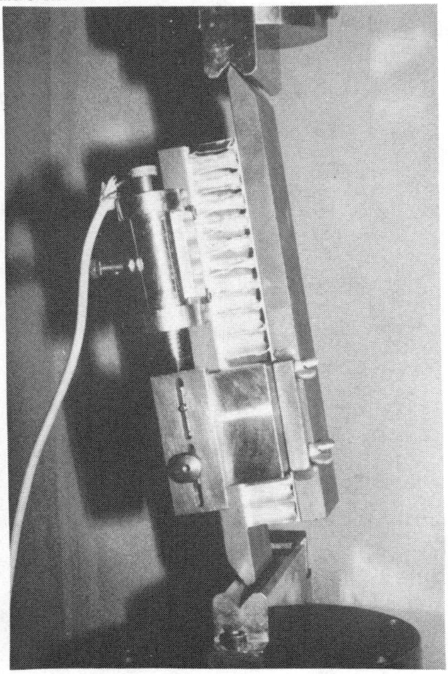

Figure 6

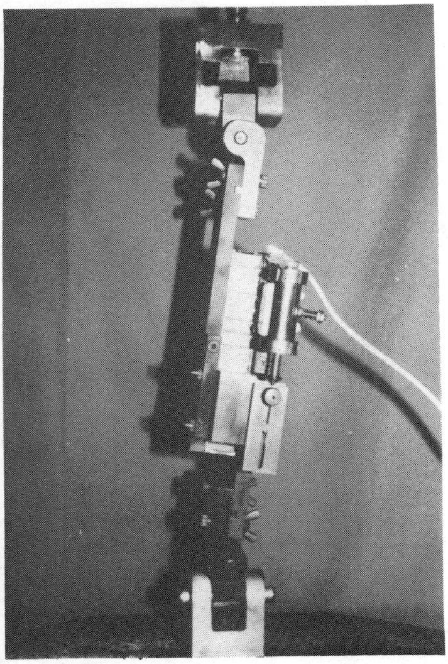

Figure 7

The plate shear test results will give slightly lower shear strengths than the beam shear test as the beam's facings do take a small amount of the shear load - the variation is a function of the skin thickness and core density (see Fig. 8).

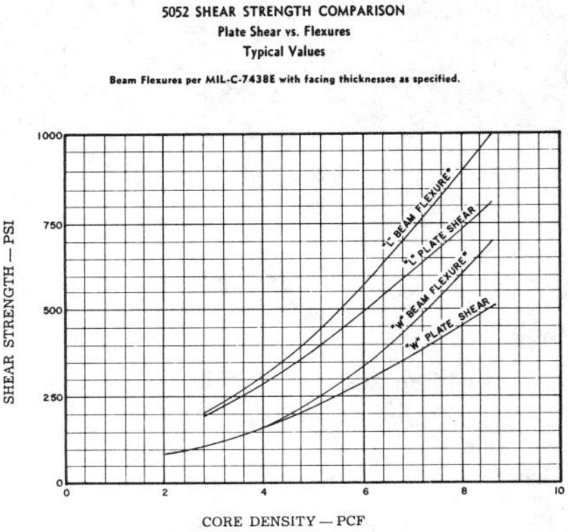

FIGURE 8

The honeycomb shear strength varies with core thickness due to the fact that the failure mode is shear buckling. Figure 9 shows the variation of the normally tested thicknesses (0.625" for aluminum core, 0.500" for fiberglass and nomex cores).

Basically what this means is a 2" thick core does not have twice the shear load capability as a 1" thick piece of honeycomb.

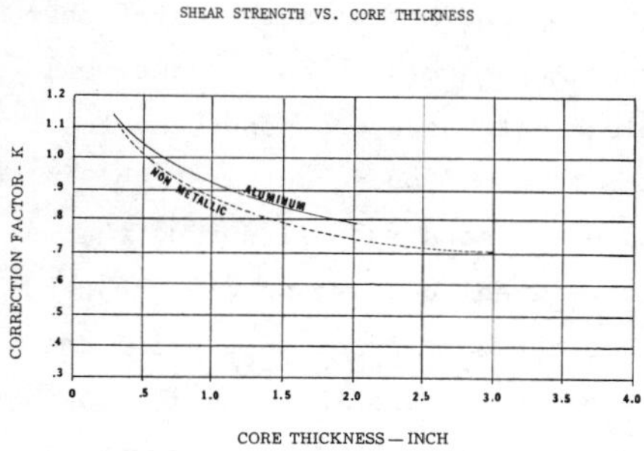

FIGURE 9

The flatwise compression strength of a sandwich can be determined from samples cut from the panel: 3" x 3" for ½" or smaller honeycomb cells and 4" x 4" for larger cells. These different specimen sizes are to minimize the edge effects of the cell walls being cut thus not having complete cells around the perimeter. In order to get good core compression mode failures and reliable test data, a self-aligning, spherical loading head should be used as shown in Fig. 10. This assures you that if the sandwich facings are not perfectly parallel then just the thicker portion is not crushed first. We usually test at a 0.020 inch per minute loading rate.

To determine the compressive modulus, a small hole is drilled through the center of the specimen and a rod attached to a transducer is inserted as shown in Fig. 10. It is normally not a good practice to use the test machine's crosshead travel to obtain the samples deformation.

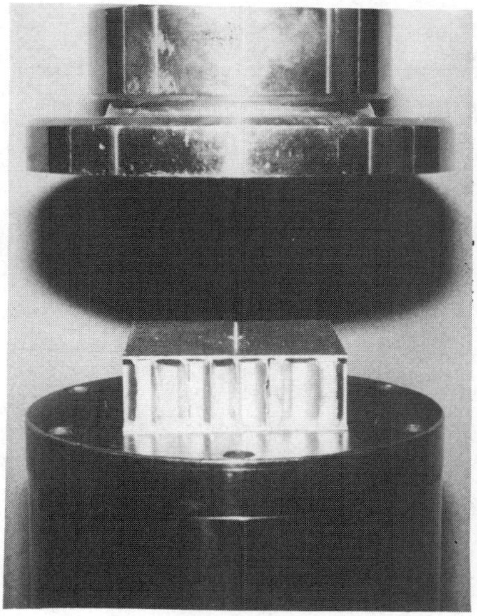

Figure 10

The edgewise compression test consists of applying compressive loads to the facings as shown below.

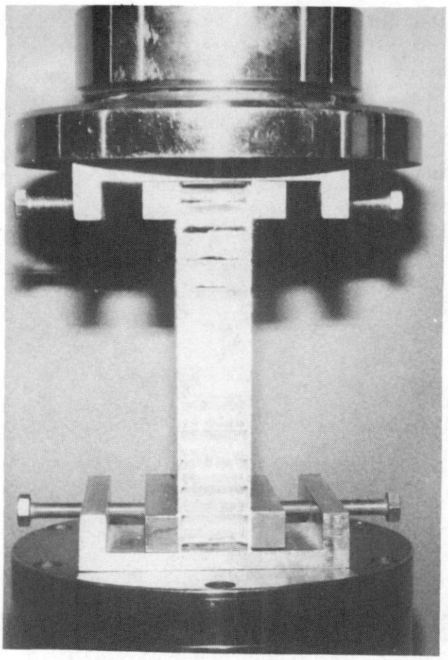

Figure 11

This test gives the column load the sandwich panel can withstand disregarding buckling. Here the specimen fails by the facings yielding or dimpling/wrinkling or delaminating from the core or shear crimping of the honeycomb. The height of the specimen must be short enough so column type buckling does not occur - generally, the unsupported height should be less than 8 times the panel thickness. If the skins are relatively thick (0.040" and up) just milling the load ends flat and parallel usually is sufficient, but on thin facings it is best to rout out about a ½" of the core and fill this area with a potting resin and then mill the ends. It is extremely important to have flat, parallel load ends with the skins adequately supported.

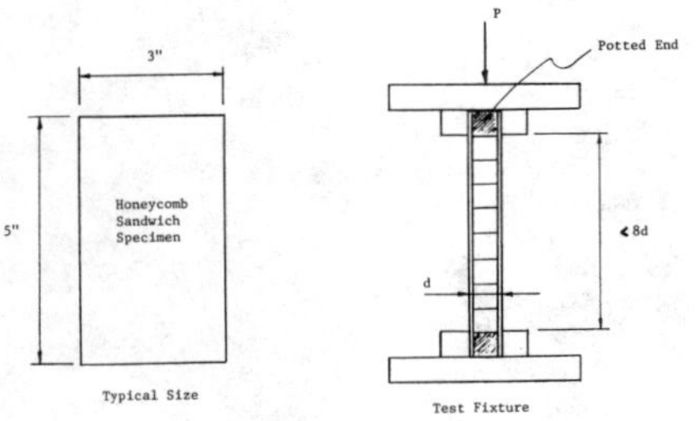

FIGURE 12

How well the skins are bonded to the honeycomb can be determined through the flatwise tensile and climbing drum peel tests. The first test consists of bonding 2" x 2" or 3" x 3" (depends on cell size) pieces of the panel to blocks and then pulling the blocks apart. The top and bottom surfaces of the specimen should be sanded and well cleaned to assure a good bond to the blocks (you do not want this adhesive bond to fail during the test). An adhesive should be used that cures at least 50°F below the temperature that the sandwich was bonded. Although sometimes this is not practical especially when you are going to run elevated temperature tests.

The test set-up is shown in Fig. 13, and the blocks should be free to pivot so the load applied to the sandwich specimen is in pure tension, and there isn't any peeling occurring - this can drastically lower the ultimate failure load. The modes of failure are the following:

(a) core tearing
(b) adhesion-to-core
(c) adhesion-to-facing
(d) cohesion of the adhesive

Figure 13

If the honeycomb tears, the only way to improve the panel flatwise tensile strength is to use a stronger core. Sometimes (heaven forbid) the core may get contaminated and cause an adhesion-to-core mode failure. Making sure the honeycomb is kept clean and not handled with bare hands should improve this problem. When an adhesion-to-facing failure happens this implies that the facings were not properly cleaned or became contaminated. A cohesive failure means that this is the best you can get out of this adhesive. You can tell this mode of failure by looking at the core and skin of the failed specimen. If there is adhesive on both, this type of failure occurred. One small point, you can increase the panels flatwise tensile strength even using the same adhesive by going to smaller cell honeycomb - this gives more bond area. To tell the other failure modes; if the facing does-not have any adhesive on it after failure then it's an "adhesion-to-facing" mode, or if the core is clean then it's an "adhesion-to-core" failure.

The climbing drum peel test should only be performed on sandwich panels with relatively thin skins; the maximum thicknesses should be approximately 0.032" for aluminum, 0.020" for steel and 0.040" for fiberglass. The reason being that the facing must be wrapped around a 4" diameter drum, and if the facing is too stiff it takes a large force just to bend the skin which causes the failure mode of the sandwich to be slightly different - the peeling of the skin from the honeycomb does not start in the desired location. Normally a 0.020" aluminum facing with 1/4-5052-7.9 honeycomb is used to evaluate different adhesives.

Test specimens are 3" wide and about 14" long as shown in Fig. 14. Usually the honeycomb "L" orientation is in the long direction, and the test set-up is shown in Fig. 15.

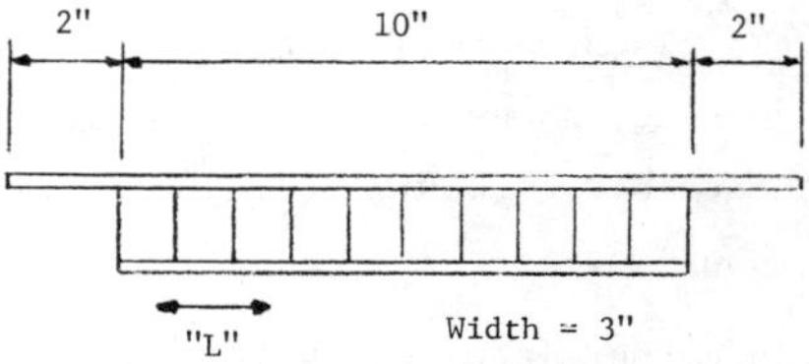

Climbing Drum Peel Specimen

FIGURE 14

Fig. 15

By pulling down on the drum straps, the drum rotates and travels upward bending the facing and peeling the facing off the core. A cross-head travel of one inch per minute is used.

Test results are expressed in torque, inch-pounds per inch (in.-lbs./in.) or inch-pounds per 3 inch width (in.-lbs./3 in.) and can include the torque to bend the facings and rotate the drum or these values can be substracted giving just the torque to peel the skin from the core. The latter method is how Hexcel reports the data.

The failure modes are similar to the flatwise tensile test: core tearing, adhesion to the facing or honeycomb, cohesive failure of the adhesive.

Both the flatwise tensile and climbing drum peel tests are good ways of evaluating the integrity of a honeycomb sandwich panel, and either test can be run after the panel has been subjected to different environments.

5. NON-DESTRUCTIVE TESTING

Many times you can not or don't want to destroy a sandwich panel to determine its integrity. Here is where non-destructive test methods are useful. There are many different non-destructive techniques that are used. Some of them are: X-ray radiography, thermal neutron radiography, ultrasonic inspection, acoustic emission, harmonic bond testers, visual inspection and the old coin tap test. A good NDT method should be able to locate, determine the size, and also classify the type of defect. Sometimes this requires using several different procedures on one panel.

Some of the panel flaws that can be detected are core-to-skin delamination areas, crushed or corroded core, water intrusion and voids in the foaming core splice adhesive.

The test equipment can be expensive to purchase and require some operator skill to obtain reliable and repeatable results. Interpreting the data may be difficult and require a bit of experience. But if the equipment is used properly it can be a very valuable tool for inspecting sandwich panels - especially in a production quality control situation.

6. SUMMARY

The basic honeycomb sandwich concepts are quite simple - the facings take the bending moment with an assumed uniformly distributed compression and tension stresses while the core takes the shear load.

Tests can be easily performed on specimens cut from the panel to obtain the pertinent mechanical properties and evaluate the sandwich's integrity. Various non-destructive test equipment are also available to perform quality control inspections.

Honeycomb panels have been successfully used for a long time by the aircraft and aerospace industries and now are starting to be used more and more in commercial applications.

Biography

Tom Bitzer is the manager of Hexcel's R&D Test Laboratory where he has been involved in the testing and design of honeycomb sandwich panels for over 14 years. He has a masters degree in Structural Engineering from the University of California in Berkeley and is a registered Civil Engineer.

13th National SAMPE Technical Conference
October 13-15, 1981

ANALYTICAL (FEM) AND TEST EVALUATION OF ADVANCED COMPOSITE ANTENNA REFLECTORS FOR COMMUNICATION SATELLITES

J. R. Badura, R. N. Gounder, S. Ino
RCA Astro-Electronics
Space Center
Princeton, New Jersey

Abstract

This presentation describes the Analytical Finite Element (FEM) Methods, in particular, modeling of complex advanced composite antenna reflector designs for communication satellites, used to evaluate the dynamic characteristics in terms of frequencies, mode shapes and dynamic response. The antenna reflector response to sinusoidal and acoustic excitation is described in terms of mode related deflections, loads and stresses. The final evaluation includes damping characteristics extracted from the conducted tests and used to finalize a design in terms of stress analysis. A brief description of dynamic tests and correlation of analytical results with test results are shown.

1. INTRODUCTION

An offset reflector of Kevlar sandwich structure was subjected to a sine vibration and acoustic survey to determine its natural frequencies, mode shapes and damping characteristics. The test was also performed to substantiate the correctness of the computer Finite Element Model developed for the prediction of dynamic characteristics and stress levels. The same techniques have subsequently been applied to design and the analysis of future flight hardware.

2. REFLECTOR DESIGN COMPLEXITY

The mechanical design of the antenna reflector shown in Figure 1 is based on a set of stringent design criteria. The choice of configuration, materials and the method of analysis is related to a variety of mechanical, electrical and environmental requirements defined for the ground, launch and space environments.

The mechanical demands on the reflector involve high dimensional accuracy in reflector surface deviations and precise control of mass properties including weight and inertias. The design must provide physical and functional compatibility with other RF antenna components, spacecraft and launch vehicle. Maintainability of shape and dimensions under thermal, radiation, dynamic and moisture environments of ground, launch and space mission is also required.

The dynamic demands impose the minimum frequency requirements to decouple antenna from the spacecraft vibratory modes during launch.

Selection of materials should ensure adequate strength, stiffness and RF transparency to meet all environments. The electrical demands require the design to support the polarizing grid elements and maintain them at defined locations and provide reference locations and means for location of the focal points.

Fig. 1. Composite Antenna Reflector

3. DESCRIPTION OF KEVLAR REFLECTOR

The antenna reflecting surface is a 50 inch focal length off-axis fed paraboloid. The aperture geometry is a 70 inch x 50 inch ellipse. The vertex of the paraboloid is located 30 inches from the center of the inner edge of the reflector in the V direction.

The reflector shell structure has the same overall geometry as the reflecting surface. It is fabricated from Kevlar cloth prepreg and consists of a honeycomb core, top and bottom face sheets shown in Figure 2 and edge closures. The honeycomb core and the edge closures are single ply Kevlar fabric. The face sheets are of two-ply construction which are precured. The cured face sheets are then bonded to the core. The shell structure is RF transparent.

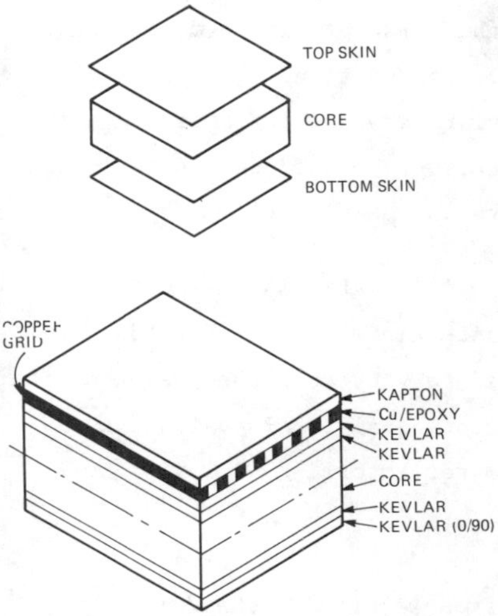

Fig. 2. Construction of the Sandwich Reflector

The antenna backframe structure is provided to strengthen and stiffen the reflector shell. It consists of ribs and joints fabricated from Kevlar fabric reinforced epoxy. The ribs are of sandwich construction using two-ply face sheets, and single ply honeycomb cores. The ribs are bonded to the reflector without any reinforcing angle between the ribs and the reflector surface. Multi-ply joint doublers and face sheet reinforcement for the ribs themselves are used in designated areas. The backframe structure is also RF transparent. Fig. 3 illustrates typical cross sections of the reflector.

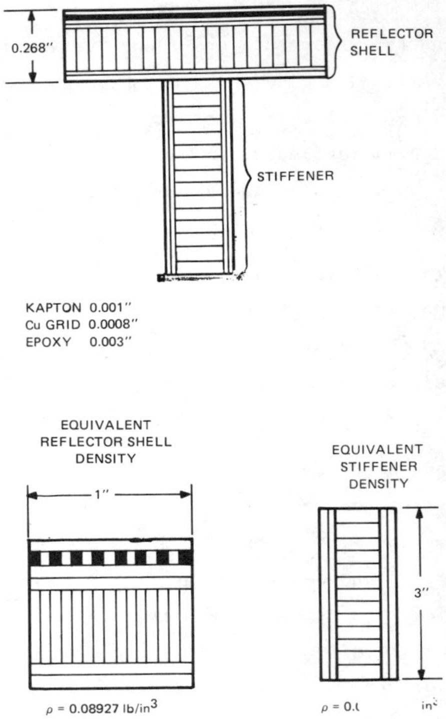

Fig. 3. Typical Reflector Cross Sections

The reflector assembly is supported on three sets of flexible legs. The legs consist of multiple-ply graphite/epoxy, 3 inch diameter tubes with Kevlar plates forming flexible top section. The flexible Kevlar laminate leg segments are adhesively bonded to the reflector ribs as shown in Fig. 4.

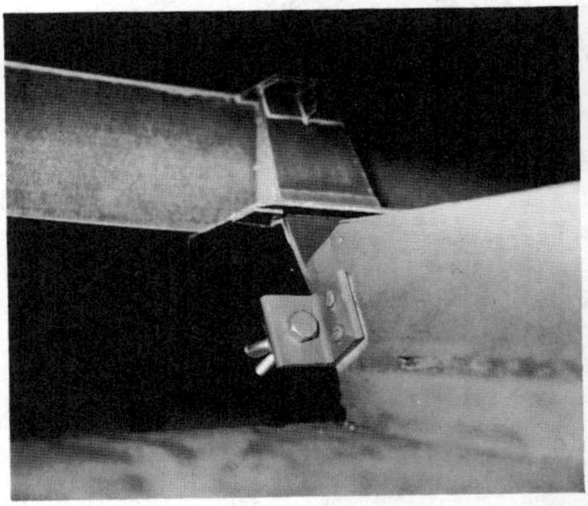

Fig. 4. Details of the Reflector Supporting Structure and Flexible Leg.

4. ANALYTICAL METHODS

Finite Element (FEM) Methods were used to represent the design and to perform analytical evaluation of the reflectors.

4.1 Selection of the Computer Program

The selection of a computer program to represent the complex geometry of the reflectors and to properly idealize the composite materials used in the construction requires a degree of judgment. Finite Element codes available commercially provide means for adequate idealization of the geometry. However, these codes do not provide features to properly represent composite design features of this nature. Proper idealization of multi-ply honeycomb construction with imbedded elements requires introduction of anisotropic or at least orthotropic properties. The materials differ in their properties with the direction of applied loads, for example tension and compression. Thus, many approximations need to be made in the course of the analysis.

Two finite element codes were considered, STARDYNE and NASTRAN. STARDYNE code was used for the reflector because of its simplicity and relatively low cost of usage. STARDYNE was considered adequate for the analysis of the reflector due to the essentially quasi-isotropic nature of the reflector. NASTRAN code was used in analyzing flight hardware to more properly represent the orthotropy introduced by the imbedded grid wires.

4.2 Material Properties

The Kevlar fabric properties used in the analysis are summarized in Table I. The laminate properties of the quasi-isotropic skins were calculated from the fabric properties using an in-house laminate program and used as inputs to STARDYNE analysis. Any contributions from the imbedded wires were neglected for the purpose of analyzing the reflectors. However, these contributions were properly included in the NASTRAN analysis of the flight hardware.

4.3 Finite Element Model

The reflector was triangulized into finite elements. The supporting ribs were modelled as off-axis beam elements. The flexible legs were modelled as beam elements. The finite element STARDYNE model (FEM) of the reflector shown in Fig. 5 consists of 109 grid points, 141 flat sandwich triangular elements and 79 beam elements. Quasi-isotropic laminate properties were specified for the face sheets of both the shell and rib elements.

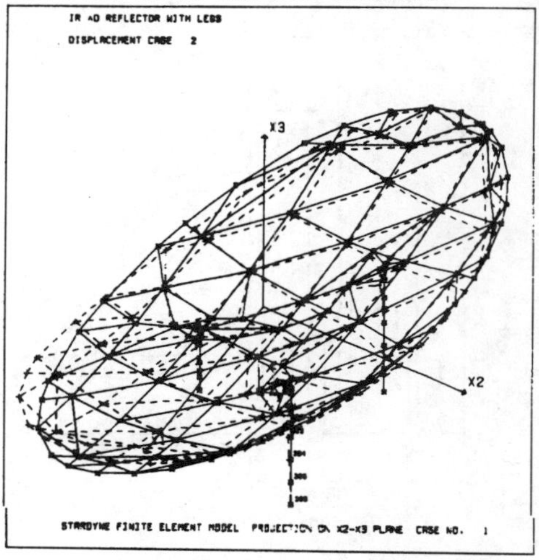

Fig. 5. FEM STARDYNE Model of the Reflector Undeflected Shape & Deflected Mode 1

TABLE I. MATERIALS DATA
KEVLAR FABRIC EPOXY MATERIAL
120 STYLE

Property	Kevlar Fabric Properties	Quasi-Isotropic Skin Laminate Properties
E_1 E_t (msi)	4.2	2.88
G_{1t} (msi)	0.3	1.17
ν_{1t}	0.12	0.14
ρ (lb/in^3)	0.048	0.048
F_{tu} (ksi)	72.1	50.0
F_{cu} (ksi)	21.1	15.0

TABLE II. COMPARISON OF NATURAL FREQUENCIES DERIVED
BY TESTS AND ANALYSIS

Mode No.	Test	Frequency Hz	Description
1	28.6	27.72	Antisymmetrical mode-Bending X_3
2	37.0	36.20	Bending horizontal- X_2
3	41.2	40.25	Symmetrical mode bending-X_3
4	51.2	50.13	Bending-horizontal-X_1
5	59.9	55.38	Second Bending-X_3
6	69.3	66.54	

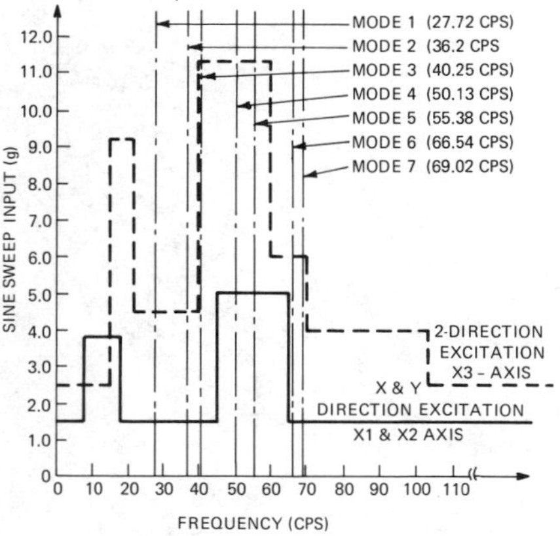

Fig. 6. Forcing Vibration Levels vs. Reflector Natural Frequencies

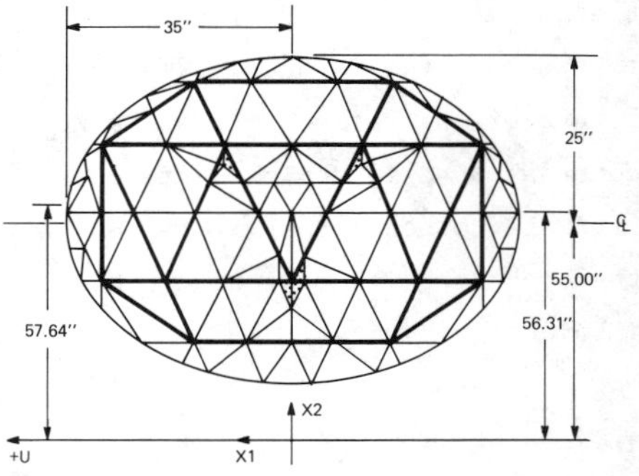

Fig. 7. Maximum Stress Areas

4.4 Analytical Results

A modal analysis was performed using the FEM model described. The natural frequencies and associated mode shapes obtained are summarized in Table II. The stress levels were also obtained from a sinusoidal excitation input shown in Fig. 6 and the margins of safety in the areas of maximum stresses were calculated. The maximum stress areas are shown in Fig. 7.

5. REFLECTOR TESTING

The reflector was subjected to an extensive test evaluation program. Sine vibration and acoustic tests were performed for the qualification level loads which were applied gradually.

5.1 Test Objectives

The reflector was subjected to sine vibration and acoustic testing with the following objectives:

- o To verify that the reflector is capable of withstanding the mission mechanical loads without failure and
- o To study the characteristics of the reflector including the natural frequencies, their mode shapes and the vibration magnification factors.

5.2 Sine Vibration Test

The sine vibration test was performed at the RCA/ASTRO-Electronics Environmental Test Facility using C-210 vibration exciter and control console. The reflector was mounted onto a rigid sine vibration test fixture and then mounted onto the vibration test facility as shown in Fig. 8.

Fig. 8. Reflector Test Set-Up

5.3 Acoustic Tests

Since the sine vibration tests were performed at the acoustic test chamber, the same set-up of the fixture and the bed table was used for the acoustic test except for removing all equipment not associated with the acoustic test other than the exciter and bed plate.

5.4.1 Sine Vibration. The reflector was subjected to three different load levels as shown below:

(1)	X-direction	0.5g
	Y- "	0.5g
	Z- "	1.0g
(2)	X-direction	1.5g
	Y- "	1.5g
	Z- "	2.5g
(3)	X-direction	full spectrum (See Fig.6)
	Y- "	" "
	Z- "	" "

This swept sine survey provides the most precise means of analyzing the dynamic characteristics. In addition, the qualification level load determines if the reflector has the

TABLE IV. MODAL DAMPING AND VIBRATION MAGNIFICATION FACTORS

Mode No.	Vibration Magnification Factor (Q)	Viscous Damping Factor (ζ)
1	13.2	.038
2	13.8	.036
3	17.0	.028
4	11.6	.043
5	9.8	.051

capacity to withstand the launch loads.

5.5 Instrumentation

The instrumented reflector test set-up is shown in Fig. 9. The measuring instruments were mainly installed at the locations determined from the analysis.

Strain gauges and rosettes were placed at the maximum expected displacement and stress locations.

A total of 13 accelerometers in addition to the control accelerometers were placed at the location of expected maximum deflections for 6 analyzed frequencies and mode shapes.

Fig. 9. Reflector Test Instrumented

5.6 Test Results

The sine vibration testings provided the reflector natural frequencies with associated mode shapes. The deflection and stress levels in the maximum stress areas were also obtained. The modal damping characteristics were deduced from the test results.

5.6.1 Natural Frequencies. The reflector natural frequencies with associated mode shapes as derived from tests are listed in Table II and compared directly with analytical predictions.

5.6.2 Damping Characteristics. The modal damping coefficients were derived from the frequencies of the maximum and minimum coincident response for the number of frequencies analyzed. The damping coefficient for the viscous damping is obtained from the equation:

$$\xi_i = \frac{1}{2}\,\frac{(f_a/f_b)_i^2 - 1}{(f_a/f_b)_i^2 + 1}$$

where ξ_i = modal damping coefficient for mode i.

i - mode number.

The clearly identified modes i.e., Mode 1 at 27.72 Hz first bending asymmetrical mode and Mode 5 at 55.38 Hz, second bending mode are excited by input in the X_1 direction. Mode 3 at 40.25 Hz which is a symmetrical bending mode is being excited by an input in vertical X_3 direction.

The modal damping factors and vibration magnification equivalent Q for these modes are given in Table IV.

7. CONCLUSION

The analytical predictions of the reflector natural frequencies are in good agreement with the frequencies obtained from the sine vibration testing. This is in-spite of the idealization of the STARDYNE FEM Model and the quasi-isotropic laminate properties assumed in the analysis. Further, the qualification level sine vibration and acoustic testing verified the adequacy of the reflector design for the intended flight environment. The vibration magnification factors obtained from the IR&D reflector testing have formed the basis for the stress analysis of flight reflectors.

8. REFERENCES

1) "Damping Applications for Vibration Control, P. J. Torvik, AMD - Vol. 38.

2) "Analytical (FEM) Evaluation of the Composite Communication Satellite Antenna Reflector Dynamics", J. R. Badura, R. N. Gounder, K. W. Tung, RCA Symposium on Mechanical Engineering, April 23, 1981.
3) "A Modal Vibration Test of the 5D-2 Satellite", D. F. Chu, RCA Technical Communications RE-26-4-5.

9. ACKNOWLEDGMENT

The authors wish to thank Dr. D. F. Chu, RCA/ASTRO-Electronics for review of the paper.

10. BIOGRAPHIES

Jack R. Badura, M.A.Sc. from Technical University of Silesia, Poland in 1965. Completed course work and passed comprehensive examination towards a Ph.D. at University of Waterloo, Canada while gainfully employed at SPAR Aerospace working on Space Shuttle Program. Employed at RCA/ASTRO-Electronics since July 1980. As a Senior Member of Advanced Structures and Composite Materials Group at ASTRO, he contributed to Satcom, SOOS and IR&D Programs. In all, he has over 15 years of Mechanical Engineering experience.

Raj N. Gounder, Ph.D. from Northwestern University in 1972. As Manager of Advanced Structures at RCA/ASTRO-Electronics, he is responsible for the application of advanced composite structures technologies to satellite hardware. In his position, he has successfully managed the design, analysis, procurement and testing of several advanced composites hardware for communications, navigational and scientific missions.

Sari Ino, received her B.S. degree in Mechanical Engineering from Columbia University in 1980. She joined RCA/ASTRO-Electronics in 1980 and has spent the past one and half years in the field of composites including analysis and testing of satellite structures.

13th National SAMPE Technical Conference
October 13-15, 1981

MATERIALS CONSIDERATIONS IN SECOND SURFACE SOLAR MIRRORS*

Victor P. Burolla
Sandia National Laboratories
Livermore, California

Abstract

Failures involving several mirror designs are described. The failures in some cases can be considered catastrophic in that the mechanical structure lost its integrity. In other cases, the failures were simply loss of reflectance of the mirror due to silver degradation. Testing of various mirror designs has produced results that are not catastrophic but have implications in terms of long term survival. The investigations into the causes of these failures are described. Apparent solutions to these failures are presented along with the description of one module design that has survived well after 4 years of environmental exposure/use. The failures seen to date have been related to materials incompatibility brought about by improper application, changes in material property due to weathering and chemical attack.

*This work is supported by the U. S. Department of Energy under Contract No. DE-AC04-76DP00789.

1. INTRODUCTION

For several years now the Department of Energy has sponsored research and development to construct a 10 megawatt solar electric power generating facility. The solar collector or reflector field for this, consists of 1800 heliostats with over 20,000 individual mirror facets attached to these heliostats. Testing and real time exposure of these mirror modules (term for each individual mirror facet) has revealed both catastrophic and gradual failures. The catastrophic failures relate to loss in structural integrity while the gradual failures relate to loss in reflectivity due to dissolution of the silver layer. The author's first involvement began in September 1978 with the discovery of mirror deterioration on several modules exposed outdoors for one year at Sandia National Laboratories, Livermore, California. Since then the failures noted have involved the structure, as well as a loss in reflectivity.

2. SILVER DETERIORATION

The problem of deterioration of the reflective layer has been dealt with extensively in the literature[1,2,3,4,5] and will not be discussed here at great length. It is due, for the most part, to water migration into the silver layer either through the mirror backing paint or in from the edges via capillary action. Certain adhesives used to bond the mirror to the remainder of the structure, have been found to greatly accelerate the deterioration rate and these could be classified as chemically incompatible. Tests to ascertain chemical compatibility are available[1], as are tests to determine the porosity of the mirror backing paint. Martin Marietta Aerospace has done a great deal of work in evaluating paints for mirrors that are applied in addition to the traditional paint as a supplemental moisture barrier. One of these paints is currently being used in the production mirror modules at the Barstow, California solar facility. All of the deterioration seen to date involves a module structural configuration that includes or permits voids immediately behind the mirror. In designing mirror modules, it therefore became necessary to integrate into the design a seal that would preclude moisture penetration to such an extent as to prevent water condensation within the voids of the module. To further compound the problem the seal had to survive for well over 10 years in order to consider the design viable for commercial use. In a typical sandwich type mirror module 1 m x 3 m x 5 cm thick, this meant a moisture tight seal for a length of more than 16 meters. Other module designs were able to reduce that length by half, but an 8 meter seal still proved to be a prodigious task. Much information was obtained from the thermal pane window industry, which eventually lead to the use of polyisobutylene as the primary sealant. In developing a sealing technology for sandwich type mirror modules, Dietrich et al.[6] determined that the only appropriate surface of the mirror to seal against was the top (unsilvered) surface as the edge was not clean enough and the paint on the back side was too porous. They were able to seal full scale modules that would provide an adequate moisture barrier for a lifetime probably in excess of 20 years. This assumes, of course, no environmental degradation of the seal. The material, therefore appeared adequate to the task, however, it has become apparent over the past few years that it is impossible to mass manufacture a high percentage (>95%) of leak free sandwich type mirror modules regardless of the quality of the sealant material. In short if the module has an internal void, moisture eventually will work its way into the void whether through or by the organic sealants. The next generation of mirror modules has for the most part recognized this problem and designed the modules to minimize or eliminate voids, thereby eliminating module sealant materials and processing problems.

3. BONDS TO MIRRORS

In May 1980, a catastrophic failure of a mirror module occurred during thermal cycling in an environmental chamber. The module design consisted of a mirror bonded to aluminum honeycomb (separated from the mirror by a coarse weave glass cloth) which in turn was bonded to a steel pan that formed a sealed structure around the back of the mirror. After several days of cyling the mirror literally blew away from the core over a diameter of about 75 cm (see Figure 1). A cursory examination of the mirror backing indicated that the paint had separated from the metallic copper layer wherever it had contacted the adhesive. A more detailed examination revealed that those delaminated areas immediately under an aluminum cell edge or a glass fiber still had paint adhering to the copper (see Figure 2). Since the coefficient of expansion of the aluminum and glass fiber more closely matched the mirror than the coefficient for the adhesive, it was thought that during thermal cycling the unrestrained

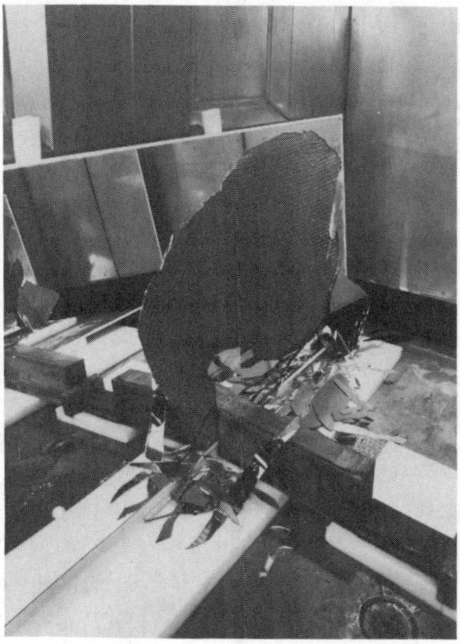

Figure 1. Photograph depicting mirror separation from mirror module core (aluminum honeycomb).

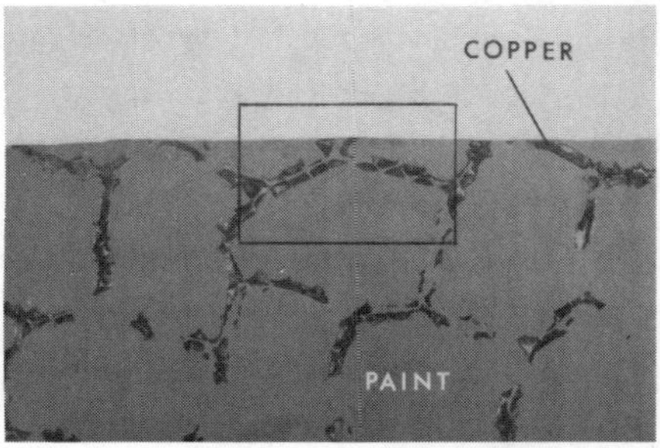

Figure 2a. Photograph of paint delamination area, indicating locations where paint still adheres to copper.

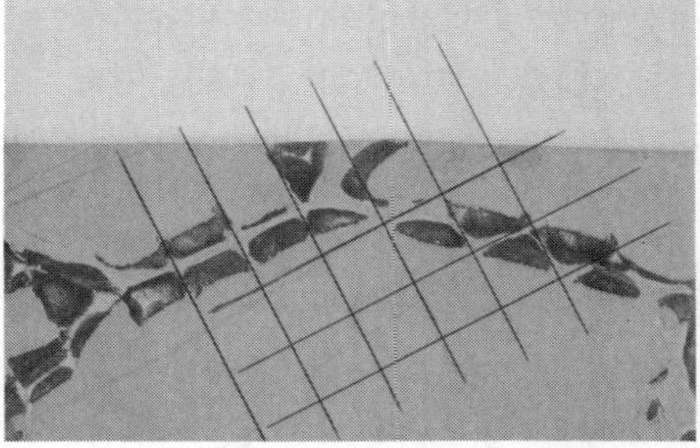

Figure 2b. Glass cloth weave pattern superimposed on fig. 2a inset.

adhesive (adhesive away from the aluminum cell edge or glass fiber) worked the paint in shear and caused it to delaminate from the copper. The shear mechanism hypothesis became more convincing after comparing the results of tensile pull and lap shear tests on the mirror backing paint. The tensile pull specimens were not successful in removing the paint over the entire bonded area. At most only 5% of the area would show copper. The lap shear tests however were always successful in removing the paint over the entire bonded area. (Most tensile specimens were run on a Sebastian Adherence Tester and all lap shear tests were accomplished by bonding a 1.25 cm x 10 cm steel strip to one end of a 10 cm x 15 cm piece of mirror with a 7.5 cm x 10 cm steel strip bonded to the other end - see Figure 3). We were able to produce very similar paint delaminations by applying thick beads of adhesive, in varying lengths and widths to the backs of mirrors and thermally cycling them. Periodically the samples would be removed and examined with polarized light and filters to determine stress distributions between the adhesive and the glass. At the start, the stresses appeared low and distributed uniformly over the entire bead. As cycling progressed, the stresses shifted to one end of the bead and became significantly more concentrated. In many instances,

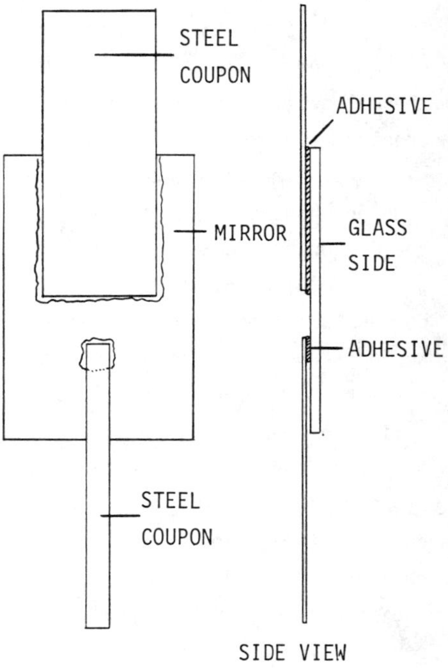

Figure 3. Diagrammatic sketch of lap shear specimens used to separate paint from copper.

after 10 cycles, the beads literally jumped away from the mirror leaving a dull copper finish. The shininess or smoothness of the copper surface varied from sample to sample and although we were able to approximate the mirror-like appearance of the copper on the failed module we were never able to duplicate it exactly. The frequency and rapidity with which the adhesives would delaminate the paint depended on the modulus of elasticity, the coefficient of expansion, the bead dimensions and the paint system on the mirror. Those mirrors with a second coat of paint seemed to be slightly less prone to failure. For all the other parameters, the higher the modulus, the higher the expansion coefficient and the larger and thicker the bead, the more complete and more rapid the delamination. For the module that failed, the probable cause of failure was a partially cured batch of adhesive (the adhesive used was a B-staged adhesive, roller coated on the aluminum honeycomb). The adhesive had not been stored at the appropriate temperature during its life and when heated, did not flow sufficiently to form a smooth tapered fillet. Instead, the adhesive bead was thick and rounded, causing higher than normal stresses. Although other phenomena such as adhesive outgassing effecting the strength of the paint to copper bond, may have been contributing factors in the failure, the basic problem is still excessive shear stress applied to the paint by the bulk properties of the adhesive (not the module design in this case). Calculating what maximum stress is allowable for the mirror paint over the entire life of the module is next to impossible since we do not have any information on how the shear strength of the paint-copper bond is effected with time and environmental conditions. If a module design requires bonding to the mirror, the bond should be made to the entire surface with an adhesive that has a low modulus of elasticity and a relatively low coefficient of expansion such that the shear stresses applied to the paint are much less than the shear strength of the paint to copper bond. The "explosive" separation of glass from the core occurred as a result of flaws in the glass surface(7) and an internal pressure. Since the module was sealed at ambient, thermal cycling to 49°C (120°F) generated internal pressures measured to be 10.0 kPa (1.5 psi). This pressure acting over a debonded region approximately 75 cm in diameter caused stresses on the order of 41 MPa (6000 psi) in the glass. With the flaws present in the surface, Mecholsky(7) has shown that this was sufficient to fracture the glass.

4. PROTECTIVE COATINGS

Since normal desert temperatures do

not rise much above 50°C (122°F) ambient, it apparently did not seem necessary to design adhesively joined components in mirror modules to survive temperatures in excess of 65°C (150°F). It was also desirable to use adhesives to bond mounting plates to the module that had excellent peel properties and could be applied with thixotropic agents enabling the adhesive to serve as a gap filler. The choice of this type of adhesive did not pose a problem until it was discovered that the galvanized steel pan to which the adhesive was the bonding agent, weathered in such a fashion as to absorb solar insolation selectively. Modules stowed face down would receive the full solar insolation on the weathered galvanized surface thereby allowing surface temperatures in excess of 88°C (190°F). Under these conditions, the adhesive lost most of its strength and structural failure occurred. The solution to this problem is probably as simple as painting the back side of the module with a relatively non-absorbing paint and/or using adhesives with better high temperature properties. In fact, this procedure was recommended by the supplier but, for unspecified reasons, never implemented. The use of unpainted galvanized surfaces and low temperature adhesives should, at all cost, be avoided.

5. ACCEPTABLE DESIGNS

To date, one mirror module design has survived with no apparent failures for at least 4 years in Albuquerque, New Mexico. It is constructed of a laminated mirror (mirror bonded to glass with polyvinylbutyral) with steel sections bonded to the glass. The lamination was accomplished using standard automobile glass laminating techniques, i.e., autoclaving of the laminate for as much as 16 hours. Although this design appears to be very promising, there is no laboratory data available that predicts the long term effects of heat and ultraviolet light on the PVB bond. Unfortunately, the most rapid weathering and degradation of full sized modules (sometimes as large as 1.5 m x 3.7 m) has to date occurred in outdoor exposure, and it may be that this is the only real test of future module designs.

6. SUMMARY

In all mirror module failures seen to date, the cause of failure was in some way material related. Either the materials were chemically incompatible with silver, were permeable to water, transferred too much stress to the mirror paint or possessed certain weathering peculiarities. All module designs with internal voids have shown the propensity for water intrusion, while a laminated mirror design using PVB in an autoclave process has survived in excellent condition.

7. CONCLUSION

The most promising design for a long lived mirror module would probably use laminated mirrors as the reflective surface, protected on the back side by a reflective paint to minimize heat buildup. The support structure, if at all possible would use mechanical means to anchor the mirror rather than adhesives. Although adhesive bonding of the support structure may prove to be more economical in the end, there is a need for long term outdoor testing of this type of design to bring to it a confidence level now associated with the laminated mirror reflector.

8. REFERENCES

(1) V. P. Burolla, Solar Energy Materials, 3 (1980) 117.

(2) J. L. Daniel and J. E. Coleman, Solar Energy Mat., 3 (1980) 135.

(3) L. R. Pederson and M. T. Thomas, Solar Energy Mat., 3 (1980) 151.

(4) J. E. Shelby, J. Vitko and R. L. Farrow, Solar Energy Mat., 3 (1980) 185.

(5) C. Q. Buckwalter and G. L. McVay, Solar Energy Mat., 3 (1980) 215.

(6) J. J. Dietrich, M. R. Boyaner and G. F. Pittinato, Solar Energy Mat., 3 (1980) 335.

(7) J. J. Mecholsky, Sandia National Laboratories Albuquerque, Memo to W. R. Delameter, dated May 12, 1980.

9. BIOGRAPHY

V. P. Burolla received his B.Ch.E. from Manhattan College in 1969 and his M.E. degree from Rensselaer Polytechnic in 1970. He has worked at Sandia National Laboratories since 1974 and has been involved in materials research and development for solar energy systems since 1976.

13th National SAMPE Technical Conference
October 13-15, 1981

RADIATION EFFECTS ON FOUR POLYSULFONE FILMS

Beatrice Santos and George F. Sykes
NASA Langley Research Center
Hampton, Virginia 23665

Abstract

Graphite fiber/polymeric matrix composites have emerged as prime candidate materials for structural applications in large space systems because of good strength, high stiffness, low density and high dimensional stability. However, the performance of these composites for long-term missions in the space radiation environment is unknown. Because most polymeric materials are known to undergo changes in physical and/or mechanical properties when exposed to radiation, a program was initiated to evaluate the response of candidate polysulfones to proton and electron radiation. The preliminary results show that both chain scission and cross-linking occur as a result of irradiation, and that a threshold for major property changes occurs at approximately 10^9 rads totaldose. Also, chemical characterization data show that a site of radiation interaction in the polymer chain is the oxygen bond. These results provide an indication of the radiation mechanisms responsible for material property changes in polysulfone materials.

Key. words: Polysulfones, cross-links, chain scission, oxygen bond

1. INTRODUCTION

Graphite fiber/polymeric matrix composite materials are prime candidates for use in building large space structures because they possess high stiffness and strength, low density and good dimensional stability(1). However, the durability of these materials in the space environment is of major concern, primarily because of the radiation interaction which could alter their attractive properties. In a Geosynchronous Earth Orbit (GEO), it is estimated that the interior dose of a 4-ply composite would be about 10^9 rads(2,3) during a 30-year use life. Fifteen to thirty years is estimated to be a cost effective lifetime of large space structures for communications and solar-collector applications. Because of the durability concern, a study was initiated to evaluate the response of selected candidate materials to the radiation environment. Specifically, 4 polysulfones of varying chemical structure were evaluated. Polysulfones were chosen because thermoplastics are excellent candidates for in-space fabrication(4) techniques and several of these materials have shown good radiation stability at low total doses(5,6).

The main purposes for conducting this investigation were: (1) to assess the radiation durability of four selected polysulfones, (2) to establish radiation interaction mechanisms with the polymer chain, and (3) to determine the dependence of radiation durability on chemical structure.

2. EXPERIMENTAL

2.1 Materials

The chemical structures of the four polysulfones which were evaluated are shown in figure 1. Three are commercially available: Polyethersulfone (300 P) is made by Imperial Chemical Industries, and P-1700 and Radel 5000 are Union Carbide products. The fourth, Bisphenol-A Hydroquinone is an experimental material described in reference 7. The ethersulfone moiety is common to all four structures, and therefore provides a means of comparison for the response of each polymer to radiation. The various additions to this ethersulfone moiety should provide some indication of the radiation durability of different groups in the chain.

2.1.1 Specimen Preparation. Solutions of 5% or 10% by weight of each polymer were made using the solvents as shown in Table I. Films were cast from the polymer solution in a low humidity, dust-free box. A mechanically driven glass plate was coated with the polymer solution and pulled at constant velocity underneath a Gardner knife edge set to obtain a film thickness of approximately 0.025 mm (one mil). Batch-to-batch film thicknesses varied from about 0.0125 mm to 0.025 mm. The variation in thickness within a test sample, however, was no more than $\pm 2\%$. After the film dried, it was placed in an oven and heated to remove residual solvent. The drying conditions are also recorded in Table 1. Verification of the solvent removal process was obtained by thermogravimetric analysis (TGA) of the film.

The films were then cut into 5.4-cm squares with eight slits. The slits were placed such that there were five 5-mm and three 2-mm wide strips of film available for characterization purposes. The 5-mm strips were for modulus value determinations and the 2-mm strips for obtaining glass transition temperatures. These strips were cut prior to radiation exposure because of the difficulties often encountered when working with specimens embrittled by high radiation doses.

The remaining film area was used for other characterization techniques such as infrared spectrophotometry (IR) and thermogravimetric (TGA) and chemical analyses. These techniques will be discussed in a later section.

2.1.2 Specimen Irradiation. The pre-cut films were irradiated at predetermined doses ranging from 10^8 to 10^{10} rads. This range encompasses the total dose expected in a thin laminate at GEO for a 30-year mission. However, these doses were obtained at an extremely accelerated rate of $2(10)^8$ rad/hr. Irradiation exposures were obtained using 1 MeV protons or electrons. At this energy, uniform deposition through a 0.025-mm film can easily be achieved by electrons. Protons of 1 MeV however, penetrate to a depth of about 0.017-mm in these polymers. Thus, protons could not achieve uniform deposition through some of the thicker films.

Before irradiation, all specimens were preconditioned for three days in a vacuum oven at 40°C. During irradiation, specimens were kept at approximately 20°C by a water cooled sample holder. Following exposure, the samples were packaged with dessicant material and stored at room temperature in vacuum until needed for testing.

2.2 Material Characterization

2.2.1 Thermomechanical Analyses. Young's modulus was obtained using an automated Rheovibron Dynamic Viscoelastometer operating at a frequency of 3.5 Hz. Consequently, the values reported may be slightly higher than those determined from static test methods. When testing specimens, the film strip was placed in the Rheovibron and several values were recorded. The film was then completely removed, re-inserted, and again several values were taken. This was done for each film strip in order to eliminate any error resulting from sample mounting effects. Each data point reported is an average of these values.

The glass transition temperature (T_g) of each specimen was determined using a Dupont Model 997 Thermal Mechanical Analyzer in the tensile mode[8], at a heating rate of 5°C/min in an air environment. The values reported here are apparent T_g's, determined by the intersection of the two tangents to the TMA curve.

2.2.2 Chemical Characterization. Infrared spectra were taken of the thin films in air using a model 990 Perkin-Elmer Spectrophotometer. The spectra for un-irradiated and irradiated films are superimposed in order to compare differences in the absorption bands of the two films.

Elemental analyses were performed on film samples to determine percent elemental composition as a function of radiation dose. This technique was used to attempt to establish the site of radiation interaction in the polysulfones.

2.2.3 Analysis of Thermal Stability. Thermal stability of the polysulfones was determined using a Perkin-Elmer Model TGS-2 Thermogravimetric System. For each test, approximately 3 mg of film sample was heated from room temperature to 700°C at 3°C/min in a nitrogen atmosphere. Prior to heating, each sample remained in the nitrogen atmosphere for 8 hours in order to remove the air from the heating chamber.

3. RESULTS

3.1 Radiation Effects on Thermomechanical Properties.

Figure 2 shows the effect of electron radiation on the modulus of the four polysulfones studied. The data points on the y-axis represent the modulus of the un-irradiated specimens. There is a definite increase in modulus for each polysulfone, the maximum increase occurring at the 10^{10} rad dose. Radel 5000 exhibits the greatest increase in modulus, which is approximately 75% above the original value, and P-1700 shows the least increase at about 25%. There appears to be a threshold for major increase in modulus at about 10^9 rads for each polymer. Before this threshold, the curves are relatively flat, and beyond it, they tend to increase more sharply. This increase is probably due to crosslinking occurring in the materials. Radel 5000 and P-1700 had some slight decreases in their modulus at low (10^8 rads) doses, which could be due to chain scission occurring as a result of irradiation.

The response of the polysulfones to proton irradiation is given in figure 3. For each polymer, the modulus decreased below its un-irradiated value. There is a substantial drop in modulus at low doses, and as the dose increases, the modulus tends to approach its original value. This suggests that chain scission predominates at low doses, and, once the threshold of 10^9 rads is attained, crosslinking begins to occur.

The effect of electron radiation on the glass transition temperature

(T_g) is given in figure 4. Here again, as in the previous data for modulus, it appears that each polysulfone undergoes some chain scission initially, as shown by the slight drop in T_g at low doses. Then, once the threshold of 10^9 rads is reached, the T_g begins to increase as a result of radiation-induced crosslinking.

As shown in figure 5, proton radiation affects T_g in a somewhat different manner. P-1700 and Bisphenol-A HQ are affected the least by protons, as shown by the slight decrease in T_g. The maximum decrease was about 6% for polyethersulfone.

3.2 Radiation Effects on Chemical Properties

In figure 6, two infrared spectra, each made from a separate film of polyethersulfone, are shown. This polysulfone was chosen as being representative of all four materials. The lower curve represents the portion from 1600 to 600 cm^{-1} frequency, of the IR spectrum of un-irradiated polyethersulfone. The upper curve represents that same portion of the spectrum of polyethersulfone irradiated by electrons at 10^{10} rads. The SO_2 absorption band occurs at approximately 1400 cm^{-1} which can be clearly seen as an absorption peak (an inverted peak) on the un-irradiated curve. However, at this same frequency on the spectrum of irradiated material, this peak is greatly diminished. Also, at around 1300 and 1150 cm^{-1} are the asymmetric and symmetric ϕ-SO_2-ϕ absorption bands, respectively. These bands can be seen on the lower curve, but have almost disappeared on the irradiated curve. Also, at approximately 1070 cm^{-1} the C-O-C absorption band, which appears on the spectrum for the un-irradiated specimen is absent in the spectrum for the irradiated polyethersulfone. The spectra shown here are the two extremes of the data actually obtained. The spectra of those specimens irradiated at lower doses resemble the spectra of the un-irradiated material, and, as the dose was increased, the spectra approached that of the curve at 10^{10} rads. All four polysulfones exhibited this same basic trend, with the SO_2, ϕ-SO_2-ϕ and C-O-C bands being the most significantly affected by the radiation. The proton data, not shown here, also followed this trend, but to a lesser extent.

In figure 7, percent change in elemental composition of P-1700 is plotted as a function of dose. As the electron dose is increased, a loss of oxygen is observed. This of course must be accompanied by an increase in some other element and here the increase is in that of carbon. This does not mean the carbon content increased, however, since there was no source of carbon atoms during irradiation. The threshold for major decrease in oxygen content occurs at 10^9 rads. This loss of oxygen could perhaps establish sites for crosslinks to occur. There is a decrease of about 24% of initial oxygen content for P-1700 at 10^{10} rads. The percent decrease in oxygen for those polymers not shown are: Radel 5000 with about a 10% decrease, Polyethersulfone with a 6% decrease, and Bisphenol-A Hydroquinone with a decrease of about 7%. This loss of oxygen is supported by the diminished absorption bands associated with oxygen in the IR data. The breaking of an oxygen bond by irradiation and subsequent combination of oxygen atoms to form oxygen molecules which could then be lost to the surroundings explains the decrease in oxygen content as well as the dimished absorption bands.

Figure 8 compares the samples of P-1700 irradiated by electrons and protons. The oxygen content decreases for the proton irradiated samples, but not to the extent of the electron irradiation samples. This does suggest, however, that similar mechanisms are involved in the two types of radiation.

3.3 Radiation Effects on Thermal Stability

Thermogravimetric analyses were performed on all irradiated polysulfone samples, but only polyethersulfone is shown in figures9 and 10, since all materials behaved similarly. It appears from figure 9 that as the electron dose is increased, the thermal stability decreases. That is, the onset of weight loss, and the total weight loss at 450°C increases with increasing dose. This suggests that chain scission occurred as a result of irradiation, and that low molecular weight fragments were produced as dose increased. However, the behavior of the polymer at higher temperatures (600-700°C) suggests that crosslinking also occurred as shown by the greater quantity of residue remaining for the sample exposed to 10^{10} rads. These data support the observation made previously that both chain scission and crosslinking are taking place. The samples irradiated by protons behaved in much the same way, but again, to a lesser extent as shown by figure 10. It seems that here, chain scission predominates, with little or no crosslinking occurring.

4. CONCLUDING REMARKS

In summary, there seem to be two basic mechanisms taking place as a result of irradiation. First, chain scission appears to predominate at lower doses up to approximately 10^9 rad for electron irradiated specimens of polysulfone.

Past this threshold, however, the second mechanism, crosslinking, seems to predominate. This is evidenced by the increase in modulus, glass transition temperature, and increased quantity of thermally stable residue at high temperatures.

A site of radiation interaction appears to be the oxygen bond. This is supported by the change in absorption bands associated with oxygen in the IR spectra. Any change such as the loss of an oxygen molecule would cause the vibrational frequency to change and thus, absorption bands would diminish. The chemical analysis data also suggest that the oxygen bond is being attacked by the radiation as the decrease in oxygen content attests. Also, this opening up of bonds may establish sites for crosslinking which is responsible for major property changes past the threshold of 10^9 rads.

The variations of chemical structure of the polysulfones seem to have little effect on the response to radiation. In all four cases, the oxygen bond seems to be the site of radiation interaction with the polymer chain.

From the data collected in this study, it appears that the threshold for major property change occurs at about 10^9 rad. Up to this dose level, which corresponds to the dose expected for a 15 to 30 year-mission in GEO, the polysulfones studied seem to be relatively stable to both electron and proton radiation.

ACKNOWLEDGEMENTS

The authors would like to thank Mr. Wayne S. Slemp of the Langley Research Center and Mr. Larry B. Fogdall of the Boeing Aerospace Corporation for the film irradiation and helpful discussion of results.

REFERENCES

1. Tenney, D. R.; Slemp, W. S.; Long, E. R.; Sykes, G. F., and Stein, B. A.: Space Environmental Effects on Composite Materials - Assessment and Research Needs. AIAA 12th Fluid and Plasma Dynamics Conf., Williamsburg, VA, July 23-25, 1979.

2. Brancato, E. L. and Allard, James G.: Summer General Meeting, Montreal, Que., Canada, June 24-28, 1957. Available from the American Inst. of Electrical Eng., 33 West 39th St., New York, NY.

3. Campbell, F. J.: U.S. Naval Research Lab. 148th National Meeting of the American Chemical Society, Sept. 1, 1964.

4. Poveromo, L. M.; Muench, W. K.; Marx, W.; Lubin, G.: SAMPE Journal, vol. 17, no. 1, Jan/Feb. 1981.

5. Brown, J. R. and O'Donnell, J. H.: J. of Polymer Science, vol. 19, pp. 405-417, 1975.

6. Davis A.; Gleaves, M .H.; Golden, J. H., and Huglin, M. B.: Die Makromolekulare Chemie, 129 (1969), 63-72, June 19, 1969.

7. Viswanathan, R.; Johnson, B. C.; Ward, T. C., and McGrath, J. E.: VPI and SU report to NASA, Resarch Grant NSG-1597, May 1981.

8. Burks, Harold D.: Jour. of Appl. Polymer Sci., p. 627-628, vol. 18, 1974.

BIOGRAPHY

Beatrice Santos is a graduate of Sam Houston State University where she earned a Bachelor of Science degree in Physics. She is currently working toward a Master of Science degree in Materials Science Engineering from The George Washington University, and is working at NASA Langley Research Center in the Environmental Effects Branch, Materials Division.

Table I

Polysulfone	Solvent	Concentration (%)	Oven Temp (°C)	Atm	Drying Time (Hr)
Polyther-sulfone	DMAC	10	225	vac	12
Radel 5000	TCE	5	200	vac	12
P-1700	$CHCl_3$	10	150	N_2	1
Bisphenol-A HQ	DMAC	10	200	vac	12

DMAC - N, N - Dimethylacetemide

TCE - Sym - Tetrachloroethane

$CHCl_3$ - Chloroform

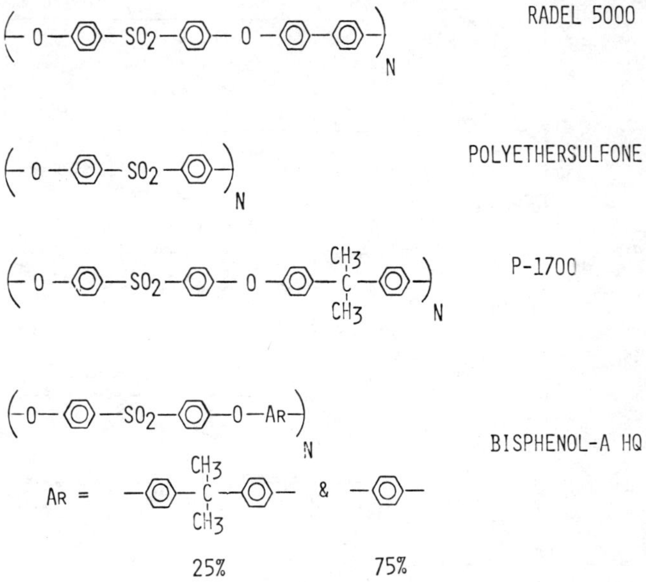

Figure 1.- The Chemical Structures of Four Polysulfones.

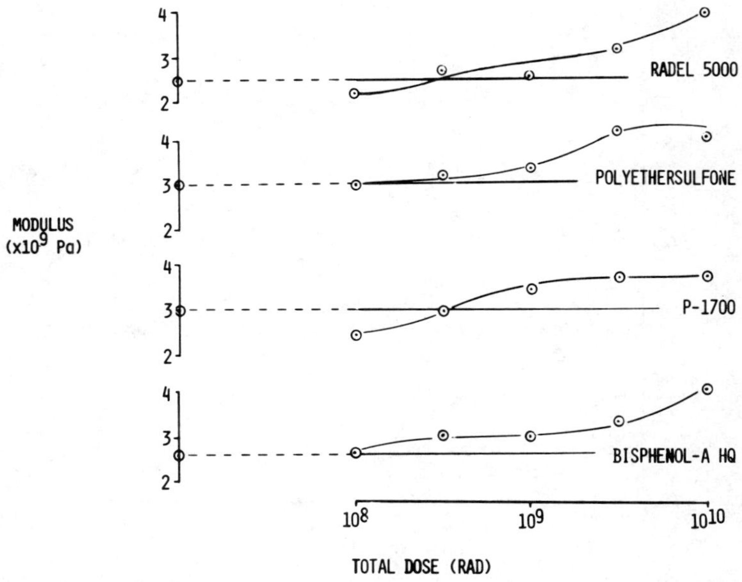

Figure 2.- The Effect of Electron Radiation on Modulus.

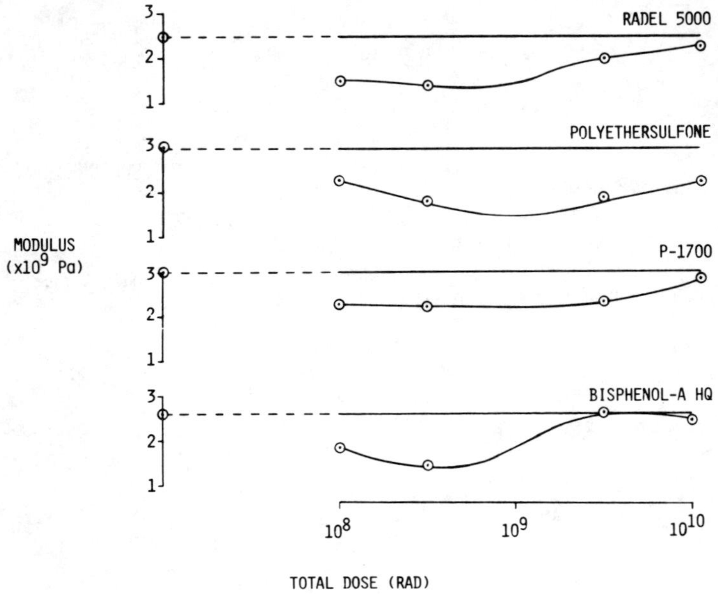

Figure 3.- The Effect of Proton Radiation on Modulus.

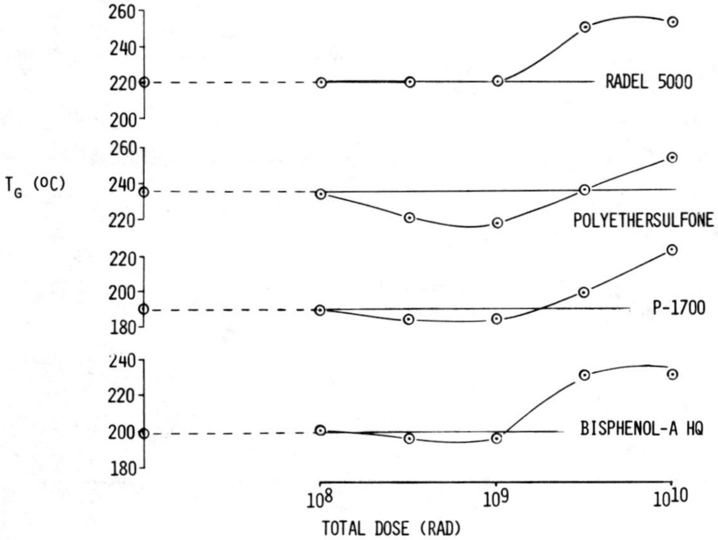

Figure 4.- The Effect of Electron Radiation on Glass Transition Temperature.

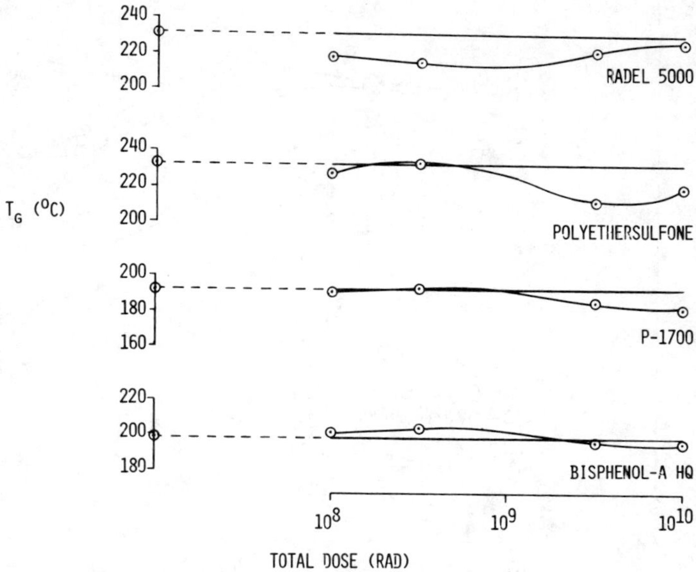

Figure 5.- The Effect of Proton Radiation on Glass Transition Temperature.

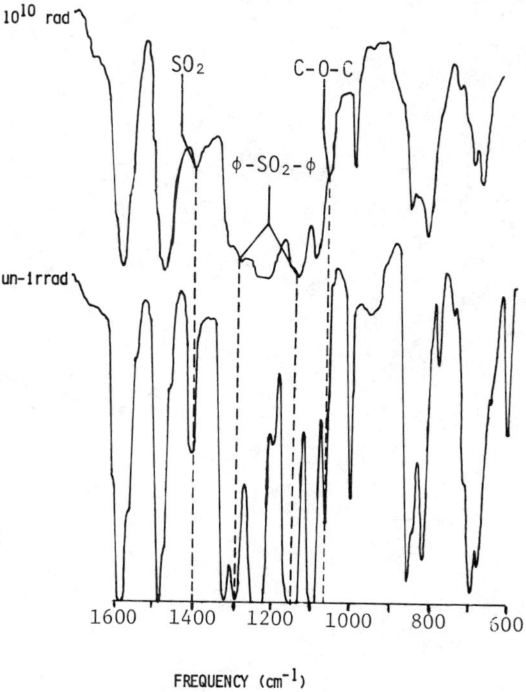

Figure 6.- The Effect of Electron Radiation on the Infrared Spectra of Polyethersulfone

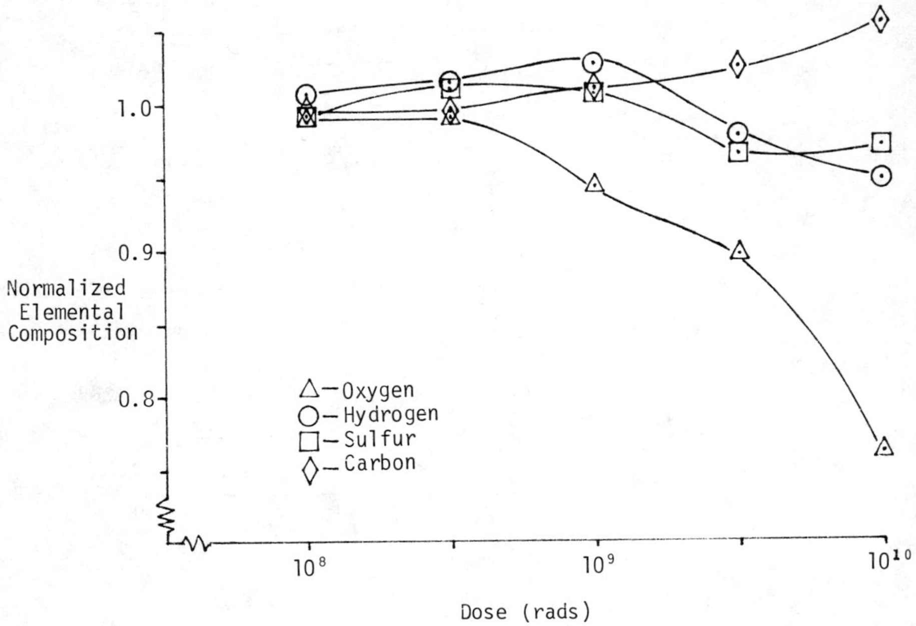

Figure 7. The Effect of Electron Radiation on the Normalized Elemental Composition of P-1700

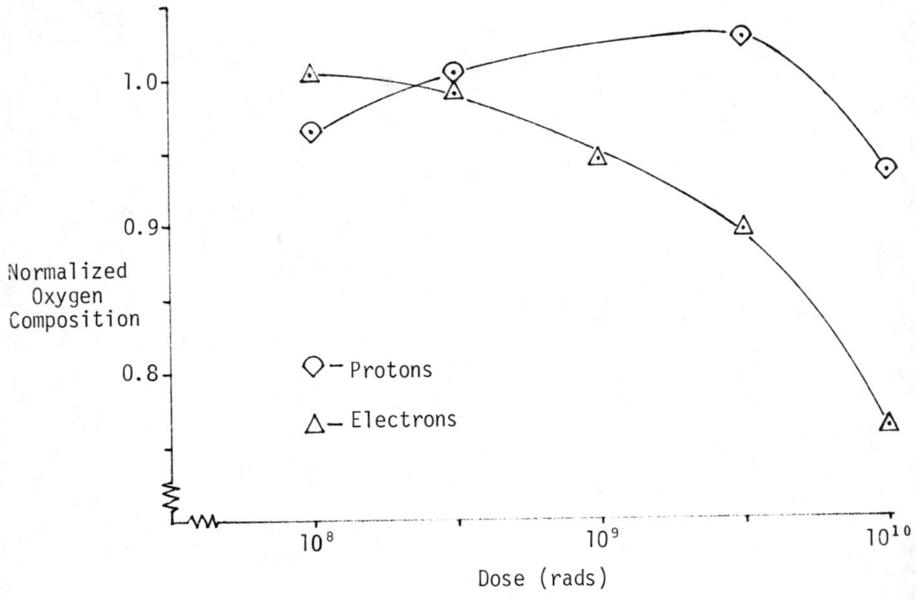

Figure 8. The Effect of Radiation on the Normalized Oxygen Content of P-1700

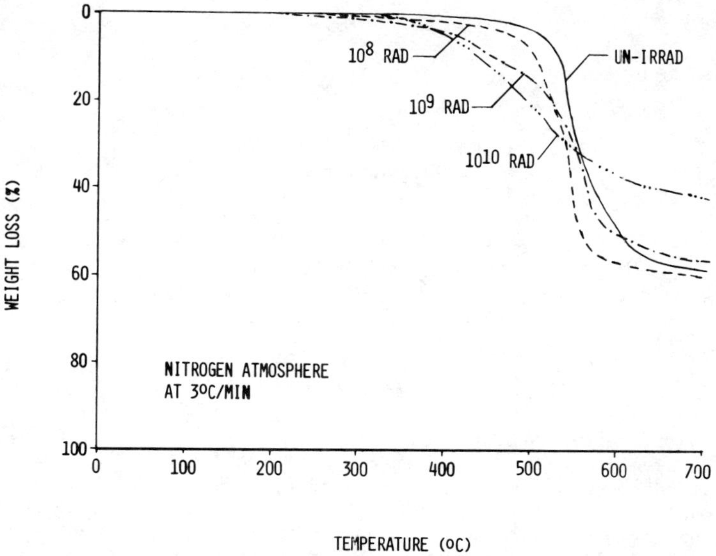

Figure 9.- The Effect of Electron Radiation on Thermal Stability for Polyethersulfone

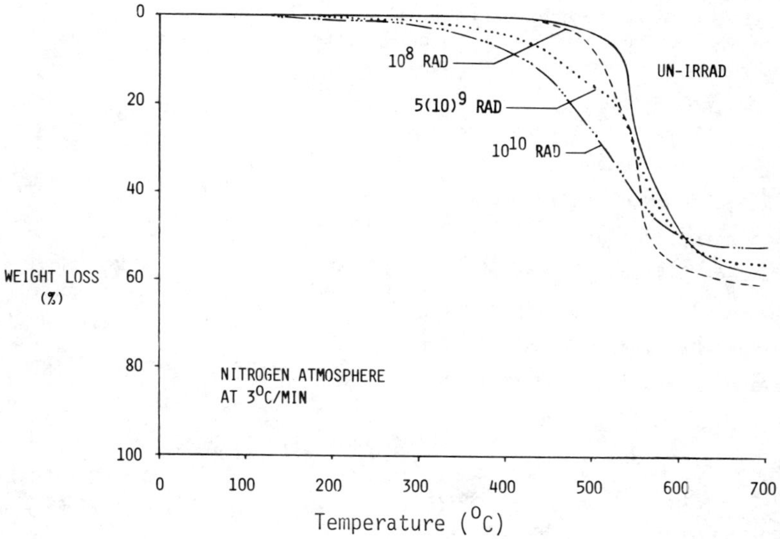

Figure 10.- The Effect of Proton Radiation on Thermal Stability for Polyethersulfone

13th National SAMPE Technical Conference
October 13-15, 1981

INFLUENCE OF SPACE ENVIRONMENT ON ADHESIVES

B. J. Mulroy, Jr., and Y. D. Izu
Materials & Processes Engineering
Space System Division
Lockheed Missiles & Space Co., Inc.
Sunnyvale, California

Abstract

A multi-year study was recently completed at LMSC concerning adhesives, processing and environmental effects on bi-material adherend bonding. The study identified: bonded-joint requirements for space vehicle hardware; identified adhesive candidates and screened candidates by evaluation of neat resins; identified and screened adherend surface pre-bond treatment; evaluated bond strengths at elevated temperatures for selection of adhesive systems for long-term environmental exposures; assessed capability of selected adhesives to cure under various temperature-time schedules representative of design and manufacturing constraints, and evaluated selected adhesive/adherend bonded joints after extended thermal-vacuum cycling exposure to simulated ten-year spacecraft orbital life space environments. The influence of simulated space environments on the bonded systems and the related materials and processes are described.

1.0 INTRODUCTION

An increase in functional and environmental performance requirements for spacecraft structures leads to more frequent design applications using advanced organic-matrix composites. Spacecraft structural designs are generally based on high strength, high stiffness and minimal weights. Operational orbital lifetimes are being increased to the seven to ten year range, necessitating increased reliability to meet these requirements. Cyclic thermal-vacuum excursions in temperature environments from -250° to +250°F may be encountered and are design requirements for certain applications. Designs of many specialized spacecrafts are concerned primarily with one-or-two of a kind type of production, and must recognize manufacturing considerations for such relatively small runs. Dimensional structural stability for optical sensors, antennas and other devices require very low (approaching zero), coefficient of thermal expansion (CTE). This low CTE design requirement necessitates that the spacecraft materials remain stable in the space environment.

The materials and processes for the spacecraft structures must complement the vehicle design. Compatible adhesive bonding techniques to provide reliable joints-to-metallic structures are needed for the effective application of the advanced organic-matrix composites to spacecraft hardware. The mismatch of CTEs between the composites and metallic adherends produces stresses in the bondline during the thermal cycling environment encountered in spacecraft operation. Appropriate materials selection, together with controlled surface preparation, bonding and curing techniques are required to achieve structurally reliable joints.

This paper presents the results of a multi-year study at LMSC on bi-material adherend bonding. The study identified: bonded-joint requirements for spacecraft design, spacecraft adhesive candidates, adhesive properties, surface bond treatments, and joint strengths under various environmental exposures. The influence of simulated space environments on bonded joints and the related materials and processes are described.

2.0 MATERIAL REQUIREMENTS AND STUDY DESCRIPTION

Because of the performance characteristics needed, the bonding materials are usually epoxy adhesives. Some typical adhesive requirements for spacecraft structural designs are:

- High glass transition temperature (Tg) or heat distortion temperature (Hd) to withstand elevated temperature in space.
- Low temperature cure to minimize the thermal stresses in dissimilar material joints in the low temperature space environments.
- Ductility to withstand mismatched CTE's of dissimilar materials.
- Capability of joining dissimilar materials such as: graphite or Kevlar composites; and magnesium or titanium metallic adherends.
- Low outgassing to prevent contamination of the critical electrical, optical or thermal surfaces.

Using these criteria, a multi-year evaluation sequence was initiated for the adhesive and bonding processes study as shown in Figure 1. The study was essentially a three-phase evaluation. Phases I and II were concerned with identification of requirements, materials selections, surface treatments, cure evaluations, and initial material-bond evaluations. Phase III was directed toward conducting a simulated, ten-year, thermal-vacuum, space environment exposure and evaluation of selected adhesive/adherend joints.

3.0 DISCUSSION

Description, evaluations and the results of the three-phase study were as follows:

3.1 <u>Phase I Adhesive Raw Material Evaluation</u>.

'New' generation adhesives as well as the old standby adhesives were

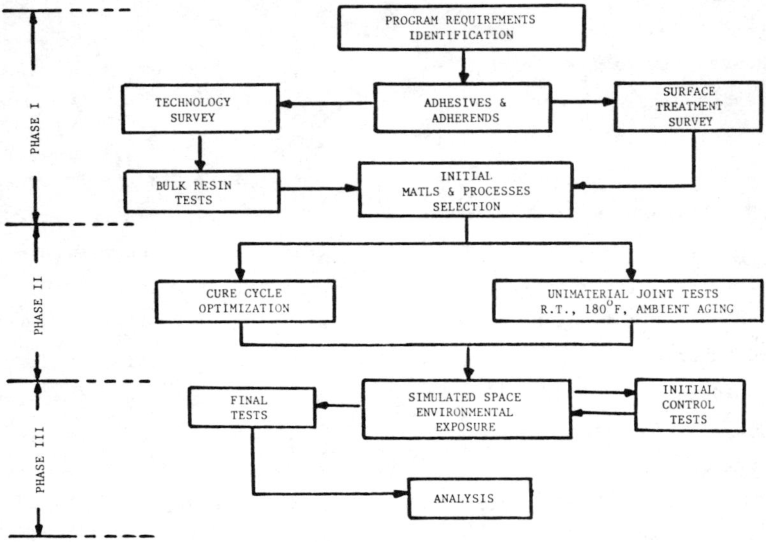

FIGURE 1
MULTI-YEAR EVALUATION SEQUENCE FOR
ADHESIVE AND BONDING PROCESSES

evaluated in Phase I. The candidates, all epoxy adhesives, were screened to those listed in Table 1. The epoxies are the primary adhesives used in spacecraft designs because of their structural, temperature capability, environmental resistance, and thermal-physical properties. These included both paste and film epoxies, which cure at room or elevated temperatures. Phase I evaluations included the determination of mechanical and physical properties of cast (paste) and laminated (film) panels. The description and test results of the evaluation were previously presented (Reference 1). Based on a combination of physical properties (vacuum outgassing, density, glass transition temperature, coefficient of thermal expansion and water absorption) and mechanical properties (tensile strength and elongation), eight adhesive systems were selected for Phase II candidates (Table 2).

3.2 Phase II Surface Preparation and Screening Evaluations.

For further screening of the selected Phase I adhesives, bonded joint configurations were prepared and tested. Bonded aluminum, lap shear joints at room temperature and at 180°F were tested. The description and results of the tests performed previously were presented (Reference 2). Based on the bonded joint test results, specific adhesive systems were selected for the bimaterial joint testing in a thermal vacuum simulated space environment.

MANUFACTURER		CURE TEMPERATURE
Adhesive Engineering	Aerobond 2143	Room Temp.
American Cyanamid	FM 53	250°F
	FM 73	250°F
	FM 400	350°F
Amicon	A 452	210°F
Armstrong	C7/W	Room Temp.
B. F. Goodrich	PL 717A	250°F
Furane	Epibond 123	150°F/R.T.
Hexcel	Hexabond 111	260°F
Hysol	EA 934	Room Temp.
	EA 9309	Room Temp.
	EA 9313	Room Temp.
	EA 9320	Room Temp.
	EA 9321	Room Temp.
	EA 9601	250°F
	EA 9602.3	250°F
	EA 9628	250°F
Narmco	Metlbond 329	350°F
	Metlbond 1133	250°F
3 M	AF 126-2	250°F
	AF 143	350°F
	AF 163K	250°F/P.C. @ 350°F
	AF 163-2M	250°F/P.C. @ 350°F
	EC 2214	250°F

TABLE 1 - ADHESIVE MATERIALS EVALUATED

CATEGORY	HIGH TENSILE	HIGH ELONG
Room-Temp Cured Paste	EA 9321/ EA 9320	EA 9313
Elevated-Temp Cured Paste	EC 2214	EC 2214
250°F Cured Film	FM 73	EA 9628/ FM 73
350°F Cured Film	M 329	AF 143

TABLE 2 - CANDIDATE ADHESIVES SELECTED

3.2.1 Adherend Surface Preparation

Achievement of maximum bond joint strengths are significantly influenced by the adherend surface preparation. Various surface preparations for both metallic and non-metallic composite substrates were selected for the evaluation.

- Chromic-sulfulric acid etch versus the Boeing phosphoric acid anodize (PAA) for aluminum substrates.
- Dow 17 anodize versus Dow 19 finish for magnesium substrates.

- Stabilized phosphate fluoride etch versus Pasa Jel[R] etch for titanium substrates.
- Nylon peel ply versus nylon peel ply plus sand blasting for epoxy composite substrates.

The results of the evaluation performed indicated: (1) PAA was the best preparation for aluminum; (2) the stabilized phosphate fluoride etch was the best for titanium; (3) Dow 17 anodize best for magnesium; and (4) peel ply plus sand blasting was the best for woven epoxy-composites with peel ply only for unidirectional composites.

3.2.2 Bonded Joint Evaluation. The adhesives selected from Phase I studies were used to prepare aluminum-aluminum and aluminum-composite bonded joints. The adherend surface preparations selected from the evaluation described in Section 3.2.1 were used. The lap shear bonded joint specimens were tested at room temperature and at 180°F. Results of these tests were previously presented (Reference 2). Based on mechanical properties achieved and processing considerations, the five adhesives listed in Table 3 were selected for the Phase III evaluation.

3.2.3 Adhesive Cure Studies. To complete the Phase II evaluation, adhesive cures were studied. Lowering of the cure cycle temperature was investigated for possible reduction of the thermal stresses in bimaterial joints. The thermal stresses in a joint are a result of the stress free (cure temperature), of the adhesive and the greatest temperature delta (-250°F) experienced in spacecraft applications.

$$\text{Thermal Stress} = \frac{E_1 E_2 \Delta T(\alpha_1 - \alpha_2)}{E_1 + E_2}$$

Where: E = Modulus
T = Temperature
α = Coef. of Thermal Expansion

The results indicated that a 200°F cure was possible with the 250°F curing films with some sacrifice in strength, but the 350°F films required a 350°F cure to achieve optimum mechanical strength performance. Based on the results, Phase III incorporated the vendor's suggested cure cycle for each individual adhesive system to achieve optimum strength properties in the bi-material joint.

3.3 Phase III Space Environment Simulations and Testing. The Phase III portion of this evaluation concluded the multi-year adhesive study. Bi-material bonded joints were thermal-vacuum cycled to simulate a ten-year spacecraft orbital lifetime. The joints were fabricated, bonded, thermal-vacuum cycled and tested to determine thermal degradation effects, if any, on bimaterial joints. The simulated ten year space environment lifetime consisted of cyclic temperature extremes of $\pm 250^{o}$F and 10^{-5} torr vacuum. A total cycle time of two hours with a twenty minute dwell at temperature extremes was used to

CATEGORY	ADHESIVES	CURE CYCLE
Room Temperature Cured Paste	Hysol EA 9320 Hysol EA 9321	Room Temperature, 7 Day Minimum
250°F-Cured Film	American Cyanamid FM 73 Hysol EA 9628	1 h, 250°F
350°F-Cured Film	3 M Company AF 143	1 h, 350°F

TABLE 3 - ADHESIVES SELECTED FOR PHASE III

expose the bi-material joints to thermal stress conditions from the simulated space environment. In this particular study, 3,650 cycles at the $\pm$ 250°F temperature extremes and 10^{-5} torr vacuum represented the ten-year lifetime.

3.3.1 Thermal-Vacuum Testing. Long term thermal-vacuum exposure to simulate space exposure on bi-material joint specimens was completed in 1980. A thermal-vacuum test chamber (Figure 2) was modified to provide the required environment and accommodate the large quantity of test specimens. An environment of -250 through +250°F and 10^{-4} torr vacuum was provided. A total cycle time of two hours with a twenty minute dwell at the temperature extremes was selected to simulate a ten-year orbital lifetime within a reasonable (one year) test period. An analysis of current and potential space vehicle missions determined the cycle

FIGURE 2: SIMULATED THERMAL-VACUUM SPACE ENVIRONMENT CHAMBER

temperature extremes. The temperature range represents a more severe total temperature environment than normally expected.

3.3.2 Mechanical Test Results. To obtain trends and determine the influence of exposure time, specimens were removed from the chamber after simulated 3 1/3, 6 2/3 and ten year exposures. The materials evaluated in the long term tests are presented in Table 4. Each of the metal adherends was bonded to each composite adherend with each of the adhesives.

METAL ADHEREND	ADHESIVE	COMPOSITE ADHEREND
Titanium (6 Al, 4V)	EA 9320 EA 9321	Kevlar/X904B HMC 330C/934 [a]
Aluminum (2024)	EA 9628	T50(PAN)/X904B [a]
Magnesium (ZK60)	FM 73 AF 143	T50(PAN)/X904B

TABLE 4 - BIMATERIAL JOINT TEST SPECIMEN MATRIX
(a) With 1581 glass interface graphite-metal-transition layer.

Lap shear bi-material joint specimens prepared from the adherend/adhesive/composite combinations of Table 4 were subjected to the thermal-vacuum cyclic exposure. Lap shear strengths of three (3) bonded joints for each combination were determined at room and at elevated temperature ($180^{O}F$). The results of the tests performed are presented graphically in Figure 3 (Room Temperature Cured Paste Adhesives EA 9321 and EA 9320), Figure 4 ($250^{O}F$ Cured Film Adhesives FM 73 and EA 9628), and Figure 5 ($350^{O}F$ Cured Film Adhesive AF 143).

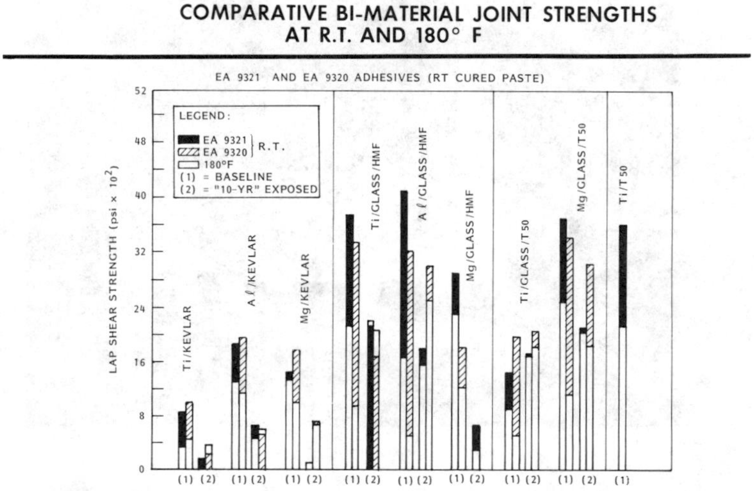

FIGURE 3 - ROOM TEMPERATURE CURED PASTE ADHESIVES EA 9321 AND EA 9320

COMPARATIVE BI-MATERIAL JOINT STRENGTHS AT R.T. AND 180° F

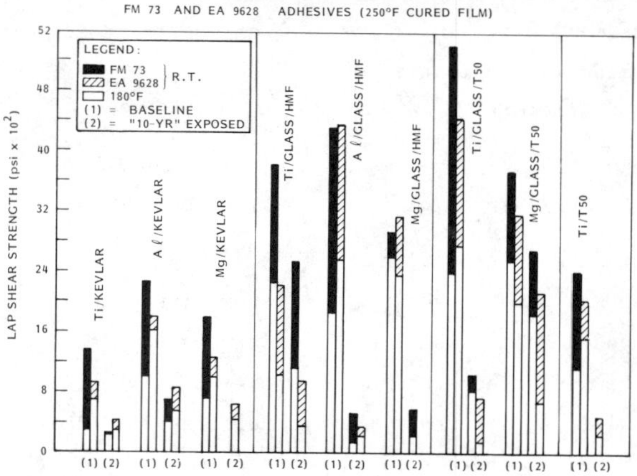

FIGURE 4 - 250°F CURED FILM ADHESIVES FM 73 AND EA 9628

COMPARATIVE BI-MATERIAL JOINT STRENGTHS AT R.T. AND 180° F

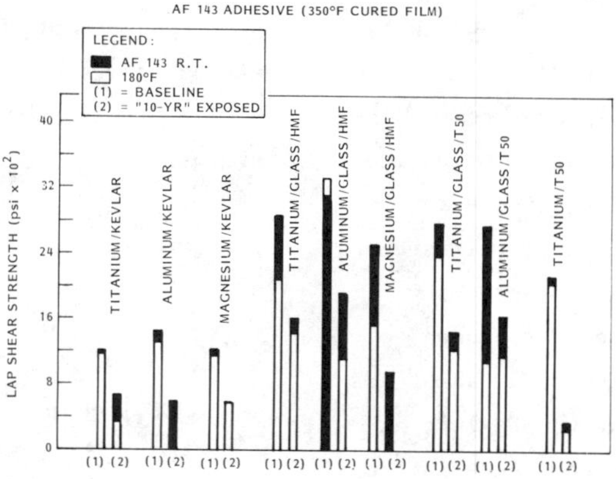

FIGURE 5 - 350°F CURED FILM ADHESIVE AF 143

Thermal-vacuum cycling of the bi-material joints caused the Kevlar adherend to delaminate with a maximum strength of 850 psi at room temperature after the simulated ten year exposure. T-50 graphite adherends were also seriously degraded when bonded directly to titanium.

This problem appeared to be resolved with the addition of fiberglass as an intermediate layer. This layer accommodated large strain differences without microcracking while serving as a galvanic inhibitor. The bond with the magnesium adherend, although utilizing the fiberglass layer concept, was also degraded. However, examination indicated failure in the magnesium-adhesive interface. This failure mode was indicative of inadequate adhesive wetting on the Dow 17 coated surface.

4.0 SUMMARY

The most significant degradation effects of thermal-vacuum cycling bi-material bonded joints were observed on the Kevlar adherend bonded to metallic substrates. The Kevlar delaminated at the adhesive interface with a maximum room temperature strength of 850 psi after the simulated ten year exposure. Delamination of the T-50 graphite composite adherend when bonded to titanium was also observed. This problem appeared to be resolved by adding a higher expansion intermediate (fiberglass) layer to the low expansion graphite/epoxy composite. It appeared that the fiberglass can accommodate large strain differences without the resin microcracking. The magnesium adherends bonded to fiberglass-graphite/epoxy adherends were degraded, but the failure exhibited itself at the magnesium interface. This is indicative of inadequate wetting by the adhesive on the Dow 17 coating for magnesium. Since the 350°F adhesive wet the surface better than the lower temperature adhesives, the 350°F adhesives are preferred for the magnesium bonding applications.

For these particular tests and environments, the failure surface was in the composite adherend. This indicated that the adhesive layer was not the design failure criteria.

For space thermal cyclic environment applications, bi-material bonded joints should have a layer of fiberglass intermediary between the composite adherend and the adhesive.

Promising magnesium-composite bonding has been accomplished by a double-adhesive process described in Reference 3. This process utilizes a primary 250°F adhesive cured to the magnesium surface which is then bonded to the composite adherend with a secondary room temperature cured adhesive. Although bonded joints from this process had not been evaluated in the simulated space environment, initial investigations have indicated excellent potential utilization for space applications.

5.0 REFERENCES

1) "Adhesive Bonding for Space Applications", B. J. Mulroy, Jr. and D. M. Mazenko; 10th National SAMPE Technical Conference, Kiamesha Lake, New York; October 17-19, 1978.
2) "Structural Adhesives for Space Systems", B. J. Mulroy, Jr. and D. M. Mazenko; Structural Adhesives and Bonding Conference; March 13-15, 1979; El Segundo, California.
3) "Adhesive Bonding Process For Space Structures", Jose L. Camahort and B. J. Mulroy, Jr.; 24th National SAMPE Symposium

3) and Exhibition; San Francisco, California; May 8-10, 1979.

6.0 BIOGRAPHY

Bernard J. Mulroy, Jr.

Mr. Mulroy is a Materials and Processes Specialist with the Advanced Technology Laboratory of the Materials and Processes Organization. Mr. Mulroy was the project leader for the Independent Development Task 'Adhesive Bonding.' He received a B.S. in Chemistry at West Liberty State College in West Virginia in 1967. He has been involved in adhesive bonding, materials testing and advanced composites since he joined Lockheed in 1967.

Y. Douglas Izu

Mr. Izu is a Senior Staff Engineer and directs the effort of the Advanced Technology Laboratory of the Materials and Processes Organization. He received a B.S. in Chemical Engineering from Stanford University in 1950. He has been involved in the development of organic and inorganic composite materials/structures since joining Lockheed in 1961.

13th National SAMPE Technical Conference
October 13-15, 1981

COMPOSITE CONTROL TUBES

Sam C. Aker and Walter G. O. Sonneborn
Bell Helicopter Textron
Fort Worth, Texas

Abstract

This paper describes the design and development of composite material push-pull control tubes to replace aluminum tubes in helicopter flight control systems. A three-year development and FAA certification program resulted in lower weight, reliable and potentially lower cost hardware that will be fielded first in the Bell Helicopter Textron (BHT) Model 214ST transport helicopter. Mechanized manufacturing methods for the production of these tubes are described.

1. INTRODUCTION

The flight control systems of most contemporary helicopters use metallic push-pull control tubes in connection with bellcranks to transfer control motions from the pilot's control sticks to the flight control actuators. In a typical 10,000-lb GW helicopter as many as 30 control tubes are used which vary greatly in length and diameter, depending on application. In military helicopters, the number can be almost twice this because of dual systems installed for ballistic survivability. For its complete fleet of helicopter models, BHT is currently manufacturing 300 different metallic tubes. Because of this large number of individual parts and a relatively high scrap factor during the swaging and subsequent threading operation at both ends of the tube, the manufacturing cost remains high. Additional cost is added by the need to apply a reliable corrosion protection system on the outsides and insides of the tubes.

A program to reduce cost was initiated in 1978 and led to the choice of filament wound composite tube construction. This paper describes the manufacturing and tooling concepts that were developed and the analyses and design support tests that guided the final design. Projected inventory simplifications, the FAA certification process, and production quality assurance procedures are also outlined together with the estimated cost and weight reductions.

2. DESIGN

2.1 Design Requirements

Helicopter control system push-pull tubes serve the function of transmitting control motions from the pilot's control sticks to the hydraulic boost actuators. Normal operating loads are 5 to 15 lbs which are needed to overcome system friction and to move the actuator servovalve. The tubes are all supported in a pinned-pinned condition, and hence are subjected to axial tension or compression. The design load according to FAR 29 is 1-1/2 times the maximum pilot's effort to cover the case of a mechanical jam in the system that a pilot is attempting to overcome. Maximum pilot effort at the grip of the stick is defined as 100 lbs and is typically multiplied by 10 to 1 through the mechanical advantages within the system.

The tube must provide for adjustment in length for the purpose of rigging the control system in the helicopter. A practical service consideration is that the tubes will be subjected to impacts from small tools. Any damage caused by such impacts that is not readily discernible must not render the tube unserviceable.

The above outlined design requirements must be met under environmental conditions from $-65^{o}F$ to $160^{o}F$ and from dry to 95% humidity. In addition, corrosion was to be eliminated. No consideration was given to lightning protection since the tubes are currently installed inside metallic airframes which are considered to give adequate protection.

2.2 Design Approach

A full set of control tubes for the BHT Model 214ST (see Figure 1) was designed in standard metallic tubing to establish a reliable baseline. Subsequently, a design for composite material tubes was developed and the two systems compared. The external dimensions of both systems remained

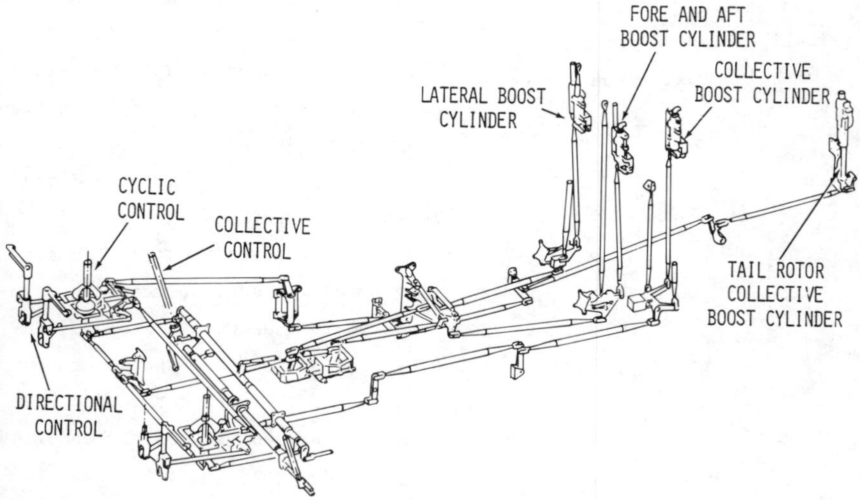

Figure 1. Control System Schematic for 214ST Helicopter.

the same to allow changeover and also to provide compatibility with other BHT models. The majority of the control tubes under consideration were critical in Euler column buckling with pinned-pinned end conditions. This condition immediately led to the choice of graphite fibers as a highly efficient structural material with high stiffness/density ratio. Also, the fibers are aligned with the longitudinal axis of the tube with the addition of a small amount of radial reinforcement to prevent local fiber buckling and to add to low energy impact resistance.

With manufacturing cost as the prime consideration, design tradeoffs were conducted to determine the lowest cost configuration. Basic tubing could be produced more easily by a helical tape wrapping technique; however, the need for separately bonded or fastened end fittings made this approach unattractive. The chosen design was a tube produced mechanically by automatic filament winding and integrally attached metallic fittings with internal threads (see Figure 2).

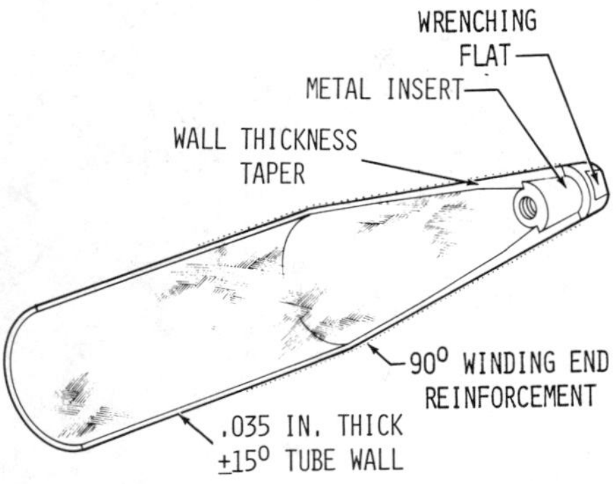

Figure 2. Cross Section of Tube End.

The winding angle in the main body of the tube is ±15 degrees relative to the lenghtwise axis, which meets strength and producibility requirements. In the tapered end section, a 90-degree overwrap was required to accommodate the local kick loads. This wrap extends 1 in. beyond the tapered section onto the cylindrical section of the tube (see Figure 3).

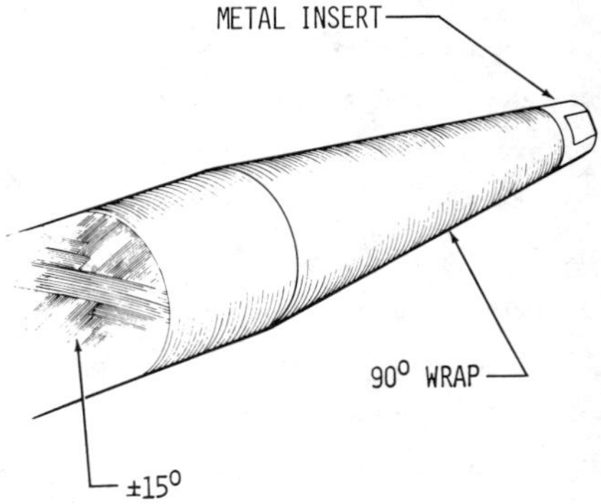

Figure 3. Gr/E Control Tube End.

A total of seven different tube diameters (as used in the metal designs) was chosen because of geometry requirements. All tubes were equipped with the same attachment fitting machined from corrosion resistant steel and threaded to 5/16 in. diameter, left- or right-hand. The fitting was wedge-shaped in order to have a tension lock for the end and contains a wrenching flat for tightening lock nuts. In addition, an end flange was included to react the compression load as well as to provide a bearing surface for the adjustment lock nut of the mating threaded end clevis. The geometry of the end fitting was designed to contain the increasing wall thickness of the end of the tube section and to conform to the tapered ends of the alternate aluminum tubes. This provided proper clearance at the existing bellcrank interfaces.

Complete tube assemblies are constructed from two semitubes, which are connected at the center by an internal sleeve that is bonded with a two-part liquid epoxy adhesive. No mechanical fasteners are used. The graphite epoxy tubes are approximately half the weight of the baseline metal tubes.

2.3 Design Support Testing

During the proof-of-concept phase of the program, a 44-in. long, 1.25-in. diameter tube was determined to be most representative of the complete set. The design ultimate load for this tube was 1275 lbs in combination with the previously stated environmental conditions.

Initial compression tests of half-length tubes revealed failure at only 800 lbs load, due to fiber buckling at the transition point from the cylindrical to the conical section. The addition of a 90° overwrap increased the compression failure load to 3000 lbs and full-length tubes were readied for test.

With compression loads applied through self-aligning bearings, the tubes typically began to bend and the load-deflection curve to flatten out at about 2200 lbs, or 172% of design ultimate. Additional compression tests showed this behavior to be repeatable. Tension loads were much higher, frequently limited by the attachment bearing capability. Tests of tubes conditioned for four weeks at 160°F and 75% relative humidity further verified the soundness of the design.

Tolerance to low-energy impact was tested by dropping a tube from an 8-ft height onto a steel rail. A clearly noticeable crushed spot near the tube mid-point resulted and the tube was then loaded in compression. Buckling occurred at 1235 lbs; i.e., 97% of design ultimate. When the load was removed, the tube self-straightened and returned to a cylindrical shape. The test was repeated twice on the same tube with the same results; thus, demonstrating a remarkable tolerance for impact damage. In addition, a compression and tension test at -67°F was performed with no discernible difference in the tube's performance.

A set of design curves for the range of sizes for graphite/epoxy and aluminum tubes is shown in Figure 4.

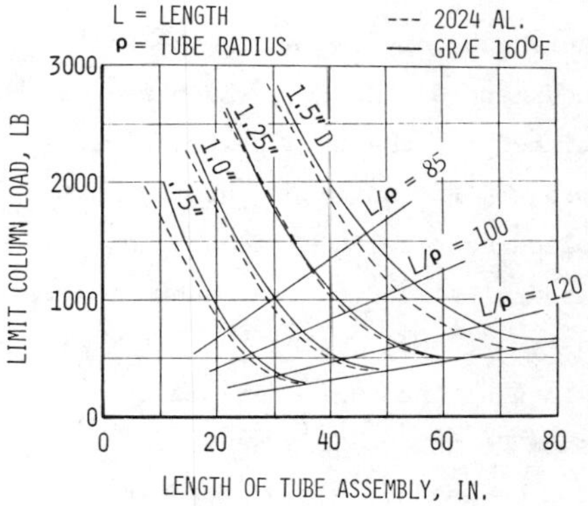

Figure 4. Limit Load Versus Length for Control Tubes.

The composite control tube program was committed to manufacturing, and the tubes are now in the process of being FAA certificated for the Model 214ST.

3. MANUFACTURING

The approach selected for quantity manufacture of the composite control tubes is described below.

The tubes are wound on a steel mandrel, which is tapered on each end. The inserts were installed at the ends of the tapered section and held in place by threaded fittings containing a stop-shoulder at the turn-around point on the insert (see Figure 5).

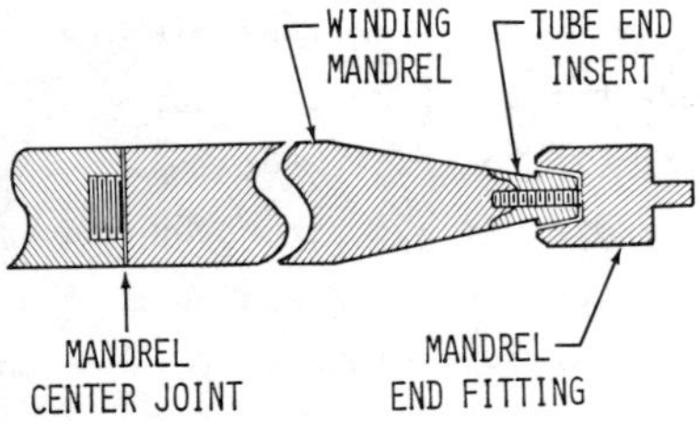

Figure 5. Winding Mandrel Detail.

After tube winding is completed, the tube is cut at the center and the two halves are pulled off the mandrel.

Curing occurs in lengthwise split and matched fixtures which are designed for use in a heated platen press. Pressure for cure is supplied by a solid silicone rubber mandrel placed in the inner diameter of the tube. Curing tools for all sizes of the tubes were made the same height so that several tools can be loaded in the press at one time. At the tapered end of the tool, a threaded plug was inserted to hold the metal end fitting in place and in alignment. The open end of the tool cavity has an adjustable end plug to trap the rubber mandrel and to adjust for shrinkage of the mandrels after repeated use. After cure, the tube sections are radiographically inspected and a dimensional check is made. The sections are then ready for final assembly. This involves cutting to length and preparing the internal diameters for bonding to the precured splice doubler tubes.

The complete assembly is placed in the assembly curing fixture and clamped in position until the adhesive has cured.

4. CERTIFICATION TEST PROGRAM

A certification test plan for all tubes was prepared and presented to the FAA for approval. The tube with the lowest design margin was selected for the plan and a series of tests of this configuration were performed. The remaining tubes will be certificated by similarity. The tests chosen established the repeatability of the manufacturing process as well as the adequacy of the design procedure and the quality control program. These tests include the various load conditions, environmental exposure levels, and load tests before and after simulated handling damage at varying levels of intensity. Table I shows the test matrix for the damage tolerance certification. Initially, all

TABLE I. TEST MATRIX FOR GRAPHITE/EPOXY CONTROL TUBES							
DAMAGE TOLERANCE							
Type of Test	Baseline Control Phase I	Number of Tests At Each Condition					
		M_1			M_2		
		V_1	V_2	V_3	V_1	V_2	V_3
Nondestructive Inspection (NDI)*	0	4	4	4	4	4	4
Static Tests							
Tension: 75°F Dry	5	2	2	2	2	2	2
Compression: 75°F Wet	5	2	2	2	2	2	2
TOTAL SPECIMENS	10	24					

M = Mass $\quad M_1$ = 7.97 oz $\quad V_1$ = 15 ft/sec

V = Velocity $\quad M_2$ = 31.42 oz $\quad V_2$ = 20 ft/sec

V_3 = 25 ft/sec

*Specimens will be subjected to NDI three times:

1. before wet conditioning
2. after wet conditioning
3. after damage

Wet specimens will be conditioned 7 days at 160°F at 95% humidity to obtain an equilibrium moisture content equivalent to 85% relative humidity.

production tubes will be subjected to a static proof load at limit load condition. It is planned to reduce the test frequency to a sampling procedure as more production experience accumulates.

The tooling system chosen (i.e., matched metal dies with internal pressure) assures a smooth exterior surface. This is not only aesthetically desirable but also for field inspection purposes. Any surface roughness occurring in service would be an indication of possible fiber damage due to mishandling.

One simple way of detecting such damage is by stroking the surface of the tube with a cheesecloth pad. Even small amounts of protruding fibers will snag the loose threads and alert the operator to potential tube damage.

5. COST CONSIDERATIONS

An analysis of costs associated with the manufacture of composite control tubes suggests that more work is needed to achieve the anticipated cost savings. Currently, low production levels, limited tooling, and the use of a general purpose winding machine show no cost savings over metal tubes. However, with improved tooling, a specialized winding machine, and increased production levels, savings in labor costs can be achieved. In addition, raw material inventory can be reduced since stocking of various diameters of metal tubing will not be required. Spares inventory is also projected to be simplified as different lengths of composite tubes can be bonded from the same stock within a short time.

6. CONCLUSIONS

1. Filament wound graphite epoxy control tubes have been developed successfully for the BHT Model 214ST helicopter.
2. Weight savings for representative medium length/medium diameter tubes are approximately 50% relative to current aluminum tube designs.
3. Inventory simplifications for spares can be achieved by the two-piece construction of the tubes.
4. Impact damage to tubes that results in strength reductions below design limits is readily detectable by visual and tactile means.

7. BIOGRAPHIES

Sam Aker is a Staff Specialist of BHT's adhesives and composite materials. He is responsible for development of bonded metal structures and development and application of nonmetallics such as reinforced plastics and honeycomb. He is assigned to the composites task group as leader in a program to develop advanced composites technology as applied to helicopters. Mr. Aker joined BHT in 1953 in the Methods and Materials Research Engineering. He participated in the development of the Model 47 bonded metal tail rotor and main rotor blades. Also, in the development of rotor systems for the UH-1 series, XV-3, and other experimental rotor systems including deicing and anti-icing systems. Mr. Aker's entire professional career has been concerned with experimental processes. He holds a BA in Chemistry from the University of Texas and has taken additional courses at Columbia University and Texas Christian University.

Walter Sonneborn is Director of the U.S. Army/Bell Helicopter Textron Advanced Composite Airframe Program (ACAP), including administration, engineering, operations, product assurance, and interface with BHT's major subcontractor, Grumman Aerospace Corporation. Mr. Sonneborn, who joined BHT in 1965, most recently was Manager, Research and Advanced Technology Projects. He was Project Engineer in 1969 for the U.S. Army AVLABS/Bell High Performance Helicopter (HPH) program during which a modified Huey set an unofficial world helicopter speed record of 316 mph, which still stands today. Mr. Sonneborn has played an important role in a wide spectrum of R&D and advanced technology assignments including the Bell Engineering Laboratories. Mr. Sonneborn holds a Dipl. Ing. (M.S. equivalent) degree from Hannover Technical University (Germany) and a M.S.E.M. degree from the University of Texas. He is currently pursuing graduate studies towards an M.B.A. degree at the University of Dallas, Irving, Texas.

13th National SAMPE Technical Conference
October 13-15, 1981

THE DESIGN, CONSTRUCTION, AND PERFORMANCE OF COMPOSITE FUSELAGE COMPONENTS FOR THE BOEING 234 HELICOPTER

Leonard J. Marchinski
Chief of Structures Technology

Robert L. Pinckney
Chief of Advanced Technology for Operations

Boeing Vertol Company
Philadelphia, Pennsylvania

1. ABSTRACT

The fuel for the commercial Boeing 234 helicopter is contained in two crash-resistant elastomeric fuel cells which are enclosed in a pair of externally mounted fuel pods. The all-composite fuel pods are attached to the airframe structure through vibration isolation devices. The composite fuel pods are of sandwich shell design employing KEVLAR and graphite fibers. Details of the design, construction, and structural evaluation of the composite fuel pods are presented, along with similar information on the all-composite cargo floor used in the commercial 234.

2. INTRODUCTION

The Boeing Vertol Company has been developing and applying composite materials to its helicopter product lines for more than 30 years. Initial U.S. Air Force contracts, "Development of a Reinforced Plastic Rotor Blade" and "Development of a Reinforced Plastic Fuselage," were awarded to the Vertol Company in 1954 for potential development and application to the H-21 helicopter. The rotor blade application appeared to offer the greatest potential and by 1964 was being actively pursued. The availability of high-quality S-glass in roving and fabric forms and its incorporation in stable preimpregnated materials by the 3M and Bloomingdale Companies made possible a whole new all-composite rotor blade concept, the Boeing Vertol Advanced-Geometry Blade. Later, the availability of high-modulus boron fibers and the subsequent fabrication of all-boron-fiber rotor blades led to full-scale flight evaluation of all-glass-fiber and all-boron-fiber rotors. A Boeing Vertol CH-47 helicopter equipped with the fiberglass blades on the forward head and the boron blades aft is shown in Figure 1. From these early beginnings, composite blade technology has grown to a full-scale production reality.

Figure 1. CH-47 Helicopter With Fiberglass Blades Forward and Boron Blades Aft

The most recent applications of composites technology at Boeing Vertol are on its Model 234 helicopter, the Commercial Chinook, shown in Figures 2 and 3. This paper deals with two major applications of composites technology in the helicopter fuselage: the long-range fuel pods and the all-composite cargo floor.

Figure 2. The Boeing 234 Commercial Chinook

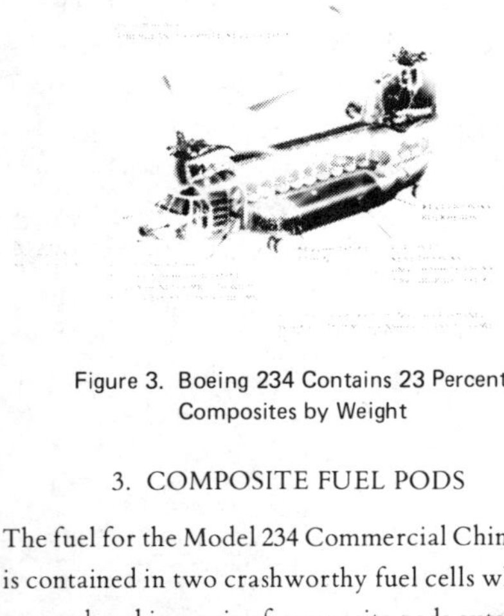

Figure 3. Boeing 234 Contains 23 Percent Composites by Weight

3. COMPOSITE FUEL PODS

The fuel for the Model 234 Commercial Chinook is contained in two crashworthy fuel cells which are enclosed in a pair of composite pods external to the fuselage, as shown in Figure 4. The tanks are supported at each end by metallic vibration isolation devices (IFIS). The vertical and lateral loads are transferred through the IFIS to the fuel tank support beams. The longitudinal loads are transferred through the IFIS to the landing gear support structure at the forward end.

The fuel tank and support beams were sized initially for stiffness to assure natural frequencies above the dominant rotor frequencies. The local structures at joints and attachments are sized primarily for emergency landing conditions; flight and water ditching conditions were also considered. The external loads are derived from the appropriate criteria. Internal forces for the fuel pod are generated by the use of a NASTRAN finite-element model analysis and classical Mc/I, VQ/I analyses are used for the fuel pod support beams.

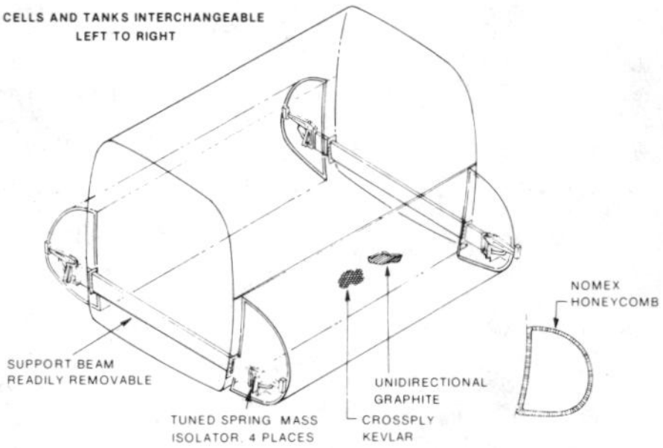

Figure 4. Composite Fuel Pods With Isolated Mounting

Strength is verified by structural tests at both room-temperature-dry (RTD) and elevated-temperature-wet (ETW) conditions for each critical structural element. The test temperature is determined by a rational thermal analysis of environmental conditions. The moisture conditioning of the test specimens was accelerated in a chamber at 120°F and 95-percent relative humidity to reach levels of moisture established by worldwide experience.

3.1 Structural Description

The pod consists of a composite sandwich shell of constant cross section along the length and closed at each end with composite flat-sandwich-plate bulkheads (see Figure 5).

The relative size of the pod is best shown in Figure 6. The capacity of the pod is over 1,000 gallons, almost 7,000 pounds. The pod shell is manufactured in three sections as shown in Figure 7, primarily dictated by the shape of the shell and the tooling requirement of release from the mold.

The shell is a sandwich with hybrid KEVLAR 49/epoxy and high-modulus graphite/epoxy faces to resist internal fuel pressures and external forces. Imbedded in the shell are four longerons located at the splices and made of unidirectional high-modulus graphite/epoxy which carry the bending loads between the bulkheads. The bulkheads are also of sandwich construction, using unidirectional KEVLAR 49/epoxy faces with some local reinforcing of graphite around the manhole door.

Figure 5. Structure of the Composite Fuel Pod

Figure 6. Scale of the Composite Fuel Pod

Figure 7. Fuel Pod Components

The two support beams span from left to right through the fuselage and are supported only at the fuselage sides. The beam is rectangular with rectangular T300 graphite/epoxy caps and KEVLAR 49/epoxy shear webs reinforced locally at bolted-joint locations with fiberglass/epoxy doublers.

All outlets to the fuel cell are designed with breakaway fittings as a crashworthy feature to preclude fuel leakage in the event of structural separation during emergency conditions.

Calculated loads are compared with test results to arrive at margins of safety for all pertinent structural elements.

3.2 Fuel Pod Support Beam

The outboard section of this beam is shown in Figure 8. The IFIS fitting attaches with two 3/4-inch pins and three 3/8-inch bolts. The beam attaches to the fuselage with six 3/8-inch-diameter bolts, through the web and reinforcing doubler pads. The beam caps and internal doublers are precured. The webs are cocured in the bonding-assembly cure cycle. Inboard of the buttline 52.4 doublers, the beam is of constant cross section.

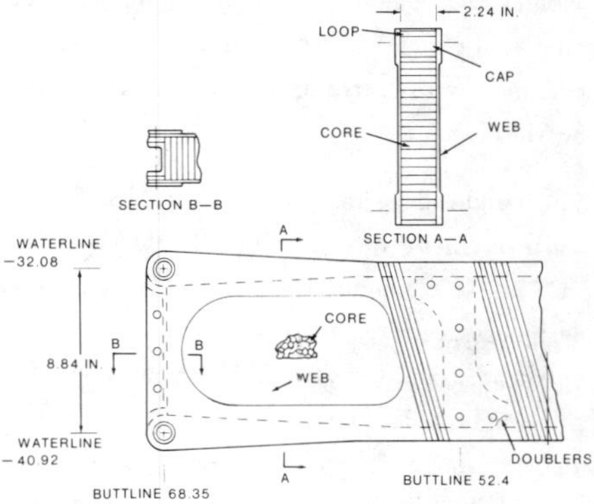

Figure 8. Outboard End of Fuel Pod Support Beam

3.3 Beam Cap Loop End

The loop end, Figure 9, consists of a titanium fitting and unidirectional T300 graphite/epoxy plies which wrap around the fittings and bond to the beam cap as shown. The titanium fitting receives 3/4-inch-diameter pins from the IFIS fitting to transfer rolling moments and lateral loads to the beam. The lower forward loop end also reacts a yawing moment due to eccentricities in the drag load path.

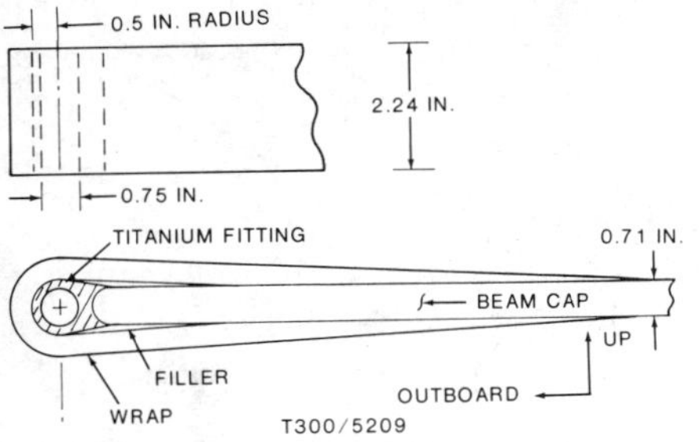

Figure 9. Loop-End Component of Support Beam

3.4 Bulkhead

The bulkheads which close the ends of the fuel pod and provide for the attachment of the IFIS support fittings are identical, front and rear. The bulkhead is of sandwich construction using primarily unidirectional KEVLAR 49 tape at 0, ±45, and 90-degree orientations for the laminate faces with a NOMEX core. A single ply of KEVLAR 49 style 285 fabric faces the sandwich to preclude fiber peeling during handling. Unidirectional T300/5209 graphite/epoxy layers are added locally around the manhole door to strengthen the sandwich plate. The bulkhead is attached to the pod shell with bolts normal to the plane of the bulkhead.

The bulkhead is analyzed with a NASTRAN finite-element model with "CQUAD" and "CTRIAL" elements. The grid points around the periphery are fixed for reactions normal to the bulkhead, fixed for vertical reactions along the sides, and fixed for lateral reactions along the top and bottom. Two loading conditions were examined, the 4-g-forward and 6-g-down emergency landing conditions. The 4-g-forward condition produces the maximum normal pressure and the 6-g-down condition produces the maximum vertical force at the IFIS fitting attachment location.

3.5 Fuel Pod Shell

The fuel pod shell is of sandwich construction using 1.5-inch-thick NOMEX core and symmetrical hybrid KEVLAR 49/HMS graphite laminates on each face. The KEVLAR 49 is woven 120 style and 285 style oriented at ±45 degrees to the longitudinal axis of the pod. The high-modulus (HMS) graphite is unidirectional tape oriented in the hoop direction (90 degrees) to resist frame bending. Some local plies of ±45-degree HMS graphite were added for dynamic tuning. The sandwich shell is shown in Figure 10 and, except for the local ±45-degree HMS plies, is constant from front to rear. Four longerons of HMS graphite provide the beam bending strength between the end supports at the fore and aft bulkheads. The pod is fabricated in three sections and spliced.

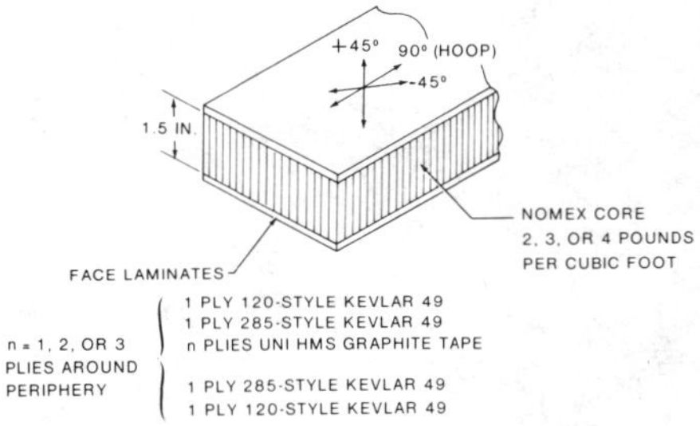

Figure 10. Sandwich Construction of Fuel Pod Shell

The fuel pod is analyzed using a NASTRAN finite-element model; a computer plot of the model is shown in Figure 11. The shell and the bulkhead are made up of "CQUAD" elements and the longerons modeled as "CONROD" elements running the length of the pod. The model is fixed at its bulkhead attachment points to the support beams. The element properties input to the program are calculated using standard laminate theory and typical material elastic properties.

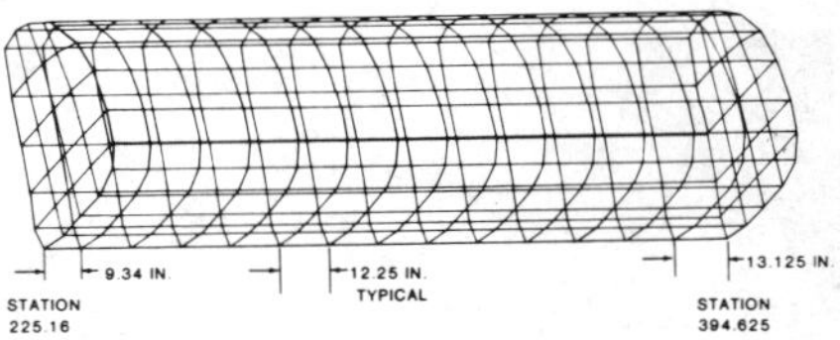

Figure 11. NASTRAN Finite-Element Model of Fuel Pod

Analyses were performed for the 6-g-down, 3-g-inboard, 4-g-forward emergency landing conditions, the flight condition, and the water landing condition. Internal pressures are generated by the accelerations for the critical conditions. Although the amount of fuel is 90 percent of the available volume, the tank is conservatively assumed to be full. The 4-g-forward condition is the most severe condition for all the "CQUAD" elements of the shell. Plate bending moments are critical for the forward half of the shell. The fuel pod is symmetrical about the center station so that it can be used for the left or right side.

Cutouts in the shell for fuel system gravity fill, probes, and sumps are located in areas that are not highly loaded. Preliminary detailed finite-element analyses of these locations have been performed and loads are not critical. Strengths have been verified by analysis and test.

3.6 Fuel Pod Shell Joint

The fuel pod shell is spliced at three longitudinal locations with a butt-joint lap-splice concept as shown in Figure 12.

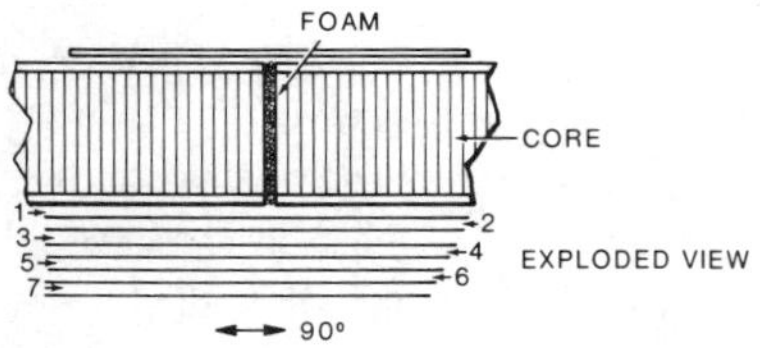

LAYER NO. 1. ADHESIVE EA 9628
LAYER NO. 2. BRIDGE — PRECURED ±45° E-GLASS
LAYER NO. 3. ADHESIVE EA 9628
LAYER NO. 4. —45° KEVLAR 49 STYLE 285
LAYER NO. 5. 90° T300/5209 GRAPHITE/EPOXY
LAYER NO. 6. 90° T300/5209 GRAPHITE/EPOXY
LAYER NO. 7. +45° KEVLAR 49 STYLE 285

Figure 12. Details of Fuel Pod Assembly Splice

3.7 Bulkhead-to-Shell Attachment

Although technical presentations or papers do not usually discuss details, it seems that the most difficult design problems in using composite materials concern detail attachments. The relatively low bearing strength of composites per se requires use of metallic inserts to distribute the loads to the laminate. The attachment of the end bulkheads to the basic shell presented such a problem. After many tradeoff studies, the method shown in Figure 13 was selected as being inexpensive, lightweight, and a reasonably easy solution to the sealing requirement.

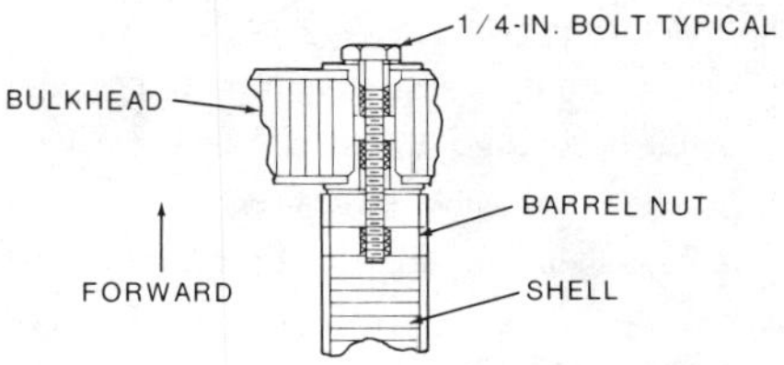

INSERT: SLEEVE SLS5163A4
PLUG SLP5163A4
(SHUR-LOK)

Figure 13. Attachment of End Bulkhead to Fuel Pod

The bulkhead is attached to the pod shell by tension bolts into barrel nuts which are mounted in solid-block inserts of chopped-fiber/epoxy. The down emergency landing condition produces maximum shear on the bolts and the forward emergency landing condition produces maximum tension.

3.8 Composite Fuel Pod — The Results

The use of composite materials in the fuel pod demonstrated the advantages of more efficient, lightweight structure and structural tailoring for dynamic purposes. Corrosion prevention, ease of repairability, and other characteristics will be demonstrated as service time accrues.

The sketch and tabulation in Figure 14 show the relative size and weight of the military CH-47C fuel pod versus the commercial Model 234. The CH-47C fuel is isolated at the bladder, whereas the Model 234 isolation includes the fuel, bladder, and structure. This evaluation shows a 36-percent reduction in the pounds of structure per gallon for the composite pod. To verify this weight saving, a study was made considering a metallic pod of the same configuration, i.e., honeycomb but with aluminum skins. That study showed the composite configuration to weigh 25 percent less than the metal.

Since the dynamic characteristics of the pod are of prime importance in the total fuel isolation system, the performance of the system deserves mention here. The curves of Figure 15 present results of a total fuselage shake test. As can be seen, without isolation the variation of natural frequency with fuel level is extreme and passes through the operating rotor speed. The isolation system not only moves the natural frequencies above the operating rotor speed, but also reduces the variation with fuel level which is only 0.25 Hertz, or about one-fourth of the nonisolated variation.

		CH-47C	234 L/R
CAPACITY	— GAL	1,100	2,100
	— LB	7,150	13,650
ISOLATED WEIGHT	— LB¹	7,150	14,580
TANK SHELL AND DOORS	— WT²	527	748
	— LB/GAL	0.56	0.36
ISOLATION SYS/IFIS AND BEAMS	— WT	137	256
	— LB/LB ISO	0.0192	0.0176

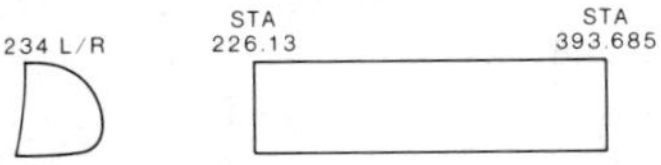

NOTES: 1. TOTAL WEIGHT SUPPORTED BY ISOLATORS
2. STRUCTURES WEIGHT ONLY

Figure 14. Weight Comparison of CH-47C and Model 234 Fuel Pods and Mounting

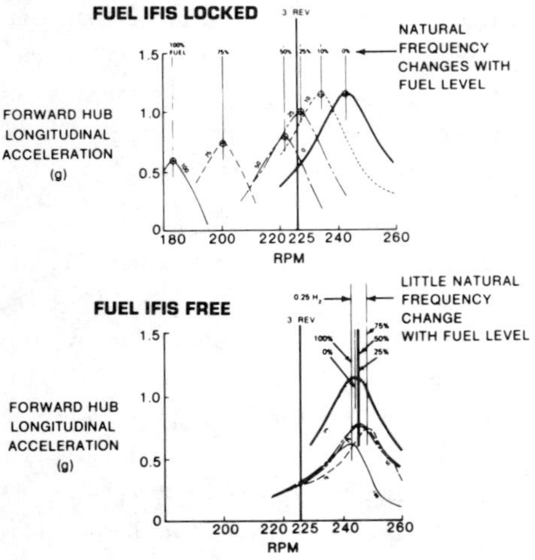

Figure 15. Vibration Reduction With Fuel Isolation System

4. COMPOSITE CABIN FLOOR

The metal floor of the CH-47 helicopter has posed a continuing corrosion problem. The floor is an assemblage of wide 30-foot-long extrusions, riveted and bolted together at relatively close pitch (hundreds of attachments), resulting in a costly assembly. The floor is isolated from the substructure by elastomeric pads, also at close pitch and an expensive installation.

The cabin floor of the Model 234 is all-composite except for the seat attachment rails and the vibration isolation units (IFIS) (Figure 16). The use of composites will eliminate the corrosion problem. The only attachments into the floor will be the seat rails and IFIS units. Thus the number of attachments will be reduced by an order of magnitude.

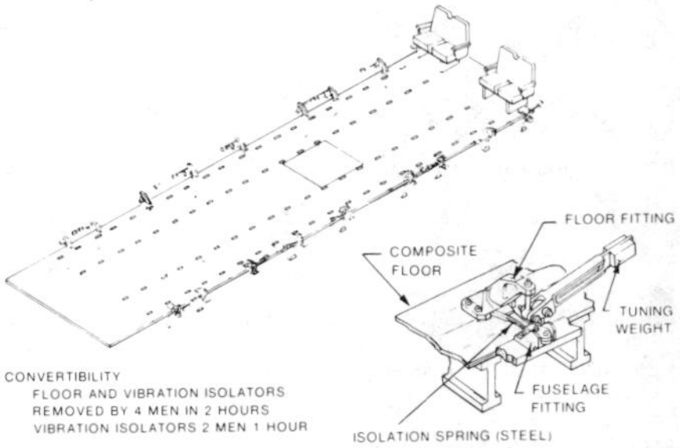

Figure 16. The Isolated Composite Cabin Floor

Similar to the pod requirements, the floor is a part of the total dynamic system; as such, its structural properties had to be matched to the total system. Again, the custom tailoring allowed by the use of composites becomes evident. Since the floor loads are transmitted longitudinally to transverse beams, stiffness and strength only in that direction are necessary. The purpose of the IFIS system is to isolate the floor from the airframe at one frequency, independent of the load carried. The system also maintains relatively constant airframe natural frequency by preventing force feedback from the fuel tanks. The vibration objectives are to provide ≤ 0.05 g vertical on the floor at the predominant excitation frequency, produced by a motion attenuation of 8 to 1.

4.1 Description of Floor

The cabin floor extends from station 121.2 to station 482.0. It consists of four composite honeycomb panels with fiberglass/KEVLAR 49 face sheets and NOMEX honeycomb core. Nine composite construction lateral beams are bonded to the lower surface of the floor panels and isolators (IFIS) at the ends of each beam attach the floor to the fuselage structure. Bolted transverse splices are located at stations 236.2, 315.0, and 392.25.

The four composite honeycomb panels are 90.0 inches wide by approximately 100 inches long. Each panel assembly includes the following common features (Figure 17):

- First (top) ply of face is unidirectional fiberglass at +45 degrees.
- Second ply of face is unidirectional fiberglass at –45 degrees.
- Third and fourth layers of face are uni KEVLAR at 0 degrees.
- One ply of EA9628 0.045-psf film adhesive is used between the face layup and the core.
- NOMEX 1/8-in. cell by 3-pcf by 1.500 ±0.005-in.-thick core
- Bottom layers are opposite order of top.
- Each floor panel has a molded rubber edge seal (channel section) which is bonded into the assembly during the cocure operation.

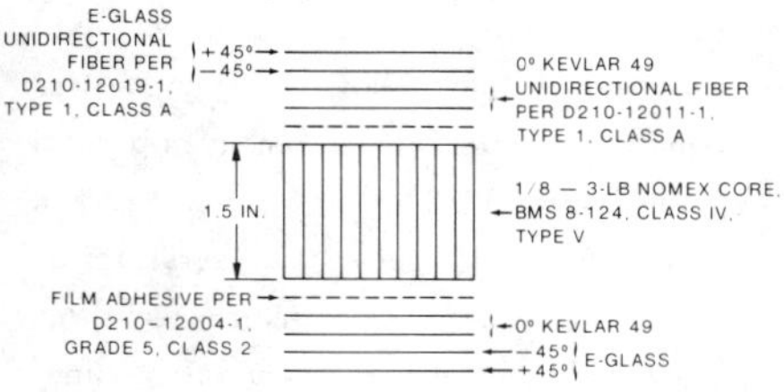

Figure 17. Sandwich Construction of Composite Floor Panels

The lateral beams located under the lower side of the floor panels consist of the following (Figure 18):

- Pultruded unidirectional KEVLAR T-caps
- Face sheets of 281 or 285 KEVLAR at ±45 degrees
- One ply each side of core of EA9628 0.045-psf film adhesive
- NOMEX core
- A machined 7050 aluminum fitting at each end for attachment to the IFIS floor isolators.

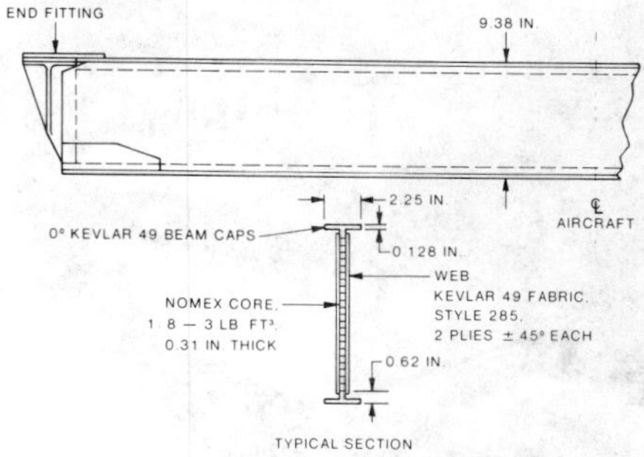

Figure 18. Composite Floor Beam Assembly

The forward floor panel runs from fuselage station 120.13 to fuselage station 235. It has two standard floor beams at station 165 and station 195 and a shallower beam at station 135. The station 135 beam is constructed using the same components as a standard underfloor beam, but is only half as deep and does not have end fittings. At the forward edge of the floor at right and left buttlines 18 the leading edge of the floor is cut back 3 inches wide by 3.25 inches deep to receive a machined heat-treated steel spring fitting for passive support. The fitting is bonded to the floor and 12-ply fiberglass doublers are locally bonded to the upper and lower face sheets.

At the aft edge of the first floor panel, SHUR-LOK SL102-type, through-bolt spacers are installed for a floor splice at station 235, on approximately 4-inch centers. The floor will have four Brownline 20415 longitudinal seat rails, at right and left buttlines 19 and 39, bolted to the floor with 1/4-inch-diameter flush-head bolts, nuts, and large-diameter washers. Holes for the seat rails will be potted in the floor during layup.

4.2 Composite Isolated Floor — The Results

Figure 19 shows a section of isolated floor before installation in the aircraft. As can be seen, the isolation units are attached to the floor. The absence of fasteners is obvious; the upper surface is unobstructed except for the seat and cargo-tiedown rails. The isolation units are installed behind the interior trim. The floor support beams which transfer cargo and seat loads to the isolation units are also structurally simple, demonstrating another benefit of using composites.

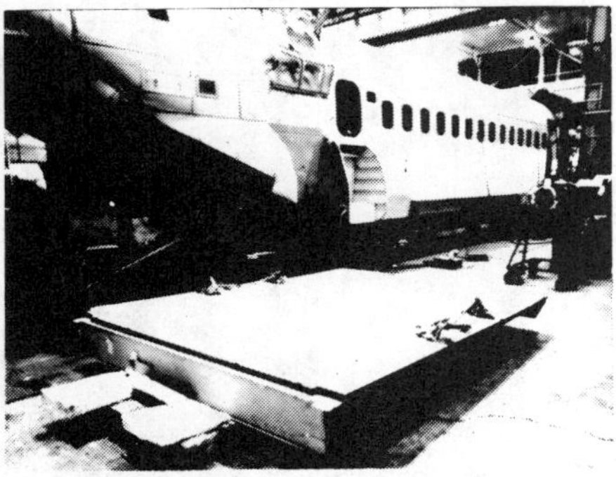

Figure 19. Composite Isolated Floor Section for the Commercial Chinook

As stated previously, composite construction was chosen due to its unique capability for structural tailoring. In the case of the isolated floor, the total dynamic system (floor structure and isolators) has achieved the objective of significant vibration reduction, as shown in Figure 20. A vibration level of 0.03 g makes the helicopter comfort level equal to that of a fixed-wing aircraft. Considering other benefits, the weight of the basic floor panel is 10 percent lighter than the metal counterpart. The weight saving would have been greater except for the added material needed to meet the stiffness requirements.

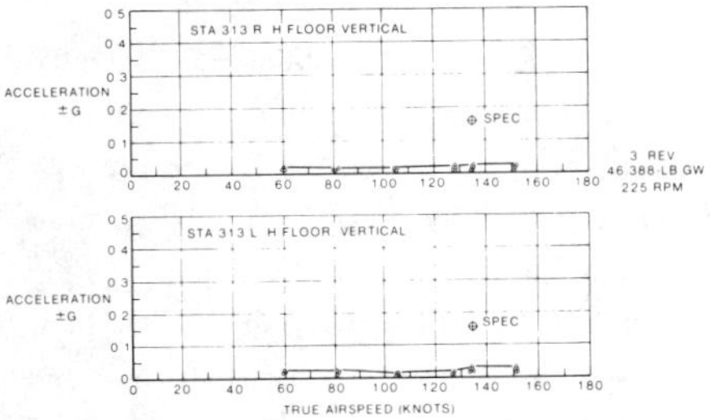

Figure 20. Cabin Vibration Levels With Isolated Composite Floor

A cost reduction is evident but can only be demonstrated by production runs. However, the lower parts count and fastener count should assure a significant cost reduction. Not previously mentioned is the impact resistance of the composite floor. In comparative tests with metal floors, the composite is better and will therefore reduce maintenance costs. All considered, the composite floor has demonstrated its benefits, guaranteeing its continued use.

5. COMPOSITE STRUCTURES CERTIFICATION

Structural verification of the use of composite materials in compliance with Federal Aviation Regulations, Part 29, was done on the Model 234 by a combination of coupon testing of the materials (part of a material specification), coupon testing of the specific laminate being used, full-scale testing of specific sections of the structure, and analysis of the full-scale structure with allowables developed from theory and test. The composite structures were shown to have the required strength over an ambient temperature range of −35°F to 120°F from dry to equilibrium moisture content. The fabrication of reproducible and reliable composite structures is assured by material specification control and manufacturing in-process control, with test verification to assure correct fabrication procedures.

Measured moisture pickup data(1) is used to determine the equilibrium moisture content resulting from exposure to a realistic environment of temperature and relative humidity. Test specimens are conditioned to the equilibrium moisture content using an environmental chamber to accelerate the absorption. The test matrices are summarized in Table I. The test temperatures are determined as a realistic case condition consistent with the loading conditions, based on maximum ambient temperatures and solar radiation effects. Detailed test plans and test reports were submitted to the FAA in a timely manner as the tests described in Table I were accomplished.

The commercial Boeing 234 helicopter was issued its FAA type certificate in June 1981.

5.1 Establishment of Static Design Allowables

The objective is to demonstrate adequacy of the structure fabricated from composite materials. The extent of testing will differ for each item depending on the test configuration and its expected load conditions, material selected, design margins, data base, and experience with similar structures.

TABLE I. SUMMARY OF TESTS OF COMPOSITE COMPONENTS

Item	Critical Load Conditions	Failure Mode	Test Substantiation — Type of Specimens RTD/ETW			
			Coupon	Laminate	Element	Component
Fuel Tank Support Beam	+6g down (crash)	Bolt bearing	–	X	–	–
	Flight cond 3	Bolt bearing	–	X	–	–
		Web shear	–	X	–	–
		Web/cap bond	–	X	–	–
		Cap tension	X	–	–	–
		Cap compression	X	–	–	–
		Cap interl shear	X	–	–	–
		Cap loop end bending	–	–	X	–
Fuel Pod	+6g down (crash)	Shell bending	–	–	X	–
	+4g long. (crash)	Splice joint bending	–	–	X	–
	±3g lat (crash)	Block insert tension	–	–	X	–
	–3g long. (crash)	Block drag reaction	–	–	X	–
		Bulkhead bending	–	–	X	–
		Bulkhead bending	–	–	X	–
		Shear on insert	–	–	X	–
		Tension on insert	–	–	X	–
		Bulkhead bending	X	–	–	–
Floor-Support Beam	150 lb/sq ft x N_z	Cap compression	X	–	–	–
		Cap interl shear	X	–	–	–
		Web shear	–	X	–	–
		Web/cap bond	–	X	–	–
Floor Panel	1,000 lb wheel load x N_z	Panel bending	–	–	–	X
	150 lb/sq ft x N_z	Beam bending	–	–	X	–
	Seat retention (6g)	Shear on insert	–	–	X	–
	2,500-lb tiedown	Tension on insert	–	–	X	–

To demonstrate proof of structure, a series of static tests was performed using coupon, laminate, element, and component tests as follows: (a) For a full-scale component, such as a complete floor panel, one test specimen is sufficient to demonstrate structural adequacy. (b) When coupons or laminates are used to demonstrate material static strength, a number of specimens are necessary to demonstrate strength properties and the allowable strength is determined statistically. (c) When elements which represent material arrangement and local geometry are tested, three specimens are tested. The demonstrated strength for elements is the bottom of scatter of the tests.

The specimens were conditioned to the appropriate moisture content and tested at temperatures consistent with the conditions which generate the maximum loads. Figures 21 through 24 show the location and type of test specimens that were used.

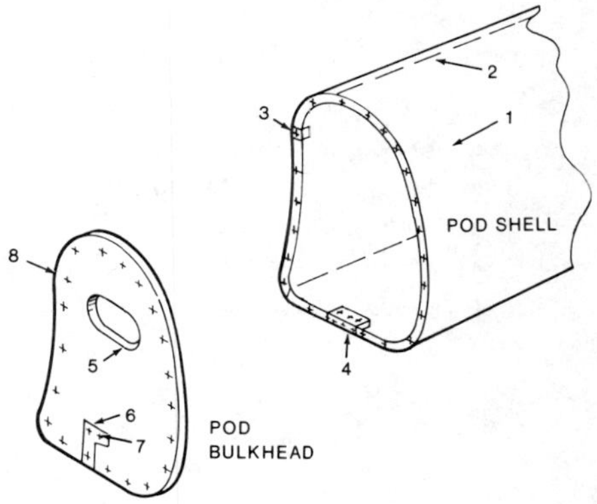

LOCATION

1 SHELL BENDING
2 SPLICE JOINT BENDING
3 BLOCK INSERT TENSION
4 BLOCK DRAG REACTION
5 BULKHEAD BENDING
6 BULKHEAD BENDING
7 ATTACHMENT, INSERT SHEAR
8 ATTACHMENT, INSERT TENSION

Figure 21. Composite Fuel Pod Tests

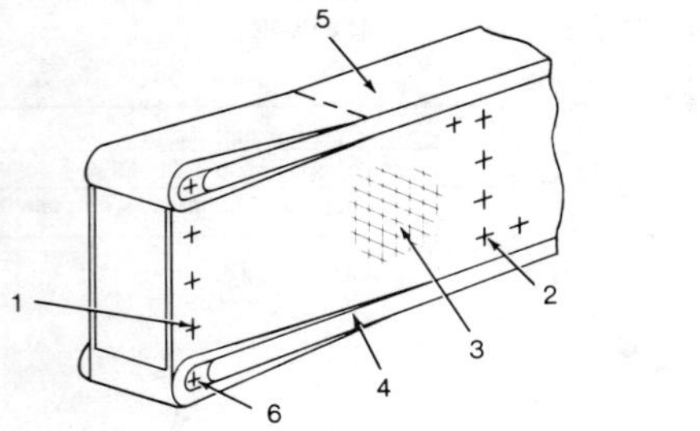

LOCATION

1 BOLT ATTACHMENT TO IFIS
2 BOLT ATTACHMENT TO FUSELAGE
3 WEB, SHEAR
4 WEB-TO-CAP BONDED JOINT
5 BEAM CAP
6 BEAM CAP END LOOP

Figure 22. Fuel Tank Support Beam Tests

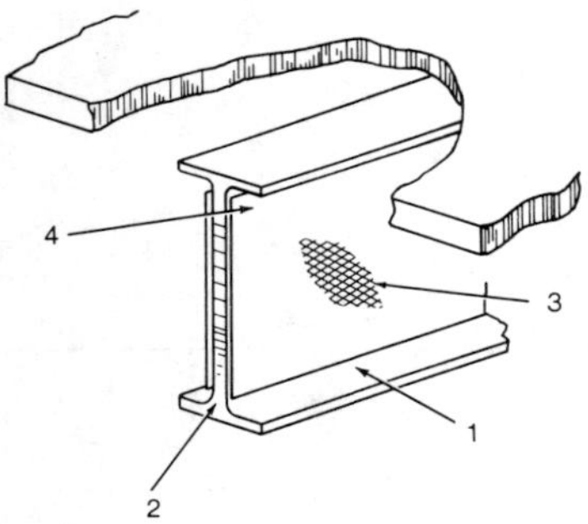

LOCATION

1 BEAM CAP, COMPRESSION STABILITY
2 BEAM CAP, INTERLAMINAR SHEAR
3 WEB, SHEAR
4 WEB/CAP BONDED JOINT

Figure 23. Floor Support Beam Tests

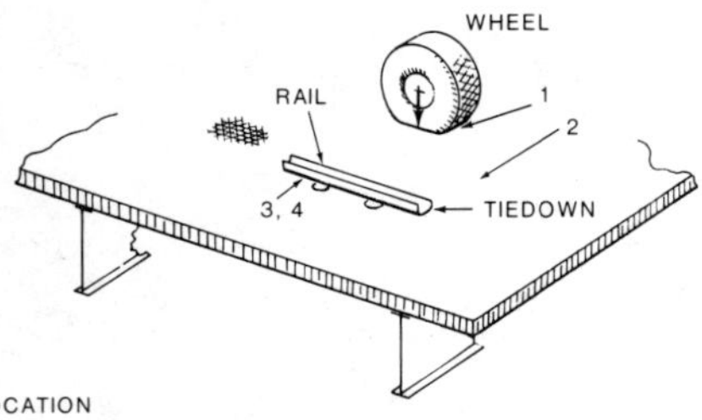

LOCATION

1 WHEEL LOAD, 1,000 LB
2 BENDING, 150 LB/FT²
3 SEAT RETENTION
4 2,500-LB TIEDOWN

Figure 24. Floor Panel Tests

6. CONCLUSIONS

In all major uses — main rotor blades, fuel pods, and cabin floor — composite materials have proven to be: (a) less costly; (b) lower weight; (c) easily tailored to requirements; and (d) simple construction.

These characteristics have been actually demonstrated and the benefits realized. The elimination of corrosion, ease of repair, and other characteristics have a high probability of being achieved. The potential for use of composite materials is only beginning to be appreciated.

7. REFERENCES

1. Pride, Richard A., ENVIRONMENTAL EFFECTS ON COMPOSITES FOR AIRCRAFT, NASA TM78716, National Aeronautics and Space Administration, Langley Research Center, Hampton, Virginia, May 1978.

8. BIBLIOGRAPHY

1. Hoffstedt, Donald J., et al, ADVANCED-GEOMETRY, GLASS-FIBER-REINFORCED PLASTIC ROTOR BLADE TEST PROGRAM, Boeing Vertol Company, Philadelphia, Pennsylvania; USAAMRDL TR71-42, Applied Technology Laboratory, U.S. Army Research and Technology Laboratories (AVRADCOM), Fort Eustis, Virginia, September 1971.

2. Hoffstedt, Donald J., et al, RESEARCH AND DEVELOPMENT OF HELICOPTER ROTOR BLADES UTILIZING ADVANCED COMPOSITE MATERIALS, Boeing Vertol Company, Philadelphia, Pennsylvania; AFML TR71-125, U.S. Air Force Materiel Laboratories, Wright-Patterson Air Force Base, Ohio, April 1971.

9. BIOGRAPHIES

9.1 Leonard J. Marchinski

Mr. Marchinski received his BS degree in Aeronautical Engineering from Notre Dame and his MS from Drexel University. His 25 years of experience in structural engineering and engineering management include work on aircraft such as the GD/Convair F-106, Martin B-57, Avro CF-105, Republic F-84, and Boeing Vertol CH-46, CH-47, YUH-61A, HLH, and 234. Formerly the vice president of a consulting engineering firm, Mr. Marchinski now directs the various activities of the Structures Technology department at Boeing Vertol. He formerly served on the board of directors of the American Helicopter Society and now chairs the AIAA Composite Structures subcommittee which he founded.

9.2 Robert L. Pinckney

Mr. Pinckney received his BS degree in Chemical Engineering from Iowa State University and later did postgraduate work in chemistry and related fields at UCLA and the University of Oklahoma. He joined Boeing Wichita in 1951 where he worked in fuels, lubricants, and corrosion protection. After later work in adhesives, sealants, and structural plastics, he came to Boeing Vertol in 1964 to perform vital tasks on the early fiberglass and boron rotor blade programs. He has since moved on to the development of low-cost techniques for fabrication of composite airframes and thermoplastic resin technology. Mr. Pinckney has been working in the field of metal-matrix composites since 1974.

13th National SAMPE Technical Conference
October 13-15, 1981

FAILURE ENVELOPES AND LAMINATE OPTIMIZATION FOR COMPOSITE MATERIALS

N. Balasubramanian
Materials Laboratory
Air Force Wright Aeronautical Laboratories
Wright-Patterson Air Force Base, Ohio 45433

Abstract

The available analytical methods are used to determine the failure surface of laminated composites. A ply-by-ply analysis is made, and the failure of each ply for a given state of stress is determined. The inner envelope of failure surfaces represents the first-ply failure (FPF) of the laminate. By superposing the FPF envelopes, the optimum laminate configuration for a given state of loading can be chosen. This is illustrated for the case of AS/3501 graphite-epoxy laminates. The results are compared with those for the other laminated composites using a simple expression for the FPF in the first quadrant of the normalized stress resultant space. A correlation between the FPF strength and fatigue strength is obtained for B-epoxy composites.

Keywords: Failure envelopes, optimization, graphite, boron, glass, kevlar, epoxy, composites, laminates, first-ply failure.

1. INTRODUCTION

In order to design efficiently with composites, their strength under combined stresses must be determined. Since a variety of laminate configurations are possible, experimental determination of the strength of all possible laminates becomes a formidable task. However, theoretical methods have been developed to accomplish this.[(1)] These methods require only the measurement of uniaxial (0 ply) strengths and modulii, the knowledge of tensor transformation relations and a clear understanding of the interaction of plies in a laminate. The aims of the present work are to use the available analytical methods[(1)] to obtain the biaxial strength envelopes of graphite-epoxy (AS/3501) composites; to make comparisons with other composites; and to illustrate the use of failure envelopes in laminate optimization (i.e. in specifying the optimum combination of ply orientations and the ply thicknesses for a given loading condition).

2. FAILURE SURFACE OF UNIDIRECTIONAL COMPOSITES

2.1 Stress Space

The failure of a unidirectional composite referred to its orthotropic axes can be described by the equation:[1,2]

$$F_{xx}\sigma_x^2 + 2F_{xy}\sigma_x\sigma_y + F_{yy}\sigma_y^2 + F_{ss}\sigma_s^2 + F_x\sigma_x + F_y\sigma_y = 1 \qquad (1)$$

This equation describes a failure surface in stress (σ) space. Only thin laminates or plane stress condition is considered, and interlaminar stresses are not taken into account. The coefficients of the equation (1) can be determined as follows. Let the strengths measured in longitudinal tensile (σ_x,0,0) and compression ($-\sigma_x$,0,0) tests be X and X' respectively. Equation (1) reduces to

$$F_{xx}X^2 + F_x X = 1 \qquad (2)$$

and

$$F_{xx}X'^2 + F_x X' = 1 \qquad (3)$$

Hence

$$F_{xx} = \frac{1}{XX'} \qquad (4)$$

$$F_x = \frac{X' - X}{XX'} \qquad (5)$$

Similarly

$$F_{yy} = \frac{1}{YY'} \qquad (6)$$

$$F_y = \frac{Y' - Y}{YY'} \qquad (7)$$

$$F_{ss} = \frac{1}{S^2} \qquad (8)$$

where Y and Y' are the transverse tensile and compressive strengths and S the longitudinal shear strength. The values for these five strengths are listed in Table 1 for the composites to be considered in this paper (graphite/epoxy, boron/epoxy, glass/epoxy, and kevlar/epoxy). The interaction term, F_{xy} in Equation (1) is difficult to determine experimentally. The value of F_{xy} is constrained by the condition

$$-1 < F^*_{xy} < 1 \qquad (9)$$

where

$$F^*_{xy} = F_{xy}/\{F_{xx}\,F_{yy}\}^{1/2} \qquad (10)$$

This condition ensures that the failure surface is a closed curve. In the present work F^*_{xy} is assumed to be -1/2 (which is the value of the interaction term in the von-Mises equation for isotropic materials). The failure surface of unidirectional graphite epoxy (AS/3501) using Eq. (1) and parameters listed in Table 2 is shown in Figure 1.

2.2 Strain Space

Equation (1) can be rewritten in terms of strains by using the orthotropic stress-strain relation, to obtain the failure surface in strain space:

$$G_{xx}\varepsilon_x^2 + 2G_{xy}\varepsilon_x\varepsilon_y + G_{yy}\varepsilon_y^2 + G_{ss}\varepsilon_s^2 + G_y\varepsilon_y = 1 \qquad (11)$$

The coefficients in Equations (1) and (11) are related by modulii (Q_{xx} etc.) as shown in reference 1. These are calculated for five composites using the modulus data in Table 3 and are listed in Table 4. The coefficients of equations (1) and (11) follow the transformation laws for fourth-order tensors. This enables us to write down the equation for the failure surface of off-axis (θ) unidirectional composites by transforming the coefficients. Additional terms are present since off-axis composites are anisotropic. The failure surface of an off-axis unidirectional composite in strain space is given by

$$G_{11}^{(\theta)}\varepsilon_1^2 + 2G_{12}^{(\theta)}\varepsilon_1\varepsilon_2 + G_{22}^{(\theta)}\varepsilon_2^2 + G_{66}^{(\theta)}\varepsilon_6^2 + 2G_{16}^{(\theta)}\varepsilon_1\varepsilon_6 + 2G_{26}^{(\theta)}\varepsilon_2\varepsilon_6 + G_1^{(\theta)}\varepsilon_1 + G_2^{(\theta)}\varepsilon_2 + G_6^{(\theta)}\varepsilon_6 = 1 \quad (12)$$

G_{11} etc. can be calculated from G_{xx} etc. by using the transformation relations given in references 1, 3, and 4.

3. FAILURE SURFACE OF MULTI-DIRECTIONAL LAMINATES

The failure surface of a laminate in strain space is obtained by superposing the failure surface of the plies, since the laminate strain is assumed to be the same through the thickness of the specimen.(1,3,4) The in-plane strength of a laminate can also be described in terms of normalized stress resultants (N^*). The stress across the laminate thickness is not constant, though strain is, because the modulus varies from ply to ply. Hence, a stress resultant (N) is obtained by integrating the ply stresses. Normalizing is done by dividing N by the laminate thickness (h). The normalized stress resultant ($N^* = N/h$) is the average laminate stress. The failure surface in terms of normalized stress resultants for a given ply in a laminate is given by

$$H_{11}N_1^{*2} + 2H_{12}N_1^* N_2^* + H_{22}N_2^{*2} + H_{66}N_6^{*2} + 2H_{16}N_1^* N_6^* + 2H_{26}N_2^* N_6^* + H_1 N_1^* + H_2 N_2^* + H_6 N_6^* = 1 \quad (13)$$

The coefficients $H_{ij}^{(\theta)}$ for the ply are calculated knowing the normalized compliances of the laminate (a_{ij}) and $G_{ij}^{(\theta)}$ for the ply:

$$H_{11} = G_{11}a_{11}^2 + 2G_{12}a_{11}a_{12} + G_{22}a_{21}^2 + G_{66}a_{61}^2 + 2G_{16}a_{11}a_{61} + 2G_{26}a_{21}a_{61} \quad (14)$$

$$H_{22} = G_{11}a_{12}^2 + 2G_{12}a_{12}a_{22} + G_{22}a_{22}^2 + G_{66}a_{62}^2 + 2G_{16}a_{12}a_{62} + 2G_{26}a_{62}a_{22} \quad (15)$$

$$H_{66} = G_{11}a_{16}^2 + 2G_{12}a_{16}a_{26} + G_{22}a_{26}^2 + G_{66}a_{66}^2 + 2G_{16}a_{16}a_{66} + 2G_{26}a_{26}a_{66} \quad (16)$$

$$H_{12} = H_{21} = G_{11}a_{11}a_{12}$$
$$+ G_{12}(a_{11}a_{22} + a_{12}^2)$$
$$+ G_{22}a_{21}a_{22} + G_{66}a_{61}a_{62}$$
$$+ G_{16}(a_{11}a_{62} + a_{12}a_{61})$$
$$+ G_{26}(a_{21}a_{62} + a_{22}a_{61}) \quad (17)$$

$$H_{16} = H_{61} = G_{11}a_{11}a_{16}$$
$$+ G_{12}(a_{11}a_{26} + a_{21}a_{16})$$
$$+ G_{22}a_{21}a_{26} + G_{66}a_{61}a_{66}$$
$$+ G_{16}(a_{16}a_{66} + a_{16}^2)$$
$$+ G_{26}(a_{21}a_{66} + a_{26}a_{61}) \quad (18)$$

$$H_{26} = H_{62} = G_{11}a_{12}a_{16}$$
$$+ G_{12}(a_{12}a_{26} + a_{22}a_{16})$$
$$+ G_{22}a_{22}a_{26} + G_{66}a_{62}a_{66}$$
$$+ G_{16}(a_{12}a_{66} + a_{62}a_{16})$$
$$+ G_{26}(a_{22}a_{66} + a_{26}^2) \quad (19)$$

$$H_1 = G_1a_{11} + G_2a_{21} + G_6a_{61} \quad (20)$$

$$H_2 = G_1a_{12} + G_2a_{22} + G_6a_{62} \quad (21)$$

$$H_6 = G_1a_{16} + G_2a_{26} + G_6a_{66} \quad (22)$$

For orthotropic laminates (e.g. cross-ply), all terms with subscripts 16 and 26 reduce to 0.

4. FAILURE SURFACE OF GRAPHITE-EPOXY

The failure surface of graphite-epoxy (AS/3501) cross-ply laminates calculated using Equation (13) are shown in Figure 2. The inner envelope represents the first-ply failure (FPF) and corresponds to the first appearance of a crack in any ply of the laminate. Failure occurs when the straight line corresponding to the loading path intersects the failure surface. For (1,0,0) loading (i.e. in tension) 90° ply fails first, but for (-1,0,0) loading (i.e. in compression) 0° ply fails first. For any direction of loading the combination of normal stresses to produce the first-ply failure or the ultimate failure can be read off from Figure 2. For all laminates simultaneous failure of 0 and 90° plies occurs at only two points in the first and third quadrants.

5. LAMINATE OPTIMIZATION

We can compare all the cross-ply laminates of Figure 2 and determine which one has the highest value of FPF strength for a specified state of loading. We can also read off the value of N^*_{FPF} for that laminate. This is done in Figure 3 by superposing the FPF envelopes of all cross-ply laminates from Figure 2 and tracing the outer envelope. For instance, $[0/90]_s$ is seen to be the optimum for hydrostatic loading. To use this information quantitatively, the ply strength data for (1,1,0) loading of cross-ply laminates are shown in Table 5. The FPF strength of $[0/90]_s$ is 347 MPa. Hence the thickness of this laminate to support a stress resultant of, say, (3.7,3.7,0) M N/m is given by

$$h = \frac{N}{N^*} = .0107 \text{ m} \quad (23)$$

The ply thickness for AS/3501 is 133×10^{-6} m (5.25 mil). Hence, the number of plies required is given by

$$n = h/h_0 = 80 \text{ plies} \quad (24)$$

The laminate required is $[0_{20}/90_{20}]_s$. The use of any other laminate would impose a weight penalty. For instance a $[0_4/90]_s$ would require a thickness of (see Table 5)

$$h = \frac{3.7}{197} = .01878 \text{ m} \quad (25)$$

or

$$n \simeq 140 \quad (26)$$

Hence a weight penalty of 75% is involved. If the loading condition were changed to, say (3,-1,0), then $[0/90]_s$ is no longer the optimum. From Table 6, it is seen $[0_4/90]_s$ is seen to be the optimum among the candidates screened. The use of $[0_4/90]$ would require a laminate thickness of .02 m (≃150 plies or $[0_{60}/90_{15}]_s$) rather than .0262 m required by [0/90] (196 plies or $[0_{49}/90_{49}]_s$) resulting in a weight saving of 23%. When loading conditions can be predetermined, the failure envelopes, such as in Figure 3, can be used for laminate optimization. If the loading conditions cannot be anticipated, design has to be conservative and the approximate first-ply failure envelopes developed in references 1, 3, and 4 may be more helpful.

6. ALL COMPOSITES - A GENERAL COMPARISON

In the first quadrant of normalized stress resultant space, laminate optimization from the point of view of FPF can be represented by the simple equation:

$$N_1^* + N_2^* = C \quad (27)$$

if we ignore the unidirectional (0° and 90°) composites. This is equivalent to approximating the maximum FPF envelope in Figure 3 by a straight line. For a given loading condition, if the optimum laminate is chosen, its thickness is given by

$$h = \frac{1}{C}(N_1 + N_2) \quad (28)$$

The value of the constant C (in MPa) for the various composites is T300/5208, 604; B/Ep, 552; AS/3501, 694; Scotchply, 172; and Kevlar-epoxy, 796. Equation (28) is plotted for graphite-epoxy, boron-epoxy, glass-epoxy and Kevlar-epoxy in Figure 4. Such diagrams are useful in comparison of composites and material substitution. Equation (28) also shows that the laminate must be optimized by considering the invariant $(N_1 + N_2)$ rather than calculating the plies required to withstand N_1 and then adding plies to withstand N_2. Similar simplifications can be used for the other quadrants as well.

7. CORRELATION BETWEEN FPF AND FATIGUE STRENGTH

Laminates can still support loads beyond the FPF, and the FPF may even relieve stresses. Yet FPF strength is a useful design parameter since damage accumulation does not occur below this value. It may be possible to correlate other mechanical properties with FPF. Figure 5 shows the correlation of FPF strength (calculated in this work) with the fatigue strength at 10^6 cycles obtained by Shockey et.al.[(5)] in the case of boron-epoxy laminates.

8. CONCLUSIONS

The failure surfaces are obtained in normalized stress resultant space and are used for laminate optimization. A simple equation is obtained for the maximum FPF envelope in the first quadrant of the stress space. This equation enables a comparison of various laminated composites and material substitution. A correlation between FPF strength and fatigue strength is obtained for boron-epoxy laminates.

9. ACKNOWLEDGEMENTS

I am grateful to Dr. S. W. Tsai for valuable suggestions. This work was done during the author's tenure of a Senior National Research Council Associateship.

10. REFERENCES

1. S. W. Tsai and H. T. Hahn, Introduction to Composite Materials, Technomic Publishing Co., Westport, CT, 1980.

2. S. W. Tsai and E. M. Wu, "A General Theory of Strength for Anisotropic Materials," Journal of Composite Materials, Vol. 5, 1971, p. 58.

3. N. Balasubramanian, "The Failure Surface of Graphite-Epoxy Laminates," Proceedings of the 36th Annual Conference Reinforced Plastics/Composites Institute, The Society of the Plastics Industry, Feb., 1981, Session 12-E.

4. N. Balasubramanian, "The First-Ply Failure Surfaces of Laminated Composites," Sixth Conference on "Composite Materials: Testing and Design", ASTM, Phoenix, Arizona, May, 1981.

5. P. D. Shockey, J. D. Anderson, and K. E. Hofer, "Structural Airframe Application of Advanced Composite Materials," AFML-TR-69-101, Vol. V, quoted in "Composite Materials Workbook" by S. W. Tsai and H. T. Hahn, AFML-TR-77-33, Air Force Materials Laboratory, 1977, p. 223.

11. BIOGRAPHY

Dr. N. Balasubramanian received his M.S. Met.E. degree from Purdue University and his Ph.D. degree from Columbia University. He is the author of 50 publications. He has been employed at Martin Aerospace, Allied Chemical Corporation, and at the National Aeronautical Laboratory of India. His interests include high-performance composites as well as inexpensive composites with natural fibers for less sophisticated applications. He is currently a Senior Research Associate of the National Research Council working at the Materials Laboratory of the Air Force Wright Aeronautical Laboratories, Wright-Patterson Air Force Base, Ohio.

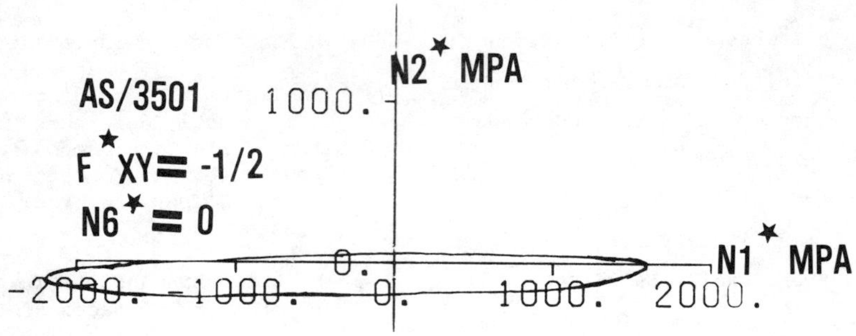

Figure 1. The Failure Surface of a Unidirectional (0°) Graphite-Epoxy (AS/3501) Laminate. The failure surface for 90 is given by a mirror image if the plane of the mirror is at 45° to the axes.

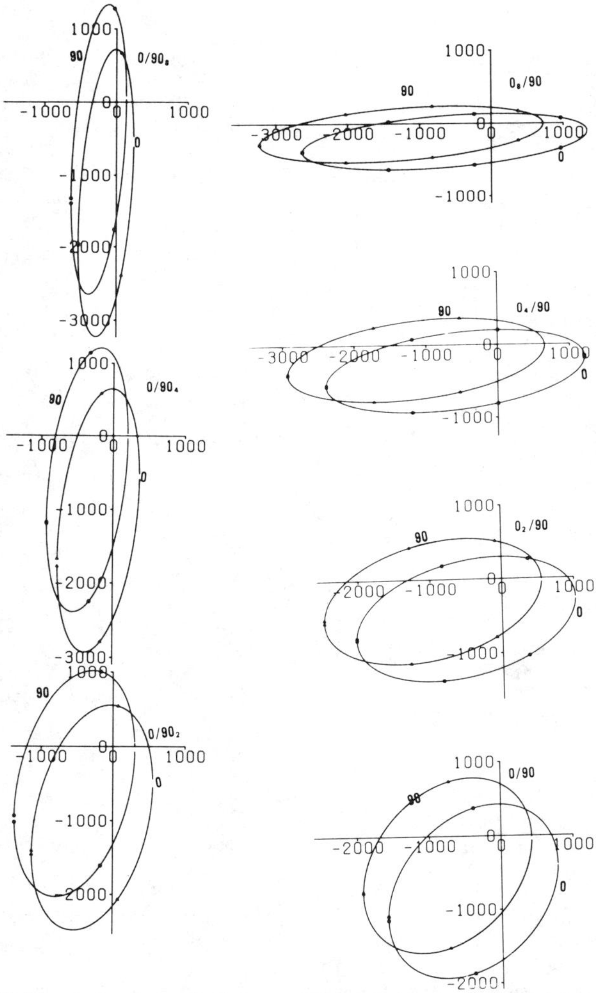

Figure 2. Failure Surfaces of Cross-Ply Laminates of Graphite-Epoxy (AS/3501) in Normalized Stress Resultant (N*) Space. The coordinates are the same as in Figure 2 but the scale is different.

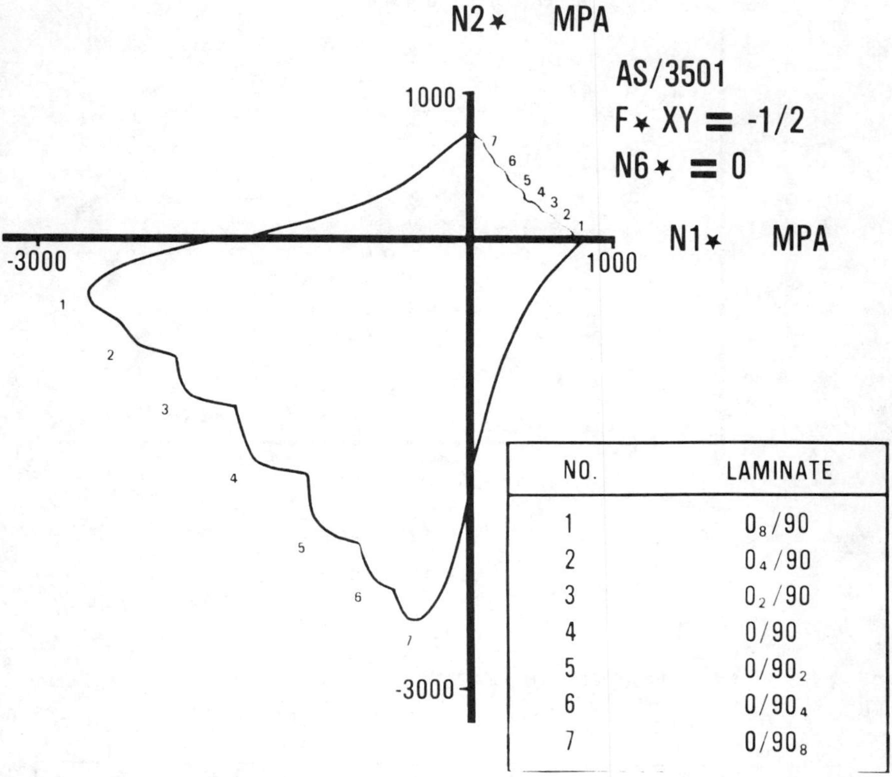

Figure 3. Laminate Optimization: Maximum First-Ply Failure Envelope of AS/3501 Cross-Ply Laminates (excluding all 0 and all 90 laminates).

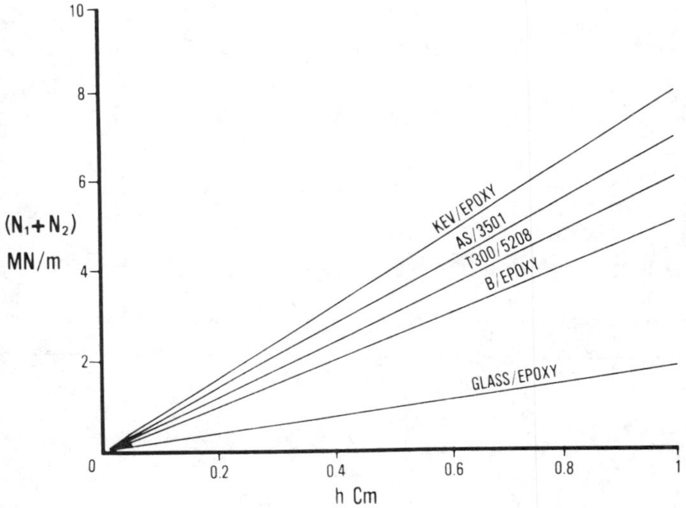

Figure 4. Comparison of Laminated Composites: Laminate Thickness Required to Support a Given Stress Resultant in the First Quadrant of Stress Space.

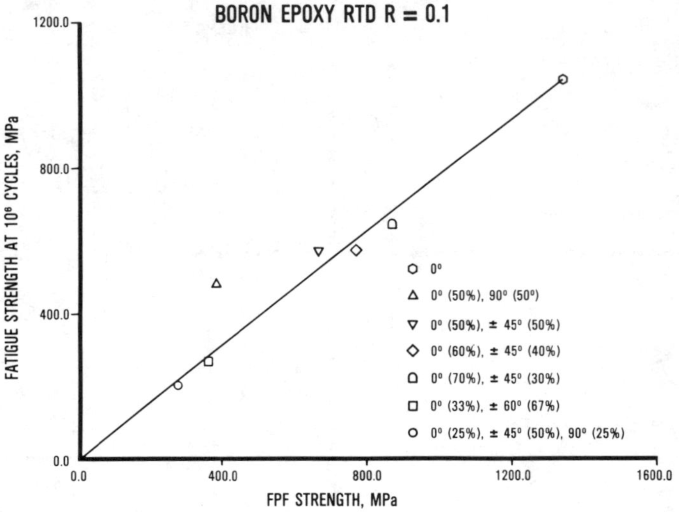

Figure 5. Correlation between Fatigue Strength and FPF Strength.

TABLE 1

STRENGTHS OF UNIDIRECTIONAL COMPOSITES IN MPa

Material	v_f	Long. tens. X	Long. comp. X'	Trans. tens. Y	Trans. comp. Y'	Shear S
Graphite/Epoxy T300/5208	0.70	1500	1500	40	246	68
Boron/Epoxy B(4)/5505	0.50	1260	2500	61	202	67
Graphite/Epoxy AS/3501	0.66	1447	1447	51.7	206	93
Glass/Epoxy Scotchply 1002	0.45	1062	610	31	118	72
Kevlar 49/Epoxy	0.60	1379	276	29.6	138	60

TABLE 2

STRENGTH PARAMETERS IN STRESS SPACE FOR UNIDIRECTIONAL COMPOSITES

$F^*_{xy} = -1/2$

Material	F_{xx} $(GPa)^{-2}$	F_{yy} $(GPa)^{-2}$	F_{xy} $(GPa)^{-2}$	F_{ss} $(GPa)^{-2}$	F_x $(GPa)^{-1}$	F_y $(GPa)^{-1}$
Graphite/Epoxy T300/5208	.444	101.6	- 3.36	216.2	0	20.93
Boron/Epoxy B(4)/5505	.317	81.15	- 2.53	222.7	.373	11.44
Graphite/Epoxy AS/3501	.476	93.48	- 3.33	115.4	0	14.50
Glass/Epoxy Scotchply 1002	1.543	273.3	-10.27	192.9	- .697	23.78
Kevlar 49/Epoxy	2.63	244.8	-25.36	277.8	-2.9	26.54

TABLE 3

MODULUS COMPONENTS OF UNIDIRECTIONAL COMPOSITES (GPa)

Material	Q_{xx}	Q_{yy}	Q_{xy}	Q_{ss}
Graphite/Epoxy T300/5208	181.8	10.34	2.897	7.17
Boron/Epoxy B(4)/5505	205.0	18.58	4.275	5.79
Graphite/Epoxy AS/3501	138.8	9.013	2.704	7.1
Glass/Epoxy Scotchply 1002	39.16	8.392	2.182	4.14
Kevlar 49/Epoxy	76.64	5.546	1.886	2.3

TABLE 4

STRENGTH PARAMETERS IN STRAIN SPACE FOR UNIDIRECTIONAL COMPOSITES (DIMENSIONLESS)

$$F^*_{xy} = -1/2$$

Material	G_{xx}	G_{yy}	G_{xy}	G_{ss}	G_x	G_y
Graphite/Epoxy T300/5208	12004	10680	-3069	11117	60.64	216.5
Boron/Epoxy B(4)/5505	10374	27646	-2988	6961	129.6	214.3
Graphite/Epoxy AS/3501	7363	7440	-1743	5821	39.22	130.76
Glass/Epoxy Scotchply 1002	1913	18881	1712	3306	24.56	198.05
Kevlar 49/Epoxy	12680	7275	-2460	1470	-172.2	141.7

TABLE 5

PLY FAILURE STRESS (MPa) FOR (1,1,0) LOADING OF AS/3501 CROSS-PLY LAMINATES

Laminate	0 ply	90 ply
$[0/90_8]$	199	135
$[0/90_4]$	273	197
$[0/90_2]$	341	280
$[0/90]$	347	347
$[0_2/90]$	280	341
$[0_4/90]$	197	273
$[0_8/90]$	135	199

TABLE 6

PLY FAILURE STRESS (MPa) FOR (3,-1,0) LOADING OF AS/3501 CROSS-PLY LAMINATES

Laminate	0 ply	90 ply
$[0/90_8]$	75	45
$[0/90_4]$	114	67
$[0/90_2]$	176	100
[0/90]	258	141
$[0_2/90]$	338	176
$[0_4/90]$	366	189
$[0_8/90]$	331	175

13th National SAMPE Technical Conference
October 13-15, 1981

DESIGN DATA FOR GRAPHITE CLOTH/LARC 160 COMPOSITES AT TEMPERATURES UP TO 600°F*

A. Wereta, Jr., J. A. Bailie, R. J. Boeddiker, and R. L. Cothern
P.O. Box 504
Sunnyvale, California

Abstract

Development of a moisture resistant 600°F advanced composite capability would provide significant benefits for certain applications in future generations of aerospace structures. The material investigated in this program is a commercially available prepreg consisting of LARC-160 addition polyimide resin on a 24 by 24 8HS Celion fabric. Adopting the philosophy that design data should be representative of manufactured hardware, the first step was to develop a cure cycle that could be readily implemented in production. A combination of analytical and empirical means were used to derive satisfactory cure cycle and layup procedures that are explained in detail. To avoid heating the working gas (nitrogen) of the autoclave to 600°F, a novel, safer method was used. Integral electric heating of the tool, controlled by a programmable temperature controller, was used to fabricate 21 by 21 inch test panels in an autoclave. This ensures tighter control of cure temperature and eliminates preheating the nitrogen before its entry to the autoclave. Panels were X-rayed and ultrasonically C-scanned prior to cutting into individual test specimens. Photomicrographs, fiber volumes, and Tg measurements were obtained for each panel. Tests included flex, short beam shear, tension, compression and in-plane shear. Test procedures are described in full. Specimens were tested both dry and after moisture conditioning at temperatures ranging from room temperature to 600°F. The results will be discussed and compared to a 350°F epoxy composite used in aerospace applications.

1. INTRODUCTION

Numerous aerospace vehicles now in production or still under development employ graphite/epoxy composites in both primary and secondary structures. Although classified as

* This work was supported by Missile System Division Independent Development Funds.

350°F systems, moisture absorption limits their upper performance levels to 300°F or less. Future generation launch vehicles may have a requirement for 600°F materials, which would increase range and/or payload by eliminating insulation. Requirements for such applications include strengths and stiffnesses comparable to graphite/epoxy composites at temperatures up to 600°F, improved moisture resistance, and reproducible processing in a manufacturing environment.

One class of resins offering high temperature capability and good processibility is the addition polyimides. Included in this category is LARC-160 [1] and PMR-15.[2] The chemistry of these resins is fairly well understood, and preliminary processing and manufacturing studies have provided enough confidence to undertake design studies.[3, 4, 5] Very little data is available on material strength, and even less on the influence of absorbed moisture on its degradation. Most of the data that is available refers to tape reinforcement rather than cloth. Continuing to develop the processing and characterization of graphite cloth/LARC-160 composites reported previously [6], innovative manufacturing technology was used to fabricate relatively large (21 in. by 21 in.) panels, to evaluate key mechanical properties under realistic environmental conditions, and to develop a preliminary data base to serve as a basis for comparison with other promising high temperature systems. The evaluation program is dedicated to the following philosophies:

(1) Integrate design, analysis material processing, and manufacturing functions early in the development effort.
(2) Understand the resin chemistry and rheology in order to develop processes that ensure optimum mechanical performance.
(3) Control the reproducibility of prepreg handling and processing in a manufacturing environment.

2. PROCESS DEVELOPMENT

LARC-160 is a NASA developed polyimide that is prepregged as a mixture of monomeric reactants in an alcohol solution.[1] The resin is first imidized to form oligomers via a condensation reaction that evolves both water and ethanol as by-products, and is then crosslinked at a higher temperature via an addition reaction to give a structurally useful material. Consequently, the process consists of two stages. In the first stage, where imidization is promoted, the reaction volatiles must be removed by a combination of diffusion and vacuum. The extent of imidization and the amount of volatiles removed influence the degree of flow during laminate consolidation. Since the condensation

and addition reactions are not completely separable, trade-offs are possible. By driving the imidization reaction to completion, the void volume is minimized, but higher pressures are required for consolidation. On the other hand, additional flow can be gained by retarding the imidization reaction but only at the expense of a higher void content. Factors contributing to the cure options include the type of reinforcement (cloth or tape), the available pressure (autoclave or press), and the part complexity (thickness and radii).
The second process stage advances the cross-linking reaction to the extent that a dimensionally stable composite is formed. Postcuring is desirable for the removal of traces of residual volatiles which act as plasticizers and lower the glass transition temperature, Tg. This can be a free-standing postcure, depending on part rigidity and allowable tolerances.

The cure cycle currently used at Lockheed Missiles and Space Company for fabricating LARC-160 panels is shown in Figure 1. It was derived semi-empirically, beginning with a NASA recommended cure cycle for tape prepreg and modifying it for cloth prepreg by a combination of hands-on experience and analytical studies addressing cure kinetics and resin rheology.[7] The process variables are shown in Figure 1.

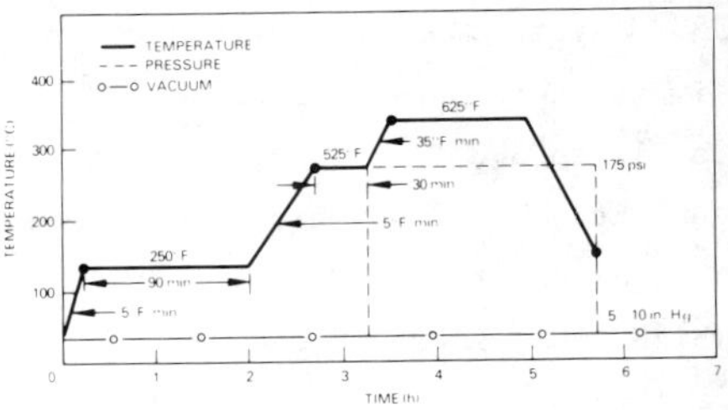

Figure 1 LARC-160 Cure Cycle

Panels were given a free-standing postcure at 620°F for 4 1/2 hours, with a one-hour hold at 550°F to prevent internal damage by expanding volatiles. Maximum oven heat-up rate was 18°F/min.

At this point, a study of the energy required to operate an autoclave at the required temperature of 600° and 200 psi was initiated. Since rather close control of heat-up rates and cure temperatures are desired, electrically heated autoclaves were appropriate. The autoclave selected for use in producing 21 by 21 in. panels to develop and verify cure cycles, was a 4 ft. diameter by 8 ft long clave. The energy required to run the circulating fan, as well as heat the inert gas, increased the cost of the development program and threatened to limit the number of panels and thereby reduce the data base. In addition, the cost of high temperature auxiliary materials, i.e., bags, sealants, vacuum couplings, etc., were considerably higher than those used for epoxy matrix materials.

A satisfactory solution was to design and fabricate an integrally heated tool, insulate both it and the panel from heat losses, and use the autoclave solely as a pressure vessel. The design was fairly simple. A 24 by 24 by 2 in. piece of molded graphite block was ground flat and parallel. Six 1 in. holes on 3 in. centers were bored through the side to allow the insertion of tubular resistance heaters. The graphite was placed in a 0.25 in.-thick steel shell, and castable ceramic was poured 2 in. thick around the bottom and all four sides. Provisions were made for a vacuum manifold around the periphery of the graphite, and for a controlling thermocouple embedded in the graphite. Vacuum ports, a thermocouple coupling and wiring ports were run through the steel box. The layups were placed on the graphite surface, a rubber caul placed on top, and standard bagging film and sealant were used to complete the layup. The maximum power consumption during heat-up is 9 kilowatts, with considerably less power required to maintain cure temperature. This power consumption is minimal compared to that of heating the autoclave in the conventional manner and circulating the heated gas.

In order to control both hold temperatures, and more importantly, heating rates, a Honeywell DCP-7700 microprocessor was installed on the autoclave panel and wired to the heating elements of the internally heated tool. This minimized panel-to-panel process variations and provided for a systematic evaluation of the key process parameters.

Although the process optimization studies are still not complete, it was decided to utilize the existing process to fabricate a sufficient number of panels for design allowables testing. The general feeling is that the credibility of the LARC-160 system for advanced aerospace structures could be established at this time, and process optimization could be aimed at improving performance, as necessary.

3. PANEL CHARACTERIZATION

In order to relate the mechanical properties of LARC-160 to laminate quality, the morphology and the glass transition temperatures of the fabricated panels were examined. The morphology was examined by a combination of techniques which included photomicroscopy, resin digestion, ultrasonic C-scans and X-ray diffraction. Glass transition temperatures (Tg) were determined by thermomechanical analysis (TMA), which measures the thermal expansion of a sample as a function of temperature.

The data related to panel morphology is shown in Table 1 and will be discussed first. Each method has its own merits and shortcomings. The resin digestion and photomicrographs are quite descriptive of small areas of a panel but may not be representative of the entire panel. On the other hand, X-ray and C-scan techniques examine the entire panel, but

but lack the definition obtainable from the small sample techniques. The results for LARC-160 indicate inconsistencies between these methods. Based on resin digestion results which show significant differences in fiber volumes and void content, one would expect large differences in panel quality between panels 051280 and 051380. However, this is not supported by any of the other techniques, nor is it indicated by the mechanical test data for tension, flex, or short beam shear where various sections of these panels were tested. Also based on resin digestion, one would expect panel 062080 to be of good quality, but C-scan, X-ray, and photomicrographs all indicate the presence of voids. Additional resin digestion data, thus sampling a greater area of the panel, is required for clarification. All of the resin digestion data and the photomicrographs were taken from samples within 1 to 2 in. from a panel's edge in order to maximize the number of mechanical specimens that could be extracted from a given panel. One might argue that the attempt to exclude edge effects may not have been successful. Note that the calculated per-ply thicknesses are not consistent with the data from resin digestion. Clearly, additional work is required to develop correlations between these morphological techniques in order to adequately describe laminate quality.

The Tg data is presented in Table 2, where as-cured and postcured Tg plus time internals between panel fabrication, Tg measurements, and mechanical testing are shown. The first observation is the relatively low Tg for both as-cured and postcured panels and the very disturbing result that some postcured Tg's were actually lower than the corresponding as-cured Tg's. Our previous experience showed a postcured Tg of 662°F or higher when using the same postcure schedule.

Since the characterization panels (ca 21 by 21 in.) used for this study were postcured intact and previous results were obtained from smaller panel sections (ca 4 by 10 in. or less), the spacial distribution of Tg's was examined. The results for Tg measurements on panel 090480 are shown in Figure 2.

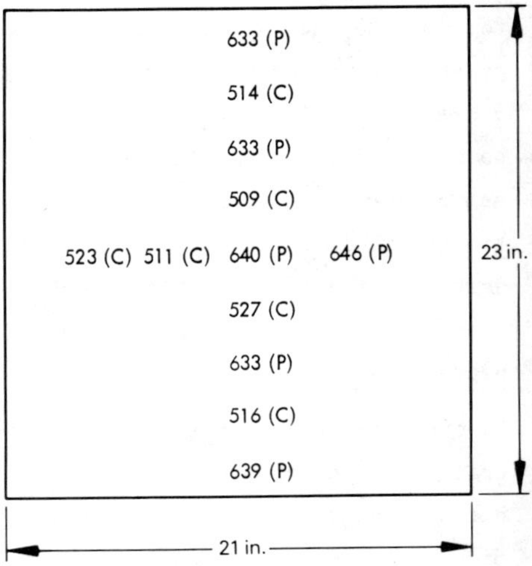

Figure 2 Distribution of Tg for LARC-160 Panel 090480

P: postcured C: as-cured

The as-cured and postcured Tg's are much closer to what was expected, and the uniformity of Tg's within the panel was quite good. These results indicate that the reason for the low Tg's was not panel size. The possibility of ambient moisture absorption causing plasticization of LARC-160 was then investigated. Sections of remnant panels already characterized were dried at 392°F for 2 hours and new Tg's were determined. As shown in Table 2 for panels 051280 and 051380, the Tg's were increased to 671° and 664°F, respectively. No additional cure is likely at 392°F, so one can conclude that desorption of a plasticizing volatile is responsible for the increase in Tg. Since the panel had only been exposed to ambient conditions, the assumption of water being the volatile appears reasonable. More dramatic evidence of the same effect was derived from a panel which was fabricated in Nov. 1979. Six days after postcure the Tg was determined to be between 644° and 662°F (two measurements). Ten months later the Tg was found to be 507°F. The effect of ambient moisture absorption on mechanical performance should be bounded by the specimens tested after saturation at 90 percent RH. Very little effect was found at 60 percent RH because of the substantial absorption of the ambient specimens exposed to 75°F/50 percent RH prior to testing (see Table 2).

A qualitative description of Tg behavior is offered in Figure 3, where time at ambient conditions accounts for moisture absorption and hence plasticization.

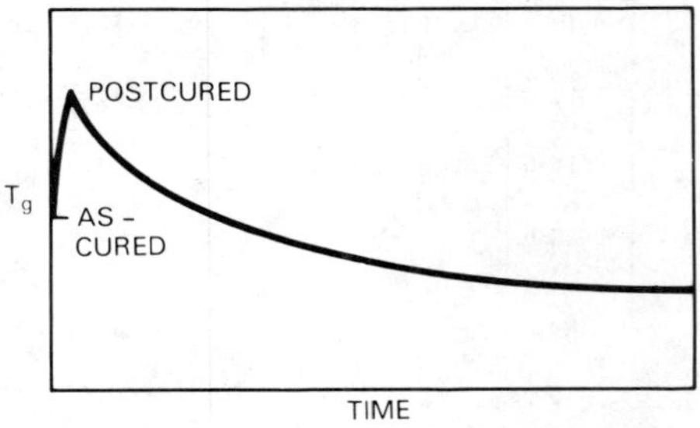

Figure 3 Tg as a Function of Time (Qualitative Description)

Quantitative data will be forthcoming. Unfortunately, the Tg's tabulated in Table 2 reflect neither the Tg at the time of testing nor the highest Tg (dry) existing immediately after postcure. These results serve to remind us that Tg is not a simple material parameter but depends on both environmental conditions and time. Consequently, the conditions under which it was obtained should be adequately specified. The highest Tg's (dry) are the best indicators of panel quality in terms of judging the extent of cure while the saturated Tg's are the best indicators of mechanical performance.

Table I

Panel No.	Fiber Volume (%)	Resin Content (%)	Voids (%)	Per Ply Thickness	Photomicrographs	C-Scan	X Ray
051280	61.4	31.0	0.79	13.1	No voids, fair compaction	Clear	Clear
051380	65.6	26.4	2.23	13.1	No voids, good compaction	Clear	Clear
061780	66.1	32.1	1.81	14.0	No voids, fair compaction	Clear	Clear
062080	65.1	32.7	1.80	12.9	Voids, excellent compaction	Dispersed regions of high scatter	Widespread porosity in resin pockets
090480	67.4	25.1	1.79	13.3	No voids, excellent compaction	Clear	Clear

fiber density = 1.78 g/cm^3 resin density = 1.30 g/cm^3

Table II

Panel	T_g As-Cured (° F)	Δt_1 [a] (days)	T_g Postcured (° F)	Δt_2 [b] (days)	Mechanical Test	Δt_3 [c] (days)
051280	541	28	518[d]	79	Tension	130
					Flex	48
					Short Beam Shear	52
051380	550	28	523[e]	79		
061780	543	28	586	60	±45° Tension	–
062080	513	25	No Postcure		Compression	86
090480	516	41	637	7	Compression	26

(a) Time interval from date of panel fabrication to date of T_g measurement.

(b) Time interval from date of postcure to date of T_g measurement.

(c) Time interval from date of postcure to date of mechanical test.

(d) T_g raised to 671° F after 2 h at 392° F.

(e) T_g raised to 667° F after 2 h at 392° F.

4. STRUCTURAL TESTING

The four tests used to obtain tension, compression, flexural and short beam shear data were all conventional and have been frequently used in composite testing. There were five replicates of each test.

Tension specimens were six ply 9.0 ± 0.25 in. long, 0.5 ± 0.01 in. wide. Glass epoxy tabs, 6 plies thick and 2.5 in. long, were chamfered to zero thickness over the half inch adjacent to the test section.

Compression specimens were the conventional 0.25 ± 0.01 in. wide, 8-ply thick with 0.5 in. test section. Two-ply graphite cloth/LARC 160 tabs were bonded with HT432 epoxy phenolic adhesive.

Flexural and S.B.S. specimens were 6-ply with spans between test rollers of 2.5 in. and 0.34 in., respectively. Conventional test facilities were used throughout, with thermal environments provided by air guns. Thermocouples attached to every specimen were utilized for test temperature control.

Heating procedures were based on assuring equilibrium temperatures prior to load application. As compression data was of greatest interest, the heating procedure for these tests will be discussed in some detail. Prior to tests at a prescribed temperature, the entire fixture was heated with loading platens and a dummy test specimen in place. An air gun temperature was set at the test temperature, until thermocouples in contact with the composite specimen reached this temperature. As an example, at 600°F this fixture conditioning took about 50 minutes. For each test, the Celanese fixture was disassembled, a specimen was installed, and the thermocouple was attached. During this, the "hot box" was closed to ensure continuous heating of the platens. Temperatures measured at the specimen side began to rise rapidly as soon as it was installed due to heat conduction from the fixture. The fixture was then reinstalled in the loading fixture and the "hot box" closed. Total time of the Celanese fixture being out of the box was under a minute. For 600°F tests, the thermocouple indicated this temperature is roughly 70 seconds after "hot box" closure. Two minutes later loading commenced.

Similar procedures were used for the other tests. Test temperature was held steady for two minutes to ensure thermal equilibrium prior to load application. This dwell time was based on response data obtained from thermal calibration specimens having thermocouples inserted in holes drilled into their midplanes. These thermal calibration specimens were not loaded structurally.

To measure strains in tension and compression tests, special commercially available clip-on relative deflection gages were used. They are based on the quadraflex principle, having strain gages bonded to a pair of arms. Eastman 610 adhesive, cured at 350°F and post-cured at 600°F was found to be satisfactory for roughly 20 tests at 600°F before rebonding became necessary.

One series of tests was conducted on specimens left in the ambient atmosphere for 16 weeks between fabrication and testing. Moisture absorption was not measured but it is estimated at roughly 0.3 percent by weight during this time. These are referred to as "ambient" specimens. A second test series were moisture conditioned at 60 percent RH at 160°F until weighing revealed no further weight gain. The weight gain in this case was measured as 0.5 percent.

5. DESIGN ALLOWABLE DATA

All of the LARC-160 specimens tested came from one 200 lb. batch of

Fiberite's HMF - 1182/78 prepreg. A single lot of Celion 3000 fibers was used to weave a 24 by 24 8-harness satin fabric devoid of any tracers. Fiber certifications were as follows: tensile strength of 476 ksi, modulus of 34.1 msi, density of 1.78 g/cc and fiber finish of 0.9 percent EP-01 resin. The prepreg had a dry resin content of 36 ± 2 percent and a volatile content of 11 ± 2 percent. Tack could best be described as moderate.

Mechanical properties for LARC-160 cloth composites in tension, Celanese compression, flex and short beam shear, are presented for "ambient" and saturated at 60 and 90 percent relative humidity conditions in Figures 4 through 7. Ambient should not be confused with dry or as-fabricated because there was moisture absorption prior to testing. With the exception of the compression data, where one as-cured panel was also tested, all other data refers to postcured properties. For comparative purposes, Fiberite 934 graphite (23 by 24 8-harness satin weave)/epoxy data from a Lockheed database is also shown.

Average tensile strengths in the warp direction range between 107 and 118 ksi between RT and 600°F and pass through a maximum around 350°F (Figure 4).

Presumably, the lower values at RT are due to internal shrinkage stresses arising from cooling below

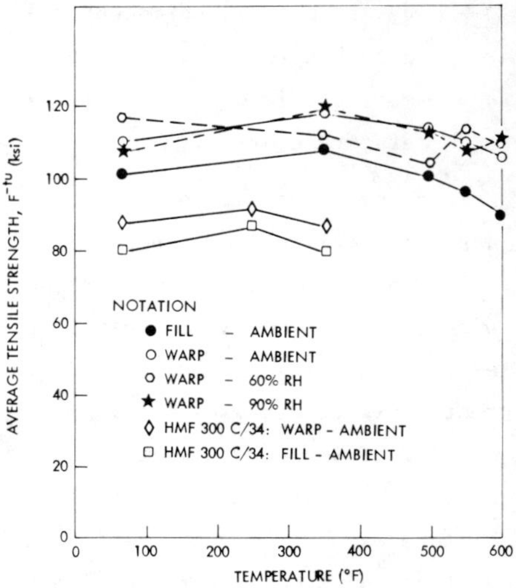

Figure 4 Tensile Strength

the cure temperature, while the lower values at 600°F are due to resin softening. The same trend is evident for the fill direction where the strengths range between 88 and 108 ksi and for the graphite/epoxy composites system designated HMF 330C/34 as well. While these values are clearly fiber dominated, it is evident that the matrix also contributes. In fact, one LARC-160 panel of poor quality (3.3 percent voids), whose results are not included in Figures 4 through 7, gave tensile strengths closer to 80 ksi throughout the same temperature range. The LARC-160 data is significantly higher than the Fiberite 934 data and this can be attributed to lower tension strengths (ca 400 ksi) for the Union Carbide T-300 fibers used in those specimens. Moisture conditioning at 60 or 90 percent RH causes no significant changes in tensile strength when compared to the ambient data. The low value of 500°F is not considered significantly different in that the error bands for the both ambient and 60 percent RH data overlapped.

Celanese compression strengths for as-cured and postcured LARC-160 specimens, conditioned at ambient, 60, and 90 percent RH, are shown in Figure 5.

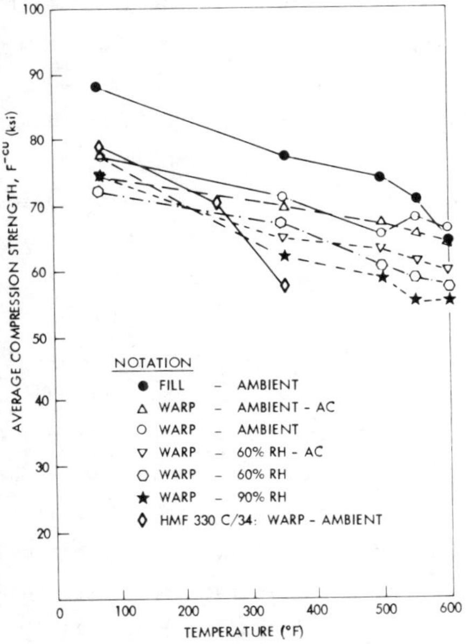

Figure 5 Compression Strength

Beginning with the ambient data, one finds postcured compression strengths in the fill direction ranging from 88 ksi at RT at 65 ksi at 600°F. Surprisingly, the warp direction strengths are lower than the fill, but little difference is found between the as-cured and post-cured strengths in this direction. As discussed in the section entitled PANEL CHARACTERIZATION, there is strong evidence for moisture absorption during ambient storage prior to mechanical testing. One might expect that the moisture absorption of the postcured specimens would exceed that of the as-cured specimens with the result that at the time of testing there might be little difference in the Tg's. Average ambient compression strengths in the warp direction range from 77 ksi at RT to 65 ksi at 600°F, thereby retaining 87 percent of the RT strength. Turning now to specimens saturated at 60 percent RH, here again, one finds no discernable difference between the as-cured or postcured compression strengths in the warp direction over the entire temperature range. There is, however, a small but definite decrease in compression strength relative to the corresponding ambient data. The magnitude of this decrease ranges from 6 percent at RT to 13 percent at 600°F. The significance of this decrease is questionable since the error bands at each temperature overlapped; error bands are not shown because of graphic complexity. After conditioning at 90% RH, a significant drop (about 15%) in compressive strength was found at temperatures between 350 and 600°F. However, there was still a 72% retention of RT strength at 600°F. The coefficients of variation for all the data were, however, quite low and ranged from 1 to 9 percent with 5 percent being typical. The failure mode in all cases was primarily shear, with some mixed-mode failures indicated by one or more "V's" apparent in the failed cross-sections. RT failures were more "broomlike" than those at 550° or 600°F. By way of comparison, LARC-160 compression strengths are as good or superior over the entire temperature range of RT-600°F to HMF 330C/34 compression strengths over the limited temperature range of RT-350°F.

Flexural strengths for postcured LARC-160 specimens are shown in Figure 6.

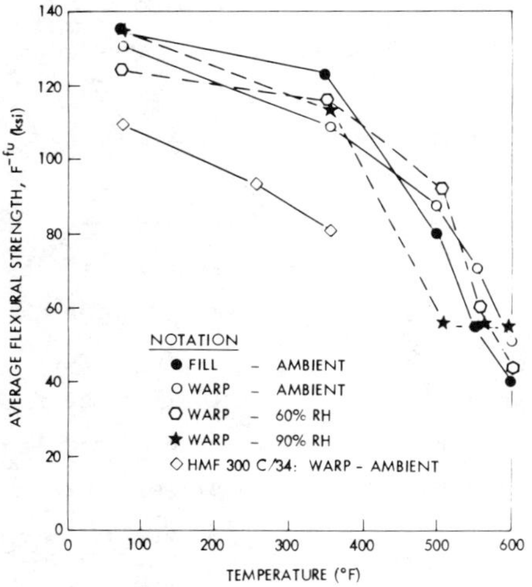

Figure 6 Flexural Strength

Warp direction strengths at RT averaged 129 ksi and dropped to 46 ksi at 600°F, exhibiting only 36 percent strength retention. Similar values were observed in the fill direction where RT strengths averaged 134 ksi and dropped to 30 ksi at 600°F. The largest difference occurred at 350°F (123 ksi in the fill direction versus 108 ksi in the warp direction).

The 60 percent RH flexural data (warp) shows no significant effect of moisture when compared to ambient data. The 90 percent RH data shows little effect of moisture at all temperatures except 500°F where a 60% reduction of strength was found relative to ambient conditioned specimens tested at the same temperature.

This result may be a consequence of rapid diffusion above Tg and the stress gradient in a flexure specimen; i. e., the wet Tg is in the vicinity of 500°F and higher temperatures quickly dry the outer plies where stresses are highest. The most striking observation for all these curves is the steep drop-off in strength above 350°F. Both flex and compression data are matrix dominant, but the strength retention was much higher in compression. The flex failure modes were difficult to determine visually, but many of the high temperature failures occurred directly beneath a loading roller. Using a fluorescent penetrant (DIB) and X-raying the specimens revealed that the majority of failures were interlaminar, sometimes occurring between rollers and other times occurring between a roller and a free edge. Improvements in both testing technique and revised specimen geometry are now under investigation.

The only flexural LARC-160 data available for comparison is zero-degree tape (Celion 3000) data taken from Rockwell International's work under NASA's CASTS Program. RT strengths averaged 243 ksi and dropped to 144 ksi at 600°F, thereby retaining 59 percent of its strength. It should be clear that the high dropoff for the cloth data is not an inherent characteristic of the LARC-160 resin, and other causes should be investigated. Since the compression properties show good strength retention at 600°F, processing was not at fault.

When compared to the HMF 330C/34 database, the LARC-160 shows superiority up to 500°F at which point the LARC-160 is equivalent. At higher temperatures, LARC-160 becomes inferior to HMF 330C/34 at its maximum use temperature of 350°F.

Short beam shear strengths for LARC-160 specimens are shown in Figure 7.

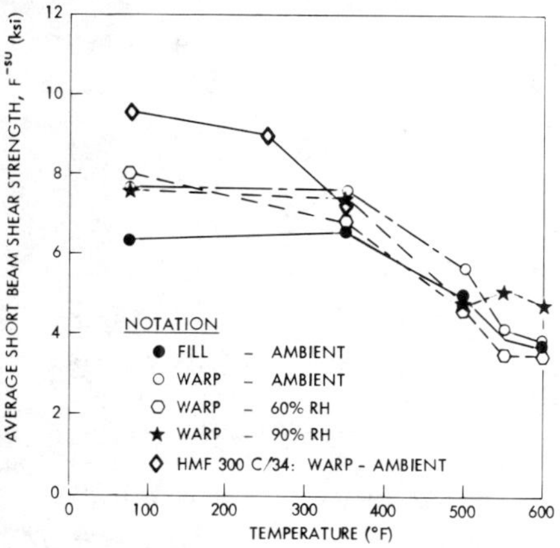

Figure 7 Short Beam Shear Strength

Warp direction strengths at RT averaged 7.8 ksi at RT and dropped to 3.6 ksi at 600°F, exhibiting a 46 percent strength retention. Strengths in the fill direction were somewhat lower at RT (6.4 ksi) but virtually identical at 600°F. Failure modes were clearly in shear with evidence of some permanent deformation at 500° and 600°F. Moisture conditioning at 60 or 90 percent RH caused no significant differences in short beam shear strength over the entire temperature range. By way of comparison, zero-degree tape short beam shear strengths (3) went from 17.4 ksi at RT to 9.2 ksi at 600°F, retaining 53 percent of its RT strength. No large discrepancy in strength retentions between tape and cloth is observed here as was the case for flexural strengths. However, the short beam shear strengths for the HMF 330C/34 system are clearly superior to those for LARC-160 over their respective temperature ranges. This may be a consequence of non-optimum processing in part, but the lower minimum viscosity during cure and the greater toughness for the epoxy are largely responsible.

In summarizing the outcome of these structural tests, the mentioned difference of temperature dependence between flex and short beam shear strengths on one hand and compression on the other deserves further attention. Similar trends have been previously noticed in unpublished graphite epoxy data, but the relative difference was not as great. This apparent anomaly gave rise to concern. At present, only a tentative explanation has been reached. The reason is believed to be the manner in which load is introduced into the test specimen. With the compression specimen, no concentrated loads are applied and smooth diffusion of the fixture load to the specimen is accomplished. In the other two test specimen types, loading rollers apply highly concentrated transverse loads. The failures at high temperature occur at the loading points. Roller marks are clearly visible on the high temperature specimens. Resin softening is known to occur at high temperatures and is responsible for the indentations. This could well lead to a local crushing of fibers underneath the rollers, which is believed to be the cause of the marked temperature dependent degradation in these two "beam" tests.

The derivation of this data and the results themselves have provided a firm basis for providing design data for this material at temperatures up to 600°F. Five replicates of every test is not sufficient to produce design curves, and cross-plied laminates are now routine and will provide firmly based design data. A separate investigation into high temperature flexural and short beam shear testing is also required.

6. CONCLUSIONS

Preliminary design data was gathered for graphite cloth/LARC-160 composites in tension, compression, flex and short beam shear tests at temperatures up to 600°F after ambient and moisture conditioning. Minimal temperature-dependent strength degradation was found in tension and compression tests. The greatest decrease appeared in the short beam shear tests. Substantial decreases in Tg were found for the ambient conditioned specimens, thereby indicating moisture absorption during the 2 to 3 months prior to testing. A need for a more quantitative measure of moisture absorption and correlation with Tg reduction was established. High temperature testing techniques were demonstrated with the result that some improvement in load introduction methods may be required for flexural and short beam shear tests in order to avoid failure at the rollers. The processing technique of using a microprocessor-controlled, integrally heated tool in an autoclave, resulted in reproducible fabrication of 21 by 21 in. composite test panels under manufacturing-like conditions. The current state of development suggests that graphite cloth/LARC-160 is a credible structural material for temperatures up to 600°F.

7. ACKNOWLEDGMENTS

The authors wish to thank K.G.Perry and S.A. Howard for their help in mechanical testing and data reduction, D.E. Lingenfelter for his help with NDE of the composite panels, and C.A. Bostwick for the measurements of glass transition temperatures.

8. REFERENCES

(1) T.L. St. Clair, and R.A. Jewell, "LARC-160: A New 550°F Polyimide Laminating Resin", 8th National SAMPE Technical Conference, Vol. 8, Seattle, WA, Oct. 12-14, 1976, p. 82

(2) T.T. Serafini, P. Delvigs, and G.R. Lightsey, "Thermally Stable Polyimides from Solutions of Monomeric Reactants". J. Appl. Polym. Sci., Vol. 16, No. 4, April 1972, pp. 905-915

(3) J.G. Davis, "Composites for Advanced Space Transportation Systems - (CASTS)". Proc of NASA Special Review of ACEE Composites Programs, Mar. 21,1979

(4) J.T. Hoggatt, W.A. Symonds, and C.H. Sheppard, "Development of Fabrication Methods for PMR-15 Polyimide Composites". Proc. of NASA Special Review of ACEE Composites Programs, Mar.21,1979

(5) J.D. Leahy, "Development of Fabrication Methods for LARC-160 Polyimides". Proc. of NASA Special Review of ACEE Composites Programs, March 21, 1979

(6) J.A. Bailie, W.C. Mace, A. Wereta, Jr., and G.D. Menke, "Development and Fabrication of Graphite Polyimide Launch Vehicle Structures". Proc. Fourth DOD/NASA Conference on Fibrous Composites in Structural Design, San Diego, CA, Nov, 1978. To be published.

(7) A. Wereta, Jr., and D.K.Hadad, "Analytical Techniques Applied to the Optimization of LARC-160 Composite Lamination", Chapter 17 in Resins for Aerospace, C.A. May, ed., ACS Symposium Series, Washington, D.C. 1980

9. BIOGRAPHIES

A. Wereta, Jr., is a Research Specialist in the Manufacturing Research Department at Lockheed Missiles & Space Company. His research centers on the rheological characterization of thermosetting resins throughout cure and the application of these techniques to quality assurance, process optimization and realtime cure monitoring of production hardware. His current responsibilities include process development and material characterization of high temperature composites and adhesives for future generation fleet ballistic missiles. Dr. Wereta is a member of SAMPE, Sigma Xi and ACS. He holds a Ph D in chemical engineering from Stevens Institute of Technology.

J. A. Bailie. Mr. Bailie is senior staff engineer in the Loads, Structures, and Dynamics Department of LMSC's Missile Systems Division. He has been project leader for an Independent Development Composite Structures and Manufacturing Program since May 1976. For the past 19 years his assignments have been involved with development to the Polaris, Poseidon, and Trident Fleet Ballistic Missiles. His training includes an apprenticeship at the de Havilland Company in

England and a Stanford University Ph D in Aeronautics.

R. J. Boeddiker is an associate engineer in the Advanced Manufacturing Technology Department at LMSC. Since joining LMSC in 1979, Mr. Boeddiker has worked extensively with high temperature polyimide/graphite composites; he is currently responsible for the fabrication of LARC-160/graphite structural test specimens.

R. L. Cothern. Mr. Cothern is responsible for the composites development group of the Lockheed Missile System Division Advanced Manufacturing Technology organization. His technical specialty is the fabrication of prototype composite components and their tooling. He is an associate manufacturing planner and has been on the staff of Lockheed for 13 years.

13th National SAMPE Technical Conference
October 13-15, 1981

AIRCRAFT COMPOSITE MATERIALS AND STRUCTURES

Samuel J. Dastin
Grumman Aerospace Corporation
Bethpage, New York 11714

Abstract

A ten year advanced composite aircraft structural history is reviewed. Composite material development is traced from fiberglass epoxy to mixed fiber polymer hybrids containing boron, graphite and Kevlar. Typical composites processing and tooling approaches utilized in the industry is given along with techniques being developed for high volume production.

1. INTRODUCTION

The reinforced plastic industry is over forty years old and is currently an established well-known commercial reality. The major material utilized in this industry is fiberglass reinforced polyester with components manufactured by hand lay-up, spray-up, vacuum bagging, autoclave molding, compression molding, pultruding, filament winding and many variations of these processes. Components manufactured from reinforced plastics demonstrate high strength to weight ratio and satisfactory life and durability. Primary structural limitations of this material compared to metals are low stiffness, bearing, and shear strengths and ductility. With the development of newer fibers such as boron, graphite and aramid in the early and mid '60s these limitations could be overcome and thus initiated a new material of construction commonly known as Advanced Composites. The aerospace industry, under sponsorship of the Department of Defense (DoD), began extensive studies in 1964 to develop and validate structural suitability and performance benefits of this new material. It was shown that many types of advanced composites can markedly improve aircraft performance but at higher costs. The newly developed fibers could be used with either organic or metallic matrices. Major development efforts during the initial study phase was with the organic matrix system utilizing epoxy resins. Currently both metal matrix composites with aluminum and titanium matrices and organic matrix composites with high temperature stable polyimide matrix are in development. Reinforcing fibers for advanced composites are primarily various types of high modulus graphite (also referred to as carbon), aramid (also referred to as Kevlar) and boron and have been well characterized and widely utilized in the aerospace industry. Current fiber development includes high strain to failure graphite, silicon carbide, boron nitride, boron carbide and several others for specialty applications. The major processing technique for the organic matrix composites is autoclave molding and the major process for metal matrix composites is diffusion bonding. These processes are being improved and automated to increase productivity and reduce labor intensity. In addition alternate processes are being evaluated to expand the industry capability and to minimize facility investment for production manufacture. The growth of advanced composites structures for aerospace applications is currently significant and expected to continue well into the 90's.

2. ORGANIC MATRIX COMPOSITE MATERIALS

Early organic matrix composite materials for aerospace applications utilized fiberglass reinforced epoxy for fairings and both fiberglass polyester and fiberglass epoxies for radomes. Development was continued with the epoxies primarily due to higher resultant strengths especially under loading that required high matrix to reinforcement adhesion. Structural evaluation of this early composite showed the need for higher laminate stiffness which was obtained with the development of continuous boron fibers. Shortly thereafter various continuous graphite fibers were also shown to overcome limitations of fiberglass reinforced plastics. Later, aramid fibers were developed that provide improved stiffness to weight ratio as compared to fiberglass reinforcements. Typical allowable unidirectional properties of boron and graphite composites are given in Table 1. The next major composite materials development was the introduction of hybrids, the use of more than one type of advanced fiber in an epoxy matrix. The basic reason for development of hybrid composites are to tailor the material for the optimum balance

Table 1. Typical Undirectional Composite Allowables in the Dry Condition

PROPERTY			BORON/EPOXY		GRAPHITE/ EPOXY TYPE T-300 OR AS-1	
			RT	350°F	RT	350°F
NO HOLES	F_1^{tu}	KSI	178	154	169	152
	F_2^{tu}	KSI	10.5	6	6.0	3
	F_1^{cu*}	KSI	480	480	162	137
	F_2^{cu}	KSI	40	12	25	15
WITH HOLES	F_1^{tu}	GROSS KSI	80	80	76	76
	F_2^{tu}	GROSS KSI	5.0	3.0	3.2	2.0
	F_1^{cu}	GROSS KSI	185	180	76	76
	F_2^{cu}	GROSS KSI	12.0	5.5	12.5	7.5
	E_{11}^t	MSI	30.3	28.3	18.5	17.7
	E_{22}^t	MSI	2.8	.9	1.6	1.1
	E_{11}^c	MSI	31.8	30.6	18.5	17.7
	E_{22}^c	MSI	2.8	.9	1.6	1.1
	G_{12}	MSI	.723	.232	.65	.20
	γ_{12}	MSI	.25	.25	.25	.25
	α_{11}	IN./IN.°F	2.5	2.5	.25	.25
	α_{22}	IN./IN.°F x 10^{-6}	13.0	22.0	15.2	20.0
	t'	IN.	.00525		.00525	
	ρ	LB/IN.^{3}	.075		.055	

NOTES: *INTERLAMINAR SHEAR MODULUS G_z IS CUT OFF ON COMP STRENGTH between performance and cost. Current organic matrix composite materials are being improved by tailoring the resin to provide greater strain to failure, higher toughness and increased resistance to thermal shock when moisture saturated.

3. ORGANIC MATRIX COMPOSITE PROCESSES

Initial organic matrix composites were manufactured by wet lay-up and either vacuum bag or press cured at low temperatures. In such processes, control of fiber volume fraction was difficult and thus resultant strengths and stiffnesses varied considerably. This variation of properties was overcome by pre-impregnating the reinforcement and using stageable epoxy resins to produce "prepregs". Subsequent development showed that pressure cures of prepregs resulted in consistent improved mechanical properties and thus the current standard process for advanced composites is autoclave molding at 50 to 100 psi at 350°F.

Currently advanced composites are in production on military aircraft and will soon be production introduced on commercial aircraft and thus processing techniques for high volume manufacture is required. To satisfy this need, various automated processes are being developed such as the Integrated Laminating Center at Grumman shown in Figure 1, automated filament winding, computer-aided manufacture with Robotics, automated pultrusion and several other high speed manufacturing methods.

4. ORGANIC MATRIX COMPOSITE STRUCTURES

The first major primary safety of flight production structure was the horizontal stabilizer of the Grumman F-14A aircraft shown in Figure 2. This structure was 18% lighter than a titanium equivalent component and has provided excellent performance in fleet operation during the past

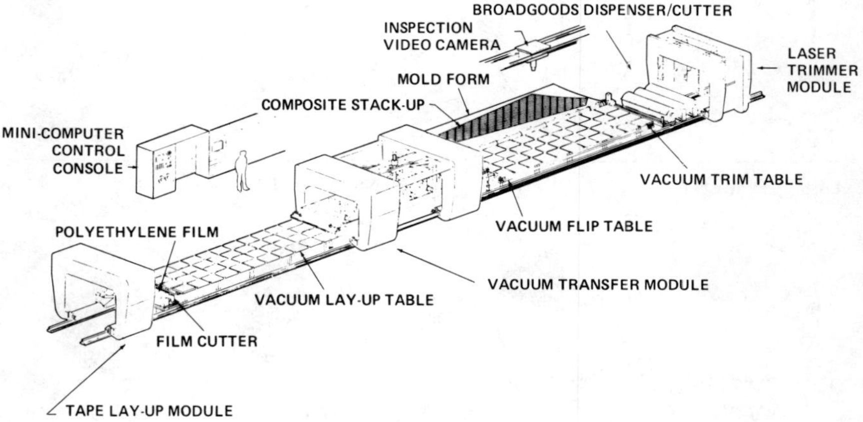

Fig. 1 Integrated Laminating Center

ten years. Composite primary structures continued to provide improved military airframes as shown by increased usage on the F-15, F-16, and F-18 as outlined in Figure 3. Presently, the AV-8B has demonstrated flight suitability with an airframe containing 26% composite weight fraction. Future V/STOL, Advanced Tactical Fighter and helicopter weapon systems will utilize composites for at least 50% by weight to achieve required performance. Growth of organic matrix composite military structures is forecasted in Figure 4 and by the year 2000 such structure will be common place and thus an established industry will have been realized.

A. AS FABRICATED

B. AS FINISHED

Fig. 2 F-14A Composite Horizontal Stabilizer

Military composite structures programs have readily shown empennage weight savings of 15 to 30% over equivalent metallic baselines as shown in Figure 5. As can be noted from the Figure, a balance between weight and cost is required for production implementation. Grumman has developed an optimal horizontal

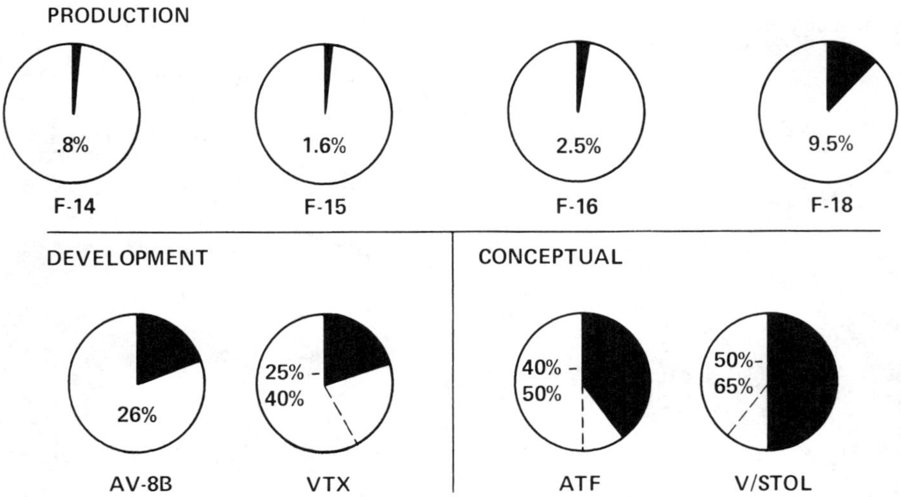

Fig. 3 Composite Structures Usage (Weight Fraction Trend)

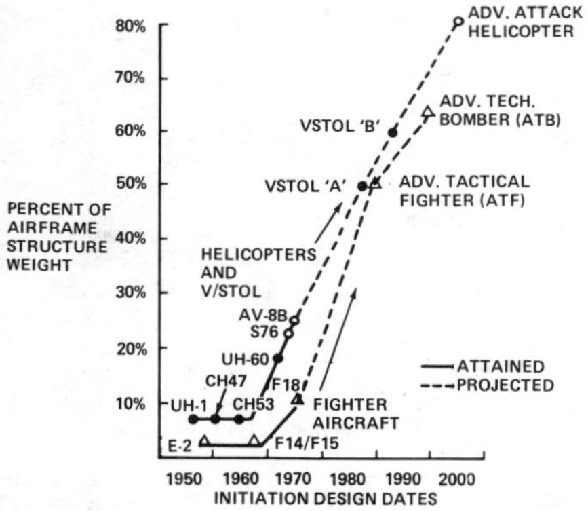

Fig. 4 Typical Growth of Organic Matrix Composite Airframe Structures

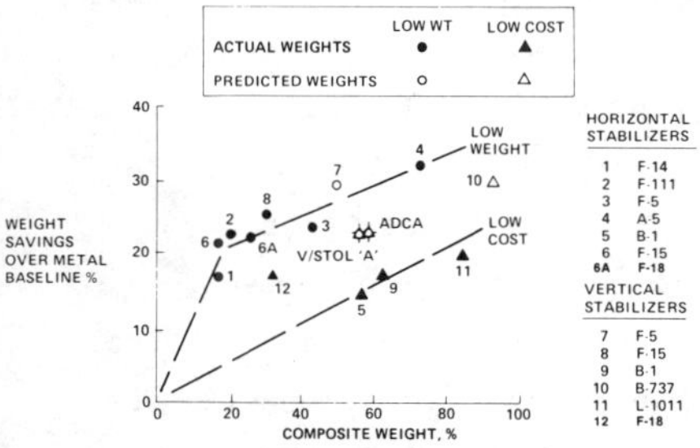

Fig. 5 Empennage Weight Savings Chart

stabilizer for the B-1 and has proven that composite structures are not size restricted and can satisfy requirements for damage tolerance in accordance with guidelines of MIL-STD-1530. Figures 6 and 7 outline the design features and certification tests successfully obtained on the structure.

Composite wing structures have been under study for several years and have shown that weight savings of 15 to 25% over equivalent metallic structures, Figure 8, are achievable. McDonnell has developed wing structures for the F-18 and AV-8B that demonstrate the flight suitability of composite wings.

Currently, composites are being evaluated in-depth for fuselage

FIRST PRODUCTION APPLICATION OF 3RD GENERATION COMPOSITE STRUCTURE (COMPOSITE COVERS & SUBSTRUCTURE)

30-FT SPAN, 17-FT ROOT CHORD

WEIGHT/AIRCRAFT = 2846 LB	
• GRAPHITE/EPOXY	48.6%
• BORON/EPOXY	7.5%
• GLASS/EPOXY	5.2%
• TITANIUM	13.3%
• ALUMINUM	18.9%
• MISC	6.9%
	100.0%
• WEIGHT SAVING OVER METAL: 500 LB, 15%	
• COST SAVING: 15–20%	

Fig. 6 B-1 Horizontal Program

- FULL-SCALE STATIC TEST STABILIZER
  - – 250° F & RT TESTS
  - – RT TO FAILURE (MOST CRITICAL STATIC CONDITION)
- FULL-SCALE FATIGUE TEST STABILIZER
  - – PROOF LOAD (± 1.1 DLL)
  - – 2-LIFE ACCELERATED, RT, FLT-BY-FLT SPECTRUM
- ENVIRON EFFECTS (TEMP, MOISTURE, REAL-TIME LOADING) ASSESSED THRU ELEMENT TESTING
- DAMAGE TOLERANCE NDI CERTIFICATION BY ELEMENT TESTING

Fig. 7 B-1 Test Program

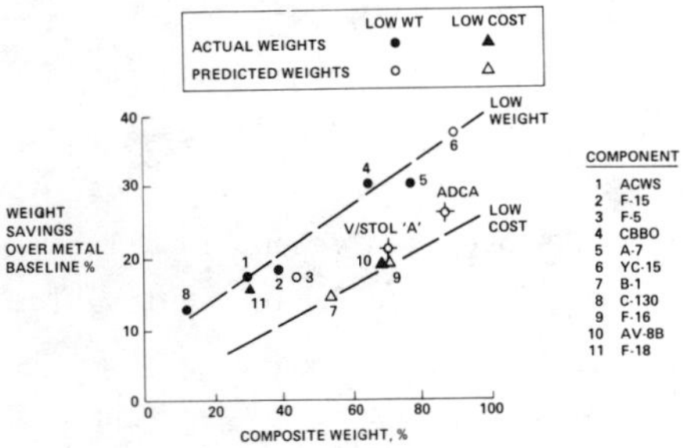

Fig. 8 Wing Weight Savings Chart

structures. Figure 9 reviews some of such programs which indicate weight savings of 12 to 28% over metallic structures. The AV-8B

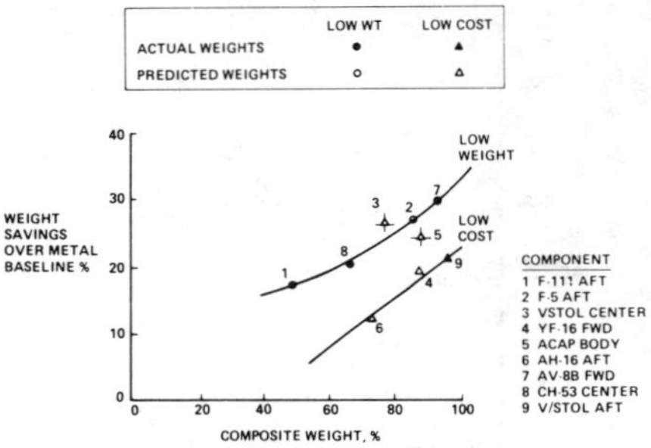

Fig. 9 Fuselage Weight Savings Chart

has shown the flight suitability of the forward fuselage and development studies at Grumman on center VSTOL fuselage are proceeding effectively. Figure 10 shows the all-composite subcomponent side panel and bulkhead which was successfully static tested to 113% of design ultimate load. Several other on-going DoD composite fuselage studies such as LTV's aft fuselage, Grumman's fuselage duct

Fig. 10 V/STOL Center Fuselage

development (Figure 11) and Northrop's wing to fuselage program will provide the needed data to overcome current concerns for fuselage structure.

Composite structures have been under serious study for the past fifteen years and have produced enough successful results to give thought to an "All Composite Airframe". Grumman's first program in this regard was an analytical study of the "Advanced Designed Composite Aircraft (ADCA)" for the Air Force Flight Dynamics Laboratory. The airframe, shown in Figure 12, indicated the use of composites for 75-80% of the

A. INTERIOR (CONTAINING STITCHING)   B. EXTERIOR (NOTE INTEGRAL CURE)

Fig. 11 ATF Fuselage Duct

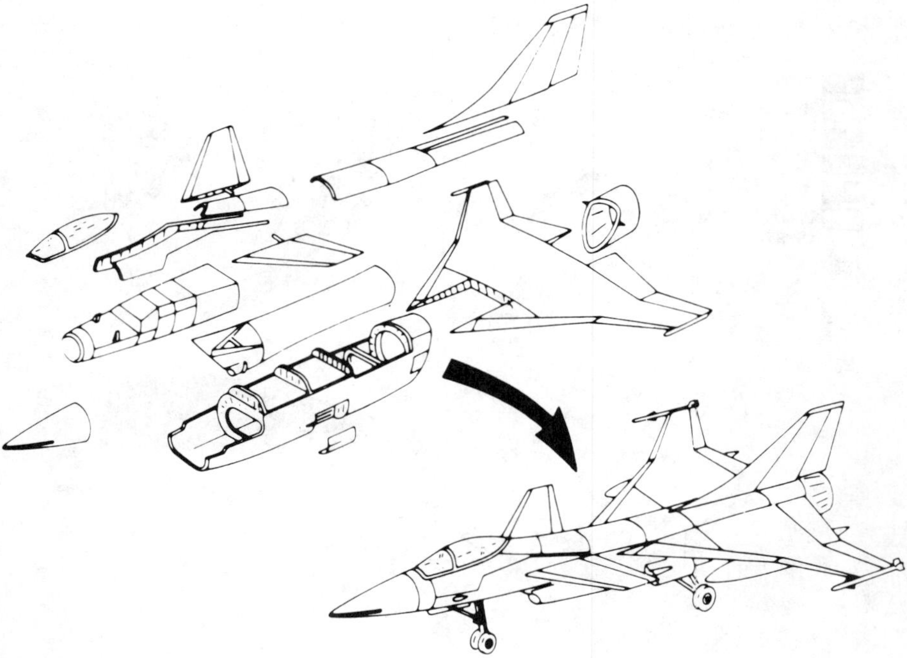

Fig. 12 Advanced Design Composite Aircraft

structures weight and resulted in an overall weapon system weight reduction of 26%. Current studies for subsonic VSTOL aircraft such as the vehicles shown in Figures 13 and 14 indicate a composite structural weight fraction to 65% with a potential system weight reduction to 20%. Currently, the U.S. Army under their Advanced Composite Airframe Program (ACAP), is evaluating an all composite helicopter structural airframe for capability to save 22% weight, 17% cost, improved ballistic survivability and lower radar cross-section than comparable metallic airframes (Figure 15).

In the commercial sector, composites are in most aircraft with expanding utility almost guaranteed. The most dynamic current program in the business aircraft sector is the "all composite" Learfan 2100. This nine passenger pressurized fuselage aircraft is contructed entirely of graphite-epoxy composite; the fuselage, wings and empennage. Successful operation and FAA certification of this aircraft will assure continued significant composite growth.

5. CONCLUSIONS

Composite materials and structures have been under development for the past twenty years and during the last 10 years have shown a steady utility growth. Growth rate for the past few years has been estimated at 8 to 10 percent with forecasts to 15% for the near future. Performance has been the major driver for the current growth and reduced

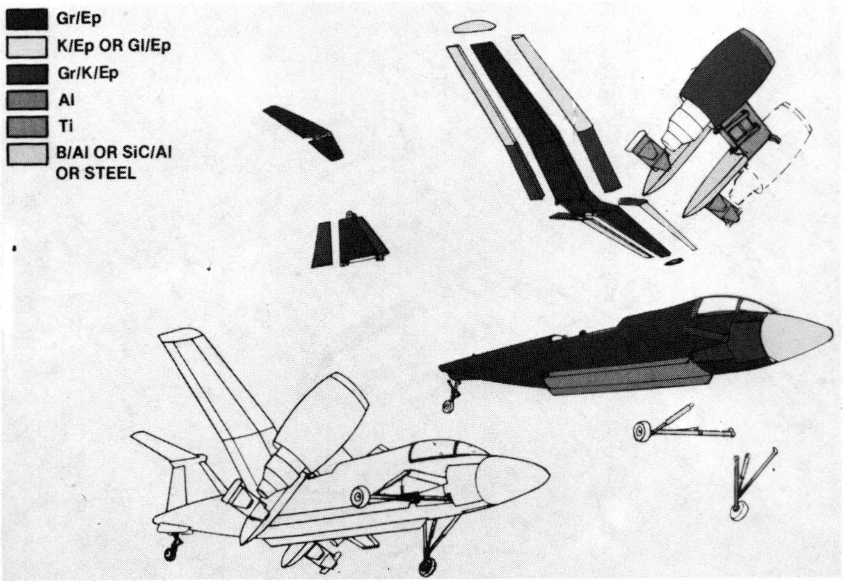

Fig. 13 Subsonic V/STOL

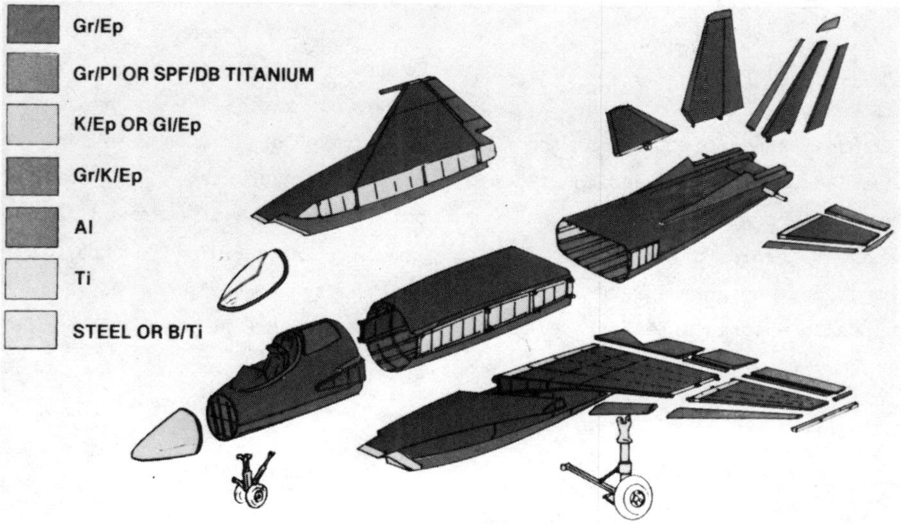

Fig. 14 Supersonic V/STOL

costs and energy consumption are the major reasons for future growth. In anticipation of the forecasted growth, many airframe manufacturers are up-grading facilities, conducting manufacturing automation studies and developing "factories of the future" to provide high volume production capability. To assure that composites are completely exploited, further exploratory and advanced development efforts are required. Some of the issues that require validated data are fatigue analysis methodology, repairability, low and high impact survivability, systems and subsystems integration, natural and hostile electrical-electromagnetic suitability, increased service temperature capability, crashworthiness and stealth effectivity.

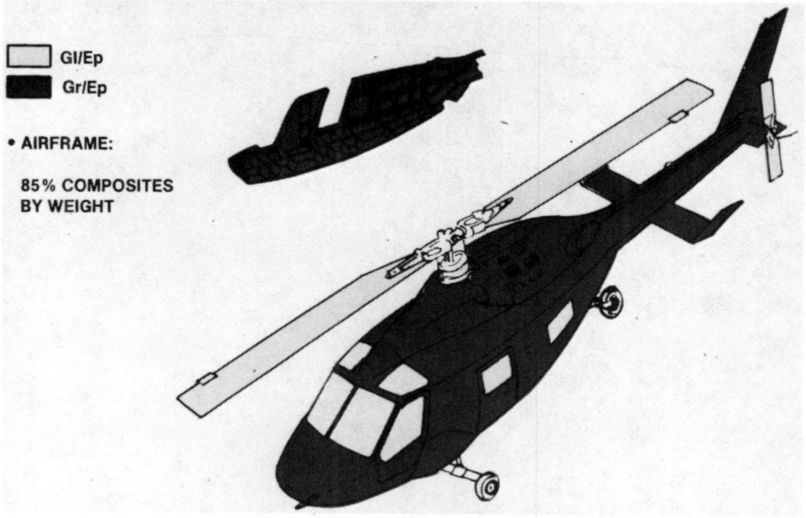

Fig. 15 Advanced Composite Airframe Program

6. BIOGRAPHY

Samuel J. Dastin

Mr. Dastin is Manager of the Advanced Composites Group of the Advanced Development Section of Grumman Aerospace Corporation. He manages a series of advanced development programs using advanced composite materials to improve the efficiency of aerospace structures. He has been involved in composite structures since 1965 and has over 27 years of experience with non-metallic structures. Mr. Dastin holds the degree of Bachelor of Chemical Engineering from the City University of New York. He is a member of SAMPE and SPI.

13th National SAMPE Technical Conference
October 13-15, 1981

A PRACTICAL DESIGN METHODOLOGY FOR THE COMMERCIAL APPLICATION OF ADVANCED COMPOSITES

James G. Claussen
International Harvester Company
Science and Technology Laboratory
Composite Materials Technology Center

Abstract

This paper discusses the design methodology developed specifically around performance critical applications which may be produced on a worldwide basis in moderate to high volumes. The significance of the initial feasibility study, in regards to "payoff" determination, is stressed. Factors to be considered during this stage, such as design trade-off and technical and economic assessments, are presented. Aspects of overall program development, component fabrication and test, and technology transfer are also addressed. The concept of "consumer value" was found to be a key factor in establishing application potential. The role of the program manager in providing an overall perspective and program continuity was also identified as being significant.

"Keywords": Composite, design methodology, commercial.

1. INTRODUCTION

Although successful application of composite technology is now common-place in the aerospace industry, the wide scale transfer of the technology to automotive and capital goods industries has yet to occur. Relatively few structural applications for composite materials have been released for production in these industries. This is the result of the high material costs and the inability to develop sufficiently low cost manufacturing techniques to counter the high material costs. The potential for composite technology to make a significant impact on the capital goods and automotive market is nevertheless great. In these industries secondary[1]

---

1. Secondary Structures: Cosmetic parts and those carrying loads imposed as a result of their own weight and/or fasteners. Examples include hoods, body panels, dash panels and fascia.

structures utilizing composite materials are currently in production. Recently, much attention has been paid to the potential of advanced composites in primary[2] structures common in the automotive and capital goods industries.

With regards to secondary structures, the use of composite materials has realized continued growth within the truck industry over the past twenty years, especially in the use of SMC hoods for medium and heavy duty trucks. As compared to their steel counterparts, composite hoods offer greater design latitude, reduced number of parts, lower tooling costs and improved corrosion resistance.

When considering primary structures, however, composites have yet to make a significant impact on either the automotive or transportation industries. Although these materials have been well proven in aerospace applications, high material costs and low-efficiency fabrication techniques appear to have constrained the transfer of this technology. In spite of a concerted effort by material suppliers and original equipment manufacturers in the passenger automotive industry, only a few applications have been released for production.

Our evaluation concluded, however, that this situation is not representative of what could be expected for the capital goods manufacturing industry. This conclusion is based upon the fact that many of these applications in truck, agricultural equipment and construction equipment industries are based upon significantly different performance criteria than the passenger automobile industry.

This is brought into perspective if one considers the concept of added "consumer value". The basis of the automotive industry's efforts to reduce the mass of the passenger car is to boost fuel efficiency. Consumer savings resulting from less frequent stops at the gas pump have been well documented. In the case of a $120,000 grain harvesting combine, (Figure 1) however, fuel efficiency is not an important incentive for lighter weight. Fleets of combines frequently operate on a continuous basis, except when being transported from field to field. Here, both the size of the crop gathering mechanism and the onboard grain

---

2. Primary Structures: Parts which, by design, carry or transfer loads as their primary function. Examples include frame rails, drive shafts and suspension components.

weight capacity can be directly related to field-time operational costs. Increased operating efficiencies resulting from mass reductions in either of these areas can add significantly to the value of the equipment. Similar examples relating decreased time in the field or increased payload capacity as a means for improving "consumer value", can be cited for construction equipment and truck applications.

Although the value of composites in these applications is now being recognized, successful implementation of these new materials would require development of a design methodology more specifically oriented toward the relevant application parameters.

This paper will present some of the key factors identified during our development of this methodology, and will suggest certain insights found useful in developing composite applications in the capital goods industry. Although many of the examples mentioned are specific to truck, agricultural and construction equipment industries, the program logic is applicable to a broad range of capital goods products.

The development of composite components for such products can be conveniently divided into three phases: initiation, validation and execution.

2. INITIATION

The initation phase originates with a product requirement that cannot be met with conventional material and design technology. The preliminary assessments determine whether there is enough potential to commit further resources. The duration of this effort is normally two to four weeks, and usually involves the interaction among product engineers, designers and the composite specialists. At this point, it is important to define as well as possible, the general objectives of the program. Application requirements, alternate or parallel programs, key timing milestones, and an estimate of resources required, must also be identified. Although selection criteria are shown in more detail in Table 1, emphasis is placed in three areas. Can the application benefit from the design capability offered by composites or are conventional materials such as steel or aluminum adequate? Can the probability of success be rated as high, or are there significant technical or economic barriers present? Finally, does the program offer potentially high paybacks in comparison to the estimated developmental costs? With regard to the last item, a global point of view must be taken when considering future implementation of the developed technology. A commitment to a specific application normally must be justified by the potential for use across several products or components.

The program manager in the industrial environment plays a similar role to that of the aerospace industry. In addition to the responsibilities outlined in Table 2, the manager must provide a sense of continuity and maintain a perspective regarding the relationship of the program team activities with respect to the overall goals. This is particularly important in longer term programs where development may span a number of years.

3. VALIDATION

The validation phase is essentially a more detailed extension of the preliminary assessments. However, whereas the early work is often limited in scope, this phase usually involves a larger group of people, representing a much broader base of expertise. As an example, Table 3 identifies groups represented on the validation team for a typical program for an application to an agricultural implement.

The duration of this phase is normally three to six months, and involves two major activities, the feasibility study and development of a detailed program plan.

3.1 Feasibility Study

The feasibility study is commonly used as the reference work when justifying the costs for a developmental program. As these expenses are often quite high, in terms of manpower, materials and capital, the need for a thorough and complete evaluation cannot be overstressed. This activity can be broken down into three major categories.

3.1.1 Economic Analysis

As far as economics are concerned, two points of view are considered. Firstly, the impact of composites on the product itself must be determined. Because composite materials are often more expensive than conventional materials, it is important to develop a clear understanding regarding the economics of the existing and/or proposed designs. Frequently, at moderate production volumes, this material penalty can be offset by significant savings on tooling costs. However, as volumes approach the high levels consistent with metal forming processes, this advantage ceases to exist. Because of this situation, it is more beneficial to consider a total system approach rather than a part-for-part substitution. Additional economic advantages are often achieved through parts consolidation, or by elimination of assembly components entirely; an example of this approach can be seen in the analysis of composite truck driveshafts, with the resulting elimination of the center bearings. The second economic factor which must be considered involves program related costs. In addition to normal manpower and laboratory expenses, these can include purchase and installation of new processing or test equipment, design software, or other special resources. As previously stated, these costs are weighed against potential payback and probability of success for the program.

3.1.2 Design Trade-Offs

The trade-offs, depending upon overall program complexity, can involve design considerations ranging from conceptual to detailed. While the preliminary design that results from this work is important, several other aspects are also significant as far as insuring that the final program outcome is relevant to the application. As an example, many parts perform subtle secondary functions, not immediately evident during the preliminary work. This is demonstrated by the truck bumper which also acts as a platform for the mechanic during engine maintainance, or the agricultural assembly which must convey an impression of "ruggedness" and "durability". Requirements such as these need to be carefully indentified and considered even when they cannot be precisely defined.

Once the desired functions have been determined, the actual operational requirement, such as environment and field loadings, must be developed. This is an area where a lot of information is usually missing, particularly that relating to structural considerations.

Where data exists, it is normally incomplete, due to the evolutionary nature of metal component development. At this point, instrumentation of an existing component and/or finite element modeling (FEM) may be beneficial in determining the operating requirements. Our experience has shown, however, that the degree of detail required at this stage, particularly in regards to FEM, should be weighed carefully against the value of the data generated. The expense associated with sophisticated analysis may provide more valuable information if applied to the detailed design of the composite part during the execution phase.

Once these tasks have been accomplished, it is then possible to evaluate the effectiveness of various design concepts. While practicality must be kept in mind, this is the opportune time for innovative thinking and one must not be prejudiced by conventional or historic designs. It will develop however, that the most viable concepts will exploit the design and manufacturing capabilities offered by composite materials, and not represent a simple part-for-part substitution. As an example, molded in ribs and bosses will replace weldments, variations in part thickness will provide localized tailoring of component strength and stiffness, and the judicious use of orthotropic material forms will provide optimized response to severe, non-uniform loading conditions. This last case is demonstrated by the part shown in Figure 2. While the general application requirements are met by the $90^{\circ} \pm \theta^{\circ}$ layers, the high vertical bending loads, imposed on this cantilevered tube, are carried by 0° fibers in the horizontal faces.

3.1.3 Technical Assessments

The final feasibility study activity concerns identifying potential technical barriers, and estimating their significance in regards to the success of the overall program. This involves evaluating the state-of-the-art for materials, design and processing, and then assessing the company's capabilities with respect to those technologies.

It can be assumed that the majority of commerical applications will utilize combinations of materials selected from the following:

Fibers	Resins
E-glass	Epoxy
S-glass	Vinylester
Graphite	Polyester
Aramid	

It was determined that sufficient published sources exist regarding properties and engineering data for these materials, to pursue initial design efforts. Where necessary, this information can be augmented by data generated on constituitive or laminate computer programs. However, as most of these data represent aerospace fabrication techniques, property verification should be addressed at the laminate and component levels, following the identification of a preliminary composite design. This will insure reliable modification of design data, consistent with the manufacturing process of interest.

In regards to manufacturing processes, emerging developments in areas such as filament winding, compression molding and resin injection, show strong potential for cost effective production. However, although many of these programs are oriented toward the production volumes associated with the automotive industry, interim developments should not be ignored in low or moderate volume situations. In many instances, the ability to address smaller, vocational market segments can justify the cost basis associated with current levels of manufacturing.

In the area of design, for non-structural applications and simple shapes, sufficient background exists to do a fairly good job with paper and pencil. However, for structural parts, especially those with complex shapes, the use of a computer is essential. A number of programs, such as ARGUS or NISA, are available with composites capability, and additional developments are underway. Currently, however, due to the "unfriendly" nature of existing software, the program manager outside of the aerospace industry will frequently find that the skilled personnel required to execute these programs often have no background in composites. Whether dealing with a structural analyst with a background in metals, or a designer with a background in aerospace composites, the program manager must maintain the perspective required for a commercial application.

3.2 Program Development

This activity covers planning and organizing the resources required over the life of a program. Although within a particular organization these procedures are often tied to some established format, the following generally applicable guidelines should be noted:

As success is dependent upon taking a team approach, adequate communications are essential. This involves periodic program review and critique by the entire team, as well as the usual written communications.

The major application timing milestones must be clearly defined. The significance of changes in deadlines, or failure to meet milestone dates, must also be addressed.

Individual tasks must be identified and assigned. Each team member should be aware of his responsibility and the impact on the overall program. Firm commitment, preferably in writing, should be obtained for the required resources.

Major program decision points should be established, and contingency plans developed.

Finally, while being rather detailed in nature, the program organization must remain flexible enough to quickly respond to new input as it occurs.

4. EXECUTION

This final program phase involves detailed design, prototyping, testing and finally production implementation. The flow of activities, commonly spanning several years, is shown graphically in Figure 3. With the exception of the initial stages, the approach shown is typical of the methodology employed in the capital goods industry for conventional material design, and will not be addressed in this paper.

The major difference between this procedure and normal industrial developmental programs is the heavy reliance on FEM. Using this analytical tool, the designer can rapidly evaluate the wide variety of material combinations, thus avoiding much of the costly and time consuming prototype effort which would otherwise be required. This is particularly true for orthotropic laminated materials, where intuitive design is of little value. Once again, however, the cost/value relationship of this technique must be kept in perspective. In general, the degree of model complexity should parallel the design concept development. A coarse model should be utilized for preliminary designs, with additional detail being added in critical areas, as the design firms up.

Once a suitable design has been developed by iterating material properties, performance requirements, part shape and laminate construction, development returns to a more familiar course, with production of the initial prototype.

5. SUMMARY

Cost effectiveness is the critical factor controlling application of new technology in the capital goods industries. This is in contrast to the aerospace industry, where performance is paramount and cost usually secondary. Application of composite technology to primary structures in the automotive and capital goods industries has been delayed by the high cost of some materials and by inefficient fabrication methods. Nevertheless, the potential remains very high. This is particularly so in capital goods due to the applicability of cost-effectiveness.

In the automotive industry, the primary rationale for composites is weight reduction to achieve fuel efficiency however, cost competitiveness is still the deciding factor. For capital goods, the broader concept of "consumer value" is used. This relates to the overall cost of equipment used by a businessman to make profits. This type of customer could tolerate initial cost penalties in the interests of net profitability. This difference in interpretation of value does require more complex value measurements during the pursuit of capital goods applications.

BIOGRAPHY

Jim Claussen received his B.S. in Chemistry from the University of Dubuque, Dubuque, Iowa in 1965. Upon graduation he joined the Corporate Materials Research Group at International Harvester and proceeded through a number of assignments involving development and application of polymeric materials. He was promoted to the position of Group Leader - Rubber and Plastics Laboratory in 1973, and was responsible for planning and coordination of research programs and technical service. He was assigned as Group Leader of the Science and Technology Composites Program in 1978. He joined the newly formed Science and Technology Laboratory, Composite Materials Technology Center in 1979 and was promoted to his present position as Senior Engineer, Application Development Section in 1980. He is a member of SPE and SAE.

TABLE 1

SELECTION CRITERIA
<u>Return/Risk</u> Paybacks divided by development cost.
<u>Consistent with Business Goals</u> Meaningful competitive advantage; leadership through product reliability and diversification; generate new/venture business.
<u>Timeliness</u> Legislative or market competition pressures.
<u>Expected Return</u> Probability of success with acceptable return.
<u>Technological Experience</u> Potential for building technological capabilities.

TABLE 2

PROGRAM MANAGER RESPONSIBILITIES
<u>Planning</u> Define objectives and select course of action.
<u>Organization</u> Implement planning; establish structure, procedure and resource requirements.
<u>Direction</u> Coordinate activities and insure proper execution of team assignment
<u>Control</u> Keep operation on intended course, provide checks and direct changes where necessary.

TABLE 3

VALIDATION TEAM	
<u>Agricultural Equipment Group</u>	<u>Science and Technology Laboratory</u>
- Product Engineering	- Electronic Controls Technology Center
- Materials Engineering	- Computer Technology and Structural Mechanics Team
- Applied Mechanics	- Composite Materials Technology Center
- Product Reliability	
- Marketing	
- Engineering Design Service	

Figure 1: 1480 Axial Flow Combine

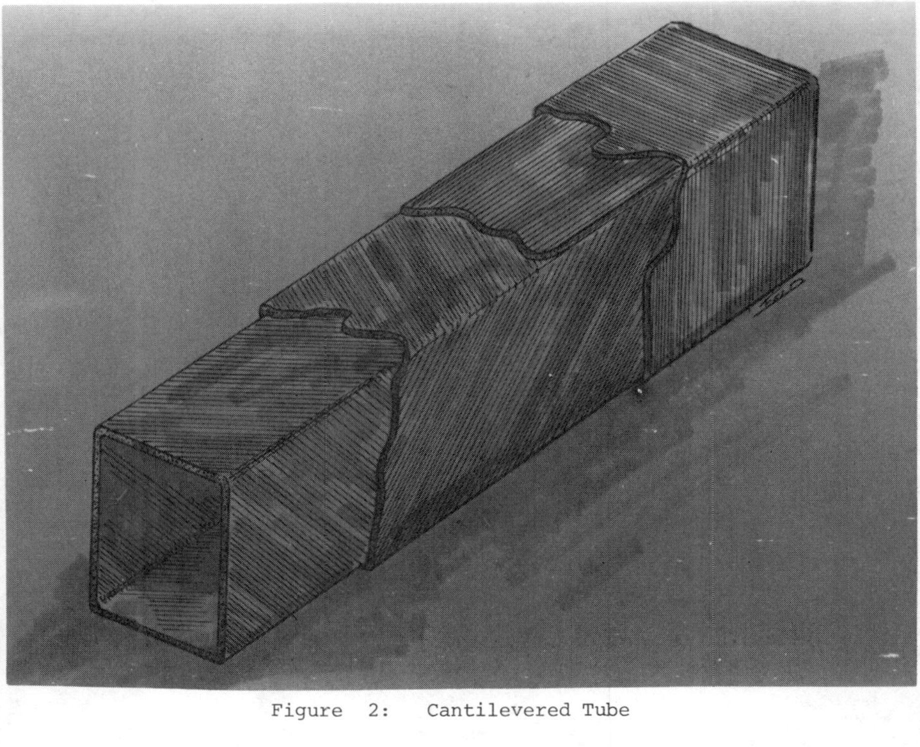

Figure 2: Cantilevered Tube

FIGURE 3

EXECUTION PHASE

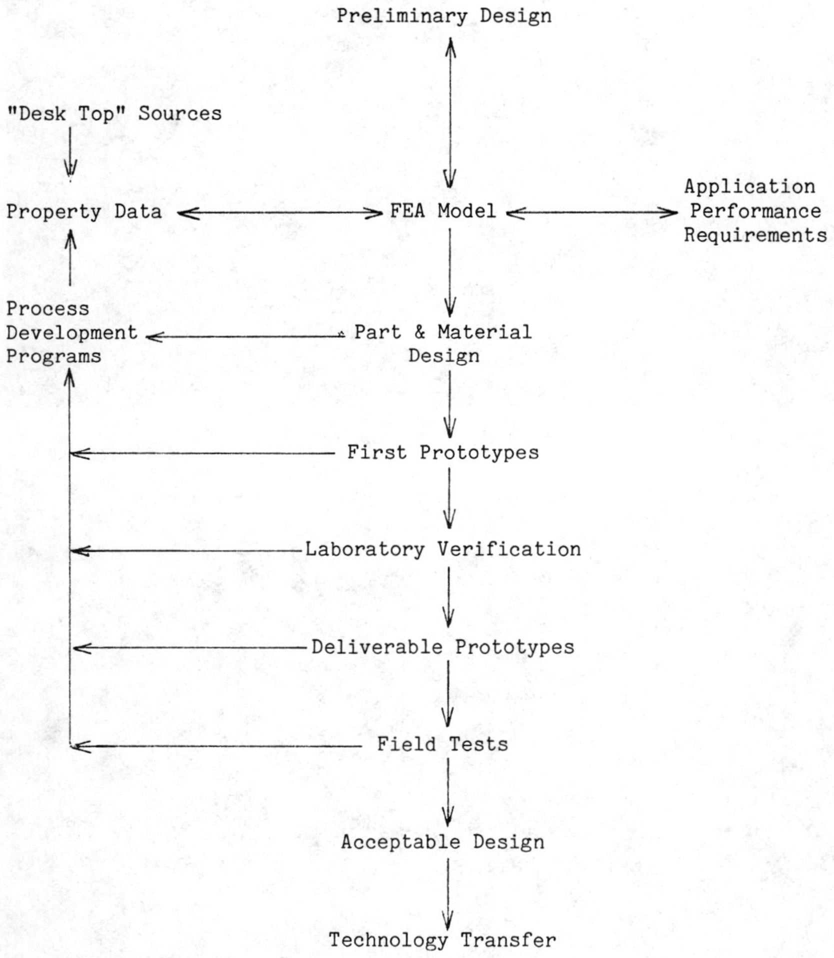

13th National SAMPE Technical Conference
October 13-15, 1981

COST/BENEFITS OF USING PREPLIED GRAPHITE EXPOXY MATERIAL IN A PRODUCTION ENVIRONMENT*

Timothy J. Bettner
Northrop Corporation
Aircraft Group
Hawthorne, California

Abstract

As the use of composite material increases both in primary and secondary aircraft structures, it becomes incumbent upon the disciplines of design, manufacturing and material systems to develop vehicles/tools which will reduce manufacturing costs. This activity is currently being undertaken by the designer in conceptualizing the next generation aircraft systems; by the manufacturing groups in their drive towards increased automation in the fabrication process; and by the material suppliers/vendors in pioneering the development of wider material forms, various fibers and resin systems, all of which enhance the manufacturability of composite material. However, the new designs, the new automation techniques and the innovative material resin systems will do little for "todays" production environment to eliminate/alleviate the inevitable manufacturing roadblocks that occur due to labor intensive operations. This is especially true of current programs, since they are basically "cast in concrete" with relation to design, to material and process specifications, and to certifications; all of which would require costly and time consuming changes. This provides a great opportunity for a system that will produce immediate results in cost reduction without causing aberrations in the existing design/manufacturing/material interfaces. To address this problem, Northrop undertook and successfully completed a program that employed preplied/crossplied material as a substitute for both woven and unidirectional materials on an existing part in a production environment. This paper addresses these accomplishments in

*This work was performed under Northrop's Independent Research and Development Program.

terms of cost reductions and physical property enhancements of preplied material relative to unidirectional and woven ones, respectively. "Keywords": preplied broadgoods, cross-plied broadgoods, prepreg, and labor standards.

1. INTRODUCTION

There are two major reasons that can be cited for emphasis on cost; the ever increasing acquisition and ownership cost of aircraft systems and, prior to the election of the current administration with their defense posture, a downward long-term trend (in constant dollars) of defense procurement funds. Thus, government and industry have been and will continually be under pressure of effectively utilizing these funds for the development of aircraft systems that meet performance goals within a cost ceiling.

Several methods have been investigated to solve this problem. One such vehicle, the Design-to-Cost concept, the implementation of which is mandatory in government contracts with industry, has proven to be an effective tool to keep cost at/or below established target levels. This concept promotes the initial integration of the cost and technical disciplies within industry and government, where the cost engineer, in addition to his traditional role of cost/economic analysis becomes more involved and sensitive to the needs and requirements of the technical groups. In addition, the design and manufacturing organizations have become increasingly aware of the cost impact of their design decisions and manufacturing plans. Another vehicle which has been employed to solve this cost dilemma has been the introduction of contemporary material systems into airframe structures. These systems include; advanced metallics, weld brazing, diffusion bonding, weld bonding, adhesive bonding, isothermal forging, hot die forging, hot isostatic pressing (HIP), superplastic forming/diffusion bonding and composite material systems. Composite materials and their fabrication is a technology that continues to find increasing applications in the aerospace industry. Initially, composites were developed and employed to exploit their inherent advantages of low density, high strength and high modulus when compared to conventional structures manufactured from traditional materials. Since the material was very expensive in developmental quantities, an increase in cost or at best no cost saving was expected for any aircraft structure in which composites were substituted for traditionally accepted material. In the early seventies, composites technology was characterized by high material cost (close to $300/lb. for graphite/epoxy) limited material width (3", 6" and 12") labor-intensive fabrication processes, and designs that directly substituted composite material for conventional metals.

Since then, through extensive research and development supported by government and industry, and through the ever increasing application of composites in both the aerospace and commercial sectors, we see today composite material costs in the \$30-\$50/lb. range despite inflation, a variety of composite material forms, e.g. 48" wide unidirectional tape, preplied, etc., improved matrices, increasing percentage of automation in composite fabrication, and development of innovative manufacturing approaches. Furthermore, innovative design now exploits the inherent advantages of composite materials, rather than merely substituting them for metal parts, i.e., through simpler designs that result in fewer detail parts and fasteners, with a corresponding reduction in tooling. Hence, we see today, more clearly than before, the life cycle cost-competitiveness of composites when applied to aircraft structures. There has also been much consternation within the composite community with regards to the transfer and implementation of technology generated in the composite research and development environment to a actual ongoing full-fledged production scenario. The cross-pollenization of this information is not easy and automatic as it should be.

2. FABRICATION PROCEDURES

In order to overcome this continued struggle with respect to technology transfer, Northrop undertook a program whose objective was to produce a current production part in the production shop without changing the design and the manufacturing process; but substituting vendor-supplied preplied material for unidirectional broadgoods or woven material. The current state of the art in composite design and fabrication is to employ single ply and/or woven material. This procedure has been driven in part by the design engineer's need to have plies drop off in order to ensure proper load distribution throughout the component. In addition, multiple layups of heterogeneous orientations, even with identical geometric ply configuration, were difficult to manufacture out of existing material systems due to the engineering requirement of not "butting" the fibers of the prepreg against each other in the layup. The problem has been basically overcome, except for the largest of structures, due to the wide material forms that are currently available from the vendor, i.e., 48" and 60" wide broadgoods. The component that was selected for this test program was an outer mold line skin for the lower leading edge extension panel of the F-18. See Figures 1 and 2. This skin has identical internal two (2) ply drop-offs which provides for an outer edge build up of 21 plies, with a four-ply web in a balanced laminate $(0^{\circ},90^{\circ},0^{\circ}, 90^{\circ}, \pm 45^{\circ} \pm 45^{\circ}, 90^{\circ}, 0^{\circ})s$. The build-up provides for a sufficiently thick edge member to accommodate fasteners and the tapered internal plies for core placement. The skin is also aluminum flame-sprayed for lightning protection and has a barrier of fiberglass between the flame spray and the first graphite preply of $0^o/90^o$. The material system employed was MMS-549 Type I, AS-3501-6, 48" wide in a cross-plied/preplied $0^o/90^o$ prepreg condition with a resin content of $42^o/o \pm 3\%$, as obtained from the vendor. In order to facilitate fabrication in the composite production facility issuance of production paper was required. This provided the capability of employing production procedures and production personnel for the fabrication of the skin with the only exception from the current production method being the substitution of preplied material for unidirectional broadgoods. However, an unexpected spillover cost surfaced in that an additional debulk cycle was added to the layup sequence in order to accommodate Northrop's material and process specifications requiring a debulk after every five (5) single plies or two (2) plies of preplied material for this skin. This currently is an arbitrary debulk cycle at Northrop for this laminate. Efforts are currently underway to modify this cycle when preplied material is used. An existing numerically controlled tape was used to cut the preplied prepreg on Northrop's Gerber cutter. This tape currently is employed to cut both a left and right hand 20 ply skin from unidirectional broadgoods. By employing this existing tape, the capability now exists to Gerber cut two left and two right hand skins when using preplied material since two plies are cut-out at the same time. This scenario reduces by one-half the relevant amount of Gerber cut time; allows for the use of the existing numerically controlled tape and also frees up valuable computer time. The savings that was generated for cutting just one preply can be magnified if preplied material is stacked, since the Gerber is capable of cutting multiple stacks of preplied. After the material was cut by the Gerber, the plies that were required to layup the right hand skin were transferred to the cleanroom, with the remaining material returned to the freezer. At the work station in the cleanroom, the composite technician placed one ply of fiberglass epoxy on the flame-sprayed layup/curing tool and subsequently stacked the ten Gerber-cut preplies to complete the layup. A three minute debulk cycle, pulling 25 inches of mercury was required after the fiberglass ply and then after every two preplied plies. Upon completion of the layup operation, the tool and layed-up skin were enveloped, bagged and transferred to the autoclave staging area. The part was then cured at a maximum autoclave pressure of 100 psi ± 5 psi and a maximum temperature of $350^{o} \pm 10^{o}$ F. Upon removal from the autoclave, the skin was stripped from the tool, flash removed and sent to the trim area. The skin was rough trimmed and transferred to Quality Control for testing.

3. TESTING PROCEDURES AND RESULTS

The test skin underwent the same rigorous NDT procedures as would be required for a F-18 production part/skin. The results of the X-Ray test exhibits no difference between the preplied, woven, and the single-ply layed-up skins, as would be expected. The C-SCAN printout exhibited areas where the skin would be categorized as either resin rich or resin starved; however, this aberation was not considered serious, and the skin passed all NDT and visual inspection procedures, i.e., it was acceptable for the next assembly sequence. At this point, the skin was pulled-out of the production loop. Since physical flow data (see Table 1) was higher for the preplied than either the average for woven or single-ply unidirection broadgoods I suspect that the prepreg (Type I 42% ± 3%) was either resin rich in $0^{o}/90^{o}$ preplied configuration, possibly due to the densification of the $0^{o}/90^{o}$ preplied material at the vendor facility, or the bleeder-ply ratio currently used for unidirectional material is invalid for preplied prepreg. This is reconfirmed in Table 2, where the fiber volume of the cured coupon is somewhat

FIGURE 1
FLAME SPRAYED LEX SKIN

FIGURE 2
LEX SKIN ON LAYUP TOOL

lower than would be expected for a laminate fabricated out of unidirectional broadgoods. Test data on a preplied prepreg with a 35% ± 3% resin content and verification of bleeder-ply ratio is probably appropriate to confirm either of these scenarios. The test data presented in Tables 1 and 2 are for

MATERIAL	INTERLAMINAR SHEAR - PSI (RT)	INTERLAMINAR SHEAR - PSI (250°)	PHYSICALS PREPREG - FLOW
• PREPLIED (0°/90°)			
- R&I 7 PLY COUPON	15,300	11,100	24.2%
- R&I 8 PLY COUPON	16,400	11,800	15.0%
- 8 PLY COUPON w SKIN	14,800	11,200	-
• UNIDIRECTION COUPON [1] 15 PLY ALL 0° ORIENTATION IN 6" DIRECTION	18,000	13,000	18.0%
• WOVEN COUPON - 8 PLY [1] COUPON, 0° IN THE 6" DIRECTION - WARP	10,000	7,000	15.5%
MATERIAL	LONGITUDINAL TENSILE - PSI	MODULUS $X10^6$	STRAIN MICRO INCH/INCH
• PREPLIED (0°/90°)			
- R&I 4 PLY COUPON	116,500	10.1	11,200
• UNIDIRECTIONAL - 15 PLY, 0°	220,000	21.1	-
• WOVEN - 8 PLY, 0° IN WARP	85,000	11.2	-

[1]CURRENTLY BEING ACHIEVED ON INCOMING F-18 MATERIAL

TABLE 1

TEST DATA FOR RECEIVING AND INSPECTION COUPONS AND COUPONS FABRICATED CONCURRENTLY WITH TEST SKIN (AVERAGE VALUES)

test coupons fabricated for initial receiving/inspection; and the process control coupon fabricated concurrently with the skin and an average of the test results currently being obtained for the 15 ply 0° coupon and the 8 ply woven coupon. These data suggest that the use of preplied 0°/90° would produce a far superior laminate when compared to an otherwise identical woven structure.

4. FABRICATION COST

A reduction in Factory Labor Standards Hours occurred in the cost elements of Gerber cut time and layup/debulk.

NOTE: At Northrop, standard hours represent "pure" labor (set-up and run time only) and do not take into consideration the elements of variance e.g. personal delays, waiting time for materials and tools, fatique, supervision etc.

SPECIFIC GRAVITY	1.578
• RESIN CONTENT	35.1
• FIBER VOLUME	56.9
• VOID CONTENT	0.093

TABLE 2

PHYSICAL PROPERTIES FOR COUPON LAMINATE - CURED

COST ELEMENT	STANDARDS FROM PRODUCTION TIME CHARGE DOCUMENT					
	WOVEN		UNID		PREPLIED	
	S.U.	RUN	S.U.	RUN	S.U.	RUN
• GERBER CUTTING STDS.			.08	.430	.08	.213
• LAYUP/DEBULK STDS.			.08	1.502	.08	.818
• ALL OTHER STDS.			.38	1.433	.38	1.433
TOTAL FACTORY LABOR STANDARD HOURS	N.A.		.54	3.365	.54	2.464
• STANDARD HOUR COST SAVINGS				+27%		
• # of plies						
- GL/EP	1		1		1	
- GR/EP	7		20		10	
• SKIN THICKNESS	.110		.110		.110	
• SKIN WEIGHT (LBS.) APPROX.	2.0		2.0		2.0	

TABLE 3

COST ANALYSIS - PREPLIED vs UNIDIRECTIONAL BROADGOODS

These standards are developed from basic stop-watch time studies performed at Northrop on the various composite fabrication and assembly operations. The reduction in standard hours that were obtained from using preplied $0^{o}/90^{o}$ material are presented in Table 3. The table shows that 27% cost saving can be achieved by using preplied rather than single ply unidirectional broadgoods. This cost saving can be considered representative of the saving that could be achieved by using preplied vs single ply whenever two or more plies have the same geometric configuration and identical play drop-offs. The standard hours presented in Table 3 were obtained from the shop production paper, which mirrors the standard data developed by Northrop's Industrial Engineers. When this labor cost saving of 27% is coupled with the cost penalty of purchasing preplied $0^{o}/90^{o}$ prepreg material over unidirectional broadgoods, see Table 5 the resultant total cost saving is approximately 20%. NOTE: In Table 3 the Gerber cut time standards for preplied material are one-half of those for unidirectional broadgoods - as expected. However, the layup/debulk time standard for the preplied material is slightly more than one-half of the unidirectional broadgoods due to one additional debulk cycle incorporated into the layup sequence. Efforts are currently underway to reduce or eliminate debulk from relatively flat laminates, since

COST ELEMENT	STANDARD ESTIMATES USING NORTHROP'S COMPOSITE COST ESTIMATING MODEL					
	WOVEN		UNID		PREPLIED	
	S.U.	RUN	S.U.	RUN	S.U.	RUN
• GERBER CUTTING STDS.	.00	.182	.00	.354	.00	.185
• LAYUP/DEBULK STDS.	.08	.557	.08	1.456	.08	.795
• ALL OTHER STDS.	.70	1.097	.70	1.123	.70	1.123
TOTAL FACTORY LABOR STANDARD HOURS	.78	1.836	.78	2.933	.78	2.103
• STANDARD HOUR COST SAVINGS		- 14% (WOVEN RUN vs PREPLIED RUN)		+ 28% (UNID RUN vs PREPLIED RUN)		
• # of plies						
- GL/EP	1		1		1	
- GR/EP	7		20		10	
• SKIN THICKNESS	.110		.110		.110	
• SKIN WEIGHT (LBS.) APPROX.	2.0		2.0		2.0	

TABLE 4

COST ANALYSIS - PREPLIED vs WOVEN vs UNIDIRECTIONAL

preplied material obtained from the vendor is already compacted, i.e., densified. When this occurs, the cost saving of preplied versus unidirectional will increase. Northrop's computerized cost estimating system for advanced composites was used to generate cost estimates for unidirectional, preplied and woven material that could be used in the fabrication of the test laminate. The primary purpose of this activity was to compare the overall cost effectiveness (labor and material) of preplied relative to woven material in addition to providing a comparison between the standards in the Production Time Change Document and those generated by the automated cost estimating system since both employ the same Industrial Engineering Time Standards. This computerized methodology estimates the recurring costs associated with the fabrication and minor assembly of composite components. The model consists of three modules; (1) Factory Labor Standards Estimating, (2) Support Functions Estimating, and (3) Cost Projection. The Factory Labor Standards Estimating Module is the foundation of this methodology. It utilizes Industrial Engineering time standard equations to calculate the "pure" labor hours associated with the fabrication operations performed on composite parts. In addition, material requirements in terms of total and net usage and scrap are calculated. This computer model is primed by a description of the parts to be fabricated; specifying its construction, the materials used and the fabrication processes to be performed. The Cost Projection Module calculates projected factory labor hours at a user-specified production unit (quantities) by the application of variance factors and improvement curve slopes to the factory labor standard hours generated by the Standards Estimating Module. It also applies the labor, material and overhead rates to arrive at total recurring cost. Estimate of unit, cumulative and cumulative average cost at specified production (quantities) are generated based upon user request. The Support Functions Estimating Module estimates recurring engineering, tooling, quality control, manufacturing engineering, and graphic services as well as support material through the application of cost estimating relationships. This fast computerized capability is characterized by the following features: estimates in detail and summary levels; sensitivity to variations in manufacturing processes and designs; cost effectiveness since less manhours are required to produce an estimate; adaptability to new processes, materials and equipment as data becomes available; and the system is excellent for cost-driver identification. The results of employing this system to generate cost estimates are presented in Table 4 which shows that a standard hour cost saving of 28% can be achieved when using preplied prepreg over unidirectional broadgoods. However, when comparing preplied material to woven for this skin, there is a standard labor hour cost penalty of 14%. This is because the skin is thickness sensitive, i.e., it requires only seven plies of woven to obtain required edge thickness whereas it takes ten plies of preplied to achieve the same thickness - 0.110. If both labor and material cost are taken into consideration, then preplied is 13% more cost effective than woven material in addition to providing a structurally better laminate and the intangible of preplied being easier to handle and layup. The differences that are evident between the standards presented in Table 3 and those presented in Table 4 are due to the standardization of manufacturing plans in the estimating system which does not account for a specialized operation such as flamespray in a generic plan. The computerized estimating system employs standard equations which were obtained from regression analysis on tabular data.

5. MATERIAL COST

There is a material cost penalty when $0^{o}/90^{o}$ preplied prepreg is ordered and employed in the fabrication of composite components of equal area density (lb/ft^2) of unidirectional broadgoods, all other things being equal. That cost penalty is estimated to be in the neighborhood of 10-15% for purchase

	UNIDIRECTIONAL BROADGOODS	PREPLIED BROADGOODS	WOVEN
MATERIAL COST PENALTY OVER UNIDIRECTIONAL BROADGOODS (BASED UPON EXISTING MATERIALS USED BY NORTHROP)	BASE	+ 10-15%	+55-65%

TABLE 5

at the vendor end-bracket. Thus, if unidirectional broadgoods can be purchased in large volumes for $40.00/pound, then this translates into paying from $44.00-$46.00/pound for equal amounts of $0^{o}/90^{o}$ preplied prepreg. This cost penalty would be different if other preplied orientations were purchased. (See Table 5). However, when $0^{o}/90^{o}$ preplied prepreg broadgoods cost are compared to the cost of equal amounts of woven material, then the cost pendulum swings in the direction of preplied material. This means that the cost advantage of preplied over woven material is between 45 and 55%. Thus, if $0^{o}/90^{o}$ preplied material were purchased for $45.00 then equal amounts of woven material would cost in the range of $65.00-$70.00/pound. Since one ply of preplied $0^{o}/90^{o}$ material can generally be substituted for one ply of woven material, except where the primary design consideration may be thickness - as it was in the case of the edge build-up of this skin, then preplied should be preferred over woven material. The fabricated skin/part will have better physical properties with far less voids with a lower net cost of material, Table 5.

6. CONCLUSION

Cost has become an increasingly significant parameter in the design, engineering and fabrication of aircraft systems. With the increased ownership costs of aircraft systems, industry is being required to build systems that generate the least cost to acquire, operate and maintain while still achieving performance goals. One of the responses to this cost reduction mandate has been the introduction of composite materials for aircraft structures. The traditional design approach for employing composites is to design the component to maximize the unique aeroelastic tailoring of composites. While this maximizes design and performance requirements, it places a constraint on production by intensifying the amount of hand labor indigenous to each part; and thus is a major cost driver. Automation is currently looked upon as a major weapon that can be employed to reduce fabrication cost. However, it will probably be sometime before production implementation, which provides the current opportunity for employing preplied material. With the introduction of $0^{o}/90^{o}$ (which can also be used as $\pm 45^{o}$ with some slight adjustment) preplied prepreg material into the manufacturing process a cost saving in labor and material can be achieved relative to unidirectional broadgoods and woven material. Additional cost savings can also be realized via the optimization of such operations as debulk cycles in all material systems. For example, a skin which is basically flat, requires five dubulk cycles prior to bagging when using unidirectional broadgoods and preplied material under current process specifications at Northrop. This may be able to be reduced especially, since preplied prepreg is compacted/densified by the vendor. It should also be remembered that the structural properties are identical, in the cured laminate, for both the unidirectional and preplied broadgoods; thus, they are substantially superior to woven material. In addition, initially, the conversion to preplied material could be limited to the replacement of woven and to two unidirection plies with the same geometric configuration and orientation that is compatible with the preplied material. In either case, there would be a negligible effect on existing engineering drawings.

7. BIOGRAPHY

Mr. Bettner is currently a member of Northrop's Integrated Computer-Aided Manufacturing (ICAM) group with responsibility for advanced composite programs. Prior to joining ICAM, he was a member of a cost technology team that provided cost/benefits data for funded contracts in composites research. In this capacity, he was also responsible for writing the cost chapter of the Second Edition of the Structural Composites Fabrication Guide. Prior to joining Northrop he was a senior price and market analyst for Ducommun Metals and Supply Company. Mr. Bettner's education includes a MBA and a MA in Economics. He also teaches at various colleges in the Southern California area.

13th National SAMPE Technical Conference
October 13-15, 1981

ENGINE INDUSTRY COST CONSIDERATIONS FOR EMERGING TECHNOLOGIES

M. A. Siegel
Manager, Material Applications
Government Products Division
Pratt & Whitney Aircraft
United Technologies Corporation
West Palm Beach, Florida

Abstract

To meet the demanding requirements for improved aircraft performance and durability at acceptable cost, a number of evolutionary and revolutionary aircraft engine materials and processes are being developed. These systems range from advanced alloys in conventional and powder metallurgy form to intermetallics, ceramics, and composites. For those systems which require fundamental changes in design, manufacturing and/or maintenance practices, studies are required to determine their cost-effectiveness prior to commitment. The pertinent cost considerations include development cost, manufacturing R&D cost, capitalization cost, production cost, and operational costs. This paper defines these considerations and the key factors which are included in application decision making.

Keywords: Engines, Aircraft, Cost, Technologies, Materials, Processes, Design, Manufacturing.

Despite the fast-moving pace of modern technology and the breakthroughs in the materials and processing areas, there is still significant resistance to change due to a variety of factors. The typical Program Manager has commitments to develop, qualify and produce his engine on a specific schedule at a given development cost, and with specified performance and life. The Project Engineer, hardened by the scars of previous problems and faced with tight schedules and budgets, doesn't want to face the risks of new technologies and prefers to rely on others to demonstrate their value. The Design Engineer has to use concepts that are demonstrated, characterized, evaluated, maintainable, repairable, and producible at a lower cost than current concepts. How then does a new technological concept find its way into a product? The answer, very simply, is with great difficulty.

While this may sound unduly harsh, it must be remembered that aircraft gas turbine development is a long and expensive process. In order to qualify a new engine or even a model change, all parts and assemblies are subjected to thousands of hours of laboratory, rig and engine testing. This extensive testing is required to ensure that high levels of safety and reliability are achieved in addition to the fundamental requirements of performance and durability. New materials must have sufficient characterization and substantiation to warrant this thorough process.

The primary concern of the decision maker who must decide whether to go with a reliable, proven material or process, or take the plunge into a sparkling new technology, can be summed up in one word — risk. He must be convinced of the benefits of a concept not only in terms of performance, durability, and reliability, but also in terms of that something more difficult to establish parameter — cost. Fig. 1 illustrates the decision maker's problem. This paper will assume that the benefits other than cost have been established, and will concentrate on the cost aspects of introducing new technologies. The Design Engineer is adequately educated and trained to evaluate the stress, temperature, vibratory mode, fatigue life and other vital statistics of a part, and can also evaluate the maintainability and reliability aspects. But the cost influences are not as approachable and manageable based upon his education and experience.

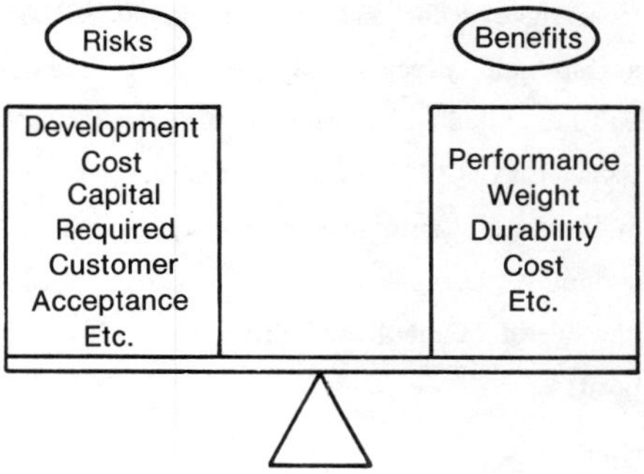

Fig. 1 The decision maker must have a balance

The cost factors in product selection cover a wide range of influence encompassing the basic cost of the product or part, the development cost, capitalization to produce the part, and operating costs, as shown in Fig. 2.

1. Development Cost — Cost To Substantiate
2. Product Cost — Production Cost
3. Capital Cost — Cost To Facilitize
4. Operating Cost — Total User Cost

Fig. 2 Major cost influences

The ability to accurately calculate these costs frequently determines acceptance of a new concept and its success in service. This is particularly true in areas of rapidly changing technology where new materials and processes transcend conventional practice.

Materials and Processes have paced the growth of the aircraft and aerospace industry. This trend is as evident today when the conventional alloys of early engines were replaced with the superalloys and titanium alloys used in current engines. One can describe an entirely new generation of materials and processes in various phases of development that are beginning to appear in the latest generation of high-performance aircraft engines. These range from the powder metallurgy superalloys that are specially processed in inert environments and fabricated in superplastic form by a process such as GATORIZING®, the intermetallics of titanium aluminum and ferrous aluminides, ceramics, the rapid solidification alloys, to the new world of metal and organic matrix composites.

This is an exciting array of new materials that can satisfy the most demanding requirements for reduced weight, higher temperature capability, longer life, or whatever is needed. The establishment of material properties is a straightforward technical effort involving the characterization of the vital properties such as tensile, compressive, fatigue, crack propagation, stress-rupture, creep or what have you. However, the establishment of producibility characteristics is another story because we are faced with significant changes, or in some cases totally different manufacturing concepts, to produce these new materials.

Let's examine the problem and several approaches. Traditionally, new concepts appeared as an engine part of a new material and were costed by establishing standard material and labor costs for purchased material and each manufacturing operation. Typically, in the past if we were changing to a new material, we found an identifiable change in material cost, and we used the traditional machining and finishing operations in our shops to produce the part. If the new material had a different machinability index, it might take longer to machine or even require a different machining operation. If there were special new operations required, they were taken into account. The change to titanium and superalloys from steel alloys is typical of this type of change. The material costs more because melting, heat treating and element content are more costly, and the forging and machining operations are similar but more difficult. This type of change was not too hard to sell because the risk was moderate, the capital investment for equipment change was low, and the confidence in cost prediction was high. The easiest way to calculate a part cost is by comparing it to a known production part and modifying the known cost for material and processing cost differences.

But what happens when we get into totally new processes? Let's take the change from conventional forging of standard ingot to GATORIZING of powder. The process for producing the starting material for forging becomes more complex, and the actual forging operation itself is significantly changed. At first glance it would appear to increase part cost significantly. The starting powder has to be more expensive than standard ingot. The material must be handled in an inert environment once the powder is compacted and in the forging operation. Special dies capable of withstanding the forging temperatures and pressures are required. Sounds like a loser, and a faint-hearted manager might have zeroed this project. But due to the ability to forge much closer to net or final shape significantly reducing input-material requirements and subsequent machining, the parts are actually lower in cost than conventional superalloy parts processed in the conventional manner.

This has been demonstrated in over 20,000 parts made for the F100 engine used in the USAF F-15 and F-16 air superiority fighters.

An additional benefit is derived in capital investment because much smaller forging press capability is needed for superplastic forging than in conventional forging. The success of this effort can be seen from the recent announcements of new GATORIZING-type presses by the leading U.S. forging companies, which will be used for titanium as well as superalloy parts. Even though this process presented formidable cost barriers, detailed cost evaluations based upon a demonstrated sequence of operations predicted lower cost for parts than by conventional forging. A typical GATORIZED titanium part, currently in production, is compared to a conventional forging in Fig. 3.

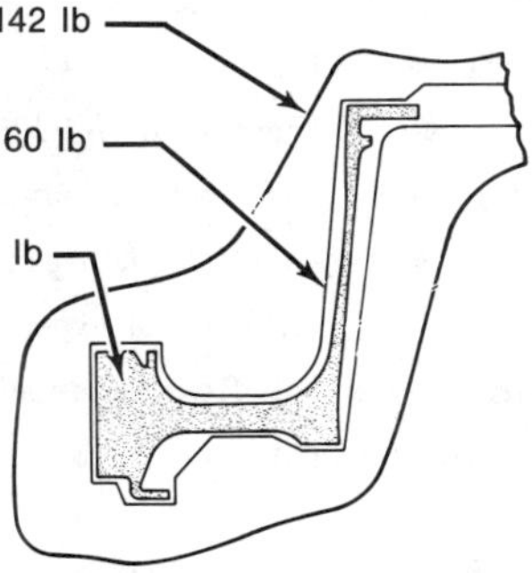

Fig. 3 GATORIZING® reduces input requirements

With product cost satisfied, what about capital requirements? In the evaluation of new processes, the same type of return on investment wickets must be passed as in any original plant construction or expansion decision. It is usually more difficult in new processes because of the lack of detailed information and experience in the process, and perhaps even difficulty in determining the plant cost when new or revolutionary processing equipment is required. On top of that there is the competition with other investment requirements and, of course, the soaring interest rates currently in effect.

If development cost, product cost, and capital cost are satisfied, we still have one more key cost parameter to satisfy — operating cost. From the customer's standpoint, product cost and operating cost are the visible cost elements. In the business of using a gas turbine engine, the operator is concerned with maintenance involving the ability to inspect and repair parts. He would obviously prefer all parts to last the full lifetime of the engine, but is faced with anticipated and sometimes unanticipated wear, erosion, corrosion, fatigue, and other failure modes that require the ability to adequately inspect and repair. His experience with previous metals and processes may not adequately prepare him for new concepts where conventional nondestructive inspection and repair procedures may not work. So another hurdle is the education and training of the user's maintenance force along with his subcontractor's people. He is also concerned about the availability and cost of replacement parts. The user's concerns are summarized in Fig. 4.

- How Long Will It Last?
- Can It Be Inspected?
- How Will It Be Repaired?
- What Is the Replacement Cost?
- Will Spares Be Readily Available?

Fig. 4 The user's concerns

Let's look at some of the problems, based on these concerns, that face us in the use of new materials.

Composite materials are receiving significant attention in the Aerospace Industry because of the weight and structural advantages available.

We see a continuing increase in aircraft applications in secondary structures, and applications being designed for primary structures due to the significant impact on weight, and therefore, fuel usage. Composites have been around a long time in the form of fiberglass, which has become the standard for the pleasure boat industry, as well as many other applications. Their usage in aircraft structures requires somewhat more sophisticated materials than fiberglass due to temperature and structural requirements. We see a plethora of materials including graphite/epoxy, graphite/polyimide, boron/aluminum, and carbon/carbon to name just a few. Significant efforts have been made, in terms of structural requirements, to characterize the many combinations and forms that are available.

Many components have been fabricated, tested, and in some cases are in daily use. Fig. 5 shows a graphite polyimide nozzle flap currently being flight tested on F100 engines in F-15 aircraft. But when we look at the cost factors, a number of problems and challenges are evident. The forecasting of product cost is difficult because of the lack of experience in similar production parts. The cost of a composite part can't be directly related to the metal part it will replace. The conventional metals fabrication processes of melting, forging, and machining are not comparable to the tape lay-up, cutting, curing, and bonding techniques required for composites. It is basically a new ball game in manufacturing, requiring new facilities and capital investment.

Operating and maintenance experience is minimal in aircraft structures for this emerging technology, but it is obvious that new inspection and repair procedures will be required compared to conventional metal structures. In

Fig. 5 Composite nozzle flap installed for flight

a radically different concept such as composites, the chicken and egg dilemma becomes evident. How can you produce composites at a competitive price without sizable production orders? Composites are competing against metal parts with considerable production experience and learning. In this case, the benefits may have to justify an initially higher cost until production learning impacts part cost. Fig. 6 is a hypothetical picture of this situation. Learning is a function of experience, trial and error, and a shift from handmade to true production-type parts.

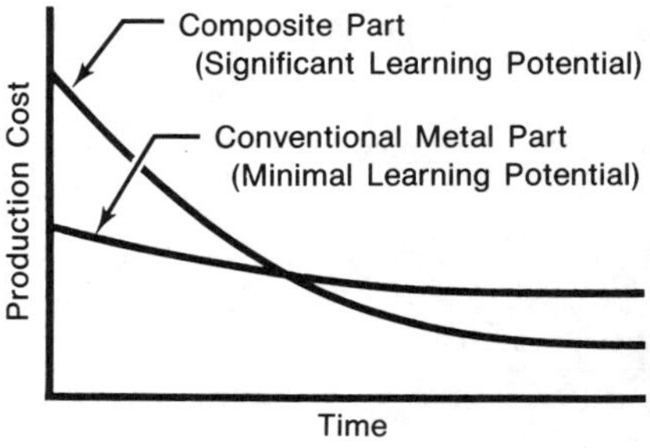

Fig. 6 Emerging technologies require learning to be competitive

The other emerging materials and process technologies face similar problems that depend upon their specific characteristics. An analysis of the cost factors is vital to ensure any chance of success. A systematic approach involving the following steps will help in establishing a basis for cost:

1. *Development Cost* — What is required to characterize the materials in all facets of anticipated operation? What testing is required to substantiate and demonstrate that the required properties can be met? It is important at this time to ensure that *all* required properties are identified for specific applications. What temperatures, stresses, vibration mode, etc., will the part see? What type of environment is anticipated?

2. *Product Cost* — As was stated earlier, comparison to existing production parts and adjusting cost for differences is one way to obtain initial cost. However, that comparison may not be available and then one must use the utmost ingenuity until some initial manufacturing experience and costing is accomplished. Comparison to nondirectly related parts of the same material may be possible, or a step-by-step projected production manufacturing process must be established and assessed.

3. *Capital Investment Requirements* — An assessment of factory equipment required to produce and inspect parts of new materials is obviously important. Without a production base, parts cannot be made regardless of the benefits achievable. Will subcontractors respond to the potential market for a new materials concept? Will your company rank an investment in this technology high enough to commit capital?

4. *Operating Costs* — A key influence is the acceptability of a new technology by the ultimate user based upon his perception of maintenance and replacement costs. His requirement for lowest life cycle costs will balance the benefits achievable in critical cost factors such as fuel cost against the costs of part inspection, repair and replacement.

In summary, the path to succesful implementation of new technologies in aircraft engines has never been an easy one. The program manager, project engineer, and design engineer had to be convinced that the concept provided assured performance, durability, and reliability benefits at an affordable cost. Today's new material technologies involve more significant and fundamental changes than ever and will, therefore, require more discipline in identification of value determination for all aspects of cost early in the game. The Design Engineer is adequately educated and trained to evaluate the stress, temperature, vibratory mode, fatigue life and other vital statistics of a part, and can also evaluate the maintainabiity and reliability aspects. But the cost influences are not as approachable and manageable based upon his education and experience.

13th National SAMPE Technical Conference
October 13-15, 1981

"MANUFACTURING COST TRADE-STUDIES IN AVIONICS"

Bryan R. Noton
Battelle's Columbus Laboratories
Columbus, Ohio

John G. Vecellio
Avionics & Missiles Group
Rockwell International
Cedar Rapids, Iowa

Robert Remski
Honeywell, Inc., AvD
Minneapolis, Minnesota

Abstract

A "Manufacturing Cost/Design Guide (MC/DG) for Avionics" is under development for the Computer Integrated Manufacturing Branch (AFWAL/MLTC), Materials Laboratory, WPAFB, Ohio. The MC/DG is being developed by a team; Battelle's Columbus Laboratories (BCL) as prime contractor, Honeywell, Inc., and Rockwell International's Avionics and Missile Group. The data analysis methodologies established for the "MC/DG for Airframes" are being used. General and detailed ground rules are developed, which include specifications of electronic discrete parts, materials, assemblies, manufacturing methods, etc. Examples of assemblies are power supply, hybrids and chassis. Examples of discrete parts are substrates and printed wiring boards. The paper presents an overview of the aerospace electronic design processes, indicating how the MC/DG will be utilized to conduct trade-studies at the conceptual design phase, as well as the detailed design phases for the circuitry and also the chassis. The paper concludes with an overview of the numerous trade-studies and data requirements to achieve affordable performance in avionic systems. This data must be presented to designers so that the cost-trade studies can be conducted within schedule limitations.

INTRODUCTION

The need to arrest and reduce costs at all levels of the aircraft system life-cycle is becoming increasingly important. Qualitative and quantitative data and other information on cost-drivers useful during the design, manufacturing, operation, and maintenance of aircraft systems are essential. This is particularly so, because, at the same time, the number of new aircraft types being developed is reducing while there is a need for increased performance meaning reduced weight, better quality, lower ownership costs, but with less energy consumption. The customer must therefore be provided with performance which is affordable.

Cost-drivers are the result of advancements in technology. For example, new electronic systems or airframes utilizing advanced composites or superplastic formed/diffusion bonded (SFP/DB) titanium require new developments in manufacturing technology, if through their application, requirements for increased performance are met while at the same time the aerospace industry remains competitive. This requires the development of new manufacturing technologies. However, this must not be for the sake of technological sophistication but to reduce costs with both emerging and conventional manufacturing processes for castings, sheet metal, and composites, which are used for airframes, avionics, and other subsystems. In meeting the mandatory objective of providing affordable performance, we find that there is a dilemma in manufacturing. The availability of the skilled work force is reducing, labor and equipment costs are increasing, as is the need for the higher performance, reduced weight, and more stringent specifications. To alleviate these problems, the most promising avenue of development is manufacturing sophistication, e.g., computer-aided manufacturing (CAM), robotics and adaptive process control.

Strong interaction between design and manufacturing is essential to achieve this required advancement in manufacturing sophistication and refinements. The "Manufacturing Cost/Design Guide" (MC/DG) Data Development Program for airframes and avionics, a follow-on of earlier contracts (F33615-75-C-5194 and F33615-77-C-5027) provides an unprecedented opportunity for the designer to study a larger number of alternative design configurations for airframes and avionics to achieve the lowest manufacturing cost.

The MC/DG identifies the cost-drivers over which the designer has control and which he can trade back for performance once the basic performance requirements of the system have been exceeded. The MC/DG also provides information to promote interaction between manufacturing and design, for example, regarding alternative facilities due to shop loading requirements. While the designer is principally interested in the lowest cost process in the manufacture of airframes, avionics, or other subsystem discrete parts, when communicating with manufacturing, the principal discussions will revolve around the alternative methods to produce a certain part or an assembly, and therefore, the MC/DG man-hour information is presented in basically three forms. These are lowest-cost processes for the designer, manufacturing methods for multiple discrete parts, and multiple manufacturing methods for single discrete parts.

The current contract requires the development of MC/DG sections for TI&E* (Phase I), castings (Phase Ia), certain emerging technologies (Phase Ib), and also electronics (Phase II); the subject of this paper.

The MC/DG is intended for utilization throughout the aerospace avionics design process. The design process is depicted in Figures 1 & 2 for the conceptual and detail design (curcuit and mechanical) phases. The utilization of MC/DG formats (CDE and CED) is shown on each figure.

A typical cost breakdown for an aircraft system is shown in Figure 3.

*TI&E: Test, Inspection & Evaluation

THE MC/DG TEAM

This paper covers work performed under Contract No. F33615-79-C-5102, by the Battelle's Columbus Laboratories (BCL)/Electronics Industry Coalition for the Computer Integrated Manufacturing Branch, Materials Laboratory (AFWAL/MLTC), Air Force Wright Aeronautical Laboratories, AFSC, Wright-Patterson Air Froce Base, Ohio 45433.

A. USAF TECHNICAL DIRECTION

This program is administered under the technical direction of Capt. Richard R. Presion, AFWAL/MLTC.

B. MC/DG COALITION

BCL is the prime contractor on the MC/DG Data Development Program. The Program Manager is Mr. Bryan R. Noton, Design/Manufacturing Interaction Project Office, BCL. The prime contractor is supported by the following airframe and elec-- tronics industry subcontractors:

Aerospace Industry Subcontractors	Program Managers
General Dynamics Corporation, Fort Worth Division	P. M. Bunting
Grumman Aerospace Corporation	A. J. Tornabe
Honeywell, Incorporated	R. Remski
Lockheed-California Company	J. F. Workman
Northrop Corporation, Aircraft Group	A. P. Langlois

Aerospace Industry Subcontractors	Program Managers
Rockwell International Corporation, North American Aircraft Division	R. A. Anderson
Rockwell International Avionics & Missiles Group, Collins Avionics Division	J. G. Vecellio

MC/DG VOLUME CONTENTS - AVIONICS

The contents for the MC/DG Volume, "Manufacturing Technologies for Electronic Fabrication, Assembly and Test" are subdivided (Table 1) into the four broad catagories which follow:

I. Procured Items
II. Detailed Fabrication
III. Assembly
IV. Test, Inspection and Evaluation (TI&E)

Selection Aids for avionic manufacturing data are available. The first is a selection-tree (Figure 4) similar to that included in Report No. AFWAL-TR-80-4115 and also, a second type, which indicates the formats required for designer guidance at the conceptual design phase when addressing specific design parameters, is shown in Table 2. The Avionic Team Members emphasize the importance of the leverage existing at the conceptual design phase to minimize cost.

AVIONIC DESIGN OBJECTIVES

In the design process for avionics, a series of systems parameters or objectives must be considered at the conceptual and preliminary design phases. An overview of the major system design parameter follows:

MAJOR SYSTEM DESIGN PARAMETSRS

1. Reliability → Goal MTBF
2. Maintainability → Goal MTTR
3. Cost → Cost Bogie or DTC Goal
4. Environmental → Quality Levels
   a. Temperature
   b. Vibration
   c. Shock
   d. Radiation
   e. Altitude
5. Operation/Mission Requirements
   a. Fault Tolerance
   b. Power-on Cycles
   c. Mission/Flight Critical
6. Mounting Configuration
7. Vehicle Power Requirements
8. Location in Vehicle
9. Volume
10. Weight
11. Interface Requirements
12. Field Maintenance Concept
13. Build Schedule
14. Part Lead-Time
15. Design Schedule
16. Design Costs

Examples of typical trade-studies conducted at the conceptual and detail design phases follow in Appendix A. Conceptual design trades are A-E and for detail design, F-J.

MC/DG SECTION DEVELOPMENT FOR ELECTRONICS FABRICATION AND ASSEMBLY

The objectives of the effort to

develop an "MC/DG for Electronics Fabrication, Assembly and Test, Inspection and Evaluation (TI&E)" are identical to those for the "MC/DG for Airframe Discrete Parts and Assemblies."

Primary tasks in developing the "MC/DG for Electronics" include the following:

- Develop glossary to improve communication
- Develop preliminary outline of contents for MC/DG
- Develop general and detailed ground rules
- Prepare sketches for parts and assemblies
- Prepare operational sequences for fabrication and assembly processes
- Specify detailed data development procedures and prepare data collection forms
- Conduct pilot run for manufacturing man-hour development for various lot sizes.

The glossary, which is being continuously updated, is included in all reports. To develop the material and manufacturing man-hour data, a series of data collection tables were developed.

MANUFACTURING DATA PRESENTATION TO DESIGNERS

As was previously the case for the "MC/DG for Airframes," the manufacturing man-hour data for materials, discrete parts, assembly and test, inspection and evaluation (TI&E) are presented in two ways. Firstly, cost-driver effects (CDE), and secondly, cost-estimating data (CED) are shown. The objectives of the CDE and CED methodologies are:

- To develop a simple approach for the use of formatted data by designers to achieve lowest manufacturing costs during all design phases (CDE and CED)
- To provide qualitative cost guidance to the designer to assure lowest manufacturing cost (CDE)
- To provide the designer with the capability through quantitative guidance to perform simple trade-offs on manufacturing costs (CED).

The CDE cost relationships, providing qualitative information, have the following objectives:

- Identify cost-drivers that increase the manufacturing cost of the design
- Show relative effects of cost elements over which designers have control
- Motivate designers to reduce the impact of the cost-drivers by designing around them.

Using the CDE approach, the designer should realize the lowest cost while satisfying the performance requirements.

The CED cost relationships, providing quantitative information, have the following objectives:

- Provide designers with manufacturing man-hour data to allow trade-offs to be quickly performed to achieve comparative costs for candidate structural and avionic configurations
- Motivate airframe and avionic designers to conduct trade-offs through the use of designer-oriented formats and manufacturing man-hour data in the MC/DG.

ELECTRONIC COST-DRIVERS

The BCL/Electronic Industry Coalition has identified cost-drivers in electronics/avionics. Examples of these cost-drivers are listed below. It will be noted that these include fabrication, assembly and test, inspection and evaluation (TI&E).

Electronic Cost-Drivers

- Single Source/Proprietary Items
- Insufficient Test Points (Shorten Fault Isolation Testing)
- Unrealistic Tolerances
- Requirement for Use of Special Equipment/Skills/Facilities
- Back-to-Back Card Modules
- Bussing
- Shielded Wire
- Unique Wire Gages
- Solder Sleeve Terminations
- Flextapes
- Solder-Cup Connectors
- Chassis Mounted Components
- Multilayer Printed Wiring Boards
- Components Unable to Be Auto-Inserted
- Multirow Interface Connectors
- Masking, Unmasking, and Touch-Up
- Special Tools for Insertion or Extraction
- Multiple Range of Fasteners
- Special Punches
- Excessive Hand-Wiring
- Castings
- Transformer Clearances.

AVIONIC FORMAT DESIGN CRITERIA

The designer-oriented formats and methodologies reported in the model of the MC/DG (AFML-TR-76-227) are being used as the basis for format development for the "MC/DG for Electronics Fabrication, Assembly and TI&E."

Designer surveys have been conducted and the feedback received on the MC/DG was as follows:

- Must be simple whenever possible
- Must not be time consuming to use in the design process
- Complicated calculations should be avoided
- Manufacturing data are urgently needed but with designer orientation
- No single aerospace company can provide all manufacturing cost data required due to varying expertise

- Designers are more concerned that it is the lowest cost rather than what it costs, i.e., qualitative comparisons are important.

The MC/DG team has agreed that the designer-oriented formats must meet the following criteria:

- Emphasize cost-drivers
- Be simple to use
- Use designer language
- Instill confidence
- Be economical
- Be accessible
- Be maintainable.

ELECTRONIC INTERCONNECT, INSERTION AND SOLDERING PROCESSES DATA DEVELOPMENT

The interconnect, insertion and soldering processes for avionics includes several materials, part types and also manufacturing methods. Some processes are accomplished by hand, while others are semi-automated and fully automated.

Designer guidance sheets were initially prepared for interconnect between components, insertion, and soldering. Examples of such a format is included here (Table 4). It will be noted that these formats resemble that earlier developed for sheet-metal designer-influenced cost elements (DICE-0).

For the insertion and soldering processes, each part type is listed in the MC/DG, e.g., resistors, capacitors, transistors, integrated circuits, and coils. For the interconnect between components, various materials are included, e.g., Teflon, polyimide and ceramic (Table 4). The actual man-hours have been subsequently calculated for each process. Other formats are shown in Figures 5 to 10 for cost analysis.

INTEGRATED EXAMPLE UTILIZING MC/DG FORMATS FOR ELECTRONICS

Utilizing the avionic designer's cost worksheet shown in Table 3, a number of trade-studies have been conducted on an aircraft glideslope-to-indicator interface assembly. An example of the baseline cost of material is indicated below. The worksheet, completed by the designer, enabled a cost savings to be achieved compared to the baseline interface assembly.

BASELINE INTERFACE ASSEMBLY

LIST OF MATERIAL REQUIREMENTS

RF ASSEMBLY

Quantity	Type	% Total Electrical Cost
1	IC Amplifier	59
2	Transistor	15
2	Diodes	3
1	Coupling Cap	1
1	Variable Cap	3
1	Variable Resistor	7
10	Resistors	3
1	Crystal	9
		100%

COST-DRIVERS

Many papers and reports have been written and conferences have been held focussed on the subject of "cost-drivers." In most cases, however, the emphasis has been placed on a specific product, component or manufacturing process or method.

The following is representative of the cost-drivers identified in the electronics section of the MC/DG.

Cost-drivers exist in the fabrication, assembly, test, and installation of electronic components and products. For example, in most cases avionic equipment is installed by the airframe manufacturer. Consequently, the interconnecting hardware, such as wire harness, is designed by the airframe manufacturer. Avoidance of cost-drivers in installation is the responsibility of both the designer of the equipment and the designer of the installation.

INSTALLATION COST-DRIVERS

- Connectors or Plugs: Both mechanical and solder type pins are primarily assembled into the connector by hand. Some larger connectors (up to 200 pins) depend upon the operator inserting the correct pin or wire into the correct connector position by reading a small letter or number. The very small miniature connectors also depend upon a high degree of manual dexterity.
- Number of Wires: Although manual advancements such as printed circuit boards, ribbon wiring, etc., are available, most airframe manufacturers use conventional wiring (an advanced fighter uses approximately 10 miles of wiring with harness sizes up to 3 inches in diameter).
- High Density Areas: Installation of harness and equipment in confined areas such as cockpits and dedicated compartments.
- Type of Harness: Trade-off selections are conventional wire, braided, molded or hard harness.
- Harness Fabrication: Large harness with numerous breakouts vs. a number of smaller simpler harnesses.
- Temperature Controls: Requirements for air-conditioning.
- Shielded Wire Preparation: Primarily a hand operation, vs. hand and automatic wire strippers for conventional wires.
- Potting Requirements: For corrosion and moisture protection.

FABRICATION AND ASSEMBLY COST-DRIVERS

- Tolerances: Circuit boards, capacitors, resistors, controls, microwave components, etc.
- Conventional wire harness vs. ribbon harness or printed circuit boards.
- Multi-piece racks and boxes vs. one formed from a single

piece of metal, plastic or composites.

- Silk screening and engraving of panels vs. decals, stampings, etc.
- Connectors: Standard cables, backshells, etc., provide poor producibility interfaces.
- Cable lacing or tieing vs. cable straps.
- Manual point-to-point chassis wiring.
- Manual insertion of pins in printed circuit boards and connectors.
- Contract requirements for multi-level, sub-system, and system testing vs. environmental testing of complete system.
- Mil-spec vs commercial components and parts.
- Sheet metal: These cost-drivers are identified in the MC/DG sections related to this technology.
- Test, Inspection and Evaluation (TI&E): These cost-drivers are identified in the TI&E section of the MC/DG.
- Shock and anti-vibration mountings: Requirement to withstand high "g" impact.
- Wire identification: Stamping on wires, sleeving or color-coding wire.
- RFI shielded cables (RED signal, breeze type): Difficult to handle because of size and weight.
- High rejection rate of PCB's due to weave exposure, blistering, measling and delamination.
- Soldering problems: Due to oxidization of leads during storage and prior to forming and insertion.
- High circuit density: Results in thinner conductor lines, and the reduced spacing contributes to higher probability of electrical shorts or opens.
- Limitations on semi-or automatic wire wrapping: Due to requirements for twisted pairs, coaxial cables, shield wire, etc.
- Environmental protection: Against moisture and ionic contaminates such as salt.
- Lot Sizes: Selection of the correct lot requires considerable evaluation of a number of variable factors. As a general rule, the larger the lot size, the lower the cost per part.

The design engineer must not only be able to recognize cost-drivers and avoid them in the design, but he must also have the capability to evaluate optional manufacturing methods and perform trade-offs between optimum performance and affordable manufacturing costs. The MC/DG also provides this capability.

ADDITIONAL BENEFITS OF THE ELECTRONICS SECTION OF THE MC/DG

- Promotes interaction between design and manufacturing engineering, resulting in a team approach to resolve Design-to-Cost problems.
- Provides a data base for an in-house Design-to-Cost (DTC) data bank that can be easily modified to reflect specific in-house cost experience.
- Provides an industry "benchmark" or reference base for manufacturing costs related to manufacture of electronic assemblies.
- Due to the commonality of electronic components, will provide savings in all electronic-related products.
- Provides an excellent training vehicle for university students and inexperienced engineers.
- Alerts the design engineer to the manufacturing cost-drivers over which he has control.
- The Electronics Volume of the MC/DG will be available as pocket and desk hard-copy versions and as a computerized system (MC/DS), similar to previous sections of the MC/DG.
- The MC/DG can be easily maintained and updated as new cost information, or manufacturing technologies are developed to improve costs. Thus, the MC/DG should not become obsolete.

APPENDIX A

Conceptual Design Trade-Offs

A. Standard Circuits vs. New Technology
- Cost
- Weight
- Size
- Reliability
- Maintainability
- Probability of Availability
- Factory Capability of Handling
- Multiple Source (for Candidate Circuits)

B. One Box Vs. Two Boxes vs. N Boxes
- Vehicle Configuration
- Redundancy Level Desired
- Vulnerability Level Desired
- Identical Configurations
- Maintenance Philosophy
- Common Functions
- Box Reliability
- Weight

C. Common Functions Vs. Unique Functions
- Multi-Mode Computational Capability
- Reliability
- Space Availability
- Data Transmission
- Fault Isolation Capability
- Flight/Mission Critical

D. Analog Vs. Digital Processing
- Interface
  - I/O
  - Required Conversions
- Part Availability

- Part Costs
- Reliability
- Hardware/Software Integration
- Number of Functions
- Operational Definition
- Test Costs

E. Ambient Vs. Controlled Temperature
- Ambient Termperature
- Power Dissipation
- Heat Rise
  - In Box
  - In Installation
- Parts for Thermal Environment
  - Cost
  - Reliability
- Air Flow
- Controlled Temperature
- Redundancy Level - System
- Redundancy Level - Conditioning Systems

<u>Detail Design Trade-Offs</u>

F. Interconnect: Hard-Wire Auto Wire Wrap Vs. Wire Lay-Up Vs. Printed Circuit Board (PCB)
- Quantity
- Size
- Assume Production (Firm Design)
- Density
- Non-Recurring Costs
- Recurring Costs
- Repairability

G. Auto-Insertion Vs. Hand Insertion
- Quantity of Assemblies
- Type of Component
- Number of Axes - Component Orientation
- Number of Boards per blank
- Footprint
- Lot Sizes
- Assume Production (Firm Design)

H. Adding Cuts/Jumpers Vs. Redesignin Printed Wiring Board
- Quantity of Cuts and Jumpers
- Cost per Cut & Jumper (Labor)
- Cost of Re-Design of PWA
- Volume of Assembly Remaining

I. Soldering: Automatic Wave Vs. Vapor Phase Vs. Manual
- Density
- Type of Component
- Type of Lead Form

J. Structural Plastic Vs. Metal Parts for Chassis
- Temperature
- Weight
- Environment
- Quantity
- Cosmetics
- Specification Requirements
- Repairability
- Labor Costs
- Tooling Costs
- Life Expectancy

TYPICAL DESIGN FLOW FOR AEROSPACE AVIONICS INDUSTRY

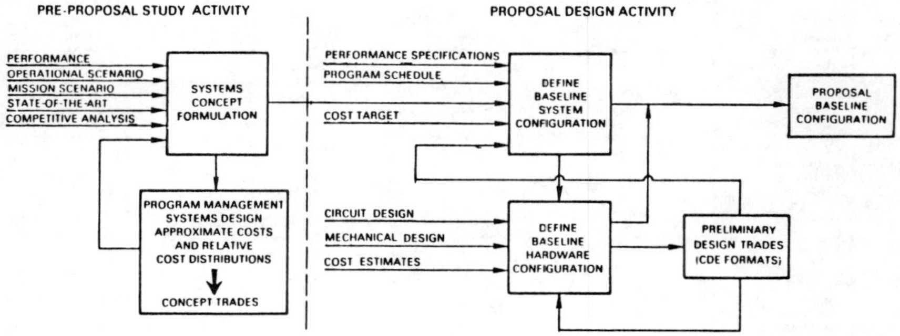

Figure 1

TYPICAL DESIGN FLOW FOR AEROSPACE AVIONICS INDUSTRY

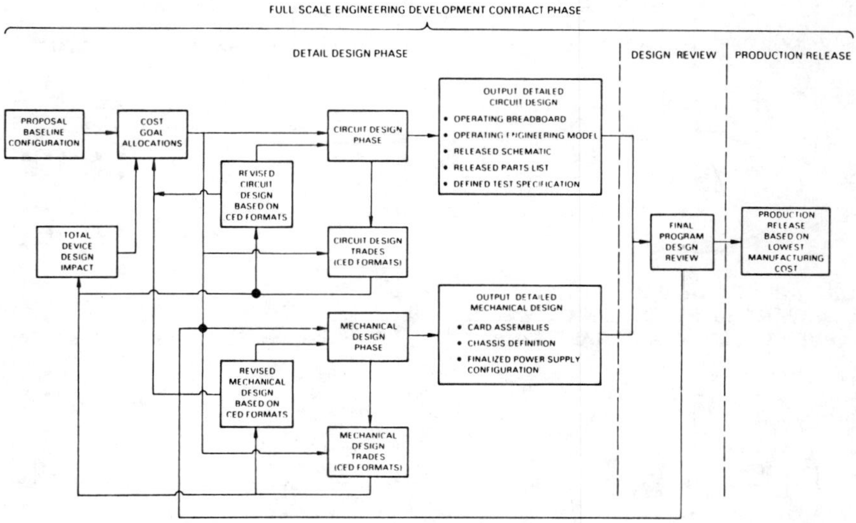

Figure 2

MC/DG VOLUME CONTENTS: "MANUFACTURING TECHNOLOGIES FOR AEROSPACE ELECTRONIC FABRICATION, ASSEMBLY, AND TEST"

I	II	III		IV
PROCURED ITEMS	DETAIL FABRICATION	ASSEMBLY		TEST, INSPECTION, AND EVALUATION (TI&E)
Schematic Parts • Electronic • Electromechanical Interconnect Parts • PWB • Connectors • Signal Interconnect – Wire – Coaxial – Flex Cable – Ribbon Cable – Wave Guides Hardware • Mechanical • Electrical Fabricated Parts • Castings • Sheet Metal • Plastic Parts • Powdered Metal • Machined Parts • Extrusions Emerging Technologies • VLSI • Displays (i.e., liquid crystal, electroluminescent, etc.) • Chip Carrier • Fiber-optics • Multilayer Flex Circuit	Metallic • Machining • Forming • Chem-milling Nonmetallic • Molding • Laminating • Machining • Forming Treatment/Coating/Marking • Anodize • Paint • Silk Screen • Plating Emerging Technologies • Laser Welding • Net Shape • Robotics	Mechanical Assembly • Connectors • Hardware Component Assembly (Pre-Wave and Post-Wave) • Automatic • Manual • Semiautomatic Cleaning Solder • Hand • Flow • Vapor Phase • Oven • Induction • Infrared Sheet Metal/Standoff Assembly (Hard Wiring) • Standoff Insertion • Part Assembly • Interconnect Cable/Wire Harness Assembly • Coaxial • Wire Bundles • Ribbon Cable	Hybrids Other Assembly Techniques • Multi-Wire • Stitch Wire • Wire Wrap Chassis Assembly Final Equipment Assembly • Mechanical • Electrical Post-Assembly Processes • Conformal Coating – Masking – Spray/Dip/Brush • Sealing Potting Adhesives Emerging Technologies • Chip Carrier • Chip/Epoxy • Robotics • Laser Soldering	Card/Module Level Test • Automatic • Manual • Calibrate Testability Burn-In/Screening Test • Automatic • Manual Device/Equipment Test • Automatic • Manual • Calibrate Emerging Technologies • Automated Assembly Handling • Improved CAD/CAM • CAT • Automatic Optical Comparison • Voice Control • Laser Inspection • Video Control Inspection

Table 1

COST APPORTIONMENT FOR AN ADVANCED SUPERSONIC AIRCRAFT

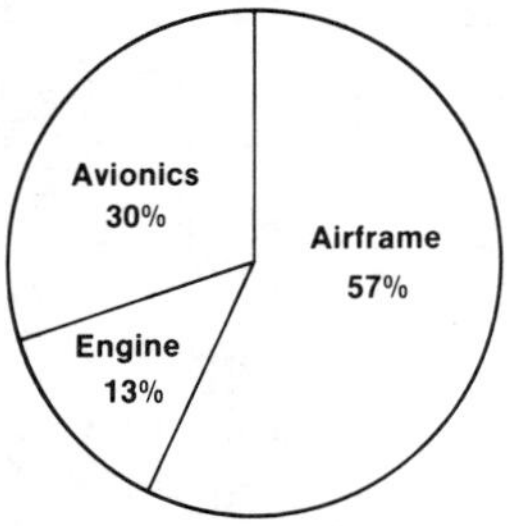

COST APPORTIONMENT FOR AN ADVANCED ELECTRONIC SURVEILLANCE AIRCRAFT

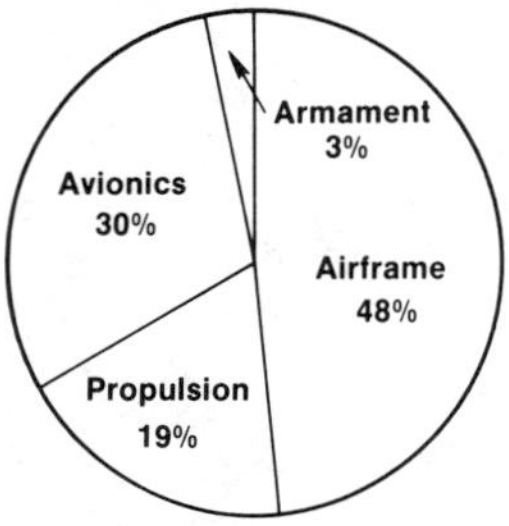

Figure 3

CONCEPTUAL DESIGN PHASE TRADE STUDY DESIGNER FORMAT SELECTION CHART

System Design Parameters	Parameter Known	Parameter Required For Format CDE				
		I	II	III	IV	V
Reliability		●	●	●	●	●
Maintainability		●	×	○	●	●
Environmental		○	●	○	●	○
Density		●	×	●	×	○
Part Costs		●	×	○	×	×
Factory Special Handling		●	○	○	○	○
Assembly Cost		●	×	×	×	×
Test Cost		●	×	×	×	●
Aircraft Configuration		○	●	×	×	●
Redundancy		×	●	×	●	●
Maintenance Concept		×	●	○	×	●
Interface		×	●	●	●	×
Partitioning, Functional		×	●	×	●	×
Fault Isolation		×	×	●	●	●
Mission Length		○	○	●	○	○
Part Count		×	○	×	×	×
Number of Functions		○	○	○	●	×
Vulnerability Levels		○	●	×	×	×

● = Required Data
× = Secondary Data
○ = Not Used

Table 2

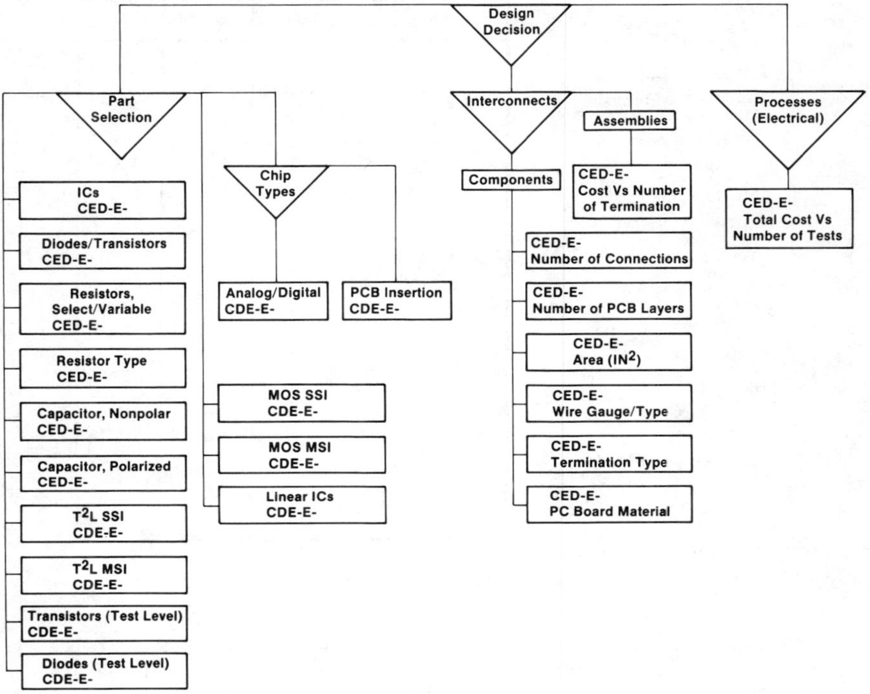

Figure 4 Format Selection Aid for Detail Design Phase

Baseline Configuration: Interface Assembly LRU
Indenture Level: RF Module Subassembly SRU

Baseline Manufacturing
Cost Allocation: $________

DESIGN DECISION	1	2	3	4	5	6	7
Part Selection - Screening Piece Part Type	% Total Electronic Part Cost	Baseline Approach	Reference CED No.	Study Approach	Ratio: Baseline to Study Approaches from CED	Cost Factor (%) 1 ÷ 5	Cost* Change (%)

Total: Cost Impact - Baseline Manufacturing Cost ($) x Cost Change Total (%) =
New Cost - Baseline Manufacturing Cost ($) - Cost Impact =

* Notes: Positive Number = Cost Savings
Negative Number = Cost Increase

Remarks____

By:________________
Date:________________

Table 3 MC/DG Designer's Cost Worksheet

PROCESS	Polyimide	G-10	Teflon	Ceramic	Multi-Wire
Part Package Suitability					
Conventional Leaded	A	A	S	S	S
Beam Ribbon	S	A	S	A	N
Leadless	S	A	N	A	N
Insertion/Attachment					
Hand	A	A	S	A	S
Semi-Auto	A	A	S	A	S
Auto	A	A	S	A	S
Soldering					
Hand	A	A	S	A	S
Wave	A	A	S	N	S
Vapor	S	S	S	A	N
Infrared	S	S	N	S	N
Laser	S	S	N	N	N

A Applicable

S Applicable (may require special processing/equipment)

N Not applicable

Table 4 Avionic Designer Format Showing Process/Material Considerations for Interconnect Between Components

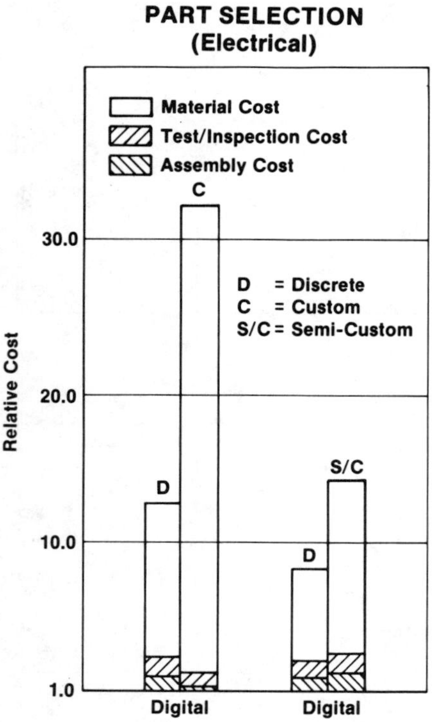

Figure 5

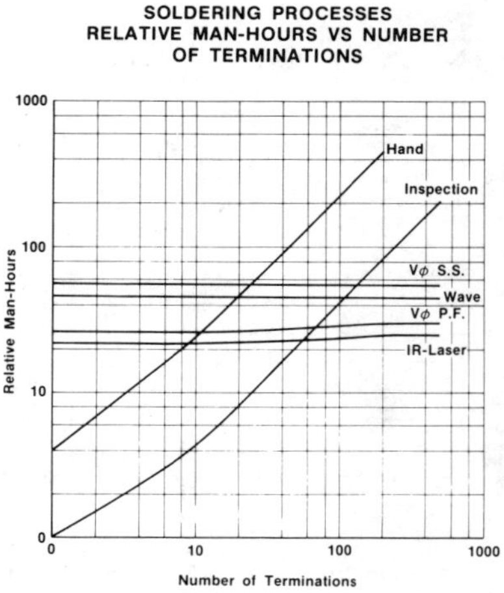

Figure 6

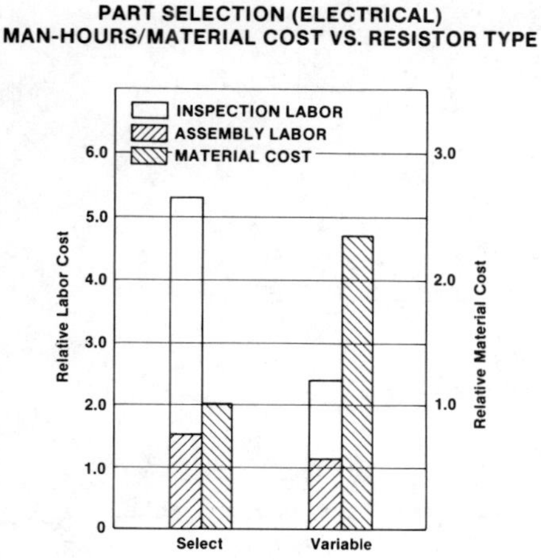

Figure 7

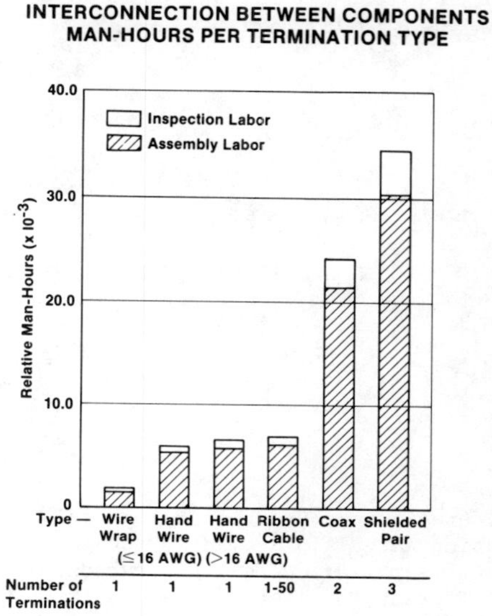

Figure 8

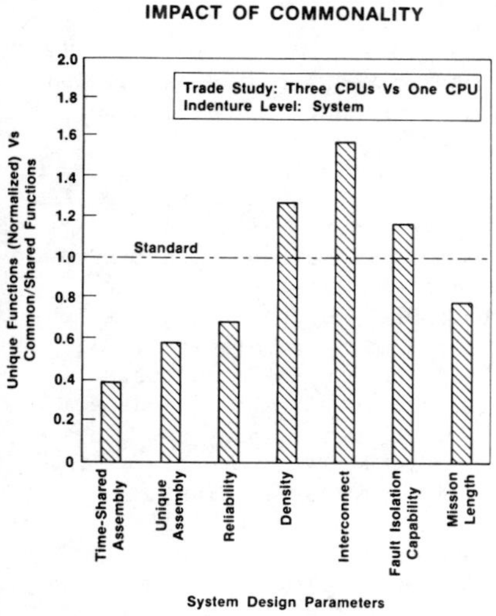

Figure 9

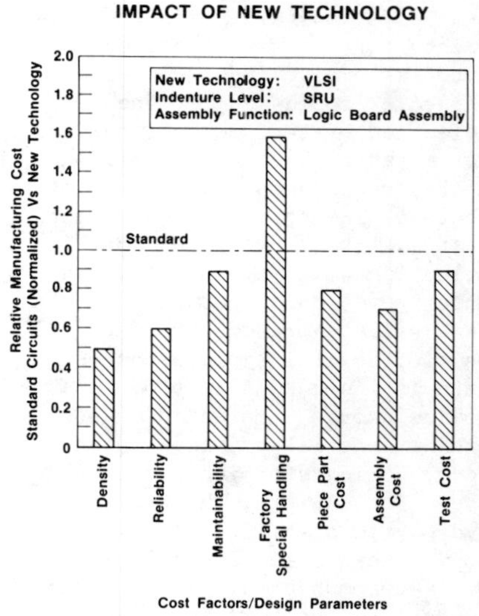

Figure 10

13th National SAMPE Technical Conference
October 13-15, 1981

SELECTION OF ENCAPSULANTS FOR NAVY ELECTRONIC SYSTEMS*

Doris S. Lequin, Naval Surface Weapons Center
Silver Spring, Maryland 20910

ABSTRACT

There is an abundance of encapsulating materials available today, with suppliers offering hundreds of different formulations, and each manufacturer claiming his material to be the "best". The dilemma of the user is to determine which of these encapsulants is really suitable for his application. It is the intent of this paper to present some evaluation techniques to aid in the selection of liquid casting resins for encapsulation of electronics. New ASTM guideline documents for using test methods in material evaluations will be discussed. Procedures for the selection of test methods needed for a particular set of environmental requirements will be presented. The type of requirements that must be known prior to this selection will be outlined. The interpretation of pertinent test data will be presented. The analysis of several typical encapsulants is included as a tool for the novice user. Some encapsulants exhibit special problems which can occur during storage or extended use. For example, the problem of hydrolytic stability of certain polyurethanes and some flexible epoxy compounds has been of great concern to the Navy in recent years. How to predict long-term stability of these materials is discussed. In summary, the knowledge of the expected service environment for encapsulated Navy electronics is critical to the selection of a proper encapsulant. Knowledge of the properties and long-term stability of the encapsulant allows the engineer to match this data to the expected service requirements. No utopia answer is available to meet every need, but proper evaluation of encapsulants can narrow down the numerous selections offered to the engineer by industry.

Keywords: Encapsulants, Epoxies, Polyurethanes, Hydrolytic Stability

1. INTRODUCTION

The environments that encapsulated electronics must survive can be very severe and unique. For example, static environmental atmospheres often vary from 70°C to -55°C, yet, after extended storage periods at these conditions, hardware must remain operable. Furthermore, during transportation or operation, hardware sees extreme dynamic environments. These include vibration, acceleration, spin, water entry and land impact, as well as possible radiation exposure. Plastics encapsulants are often selected because they have a low density, they withstand mechanical and thermal shocks well, they have good electrical characteristics, and provide a unique combination of properties.

---

*This paper has been adapted from reference (1).

Temperature and pressure sensitive electronics have made the use of room and low temperature curing liquid encapsulants most desirable. However, there is an abundance of encapsulating materials available today, with suppliers offering hundreds of different formulations, and each manufacturer claiming his material to be the "best". In addition, each hardware item presents a different set of requirements, and as a result, no single liquid casting resin can be used for every application. The dilemma of the user is to determine which of these encapsulants is really suitable for his application. It is the intent of this paper to present some evaluation techniques to aid in the selection of liquid casting resins for encapsulation of electronics.

2. AVAILABLE TEST METHODS

Numerous test methods are used to evaluate liquid encapsulants. The ASTM F-1 Committee on Electronics requested F-1.05 on Encapsulation to research available test methods used in industry today. The result of this study has been documented by ASTM F635-80.[2] This standard guide lists twenty-eight properties that can be measured. The significance of each property as it relates to liquid encapsulants evaluation is documented. Since Navy requirements are unique, a few additional in-house test methods are discussed in this paper. All methods mentioned in this paper are listed in Table I.

TABLE I. TEST METHODS

TESTS PERTAINING TO POLYMERIZATION	
Gel Time	Mol-Rez Gel Time Meter
Peak Exotherm	ASTM F 542 Determining Exothermic Temperature of Encapsulating Compounds
Spiral Flow	Gravity Flow into Spiral Mold
Volumetric Shrinkage	Calibrated One Inch Cube Molds
Initial Polymerization and Long-Term Stability	CCMD (Reference [3])
TESTS PERTAINING TO CURED PROPERTIES	
Specific Gravity	ASTM D 792 Specific Gravity and Density of Plastics by Displacement
Ash Content	ASTM D 2584 Ignition Loss of Cured Reinforced Resins
Hardness, Shore D	ASTM D 2240 Indentation Hardness of Rubber and Plastics by Means of a Durometer
Hardness, Rockwell M	ASTM D 785 Rockwell Hardness of Plastics and Electrical Insulating Materials
Izod Impact Strength	ASTM D 256 Impact Resistance of Plastics and Electrical Insulation Materials
Thermal Shock	ASTM D 1674 Ploymerizable Embedding Compounds Used for Electrical Insulation
Embedded Thermometer	ASTM F 135 Embedment Stress Caused by Casting Compounds on Glass-Encased Electronic Components
Corrosivity Index	Method 7071 of Fed. Test Method Std. No. 406
Water Absorption	ASTM D 570 Water Absorption of Plastics
Water Vapor Transmission Rate	Method 7032 of Fed. Test Method Std. No. 406
Hydrolytic Stability	ASTM F 74 Determining Hydrolytic Stability of Plastic Encapsulants for Electronic Devices

3. EVALUATION OF ENCAPSULANTS

Evaluation of encapsulants can be divided into three areas: Polymerization, initial cured properties, and changes with long-term storage.

3.1 Polymerization Properties

Data needed to evaluate the polymerization properties begins with the determination of the peak exotherm temperature and the time to reach that temperature. Peak exotherm temperatures developed during polymerization are dependent, in part, on the chemical reaction taking place, the total volume of the material, and whatever heat sink may be present. Some systems can reach temperatures high enough to cause the material to foam or even char. The peak exotherm is obtained by inserting a thermocouple into the center of the resin and monitoring the temperature as the liquid resin becomes a solid. Selection of the curing agent or hardener for any particular resin system will determine the chemical reaction and the heat of reaction to be expected, as shown in Fig. 1. The volume of resin and the heat sink available also affect the exotherm temperatures encountered by a single resin system as shown in Fig. 2. Some other polymerization properties are helpful in the selection of proper encapsulants. The pot life, or working time, is slightly shorter than time to peak exotherm, and is used as an indication of the time available for miximg and pouring the resin into the hardware. It is easily determined by a Mol-Rez Gel Time Meter. The ability of liquid casting resins to gravity-flow into units is estimated by the distance a 100 ml volume of resin travels from the transfer pot into a spiral flow mold. Volumetric shrinkage during polymerization includes both chemical crosslinking shrinkage and thermal shrinkage and is measured with calibrated one-inch cube molds.

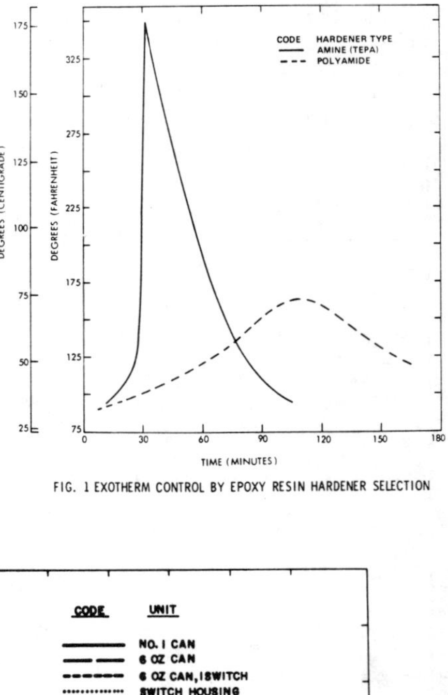

FIG. 1 EXOTHERM CONTROL BY EPOXY RESIN HARDENER SELECTION

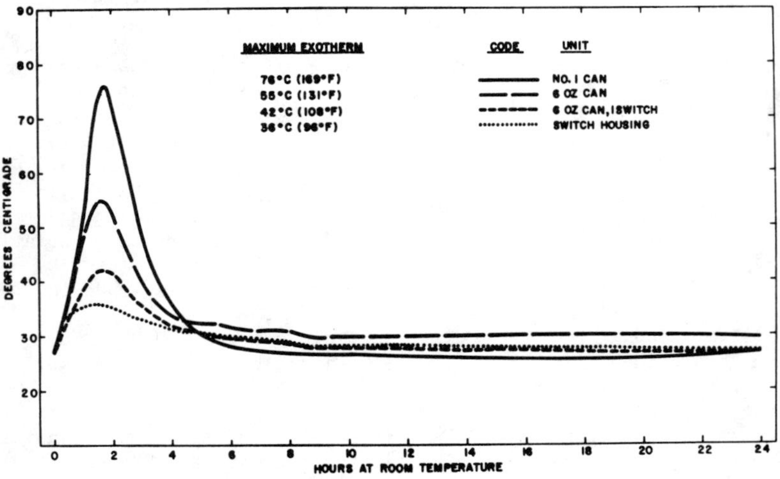

FIG. 2 EFFECT OF VOLUME AND HEAT SINK ON EXOTHERM TEMPERATURES

3.2 Cured Properties

The use of pressure and temperature sensitive electronic components has made room or low temperature curing resin systems most desirable because of their low shrinkage characteristics. The volume shrinkage cubes are also used to obtain specific gravity, hardness, and ash content of the cured resin. The effect of thermal expansion properties on the encapsulant during thermal cycling is evaluated by thermal shock tests using the hex bar insert. The ability of the electronics to withstand thermal cycling without changing their properties or cracking their glass housings is determined by the embedded thermometer test. Stresses as high as 4000 psi or more have been measured at -40°C as plotted in Fig. 3. These stresses are often relieved by the use of circuit board coatings, or the selection of flexible resins for the specific application.

Many times an epoxy resin is used to provide a moisture barrier for components exposed to the atmosphere. The water vapor transmission and water absorption characteristics of the resin are important, especially during long-term storage. If moisture is absorbed into the resin, it is important to determine if any corrosive materials are developed which, in turn, could corrode electrical contacts, etc. The water extract conductance test measures potential corrosiveness of the encapsulants. Most water absorption values for epoxy resins are less than two percent over a six month test period, but will increase with greater flexibility of the resin.

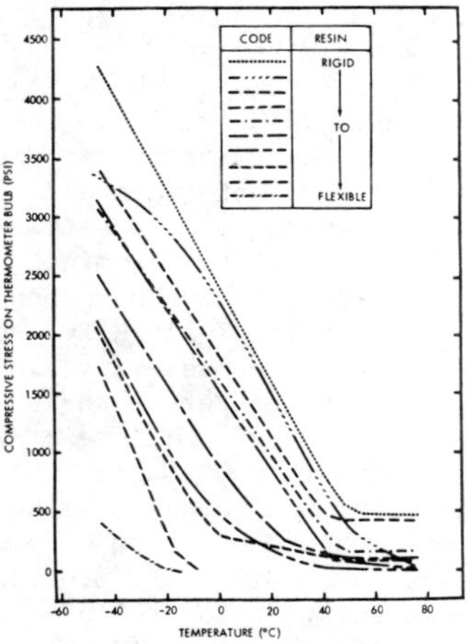

FIG. 3 EMBEDMENT STRESSES DEVELOPED IN CAST EPOXY RESINS

Test results for six encapsulants are given in Table II. Those test values which are outside of acceptable limits are circled. For example, in Section A, a 28-minute gel time is not recommended. For most applications, the working life, or the pot life, required would be at least 45 minutes. This gives sufficient time for mixing and pouring of the resin into hardware components. The peak exotherm for two of the encapsulants is over 300 degrees. Even with the available heat sink in electronic units, a peak exotherm of this range would probably cause excessive heat in actual hardware. Spiral flow values under ten inches do not flow readily into areas to be encapsulated, and values over 27 inches often leak out of housings or flow into unwanted areas. Volumetric shrinkage, which includes both polymerization crosslinking and thermal shrinkages, may lead to cracking of electronics or the encapsulant. In order to minimize this effect, resins selected should have volumetric shrinkages below four precent. In Section B of Table II, the specific gravity of one system may be too high for the application's weight tolerance. Comparison of the Shore D and Rockwell M hardness values indicate that one system has very little flexibility available to relieve stresses in the encapsulant. This is confirmed by failure of this resin to pass the thermal shock test. It would seem that selection of a material could depend greatly on flexibility, with the more flexible materials desirable for relieving harmful stresses. Actually, other factors must be considered. For example, the higher the flexibility, the higher the thermal expansion of the polymer, which would increase the stress if the mass is fully confined.

TABLE II. ENCAPSULANT EVALUATION

Property	Casting Resin System A	B	C	D	E	F
A. Properties Pertaining to Polymerization:						
1. Mol Rez gel time, 100 ml, min.	65	28	45	113	94	59
2. Peak exotherm, 150 ml, embedment test, °F	103	121	124	97	101	103
3. Peak exotherm, 300 ml, No. 1 can, °F	248	311	332	220	256	246
4. Spiral flow, inches	15.9	22.1	8.8	25.5	15.2	5.0
5. Volumetric shrinkage, percent	3.91	4.57	3.98	3.82	3.89	4.47
B. Properties of Cured Compound:						
1. Specific Gravity	1.84	1.62	1.42	1.33	1.94	2.27
2. Ash content, percent	56.2	49.6	38.0	28.6	59.3	68.6
3. Hardness, Shore D	88	92	88	70	88	90
4. Hardness, Rockwell M	55	96	64	--	64	76
5. Izod Impact Str., ft.lbs/in. notch	.269	.235	.285	.165	.264	.374
6. Thermal shock, 10 cycles, 70°C to -54°C	Passed	Failed Second Cycle	Passed	Passed	Passed	Passed

The addition of fillers reduces thermal expansion but also reduces the flexibility and increases the viscosity, thus making application more difficult. All such factors must be considered in compromise in the selection of liquid casting resins or encapsulants.

3.3 Changes With Long-Term Storage

Some encapsulants exhibit special problems which can occur during storage or extended use. For example, the problem of hydrolytic stability of certain polyurethanes and some flexible epoxy compounds has been of great concern to the Navy in recent years. These materials are cured initially, but during storage gradually revert to the liquid state. Hydrolytic stability may be determined by exposing cured pucks of the encapsulant to 95% relative humidity at either 97°C, 85°C or 71°C. The hardness of the pucks is measured at regular time intervals and plotted as hardness versus time. (See Fig. 4.) Visual examination of the pucks indicate that some materials completely liquify, some soften into gummy masses, while others soften but retain their original geometry. Failure of the test samples is reported as the time required for the hardness value to read less than one on the durometer. These failure points are plotted on an Arrhenius graph of failure time versus logarithmic reciprocal-

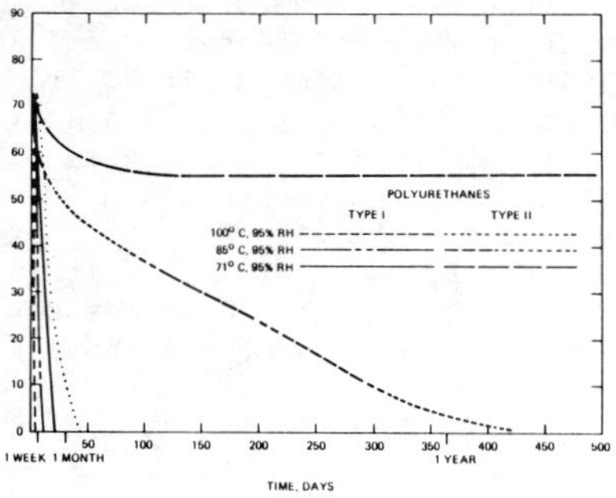

FIG 4 HARDNESS VS. TIME MEASUREMENTS FOR HYDROLYTIC STABILITY TESTS

temperature. Straight lines through the plotted points are calculated by the method of least squares, and then extrapolated to 23°C as shown on Fig. 5. This provides an indication of the encapsulant's hydrolytic stability in service use over a period of several years.

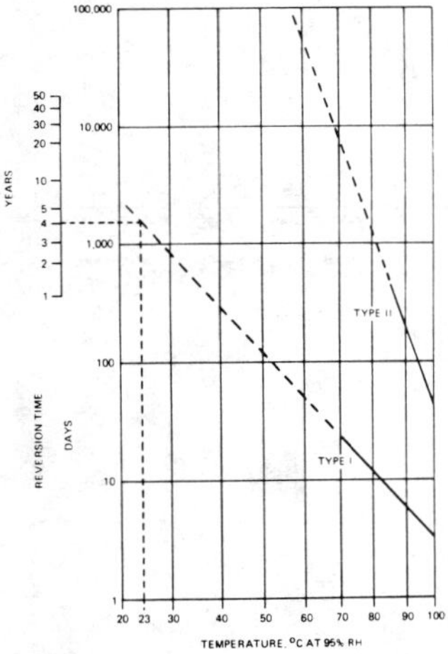

FIG. 5 HYDROLYTIC STABILITY OF POLYURETHANE ENCAPSULANTS

In the laboratory evaluation of electronic components, test programs are of necessity conducted within a very short time following the potting process. This raises a question as to how well the encapsulants have been polymerized or cured prior to testing. Prototype hardware may be potted for months prior to testing, whereas, ultimately, the encapsulated units may be in storage for many years. Therefore, in order to evaluate encapsulants and to predict service and storage life, pertinent data on the extent of polymerization and the stability of the cured compounds are necessary. The principal test equipment employed in the initial polymerization and long-term stability studies is the Continuous Current Monitoring Device (CCMD).[3] The test apparatus consists of a sample cell containing 10 ml of resin cured between two electrodes (Fig. 6). The conductivity of the resin during polymerization is measured, and the volume resistivity calculated and plotted.

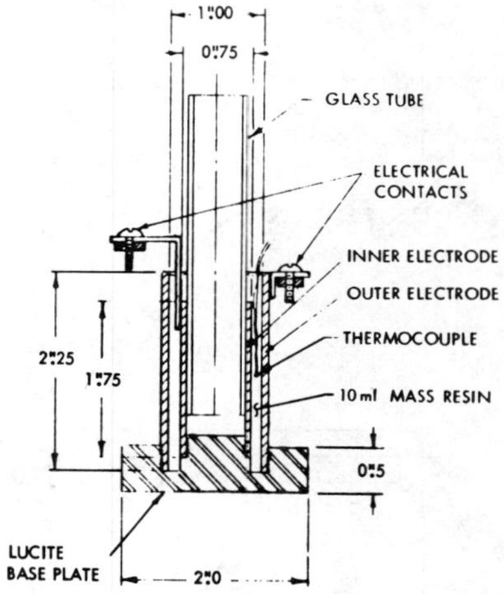

FIG. 6 10 ml SAMPLE CELL WITH ISOTHERMAL ELECTRODES

The magnitude of the plotted data is indicative of the degree of flexibility obtained during polymerization (Fig. 7). Completion of the initial polymerization process is indicated by the flatness of the curve after 12 hours. The cured test specimen is subsequently heated to 71°C, removed from the oven, and the conductivity measured as the unit cools to about 45°C. This data is used to obtain the temperature dependence of the resistivity shown by Fig. 8. The slope of this plotted data provides the activation energy for the electrical conduction process (E_c). This E_c value is used as an index of the degree of crosslinking within the polymer structure.

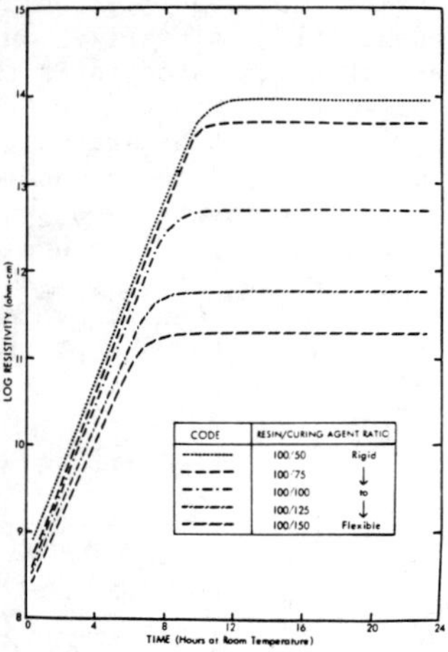

FIG. 7 INITIAL POLYMERIZATION STUDY OF AN EPOXY RESIN INDICATING VARIATIONS IN RESIN FLEXIBILITY

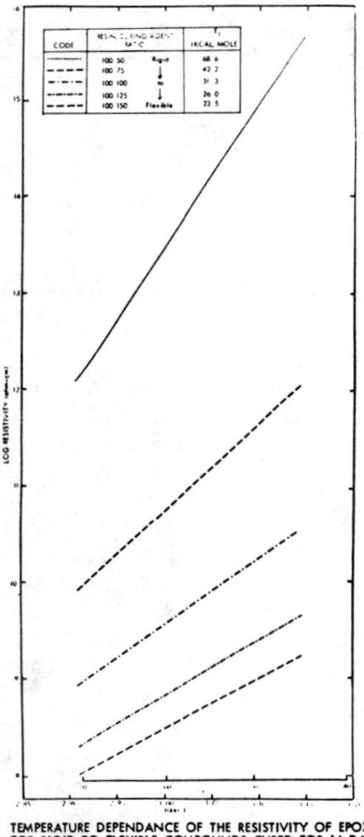

FIG. 8 TEMPERATURE DEPENDANCE OF THE RESISTIVITY OF EPOXY RESINS FOR RIGID TO FLEXIBLE COMPOUNDS CURED FOR 14 DAYS AT ROOM TEMPERATURE

Since the test is nondestructive, cooling curves may be determined at any interval of time desired. The analysis of this superimposed data is indicative of the long-term stability of the encapsulant. Fig. 9 shows the data for an encapsulant that stabilized within two months, and Fig. 10 indicates a stable material within 24 hours.

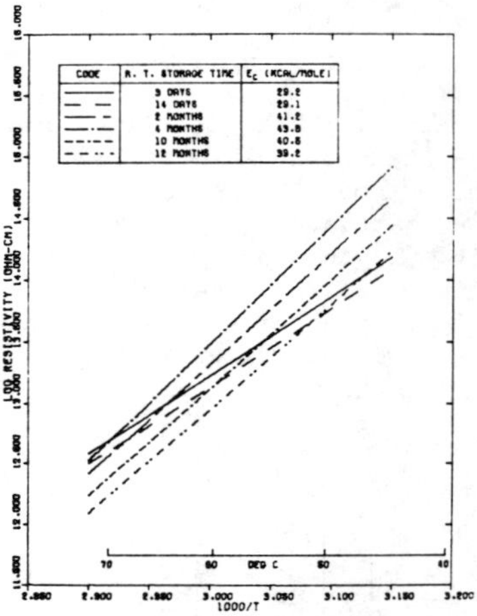

FIG. 9 LONG-TERM STABILITY STUDY OF AN EPOXY RESIN COMPOUND WHICH REACHED STABILITY WITHIN TWO MONTHS

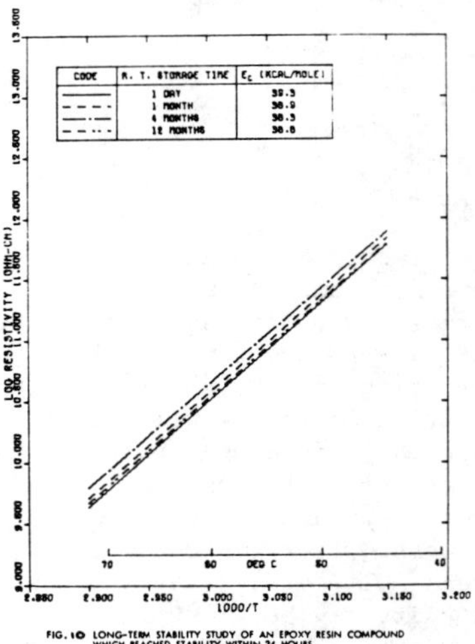

FIG. 10 LONG-TERM STABILITY STUDY OF AN EPOXY RESIN COMPOUND WHICH REACHED STABILITY WITHIN 24 HOURS

4. SUMMARY

In summary, the knowledge of the expected service environment for encapsulated electronics is critical to the selection of a proper encapsulant. Knowledge of the properties and long-term stability of the encapsulant allows the engineer to match this data to the expected service requirements. No utopia answer is available to meet every need, but proper evaluation of encapsulants can narrow down the numerous selections offered to the engineer by industry.

5. REFERENCES

1. D. S. Lequin, "Choosing Liquid Encapsulants for Electronics," Journal of Elastomers & Plastics, Vol. 13 (Jan 1981), p 3-18.

2. "Standard Guide for Selection of Tests for Material Properties of Liquid Thermoset Encapsulating Compounds for Electronic and Microelectronic Encapsulation," ASTM F 635-80, ASTM Standards, Part 43, American Society for Testing and Materials, Philadelphia PA 19103 (1980).

3. D. S. Lequin, R. D. Ezell, "Evaluation of Potting Compounds by Electrical Resistivity Techniques Using the Continuous Current Monitoring Device (CCMD)," NSWC Technical Report NOLTR 73-10, Naval Surface Weapons Center, Silver Spring, MD 20910 (23 Feb 1973).

6. BIOGRAPHY

Mrs. Lequin is a materials engineer in charge of the Plastics Engineering and Services Group of the Nonmetallic Materials Branch at the Naval Surface Weapons Center in Silver Spring, Maryland. She serves as a plastics consultant in-house and throughout the Navy in the use of all types of plastics and elastomers. She has received the U.S. Navy's Meritorious Civilian Service Award twice, in 1964 and again in 1976, for her contribution to many Navy projects. Mrs. Lequin was graduated from Mary Washington College of the University of Virginia. She is a member of the Society of Plastics Engineers, American Society for Testing and Materials, and American Scientific Affiliation. She has authored over 45 technical reports and numerous material specifications and ASTM test methods.

13th National SAMPE Technical Conference
October 13-15, 1981

HIGH-SPEED ENCAPSULATION OF ORDNANCE ELECTRONIC CIRCUITRY

Allan B. Goldberg
US Army Electronics Research
and Development Command
Harry Diamond Laboratories
Adelphi, MD 20783

Abstract

Ordnance electronic circuitry, components, and housings containing electronic assemblies are usually encapsulated with plastics to provide electrical insulation, mechanical support, and protection from environmental effects and handling abuses. The encapsulation is particularly critical for weapon projectiles that must survive very high accelerations during firing, and, in some cases, function after impact. In the past, we have used conventional liquid resin systems, especially epoxies, silicones, and polyurethane foams and elastomers. These materials usually require mold dwell times of 2 to 24 hr, resulting in increased production costs due to capital costs for equipment and fixtures, energy consumption, and space requirements. This paper will center on our efforts to identify encapsulating materials that will allow the processing of 100,000 circuits/mo with mold dwell times as short as 1 min. Most of the work was done on two contracts involving state-of-the-art review, processing, and mechanical property testing, and prototype molding of representative discrete or hybrid circuits. The remaining part of the work, consisting of environmental and electrival testing and air-gun firing of molded circuits, was done by HDL. Honeywell's Defense System Division considered 72 commercially available products with emphasis on pourable liquid encapsulants. The list was narrowed to 21 for laboratory tests and then to 10 for molding trials. A final recommendation of 5 products included novel systems that take advantage of savings identified in an analysis of the cost drivers in circuit encapsulation. Springborn Laboratories investigated the possibility of injection or transfer molding of complete hybrid circuits.

1. INTRODUCTION

The Harry Diamond Laboratories (HDL), as part of the Electronics Research and Development Command, is involved with the development of electronic fuzing as part of our responsibility for intelligence, surveillance, and target acquisition. In most of these applications, the electronic circuitry must function with 96 to 99% reliability after long-term storage up to 25 yr, without prior testing. This places severe challenges on the polymeric materials used for environmental and structural protection, for their failure could lead to system failure. Operating environments may include spinning, balloting (side-to-side slapping in worn gun barrels), and setback forces of 50 to 100,000 g, with spin rates of 0 to 1000 rps, at temperatures of -48°C(-55°F) to 71°C (160°F) and humidities from 0 to 100%. The encapsulating plastics (or in some cases, thick conformal coatings) must be processable under conditions that do not damage sensitive electronics. They must survive thermal shock and rough handling during storage and transportation. Further, although the actual cost of the polymeric materials is only a small part of the overall cost of the electronic assembly, processing costs and energy savings are a necessary consideration. Generally desirable properties for the materials that we use are given in Table 1.

Cost reduction and increased

Table 1. Desirable Properties of Encapsulants.

Property	Value
Mechanical	
Shore hardness	60-90A;60-90D
Thermal shock, MIL-I-16923G (hexbar, 10 cycles between 130°C and -55°C)	Pass
Embedment stress, ASTM F135	Low or none
Tensile strength, 71°C, psi	$\geq$ 400
Ultimate elongation, -40°C, %	$\leq$ 10
Electrical	
Volume resistivity, ohm-cm	$\geq 10^{13}$
Dielectric constant, 60 Hz	$\leq$ 5.0
Dielectric constant, 1 MHz	$\leq$ 4.5
Dissipation factor, 60 Hz	$\leq$ 0.04
Dissipation factor, 1 MHz	$\leq$ 0.05
Aging	
Hydrolytic stability, MIL-I-16923G, 120 days at 71°C/95% rh	Pass
Water absorption, MIL-I-16923G, 24 hr	$\leq$ 0.2%
Air aging	Retain functional properties for 25 yr
Noncorrosive to copper, aluminum, steel	
Repairable or removable	
Transparent	
Nontoxic, no health hazard	

production rates continue to be of interest. With fuzes and other assemblies being produced in quantities as high as 100,000/mo, a great deal of capital, floor space, and energy can be required to encapsulate with the conventional liquid epoxies, silicones, and polyurethanes (foams and elastomers) we now use. Such products usually require elevated temperature cures of from 2 to 24 hr. Since new products are frequently being introduced, we have attemped to advance the state of the art in ordnance electronic circuit encapsulation by using either newer more advantageous pourable liquid encapsulants, or new techniques such as liquid injection molding (LIM), reaction injection molding (RIM), and transfer molding. Our goal is to be able to remove encapsulated circuits from molds or fixtures in from 1 min to 1 hr. The effort involved work at HDL and, under contract, at Honeywell's Defense Systems Division and at Springborn Laboratories.

2. STATE OF THE ART

The encapsulants used most often for our current applications are listed in Table 2. All are castable liquids. Most of our high production quantity fuze electronics

Table 2. Examples of Encapsulants.

Product	Supplier	Use	Features
Isofoam PE-6S	Witco	M734 Multi-Option Mortar Fuze	Low-density polyurethane foam with high expansion factor; low dielectric constant and loss; cure 1 hr at 65°C.
Isofoam PE-10S	Witco	XM445 Multiple Launch Rocket System Fuze	Medium-density polyurethane foam; cure 1 hr at 65°C.
Isofoam PE-18S	Witco	M732 Artillery Fuze	High-density polyurethane foam; relatively low dielectric constant and loss; cure 1 hr at 65°C.
C9-R246/H-R248 R1017/H4003 Y921-A/Y921-B (Honeywell Spec MH20278P)	Hysol Epic Resins Sterling Varnish	M724 Electronic Artillery Time Fuze	Filled, rigid, epoxy; low thermal expansion coefficient; cure 16 hr at 65-79°C.
Sylgard 184	Dow Corning	M817 Target Detecting Device for Chaparral Missile	Repairable silicone; transparent, no embedment stress; cure 4 hr at 65°C.
Polymercast V356-HE80	US Polymerics	M735 Fuze for 8-in Nuclear Projectile	Polyurethane;good aging stability;used as thick conformal coating;cure 2 hr at 82°C plus 3 hr at 110°C

are encapsulated in a polyurethane foam. The foam has good electrical properties, with low dielectric constant and loss factor. It is lightweight, but strong enough to withstand forces up to 25,000 g. Since it goes into the cavity containing the electronic circuitry as a free-flowing liquid under essentially no injection pressure, components are not dislodged. The expansion of the foam produces little stress and guarantees complete filling of the cavity. Different densities of the foam are used for different applications, but in all cases the foam is overpacked. Thus, nominal 18 lb/ft^3 (pcf) foam ends up as 20 to 24 pcf foam in the M732 arillery fuze, where we encapsulate discrete component circuitry on standard epoxy/glass printed circuit boards. In the M734 multi-option mortar fuze, the hybrid circuits on ceramic substrates are encapsulated in a nominal 6 pcf foam. Although these foams are designed to cure relatively quickly at room temperature, we have found that the small masses we normally use (5-20 g) do not generate enough exotherm to produce full cure. In order to guarantee dimensional stability, we recommend preheating the electronics to 65°C or curing the system for 1 hr at 65°C.

For especially high shock environments, a rigid, 75% silica-filled epoxy has been very successful. The discrete components are held very firmly in position so that no leads are broken and no components are damaged. The high filler level results in a low coefficient of thermal expansion, minimizing stress to encapsulated components on the thermal cycling. The disadvantages of this material are the expensive equipment needed to process it during production and the 16 hr cure time at 65-79°C. The latter requires numerous ovens and ties up fixtures for long times.

3.1 Encapsulation with Pourable Liquids

Our first contract* was awarded to Honeywell's Defense Systems Division for the "Evaluation Selection of Encapsulating Plastics for Ordnance Electronic Assemblies." Although the program considered such innovative processes as LIM, RIM, and transfer or injection molding of solids, it concentrated on identifying liquid, pourable encapsulants with distinct processing advantages over those we are now using. After drawing up a specification based on the desirable properties itemized in Table I, Honeywell contacted 90 material manufacturers and suppliers who recommended 72 products. This list was narrowed to 21 products. Eliminations were made on the basis of commercial availability, compatibility with the electronics, rapid demold time, and uniqueness of the material. The materials included the most promising product(s) in each of nine categories: (1) chemically blown foam, (2) two-component syntactic foam, (3) one-component

* Contract DAAK21-79-C-0017

syntactic foam, (4) two-component filled epoxy, (5) one-component filled epoxy, (6) semi-rigid urethane (Shore D > 70), (7) Urethane elastomer, (8) transfer moldable epoxy, and (9) injection moldable thermoplastic urethane.

These 21 materials were tested extensively in the laboratory using appropriate ASTM and military specifications. The tests were selected to determine (a) processibility, (b) mechanical and thermal properties, and (c) stability on aging.

The laboratory test results were used to select 10 materials for the next phase involving encapsulation of test circuits. Two circuit types were used. The first, Fig. 1, was a thick-film hybrid circuit with small discrete components including tantalum and ceramic capacitors and pre-coated chip-and-wire active components. These were actual mortar fuze amplifier circuits and could only be tested for total circuit function.

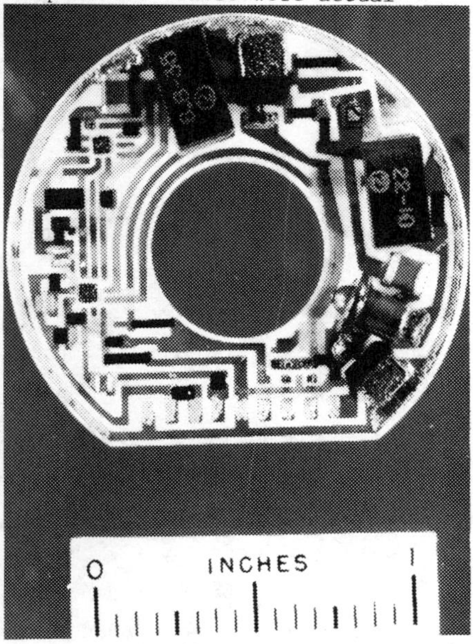

Figure 1. M734 Amplifier Circuit.

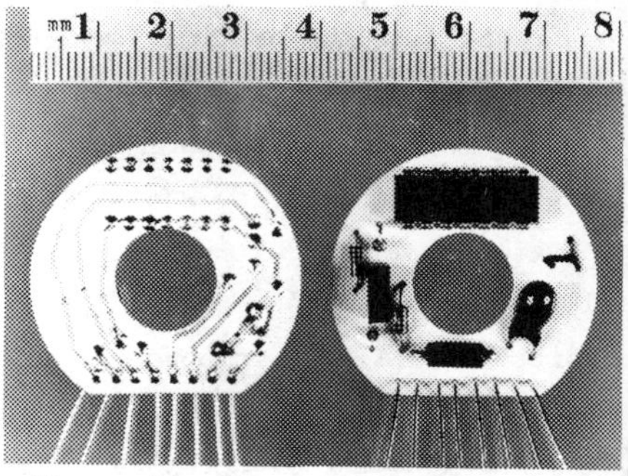

Figure 2. Discrete Component Test Circuit.

The second test circuit, Fig. 2, designed to fit in the same mold as the first circuit, contained discrete components that typically suffer problems during or after encapsulation. On the epoxy/glass board were placed such fragile comcomponents as glass diodes, small resistors, and a choke, all close to the sharp corners of a molded capacitor chosen to serve as a crack generator. A liquid tantalum capacitor was included for its heat sensitivity, and a 14-pin molded DIP was meant to restrict flow of the encapsulant. Each component except the DIP could be individually tested electrically.

After each of the 10 materials had been used to encapsulate 5 circuits of each type, the final 5 products were chosen. These products were then used to mold an additional 5 circuits of each type. The molded circuits were subjected to temperature and humidity cycling per MIL-STD-331A, Test 105.1. During this time they were exposed to two 14-day periods consisting of nine cycles between the extremes of 71°C and 95% rh and -54°C, with additional storage at 71°C and 95% rh and at -62°C. The second set of encapsulated circuits, prepared from the final list of 5 products, was subjected to a modified cycle in which the lowest temperature was limited to -40°C, since this had been the low-temperature criterion used during selection. In either case, the circuits were tested electrically after one week of exposure and again after four weeks. All discrete circuits and surviving hybrid circuits were then fired at room temperature in a 4-in. air gun to simulate actual use. Discrete circuits were subjected to 15,000 g and hybrids to 12,000 g.

Laboratory test results are presented in Table 3 for the final 5 materials. Electrical tests results are given in Tables 4 and 5.

3.2 Cost Analysis

Early in the program, we realized that shortening the dwell time in

Table 3. Laboratory Test Results.

Product	Coro-foam 589	PR2036	Stycast 1495/ Catalyst 9	EO-0029	ISP-100
Type	Chemically Blown Urethane Foam	Epoxy Syntactic Foam	Filled Epoxy	Single-Component Filled Epoxy	Semi-rigid Urethane
Supplier	Cook Paint & Varnish	Formulated Resins	Emerson & Cuming	Hysol	Dow Chemical
Mix Ratio	1:1	2:1	100:9	N/A	27:10:0.05
Demold Time	15 min	45 min	30 min	3 hr	10 min
Cure Temperature	23°C	65°C	65°C	74°C	23°C
Postcure	23°C	23°C	23°C	13 hr/74°C	23°C
Peak Exotherm	----	163°C/150ml	47°C/150ml	149°C/150ml	129°C/150ml
Density/Specific Gravity	16 lb/ft^3	0.84	1.87	1.73	1.13
Glass Transition Temperature	81°C	76°C	90°C	89°C	68°C
Coefficient of Thermal Expansion	----	29×10^{-6}/°C	36×10^{-6}/°C	37×10^{-6}/°C	82×10^{-6}/°C
Compressive Strength(23°C)	750psi	----	----	----	----
Flexural Strength(23°C)	---	7,100psi	12,000psi	16,000psi	9,400psi
Thermal hock per MIL-I-6923G	---	Pass	Pass	Pass	Pass
Lap Shear Adhesion/Aluminum	--	2,600psi	1,900psi	1,820psi	700psi
Hydrolytic Stability per MIL-I-6923G		Pass	Pass	Pass	Pass

the mold is not the only way to reduce the cost of circuit encapsulation. Each category of encapsulant has special storage, processing, or tooling requirements that impact the cost for their use. For example, a single component material requires less complex processing equipment and eliminates the need for several inspection tests, but has a longer cure time, special storage conditions, and generally higher cost per pound.

If certain assumptions are made, then the recurring and nonrecurring costs associated with the use of each category of material can be quantified, and the major factors affecting the cost of circuit encapsulation can be identified. These assumptions were: (a) Each category of material would function satisfactorily. (b) The production rate is 100,000 units/mo on a single shift

Table 4. Electrical Performance of Discrete Circuits.

(Number Passed/Number Tested)				
Encapsulant	After Molding	After 4 Weeks T&H(a)	After 4 Weeks T&H(b)	After 15,000 g Shock
Coro-foam 589	10/10	5/5	3/5	9/10
PR2036	10/10	3/5	4/5	10/10
Stycast 1495/ Catalyst 9	10/10	4/5	4/5	9/10
EO-0029	10/10	5/5	5/5	10/10
ISP-100	10/10	5/5	5/5	10/10
Isofoam PE-18S	5/5	5/5	---	5/5

(a) MIL-STD-331A Test 105.1.
(b) MIL-STD-331A Test 105.1 modified to -40°C.

Table 5. Electrical Performance of Hybrid Circuits.

(Number Passed/Number Tested)				
Encapsulant	After Molding	After 4 Weeks T&H(a)	After 4 Weeks T&H(b)	After 12,000 g Shock
Coro-foam 589	10/10	0/5	0/5	---
PR2036	9/10	2/5	2/5	3/4
Stycast 1495/ Catalyst 9	9/10	1/5	2/5	2/3
EO-0029	10/10	2/5	0/5	1/2
ISP-100	9/10	0/5	0/5	---
Unpotted	4/5	---	0/5	---

(a) MIL-STD-331 Test 105.1.
(b) MIL-STD-331 Test 105.1 modified to -40°C.

of 8 hr for a 5-day week. The work year contains 240 days. (c) The production facility has already been set up and debugged (no start-up costs). (d) Floor space is $\$1/ft^2/mo$. (e) Factory labor is $25/hr while engineering and quality assurance cost $35/hr. An 82% efficiency factor is used. (f) Each unit requires 17 cm^3 of encapsulant. The actual weight of encapsulant needed must take into account both the density and material losses due to handling and scrap. (g) Recurring costs include those for materials, molds, labor, floor space, and energy. Nonrecurring costs cover equipment, permanent tooling, and facilities.

A detailed discussion of the cost analysis is beyond the scope of this paper, but has been reported elsewhere.* However, the following general conclusions regarding annual costs can be given here: (a) As a category, chemically blown foams should be the least expensive materials and filled epoxy systems the most. (b) With a 1 hr demold time, integral potting cups that become a part of the final assembly could reduce tooling costs by as much as 58% ($320K) when compared to matched-metal molds with replaceable aluminum die-cast cavities.

*S.J. Price, R.E. Isliefson, and R.J. Ryan "Evaluation Selection of Encapsulating Plastics for Ordnance Assemblies: Final Report," Harry Diamond Laboratories, HDL-CR-81-017-1, May 1981.

Matched metal molds with plastic inserts could save 33% ($184K). (c) With a rigid epoxy, reducing the cure time at 65°C from 16 hr to 1 hr would save 8% ($94K), primarily in ovens and floor space. (d) For a chemically blown foam, eliminating the need for a 1 hr oven cure would save 18% ($132K), mostly in labor, ovens, and floor space. (e) Selecting a two-component, rigid, filled epoxy so as to avoid the need of precoating all or part of the circuit with a stress-relieving silicone conformal coating could save 22% ($239K), mostly in labor, materials, and equipment. (f) If two rigid epoxies requiring cure cycles of 16 hr at 65°C are compared, the one-component system will cost 9% ($10K) less than the two-component system. Savings are mainly in material costs, equipment, labor and floor space. (g) An interesting comparison can be made between liquid casting with a two-component epoxy and transfer molding with a solid epoxy. If a 15 min demold time at 107°C is used with the latter material, encapsulation will end up costing 16% ($172K) ***more***. If the transfer molding can be completed in 1 min at 150°C, a 42% ($448K) savings will result.

3.3 Encapsulation by Injection and Transfer Molding

During the Honeywell study, a limited effort was made to transfer mold circuits in an epoxy. Questions were raised regarding the circuit's tolerance to pressures of the incoming material wavefront, to short-term high-temperature exposure, and to mechanical design. The use of heat sinks, coatings, or prepackaging was considered on this program. However, available funds on this program did not permit an extensive investigation of these topics.

Still, the concept of injection or transfer molding was deemed to have merit. Subsequently, a different contract* under the Army's Manufacturing, Materials, and Technology Program was awarded to Springborn Laboratories for the "Development of Low-cost Molded Packaging for Hybrid Electronics." The objectives were to identify and evaluate suitable materials and processes for the molding of 100,000 hybrid circuits a month, and to develop design guidelines for circuits, to improve their moldability, and for molds and molding techniques.

Forty resin suppliers were contacted for information on epoxies, silicones, polyurethanes, and polyesters that could be used with transfer molding, LIM, machine casting, or RIM. Approximately 30 products were considered by comparing coefficients of thermal expansion (preferably less then 30×10^{-6} in./in./°C), molding pressures, and demold times and temperatures. Discussions with HDL led to a list of 10 materials for which additional data were obtained by laboratory testing. Because of the concern about excessive heat, exotherm studies were done on masses approximating those required for encapsulation of M734 amplifier circuits. These studies showed no exotherms that would be considered alarming. Additional mechanical and electrical tests were run to fill in gaps in the manufacturer's data. Because of these test results and our desire to consider each of the possible processes, four materials/process sets were selected, as listed in Table 6.

As this paper is being written, molding trials are in progress. When enough experience has been gained to successfully mold the hybrid circuits (the M734 amplifier circuit pictured in Fig. 1), then the most promising material and process will be used to demonstrate circuit encapsulation under a production situation. In addition, successfully molded circuits will be tested at HDL as done earlier on the liquid encapsulation program. Electrical tests will be made after encapsulation, again after temperature and humidity cycling, and finally after 12,000 g shock in an air-gun firing.

4. DISCUSSION

The work by HDL and its contractors very strongly indicates that there are superior materials and processes to the castable encapsulants we are now using. The cost analysis has quantified the savings that could be expected by reducing or eliminating oven curing,

*Contract No. DAAK21-80-C-0115.

by using a single-component system, or by designing so as to incorporate integral molds. Five commercially available encapsulants have been identified that take advantage of these cost savers and that have shown great promise during laboratory tests. In particular, the single-component filled epoxy, Hysol EO-0029, although still requiring a lengthy oven cure, would eliminate a number of quality assurance steps and would be much more amenable than a two-component system in a high production rate line. Another very attractive material is the Dow ISP-100, which is a very low viscosity liquid for about the first 30 sec after mixing and then becomes solid within 1-2 sec. The Cook Coro-foam 589 was destined for use in place of our present foam system, but Cook found it necessary to reformulate. They are now supplying Cook Coro-foam 599, which is different enough chemically (although still an MDI-based system) that we feel additional tests are needed.

Although the five final products did well in protecting discrete component circuits, they did not do so well with the hybrids. In an effort to take a giant step forward and eliminate a fabrication step, no protective stress-relieving conformal coating was used on the hybrid circuits at Honeywell. The difference in coefficients of expansion between the encapsulants and ceramic substrate caused cracking of the substrate and catastropic electrical failure. It is important to note that

Table 6. Processes and Materials for Encapsulation of Hybrid Electronic Circuits.

Process	Material	Supplier	Attractive Features
Reaction Injection Molding	CPR2164-75A	Upjohn	Demoldable in 2-3 min at 50-75 psig; urethane.
Transfer Molding	PTX-1200	GE	Very low thermal expansion (7-10x10^{-6}in./in./°C);short molding cycle (2 min at 300°F); epoxy.
Liquid Injection Molding	Eccomold/ LTM-2850	Emerson & Cuming	Low thermal expansion (24x10^{-6}in./in./°C); short molding cycle (1-3 min at 250°F at 40-100 psig); liquid epoxy.
Liquid Injection Molding	LIM-2600	GE	Fast cure (30-40 sec at 275°F); moderate pressure (200 psi); silica-filled silicone.

unencapsulated circuits also failed, due to lifted pads and broken leads. Thus, the encapsulants did apparently provide some protection for the circuits. Conformal coatings are very much a concern on the Springborn program.

The satisfactory performance of a commerical product that led to its selection on either testing program may not result in our using it in an actual ordnance system. Final approval comes only after developmental use of the encapsulant in fuzes, etc., that are field tested.

Further because of the ever-changing nature of the marketplace, we do not claim these to be the only products that can serve our needs, but they are the only ones we have been able to evaluate. We still feel confident that the risk of replacing our established encapsulants with these new materials and processes is a reasonable one that will yield significant cost savings without endangering system reliability.

5. ACKNOWLEDGEMENTS

The work at Honeywell was performed primarily by Stephen J.Price, as principal investigator, Robert Isliefson, and James Ryan. At Springborn Laboratories, the principal investigator was Robert Diehl. Also participating there were Stephen King and William Holley,Jr. Their cooperation in this presentation is gratefully acknowledged.

6. BIOGRAPHY

Allan B. Goldberg is a materials engineer in the Microelectronics, Materials, and Reliability Branch of the Harry Diamond Laboratories. He joined HDL as a summer student in 1963, and has been a full time employee since 1968. He is currently responsible for the evaluation, selection, specification, and correct usage of polymeric materials for specific military electronics applications. He received an A.B. Degree in chemistry from The George Washington University in 1965, an M.S. Degree in organic chemistry from The Ohio State University in 1968, and an M.S. Degree in polymeric materials from the Polytechnic Institute of Brooklyn in 1972.

13th National SAMPE Technical Conference
October 13-15, 1981

AN EVALUATION OF SEVERAL POLYMERS FOR HIGH DENSITY HUMIDITY PC COATINGS

H. E. Frankel, C. Ray, and F. P. Casto
Fairchild Space and Electronics Company
Germantown, Maryland

Abstract

As a result of a corrosion problem that had arisen, a narrowly focused, short-range investigation was undertaken to determine the imperviousness of several classes of polymeric fluorocarbons and two urethanes. Preparation and cure cycles used were those as recommended by the manufacturer and each board thus coated was subjected to a high-altitude, high-humidity temperature cycling profile exceeding the requirements of MIL-STD-810-C, Method 518, Environmental Test Methods. Limited results, based upon thickness measurements and three different observers' analyses, indicate that, under the conditions extant, the best protection was that offered by the epoxy and the poorest by the urethanes. Other significant observations include the attack of baseboard material and solder joints by one of the fluorocarbons. Analyses of breakdowns, i.e., penetration progression, as well as the micrography and morphology of fluorocarbon attack are presented.

1. INTRODUCTION

The prime objective of this short-range investigation was to up-grade material(s) and procedure(s) relevant to one of the most severe environments to which military aircraft are subjected; namely, rapidly changing temperatures at seacoast locations (especially where the atmosphere may be laden with industrial or fossil fuel by-products).

The coating used for printed wiring boards at Fairchild Space and Electronics Company (FSEC) is a urethane that, heretofore, generally afforded excellent protection against the elements. However, as a result of our customer imposing new, more stringent requirements, especially with regard to rapidity of high-altitude, high-humidity cycling, it became clear that only either a new type of coating or a much thicker urethane afforded any relief.

Accordingly, a simple qualitative corporative experiment was devised to evaluate the single parameters of imperviousness to moisture. The only criteria used was that of visual observation, i.e., did corrosion in any of its manifestiations occur.

2. HISTORICAL PERSPECTIVE

External growth emanating at closely spaced circuitry where a voltage potential is present causes a high resistance path to form during the plus temperature humidity cycle of the test. This problem condition continues to accelerate until the unit experiences an electrical problem thus terminating the test.

All units and samples had been exposed to environmental conditions in excess of the intended specified test requirements for this test. The two main areas of concern were rapid temperature changes during cycling and contamination in the test chamber and/or unit.

All companies contacted which are performing this type test plus companies supplying military electronic equipment coating materials indicate the use of one test chambe or at least simulate a one-chamber temperature change during cycling, also insurance that both the chamber and unit are as clean as possible just prior to and during the test.

It is a proven fact that moisture and contaminants have a more pronounced effect on surface resistivity than on volume resistivity. Whereas it may take several days or weeks for volume resistivity to change when subjected to humid or contaminating environments, surface effects are almost immediate. Drastic reductions in surface resistivity occur through contamination even for coatings that have very high initial values.

3. PROGRAM

Mock-ups of the boards, as well as several test coupons, were prepared; Figure 1 depicts the test parameters involved and a qualitative description of observations. In analyzing the data, it is perhaps easier to group the results - basically, corrosion/no corrosion - to indicate if any trends have developed. If one compares specimens A with B and E with F, one can conclude that two parameters, drag vs. hard soldering and "super-clean" (a proprietary PCI process) vs. regular clean (FSEC), are not operative in these tests. The one variable that seems to be pervading is that of the thickness of the coating, i.e., single conformal coat (cc) as opposed to a double cc. It is interesting to note that out of seven single-coated specimens, seven showed the presence of corrosion and only three out of seven for the double coated samples. Furthermore, it should be pointed out that half of the single cc test pieces had coatings that were significantly below the minimum of 1 mil set by the applicable military specification (ranging from 10% to 40% lower). The other half met the spec - but by the barest minimum. It should be emphasized that in the interest of proceeding expeditiously with the testing, it was decided to let the below minimum thickness coatings go through rather than reject and recoat per our established FSEC procedures. Thus, this provided the opportunity to determine the effect of a below-minimum spec thickness. It can be concluded that two coatings are certainly better than one - but even the two may not be wholly adequate for total elimination of corrosion, as witnessed by specimens C, G, and K.

The conformal coating used is PC 29M which is a two-part polyurethane system, produced by the Hysol Division of the Dexter Corporation. Data on urethanes indicate that its water absorption characteristics are among the highest of embedding resins and concomitantly their volume resistivity properties the largest, 0.65% by weight and 2×10^{12} ohm-cm, respectively (Handbook of Electronic Packaging, Tables 32 and 33, Pages 7-94). These values are substantiated by Section 3.2 of Hysol's Bulletin E4-815E whereby the insulation resistance is 1.5×10^6 megohms before and 5×10^3 meghoms after moisture exposure. Additional confirmation was received from Seymore Ponemone (Westinghouse) in that there was a thousand-fold decrease in resistivity after humidity testing for ten days.

In addition to their hygroscopic properties, the permeabilities of coatings are of great importance and are expressed as m, the amount of moisture passing in a time (t) through a section of the surface (s) of a layer of material of thickness (h) when the water pressure on one side is p_1 and the other p_2. Thus,

$$m = \frac{k(p_1-p_2)\ st}{h}$$

and is wholly analogous for the conduction of an electric current through the body, where $(p_1 - p_2)$ is similar to the difference in potential, m/t = the current and k/Kt is equivalent to the resistance. The coefficient, K, which is analogous to the conductivity, is the moisture permeability of the material,

$$\text{or } K = \frac{mh}{\Delta Ets}$$

Thus, as h becomes thinner, K increases. All this equation does is to put into perspective the importance of thickness (h) in resisting water penetration. Most, if not all of the corrosion, is on the surface of the cc, and thus we now have a mechanism for migration of the various ions through the "water-laden" urethane coating in which conductivity has been further enhanced by the water being moderately acidic, especially in a "power-on" condition.

As a result of the preceding, the following experiment was performed to determine the effectiveness of our present conformal coat in comparison to other coatings. Six types of coating were used in experiment: PC-29 and 1A33, polyurtheane; 12007 epoxy; L-4739 and L1977, fluoropolymers; and 1B31, acrylic. Three sample boards were used for each coating. All coatings were brushed on in exactly the same manner to provide uniformity in application and delete one variable. After brushing, each sample was allowed fifteen (15) minutes to settle before curing. Curing was followed by the manufacturer's recommendation of time and temperature (See Table 1). Thickness measurements of the sample boards were made before the calculation of the average thickness (see Table 2). Resistance measurements were also made after coating and after testing. The reason for doing this was to insure that equal amounts of current were flowing through each sample.

The samples were subjected to rapid changing T.A.H. (Temperature, Altitude, and Humidity) conditions to stress the coatings as much as possible in a limited time frame (See Figure 2). Two complete cycles of the T.A.H. test were completed and a visual examination of the test samples was made. From this examination it was deduced that 1A33 and L-4739 were breaking down and allowing arcing. All samples showed some sign of contamination. The most noticeable were 1A33 and L-4739. A microscopic visual examination was made and the results noted as detailed in Table 4.

4. RESULTS

The results of this experiment were encouraging but also raised questions for future experiments. The L-1977 coating appears to be extremely good, even with the bubbly, imperfect coverage. Future experiments should be performed to determine the exact methods necessary to apply a more perfect coating and also to determine the properties of that coating.

The L-4739 also needs more experimentation to determine the effect of thickness. The average thickness of this coating was under .0001 inches during the present experiment. There appears to be a correlation between thickness and properties. The best coating was the thickest and the worst was the thinnest. More experimentation is necessary to determine if a correlation does exist. Although this experiment was most encouraging, it has no conclusive ending without future experiments and studies.

Group	Specimen		cc thickness, mils	Super Clean One cc	Super Clean Two cc	Regular Clean One cc	Regular Clean Two cc
I	A	F	.08	X	–		
		B	.06				
	B	F	1.2 mils	X			
		B	1.3		–		
	C	F	1.7 + 1.3	–	X		
		B	1.8 + 1.7	extremely small			
	D	F	0.9 + 0.8				
		B	0.6 + 1.9		0		
II	E	F	0.8			X	–
		B	0.6				
	F	F	1.2			X	
		B	1.3				
	G	F	1.7 + 1.3				X
		B	1.8 + 1.7				
	H	F	0.9 + 0.8				
		B	0.6 + 1.4				0
III	J	F	1.2				
		B	1.3			X	
	K	F	1.7 + 1.3				X
		B	1.8 + 1.7				
IV	L	F	1.2 (solder mask)			0	
		B	1.3 (solder mask)				
	M	F	1.7 + 1.3 (solder mask)				
		B	1.8 + 1.7 (solder mask)				0
V	R	F	0.8			X	
		B	0.6				
	N	F	0.9 + 0.8				?
		B	0.6 + 1.9				
	T	F	0.8			X	
		B	0.6				
	P	F	0.9 + 0.8				
		B	0.6 + 1.9				0

F = Front of Board   B = Back of Board

Two readings indicate the thickness of each individual coating.

X = Corrosion noted   0 = No corrosion evident

Figure 1. Evaluation of Processing Variable

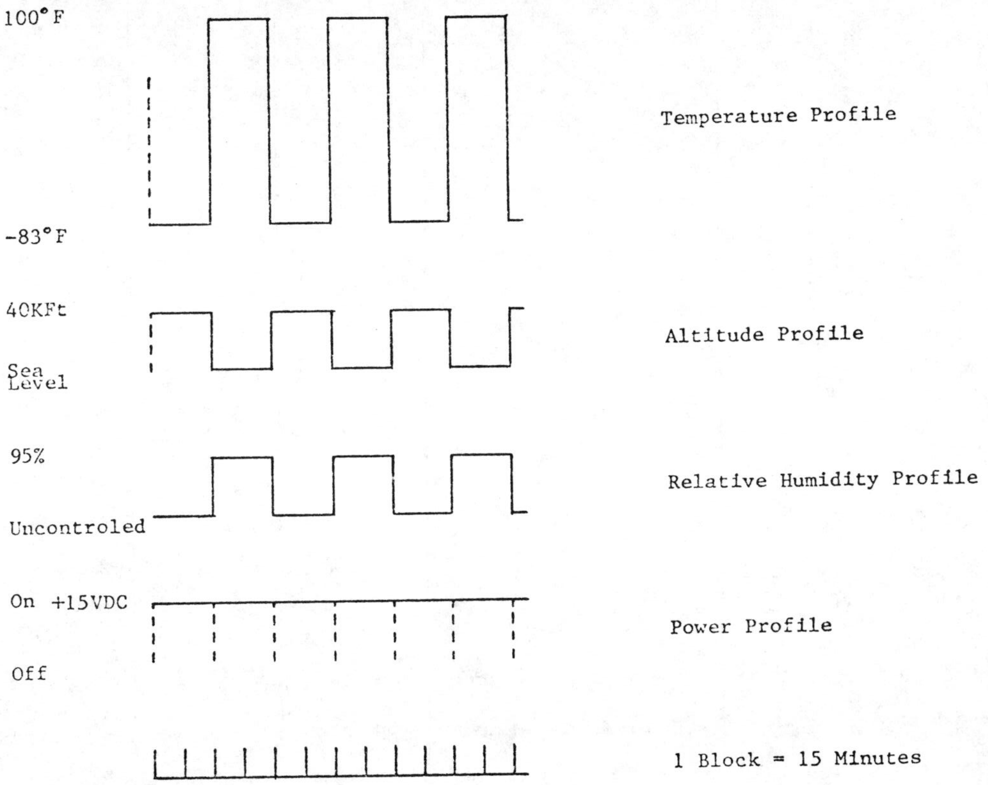

Figure 2. Temperature, Altitude, and Humidity Profile (TAH) to which Conformal Coatings were Subjected

Table 1. Materials Evaluated

Coating Type	Board S/N	Curing Time	Curing Temp.	Observations
1A33	001 005 009	4 Hrs.	170°F	Coating was soft.
1B31	012 015 016	½ Hrs.	170°F	Coating was soft.
PC29	002 003 011	4 Hrs.	170°F	Coating was hard.
PC12007	004 007 018	4 Hrs.	170°F	Coating was extremely hard.
L-4739*	006 013 014	½ Hrs.	200°F	Coating was hard.
L-1977*	008 010 017	½ Hrs.	200°F	Coating was very soft and bubbly. Either the application or the curing was not correct.

*Experimental fluorocarbons supplied through courtesy of 3M Company St. Paul, Minn.

Table 2. Thickness Measurements

S/N	C Before	C After	D Before	D After	E Before	E After	Both Sides (AVERAGE CONFORMAL COATING IN INCHES)	Single Side in MILS.
001	.0587	.0609	.0586	.0605	.0390	.0608	.0020	1.0
002	.0590	.0597	.0592	.0601	.0590	.0599	.0008	0.4
003	.0591	.0610	.0591	.0609	.0596	.0607	.0016	0.8
004	.0592	.0620	.0591	.0630	.0592	.0618	.0031	1.6
005	.0579	.0595	.0592	.0606	.0580	.0601	.0017	0.9
006	.0582	.0581	.0587	.0588	.0582	.0582	.0000	0.0
007	.0571	.0594	.0592	.0620	.0576	.0604	.0026	1.3
008	.0577	.0590	.0592	.0606	.0581	.0590	.0012	0.6
009	.0556	.0567	.0588	.0609	.0555	.0567	.0015	0.8
010	.0557	.0563	.0578	.0593	.0558	.0565	.0009	0.5
011	.0564	.0587	.0566	.0577	.0579	.0096	.0017	0.9
012	.0595	.0607	.0561	.0576	.0576	.0590	.0014	0.7
013	.0589	.0592	.0587	.0591	.0588	.0588	.0002	0.1
014	.0599	.0598	.0587	.0587	.0592	.0592	.0000	0.0
015	.0601	.0618	.0588	.0600	.0579	.0590	.0013	0.7
016	.0595	.0610	.0589	.0604	.0595	.0607	.0014	0.7
017	.0590	.0602	.0587	.0600	.0591	.0611	.0015	0.8
018	.0591	.0613	.0586	.0625	.0592	.0620	.0030	1.5

Table 3. Visual Observations of Various Conformal Coatings

S/N	MICROSCOPIC EXAMINATION			ANALYSIS AND REMARKS
1A33	Minor copper migration	1		
	Heavy tin-lead migration	1		
1A33	Heavy copper migration	4	3	
	Heavy tin-lead migration	4	4	
1A22	----- copper migration	3		
	----- tin-lead migration	3		
1B31	----- copper migration	3		
	Minor tin-lead migration	1		
1B31	----- copper migration	3	3	
	----- tin-lead migration	3	2	
1B31	----- copper migration	3		
	----- tin-lead migration	3		
PC-29M	Heavy copper migration	4		This coating appears
	Light tin-lead migration	2		to allow copper
PC-29M	Heavy copper migration	4	4	migration but retards
	Light tin-lead migration	2	2	tin-lead migration.
PC-29M	Heavy copper migration	4		
	Minor tin-lead migration	1		
PC12007	Light copper migration	2		This coating appears
	No tin-lead migration	0		to be extremely good
PC12007	Minor copper migration	1	1	at protecting
	Minor tin-lead migration	1	0	circuitry
PC12007	No copper migration	0		
	No tin-lead migration	0		
L-4739	Minor copper migration	1		This coating appears
	Heavy tin-lead migration	4		to allow tin-lead
L-4739	Minor copper migration	1	1	migration but retards
	Heavy tin-lead migration	4	4	copper.
L-4739	Light copper migration	2		
	----- tin-lead migration			
L-1977	Light copper migration	2		Even with the bubble
	Minor tin-lead migration	1		problem, this coating
L-1977	Light copper migration	2	2	is better than four
	Light tin-lead migration	2	1	of the others.
L-1977	Minor copper migration	1		
	Minor tin-lead migration	1		

LEGEND:

NO	=	0
MINOR	=	1
LIGHT	=	2
-----	=	3
HEAVY	=	4

13th National SAMPE Technical Conference
October 13-15, 1981

COMPOSITES FOR COMMUNICATIONS SATELLITE WAVEGUIDES

C. K. H. Dharan
Communications Satellite Corporation
Palo Alto, California

Abstract

Graphite and aramid composites are attractive for space microwave applications such as waveguides because of their high specific stiffness and low coefficients of thermal expansion. For successful applications, however, they must be able to survive long-term thermal cycling characteristic of a space environment. In this paper, the response of metallized graphite epoxy composites to thermal cycling is described. An analysis of the stress state in the plating shows that it undergoes significant fully reversed plastic deformation during each cycle. Using low-cycle fatigue concepts, it is shown that thermal fatigue of the composite substrate occurs significantly earlier than thermal fatigue of the metal plating. The failure mechanism is that microcracks generated in the composite substrate propagate into and cause failure of the plating. This finding is essentially corroborated by experimental observations of metallized graphite epoxy composites. Woven and unidirectional substrates and the effect of an interface layer are evaluated experimentally. The important influence of prepeg processing parameters is briefly discussed.

Key words: composite materials, graphite-epoxy, metallization, thermal cycling, fatigue, microwave components, waveguides.

1. INTRODUCTION

Graphite and aramid reinforced composites possess, in addition to high specific strength and stiffness, low coefficients of thermal expansion. They are thus ideally suited for spacecraft microwave applications where dimensional stability over large temperature excursions is a prime requirement. Such applications include refectors, feeds, antenna support structures and waveguides. In the INTELSAT V spacecraft, for example, nearly all antenna components are constructed of graphite and aramid composites. Recently introduced satellite comunications techniques place additional demands on the dimensional stability of microwave components. Frequency re-use, for example, uses polarization isolation to double the number of communications channels for a given frequency bandwidth. On INTELSAT V, frequency re-use at C-band frequencies (6/4 GHz) is achieved by employing left-hand and right-hand circular polarizations. Discrimination (or isolation) between the two polarization signals is critically dependent on accurate maintenance of dimensions over the thermal gradients and temperatures typically encountered in space (-180 C to 100C.)

Due to the heterogeneity of composite materials, however, both fiber and matrix respond differently to temperature changes resulting in induced stresses which, when allowed to relax by microcracking or viscoelastic processes, produce net displacements which in turn affect the overall dimensional stability of the component. In addition, metallization, which is done to minimize electrical losses in microwave components, provides an additional source for induced thermal stresses due to CTE differences between the plating and the graphite-epoxy substrate. Thus, although (elastic) laminated plate theory predicts low CTE's (0.1 - 2 PPM/C) which, for a given fiber, fiber volume fraction and resin, is dependent only on lay-up order, in practice, non-linear effects, induced by thermal cycling can significantly alter the dimensional stability of the composite structure. Successful application therefore depends upon careful control of process variables and an understanding of the non-linear reponse of the composite material. In the INTELSAT V satellite, metallized graphite-expoxy composites have been used successfully in several critical applications. For example, there are over 40 meters of metallized graphite epoxy waveguide runs in the satellite.

This paper examines some of the factors affecting the application of metallized composites to satellite communication waveguides particularly for long-term performance. After defining the thermo-mechanical stress state in the plating and the composite substrate, the effect of long-term thermal cycling will be examined. Factors such as low-cycle fatigue, woven vs. unidirectional composites, prepreg process variables and flexible interface layers will be evaluated in terms of their effect on the overall thermal cyclic response of composite waveguides.

2. MATERIALS AND PROCESSES

Composite waveguides can be manufactured with a variety of graphite fibers; resin matrices are limited, however, to those that pass low outgassing requirements for spaceflight use. In this investigation, the fibers examined are Thornel 300 (Union Carbide), HMS (Hercules) and GY-70 (Celanese) all obtained as prepregs with Fiberite's 934 epoxy resin. Specimen geometry, shown in Figure 1, consists of a square or rectangular graphite-epoxy tube internally plated with an electrolytically deposited copper layer, 12-13 microns thick (0.0005 in). Bonding of the copper layer to the graphite-epoxy substrate occurs at the cure temperature of the resin (175 C). Subsequent cool-down to room temperature results in a residual tensile stress in the higher CTE metal layer.

3. ANALYSIS

An elastoplastic analysis conducted in an earlier investigation obtained the strain-history of the plating from cool-down (from the cure temperature) and through thermal cycling between -180 C and 100 C (1) The Ramberg-Osgood equation below models the elastoplastic response of copper (2).

$$\epsilon = \frac{\sigma_1}{E_1} + \frac{3}{7}\frac{\sigma_1}{E_1}\left(\frac{\sigma_1}{\sigma_{.7}}\right)^{n-1}$$

where σ_1 is the uniaxial stress, n is a shape factor and $\sigma_{.7}$ the stress at which the secant modulus is 0.7 E_1. The results are shown in Figure 2 which depicts the strain-temperature history of the copper plating from cure (point 0) to -180 C (B) to 100 C (E). It can be seen that the copper yields in tension during the very first cool- down from the resin cure temperature to room temperature. Subsequent temperature cycling causes the plating to yield alternately in compression and tension during each cycle. The plating, therefore, undergoes low cycle thermal fatigue.

Next, an estimate of the number of cycles to failure can be obtained by using the following strain-life relation (3):

$$\frac{\Delta\epsilon}{2} = \frac{\sigma_f'}{2}(2N_f)^b + \epsilon_f'(2N_f)^c$$

Where $\Delta\epsilon$ is the cyclic strain range, N_f is the number of, cycles to failure, and σ_f', ϵ_f', b and c are material constants. Since fatigue data on thin electroplated copper films are not available, one can use, for the purposes of this estimate, fatigue data obtained on OFHC bulk copper (4). From this data, σ_f' = 620 MPa (90 ksi), ϵ_f' = 0.48, b = -.149 and c = -0.534. Substituting these values in equation (2), one obtains the number of cycles to failure, N_f = 50000 cycles for the predicted cyclic strain range (Figure 2) of 0.00416. The analysis predicts low-cycle fatigue failure of the plating after 50000 thermal cycles between -180 C and 100 C.

There are several major assumptions implied in the foregoing analytical model, the most serious of which is the assumption of homogeneity of the graphite-epoxy substrate. While 0/90 laminates are quasi-isotropic in CTE as predicted by laminated plate theory, on a micro-scale, composites are heterogeneous and are prone to localized thermal fatigue processes which result in microcracking. Using room temperature mechanical fatigue data for graphite-epoxy (5) and the method outlined above, one does indeed obtain the result that microcracking in the resin matrix can occur after only N_f = 400 thermal cycles. Thus the graphite epoxy substrate itself becomes discontinuous as a result of thermal cycling-induced microcracks which can then propagate into the plating resulting in failure in the plating well before the 50000 cycles predicted earlier.

To verify which particular failure theory is applicable, samples of candidate metallized graphite-epoxy waveguides were cycled between -180 C and 100 C, and surface morphology observed using scanning electron microscopy.

4. THERMAL FATIGURE FAILURE MODES

In all cases, changes in plating surface morphology become evident with the first exposure to the low temperature limit. That is, although the negative temperature differential from cure temperature to room temperature is a significant fraction of the total cyclic temperature range, little change in the plating surface is seen initially during cool-down from cure to room temperature. This is because the plating is biaxially stretched as predicted by Figure 1 (initial room temperature point is between A and G.) However, after exposure to the first cold cycle, subsequent heating back to room temperature causes the plating to go into compression (room temperature point is now between D and E.) If there are any voids or imperfections in the graphite-epoxy substrate, the copper plating will buckle into these imperfections resulting in a change in its surface morphology. This is particularly true for woven graphite-epoxy substrates where voids often exist at the weave interstices.

4.1 Woven vs. Unidirectional Graphite-Epoxy Substrates

In woven graphite-epoxy substrates, a regular pattern of localized depressions in the plating was soon evident as thermal cycling progressed. These depressions correlated with the weave interstices where voids were present due to overbleed of the resin during cure. However, no cracks in the plating were observed until 500 - 1000 cycles had elapsed, at which point localized cracking of the plating occurred, usually at the center of the depression. Figure 3 shows a typical depression of the plating and a localized arrested crack at its center. The specimen is woven Thornel 300 after exposure to 1000 thermal cycles. Figure 4 is another micrograph showing similar plating cracking at a weave intersection. Unidirectional graphite-epoxy substrates, however, showed cracks in the plating at a considerably shorter number of thermal cycles, usually between 100 and 500 cycles. The cracks were also not localized but were extensive and followed the fiber direction of the unidirectionally reinforced graphite-epoxy ply immediately adjacent to the plating. Figure 5 shows a typical example (HMS/934). Similar features were reported by Camahort et al, for non-metallized graphite epoxy laminates after 25 thermal cycles between -196 C and 100 C [(6)].

4.2 Prepreg Process Variables

It is evident that the prepreg resin cure schedule and the arrangeement of bleeder plies in the vacuum bag process play an important role in provididng a well consolidated substrate. The 934 resin system has a well-known viscosity-temperature-time characteristic. Direct heating (i.e., without staging) to the cure temperature causes a rapid decrease in viscosity before the gel state is reached often resulting in over-bleeding of the resin. The result is a substrate that has a large number of voids and is poorly consolidated. Such a substrate responds during thermal cycling, as we have seen, with early cracking of the plating. At the other extreme, over staging the resin will result in a part with interlaminar voids because the autoclave pressure is not sufficiently high to consolidate the laminate. A recent study has shown a close correlation between interlaminar voids and sites for microcracking using nondestructive evaluation techniques such as ultrasonic C-scan and acoustic emission (7). In any event, accurate process control is essential to produce dimensionally stable and thermal cycling resistant graphite-epoxy components.

4.3 Effect of a Flexible Interface

An interfacial layer of resin between the plating and the composite substrate can be thought to be advantageous in isolating the plating from imperfections (voids, fiber concentrations or resin-rich zones) in the composite substrate. Under thermal cycling, the flexible interfacial layer can allow the plating to "float" and thus expand or contract more freely and with less constraint from the low CTE graphite-epoxy substrate. There are several flaws to this argument, however, chief of which is the fact that polymerinterfacial layers have very high CTE's (30-60 PPM/C.) In addition to increasing the overall CTE of the structure, the interfacial layer actually places the metallization in a greater state of stress during thermal cycling leading to a greater tendency toward buckling. Another factor is the lower modulus of the interface (relative to the modulus of the composite substrate) which makes it less effective as a support to the plating when the plating undergoes a compressive state of stress. This situation is analogous to the buckling stability of a beam on an elastic foundation. Reducing the modulus of the substrate (the foundation) by substituting a polymer interface layer for the graphite-epoxy composite results in the plating (the beam) buckling at a lower compressive stress level.

These effects appear to be borne out by experimental data. Figure 6 shows a woven Thornel 300/934 plated specimen with an interfacial layer showing that extensive buckling of the plating has occurred. Once again, such features become visible only after 500 - 1000 thermal cycles have elapsed. Figure 7 is the same specimen at higher magnification showing that a crack in the buckled plating has been initiated.

5. SUMMARY

High specific stiffness and low coefficient of thermal expansion (CTE) make graphite fiber composites attractive for waveguides for communications satellites. This paper examines the factors affecting the performance of composite waveguides during the thermal cycling that is representative of space environments. Because composite structures are laminated, and because of their inherent heterogeneity, internal stresses are introduced whenever the temperature is changed. In fact, the material is in a constant state of residual stress at room temperature as a result of cool-down from the temperature at which the laminate is cured. In addition, since composite waveguides are metallized, the plating also undergoes cyclic stresses because of the difference in CTE between it and the composite lamina substrate.

The stresses and failure modes in metallized composite waveguides which have undergone thermal cycling are examined. Analysis of the stress state in the plating shows that it experiences significant plastic deformation during each cycle. The deformation is fully reversed, the plastic strain attaining its maximum tensile value at the low temperature limit and its maximum compressive value at the high temperature point in the thermal cycle. Using low cycle fatigue concepts, and assuming the composite substrate to be homogeneous, an estimate is made of the number of cycles required for thermal fatigue failure in the plating to be initiated. However, when the same approach is applied to the substrate, the number of cycles required to initiate thermal fatigue failure in a lamina is considerably less. This means that composite thermal fatigue failure in the substrate is the mechanism that initiates failure in the plating. This finding is essentially corroborated in experimental observations of the effect of thermal cycling on metallized composites.

In unidirectional composite substrates, thermal cycle induced failure is characterized by long cracks in the plating that extend along the fiber direction of the lamina that is adjacent to the plating layer. These cracks originate in the unidirectional composite lamina and propagate into the plating where they become visible. In metallized woven composites, however, thermal fatigue failure is characterized by initial local buckling of the plating into voids or resin shrinkage zones at the weave interstices. Additional thermal cycling results in the initiation of cracks in the plating, usually at the center of the buckled depression. Unlike in unidirectional substrates, these cracks do not propagate but are instead confined locally to the buckled region. A woven substrate thus provides a crack arresting feature; it has the disadvantage, however, of developing a wavy surface into the plating after thermal cycling which may increase radio frequency loss in some applications. It is evident from the results of this investigation, that prepreg processing parameters play a crucial role in determining the response of metallized composites to thermal cycling. The cure temperature of the resin establishes the magnitude of the residual thermal stress after cure, and the time-temperature cure schedule influences the degree of voids and resin-rich zones. In woven composites, for example, over-bleeding during cure can result in lack of resin at the interstices of the weave which can then initiate failure of the plating. A polymer interface layer between the plating and the composite substrate, while intuitively attractive because it can serve to isolate the plating from the inhomegeneities of the composite, is found to initiate extensive plating buckling after several thermal cycles. This behavior is attributed to the high CTE and low modulus of the interface layer which causes the plating to buckle at a lower compressive stress level.

6. REFERENCES

1. C.K.H. Dharan, Proc. Second International Conference on Composite Materials, pp. 735-749,

AIME, New York, 1978.

2. W. Ramberg and W. R. Osgood, "Description of Stress-strain Curves by Three Parameters", NACA Technical Note 902, 1943.

3. Manual on Low Cycle Fatigue, ASTM, STP 465, American Society for Testing and Materials, Philadelphia, 1969.

4. A.J. Nachtigall, "Strain-Cycling Fatigue Behavior of Ten Structural Metals in Liquid Helium, in Liquid Nitrogen and in Ambient Air," NASA Technical Note D-7532, 1976.

5. C.K.H. Dharan, "Fatigue Failure in Graphite Fibre and Glass Fibre - Polymer Composites", J. Materials Science, 10, pp. 1665-1670, 1975.

6. J. L. Camahort, E. H. Rennheck and W. C. Coons, "Effects of Thermal Cycling Environment on Graphite/Epoxy Composites", Environmental Effects on Advanced Composite Materials, ASTM STP 602, American Society for Testing and Materials, Philadelphia, pp. 37-49, 1976.

7. T.J. Delacy and C.K.H. Dharan, "Nondestructive Characterization of Stability Mechanisms in Advanced Composites", Proc. AF/DARPA Review of Progress in Quantitative NDE, University of Colorado, Boulder, August 2-7, 1981.

BIOGRAPHY

C.K. Hari Dharan is Manager of the Product Design Department in the Spacecraft Engineering Division of Communications Satellite Corporation in Palo Alto, CA. There he is involved with the development of mechanical systems, structures and materials for communications satellites. He was graduated from the Indian Institute of Technology, Bombay in 1964 and earned his M.S. and Ph.D from the University of California, Berkeley. In 1968, he joined the Scientific Research Laboratories of Ford Motor Company, Dearborn, MI, where he initiated Ford's first in-house research program on composite materials and later developed automotive applications for composites. In 1975, he became Senior Staff Scientist at Ford Aerospace and Communications Corporation in Palo Alto where he developed processes for advanced composite microwave components for communications satellites. He is a member of the editorial board of the SAMPE Journal, and was Technical Chairman of the 1981 Advanced Composites Symposium sponsored by the Northern California Chapter of SAMPE. He is the author of forty publications and is a reviewer for several engineering journals.

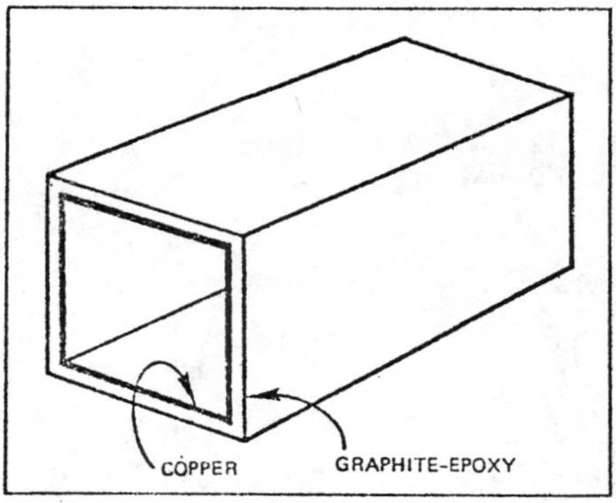

Fig. 1 Specimen geometry

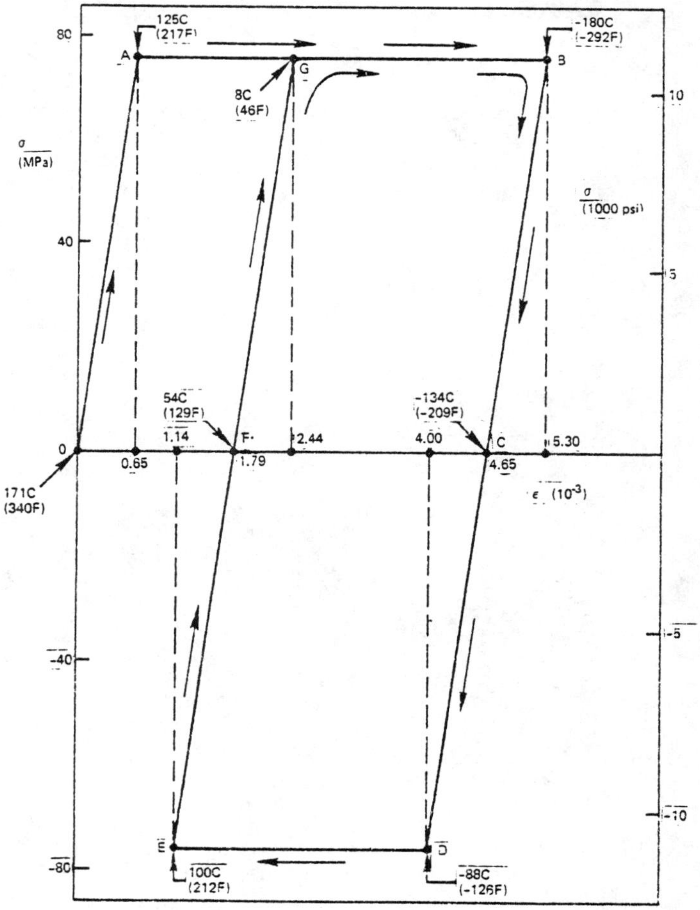

Fig. 2 Stress-strain history of plating from bonding temperature through one thermal cycle (-180 C to 100 C). Yield stress= 76 MPa

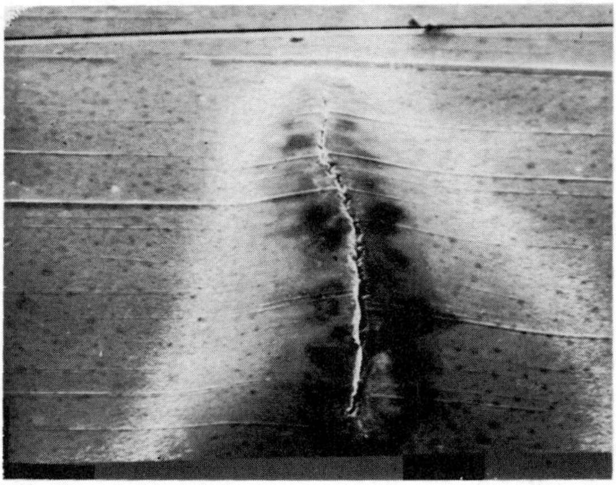

Fig. 3 Plating failure at buckled depression formed after 1000 thermal cycles. Substrate was woven Thornel 300/934. ——— = 50 microns.

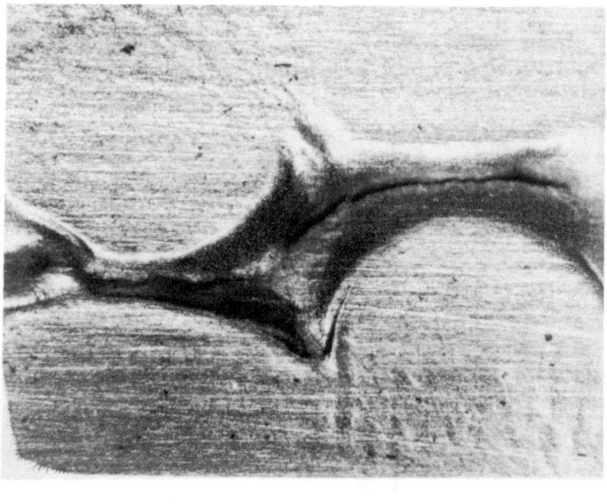

Fig. 4 As in Fig. 3 except substrate is a finer weave. ——— = 50 microns.

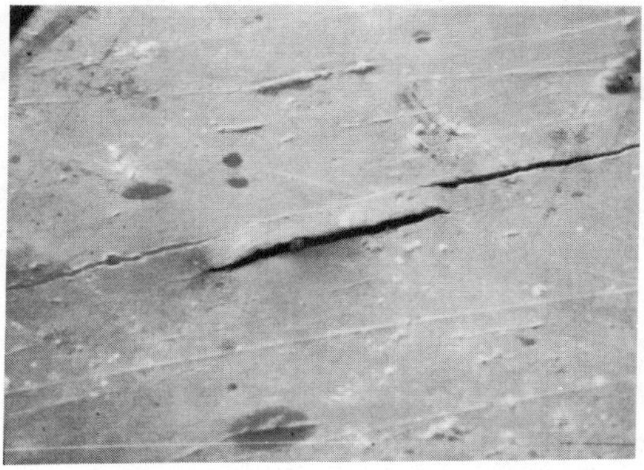

Fig. 5 Thermal cycle induced plating cracks in metallized laminated unidirectional HMS/934. ——— = 50 microns.

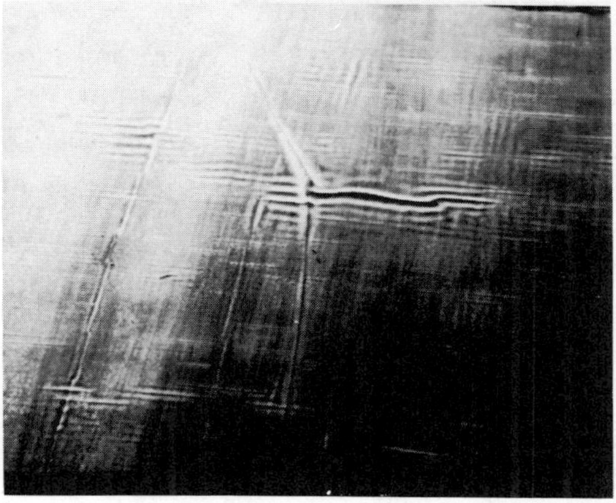

Fig. 6 Thermal cycle induced buckling of plating over an interface layer. ——— = 500 microns.

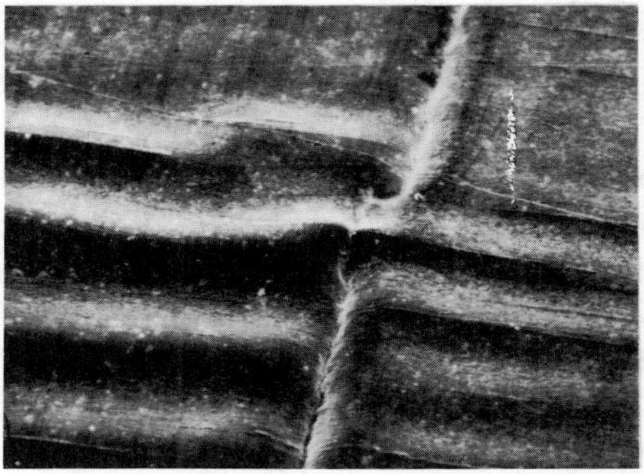

Fig. 7 As in Fig. 6, at higher magnification, showing crack in plating. ——— = 50 microns.

13th National SAMPE Technical Conference
October 13-15, 1981

EFFECT OF WATER CONTAINED IN BOND PLY ON MLB FABRICATION

R.S. Dudinyak, S.E. Susman
Engineered Materials Department
American Cyanamid Company
Saugus, California

ABSTRACT

Production of a multi layer printed circuit board (MLB) for a military application was complicated by blisters which appeared during post baking of the MLB's. The blisters were aesthetically as well as functionally detrimental to this high reliability capacitance controlled board. An extensive program was undertaken to identify the cause of and solution(s) to the problem. A variety of analytical techniques including differential scanning calorimetry (DSC), thermogravimetric analysis (TGA) and moisture emission analysis (MEA) were employed in this work. It was concluded that volatile matter in the prepreg caused the blistering. Moisture was by far the major contributor to the volatiles content, although diverse other species in lesser amounts were identified. As a consequence of this program several recommendations and changes in manufacturing procedures were made regarding bond ply packaging and storage following manufacture and customer receipt as well as pre-lamination conditioning of the bond ply. Adherence to these recommendations has led to virtual elimination of this problem.

1. INTRODUCTION

In principle the production of multi layer circuit boards can be thought of as a bonding and encapsulation process. In practice etched circuit boards are stacked atop each other in precise registration with layers of B-staged resin/glass (bond ply) placed between the individual boards as in Figure 1. The layup is then bonded under heat and pressure. Holes drilled through the board have their interiors plated with metal, thus serving to connect the individual boards electrically. The functions of a good bond ply are several and require that it:

A. be a good adhesive
B. flow sufficiently to fill in the etched portions of the artwork

C. possess a degree of flow control sufficient to give consistent thickness

D. give a high quality uniform dielectric layer.

Inasmuch as the bond ply is used in one of the final stages of MLB production, it must perform reliably since considerable time and money have by this time been expended in fabrication of the individual boards. Imperfections such as non flow, voids or thickness variations can adversely affect the electrical performance of the board. A good case in point is capacitance, since this property is closely related to the thickness of the dielectric separating the conductors. This paper describes a case in which blisters were found in the MLB's following post bake. Since these boards were of a close capacitance tolerance type and were intended for high reliability military use, they were rejected at considerable expense and consternation on the part of the fabricator.

2. DISCUSSION

A typical MLB production cycle would involve placing the stack in Figure 1 between machined flat caul plates with one or more layers of Kraft paper between the the stack and caul plates (to insure uniform pressure and control heat input). The plates/stack is placed in a cold press and heated rapidly (20°F/min.) while pressure in the range of 400-500 psi is applied. The laminate is held for about 60 minutes at 350°F and the press platens are rapidly cooled with water while pressure is maintained. A post bake is frequently used prior to further processing of the board.

The onset of blistering in this case occurred early in a 6 hour x 325°F. post bake used prior to drilling. Initially there appeared to be little or no correlation between bond ply lots, press cycles etc. relative to the extent and size of blisters observed. Cross sectioning of the laminates showed the blisters to be both within the bond ply layers as well as at the copper/bond ply interface. The former case appeared, however, to predominate.

Since blistering was first observed during the unrestrained post bake, it was reasonable to suspect that vaporization of some volatile material was the source of the problem. This was confirmed by carrying out the post bake in a press under laminating pressure. In one case the pressure was released just before the laminate was cooled, while in another the pressure was maintained throughout the post bake and cool down. The results were as expected; blistering was observed in the first case but not in the second.

It then became of immediate interest to identify the volatiles in the blisters to provide guidance in seeking a solution to the problem. A combination gas chromatograph/mass spectrometer (G.C./M.S.) was used in this work.

Table I presents a qualitative and semi quantitative analysis of 5 samples selected from a large data base. No attempt was made at this point to place absolute numbers on the amounts of the observed components. The table reports only relative concentrations based on G.C. peak heights. Water is present in adsorbed/absorbed form in all dicyandiamide hardened bisphenol A epoxy resin matrices. MEK and acetone are the primary resin carrier solvents used in the impregnation process. Carbon dioxide is atmospherically present and may also result from ketone or dicyandiamide decomposition under certain conditions. Volatile aromatics may be epoxy backbone decomposition products or contaminants in raw materials used to prepare the coating. The presence of methyl methacrylate is at present unexplainable.

Finding water as a major component of the blisters held forth the possibility of a twofold problem. Obviously any residual water not removed in the MLB lamination would readily vaporize during the post bake and cause the observed blisters. In addition, it is known that water is capable of inhibiting the degree of cure and hence the thermal stability of dicyandiamide cured epoxy resins. Reduced degrees of cure could potentially increase the concentration of volatile fragments of the board on exposure to heat and thus aggravate even further the blistering problem.

To assess the degree of cure of the bond ply as well as cured laminates differential scanning calorimetry (D.S.C.) was used. Since the presence of water in laminate blisters had been established, it was decided to scan the bond ply in both the "as received" condition as well as after vacuum/chemical dessication followed by controlled low humidity storage. No difference greater than 2°C in Tg was observed in any of the laminates whether blistered or not. The Tg of the bond plies cured during the scanning process varied by only 1-3°C, a difference considered to be insignificant. The shape of the scans in terms of melt onset (endotherm) and exotherm onset and peak temperature (170-180°C) were virtually identical for both "as received" and conditioned samples. It was concluded that the cure mechanism was not grossly affected by uncontrolled storage and predicted that thermal stability would not be greatly affected.

Quantitative data on moisture content of various bond plies was generated using a DuPont Model 902 Moisture Emission Analyzer (M.E.A.). In this instrument a weighed sample is placed in a chamber and swept by a dry stream of nitrogen. The moisture passes over a pair of platinum electrodes coated with phosphorous pentoxide. Absorption of the water by P_2O_5 causes an electric current to flow and the current is measured in dimensionless units called "counts". The counts are quantized using a known quantity of water to calibrate the instrument.

Table II presents quantitative moisture contents on the bond ply samples which were used to prepare the laminates analyzed by G.C./M.S. and presented in Table I. As in the D.S.C. work the samples were analyzed in "as received" and dessicated/controlled storage conditions. The results of this analysis are quantitatively significant and consistent. Dessication and controlled storage are seen to give a 2x-4x decrease in retained moisture. It should be noted that the sample weights used in this study correspond roughly to about 4 in^2 of bond ply.

Back calculation from the data of Table II indicates that the absolute amounts of water retained in the bond ply are really quite small in both "as received" and conditioned states. Indeed, it might be questioned whether such quantities of water, even when completely vaporized, would be sufficient to cause observable blisters. The answer to this question was demonstrated dramatically by preparing MLB's using conditioned and uncontrolled storage bond ply. Upon post cure the latter gave a very high incidence of blistering; while the former was, in most cases, blister free or at worst very greatly reduced in the number and size of blisters. During this phase of the study the effects of retained volatiles in the inner etched boards (etching, stripping and rinse solvents) were shown to be of importance in blistering. These effects, however, were variable and frequently irreproducible.

3. CONCLUSIONS

The vaporization of retained moisture in bond ply has been shown to be a major cause of the formation of blisters during MLB fabrication. The key to prevention or minimization of the blister problem is to limit exposure of the bond ply to moisture as much as possible.

Immediately after production the bond ply must be heat sealed in a polyethylene bag containing a dessicant. During storage, humidity should be limited to 50% maximum. For pre-lamination conditioning we recommend dessication in a sealed chamber containing $CaSO_4$ for 24 hours minimum (vacuum is desirable, but optional). Portable dehumidifiers have been found to be effective in storage areas.

4. BIOGRAPHY

Ronald S. Dudinyak was graduated

from the Pennsylvania State University with a Ph.D. in organic chemistry. He has been employed by the Fabrics & Finishes Department of E.I. DuPont de Nemours & Co. as a staff chemist for the past 15 years and presently is on loan to American Cyanamid Company. His work has centered about the development of polymers for composites, adhesives and coatings. He has authored a number of technical papers dealing with composites and holds U.S. patents in the area of polymer chemistry.

S.E. "Gene" Susman is a graduate of San Diego State, holding the degree of Master of Science in Chemistry. He has been involved with the research and development of polymers for structural application in composites, adhesives, sealants, coatings and non-metallic core for 28 years. He is presently responsible for the development of Advanced Composite Systems for the Engineered Materials Department of American Cyanamid. He is the author of many technical reports, articles, papers, and holds several patents in the field of structural plastics.

FIGURE 1

Typical MLB Layup

Copper Foil

Bond Ply (number depends on desired thickness)

Etched Circuit Board

Bond Ply

Copper Foil

TABLE I

Blister Content Analysis

By G.C./M.S.

Sample	1	2	3	4	5
Components	MEK*	H_2O	H_2O	H_2O	H_2O
	Acetone	CO_2	MMA	MMA	MMA
(decreasing	H_2O	MMA**	CO_2	Toluene	MEK
relative	Toluene	Toluene	Toluene	Aliphatics	CO_2
concentration)	CO_2	Aliphatics	Acetone	CO_2	Acetone
	---	---	---	Acetone	Toluene

*MEK - Methyl Ethyl Ketone

**MMA - Methyl Methacrylate

TABLE II

Moisture Emission Analysis

Sample		1	2	3	4	5
"As Received" Condition	Sample Wt. (g)	0.71400	0.62512	0.60330	0.72500	0.54381
	% $H_2O \times 10^4$	2.7	2.0	2.8	1.4	2.1
Dessicated Condition	Sample Wt. (g)	---	0.51767	0.57375	0.73600	0.58031
	% $H_2O \times 10^4$	---	0.7	0.7	0.4	1.2

13th National SAMPE Technical Conference
October 13-15, 1981

TRANSITION of AEROSPACE ADHESIVE BONDING TECHNOLOGY from R&D to OPERATIONAL USE

Herbert S. Schwartz
AFWAL/Materials Laboratory/MLBC
Wright-Patterson Air Force Base, Ohio

Abstract

Early applications of adhesive bonding technology in aircraft structures exploited intrinsic structural advantages such as long fatigue life and high stiffness-to-weight ratio. Temperature requirements were moderate since flight was principally subsonic and low supersonic. Subsequently, high temperature polymers such as polyaromatic heterocyclics and formulated adhesives incorporating them were developed for use up to 700°F. However, only limited application was made of these adhesives because of processing complexity and limited systems needs. Service experience during the 1960s revealed deficiencies in durability (moisture resistance) of the then current epoxy based adhesive bonding technology for aluminum structures. This led to development of high durability bonding technology which transitioned to production use in the mid-1970s. Current emphasis in bonded structures is on reliable performance plus economical production. Future transition of new adhesive bonding technology will depend on how well these criteria are satisfied, as well as the extent of use of advanced composites in lieu of adhesive bonded aluminum structures.

"Keywords"

Adhesive bonding materials and processes, adhesive bonding applications, adhesive bonded aluminum, adhesive development, aircraft structures, technology transition.

1. INTRODUCTION

Starting shortly after World War II, the use of adhesive bonding to fabricate metallic airframe structures for both civil and military aircraft began to be significant and it has progressed to a high level of utilization by the airframe industry.

The initial reasons for using adhesive bonded assemblies in aircraft structures were primarily their high structural efficiency (e.g. high stiffness-to-weight and long fatigue life), design simplicity, and manufacturing economy. For example, the extensive use of sandwich construction is based on these advantages.[1] Subsequently, adhesive bonding has been used in laminated construction [2, 3] to exploit its excellent damage tolerance capability (crack propagation resistance) and in integral fuel tanks to exploit combined sealing and load transfer capabilities.[4]

In order to exploit the intrinsic design and structural efficiency advantages of adhesive bonding in aircraft structures, it is necessary that the materials, manufacturing processes, and quality assurance procedures be appropriate and adequate to meet the requirements. These requirements are based not only on technical considerations, but also on economic, managerial, and geopolitical factors. The materials, processes, and quality assurance procedures to meet these requirements have evolved during the past three-and-a-half decades through the combined research and development efforts of industry and government agencies. As is the case for most technologies, only a portion of the research and development efforts which were initiated have been successfully transitioned to utilization in production hardware. Even though a research and development program achieves its technical objective, there are additional criteria it must meet in order to transition to production use in a timely manner. The extent and timeliness of the transition depend not only on technological and economic factors, but also on psychological factors, such as the degree of confidence that the potential user has in the technology.

In this paper, a chronological review is presented of important developments in the materials, processes, and quality assurance aspects of aerospace structural adhesive bonding, and the degree of success in transitioning these developments into use. The motivations for initiating the research and development and the reasons for successful transition or lack of transition will be discussed. Technology which has recently been developed or is in progress and has potential for transition to use in the future will also be discussed.

2. DISCUSSION

For discussion purposes, adhesive bonding technology developments have been grouped into the following five time periods or "eras."

(1) Mid-1940s to mid-1950s
(2) Mid-1950s to mid-1960s
(3) Mid-1960s to mid-1970s
(4) Mid-1970s to present
(5) Future

2.1 Mid-1940s to mid-1950s

According to deBruyne, the deHavilland "Dove," built in 1946 in Great Britain was the first all metal aircraft designed for adhesive bonding. [5] It used bonded stringers in the fuselage and wing as well as bonded local reinforcements. The adhesive used was Redux, a Formvar modified phenolic material previously developed by deBruyne in 1941. The motivation for using adhesive bonding was higher structural efficiency and lower manufacturing cost than for riveted structure.

Further application of the Redux type bonding technology was begun by the Fokker Company in the Netherlands starting shortly after World War II. In addition to using bonded stringers, they also used stepped laminated construction to increase local stiffness. [6]

Another important contribution by the Fokker Company was their recognition of the need to properly surface treat and prime the aluminum adherends in order to achieve long term durability (moisture resistance) of the bonded joints. According to R. J. Schliekelmann of Fokker, as early as 1953 they decided to use chromic acid anodizing preceded by a chromic sulfuric acid etch. They used an unsealed chromic acid anodize followed by a sprayed phenolic resin primer. Another development by Mr. Schliekelmann in the early 1950s was the Fokker Bond Tester, an ultrasonic nondestructive inspection instrument which assesses the cohesive bond quality of adhesive bonded joints based on resonance principles. The Fokker Bond Tester is especially useful for detecting excessive porosity in bond lines. It is therefore especially useful in testing bonded joints made with phenolic based adhesives such as Redux, which can have porosity due to evolution of water during cure.

In the United States, structural adhesives development in the early and mid-1950s was based on epoxy, epoxy-phenolic, and elastomer modified phenolic resins. The epoxies were formulated into two-part room temperature curing pastes and one-part heat curable films. Their advantages over the phenolics were that they were 100% solids, had higher strength, would cure at lower temperature and required less pressure to fabricate bonded joints and structures. An epoxy-phenolic formulation was developed (originally known as 422J) which had good strength at temperatures up to 400°F, and was used to fabricate structural components used on the U.S. Air Force's B-58. Nitrile rubber modified phenolic adhesives were also developed which had high peel strength and high strain-to-fracture at room temperature. One formulation of this type, known as AF-10, was used to "sealbond" integral fuel tanks on USAF's F-102 and F-106 aircraft and on the Convair 880 and 990 commercial aircraft. These nitrile phenolic adhesives provided excellent corrosion protection to the structures as well as very effective sealing.

During this era, the adhesives were used in conjunction with other elements of adhesive bonding technology that were considerably less than optimum from a durability standpoint. Bonded structures made with phenolic based adhesives had better durability than those made with epoxy adhesives. The poorer durability of joints made with the epoxies of this era was not due solely to deficiencies in the epoxy adhesives but involved additional interrelated aspects pertaining to adherends, surface treatments, primers, and aluminum honeycomb core. However, as is discussed later in this paper, very great improvements in durability of epoxy based adhesive bonded structures were made during the late 1960s and early 1970s.

2.2 Mid-1950s to mid-1960s

During the mid-1950s to mid-1960s it was anticipated that the "trend curve" of increasingly higher flight velocities for manned military aircraft would extend to Mach 3 and even beyond into the hypersonic regime. At these speeds aircraft skin temperatures could reach $600^{o}F$ for Mach 3 and progress up to $1000^{o}F$ for the lower end of the hypersonic regime. The adherends of interest were titanium in the $600^{o}F$ - $700^{o}F$ range and stainless steel for the $700^{o}F$ - $1000^{o}F$ range. Based on these anticipated requirements, research and development was conducted primarily by the U.S. Air Force on adhesives which would have acceptable structural performance in this temperature range.

For use up to $700^{o}F$, polymeric adhesives of the polyaromatic heterocyclic types were developed. These included polyimide (PI), polybenzimidazole (PBI), polybenzothiazole (PBT), and polyphenylquinoxaline (PPQ).[7, 8] When properly processed and fabricated, adhesive bonded joints made with formulated versions of these resins had good structural properties at $700^{o}F$ in air after tens of hours exposure at $700^{o}F$. Some polybenzimidazole formulations had useful structural properties at temperatures up to $1200^{o}F$. The principal problem with these resins and the adhesives derived from them was the difficulty in processing and fabricating bonded joints made with them. The two factors which caused the difficulties were liberation of volatiles during cure and poor flow characteristics. Complicated, long, stepwise cure cycles requiring hold periods at successively increasing temperatures up to $700^{o}F$ and in some cases even higher were required. One exception to this complexity was the PPQ material which was fully polymerized into a high temperature thermoplastic adhesive. It required only sufficiently high temperature and pressure to cause sufficient flow to achieve bonding. PPQ adhesive had exceptionally high toughness compared to other high temperature resistant adhesives. However, due to its thermoplastic nature, its use was limited to temperatures no higher than about 500°F.

Of the polyaromatic heterocyclic adhesives described above, the one generic type which has been used to the greatest extent and is being further developed is the polyimide. The principal reasons for this are (1) the thermo-oxidative stability of the polyimide; (2) ability to achieve polymerization or chain extension from low molecular weight forms without liberating volatiles during the bonding operation and (3) ability to obtain adequate flow by using low molecular weight prepolymers.

For use up to 1000°F, inorganic, nonpolymeric adhesives were developed, primarily under sponsorship of the U.S. Air Force.[9, 10] These adhesives were of two generic types. One type was analogous to porcelain enamel and the other type, which had both ceramic and metallic constituents, was like a ceramic modified braze alloy. This technology was developed to the stage where excellent bond strengths were obtained in stainless steel joints at temperatures as high as 1000°F. However, the requirement for high temperature furnaces and high temperature resistant fixturing to fabricate the bonded joints was a disadvantage. Although this work was technically successful, as far as is known it was not utilized to fabricate production parts for aircraft structures. A principal reason was the stigma of low strain-to-fracture generally associated with ceramic materials, even though these specific inorganic adhesives had adequate strain capability.

2.3 Mid-1960s to mid-1970s

The mid-1960s to mid-1970s was the era of the development and introduction to production of "improved durability" adhesive bonding technology for aluminum. This included the phosphoric acid anodize surface treatment[11]; chromated adhesive bonding primers; high toughness, moisture resistant, 250°F curing nitrile rubber modified epoxy film adhesives; corrosion resistant aluminum honeycomb core; and the use of bare rather than clad sheet material. This "new" compared to the "old" technology is shown in flow diagram form in Figure 1.

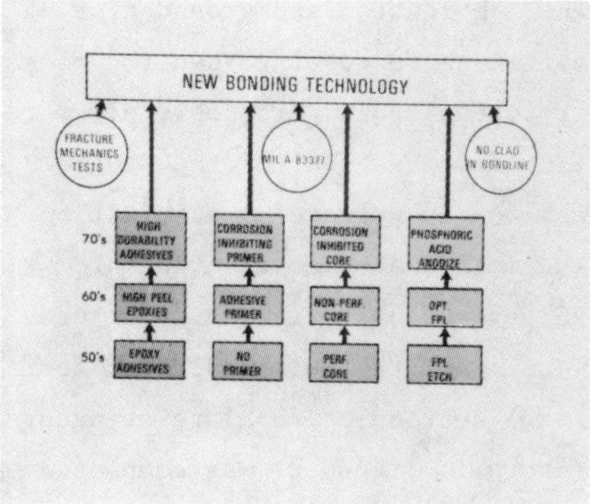

Figure 1. Evolution of Improved Durability Adhesive Bonding Technology

The impetus for introducing this new technology was the service experience by both commercial aircraft operators and the military services indicating that de-bonding and corrosion in bonded structures due to moisture ingress was occurring to a greater extent than could be tolerated. This new technology has been implemented by a major portion of the aerospace adhesive bonding fabricators. Service experience to date confirms predicted improvements in durability of the new over the old technology.

After the development of the new improved durability bonding technology, the U.S. Air Force initiated the Primary Adhesively Bonded Structure (PABST) program in February 1975 with the Douglas Aircraft Co., Long Beach, California, as the contractor.[12] The purpose of the program was to validate the concept of using this new technology to bond primary fuselage structural members together instead of using rivets. The program culminated in the fabrication and ground test of a full scale fuselage section of a prototype medium transport aircraft (YC-15). The goal of the program was to demonstrate an acquisition and maintenance cost saving of 20 percent over riveted structure. Testing of the full scale (42 foot long) section of fuselage structure (shown in Figure 2) was completed

Figure 2. "PABST" Adhesive Bonded Fuselage Structure in Test Fixture

in 1979. In order to test the damage tolerance of the structure, flaws (saw cuts) were deliberately made in the fuselage skin prior to pressure cycling fatigue testing. The structure demonstrated a crack arrestment and containment capability not found in riveted structure.

During this era, the weldbonding process, which is a combination of spotwelding and adhesive bonding, was developed and evaluated for aluminum. In weldbonding, as developed in the U.S., adhesive is placed between the faying surfaces of adherends, such as between the flanges of a stiffener and a sheet to which it will be bonded. Spot welds are then made at intervals to hold the assembly together after which it is placed in an oven to cure the adhesive. The expected advantages of weldbonding are superior fatigue properties to spotwelding or mechanical fastening and lower manufacturing cost than autoclave bonding.

Early weldbonding processes could not tolerate the use of corrosion inhibiting primers on the aluminum adherends because the primers insulated the aluminum and prevented attainment of good quality weld nuggets. This led to concerns about durability or moisture resistance of weldbonded structures which in turn led to development of special, improved durability surface treatments for weldbonding. This resulted in durability improvements over the original weldbonding surface treatment, but the durability was not quite as good as that achieved with the PABST technology.[(13)] Weldbonding has been transitioned to production to make fuselage panels for the A-10 aircraft. The production experience and service experience obtained with these panels will probably be a major factor in decisions concerning future applications of this technology.

2.4 Mid-1970s to Present

Starting in the mid-1970s, geopolitical and economic pressures exerted a strong influence on the objectives of aerospace oriented R&D in adhesive bonding technology. Considerations of Energy, Ecology, and Economy (which I have designated pseudomathematically as the E^3 parameter) and quality assurance aspects of Reproducibility and Reliability (R^2) were considered equal in importance to Performance Payoff (P^2). The E^3 and R^2 parameters provided the impetus for several productive R&D efforts. One of these was the development of an electrodeposited corrosion inhibiting adhesive bonding primer.[(14)] A schematic diagram of the process for applying the primer is shown in Figure 3.

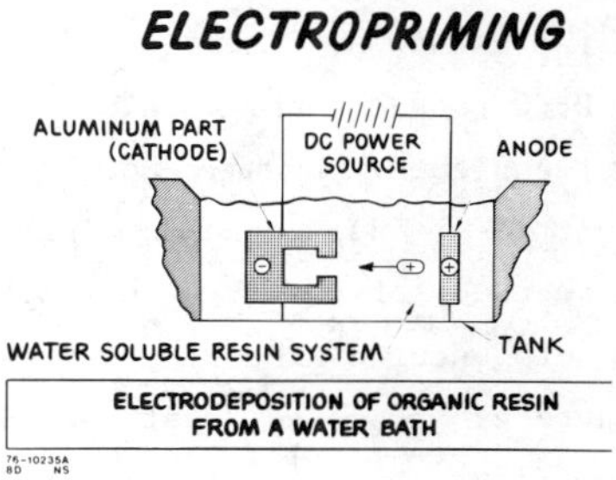

Figure 3. Schematic of Electropriming Process

This primer does not contain chromates, is deposited from an aqueous dispersion, and provides a primer film of highly uniform and controllable thickness. The performance of this primer, when applied to phosphoric acid anodized aluminum and subsequently bonded, gave joint strengths and durability at least as good as the best chromate containing, spray applied adhesive bonding primer.

Emphasis on improved quality assurance for both the materials and processes used in adhesive bonding has led to new developments which have been validated on a laboratory scale and in some cases have been transitioned to a production environment. Examples are the control of uniformity of production batches of adhesives by chemical composition analysis, rheological measurements and dynamic mechanical analysis; the use of ellipsometry (a non-contact optical technique) to assure proper thickness and cleanliness of the oxide layer formed by phosphoric acid anodizing; and the use of A. C. dielectric and D. C. conductance cure monitoring techniques. Dynamic mechanical properties (modulus and damping) can be used for quality assurance for both uncured adhesive prepreg (Figure 4) and cured "neat" adhesive (Figure 5).[15]

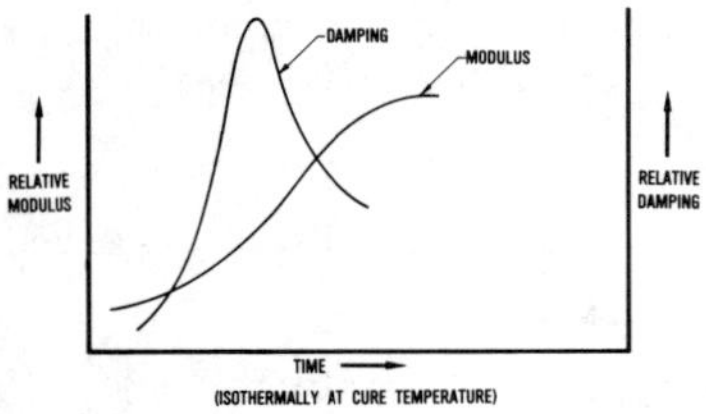

Figure 4. Dynamic Mechanical Properties of Adhesive Prepreg During Isothermal Cure

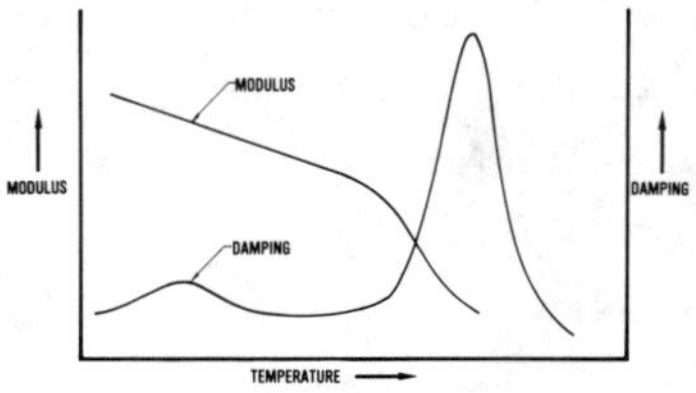

Figure 5. Dynamic Mechanical Properties of Cured Adhesive versus Temperature

Another area of bonding technology in which significant advances have been made in the last few years is that of room temperature curing adhesives. From an economic standpoint, room temperature curing adhesives are attractive to use. Room temperature curing epoxy adhesives have been used successfully for three decades. However, there are some disadvantages in using room temperature curing epoxies such as limited pot life, limitations on elevated temperature capability and environmental durability, and the possibility of mixing incorrect ratios of curing agent with resin. However, two other generic classes of room temperature curing adhesives, the anaerobics and acrylics, have been developed and improved during the past ten years to the point where they do not have most of these disadvantages. The brittleness problems and elevated temperature limitations of early versions of these adhesives have been improved upon by new chemical modifications. One type of use for these adhesives could be in splice joints, in conjunction with mechanical fasteners. This would help reduce stress concentrations around the fasteners, increasing fatigue life, and would also act as a seal against moisture, thereby reducing corrosion damage. For the acrylics, one drawback which still needs to be improved upon is the high volatility of acrylic monomer in some formulations, which creates a disagreeable odor.

One interesting area of research recently initiated by the U.S. Air Force, has, if successful, the potential for greatly improving the reliability and resistance to fracture of adhesive bonded joints. In many cases, fracture is initiated by cleavage and peel forces (which are perpendicular to the plane of the adherends) even though the applied loads are nominally in the plane of the adherends. This is illustrated for the simple overlap joint shown in Figure 6.

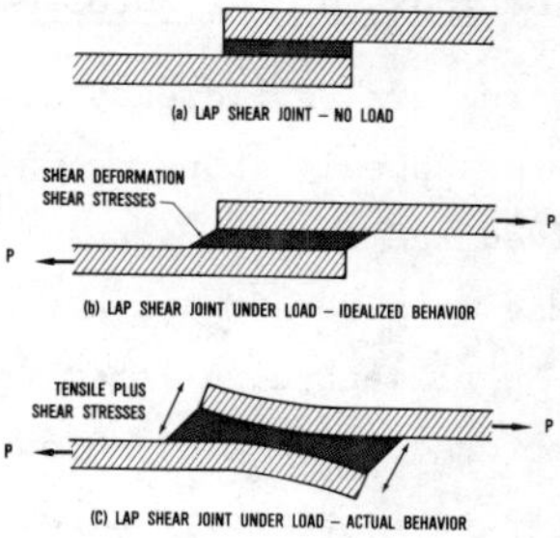

Figure 6. Induced Tensile Stresses in Lap Shear Specimen Due to Eccentric Loading

The objective of the research referred to is to improve the cleavage and peel strength of bonded joints by reinforcing the adhesive in the direction perpendicular to the adherends with polymeric fibers which are grown on anodized aluminum adherends from solution prior to bonding. A schematic diagram illustrating the process is shown in Figure 7.

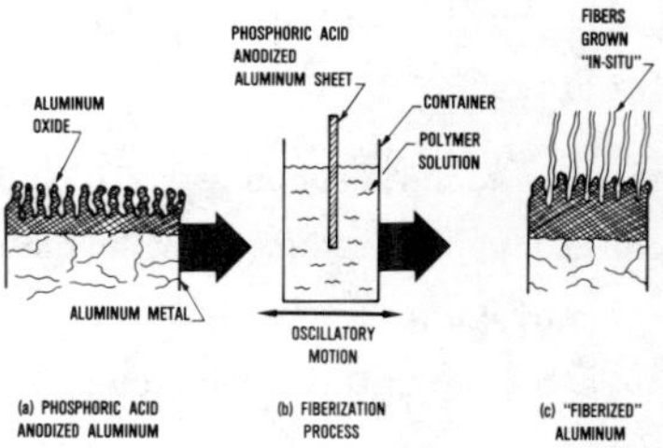

Figure 7. In Situ Fiberization on Phosphoric Acid Anodized Aluminum

The basic technology for growing these type of fibers was reported at a previous SAMPE meeting.[16]

2.5 Outlook for the Future

The structural efficiency advantages of adhesive bonding which provided the initial impetus for its use in aircraft structures have been supplemented in later years by improved durability materials and processes and by demonstrated structural performance and damage tolerance in a large primary aircraft structure, i.e. the "PABST" fuselage. These are considered to be positive factors for the increased use of adhesive bonding in aircraft structures. However, additional improvements in manufacturing, for example, more automated, less labor intensive manufacturing approaches are needed to reduce costs and increase reproducibility and reliability. Another area which needs to be emphasized is the structural analysis and design of bonded joints. The publications of Hart-Smith on this topic provide useful, practical guidelines.[17, 18]

Although adhesive bonded aluminum technology has made major advances in recent years, there is formidable competition from advanced composites (e.g. graphite and Kevlar reinforced plastics) for use in future aircraft structures. We can refer to this pseudomathematically as the C^2 (Composites Competition) parameter. Those composite structures that will have integrally fabricated stiffeners and those that will have skins cocured with honeycomb without using a separate layer of adhesive represent competition. However, it would be expected that many composite structures will be "built-up" rather than integrally fabricated in order to simplify tooling, facilitate inspection, and reduce the cost of scrapped defective parts. This would offer the possibility of adhesive bonding or bonding in conjunction with mechanical fastening and therefore presents new opportunities.

Precise predictions for Technology Transition (T^2) of adhesive bonding developments to production use are not possible. However, again employing pseudomathematics, we can arrive at the general form of a relationship for predicting the probability of transitioning new adhesive bonding technology into production hardware. This relationship is as follows:

$$T^2 = \frac{E_x}{E_m}\left(\frac{P^2\ E^3\ R^2}{(C^2)}\right)$$

where: T^2 is Technology Transition

E_x is the Experience Factor

E_m is the Emotional Factor

P^2 is Performance Payoff

E^3 is Energy, Ecology, Economy

R^2 is Reproducibility, Reliability

C^2 is Composites Competition

3. REFERENCES

1. MIL-HDBK-23, Part I, Composites Construction for Flight Vehicles. Part I: Fabrication, Inspection, Durability and Repair. October 1959.

2. R. J. Schliekelmann, "Operational Experience with Adhesive Bonded Structures," AGARD Lecture Series No. 102, Bonded Joints and Preparation for Bonding, Dayton, Ohio, October 1979.

3. R. J. Schliekelmann, "Adhesive Bonding the Fokker-VFW F-28 Fellowship," Fokker-VFW Brochure, February 1973.

4. W. M. Scardino, et al, "Adhesive Sealing - A Fuel Leak Deterrent," Proceedings of the 10th National SAMPE Technical Conference, Kiamesha Lake, New York, October 1978, pp 575-586.

5. N. A. de Bruyne, "Pioneering Times," Conference on Adhesives for Industry, Technology Conferences, El Segundo, California, June 1980, pp 1-31.

6. R. J. Schliekelmann, "30 Years Experience with Primary Adhesive Bonded Structures," Conference on Adhesives for Industry, Technology Conferences, El Segundo, California, June 1980, pp 32-63.

7. H. H. Levine, "Research and Development of High Temperature Resins for Structural Laminates and Adhesives," ASD-TDR-63-498, May 1963.

8. J. H. Freeman, et al, "Development of Polyimide Resin Based Structural Adhesives and Laminates for Long Time High Temperature Use," AFML-TR-65-188, August 1965.

9. R. Forlano, et al, "Elevated Temperature Resistant Ceramic Structural Adhesives," WADC-TR-55-491, Part VII, October 1962.

10. O. Johnston and W. Robbins, "Research on Inorganic High Temperature Adhesives for Metals and Composite Constructions," WADC-TR-59-113, Part III, May 1962.

11. A. W. Bethune, "Durability of Bonded Aluminum Structure," SAMPE Journal, Vol. II, No. 3, July/August/September 1975.

12. E. W. Thrall, Jr., "Primary Adhesively Bonded Structure Technology (PABST)" AIAA Journal of Aircraft, Vol. 14, No. 6, June 1977.

13. R. G. Hocker, "Improved Durability for Weldbonded Aluminum Structures," Proceedings of the 26th National SAMPE Symposium and Exhibition, Los Angeles, California, April 1981, pp 652-663.

14. S. L. Diener and S. J. Mels, "Electrodeposited Corrosion Inhibiting Primers," Proceedings of the 11th National SAMPE Technical Conference, Boston, Massachusetts, November 1979, pp 759-769.

15. D. J. Crabtree, "Dynamic Mechanical Analysis (DMA) for Quality Assurance of Epoxy-Based Adhesives and Composite Materials," Proceedings of the 12th National SAMPE Technical Conference, Seattle, Washington, October 1980, pp 614-636.

16. R. E. Kelchner and L. B. Keller, "In Situ Grown Fiber Reinforcement

for Improved High Voltage Component Encapsulants," Proceedings of the 26th National SAMPE Symposium and Exhibition, Los Angeles, California, April 1981, pp 444-461.

17. L. J. Hart-Smith, "Further Developments in the Design and Analysis of Adhesive-Bonded Structural Joints," ASTM Symposium on Joining of Composite Materials, Minneapolis, Minnesota, April 1980.

18. L. J. Hart-Smith, "Adhesive Bonding of Aircraft Primary Structures," SAE Aerospace Congress and Exposition, Los Angeles, California, October 1980.

4. BIOGRAPHY

Herbert S. Schwartz is the Focal Point for Adhesive Bonding Technology, AFWAL/Materials Laboratory, Wright-Patterson AFB, Ohio. In addition to his current and recent activities in adhesive bonding, he has also worked in structural and ablative plastics and composites, transparent plastics, composite armor, and the effects of special environments such as high intensity thermal radiation and lightning strikes on nonmetallic materials. He has a number of publications and several patents in these fields. Mr. Schwartz received a B. E. degree in 1949 from Vanderbilt University and a M. CH. E degree in 1951 from the University of Louisville.

13th National SAMPE Technical Conference
October 13-15, 1981

RAPID CURING MODIFIED ACRYLIC ADHESIVES

Herbert E. Harris II
Hysol Division, The Dexter Corporation
Pittsburg, CA

Abstract

The modified acrylic adhesives that achieve fixturing strength within three minutes have been improved to achieve fixturing strength within 30 seconds. The improved adhesive bonds a variety of substrates. Bond strengths are maintained after environmental exposure to various fluids. Applying the activator to the substrate surfaces limits the bond line thickness to approximately 15 mils. This limitation can be overcome by mixing the activator with the base resin.

1. INTRODUCTION

The modified acrylic adhesives have processing advantages over other adhesives. No mixing is required, and therefore, no pot life need be considered. The curing agent (or activator) and base resin are applied separately to the substrate surfaces. They attain fixturing strength rapidly, and they cure at room temperature.(1) The modified acrylic adhesives have good environmental resistance when exposed to fluids. They demonstrate cold, ambient, and elevated temperature properties suitable for aerospace and industrial uses. Some industrial examples are: magnet bonding in speaker and electric motor assemblies, automobile brackets and upholstery, computer consoles, vinyl trim and gap filling (eg. fairings).

The gap filling application involves a two part acrylic adhesive, which can be sanded or abraded after a 30 minute cure.

A recently formulated modified acrylic adhesive, designated EA 9447, has a fixturing time of 30 seconds compared to several minutes for earlier versions. This paper is concerned with this rapid fixturing adhesive (EA 9447) and how it compares with several other modified acrylic and epoxy adhesives. The epoxy adhesive

(EA 907) is included to contrast the properties of these two different classes of adhesives.

2. SOME ASPECTS OF THE CHEMISTRY

The principal reason for the properties possessed by the rapid fixturing modified acrylic adhesive is the presence of a prepolymer from which graft polymerization occurs. This prepolymer is a sulfonylchlorinated polyethylene rubber[3]. The acrylic monomer polymerizes from the sites of the sulfonylchloride groups on the polyethylene backbone, by a free radical mechanism. The reactive polyethylene backbone is chemically incorporated into the polymerized acrylic network. In other systems, the prepolymer is non-reactive and crosslinking is not possible. The graft polymerization process influences basic polymer properties such as surface wetting, elongation and ultimate strengths. Figure 1 illustrates the grafted structure.

The density of the sulfonylchlorinated sites, the molecular weight of the polyethylene backbone and the type of acrylic monomer used are some of the more important considerations governing the properties of the cured adhesive.

3. EXPERIMENTAL

All aluminum substrates, except those used in environmental exposure studies, were 2024 T-3 alclad aluminum with chromic acid etch[4]. Those used in environmental exposure studies were phosphoric acid anodized 2024 T-3 alclad aluminum [4]. The oiled steel substrates were prepared by brushing oil onto non-cleaned steel. All other substrates were solvent wiped (MEK).

Tensile shear specimens were used for all test data. The specimens bonded with acrylic adhesives were cured 24 hours at room temperature. Those bonded with epoxy adhesives were cured 7 days at room temperature. All specimens were given a minimum 10 minute equilibration soak at test temperature and tested according to Federal Specification MMM-A-132. Fixturing strength refers to time of cure at which the specimen, by moderate hand force, cannot be pulled apart (about 30 psi). The procedure for preparing the acrylic adhesive specimens was as follows: The activator was applied to both substrates. A bead of resin base was then applied to either substrate, with subsequent mating of the substrates. Substrate positioning must be accomplished rapidly to prevent bond rupturing motions which adversely affect ultimate strengths.

4. STRENGTH WITH VARIOUS SUBSTRATES

The acrylic adhesives, EA 9446 and EA 9447 bond a wide variety of substrates (Table I). This ability is similar to that of the versatile epoxy adhesive EA 907. The acrylic adhesive EA 9446 has overall superior strength on metal substrates compared to EA 9447. On non-metallic substrates (eg. wood, nylon, PVC) only small differences exists between EA 9446 and EA 9447.

When comparing oily vs solvent wiped steel, the bonding strengths of the acrylic adhesives EA 9446 and EA 9447 are reduced less than 10%. However, the strength of the epoxy adhesive EA 907 is reduced 40%. The acrylic adhesives apparently are able to accomodate non-polar contaminated surfaces more readily than epoxy adhesives.

5. FIXTURING TIME VERSUS STRENGTH

Rapid fixturing time in many applications has become more important than cured strength properties. The plot of fixturing time versus strength (Figure 2) shows that the epoxy adhesive EA 907 possesses no strength for the first 60 minutes. However EA 9446 has almost 50% of its strength in 30 minutes and EA 9447 has almost 50% of its strength in 15 minutes. Fixturing time is substrate dependent. In general, a 5-15% increase in fixturing time is observed for the acrylic adhesive on non-metallic substrates.

6. FIXTURING TIME AND ENVIRONMENTAL EXPOSURE

The ability of the modified acrylic adhesives to resist deterioration from environmental exposure makes them very useful for industrial applications. The strengths of the acrylic adhesives EA 9446, EA 9447 and LR 100-405 are only slightly affected by exposure to a variety of fluids. For example, virtually no strength reduction occurs after exposure to 98% relative humidity for 30 days, and 5% salt spray for 30 days. (Table II, data for the Hysol epoxy adhesive EA 9320 is given for comparison.) A noteworthy point is that a rapid curing adhesive such as EA 9447 does not sacrifice environmental exposure properties. The good retention of physical properties under adverse environmental exposures places the modified acrylic adhesives in the structural engineering material class.

7. PROPERTIES AT VARIOUS TEMPERATURES

The acrylic adhesives EA 9446, EA 9447 and LR 100-405 shows a greater reduction in strength than a typical epoxy adhesive both at low and at elevated temperatures (Figure 3). At -55°C, EA 9446 is reduced 7 MPa, EA 9447 is reduced 5 MPa, while the epoxy is reduced 2.5 MPa. All three acrylic adhesives demonstrate a Tg of approximately 70°C. This Tg explains the acrylic adhesives lower elevated temperature properties when compared to epoxies (Figure 3). At 80°C, EA 9447 possesses an adhesive strength of 3 MPa, while EA 9446 and LR 100-405 possess strengths of 15 MPa and 14 MPa respectively.

8. BONDLINE THICKNESS

Fixturing strength versus bondline thickness is illustrated in Figure 4. Fixturing strength for EA 9447 is achieved in 30 seconds with 2 mil bondlines and 3 minutes with 10 mil bondlines. EA 9446 achieves fixturing strength in 3 minutes with 2 mil bondlines and 5 minutes with 10 mil bondlines.

Figure 5 contrasts the ultimate strength versus bondline thickness for these adhesives. The optimum bondline thickness for the acrylic adhesives is 6-10 mils. Figure 5 also points out a practical bondline limit of 13-17 mils for the acrylic activators. At this thickness, EA 9447 still possesses 7 MPa of tensile shear strength on aluminum substrates (activator applied to both surfaces).

9. THICK BONDLINES AND POTTING COMPOUNDS

As mentioned above, a practical average bondline limit of 15 mil exists for surface activated acrylic adhesive systems. Apparently the practical limitation of activator migration is approximately 15 mil. The acrylic LR 100-405, a two part system overcomes this limitation. Interestingly no stoichiometric relationship exists between activator and acrylic monomer so careful monitoring of mix ratios is not necessary.

An additional advantage to the structural acrylic adhesives is that the material cures and is sandable in 30 minutes.

10. CONCLUSION

The rapid fixturing time acrylic adhesives EA 9447 and EA 9446 bond to a variety of substrates such as oiled steel, PVC plastic and wood. These adhesives, also, possess good environmental stability. For example, virtually no strength reduction occurs after exposure to 98% relative humidity for 30 days. With the surface applied activators and resin base, no mixing is required. Presently, surface application does limit the bondline thickness; however, future research efforts are expected to overcome this limitation. The acrylic adhesive LR 100-405 overcomes this limitation by mixing the activator with the resin base.

Acrylic adhesive applications are expected to expand into many areas now occupied by epoxy adhesives, primarily due to their shorter processing time and lower energy requirements.

ACKNOWLEDGEMENT

Acknowledgement is given to Robert R. Adler and to Larry M. Doweidt for their assistance in this work.

REFERENCES

1. C. L. Mahoney, "Structural Adhesives for Rapid Cure Applications", Parts I and II, Adhesive Age, 10 (22):30, 12(22):26.
2. Structural Adhesives EA 9446, EA 9447, LR 100-405, EA 907 AND EA 9320, Hysol Division, The Dexter Corporation
3. Adhesive Age 8 (23):35
4. "Preparing the Surface for Adhesive Bonding", Bulletin G 1-600, Hysol Division, The Dexter Corporation.

BIOGRAPHY

Herbert E. Harris II is employed by Hysol Divison of The Dexter Corporation as a Development Chemist in Industrial Products, which include acrylic, epoxy, and urethane adhesives. He graduated with B.S. and M.S. degrees in Physical Chemistry from California State University, Hayward, California.

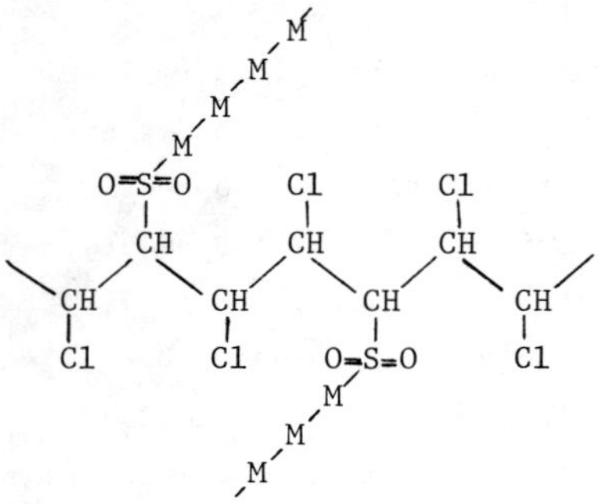

Figure 1. Structure of graft polymerized sulfonylchlorinated polyethylene. M is monomer that is polymerized.

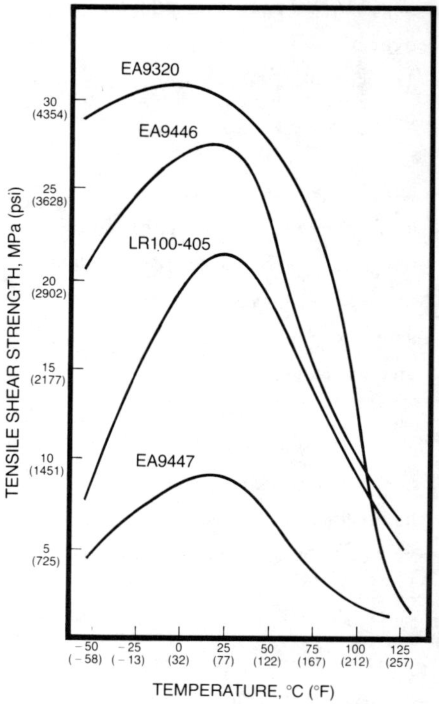

FIGURE 3. Tensile shear strength with change in temperature. 2024 T-3 clad aluminum, chromic acid etch.

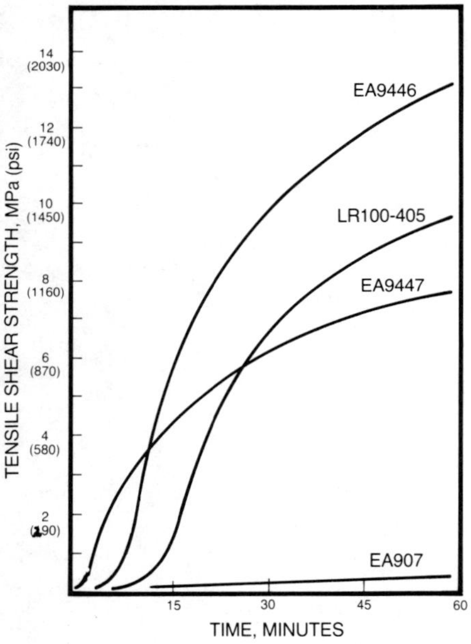

FIGURE 2. Tensile shear strength increase with time at room temp., 2024T-3 clad aluminum, chromic acid etch.

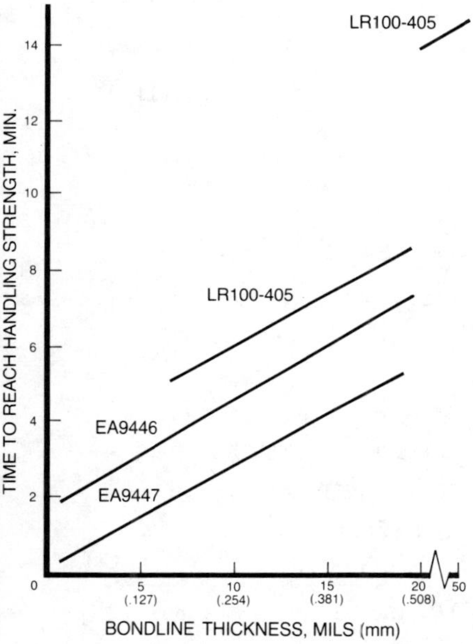

FIGURE 4. Time to reach handling strength with change in bondline thickness. Tensile shear specimens, clad aluminum, chromic acid etch.

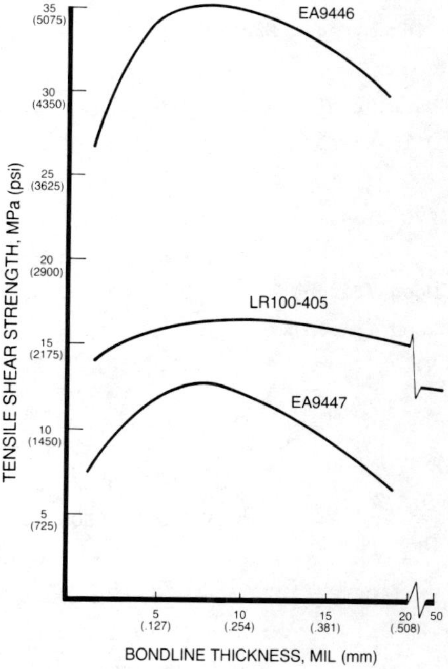

FIGURE 5. Tensile shear strength with change in bondline thickness, clad aluminum with chromic acid etch.

TABLE I

TENSILE SHEAR STRENGTH WITH VARIOUS SUBSTRATES, TESTED AT ROOM TEMPERATURE

SUBSTRATES	EA 9447 (Acrylic) MPa (PSI)	EA 9446 (Acrylic) MPa (PSI)	LR 100-405 (Acrylic) MPa (PSI)	EA 907 (Epoxy) MPa (PSI)
Aluminum (Clad. 2024 T-3, Chrom. Acid Etch)	9.0(1300)	28(4000)	21(3000)	23(3300)
Bare Steel	7.6(1100)	15(2200)		13(1900)
Oily Steel	7.3(1050)	14(2000)		7.6(1100)
PVC Plastic	3.4(500)	Sub. Fail.		--
Wood (Birch)	Sub. Fail.	Sub. Fail.		--
Wood (Maple)	--	--		9(1300)
Nylon	1.7(250)	1.7(250)		--

TABLE II

TENSILE SHEAR STRENGTH AFTER ENVIRONMENTAL EXPOSURE, PHOSPHORIC ACID ANODIZED 2024-T3 ALCLAD ALUMINUM

ENVIRONMENTS	EA 9447 (Acrylic) MPa (PSI)	EA 9446 (Acrylic) MPa (PSI)	LR 100-405 (Acrylic) MPa (PSI)	EA 9320[2] (Epoxy) MPa (PSI)
Control	9.0(1300)	27.5(4000)	21.4(3100)	32.2(4700)
49°C(120°F), 98% R.H., 30 Days	13.1(1910)	26.0(3780)	23.0(3330)	32.0(4650)
25°C(77°F), Tapwater, 30 Days	9.0(1310)	26.9(3880)	22.3(3230)	30.5(4440)
25°C(77°F) SkyDrol 500A, 30 Days	6.4(930)	24.3(3525)	15.2(2200)	31.0(4510)
41°C(105°F), 5% Salt Spray, 30 Days	9.5(1380)	28.0(4065)	23.2(3360)	29.9(4340)
25°C(77°F), JP-4, 7 Days	13.1(1910)	27.5(4000)	23.0(3320)	34.3(5000)

13th National SAMPE Technical Conference
October 13-15, 1981

CORRELATION OF SURFACE CHARACTERIZATION OF PHOSPHORIC ACID ANODIZE OXIDE WITH PHYSICAL PROPERTIES OF BONDED SPECIMENS

Kent K. Knock, Melvin C. Locke
The Boeing Company, Seattle, Washington

ABSTRACT

Phosphoric acid anodization (PAA) has proven to be a highly reliable surface preparation for aluminum bonding applications. This study has placed the process control limits of PAA on a sound scientific basis. Scanning electron micrographs, microtome/transmission electron micrographs, chemical data and Auger/ESCA spectroscopy were used to distinguish between optimal and less desirable oxides. These analytical results have been correlated with physical bondment characteristics such as dry/wet peel strength and environmental crack extension resistance. The result is firm establishment of PAA process control limits which were initially established by physical bondment testing only.

INTRODUCTION

Over the last decade significant increases in the durability of adhesively bonded aluminum aircraft structure have been achieved. One of the factors involved has been the introduction of Boeing's phosphoric acid anodize (PAA) process. The older chromic-sulfuric acid (FPL) etch surface preparation was sometimes difficult to control and produced bonds with poor resistance to attack by moisture (ref. 1). Typical failures of those bonds occured by relatively rapid delamination of the bond near the polymer-substrate interface followed by corrosion of the debonded metal. The dry/wet modified floating roller peel (ASTM D3167-T73) test was therefore chosen for this work to simulate in-service bond failures. Some related double cantilever beam data will also be presented.

Since the introduction of PAA, Boeing has not experienced any in-service bond delaminations attributable to poor durability of bonds produced using PAA. Boeing has, however, continued research on the PAA process with particular emphasis on oxide characteristics. This work was undertaken to define the differences between 'good' oxides produced by proper processing and some 'bad' oxides formed by processing outside specification limits. The 'bad' oxides form water sensitive bonds. The differences are presented in this paper. In addition to illustrating the results of improper processing, these results shed some light on the mechanism by which PAA produces excellent bond durability. This information should be useful in the development of new surface preparations for other metals as well as PAA process control.

EXPERIMENTAL

Based on previous studies (refs. 2 and 3) we chose two alloys and process conditions for primary emphasis:

1) 2024 clad anodized at 15 volts and 62°F.
2) 7075 bare anodized at 6 volts and 90°F with approximately 2 min. delay between current shut down and removal of the part from the anodize bath.

The first condition was known to produce an easily visible color on the dry anodized part. The color verifies the presence of PAA oxide but under these particular conditions a bond is produced which is sensitive to attack by water under peeling stress. We have previously presented a detailed comparison of properly and improperly anodized 2024 clad aluminum substrate (refs. 4 and 5).

The second condition was known to produce a marginal oxide density on processed parts. Little if any color was visible on parts processed under these conditions. The lack of color indicated a thin or nonexistant oxide layer after processing. Both of these conditions were compared to that produced on parts processed within specification conditions with each alloy.

Dry/wet peel specimens were prepared per the Boeing PAA specification (BAC 5555) except for the noted deviations. The process is briefly summarized in Figure 1. Each peel panel was assembled from a backing sheet (0.040 inch thick) and a peeling sheet (0.020 inch thick). After bonding the panels were sheared into 1 by 14 inch strips for peeling. The primer used was American Cyanamid BR127. The adhesive used was either American Cyanamid FM 73, 5 mil or 3M AF-126, 5 mil. Samples were peeled using a modified floating roller peel test shown in Figure 2. Liquid was applied during peel to produce the 'wet' portion of each peel strength curve.

Samples displaying failures which appeared visually to be between the adhesive and oxide were examined using scanning electron microscopy, (SEM), or microtome/transmission electron microscopy (TEM). Samples of the unbonded 'good' and 'bad' oxide were also studied. Fractured oxide specimens were prepared for SEM analysis by sawing a notch into the back of each sample, immersing in liquid nitrogen and breaking by bending toward the notch. The resulting specimen displays fractures in the oxide normal to the metal surface.

RESULTS

The two surface preparation conditions previously described produced apparent interfacial bond failures as expected. The 'wet' peel values were significantly lower than the dry peel values. This is represented schematically in Figure 3. All the controls (processed within specification limits) produced failures which were cohesive within the adhesive under both wet & dry conditions.

Figures 4 and 5 are microtome/TEM micrographs of 2024T3 clad anodized at 62°F ('bad') and 77°F ('good') respectively. Figures 6 and 7 are SEM micrographs of the same oxides. The differences in these two oxides have been discussed in detail by Knock & Locke (ref. 5). The conclusions of that analysis are represented by Figure 8, which is a schematic summary of the differences between the two oxide structures. The 'good' oxide (produced at 77°F) was found to have the following characteristics:

1) Larger pore/cell wall size ratio.
2) More pronounced tapering of the cell walls toward the oxide polymer interface.
3) A less well defined 'plane' at the interface between oxide and polymer.
4) Less oxide surface parallel to the metal surface and therefore to any potential openning mode crack.

Figure 9 is a TEM micrograph of a 7075 bare peel specimen which showed a 'slick' failure in wet peel. The anodize conditions were 90°F at 6 volts with approximately 2 minutes rinse delay. Notice that no oxide characteristic of PAA was present on the sample surface. Figures 10 shows the fractured oxide of a failed specimen produced under the same conditions. Notice that although present, very little oxide remains on the metal. For comparison Figures 11 and 12 are SEM micrographs of unbonded 7075 bare oxides produced at 6 and 15 volts at 90°F. As would be expected, the oxide formed at higher voltage is thicker while the lower voltage material is thinner and apparently more fragile. The 90°F, 15 volt anodized specimens failed cohesively in the adhesive layer under wet peel. Notice that the amount of oxide present on the failed 6 volt specimen is considerably less than that on the 6 volt unbonded specimen. Apparently a portion of the very thin oxide layer originally present on the surface was broken off and remained on the adhesive side of the failure. No analysis was done on the adhesive side of the failure to confirm this.

For comparison, specimens of 2024 bare were prepared using the 77°F, 15 volt, 1 1/4 minute rinse delay conditions. Figure 13 is a fractured oxide SEM micrograph of the resultant oxide. The oxide is more substantial than that shown in Figure 12. The 2024 bare specimens also failed cohesively within the adhesive under wet peel.

To test the relevance of the wet peel test to environmental conditions likely to be encountered, double cantilever beam (DCB) specimens were fabricated and tested. The specimens were of the 1/2 inch thick adherend type previously described by Marceau & Scardino (ref. 6). The specimens were exposed for a year at McDonnell Douglas' El Sequndo beach site. The site provides a marine environment with industrial and automobile generated air pollution. The results of that test are presented in Table 1. Notice the good correspondence between the types of conditions which produce wet peel failures and those with large crack growth in this experiment. It should be noted, however, that less polluted beach site exposure has not been found to produce large crack growths in one year except in the case of 7075 clad specimens.

DISCUSSION AND CONCLUSIONS

This work has defined two distinct PAA/alloy conditions which produce the so-called slick failure mode in wet peel. An aggressive environmental beach site exposure has confirmed the correspondence between wet peel failures and reduced environmental durability. One condition which has been found to produce water sensitive bonds is low temperature (62°F) anodization of 2024 clad (and presumably other materials with relatively pure aluminum surfaces). This condition must be carefully guarded against because visual examination of the anodized surface shows good "color" which confirms anodization. The surface produced, however, has a microstructure which is not capable of effective load transfer under wet conditions.

The second condition is anodization of 7075 bare aluminum at high temperatures (90°F) and low voltages (6V). Apparently the dissolution rate of the oxide formed is rapid enough that insufficient porous oxide remains after anodization to form a stable polymer-oxide composite. The result is that under wet peeling conditions of the oxide structure apparently breaks resulting in slick wet failure.

Fortunately the conditions which produce the undesirable wet peel 'slick' failure mode (out-of-specification temperature, voltages and rinse delays) are easily avoided by proper process control.

By comparing the oxide structures which produce the 'slick' wet peel failures with those that do not it is possible to define several 'good' oxide characteristics which contribute in varying degrees to the exceptional adhesive bond environmental durability on properly prepared PAA. Those characteristics include:

1) The walls of the cells in the oxide must be thin enough to allow gradual load transfer into the polymer from the oxide. No well defined plane exists at the polymer-oxide interfaces.
2) The tapering of the cell walls toward the outer extremity of the oxide also contributes to gradual load transfer between the oxide and polymer.
3) The extremely thin cell walls at the upper extremity of the oxide precludes a large oxide/polymer interfacial area parallel to openning mode cracks. This makes displacement of sensitive oxide-polymer bonds by water in the peeling crack more difficult.
4) The oxide structure must, however, be substantial. It must provide a large surface area (i.e., have sufficient depth). The cell walls, while being thin enough to meet the above stress transfer requirements, must be thick enough to withstand stresses which may be encountered in service.

ACKNOWLEDGEMENTS

The authors wish to acknowledge the work of Mr. Gene Ledbury in performing the scanning electron microscopy and Mr. Ron Radecky and Mr. Ray Shannon of the McDonnel Douglas company for monitoring the beach exposure of the DCBs.

REFERENCES

1. A.W. Bethune, "Durability of Bonded Aluminum Structure," SAMPE Journal, Vol. 11, No. 3 July/Aug/Sept 1975 pp 4-10.

2. R.W. Benoliel, "Durability of Adhesive Bonded FPL Etch and Phosphoric Acid Anodized Test Specimens," 10th National SAMPE Technical Conference, Vol. 10 pp 873-888, Oct 1978.

3. K.M. Harriman and M.C. Locke, "A Cyclic Test for Environmental Durability Evaluation of Bonded Honeycomb Structure," 10th National SAMPE Technical Conference, Vol. 10 pp 138-148, Oct. 1978.

4. J.C. McMillan, "Surface Preparation - The Key to Bondment Durability," AGAARD Lecture Series No. 102 pp 7-1 to 7-30, March 1979.

5. K. Knock and M. Locke, "A Case Study of Water Sensitivity of the Polymer Coating - Substrate Interface for Two Aluminum Surface Preparations," presented to The Electrochemical Society, 159th Meeting, Minneapolis, Mn, May 13, 1981. To be published in the proceedings volume edited by K.L. Mittal.

6. J. A. Marceau and W. Scardino, "Durability of Adhesive Bonded Joints," 1975, AFML-TR-75-3.

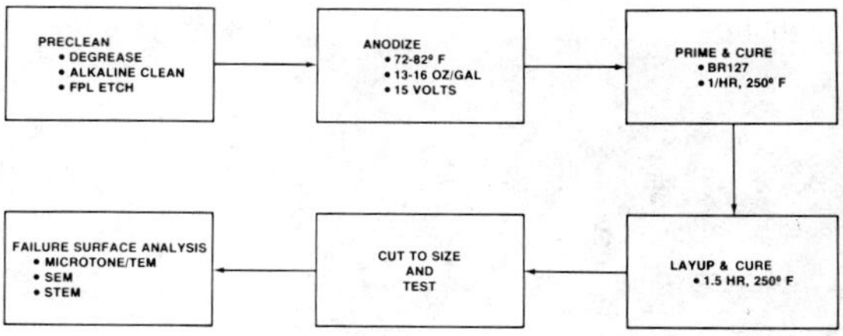

Figure 1. Process summary for fabrication, testing and analysis of peel specimens.

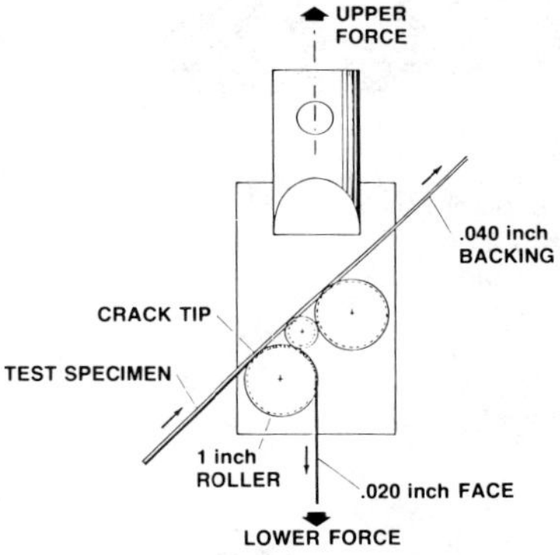

Figure 2. Modified floating roller peel apparatus.

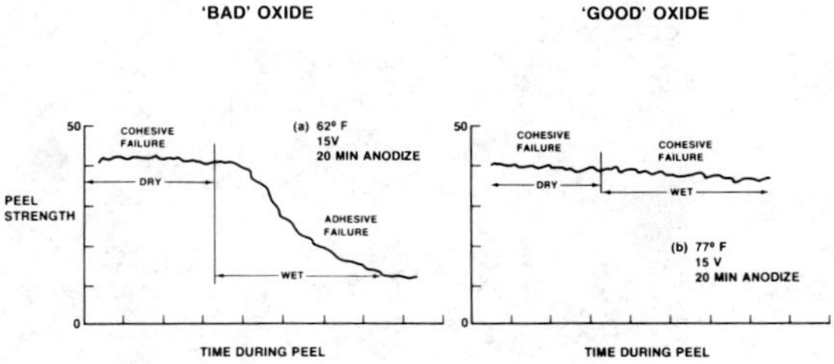

Figure 3. A schematic representation of dry/wet peel strip charts.

Figure 4. Microtome/TEM micrograph of 2024 clad anodized at 62°F. Approx. magnification 80,000X.

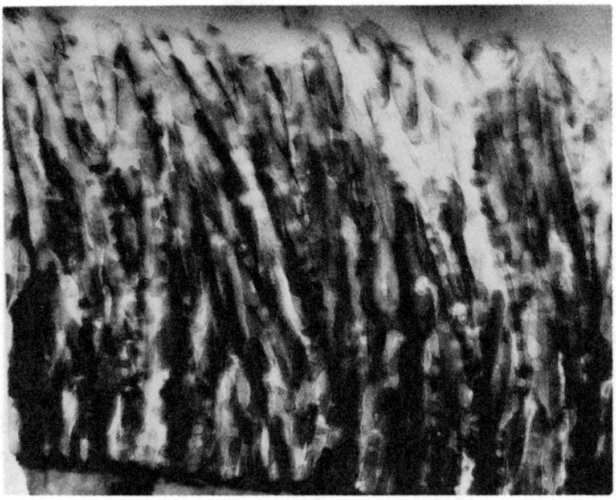

Figure 5. Microtome/TEM of 2024 clad anodized at 77°F. Approx. magnification 80,000X.

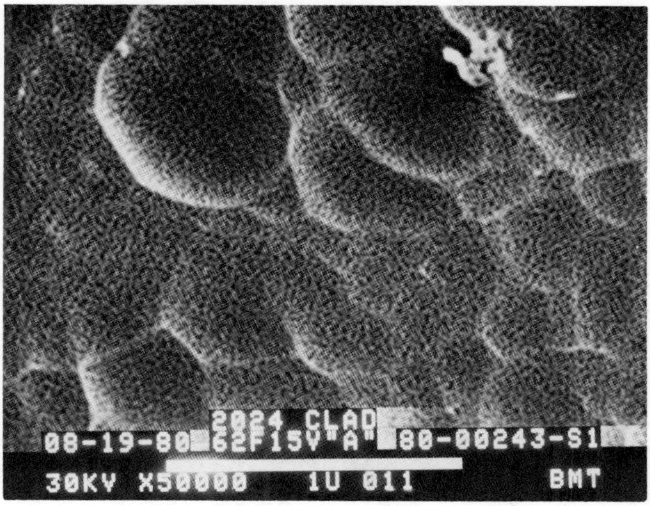

Figure 6. SEM micrograph of 2024T3 clad anodized at 62°F("Bad"). Approx. magnification 38,000X.

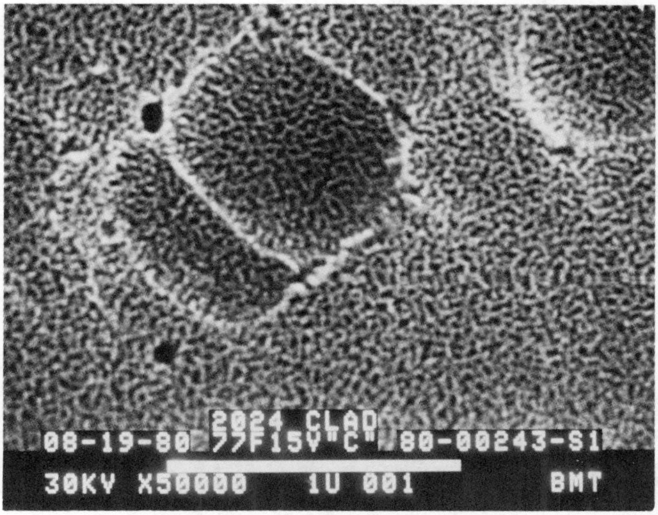

Figure 7. SEM micrograph of 2024T3 clad anodized at 77°F ("Good"). Approx. magnification 38,000X.

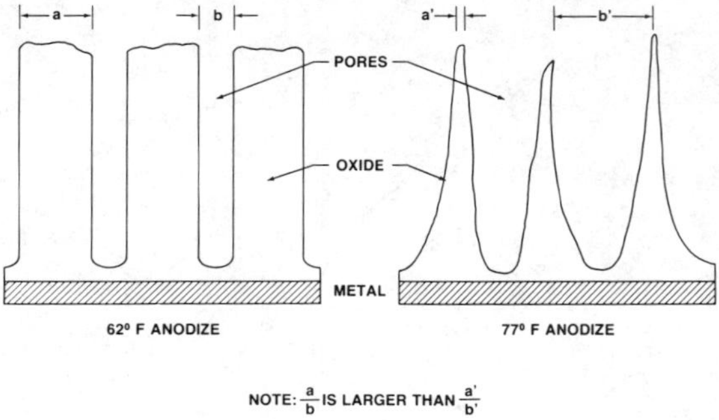

Figure 8. Schematic representation of "Bad" (left) vs. "Good" oxide.

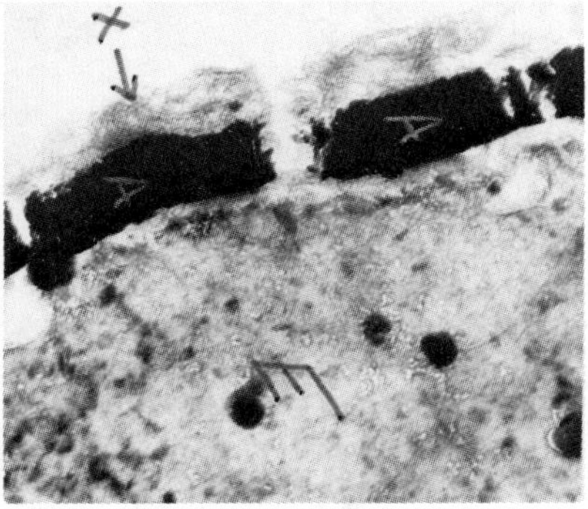

Figure 9. Microtome/TEM micrograph of failed wet peel 7075 bare anodized at 90°F, 6 volts. A: Au/Pd marker layer applied <u>after</u> peel; E: Al metal; X: Artifact or contamination. Approx. magnification 80,000X.

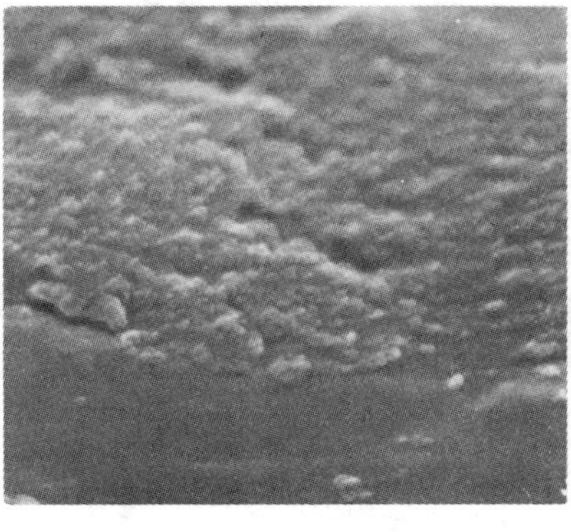

Figure 10. A fractured oxide SEM micrograph, 7075 bare anodized at 90°F, 6 volts. Approx. magnification 38,000X.

Figure 11. SEM fractured oxide micrograph of 7075 bare anodized at 90°F and 6 volts. Approx. magnification 38,000X.

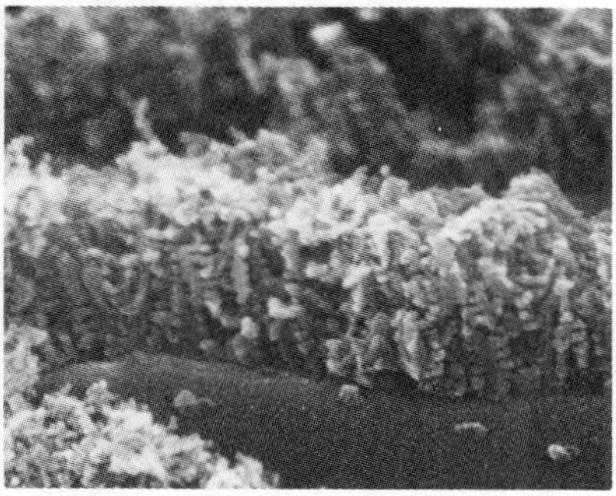

Figure 12. SEM fractured oxide micrograph of 7075 bare anodized at 90°F and 15 volts. Approx. magnification 38,000X.

Figure 13. SEM fractured oxide micrograph of 2024 bare anodized at 77°F and 15 volts. Approx. magnification 38,000X.

Table I
Double Cantilever Beam Results

Anodize Conditions		Alloy			
Temperature	Voltage	7075 clad	7075 bare	2024 bare	2024 clad
60°F	10 Volts	Delam.	Pass*	Pass	Fail
	15 Volts	Delam.	Pass*	Pass	Fail
75°F	10 Volts	Delam.	Pass	Pass	Pass
90°F	10 Volts	Delam.	Delam.	Pass*	Pass*
	15 Volts	Fail	Pass*	Pass	Pass*

The specimens were exposed for 363 days at McDonnell Douglas' El Segundo, California beach exposure site. Crack extensions of less than 3 inches passed; more than 3 inches failed; "Delam." indicates complete delamination of specimen during test; * indicates no detectable crack growth.

13th National SAMPE Technical Conference
October 13-15, 1981

CORRELATION OF DYNAMIC MECHANICAL MODULUS WITH ADHESIVE BOND STRENGTHS

Georgi T. Beckwith
David J. Crabtree
Northrop Corporation
Hawthorne, California

Abstract

A rapid and inexpensive instrumental technique, Dynamic Mechanical Analysis (DMA), can be used to reduce the number of mechanical bonding tests that are currently used to characterize an adhesive's lap shear strength and temperature aging effects. DMA experiments yield both the dynamic modulus of elasticity of adhesives and their mechanical damping, or energy dissipation, chacteristics. This paper presents the DMA dynamic moduli and mechanical damping data obtained on six aerospace structural epoxy film adhesives: FM-73 and FM-300 (American Cyanamid); AF-3101 and AF-147 (3M); EA 9628 and EA 9649 (Dexter-Hysol). A direct correlation between lap shear strength and dynamic modulus from room temperature to 177°C was obtained. In addition, the temperature aging effect and mechanical properties of two adhesives, AF-147 and EA 9649, that were exposed to 177°C for 2000 hours has been related to changes in their mechanical damping peaks. Therefore, it is concluded that the time-consuming and expensive mechanical testing of bonded specimens for obtaining strength/temperature profiles of adhesives can be significantly reduced by using DMA.

1. INTRODUCTION

A thorough characterization of an adhesive's service temperature range is required when evaluating new adhesives and is highly desirable in quality assurance (QA) evaluations of production lots of adhesive. The usual practice for determining an adhesive's service range is to measure lap shear strengths over a broad temperature range. The data are obtained by fabricating adhesive bond lap shear specimens and then destructively testing them at several temperatures per ASTM specifications. The results of this mechanical testing are dependent on a multitude of variables that exist in the fabrication of bonded specimens.

These adhesive bond variables include adherend surface preparation, bondline thickness, joint length and width, flaws in the adherend, and perhaps, even the bonding technique of the fabricator. A large number of bonding specimens are required to minimize the effects of these variables.

These destructive tests are costly in both bonding materials and labor dollars. The typical autoclave fabrication procedure requires 11 person hours to prepare 20 lap shear specimens. Testing 5 lap shear specimens each at four temperatures (e.g., -55°C, 24°C, 82°C, and 121°C) using ASTM test procedure requires approximately 12 person hours. These tests are expensive for routine QA testing and make the evaluation of new adhesives a costly procedure. An inexpensive, rapid test method is needed which can screen the mechanical properties of an adhesive over a broad temperature range.

The purpose of this paper is to demonstrate that Dynamic Mechanical Analysis (DMA) is a quick, reliable, and inexpensive test for determining an adhesive's service temperature range. DMA experiments yield two useful physical characteristics of polymers as a function of temperature change that are helpful in characterizing adhesives: 1) dynamic modulus of elasticity, and 2) mechanical damping or energy dissipation.

Figure 1 shows a typical DMA scan of an adhesive. The dynamic modulus of elasticity of a polymer is a measure of its resistance to deformation when a sinusoidally oscillating stress is applied to it. There are three regions of viscoelastic behavior of a polymer that are temperature dependent and can be defined on a DMA curve: 1) the glass region, 2) the glass-rubber transition region, and 3) the rubber region.

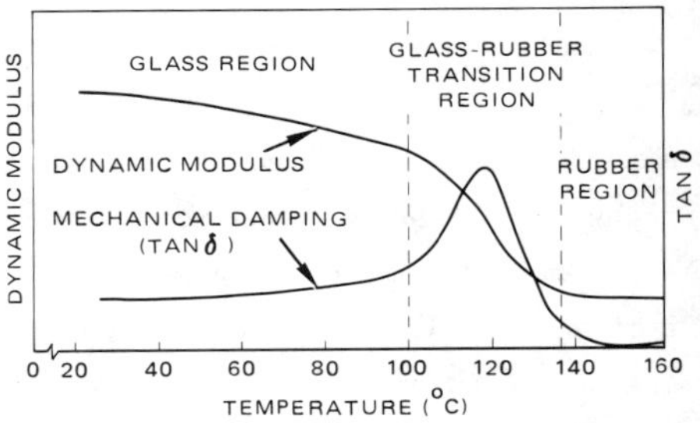

FIGURE 1. DMA CURVE OF A TYPICAL ADHESIVE

The glass region is the primary useful region for structural adhesives. In the glass-rubber transition region, a polymer has a progressively decreasing ability to carry load and creep within the bond becomes a serious problem. A structural adhesive can be used throughout the glass region. It may be possible to use it in the initial portion of the glass-rubber transition region. Careful consideration must be given to creep when selecting the maximum use temperature. In the rubber region, the adhesive is no longer structurally useful.

The mechanical damping measurement (tan δ) of a polymer is the ratio of its viscous loss of energy to its stored elastic energy. This measurement detects molecular relaxation processes. The structure of the polymer's molecular network determines the mechanical damping properties. Mechanical damping reaches a maximum in the glass-rubber transition region.

A DMA adhesive specimen can be autoclave cured and machined for testing in approximately 4 hours. The DMA instrumental analysis of an adhesive's viscoelasticity changes over the temperature range -55°C to 200°C takes approximately 30 minutes of operator time.

This paper presents the DMA dynamic moduli and mechanical damping data that was obtained on six aerospace structural epoxy film adhesives: FM-73 and FM-300 (American Cyanamid); AF-3101 and AF-147 (3M); EA 9628 and EA 9649 (Dexter-Hysol). A correlation between lap shear strength and dynamic modulus over the temperature range 24°C to 177°C was obtained. In addition, the temperature aging effect on lap shear strength of two adhesives, AF-147 and EA 9649, has been related to changes in their DMA mechanical damping peaks. These two adhesives were exposed to 177°C for 2000 hours.

2. EXPERIMENTAL CONDITIONS

2.1 Adhesive Bonding

Both thin and thick adherend lap shear specimens were fabricated. Both types of specimens had a 0.5-inch overlap. The thin adherends were made with 0.063-inch thick 2024-T3 alclad aluminum finger panels (see Figure 2). The finger panels were surface-treated using the FPL etch or the 10-volt phosphoric acid anodize methods. The FPL etch was used for the FM-73/BR-127, AF-3101/EC-3960, and AF-147/EC-3917 adhesive/primer systems. The 10-volt phosphoric acid anodize was used on the EA-9628/BR-127, EA-9649/EA-9205, and FM-300/BR-127 adhesive/primer systems. The thin adherend finger panel bonds were fabricated with a bonding tool that was designed for a 0.5-inch adherend overlap.

Thick adherend specimens were fabricated with 0.25-inch thick 2024-T81 bare adherends that were 10-volt phosphoric acid anodized. Thick adherend specimens were prepared with the AF-147/EC-3917 and EA-9649/EA-9205 adhesive/primer systems. Specimens of 1" x 6" size were cut from bonded 6" x 6" panels and machine notched for a 0.5-inch overlap (see Figure 2).

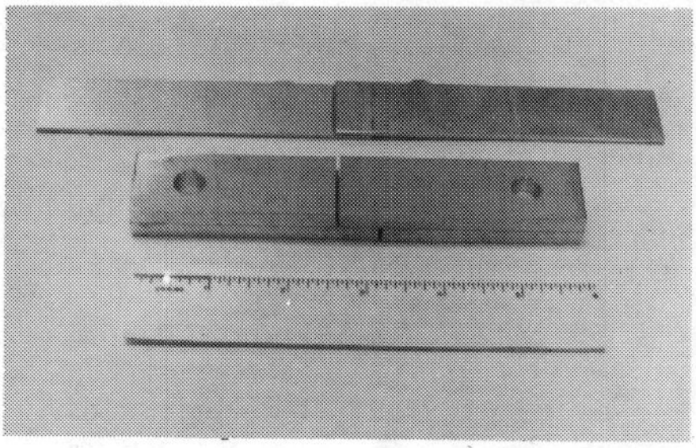

FIGURE 2. LAP SHEAR SPECIMENS — THIN ADHEREND (.063") AND THICK ADHEREND (.250")

Surface preparation, primer application, and adhesive cure parameters were per vendor specifications. Mechanical testing was done per the Northrop IT-34 specification.

2.2 DMA

DMA measurements were made using the DuPont Model 981 Dynamic Mechanical Analyzer. Specimen size was 24 mm by 12.7 mm (see Figure 3). All specimens were milled to a thickness of 1.3 mm.

FIGURE 3. DMA SPECIMEN

3. RESULTS AND DISCUSSION

3.1 Dynamic Modulus and Lap Shear Strengths

The DMA dynamic modulus and lap shear strength profiles over a 24°C to 200°C temperature range for six adhesives are shown in Figure 4. Each adhesive's lap shear curve is superimposed on its DMA dynamic modulus curve. The adhesive lap shear strength profiles were obtained from testing bonded specimens at several temperatures.

It is significant that each adhesive's lap shear strength profile is similar in shape to its dynamic modulus curve. This similarity shows that there is a good correlation between an adhesive's change in modulus and its ability to sustain a load in a bondline at specific temperatures. The three main regions of the dynamic modulus curve (glass, glass-rubber transition, and rubber) are evident and can be correlated with the strength characteristics of the adhesives. In the glassy region, where the modulus is high, each adhesive is able to sustain a load adequate for a structural adhesive. The dramatic drop in adhesive strength in the glass-rubber transition region is related to the reduction in the modulus.

The 121°C cure adhesives (FM-73, EA-9628, AF-3101) are shown in Figures 4A, 4B, and 4C. The similarities in the shapes of the moduli curves of these three adhesives show that they can each sustain a structural load in the same service temperature range.

The 177°C cure adhesives (FM-300, AF-147, EA-9649) are shown in Figures 4D, 4E, 4F. FM-300 has a dramatic drop in modulus within the transition region of the DMA curve. The FM-300 adhesive's bond strength is similarly dramatically reduced in this region. AF-147 has a more gradual negative slope with a wider transition region in its dynamic modulus profile. This wider transition region for AF-147 is duplicated by its gradual reduction in lap shear strength. The EA-9649 adhesive has a dynamic modulus profile

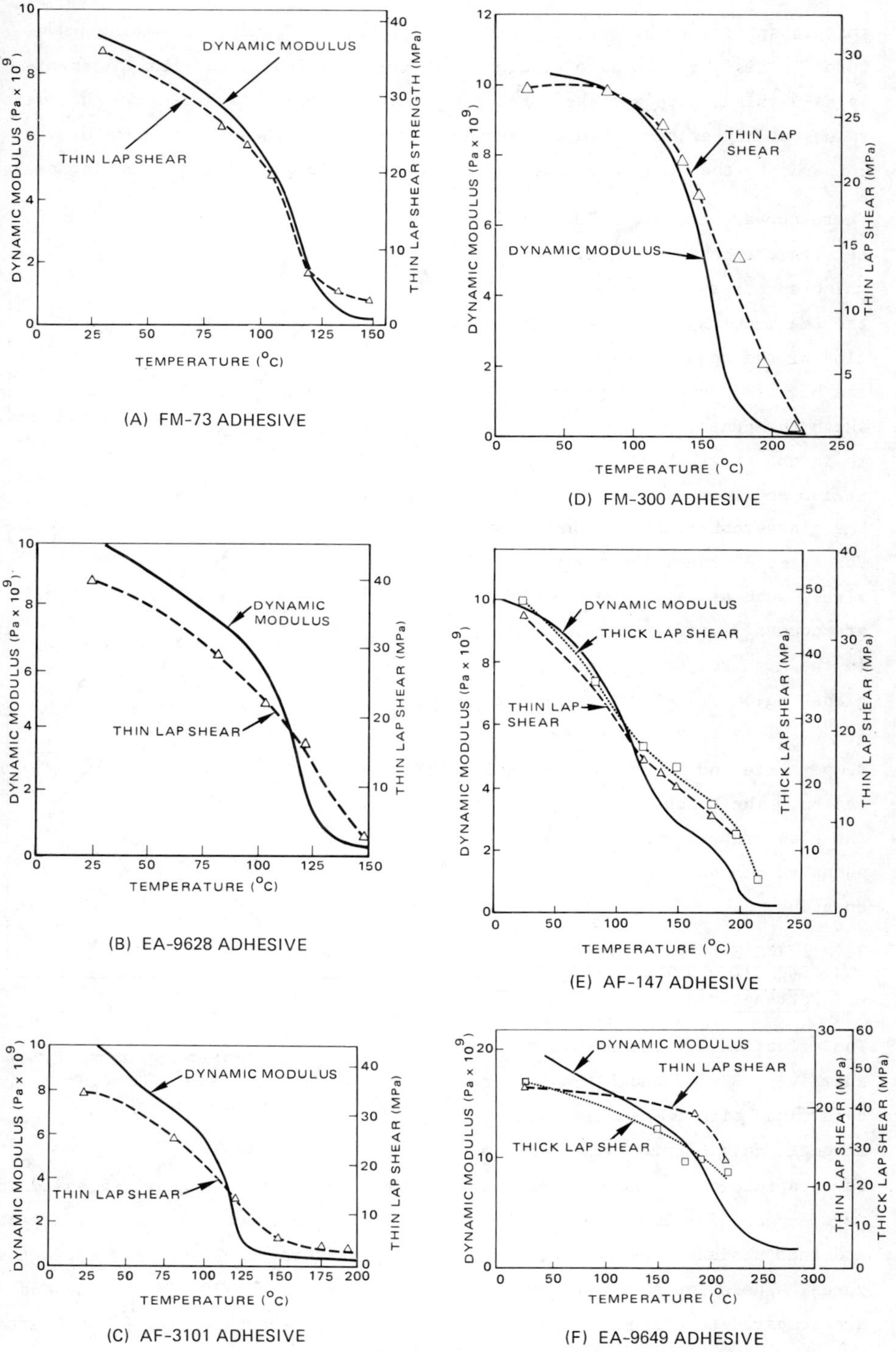

FIGURE 4. DYNAMIC MODULUS AND LAP SHEAR STRENGTH VERSUS TEMPERATURE FOR SIX ADHESIVES

that is shifted to higher temperatures. This shift is also apparent by EA-9649 adhesive's higher strength values at higher temperatures, when compared to the other adhesives.

There appears to be a relationship between the maximum temperature at which an adhesive can be used and the DMA modulus curve. An examination of the moduli curves shows that the highest service temperatures at which the adhesives are being used is in the extreme end of the glass region and in the initial portion of the glass-rubber transition region. For example, the 121°C curing adhesives, such as FM-73 and EA-9628, are commonly used at 82°C. This is in the high temperature end of the glass region. EA-9649 is used at 177°C. This is at the extreme high temperature end of the EA 9649 glass region. The highest temperature in the glass-rubber region at which an adhesive can be used is an open question.

3.2 177°C Exposure Effects on Dynamic Modulus and Lap Shear Strengths

The effect of 177°C exposure on two adhesives, AF-147 and EA-9649, were determined with DMA and lap shear strength data. Thick adherend lap shear specimens of these adhesives were exposed for 2000 hours to 177°C and then tested at several temperatures. These lap shear strengths are compared to strengths that were obtained with unexposed specimens in Figures 5 and 6. Damping peaks were also obtained with DMA specimens of AF-147 and EA-9649 that were exposed to the same temperature conditions. These DMA damping peaks are shown in Figures 7 and 8.

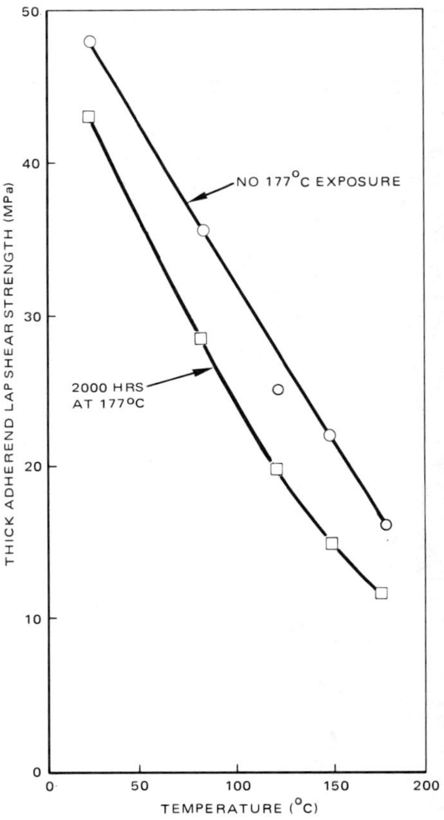

FIGURE 5. AF-147 LAP SHEAR STRENGTH REDUCTION AFTER 2000 HOURS EXPOSURE TO 177°C

The lap shear strengths decreased at all temperatures. At the 177°C test temperature, the lap shear strengths decreased about 30% for AF-147 and decreased about 50% for EA-9649 after 2000 hours exposure.

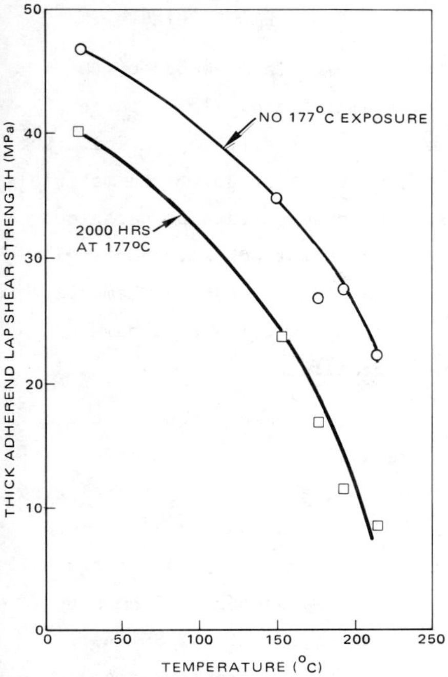

FIGURE 6. EA-9649 LAP SHEAR STRENGTH REDUCTION AFTER 2000 HOURS EXPOSURE TO 177°C

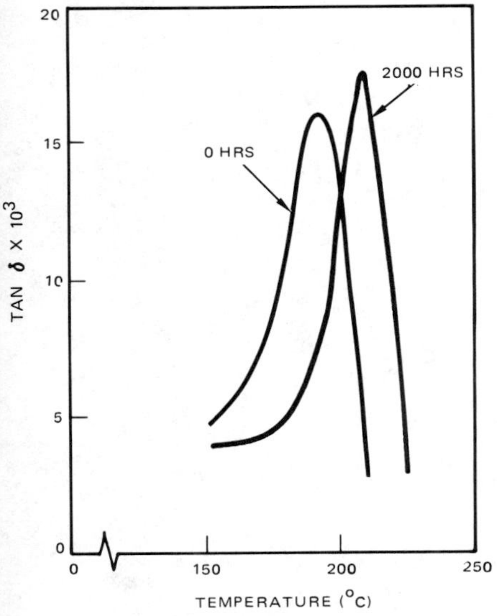

FIGURE 7. AF-147 DAMPING PEAK SHIFT AFTER 2000 HOURS EXPOSURE TO 177°C

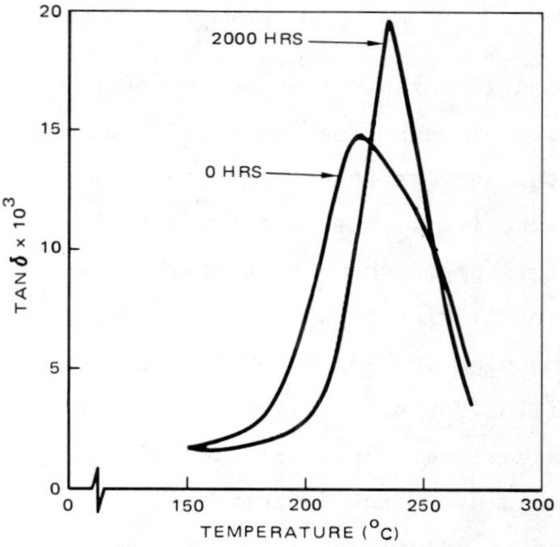

FIGURE 8. EA-9649 DAMPING PEAK SHIFT AFTER 2000 HOURS EXPOSURE TO 177°C

Changes also occurred in the mechanical damping curves. The peaks are not only shifted to higher temperatures but the intensities of the peaks are increased and the peak widths are decreased. These changes are indicative of major changes occurring in the polymer network due to the temperature exposure. The reduced width of the 2000-hour damping peak indicates that there are fewer relaxation processes occurring in the exposed adhesive as compared to the unexposed adhesive. The increased temperature of the 2000-hour peak indicates that the glass transition temperature has increased crosslinking. The increased cross-linking could be caused by reactions of free radicals that are formed in the polymer chains due to thermal degradation. This embrittles the polymer and thus lowers the shear strength.

4. CONCLUSIONS

DMA is a rapid inexpensive instrumental technique for characterizing the service temperature range of adhesives. The data presented in this paper show that there is a correlation between the dynamic modulus and lap shear strength profiles for six epoxy structural film adhesives. Thus, it is possible to obtain a quick determination by DMA of an adhesive's strength/temperature profile. This can aid in the selection of adhesives for a needed temperature application, and in the QA analysis of production lots. There is also a relationship between 177°C aging effects on adhesives and temperature shifts of their damping peaks.

DMA analysis can therefore reduce the number of time-consuming bonding tests that are currently used to obtain strength/temperature profiles of adhesives. Assuming three DMA samples are run to define the adhesive properties, then a 4 to 1 time saving can be realized in using DMA rather than lap shear testing.

5. ACKNOWLEDGEMENT

The technical contributions of Ms. Elizabeth Park in making DMA measurements is greatly appreaciated.

6. BIOGRAPHIES

Ms. Beckwith is a member of the Nonmetallic Materials Research Department at Northrop in which she is investigating polymeric materials and improved surface preparation techniques for adhesive bonding. She has an MA degree in Chemistry from the University of Southern California.

Mr. Crabtree is a member of the Nonmetallic Materials Research Department at Northrop specializing in the study of non-metallic materials used in aerospace structures. He has an MS degree in Chemistry from the University of Southern California. Currently he is active in developing methods for making compositional analyses of adhesive and composite resins.

13th National SAMPE Technical Conference
October 13-15, 1981

STUDIES OF THE EFFECT OF CURED PRIMER THICKNESS ON THE PEEL STRENGTH OF METAL TO METAL BONDS

Alphonsus V. Pocius
Adhesives, Coatings and Sealers Division, 3M
St. Paul, MN. 55144

Abstract

A detailed study of the effect of primer thickness on the peel strength of metal to metal bonds is presented. The floating roller peel strength is inversely proportional to the primer thickness while the T-peel strength is unaffected by primer thickness. Various other parameters such as angle of peel, rate of peel, temperature, adhesive thickness and scrim were also examined. The data was used to develop a phenomenological description of the peel process in the presence of highly crosslinked primer.

1. INTRODUCTION

Corrosion inhibiting primers (CIP's) are used extensively in modern aircraft structure. The many functions of a CIP include corrosion inhibition in the bondline of adhesively bonded structure, protection of the carefully prepared aluminum surface before the bonding step and provision of a corrosion inhibiting coating outside the bondline. As a result of this multipurpose usage, the demands on the CIP are great. It must contain enough corrosion inhibiting pigment to provide adequate protection over extended periods of time; but the pigment level must be low enough not to interfere with the physical properties of the cured primer. The CIP must be curable so that it provides protection before bonding but it must also be reactive enough after cure to easily bond to the adhesive during the adhesive cure. Finally, the CIP must be crosslinked enough during cure to provide jet fuel and hydraulic fluid resistance but yet be flexible enough to remain intact during stressing of the adhesively bonded structure.

In this report, I describe a detailed study of a phenomenon that results from the fact that modern CIP's are the best compromise to fit the above mentioned requirements. The phenomenon is the effect of the thickness of the cured primer on the peel strength of metal to metal bonds. It is widely known that if the cured CIP is too thick, there can be a reduction in the metal to metal peel strength. We have made a detailed study of the primer thickness effect on one of our presently used materials, EC-3961 B/A. The results lead to a phenomenological picture of the peeling process in the presence of a cured CIP. This study shows that CIP brittleness is not the entire cause of the primer thickness effect, but that the sample geometry also plays a major role.

2. EXPERIMENTAL

The metal used throughout this study was commercially obtained 2024T-3 bare alloy. The majority of the data was taken on "optimized FPL etched" aluminum, but some was obtained on H_3PO_4 anodized and CrO_3 anodized aluminum. The optimized FPL etch and H_3PO_4 anodize procedures are described by McMillan (1). The CrO_3 anodization was carried out in a glass jar at 104°F with a CrO_3 concentration of 11 oz./gallon of distilled H_2O. The voltage schedule was 5V/min. to 40V and then 30 minutes at 40V. The aluminum was alkaline degreased (Oakite 164) and FPL etched before either of the anodization processes.

The adhesive used in this study was AF-163-2, manufactured by 3M. The adhesive was used in the unsupported (U) form as well as the knit (K) scrim supported form. In most cases, the adhesive weight was 0.06 lbs./ft.^{2}. The primer used was EC-3961 B/A, also manufactured by 3M. This primer is a two part, 250°F curing primer that has been developed for roll coating application. After application the primer was air dried for 1/2 hour and then cured at 250°F for an hour. The primer thickness was measured by means of a Permascope. The adhesive cure cycle consisted of a 4-5°F/min. rise rate to 235°F followed by 90 minutes at 235°F. The applied pressure was 35 psi. A vacuum was drawn on the samples during the initial portion of the cure cycle in an attempt to remove air entrapped in the bondline.

The peel results were obtained on Instron testing machines equipped with Missimers environmental cabinets. The samples equilibrated for 1/2 hour before testing. The cross head speed was 6"/min. unless otherwise specified. The floating roller peel tests were done on a jig similar to the one shown in Fig. 1a. The T-peel tests as a function of angle of peel were done on an Instron in the usual manner except that the angle of peel was varied by means of the jig shown in Figure 1b.

3. RESULTS AND DISCUSSION

There are many mathematical relationships in the literature that are used to describe the peel strength of adhesive bonds. Likely the most widely known theory is that developed by Kaelble while he was at 3M (2). However, all of the early peel strength theories were based upon the theory of beams on an elastic foundation. As a result of the application of this theory, it is found that the peel strength should be directly proportional to the thickness of the adhesive layer. (3-5). If we assume that primer thickness can be directly additive to adhesive thickness, we find that the data in Figure 2 does not follow the results of the peel strength theories. Figure 2 shows the -67°F floating roller peel strength as a function of primer thickness. Obviously, the peel strength decreases as a function of primer thickness.

An analysis of all of the data taken in this work indicates that the functional dependence of peel strength on primer thickness in the floating roller peel test is as follows:

$$\frac{P}{W} = \frac{Co}{1 + ba}$$

where P is the peel force, W is the width of the bond, Co is the floating roller peel strength at 0 primer thickness, a is the primer thickness in mils and b is a proportionality constant. The lines drawn through the data in all of the figures in this report are based upon the above equation. The data in some cases show scatter from the line generated from the above equation, but in general, the data follow along well. The largest amount of scatter occurs at small primer thicknesses.

The first set of figures shows the effect of various bond parameters on the primer thickness effect. Figure 2 shows the effect of having a scrim present in the bondline. The effect of the scrim is to flatten out the primer thickness function, i.e., decrease b. The failure surface at moderate primer thicknesses (less than 0.25 mils) shows the pattern of the scrim on the thin adherend. Figures 3 and 4 show the effect of cured adhesive thickness on the -67°F and 75°F floating roller peel strength, respectively. The functional dependence in Figure 3 is essentially the same as the one shown in in Figure 2. The effect of adhesive thickness is very slight as may be apparent from Figures 2 and 3. However, there does seem to be a tendency for a thicker adhesive layer to worsen the primer thickness effect. Figures 5 and 6 show the effect of thin adherend thickness on the primer thickness function. Here some major effects are observed. At -67°F (Figure 5) and 75°F (Figure 6), we find that the thicker thin adherends give a higher peel strength initially (expected from Kaelble's analysis) but also show a small smoothing effect on the primer thickness function. The value of b is not substantially changed from one thin adherend thickness to another. The most interesting effect encountered in this work was the effect of surface preparation on the primer thickness function. It has certainly long been known that surface preparation can affect peel strength (6), but that finding was reserved for unprimed adherends. In Figure 7 we show the -67°F floating roller peel strength as a function of primer thickness and surface preparation. There is an obvious difference between the surface preparations with CrO_3 anodization showing the most severe effect. The surface preparation effect on the primer thickness function goes in the order of the thickness of the oxide (i.e. FPL etch is thinner than H_3PO_4 anodic oxide is thinner than CrO_3 anodic oxide) with the FPL etch giving the best performance. At this time we have no explanation for this phenomenon, but it is a matter of continuing concern for us.

More recent work on the phenomenon of peeling rupture has lead to the inclusion of viscoelastic considerations of the adhesive bondline (7-8). The original theory was based on the theory of beams on an elastic foundation and did not take into consideration the viscoelastic properties of most adhesives. We investigated viscoelastic effects on the primer thickness effect by varying the peel rate (crosshead speed) and the temperature of the environment. This data is shown in Figures 8-10. Figure 8 shows that if the primer thickness is kept low (i.e., approximately 0.04 mils) in the temperature range -67°F to 120°F there is no effect of the primer on the peel strength no matter the rate of peel or the temperature. At ~0.28 mils of primer, we see a very interesting dependence (Figure 9). At high peel rates at 90°F, we get low peel strengths and failure that is both in the primer and in the adhesive. At higher temperatures and lower peel rates, the failure goes entirely into the adhesive while at 140°F, the failure is always in the adhesive, no matter the peel rate. At ~0.44 mils of primer at 90°F, the failure is always cohesive in the primer, no matter the peel rate. However, at 120°F, a transition takes place at a peel rate of about 0.5 inches/min. in which the failure goes into the adhesive. A similar effect is seen at 140°F. At 160°F, there is no dependence on peel rate. The sharpness of the peel rate dependent transition in the 0.28 mil and 0.44 mil cases indicates the presence of a primer thickness dependent viscoelastic process.

All of the data discussed above has been concerned with floating roller peel specimens. It was of interest to see if the phenomenon carried over into other types of specimens. The specimen design most different from the floating roller peel specimen is the "T-peel" specimen. T-peel specimens were made up using various adherend thicknesses and primer thicknesses out to 1 mil. Figure 11 shows that T-peel specimens do not display any primer thickness dependence. However, this lack of sensitivity to primer thickness is only observed if the 90° "T" angle is maintained. If the 90° angle is not maintained, substantial effect are seen. E.g. in Figure 11, one specimen was mis-prepared having mismatched adherend thicknesses. The peel angle could not be maintained at 90° and substantially lower peel strengths resulted.

This effect of peel angle piqued our interest and a study was made of the dependence of "T"-peel strength on the angle which the "leg" of the T makes with the "bar" of the T as shown in Figure 1b. The first set of data of this sort is shown in Figure 12. This data was taken at 75°F and shows that at thin primer thickness a peel angle of 50° or less is necessary to decrease the peel strength. However, at thick primer levels, the "critical peel angle" was at just under 90°. At very low angles for all primer thicknesses, the peel force fluctuated in a slip-stick manner as in the case of unstable crack propogation through a brittle material. The failure surfaces also changed as a function of angle. At high angles the failure was cohesive in the adhesive. At intermediate angles at high primer thicknesses the failure was cohesive in the primer and at low angles at any primer thickness, the failure was at the thin adherend surface. We can make a correlation of primer thickness with this critical peel angle and this is shown in Figure 13. A generally linear function results. This indicates that the angle of peel controls the plane in the bondline through which the crack propogates. Further work of this sort was done at -67°F and 120°F and these results are shown in Figures 14 and 15. At -67°F only the thinnest primer thicknesses show critical angles less than 80°, at 120°F only the thicker primer levels show high critical angles.

We sort out the data shown in Figures 2-15 with the following phenomenological picture. The angle of peel that the thin adherend makes with the other adherend governs the plane in the bondline through which the crack wishes to propogate. At room temperature or lower temperatures both the adhesive and the primer do not display much viscoelastic character. At these temperatures the peel force depends upon whether the crack is propogating through the primer (thick primer) or through adhesive (thin primer). If one assumes that the peel angle governs this plane of crack propogation, then it suggests that there should be a critical angle (critical distance of the plane of crack propogation from the thin adherend) and this is observed. This suggestion of a critical primer thickness (corresponding to the critical peel angle which in turn corresponds to the plane of crack propogation) is supported by the knit scrim data shown in Figure 2. The scrim apparently pushes some adhesive into the primer layer during cure. Then, as the crack propogates in the plane specified by the sample geometry and angle of peel, it encounters regions of adhesive as well as primer. This is shown by the smoothing of the primer thickness function as well as the failure surface (described above). The viscoelastic response comes about only at higher temperature. At these temperatures the primer possibly has enough viscoelasticity to change the stress distribution to throw the maximum stress into the adhesive. This viscoelastic feature of the primer thickness function as well as the effect of surface preparation will be the subject of future study.

The conclusion we may draw is as follows: Test piece geometry plays a major role in the primer thickness function. The 45° floating roller peel is especially suceptible while the T peel shows no effect. It is expected that climbing drum peel would show an intermediate effect. This effect of test piece geometry might overshadow primer properties if it is not taken into account.

4. LITERATURE

1. J. Corey McMillan, in "Bonded Joints and Preparation for Bonding", AGARD Lecture Series

No. 102, Harford House, London, 1979, Pages 7-1 to 7-30.

2. D. H. Kaelble, Trans. Soc. Rheology, III, 161 (1959).

3. J. J. Bikerman, J. Appl. Phys., 28, 1484 (1957).

4. J. L. Gardon, J. Appl. Poly. Sci., 7, 643, (1963).

5. A. N. Gent and G. R. Hamed, Polymer Sci. Eng., 17, 462 (1977).

6. P.F.A. Bijlmer, Metal Finishing, December, 1971, pp. 34-39.

7. D. L. Hunston, J. L. Bitner, J. L. Rushford, J. Oroshnik, W. S. Rose, J. Elast. Plast., 12, 133 (1980).

8. J. A. Clarke, "Polymer Viscoelasticity: Relation to Peel Strength in Structural Adhesives" in Recent Advances in Adhesion, Edited by L. H. Lee, Gordon & Breach, N.Y., 1973.

5. BIOGRAPHY

Alphonsus V. Pocius received his B.A. in chemistry from Knox College in Galesburg, IL in 1970 and his Ph.D. in physical chemistry from the University of Illinois in Champaign-Urbana in 1974. He joined 3M in 1974 and his work has included polymer characterization, surface chemistry and corrosion research.

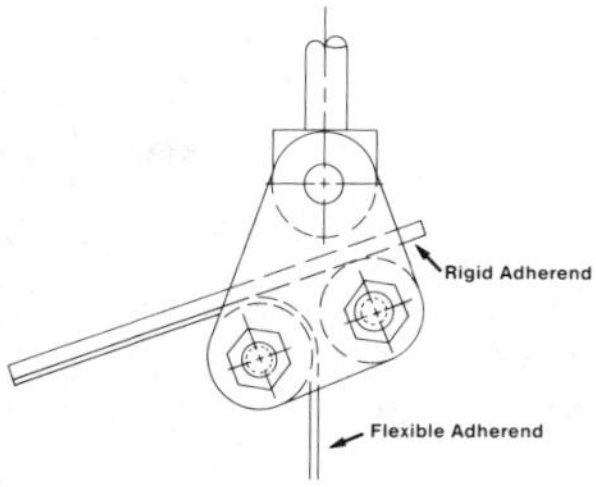

Figure 1a

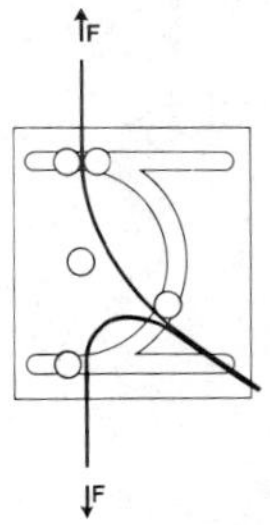

Figure 1b

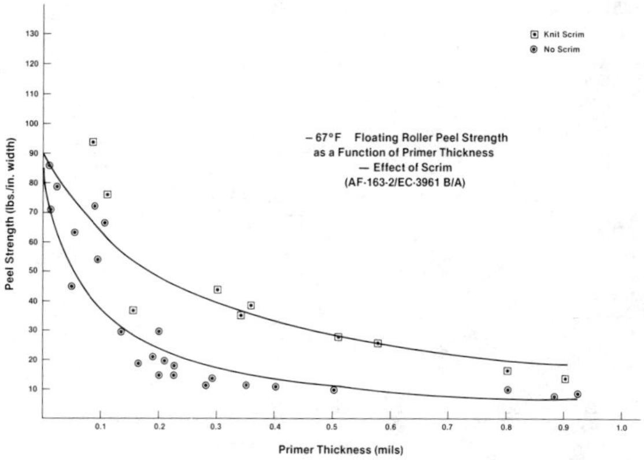

Figure 2

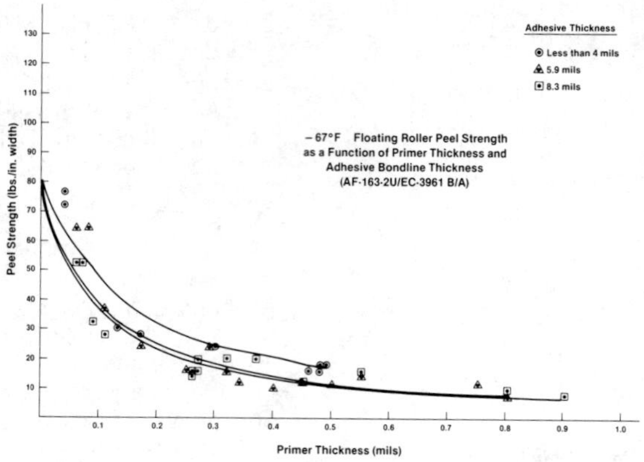

Figure 3

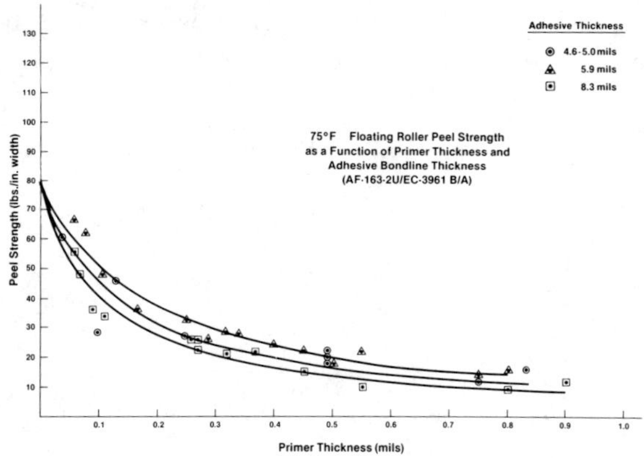

Figure 4

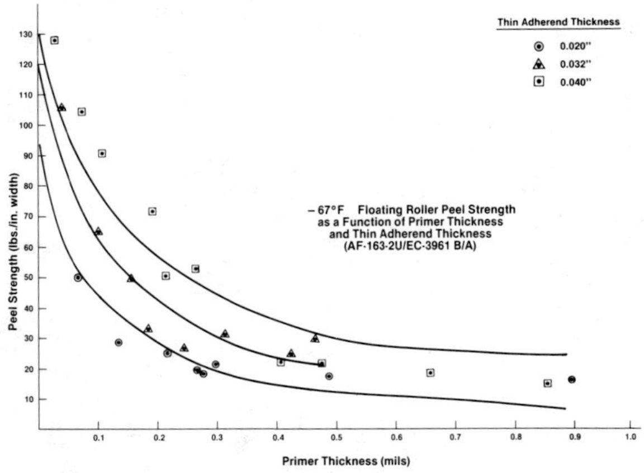

Figure 5

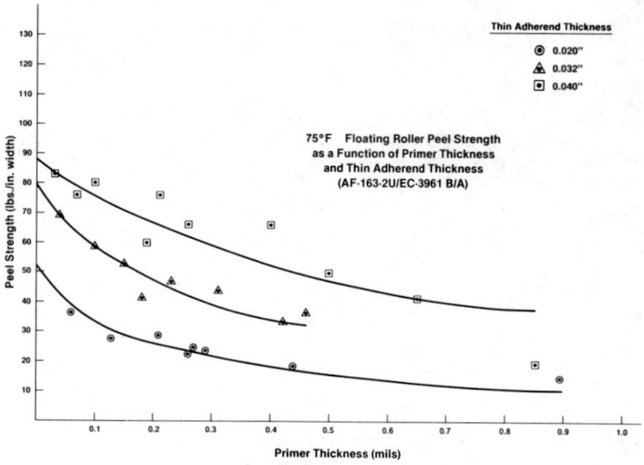

Figure 6

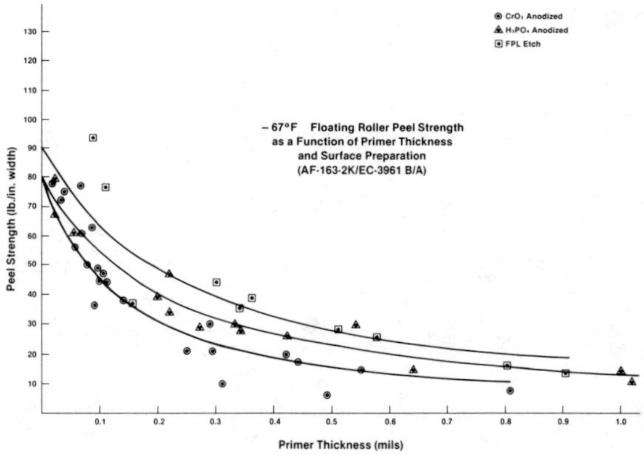

Figure 7

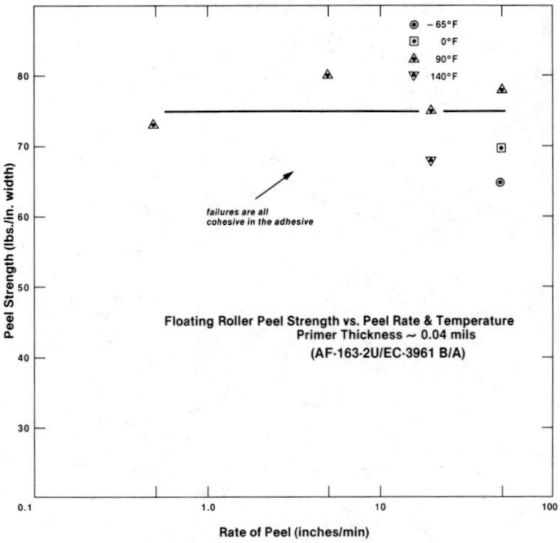

Figure 8

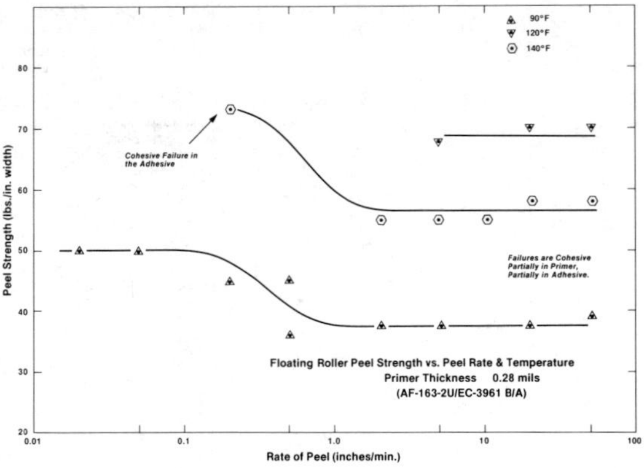

Figure 9

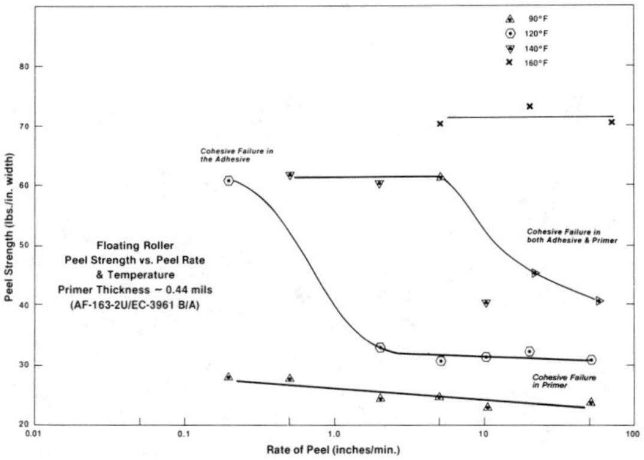

Figure 10

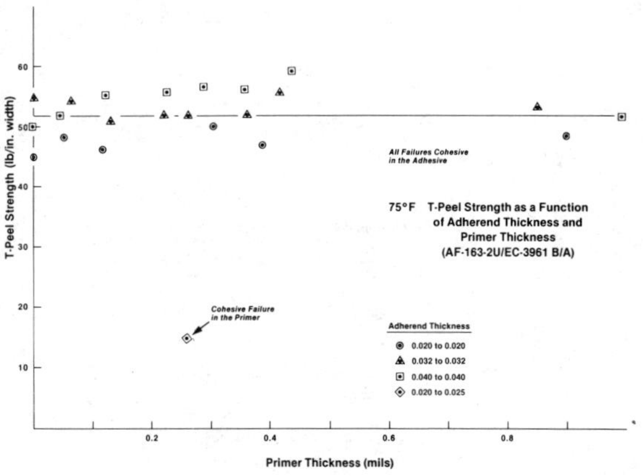

Figure 11

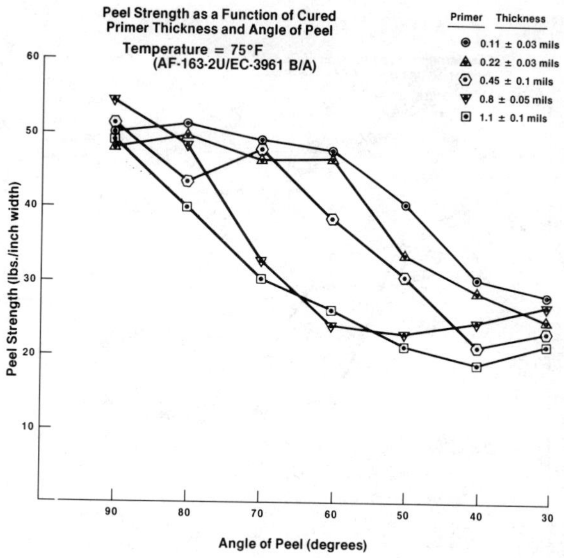

Figure 12

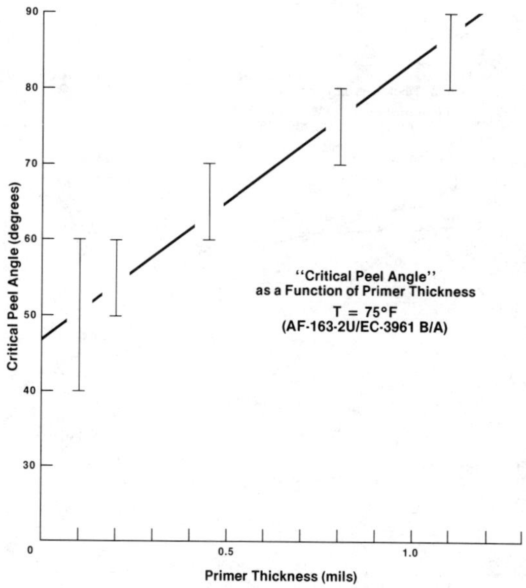

Figure 13

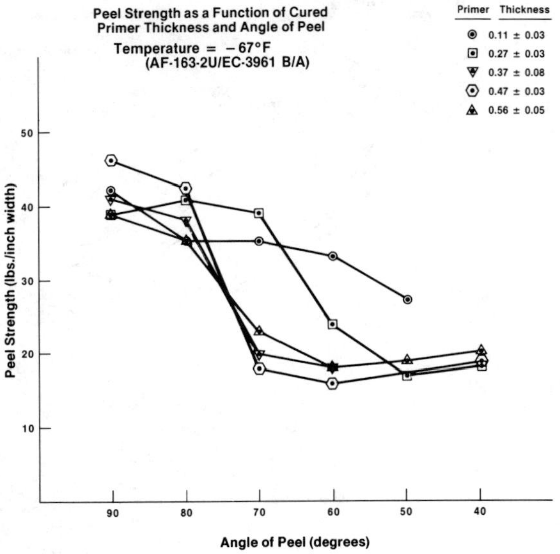

Figure 14

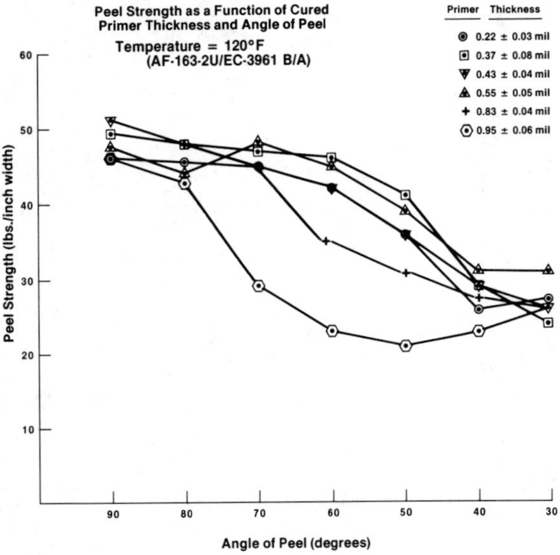

Figure 15

13th National SAMPE Technical Conference
October 13-15, 1981

EFFECTS OF NADIC ESTER CONCENTRATION AND PROCESSING ON PHYSICAL AND MECHANICAL PROPERTIES OF PMR/CELION 6000 COMPOSITES

Frances I. Hurwitz
and
J. Daniel Whittenberger
National Aeronautics and Space Administration
Lewis Research Center
Cleveland, Ohio 44135

ABSTRACT

PMR polyimide resin is prepared by reaction of 4,4'-methylene-dianiline (MDA), a dimethyl ester of 3,3',4,4'-benzophenonetetracarboxylic acid (BTDE) and a methyl ester of 5-norbornene-2,3,-dicarboxylic acid (NE). The stoichiometry of the reactants controls the length of imide prepolymers formed during *in situ* polymerization. The monofunctional NE groups serve as chain terminators and as crosslinking sites. Varying the concentration of NE in the monomer reactant mixture theoretically would be expected to alter the average distance between crosslinks, crosslink density, number of unreacted amine moieties and resin flow prior to gelation.

The influence of variation in NE stoichiometry on physical and mechanical properties of PMR polyimide/Celion 6000 graphite fiber composites is presented. Four stoichiometries, each processed by two different molding cycles, were studied. No general trends in the glass transition temperature (T_g), or in tensile, flexural or interlaminar shear properties are observed in the as-fabricated laminates with NE concentration. Mechanical properties do differ with processsing procedure. Larger differences among the laminates as evidenced by weight loss, residual flexural modulus and microstructural changes become apparent following exposure at 316°C (600°F) for 1000+ hours.

Keywords: composite materials; PMR; graphite fiber; mechanical properties; elevated temperature; nadic ester; processing; glass transition.

1. INTRODUCTION

The PMR polyimide resin system developed at NASA Lewis Research Center[1-2] is based on the *in situ* reaction of three monomer components on the reinforcing fiber: (a) a diamine, (b) a diester-diacid and (c) a monofunctional ester-acid, to form endcapped imide prepolymers. The monofunctional reactant serves as the site of an addition reaction to yield a thermally stable crosslinked matrix. The chemistry of the PMR system is summarized in Figure 1.

The present study explores the effects of variation in the concentration of monofunctional molecules on physical and mechanical properties of PMR/Celion 6000 composites. The stoichiometry of the reactants controls the molecular weight distribution of the imidized prepolymer[3], and hence the average distance between crosslinking sites. Varying the stoichiometry of the reactant mixture would be expected to influence the average molecular weight of the prepolymer as well as the average crosslink density, and therefore both the processability and properties of the matrix resin[4-6]. As the average chain length of the prepolymer changes, so does the viscosity of the resin prior to gelation[7]. In view of this potential variation in flow characteristics, all resins were processed using two different cycles.

The glass transition temperature T_g, thermo-oxidative stability, tensile and flexural strengths and moduli, and interlaminar shear strength were determined for each resin chemistry and processing method. Thermo-oxidative stability and interlaminar shear strength would be expected to be the most sensitive to resin chemistry, with tensile and flexural properties being more fiber dominated and exhibiting only slight matrix sensitivity at room temperature[8]. However, tensile and flexural properties also could be influenced by the fiber-matrix interface[9].

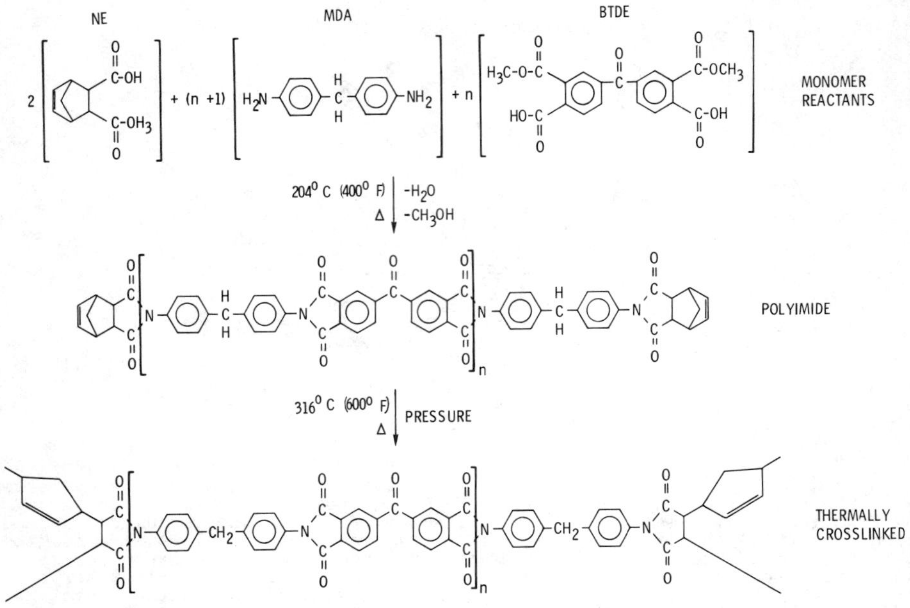

Fig. 1. Chemistry of the PMR polyimide system.

2. EXPERIMENTAL PROCEDURES

2.1 Resin preparation

The monomethyl ester of 5-norbornene-2,3-dicarboxylic acid (NE) (Burdick and Jackson lot 6858), 4,4'-methylenedianiline (MDA) (Eastman lot A6X) and 3,3',4,4'-benzophenonetetracarboxylic dianhydride (BTDA) (Aldrich lot 12 28 37) were purchased. The dimethyl ester of 3,3',4,4'-benzophenone-tetracarboxylic acid (BTDE) was synthesized by refluxing a suspension of 100 g of BTDA in 140 g of methanol for 2 hours after the material went into solution (approximately 3 hours total) to yield a BTDE solution which was 50 weight percent methanol. The BTDE not used immediately was stored under refrigeration to retard the formation of higher order esters[10]. Any refrigerated BTDE not used within 30 days was discarded.

The PMR resins were prepared utilizing a fixed molar ratio of MDA to BTDE of 3.087:2.087; the nadic ester (NE) concentration was varied from 1.25 to 2.00 moles in 0.25 mole increments. The 3.087 MDA: 2.087 BTDE: 2 NE ratio is that of PMR 15[4] and represents a "balanced" stoichiometry, i.e. the number of amine moieties equals that of the ester-acid groups; the other stoichiometries represent a "deficiency" in ester-acid functionality (Figure 1).

2.2 Composite fabrication

All laminates were compression molded. Unsized Celion 6000 graphite fiber tows (Celanese lot HTA-7-9431) were wound on a mandrel at 12 turns per inch. The monomers were applied to the fiber as a 50 w/o methanol solution. Approximately 65 g of solution were used for each 34.5 g of fiber. The prepreg was air dried to reduce the solvent content, removed from the drum and cut into 76.2 x 203.2 mm (3 x 8 in) plies. It was then stacked unidirectionally 8 plies thick in a preforming mold and staged at 204°C (400°F) for 1 hour.

Two processing procedures were used for each resin chemistry: (1) The staged prepreg was placed into a cold matched metal die and inserted into a press preheated to 316°C (600°F); or (2) The staged prepreg was placed in a mold preheated to 232°C (450°F) in the press, and the mold returned to the press. In both processing procedures contact pressure was applied until the temperature of the mold (as measured by a thermocouple inserted in the rail) reached 232°C (450°F), at which time 3.45 MPa (500psi) molding pressure was applied. The temperature was increased to 316 C (600°F) at an average rate of 4°C (7.2°F)/min, and both pressure and temperature were maintained for 2 hours. The heater was turned off and pressure maintained until the laminate had cooled to 232°C (450°F), at which time it was removed from the press. All composites were then postcured in a convection air oven at 316°C (600°F) for 16 hours. Two laminates were prepared by each fabrication process, for a total of four panels of each stoichiometry.

2.3 Composite characterization

Prior to testing all laminates were examined for gross defects using an ultrasonic C-scan technique. Standard materiallographic procedures were used to investigate each panel for microporosity and uniformity of composite structure.

The glass transition temperature (T_g) of each composite was determined using a thermomechanical analyzer (TMA) to measure the change in the rate of expansion of the laminate with increasing temperature. Measurements were made at a heating rate of 5°C (9°F)/minute.

Tensile properties were determined at 316°C (600°F) for unidirectional specimens having nominal dimension of 9.5 x 203 x 1.5 mm (.375 x 8 x .060 in) using the gripping procedure and test methods described in reference 11. For each resin chemistry and molding cycle, nominally four specimens from the same laminate were tested.

Flexural specimens, nominally 6.35 x 63.5 x 1.5 mm (.25 x 2.5 x .060 in) were tested in three point bending utilizing a fixed span of 50.8 mm (2 in) and center loading rate of 1.27 mm (.05 in)/min in accordance with ASTM standard method D790-66. Because of variations in test specimen thickness, the span to depth ratio ranged from 30:1 to 37:1. Tests were conducted at both ambient temperature and 316 C (600°F). Interlaminar shear strength (ILSS) was measured on 6.35mm (.25in) wide specimens at a span to depth ratio of 5:1 following ASTM standard method D2344-72 at both ambient temperature and 316°C (600°F). Flexural and ILSS values reported are the average of nominally six tests from two different laminates for each resin chemistry and molding process. Baseline data for cold molded PMR 15 represents the average of a total of twelve to fifteen tests of samples taken from at least three laminates.

Thermo-oxidative stability was evaluated by following changes in weight and in the 316°C (600°F) flexural strength and modulus of samples after exposure at 316°C (600°F) for up to nominally 1000 hours. Test coupons, the size of a flexural specimen, were placed in a convection oven having an air exchange rate of 100 cm^3 (6.1 in^3)/min. Weight measurements were made at time intervals of approximately 250 hours. An average of three samples of each chemistry were tested in flexure at 316°C (600°F) after each exposure interval.

TABLE I

Glass transition temperature (T_g) of laminates as a function of NE concentration and fabrication procedure.

	Cold Mold		Hot Mold	
Moles NE	T_g	std dev	T_g	std dev
	°C			
2.00	347	4.7	345	2.5
1.75	346 362	6.2 3.5	360	0.7
1.50	351	6.3	352	3.2
1.25	345	1.9	355	6.5

3. RESULTS

All resin chemistries and processing methods yielded void free laminates, as determined by C-scan and optical microscopy of polished cross-sections. Laminates ranged in thickness from 1.346 to 1.701 mm (.053-.067 in), with a mean thickness of 1.524 mm (.060 in). On average, the "hot mold" processing produced greater resin flow, as evidenced by slightly thinner composites, than did the "cold mold" fabrication procedure.

The T_g data are summarized in Table I. For all but the laminates cold molded with 1.75 moles NE, there is no statistically significant variation between laminates of the same chemistry and fabrication cycle, and the data given are average values. In the case of the 1.75 mole NE cold molded laminates a difference was found, and the average for each laminate is presented. This difference for the 1.75 mole NE material cannot be explained at this time. No consistent trend in T_g is seen with either NE concentration or fabrication procedure.

Tensile and flexural strengths and moduli were normalized for variation in fiber loading to a nominal laminate thickness of 1.524 mm (.060 in), or approximately 51 v/o fiber, utilizing the relationship that, for composites made from the same fiber and having the same number of plies, fiber volume fraction is proportional to 1/t[12]. Tensile properties of the as-fabricated laminates tested at

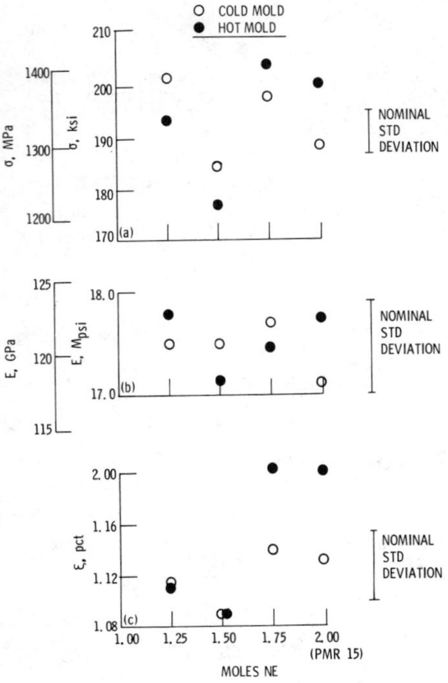

Fig. 2. Tensile properties of composites as a function of moles of NE in the resin. (a) Tensile strength; (b) tensile modulus; (c) strain to failure. All measurements were made at 316°C (600°F). Points represent an average of nominally four tests.

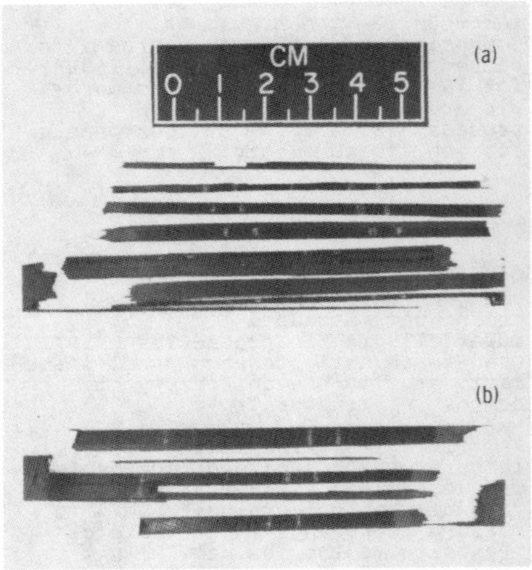

Fig. 3. Typical tensile fractures. Both laminates contain 1.75 moles NE. (a) Cold molded; (b) hot molded.

316°C (600°F) are summarized in Figure 2. No general trend is observed as a function of variation in NE stoichiometry or molding procedure. A pairwise comparison of laminates with the same chemistries processed by different molding cycles shows a statistical difference in strength only between the hot and cold molded laminates containing 2.0 moles of NE, the standard PMR 15 chemistry. There is no statistical difference among any of the measured tensile moduli. A slight increase in strain to failure is noted for the hot molded laminates having 1.75 and 2.0 moles NE as compared with their cold molded counterparts. Typical tensile failures for the composites containing 1.75 moles NE are depicted in Figure 3. The failures for other resin stoichiometries were quite similar to these, and to those of previously tested PMR 15 laminates[11].

The results of three point bending tests at both ambient temperature and 316°C (600°F) are presented in Figure 4. At room temperature all materials failed in tension at stresses > 1760 MPa (255 ksi); however at elevated temperature failures occurred in compression at much lower stresses [< 1185 MPa (172 ksi)]. Comparison of hot and cold molded laminates with the same stoichiometries again shows some differences for the 1.75 and 2.0 mole NE chemistries, with the hot molded composites being somewhat stronger than the cold molded ones. However, there is no overall trend with either stoichiometric variation or processing cycle. The flexural moduli of the laminates containing 1.75 and 2.0 moles of NE are slightly greater for the hot molded than for the cold molded composites tested at room temperature (Figure 4A), but there is no statistical difference in moduli among any of the laminates tested at 316°C (600°F) (Figure 4B).

Figure 5 gives the interlaminar shear strength (ILSS) at both room temperature and 316°C (600°F). At ambient temperature the load at failure was used to calculate the shear strength. Since all samples tested at elevated temperature exhibited plastic behavior, and the shape of the load-time curves varied with resin composition, it was difficult to determine a consistant endpoint. Therefore, a yield stress calculated from a point on the loading curve corresponding to 1.27 mm (.050 in) crosshead travel beyond the proportional limit (the point at which the behavior ceases to be elastic) was chosen as the ILSS (Figure 6). At room temperature a good deal of variation in strength is noted (Figure 5), but no overall trends can be found. At elevated temperature no significant differences among the laminates are

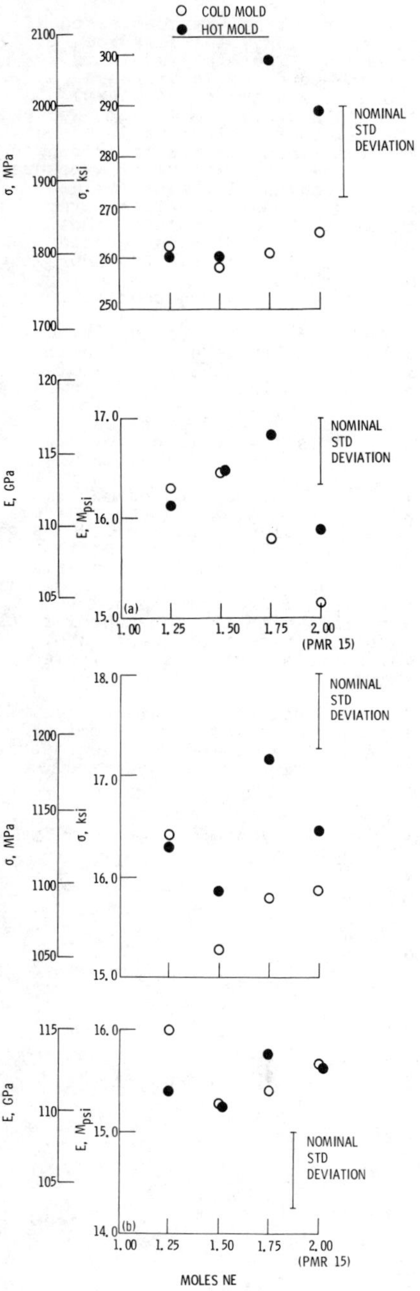

Fig. 4. Flexural properties as a function of moles of NE. (a) Room temperature; (b) 316°C (600°F). Each point is an average of nominally six tests. Baseline data for cold molded PMR 15 is the average of twelve tests for three laminates.

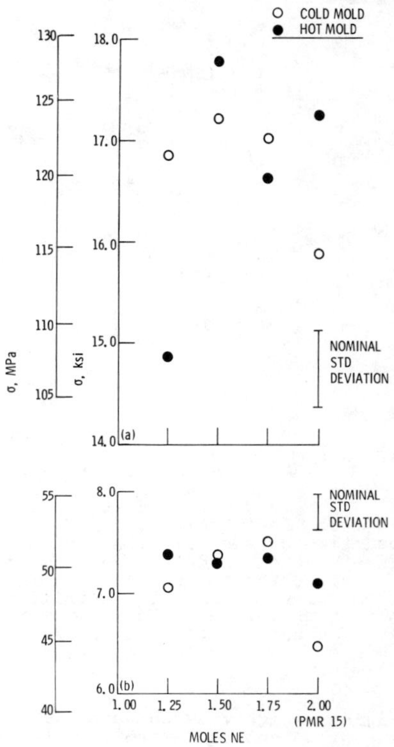

Fig. 5. Interlaminar shear strength as a function of moles NE. (a) Room temperature; (b) 316°C (600°F). Each data point is the average of six tests. Baseline data from PMR 15 represents an average of twelve tests of three laminates.

seen, with the exception of a small difference in hot and cold molded PMR 15 (2.0 mole NE composition).

While variation in the room temperature and 316°C (600°F) mechanical properties of the as-fabricated laminates with resin chemistry is relatively slight, more significant differences are seen following isothermal exposure at elevated temperature. Weight loss for exposed laminates at 316°C (600°F) as a function of exposure time is shown in Figure 7. Although there are no generalized trends with either stoichiometry or processing, considerable variation in weight loss is observed by 1050 hours. Pairwise comparison of laminates having the same stoichiometry but processed by the two different molding procedures reveals little difference for the 1.25-1.75 mole NE chemistries. However, a substantial difference is observed between hot and cold molded PMR 15 (NE = 2 moles); the hot molded material retained 95.5 percent of its

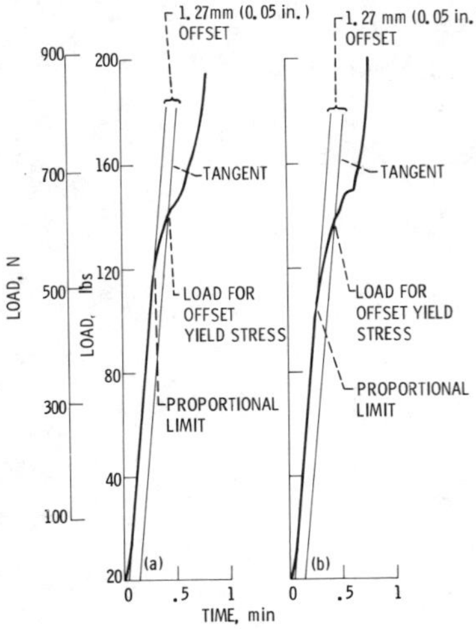

Fig. 6. Representative ILSS load-time curves illustrating the offset method used to determine load in shear strength calculations for two samples (a) and (b) exhibiting different yielding behavior. Tests were conducted at 316°C (600°F) at a crosshead rate of 1.27 mm (0.05) in)/min.

original weight, as compared with 89 percent weight retention for the cold molded material.

For resins with decreased NE concentrations, the flexural strength (Figure 8A) exhibits a general tendency to increase on short term aging (<500 hours) and then to decrease on longer exposure. PMR 15, on the other hand, nominally retains its as-fabricated strength even after 1000+ hours of exposure. The strengths of the hot molded materials generally are greater than those of the cold molded. Hot molded PMR 15 possesses greater residual strength at 1000 hours than any of the other chemistry/processing combinations.

The flexural modulus (Figure 8B) also exhibits a small initial increase at about 250 hours of exposure, which is then followed by a continous decrease. While the moduli of the as-fabricated composites are quite close in value, after 1000 hours a significant spread is observed. In general, the moduli of the hot molded materials are slightly greater than that of the cold molded, and the modulus of hot molded PMR 15 is higher than that of any of the other materials examined. The difference in moduli between hot and cold molded materials with the same chemistries is slight with the exception of PMR 15, where a significant difference exists.

Microstructural evaluation of the samples exposed for nominally 1000 hours revealed that degradation of the laminates has occurred

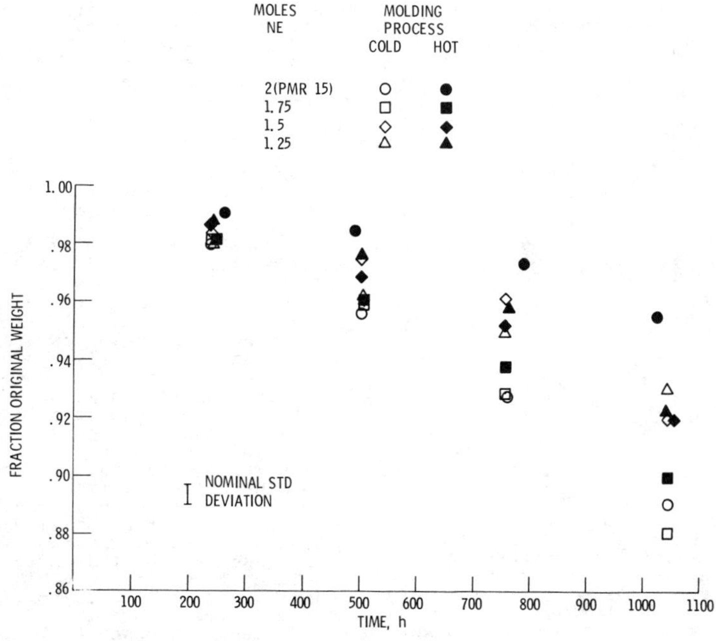

Fig. 7. Effect of isothermal exposure at 316°C (600°F) on laminate weight.

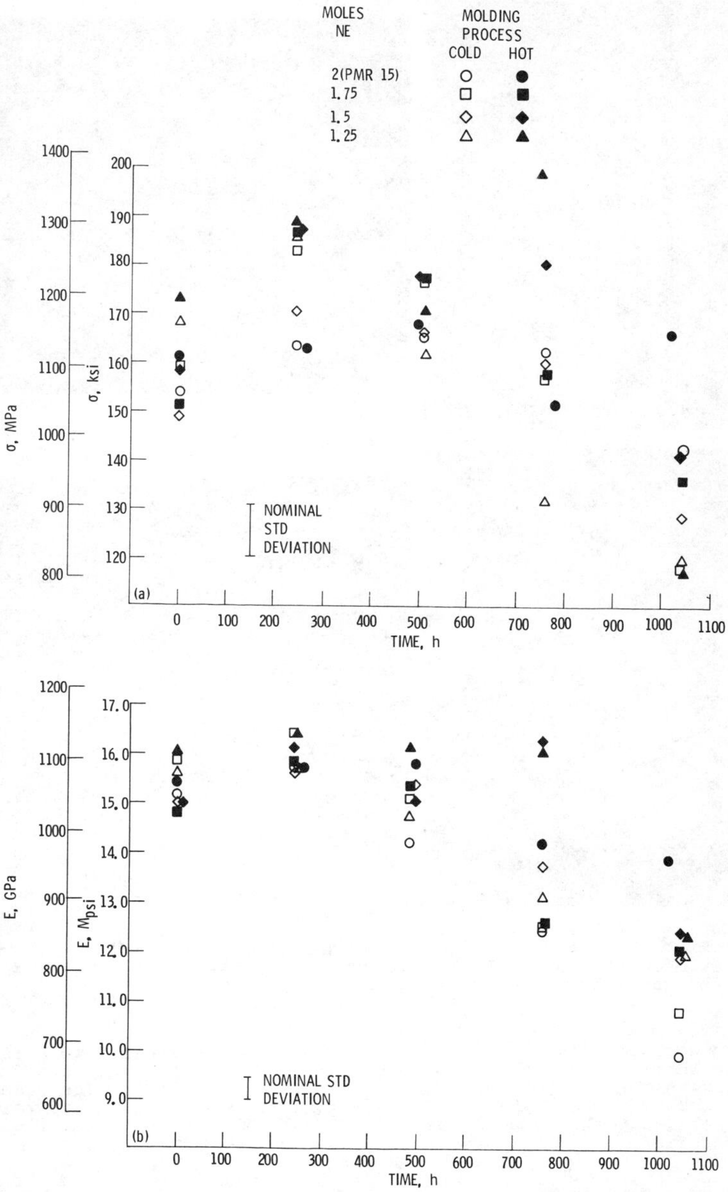

Fig. 8. Effect of isothermal exposure at 316°C (600°F) on residual flexural properties. (a) Flexural strength; (b) flexural modulus.

(Figure 9). This is evidenced by changes in the appearance of the composite near the exposed surface. Surface connected cracks were noted in many of the laminates, as was a decrease in the diameter of fibers in the oxidized region. The appearance of the "core" region of the composite is the same as that for unaged material. Comparison (Figure 9) of cross sections of hot and cold molded PMR 15 exposed for nominally 1000 hours shows the depth of penetration to be substantially greater in the cold molded than in the hot molded material, thus accounting for the large differences in residual flexural moduli after 1000 hours of exposure. The large scatter in these flexural data can be explained by variations in the thickness of the undamaged "core"

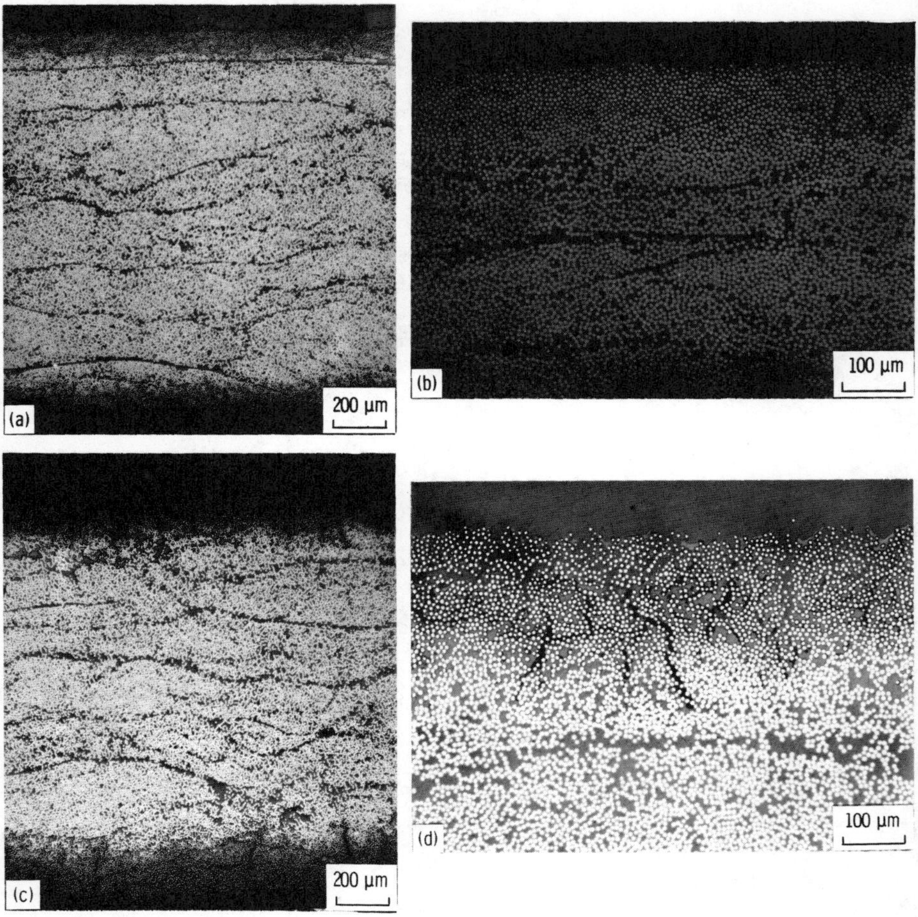

Fig. 9. Microstructure of hot and cold molded PMR 15 exposed at 316°C (600°F) for nominally 1000 hours. (a) and (b) Hot mold; (c) and (d) cold mold process.

region among laminates with differing chemistries and processing, since the residual mechanical strength must be attributable to the undamaged region of the material.

4. DISCUSSION

Variation in the concentration of monofunctional NE is assumed to produce alterations in (1) average crosslink density and (2) flow prior to gelation and hence the processability of the resin. Theoretically, changes in the crosslinked network should influence the T_g of the laminate. However, on the basis of the TMA data the net effect appears to be that the T_g of the composites is relatively unchanged by the variation in NE stoichiometry. As the nadic concentration is decreased, amine moieties are present in excess; these might react to form another type of crosslink. The occurrence of a second crosslinking reaction would explain the lack of change in T_g with NE variation.

In addition, since the nadic group has been shown to be one site of oxidation[13-15], variation in nadic concentration also would be expected to be accompanied by differences in thermo-oxidative stability of the resin. Again, after exposure at elevated temperature the laminates showed differences in weight loss and residual mechanical properties with NE concentration, but no generalizable relationship is observed. As the chemistry and crosslinked network structure is altered, the fiber-matrix interface also might be changed. These effects could compete with one another in translating molecular changes into observed mechanical properties. The net result is that no consistent trend with stoichiometric variation of NE is observed, although the T_g, flexural, ILSS and 316°C (600°F) tensile properties of some of the as-fabricated laminates are altered relative to PMR 15. Indeed, the largest differences in mechanical properties seen in this study are for the PMR 15 resins processed by two different fabrication cycles. Hence, at least in the case of unidirectional composites, rheological properties and/or thermal history seem to outweigh the difference in chemistry.

Only slight differences between as-fabricated hot and cold molded laminates are observed in tensile strength and strain to failure at 316°C (600°F), and in flexural strengths measured at room temperature and 316°C (600°F). While interlaminar shear strengths exhibit considerable variation among laminates tested at room temperature, there are no overall trends with NE content. At elevated temperature, all resins are plastic under load, and no ILSS differences among laminates are observed. However, larger differences are apparent in the weight loss, residual flexural modulus and microstructure after 1000+ hours of exposure at elevated temperature. Since greater resin flow occurs for the hot mold process relative to the cold mold cycle, it is possible that the resulting crosslinked network differs for laminates produced by the two procedures. It also is possible that lower molecular weight material is trapped as crosslinking proceeds in the cold mold process, and leads to decreased thermo-oxidative stability as well as the observed mechanical property differences.

The thermo-oxidative stability data for PMR 15/Celion 6000 presented in this work show a greater decrease in weight and mechanical properties at 1000 hour exposure times than that presented earlier[16]. This reflects the 25-30 percent difference in the thickness of the samples exposed at elevated temperature rather than a substantive disagreement between the two studies. Since our microstructural analysis reveals a damaged layer at the exposed surface, the same depth of penetration would represent a much larger percentage change in the thinner laminates used in this work.

5. CONCLUSIONS

Based on a study of the effects of variation in nadic ester (NE) concentration on properties of the unidirectional PMR/Celion 6000 system, several conclusions can be drawn: (1) Resins containing 1.25-2.00 moles of NE can be processed into void free unidirectional composites having mechanical properties comparable to those of as-fabricated PMR 15 laminates; (2) Larger differences in mechanical properties are introduced by processing changes than by stoichiometric variation within the range studied; and (3) The mechanical properties of the composites following exposure in air at 316°C (600°F) depend upon the thermal history of the resin, as well as on matrix stoichiometry.

6. REFERENCES

1. Serafini, T. T., Delvigs, P. and Lightsey, G. R., "Thermally stable polyimides from solution of monomer reactants," J. Appl. Polym. Sci. 16, 905-916, 1972.

2. Serafini, T. T., " Processable high temperature resistant polymer matrix materials," International Conference on Composite Materials vol. 1. E. Scala, ed. AIME, New York, 1976, pp. 202-220. (Also NASA TM-X-71682, 1975.)

3. Odian, G. G., Principles of Polymerization McGraw-Hill, New York, 1970, pp. 69-83.

4. Serafini, T. T. and Vannucci, R. D., "Tailor making high performance graphite fiber reinforced PMR polyimides," presented at the Reinforced Plastics - Milestone 30; Proceedings of the Thirtieth Annual Reinforced Plastics Composites Conference, Society of the Plastics Industry, Inc., New York, 1975, pp. 14E 1-5. (Also NASA TM-X-71616.)

5. Hurwitz, F. I., "Influence of excess diamine on properties of PMR polyimide resins and composites," Materials 1980 Proceedings of the Twelfth National SAMPE Technical Conference, Society for the Advancement of Material and Process Engineering, Azusa, CA, 1980, pp. 517-530. (Also NASA TM 81580.)

6. Vannucci, R. D. and Alston, W. B., "PMR polyimides with improved high temperature performance," Proceedings of the Thirty-first Annual Reinforced Plastics/ Composites Conference, society of Plastics Industry, Inc., New York, 1976, Section 20A, pp. 1-8. (Also NASA TM-X-71816.)

7. Billmeyer, F. W., ed., Textbook of Polymer Science, 2d ed., Wiley, New York, 1971, p. 229.

8. Chamis, C. C., Hanson, M. P. and Serafini, T. T., "Criteria for selecting resin matrices for improved composite strength," Reinforced Plastics - Ever New; Proceedings of the Twenty-eighth

Annual Conference, Society of the Plastics Industry, Inc., New York, 1973, section 12C, pp. 1-12. (Also Modern Plastics 50, 90-106, 1973 and NASA TM-X-68166.)

9. Harris, B., "The strength of fibre composites," Composites 3, 152-167, 1972.

10. Lauver, R. W. and Vannucci, R. D., "Characterization of PMR polyimides - correlation of ester impurities with composite properties," Enigma of the Eighties: Environment, Economics, Energy; Proceedings of the Twenty-fourth National SAMPE Symposium and Exhibition, Book 1, Society for the Advancement of Material and Process Engineering, Azusa, CA, 1979, pp. 522-532. (Also NASA TM 79068.)

11. Whittenberger, J. D. and Hurwitz, F. I., "Elevated temperature tensile properties of PMR composites," to be submitted to Polymer Composites.

12. Whittenberger, J. D. and Hurwitz, F. I., "On determination of fiber volume fraction in composite materials," to be submitted to J Test Eval.

13. Alston, W. B., "Characterization of PMR 15 polyimide resin composition in thermo-oxidatively exposed graphite fiber composites," Materials 1980 Proceedings of the Twelfth National SAMPE Technical Conference, Society for the Advancement of Material and Process Engineering, Azusa, CA, 1980, pp. 121-137. (Also NASA TM 81565.)

14. Jewell, R. A. and Sykes, G. F., "Thermooxidation of methylene bridging groups in polyimides," in Chemistry and Properties of Crosslinked Polymers, S. S. Labana, ed. Academic Press, New York, 1977, pp. 97-106.

15. St. Clair, T. L. and Jewell, R. A., "Solventless LARC-160 polyimide matrix resin," Selective Application of Materials for Products and Energy, Proceedings of the Twenty-third National SAMPE Symposium and Exhibition, Society for the Advancement of Material and Process Engineering, Azusa, CA, 1978, pp. 520-527. (Also NASA TM 74994.)

16. Vannucci, R. D., "Properties of PMR polyimide composites made with improved high strength graphite fibers, Materials 1980; Proceedings of the Twelfth National SAMPE Technical Conference, Society for the Advancement of Material and Process Engineering, Azusa, CA, 1980, pp. 15-30. (Also NASA TM 81557.)

7. BIOGRAPHIES

Dr. Frances Hurwitz is a polymer physical chemist in the Polymer Matrix Composites Section of the NASA Lewis Research Center. Her current research interests include characterization and structure/property relationships of polymeric resins for composites used at elevated temperature.

Dr. J. Daniel Whittenberger is a physical metallurgist in the Materials Development Section of the NASA Lewis Research Center. His research interests include characterization of high temperature mechanical properties of superalloys and nickel-based oxide dispersion strengthened alloys and properties of intermetallics, as well as mechanical properties of composite materials.

13th National SAMPE Technical Conference
October 13-15, 1981

THE PROPERTIES OF NOVEL BISIMIDE AMINE CURED EPOXY/CELION 6000 GRAPHITE FIBER COMPOSITES

Daniel A. Scola and Ruth H. Pater*
United Technologies Research Center
East Hartford, CT 06108

Abstract

A series of composites consisting ov novel bisimide amine cured epoxy resins reinforced with Celion 6000 graphite fibers were evaluated. The novel curing agents are represented by the formula:

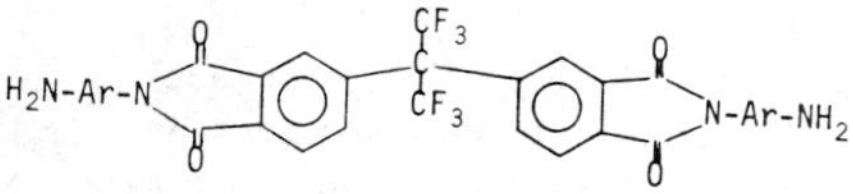

The processing characteristics of the composite systems were determined. The moisture absorption characteristics, glass transition temperature (150° and 177°C) shear and flexural properties of composites in the "as fabricated" condition and in the "wet" condition were determined. Two of the bisimide cured epoxy/Celion 6000 systems were found to have superior properties compared to state-of-the-art graphite epoxy systems.

1. INTRODUCTION

State-of-the-art epoxy resins are finding wide application as matrices in advanced fiber reinforced resin composites which are being used to fabricate primary and secondary structural components. However, the brittle nature and poor char forming characteristics of highly crosslinked state-of-the-art epoxy resins dictates that changes in molecular structure are necessary to improve their toughness, moisture resistance, and increase char forming characteristics. Toughness is required for resin matrices used in composite aerospace structures where high strain-to-failure, durability, and reliability are of prime importance. Moisture resistance is required to prevent moisture induced elevated temperature strength loss. High char yield resins are required to minimize the release of graphite fiber in the event the graphite/epoxy composite system is subjected to a fire-impact event, or some combination of fire and accident that could release graphite fibers.

The overall objective of this study was to further develop the bisimide amine cured epoxy resins first reported by investigators at NASA-Lewis[1].

This paper presents properties of the neat resins and composites made with these novel epoxy resins and Celion 6000 graphite fibers.

2. EXPERIMENTAL

2.1 Monomers and Resin Compositions. The monomers used in this

*Present address: NASA-Lewis Research Center, Cleveland, OH

study are listed in Table I. Tetraglycidylmethylene dianiline was purchased from a commercial source under the trade name Araldite MY 720[2]. The diamines, 4,4'-methylenedianiline (MDA), 4,4'-sulfonyldianiline (DDS), 4,4'-oxydianiline (ODA) and p-phenylene diamine (PDA) were obtained from commercial sources. The fluoro compound, 4,4'-hexafluoroisopropylidene-bis (phthalic anhydride) (6F), was isolated from DuPont's NR-150B2 polyimide resin solution according to the procedure of Alston[3].

The bisimide amine hardeners were prepared by refluxing the aromatic diamine (0.48 mole) and 6F (0.24 mole) in N-methyl-pyrrolidone (350 ml) for 4 hrs according to a published procedure[1,4]. The cooled reaction mixture was poured onto an ice-water mixture (500 ml), and the solid was filtered, washed with distilled water (5 x 100 ml), and dried in vacuum at 212°F to yield bisimide amine (6F-aromatic diamine) in 95-99% yield.

The compositions of nine resin systems studied are given in Table II.

2.2 <u>Preparation of Homogeneous Resin Powders</u>. Except for the control resin system (c) MY 720/DDS homogeneous resin powders were prepared by dissolving the epoxy resin (MY 720) and the curing agents in methylethylketone, and concentrating the mixture in a hood at room temperature to a solid powder. Because of the high reactivity of imide epoxies, residual solvent was removed in vacuum over a 2 hr period. The homogeneous powders prepared in this manner were used for the TGA and resin fabrication studies.

2.3 <u>Resin Fabriation</u>. Tensile specimens were prepared from resin C by pouring a solution of the resin into an open mold (ASTM D638-68, Type I size) and subjecting the specimens to the cure cycle listed in Table III. For IME-2, -3, -4, -6, -7, -8, a compression molding technique was utilized. The specimens were compression molded according to the details in Table III. For IME-1 and IME-5 no pressure was used to fabricate specimens. In addition to 1" diameter resin disks, 1 1/2" x 2.0° x 0.25" resin specimens of IME-1 and -5 were prepared.

2.3.1 <u>Resin Thermal Properties</u>
A thermomechanical analyzer (TMA) and a thermogravimetric analyzer (TGA) were used to determine the glass transition temperature (Tg) and weight losses in air as a function of temperature for each resin specimen.

The aerobic char yield of each resin sample was determined in air by the TGA technique up to a temperature of 800°C. A high temperature muffle furnace was also used for determination of aerobic char yield. The sample was placed in the oven at 800°C for a period of three (3) minutes. The char yield was determined as follows: % resin char yield = wt char after burning X100/wt resin before burning.

2.3.2 <u>Resin Mechanical Properties</u>
Tensile specimens of resin C were tested according to ASTM D638-68. Miniature tensile specimens of IME-1 and -2 were cut from 1 1/2" x 2" x 1/4" specimens. Final dimensions of these specimens were 2" long x 1/4" wide, having a 1/8" reduced section in the 1/2" wide gage section. All samples were tensile tested at a crosshead speed of 0.05 in./min.

2.4 <u>Celion 6000/Epoxy Tape and Composite Fabrication</u>.

2.4.1 Fabrication of Prepreg

Prepregs were fabricated by drum winding epoxy sized Celion 6000 graphite fibers followed by brush application of a solution of the resin in acetone to the dried fiber, calculated to yield a composite with a fiber volume of 60 ± 2%.

A 10% excess of resin was applied to account for loss in transfer and resin bleed-out during laminate fabrication. After application of the resin solution to the dry tape, the tape was allowed to stand at room temperature for 2 hrs. For resin C, IME-1, -2, -3 and -5, trace solvent was removed at room temperature in vacuum for 2 hrs to prevent advancement of the resins. The tape was cut in plies and stacked unidirectionally in a mold for the compression molding process.

2.4.2 Fabrication of Composites by Compression Molding Procedure

An open ended mold was used to follow the gelation point by probing the laminate as a function of time at specific temperatures. The composite fabricated and the conditions used in the fabrication process are listed in Table IV. These composites were molded to a constant fiber volume for better control of the final composition (fiber 60 vol %, resin 40 vol %).

2.5 Composite Characterization.

2.5.1 Physical Properties

The density of each composite was determined by the liquid displacement technique. Fiber, resin and void volumes were determined by the acid digestion method. Optical micrographs were determined on polished cross sections of 1" wide specimens. The same equipment described in Section 2.3.1 was used to determine the thermal properties of composites.

2.5.2 Char Forming Properties

The char yields were determined in air by the TGA technique up to a temperature of 800°C. A high temperature muffle furnace was also used to determine aerobic char yield. The sample was placed in an oven at 800°C for a period of three (3) minutes. The following calculation was used for composites charred in the muffle furnace.

$$\text{\% resin char yield} = \frac{\text{Wt char after burning}}{\text{Wt resin before burning}} \text{X100}$$

Wt resin before burning was determined by H_2SO_4/hydrogen peroxide digestion method[5]. The fiber containment measurements were carried out on muffle furnace charred composite specimens immediately after removal from the 800°C furnace. The charred specimens were suspended between two steel supports at the bottom of a 1/8" thick x 2" diameter x 15" long steel tube. A 100 gram weight was released to impact the hot charred specimen. The weight of the impacted debris was determined. The composition of the composite (fiber weight) was also determined on another sample by the H_2SO_4/hydrogen peroxide method before and after burning and the fiber containment was calculated as follows:

$$\text{\% fiber contaiment} = \frac{\text{Wt fiber after burning}}{\text{Wt fiber before burning}} \text{X100}$$

2.5.3 Mechanical Properties

Interlaminar shear strength test. The short beam shear test was conducted at a span-to-depth ratio of 4:1. The loading nose was 1/4" diameter and the support pins were 3/16" diameter. Crosshead speed was 0.05"/min. Nominal specimen dimensions were 1/4" wide by 1.2" long x 0.1" thick.

Flexural Test. A three point bend test at a span-to-depth ratio of 20:1 was conducted on composite specimens. Crosshead speed was 0.05 in./min. Specimen dimensions were nominally 1/4" wide x 2.2" long x 0.1" thick.

3. RESULTS AND DISCUSSION

3.1 Resin Properties

3.1.1 Physical Properties

To determine moisture absorption characteristics of each resin system, specimens (∿ 1/4 section of a 1" diameter specimen) were subjected to 87% RH, 82°C for approximately 8 days. The results of the moisture gained due to the exposure are listed in Table V. A comparison of the moisture absorbed by the IME resins with a typical epoxy under the same condition shows that the IME resins absorb considerably less moisture than the control resin and typical epoxy resins.

Table V also shows the glass transition temperatures of the neat resin specimens. Two thermal transitions were observed in the TMA thermograms of all resin systems, except IME-5, which exhibited a single transition. The first Tg varied from 184° for the control to 211°C for IME-8 sample, and the second Tg ranged from 212°C to 227°C. Contrary to the work published previously, replacement of DDS curing agent in the control resin with a bisimide amine does not lower the glass transition temperature of the resin, even though the bisimide amine curing agent contains the flexibilizing unit

$$-\underset{CF_3}{\overset{CF_3}{\underset{|}{\overset{|}{C}}}}-$$

in the backbone of the polymer network.

The aerobic char yields of the resins were determined by two techniques: (1) exposure in a muffle furnace at 800°C for 3 min, and (2) TGA weight loss up to a temperature of 800°C in air. The data listed in Table V shows that the char yields of the IME resins are considerably greater than that experienced with the control resin in complete agreement with the report published[1].

3.1.2 Tensile Properties

Because of the difficulties encountered in fabricating resin specimens from the IME resin systems, miniature resin specimens 2" long x 1/4" wide, having a 1/8" reduced section in the 1/2" gage length were fabricated for IME-1 and -5. Standard ASTM resin specimens were prepared for the control resin system. The tensile properties of these resins are listed in Table VI along with state-of-the-art epoxy resin (Ciba-Geigy 6350) for comparison purposes. Based on strength and strain-to-failure, resin system IME-1 containing 6F-DDS appears to be tougher than the control resin or IME-5 with 6F-DDS and DDS.

3.2 Composite Properties

3.2.1 Physical Properties

Table VII lists the physical properties of compression molded 9 ply composites (6" x 10" x ∿ 0.100").

Using precut flexural specimens, the equilibrium moisture absorption varied from 0.3 to 0.7 weight percent after exposure for approximately 8 days at 82°C at 87% relative humidity. Like the neat resin specimens mentioned previously, the composites based on the IME epoxy resins absorbed considerably less water than did the control sample.

As shown in Table VII, all of the composites exhibited two glass transition temperatures as observed for the neat resin specimens discussed previously. The first Tg's of the composites (Table VII) compared favorably with the first Tg's of their corresponding neat resin samples (Table V). However, the second Tg's of the composites are significantly higher than the second Tg's of the corresponding resins.

3.2.2 Char Forming Properties

Two techniques were utilized to determine the char forming characteristics of the Celion 6000 epoxy resin composites. The first technique utilized the TGA method, whereby the weight loss as a function of temperature was determined up to a temperature of 800°C in air. In the second technique, the char yield was determined after a 3 min exposure in a muffle furnace at 800°C. The results of the tests listed in Table VIII show that about all IME systems exhibited good char forming capabilities, with IME-1 and -3 exhibiting better char yield than the other resin systems. For the composites investigated, the fiber containment yield of the composites was essentially 100%, showing that the char encapsulated the fiber, thereby preventing fiber release.

3.2.3 Mechanical Properties

3.2.3.1 Shear Properties. The shear strength properties of each composite system in the dry and humidity exposed conditions are listed in Table IX. In general, the dry shear strengths at RT (11 to 16 ksi), at 150°C (6 to 11 ksi) and at 177°C (5 to 9 ksi) exhibited by these composites are considered quite good for new systems. A 24 hr post-cure at 204°C caused considerably improvement in the dry 177°C shear strengths. Therefore, postcuring is necessary in order for these composite systems to have a 177°C temperature capability. The effect of moisture on RT and elevated temperature shear properties can be seen by comparing dry strength with wet strength. Composite systems with resins IME-1 and IME-5 exhibited the best wet strength retention of the composite systems investigated.

3.2.3.2 Flexural Properties

The flexural properties of each composite system in the dry and humidity exposed conditions are listed in Table X. In general, the room temperature flexural properties exhibited by these experimental composites are considered acceptable relative to state-of-the-art epoxies. However, only composites containing the novel resins IME-1 and IME-5 exhibited acceptable 150°C and 177°C flexural strengths relative to commercial Celion 6000/epoxy resin composites. Postcuring of the composites for 24 hrs had no apparent effect on the RT flexural properties, but caused a large increase in the 177°C flexural strengths.

The effect of moisture on the flexural properties at RT and elevated temperatures can be seen by comparting the dry flexural strength data with the wet flexural data for the corresponding composites. Similar to what was observed in the shear strength data, examination of the flexural data reveals that composites containing novel resins IME-1 and -5 show the least moisture effects on the RT, 150°C, and 177°C flexural properties among all of the composite systems investigated.

4. SUMMARY OF RESULTS

As reported previously[1], the use of bisimide amine hardeners to cure epoxy resins causes a significant improvement in the char forming characteristics of the cured resin. Compared to epoxy resins cured with a conventional amine, such as 4,4'-diaminodiphenylsulfone, the bisimide cured epoxies exhibit about a twofold improvement in the aerobic char yield at 800°C.

The strain-to-failure of the MY 720/6F-DDS resin system was improved 25% by replacement of DDS with 6F-DDS (1.25% versus 0.99%).

The levels of moisture absorption exhibited by the bisimide amine cured epoxies were considerably lower than those exhibited by state-of-the-art epoxies (0.5 wt% versus 2.0%).

The processing characteristics of the bisimide amine cured epoxy systems are similar to conventional aromatic amine cured resins.

Two of the composite systems Celion 6000/IME-1 and Celion 6000/IME-5 were found to have superior properties compared to the control and to the other Celion 6000/IME composite systems. In the dry condition, IME-1 and IME-5 exhibited shear strengths of 10.5 ksi at 150°C compared to 9.86 ksi for the control and 6.80 ksi or less for the other IME composites. Similarly, the dry flexural strengths of IME-1 and IME-5 are 128 ksi and 153 ksi, respectively, at 177°C, while for the control system the flexural strength is 120 ksi and considerably less for the other IME composite systems.

This was further demonstrated by the excellent values for the shear and flexural strength at the elevated temperatures (150°C and 177°C) after exposure to moisture. For example, at 150°C, IME-5 exhibits a shear strength of 7.27 ksi while the control composite shear strength is 5.92 ksi. At 177°C, IME-1 exhibits a shear strength of 5.00 ksi while the control shear strength is 4.09 ksi. At 177°C, the flexural strength of IME-5 is 124 ksi while that of the control composite is 106 ksi.

These results suggest that the use of bisimide amine hardeners is a promising approach for improving toughness, moisture resistance and char yield of epoxy resins used as matrices for advanced composites.

5. REFERENCES

1. Serafini, T. T.; Delvigs, P.; and Vannucci, R. D., National SAMPE Tech. Conf., Boston, MA, SAMPE, Vol. 11, p 564, 1979.

2. Araldite MY 720, Technical Product Data, Ciba-Geigy Corp.

3. Alston, W., NASA Lewis Research Center, private communication.

4. Scola, D. A. and Pater, R. H. Final Report NASA CR-165229, Nov. 15, 1980.

5. Haynes, W. M. and Tolbert, T. L., J. Composite Materials, Vol. 3, p 709, October 1979.

6. Augl, J. M. and Bergen, A. W., Tech. Report NSWC/WOL/TR 76-7, Sept. 23, 1975.

7. Augl, J. M., NSWC TR 79-39, March 30, 1979.

8. Augl, J. M., NSWC/WOL TR 76-149, Feb. 3, 1977.

6. BIOGRAPHIES

DR. DANIEL A. SCOLA

Dr. Daniel A. Scola received his B.A. degree from Clark University in 1952 and an M.A. degree in Chemistry from Williams College in 1954. His graduate studies were continued at the University of

Connecticut where he received a Ph.D. in Chemistry in 1958. Dr. Scola was associated with Monsanto Research Corp. in Everett, Mass. for six years and as a Group Leader, supervised the synthesis of various types of organic compounds and conducted catalytic hydrocracking studies of polyphenyl hydrocarbons. He was associated with Norton Company for about one year during which time he studied the interaction of phenolic resins with constituents of organic bonded abrasive wheels. Currently, Dr. Scola is a Senior Materials Scientist at United Technologies Research Center where he is responsible for conducting research in advanced composite materials including interfacial bonding studies, synthesis and evaluation of resin systems and composite materials evaluation.

DR. RUTH H. PATER

Dr. Ruth H. Pater received a B.S. degree in chemistry from Tamkang College, an M.S. degree in Organic Chemistry from Southeastern Massachusetts University, and a Ph.D. in Organic Chemistry from Brown University in 1977. For two years, she was associated with the Materials Sciences Laboratory of United Technologies Research Center as a research scientist. Her research interests include composite surface analysis, interface studies, and epoxy and polyimide resin synthesis and characterization as well as composite fabrication, testing and evaluation. Dr. Pater has published in areas of advanced composite technology encompassing the resin matrix, the interface and fiber surfaces.

Presently, she is a research scientist at the NASA-Lewis Research Center which she joined in 1980. At NASA, Dr. Pater is conducting research in advanced fiber reinforced composites with emphasis on high temperature resin matrices.

7. ACKNOWLEDGEMENT

This work was supported by NASA Lewis Research Center. Dr. Tito L. Serafini was the NASA Program Manager.

TABLE I

Structures of Hardeners and Epoxy Resin

Structure	Designation
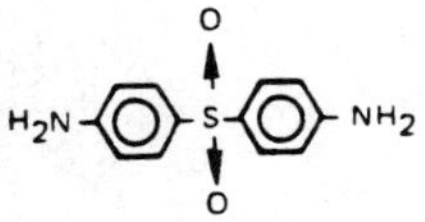	DDS
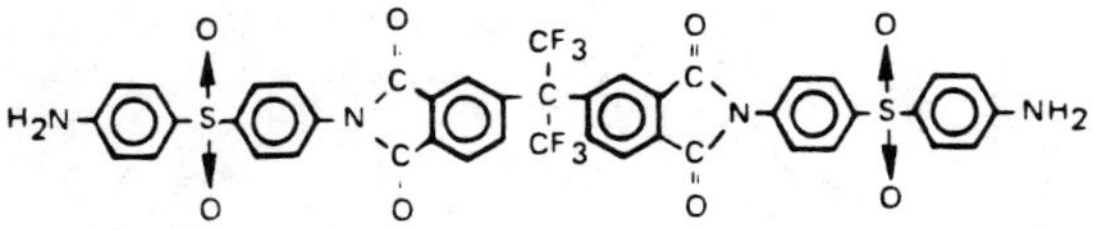	6F-DDS
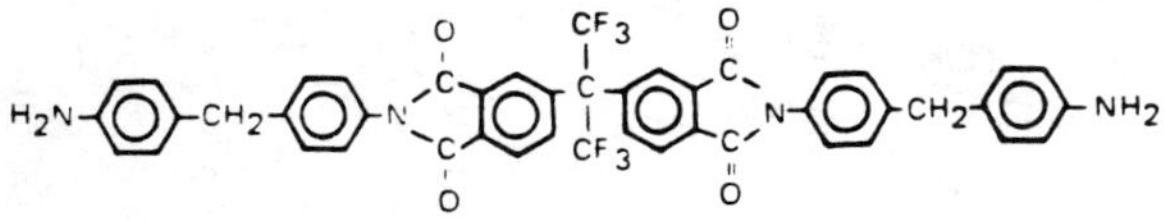	6F-MDA
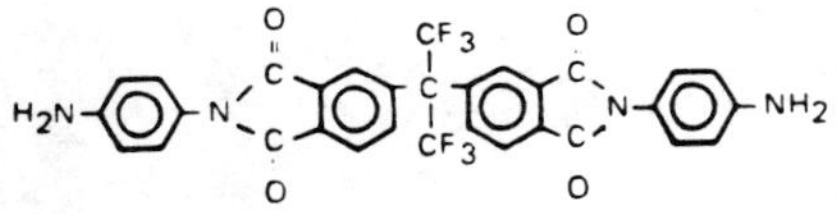	6F-PDA
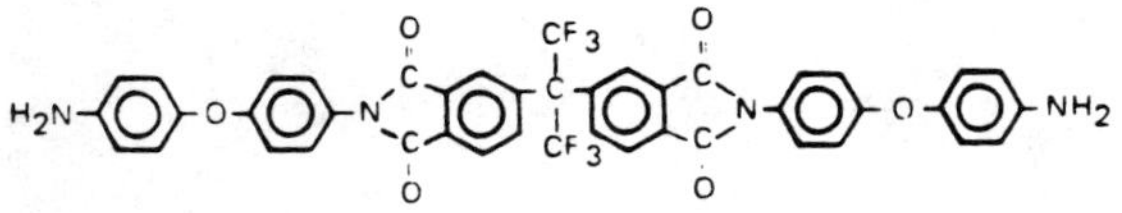	6F-ODA
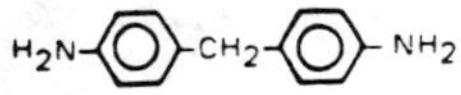	MDA
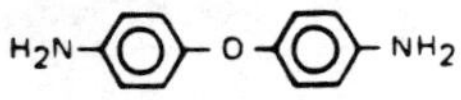	ODA
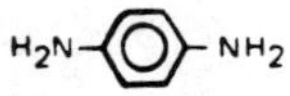	PDA
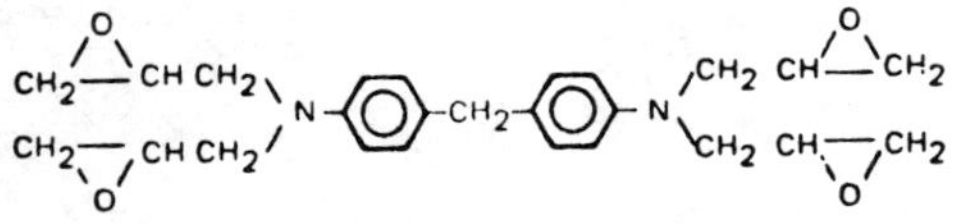	MY 720

TABLE II

Chemical Compositions of Neat Resin Specimens

Resin System No.	Resin System Epoxy/hardener(s)	Mole Ratio of Epoxy/hardener(s)
Control	MY 720/DDS	0.012/0.012
IME-1	MY 720/6F-DDA	0.0064/0.0064
IME-2	MY 720/6F-MDA	0.0069/0.0069
IME-3	MY 720/6F-ODA	0.0069/0.0069
IME-4	MY 720/6F-PDA	0.008/0.008
IME-5	MY 720/6F-DDA/DDA	0.0084/0.0042/0.0042
IME-6	MY 720/6F-MDA/MDA	0.009/0.0045/0.0045
IME-7	MY 720/6F-ODA/ODA	0.0088/0.0044/0.0044
IME-8	MY 720/6F-PDA/PDA	0.0104/0.0052/0.0043

TABLE III

Processing Parameters to Fabricate Epoxy Resin Specimens

Specimen No.	Initial Sample Preparation	Cure Cycle[1]
Control resin MY 720/DDS	Mix MY 720 + DDS, heat to 150°C, pour liquid in preheated (150°C) mold	150°C/1 hr + 177°C/2 hrs + 204°C/2 hrs.
IME-1	Removed solvent at 70°C, 2 hrs in vacuum to produce resin powder	70°C vacuum, raise T to 150°C, hold 1/2 hr, raise T to 180°C/2 hrs, raise T to 204°C, hold 204°C/2 hrs.
IME-2	Same process as IME-1	Compression molded. Placed in preheated press at 180°C, consolidated sample at 100 psi and removed pressure, flow occurs at 102°C, raise T to 150°C, gelation occurs immediately, applied pressure (100 psi), raised T to 180°C, hold 180°C/2 hrs, raise T to 204°C, hold 204°C/2 hrs
IME-3	Same as IME-1	Same as IME-2
IME-4	Removed solvent at RT in vacuum over 72 hr period to produce resin powder	Same as IME-2
IME-5	Same as IME-1	50°C in vacuum, raise T to 150°C, hold 1/2 hr, release vacuum, hold 150°C 1 hr, raise T to 180°C, hold 180°C/2 hrs, raise T to 204°C, hold 204°C/2 hrs
IME-6	Same as IME-4	Same as IME-2
IME-7	Same as IME-4	Same as IME-2
IME-8	Same as IME-4	Same as IME-2

[1] All samples were postcured at 204°C for 24 hrs

TABLE IV

Celion 6000 Resin Composites Compression Molded
(15.24 cm x 25.4 cm x 0.25 cm)
9 plies

Composite No.	Processing Parameters
33-C	RT→148°C, contact P (1.7°C/min) then at 148°C application of press. (1.38 MPa) to STOPS 148°C→177°C (0.8°C/min), hold, 177°C/2 hrs (pressure) + 204°C/2 hrs (air circulating oven)
20-IME-1	RT→150°C, contact P vacuum (1.4°C/min) at 150°C application of press. to STOPS, 150°C→177°C (1.7°C/min), hold 177°C/2 hrs (press. 1.38 MPa) + 204°C/2 hrs (air circulating oven)
32-IME-2	RT→136°C, contact P (2.0°C) at 136°C application of press. (1.38 MPa) to STOPS, 136°C→177°C (1.1°C/min), hold 177°C/2 hrs (press. 1.38 MPa) + 204°C/2 hrs (air circulating oven)
28-IME-3	RT→126°C, contact P (2°C/min), at 136°C application of press. (1.38 MPa) to STOPS, 126°C→177°C (1.4°C/min), hold 177°C/2 hrs (press. 1.38 MPa) + 204°C/2 hrs (air circulating oven)
29-IME-4	RT→145°C (1.9°C/min) 0.69 MPa initially, release P, 145°C→150°C, (1°C/min) (contact), at 150°C application of press. (1.38 MPa) to STOPS, 150°C→177°C (0.5°C/min), hold 177°C/2 hrs (press. 1.38 MPa) + 204°C/2 hrs (air circulating oven)
24-IME-5	RT→155°C, contact P (1.8°C/min), at 155°C application of press. (1.38 MPa) to STOPS, 155°C→177°C (1°C/min), hold 177°C/2 hrs (press. 1.38 MPa) + 204°C/2 hrs (air circulating oven)
31-IME-6	RT→140°C contact P (2.0°C/min) at 140°C application of press. to STOPS, 140°C→178°C (1.5°C/min) (press. 1.38 MPa) hold 178°C/2 hrs, press. 1.38 MPa + 204°C/2 hrs (air circulating oven)
26-IME-7	RT→125°C contact P (2°C/min), at 125°C application of press. (1.38 MPa) to STOPS, 125°C→178°C (2.26°C/min), hold 178°C/2 hrs (press 1.38 MPa) + 204°C/2 hrs (air circulating oven)
27-IME-8	RT→80°C contact P (1.7°C/min), 80°C→115°C (2.5°C/min), at 115°C application of press. (1.38 MPa) to STOPS, 115°C→179°C (2.8°C/min), hold 177°C/2 hrs (press. 1.38 MPa) + 204°C/2 hrs (air circulating oven)

TABLE V

Physical Properties of Epoxy Resins[3]

Resin System	Moisture[4] Absorbed	Glass[5] Transition Temp. °C First	Glass[5] Transition Temp. °C Second	Aerobic Char Yield: Muffle[6] Furnace Method	Aerobic Char Yield: TGA[7] Method
Control	3.4	184	227	20.7	20
IME-1	2.4	196	215	39.3	39
IME-2	1.6	185	215	33.8	40
IME-3	1.4	192	220	39.9	44
IME-4	1.8	207	-	29.4	45
IME-5	1.8	192	220	33.2	40
IME-6	1.8	193	217	35.8	42
IME-7	2.0	190	212	31.3	34
IME-8	2.5	211	-	30.8	43
Narmco 5208[1]	4.5 (80% RH, 75°C)				
3501-6[2]	4.5 (87% RH, 50°C)				

1. Ref. 6
2. Ref. 7
3. Postcured at 204°C for 24 hrs
4. Saturation at 82°C, 87% RH, Wt %
5. Determined by thermomechanical analysis
6. High temperature muffle furnace for 3 min at 800°C
7. Thermogravimetric analyzer in air at 800°C

TABLE VI

Tensile Properties of Epoxy Resins[1]

Resin System	Strength psi	Strength MPa	Modulus 10^6 psi	Modulus MPa	Strain-to-Failure %
MY 720/DDS (control)	5966[2]	41.1	0.60	4.13	0.99
IME-1 MY 720/6F-DDS	7870[3]	48.7	0.55	3.76	1.25
IME-5 MY 720/6F-DDS/ DDS	6500[3]	44.8	0.65	4.82	0.99
Ciba-Geigy 6350 MY 720/Dicyan-diamide	4643[2]	29.9	0.63	4.34	0.80

1. Postcured at 204°C for 24 hrs
2. An average of five specimens
3. An average of two specimens

TABLE VII

PHYSICAL PROPERTIES OF CELION 6000/EPOXY RESIN COMPOSITES[1]

Composite No.	Density g/cc	Composition, Vol % Resin	Fiber	Void	Moisture Absorbed, wt%[2]	Glass Transition Temp. °C[3] First	Second
Control	1.53	36.1	61.4	2.6	0.69	230	275
20-IME-1	1.58	38.2	62.6	1.5	0.56	175	285
32-IME-2	1.52	40.3	57.7	2.2	0.32	175	275
28-IME-3	1.54	42.8	55.5	1.8	0.41	195	235
29-IME-4	1.56	41.8	56.9	1.3	0.57	160	240
24-IME-5	1.55	34.9	62.1	2.9	0.64	185	225
31-IME-6	1.55	42.5	57.1	0.27	0.42	185	225
26-IME-7	1.49	38.8	56.4	5.2	0.39	180	250
27-IME-8	1.52	39.8	57.4	2.8	-	180	245

1. Postured at 204°C for 24 hrs
2. Saturation at 82°C, 87% RH for 8 days
3. Determined by thermomechanical analysis

TABLE VIII

CHAR FORMING PROPERTIES OF CELION 6000/EPOXY RESIN COMPOSITES[1]

Composite System	Wt % Retention Due to Pyrolysis: TGA Method[2]	Wt % Retention Due to Pyrolysis: Muffle Furnace Method[3]	Wt % Resin Char Yield in Composite	Wt % Fiber Containment
Control	85	84	15	99
IME-1	86	87	46	101
IME-2	83	75	41	111
IME-3	85	83	47	104
IME-4	84	80	35	100
IME-5	86	84	33	95
IME-6	87	75	41	103
IME-7	83	81	44	106
IME-8	83	78	30	94

1. Postcured at 204°C for 24 hrs
2. Thermogravimetric analyzer in air at 800°C
3. High temperature muffle furnace for 3 min at 800°C

TABLE IX

SHORT BEAM SHEAR STRENGTHS[1] OF CELION 6000/EPOXY RESIN COMPOSITES[2,3] IN DRY AND HUMIDTY EXPOSED[4] CONDITIONS

Composite No.	Short Beam Shear Strength, ksi					
	Dry Condition			Humidity Exposed		
	RT	150°C	177°C	RT	150°C[5]	177°C[5]
Control	13.70	9.86	8.44	9.73	5.92	4.09
20-IME-1	14.80	10.60	9.37	10.55	5.91	5.00
32-IME-2	12.30	5.71	4.52	9.63	4.21	2.51
28-IME-3	12.20	6.80	5.79	10.15	4.12	3.31
29-IME-4	10.50	8.32	7.79	9.51	4.76	3.91
24-IME-5	16.15	10.45	7.63	13.60	7.27	4.49
31-IME-5	12.20	6.80	5.26	9.56	4.39	3.18
26-IME-7	11.20	6.80	5.95	9.61	4.90	3.80
27-IME-8	12.95	6.30	5.23	11.40	5.15	3.13

1. Span-to-depth ratio, 4/1
2. Composites were compression molded, for cure cycle see Table IV
3. Composites were postcured at 477 K (204°C) for 24 hrs
4. 87% RH at 355 K ([illegible]C) to saturation
5. after 20 min soak at temperature

TABLE X

Flexural Properties[1] of Celion 6000/Epoxy Resin Composites[2,3] in Dry and Humidity Exposed[4] Conditions

	Dry Condition						Humidity Exposed					
	RT		150°C[5]		177°C[5]		RT		150°C[5]		177°C[5]	
Composite No.	Strength ksi	Modulus 10^6 psi	Strength ksi	Modulus 10^6 psi	Strength ksi	Modulus 10^6 psi	Strength ksi	Modulus 10^6 psi	Strength ksi	Modulus 10^6 psi	Strength ksi	Modulus 10^6 psi
Control	248	15.9	160	15.4	120	16.6	216	16.4			106	14.6
20-IME-1	248	16.9	150	15.5	128	14.7	231	16.2	148	14.0	109	13.0
32-IME-2	196	15.1	74.8	14.4	43.4	9.63	216	17.1	98	15.5	52	9.00
28-IME-3	224	15.4	104	14.3	73.6	12.5	208	16.7	134	15.8	53	7.10
29-IME-4	218	14.1	113	14.9	92.5	14.4	218	16.0	105	14.2	90	12.8
24-IME-5	246	17.9	183	17.1	153	19.7	239	15.6	152	15.6	124	14.6
31-IME-6	221	15.9	106	14.7	76.2	13.6	228	18.1	119	14.1	93.5	12.7
26-IME-7	223	16.9	87.6	12.8	69.8	13.2	165	15.2	116	13.3	74.7	11.9
27-IME-8	203	14.7	125	14.0	91.7	13.1	191	15.4	84.6	15.6	90.7	12.2

1. Three point flex test at a span-to-depth of 20/1
2. Composites were compression molded composites; for cure cycle, see Table IV
3. Composites were postcured at 204°C for 24 hrs
4. 87% RH at 355 K (82°C) to saturation
5. After 20 min soak at temperature

13th National SAMPE Technical Conference
October 13-15, 1981

EFFECTS OF MOISTURE DEGRADATION: MOLECULAR STRUCTURE OF COMPOSITE RESIN SYSTEMS

W.D. Humphrey*, S.H. Liu**, N.C. Plass*, D.C. Timm**

* Brunswick Corporation
Defense Division
Lincoln, Nebraska

** University of Nebraska-Lincoln
Department of Chemical Engineering
Lincoln, Nebraska

ABSTRACT

Kevlar 49, filament wound, high performance composites are being subjected to humidity aging. Resin systems include lightweight t-butylstyrene-polybutadiene copolymer formulations and epoxy formulations. Test specimens are NOL rings and 6.7 in. diameter by 10.3 in., midcylinder failure mode, pressure vessels. Molecular distribution analysis incorporates numerical interpretation of chromatograms obtained through gel permeation chromatography. Data include weight distributions for low molecular weight extractables leached from the thermoset composite. These materials encompass unreacted raw materials, products of moisture degradation, and low molecular weight prepolymers including telomers and oligomers. The effect of aging is apparent as observed by the dynamic, time dependent results. An epoxy formulation LRF 092 contains a significantly higher level of extractables after aging. A second epoxy formulation LRF 216 is relatively stable. A third epoxy formulation EXP 117 cures to a higher extent during humidity aging. Hydrocarbon resins are observed to contain distinct distributions with respect to unreacted monomers and telomers, yielding flexible resins to rigid resins. For four systems of similar monomer and telomer distributions, short beam shear is observed to increase with a decrease in the average molecular weight of the telomer fraction. This molecular distribution correlates with changes in crosslink density.

"Keywords": Humidity-aging; Composite; Molecular-distribution; Filament-winding; Chromatography; Kevlar; Hydrocarbon; Epoxy.

1. INTRODUCTION

Resin castings and Kevlar 49 composites were subjected to controlled humidity aging tests. Resin formulations include epoxy and experimental lightweight hydrocarbon systems which are comprised principally of polybutadiene/t-butylstyrene monomers. Solvent leaching of cured materials, subject to analysis by gel permeation chromatography, yields descriptive molecular detail for the extractable fractions. These molecules are precursors for the network resin fraction, and, therefore, are descriptive, in a kinetic molecular sense, of the crosslinked material. The majority of the thermoset resin is insoluble.

Numerical analysis(1) of chromatograms identifies mass fractions and molecular weights of the solvent soluble fraction. Soluble materials observed include unreacted monomers and relatively low molecular weight prepolymers or telomers. Molecular distributions are observed to be dependent on specimen humidity aging.

Polymerizations in the presence of fiber provide distinctive molecular distributions of extractables compared to data analysis of original resin castings(2). Therefore, competing chemical reactions with the fiber and/or sizing agent exist. Previous studies(3), utilizing differential scanning calorimetry, have reported similar findings. The current fiber contains a minimum of sizing which is applied to control static electricity in manufacture.

2. RESIN AGING

Test specimens evaluated were subjected to prolonged humidity exposure at 160°F and 95% relative humidity. Samples include a) original cure materials and materials after b) twenty-one days. Future data will include materials aged to fourteen months. Specimens include filament wound NOL rings and 1/10th scale MX, 1st stage motor cases. Table 1 provides a description of the cure cycle for the several resin systems.

3. EPOXY RESINS COMPOSITES

2.1 LRF 092

Normalized chromatograms observed for the originally cured and for the 21 day aged composite are presented by Figure 1. Data of Table 1 identify this resin as an oxirane, anhydride cure, which yields intramolecular ester linkages. Results tabulated in Table 2, indicate the elution mark (or volume) of the chromatogram peak of the several monomeric species. Water produces a negative response in the proximity of 130 marks (26.0 ml of eluent). The t-amine catalyst BDMA elutes as a positive peak with a maximum also at 130 marks. The anhydride's peak occurs at 118 marks and the major component of Epon 828, diglycidyl ether of bisphenol A, is a maximum at 102 marks. A material eluting similar to the anhydride NMA was added at a peak elution volume 115. This standard significantly improved the numerical fit(1) of the original chromatogram.

A visual inspection of the chromatograms for the resin system LRF 092, coupled with the mass fractions of component species in Table 2, indicates a significant percentage of material elutes at 115 marks. This material, being similar in molecular weight to NADIC methyl anhydride, indicates that stoichiometric quantities of unreacted Epon 828 & NMA may not exist. A fiber/sizing monomeric reaction may be producing a relatively stable intermediate. Research[2] on resin castings indicates a greater fraction of unreacted Epon 828 in the absence of fiber. A closer stoichiometric balance during the cure will yield an improved shear strength, increased crosslinking.

The effect of humidity aging on the resin system LRF 092 is also apparent in Figure 1 and in Table 2, Particularly in the neighborhood of 125 marks. Moisture, which produces a negative response at 130 marks, can cause the observed shift at 120 to 130 marks. Data of Table 2 indicate an increase in moisture.

Specimen testing yields the data of Table 3. All percentages are given as the observed minus the original property divided by the original property. The cumulative absolute area of the chromatogram per gram of resin content within the composite is an indicator of the quantity of material being leached from the insoluble resin fraction. The sample, after aging, contains a significant increase in extractables, an indication of moisture degradation (hydrolysis of the ester linkage catalyzed by the acid of the anhydride).

Short beam shear of sections from the naval ordinance rings results in a significant reduction in shear strength. An efficiency for pressure vessels is reported as the product of pressure P and volume V and inversely proportional to the weight of the composite W_c. Hoop fiber tensile failure for the vessel is also reported. Tensile failure is relatively stable compared to short beam shear failure.

3.2 LRF 216

A second epoxy system, LRF 216, utilizing a different catalyst, is also reported (the conditions for the cure are reported in Table 1). Normalized chromatograms for the originally cured composite and for the composite after humidity aging are presented in Figure 2. Numerical analysis results are presented in Table 2. The sum of material eluting at 118 and 115 marks is relatively constant, but in stoichiometric excess when compared to the weight fraction for Epon 828. This excess is consistent with that discussed for the LRF 092 formulation. The negative peak at 130 marks is reported as the catalyst, but could be moisture. The test results of Table 3 indicates that LRF 216 is stable, except for the drop in short beam shear.

3.3 EXP 117

The third epoxy system utilized the catalyst EMI-24, 2 ethyl-4 methyl-imidazole, which is reported to have a high crosslinking efficiency and heat distortion temperature[(4)]. Chromatography results, Figure 3, produced results of a system whose molecular structure is significantly altered by humidity aging, resulting in a distinct shift in mass fractions (Table 3) from monomeric species to telomers. The peak at 130 marks is reported as the catalyst, but may also contain moisture. The standard at 108 marks was added to improve the numerical fit.

Data of Tables 2 and 3 indicate that aging advanced the cure, yielding fewer extractables of higher average molecular weight as indicated by relative chromatogram areas. A significant reduction in shear strength again occurred. A small reduction in hoop fiber stress was observed for the pressure vessel.

4. HYDROCARBON RESIN COMPOSITES

Major objectives for the resin component in a pressure vessel are to distribute the tensile load and to maintain dimensional stability. A potential economic advantage of a hydrocarbon system is related to a reduction in resin specific gravity. The major constitutive resin components include free radical initiators and the co-monomers. The current, proprietary resin formulations are supplied by Lockheed. One resin formulation EXP 105 is an inhouse formulation which utilizes 1-2 polybutadiene, for which information is supplied in Table 1. Formulation variables include quantities and average molecular weight of monomers and initiator. Cure variables include time and temperature.

4.1 EXP 105

Figures 4 and 5 present chromatograms observed for two resin formulations, distinguished by the average molecular weights of the polybutadiene component, EXP 105A and B. The reader will observe a significant variance in the distribution of telomers which elute at marks <105 leached from the composit. Resin A is flexible; resin B is rigid. Soluble polymeric fractions eluting before 70 marks are linear or branched polymeric materials of substantial molecular weight. The molecular distribution of telomers indicates a relatively low level of crosslinking and a relatively high average molecular weight between network segments. Therefore, the resin is relatively elastic with a low shear strength (see Table 3).

The fraction of soluble materials in formulation A is about half that of resin B, but due to refractive index/compound dependence, a direct comparison with epoxy resins should not be made. Inspection of the four chromatograms coupled with data of Table 3 indicates opposite effects with respect to aging.

Resin A contains a greater fraction of extractables after exposure, whereas resin B is stable, though molecular distribution shifts occur on both Figures 4 and 5. The negative response near 130 marks is reported as water.

4.2 EXP 110

This composite system was cured subject to a nitrogen and atmospheric blanket and a protective, surface film. Figures 6-8 and data of Tables 3 and 4 present results. Similarities in chromatograms and resultant molecular distributions are observed for specimens A and C, both as originally cured and after humidity aging (see Table 4). Chromatogram B shows relatively less t-butylstyrene elutes at 117 marks. Low molecular weight oligomers elute in the interval of 90 to 110 marks. Aging results in subtle molecular changes.

A comparison of chromatographic test results with EXP 105 shows that formulation 110 yields significantly less extractables in the region of 50-70 marks. Thus the resin is of higher crosslink density. However, the substantial increase in chromatogram area (Table 3) indicates that the cure is less than that of EXP 105A. The shear strength of EXP 110A is substantially improved, but the lower cure level results in a rubbery, plasticized material. Aging advanced the cure as observed by a reduction in extractables. The efficiency factor PV/W_c is less than that for the epoxy formulations.

4.3 EXP 112

An apparent formulation change in polybutadiene produces a distinctive bimodal distribution of Figure 9, which resembles EXP 105A, except the oligomer's molecular weight is less. Humidity aging yields a higher moisture content and a relatively stable molecular distribution on the normalized chromatogram. Aging advances the curve. Results in Table 3 are similar to those of EXP 105. Chromatographic analysis indicates a more complex formulation, see Table 4.

4.4 EXP 114

Original chromatographic data of EXP 114 resemble that of formulations 105A and 112, except that the molecular weight of the polymeric material has been reduced. Aging yields a higher moisture content, shifting substantially the t-butylstyrene peak. Chromatographic area of Table 3 indicates humidity aging advanced the cure, and also the molecular weight of telomers, of Figure 10. The short beam shear is approaching that of the epoxy systems. The pressure vessel efficiency factor is substantially improved. The initial cure, in terms of chromatographic area, is similar to EXP 112. The increase of crosslink density, which correlates with lower molecular weight of polymeric extractables, may be contributing substantially to improved shear strength. Chromatography analysis presented in Table 4 includes additional components to improve the fit.

4.5 EXP 119/120

EXP 119 formulation may have resulted in a substantial increase in polybutadiene content. The resultant resin is a rubbery cured composite. A Comparison of normalized chromatograms, Figures 11 and 12, suggests that formulation EXP 120 is exceedingly high in t-butylstyrene content relative to polybutadiene. The result, see Table 3, is a rigid resin with improved shear strength. Aging further cures the resin, reducing chromatographic area and increasing molecular weight of telomers into high molecular weight soluble polymers. The system is relatively moisture resistent at 21 days.

5. DISCUSSION

Chromatography studies have been utilized to fingerprint cured resins in an attempt to gain insight into molecular distributions of soluble resin fractions. Knowledgeable predictions can then be made for the insoluble fraction with respect to crosslinking. The ultimate objective is to correlate molecular structure with physical properties.

Originally cured hydrocarbon resins EXP 105A, 112, 114 and 120 provide data at relatively constant levels of cure and telomer distribution. Average molecular weight changes in the telomers are observed to correlate with short beam shear strength. As the molecular weight of telomers decreases, shear strength increases. Inasmuch as these molecules are precursors of macromolecules within the network structure, reductions in average molecular weight of telomers indicates increasing average molecular weight within the insoluble cured resin, i.e. higher crosslink density, higher average molecular weight. Shear strength increases with increasing crosslink density[4].

Comparison of data in press[2] for resin castings with the current observations clearly shows that Kevlar 49 fiber has a pronounced effect on cure. In epoxy formulations, the oxirane content of extractables is significantly reduced relative to the anhydride content. Competing chemical reactions effectively cause a stoichiometric imbalance, which substantially limits the development of macromolecules through crosslinking[5].

Hydrocarbon systems also display a dependence on the fiber. The unreacted t-butylstyrene content relative to telomer content is substantially greater in resin castings[2]. The fiber/sizing must be providing reactive sites which, through termination, substantially reduce the relative rate of crosslinking involving free radicals of t-butylstyrene and pendent ethylene groups of 1-2 polybutadiene. The net effect is a relative increase in telomer content.

With respect to humidity aging, authors[6-7] attribute reductions of physical properties to the plasticizing effect of water. The current research demonstrates increasing resin moisture content. Test specimens were oven dried at 150°F and stored with a desiccate until tested. However, resin molecular distributions, including telomer fractions as well as unreacted monomers, changed during the accelerated humidity aging. Some resin systems continue to cure; others degradate.

Test descriptions and results and specimen descriptions are described in detail by Humphrey, et al.[8]. Requests for this document must be referred to DCASMA, Cedar Rapids, Iowa.

6. REFERENCES

1. Yen, H.C., Monomer/Polymer Analysis of Epoxy Resin Systems by Gel Permeation Chromatography, unpublished M.S. dissertation, University of Nebraska Library, Lincoln, NE (1981).
2. Yen, H.D., C.W. Tien, D.C. Timm, W.D. Humphrey and N.C. Plass, Effects of Moisture Aging on Molecular Structure of Thermoset Resins, accepted for publication, 2nd World Congress of Chemical Engineering, Montreal, Canada (1981).
3. Price, H.L. and S.G. Gupschalf, Pultrusion of Composite Materials as a Reactive Process, accepted for publication, First International Conference on Reactive Processing of Polymers, Pittsburgh, PA (1980).
4. May, C.A. and Y. Tanaka, Epoxy Resins: Chemistry and Technology, Marcel Dekker, Inc., New York, NY (1973).
5. Flory, J.P., Principles of Polymer Chemistry, Cornell University Press, Ithaca, NY (1953).
6. Lubin, G., and R. Donohue, Real Life Aging Properties of Composites, SPI Proceedings of the Reinforced Plastics/Composites Institute, 17-E, 1 (1980).
7. J. Brandrup and E.H. Immergut, Polymer Handbook, John Wiley and Sons, New York, NY (1975).
8. Humphrey, W.D., N.C. Plass and R.A. Andersen, Resin Study for Rocket Motor Case Improvements, presented at JANAF Symposia, New Orleans (1981).

7. ACKNOWLEDGMENT

Financial support from Brunswick Corporation, Defense Division, Lincoln, Nebraska and the University of Nebraska's Engineering Research Center is appreciated.

8. BIOGRAPHIES

W. Donald Humphrey is actively involved in the field of composites, principally design analysis and fabrication of filament-wound structures. He is currently

Program Manager of IR&D and is responsible for internally funded R&D. He is a graduate of the University of Utah in Civil Engineering.

Norman C. Plass is a Senior Research and Development Engineer specializing in materials and processing development. As manager of the design engineering laboratory, he supports both the engineering and production functions with materials and processing research, testing, and evaluation. He has extensive and diversified experience in filament winding, autoclave and vacuum bag molding, compression and transfer molding, and resin system formulation. He received his B.S. degree in Physical Sciences from Capital University in 1950.

Delmar C. Timm has published extensively in the field of polymeric analysis utilizing gel permeation chromatography. The current research is a joint University of Nebraska/Brunswick, Defense Division project. Professor Timm received his doctorate in Chemical Engineering at Iowa State University in 1967. He previously worked as a process engineer in a polymer group at Esso Research and Engineering.

Shiann-Hsiung Liu is a graduate student in Chemical Engineering at the University of Nebraska. He received his undergraduate degree at Tunghai University, Taiwan. Industrial experience includes employment with the National Bureau of Standards, Ministry of Economic Affairs, Taipei, Taiwan.

TABLE 1

Resin Description

Formulations

LRF 092 Brunswick Production
- Ingredients
  - Epon 828
  - NADIC Methyl Anhydride (NMA)
  - Benzyldimethyl Amine (BDMA)
- Application
  - Polaris A-3, Skylab and Space Shuttle oxygen tanks
- Cure*
  - B-stage 8 hrs minimum-150°F
  - 170°F-2 hrs 280°F-10 hrs
  - 235°F-2 hrs

LRF 216 Brunswick/Aerojet/Thiokol
- Ingredients
  - Epon 828
  - NMA
  - ATC-3
- Application
  - MX, 2nd stage
- Cure*
  - B-stage 18 hrs minimum-140°F
  - 170°F-2 hrs 210°F-17 hrs

EXP 117 Thiokol
- Ingredients
  - Epon 828/871
  - NMA
  - EMI-24
- Application
  - Stand-off Missile
- Cure*
  - B-stage 12 hrs minimum-140°F
  - 225°F-4 hrs 280°F-5 hrs

EXP 105 Colorado Chemical Speciality
- Ingredients
  - 1-2 polybutadiene
  - t-butylstyrene
  - dicumyl peroxide
- Application
  - Brunswick (Marion) R&D
- Cure*
  - 175°F-5 hrs 250°F-3 hrs
  - 220°F-3 hrs 280°F-4 hrs

* Temperature rise may require 1-2 hours of time interval shown, temperature control ± 10°F

TABLE 2

Gel Permeation Chromatography
Molecular Distribution Analysis
Reinforced FIBER EPOXY RESIN Coupons

Material	Peak Mark	Original Weight Fractions	21 day
LRF 092			
H_2O	-(130)	.09	.11
BDMA	130	.00	.00
NMA	118	.09	.00
ADD	115	.52	.53
Epon 828	102	.04	.10
Telomers	<110	.26	.26
LRF 216			
ATC-3	-(130)	.09	.10
NMA	118	.10	.33
ADD	115	.47	.26
Epon 828	102	.12	.10
Telomers	<110	.22	.21
EXP 117			
EMI-24	-(130)	.10	.06
NMA	118	.00	.08
ADD	115	.32	.06
ADD	108	.19	.17
Epon 828	102	.04	.03
Telomers	<110	.35	.61

TABLE 4

Gel Permeation Chromatography
Molecular Distribution Analysis
Rein. FIBER HYDROCARBON RESIN Coupons

Material	Peak Mark	Orig. Mass Fractions	21 day
EXP 110A			
Water	-(130)	.06	.15
Add	120	.09	.00
t-butylstyrene	117	.29	.37
Telomers	<105	.56	.48
EXP 110B			
Water	-(130)	.04	.09
ADD	120	.02	.00
t-butylstyrene	117	.32	.30
Telomers	<106	.62	.61
EXP 110C			
Water	-(130)	.03	.15
ADD	120	.02	.00
t-butylstyrene	117	.37	.37
Telomers	<105	.58	.48
EXP 112			
Water	-(130)	.05	.12
"Epoxy/Amine"	131	.02	.00
t-butylstyrene	117	.49	.47
"Epoxy/Amine"	101	.08	.10
Telomers	< 95	.36	.31
EXP 114			
Water	-(130)	.00	.20
"Epoxy/Amine"	131	.00	.00
ADD	120	.31	.00
t-butylstyrene	117	.28	.00
ADD	106	.00	.46
"Epoxy/Amine"	101	.04	.00
Telomers	< 95	.37	.34
EXP (119)			
Water	-(130)	.04	.14
"Epoxy/Amine"	126	.02	.00
t-butylstyrene	117	.09	.05
ADD	108	.03	.04
"Epoxy/Amine"	101	.02	.03
Telomers	< 95	.80	.74
EXP (120)			
Water	-(130)	.02	.04
"Epoxy/Amine"	126	.07	.02
t-butylstyrene	117	.42	.40
ADD	108	.17	.14
"Epoxy/Amine"	101	.07	.07
Telomers	< 95	.25	.33

TABLE 3
Naval Ordinance Rings and 6.7 in. Pressure Vessels
Gel Permeation Chromatography and Physical Property Changes

Material	Extractables area/gm x 10^{-3}	Shear Strength psi	Tensile Strength psi x 10^{-3}	PV/W_c Factor in x 10^{-6}	Hoop Fiber Stress KSI
EPOXY					
LRF 092					
original	9.5	6500	388	1.09	394
21 day	+26%	-35%		-3.7%	-3.0%
LRF 216					
original	9.6	6700	398	1.00	368
21 day	+2%	-34%		0.0%	+1.9%
EXP 117					
original	11.2	6100	402	1.04	389
21 day	-33%	-43%		-1.9%	-6.2%
HYDROCARBON					
EXP 105					
A					
original	5.1	2200	420	1.25	423
21 day	+33%	-36%		0.0%	0.0%
B					
original	9.3	2400	408		
21 day	-1%	-29%			
EXP 110					
A					
original	11.7	5100	371	0.97	396
21 day	-18%	-35%		+3.1%	+5.0%
B					
original	11.3	(a)	388		
21 day	+9%	(a)			
C					
original	12.8	(a)	386		
21 day	-24%	(a)			
EXP 112					
original	6.4	2600	384	1.28	429
21 day	+11%	-46%		0.0%	0.0%
EXP 114					
original	6.2	5300	331	1.24	439
21 day	-20%	-40%		0.0%	0.0%
EXP 119					
original	20.5	(a)	392	1.10	416
21 day	-7%	(a)		-5.4%	-4.0%
EXP 120					
original	6.4	4600	327	1.23	438
21 day	+8%	-46%		0.0%	0.0%
EXP 120(R)					
original				1.29	441
21 day				0.0%	0.0%

(a) too flexible for test results

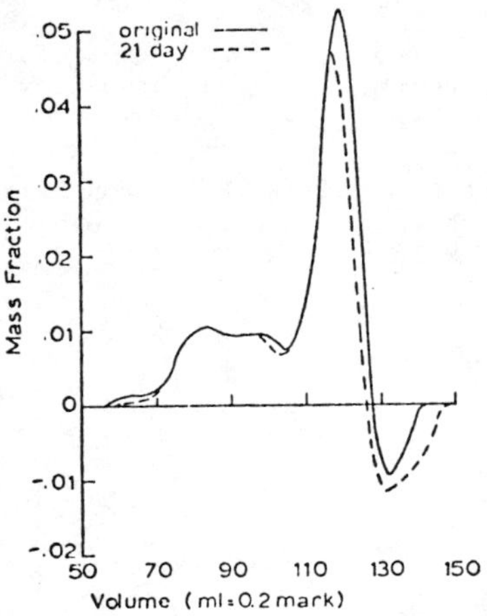

Fig. 1: Epoxy Composites LRF-092

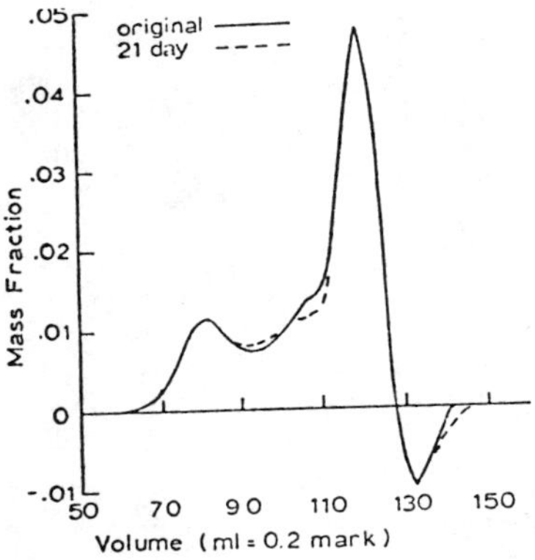

Fig. 2: Epoxy Composites LRF-216

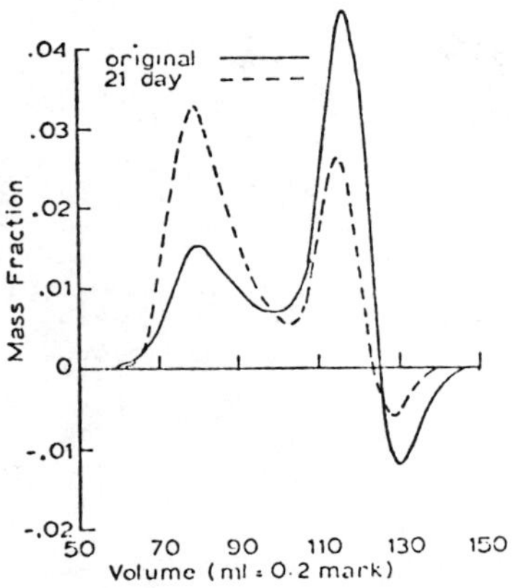

Fig. 3: Epoxy Composites LRF-117

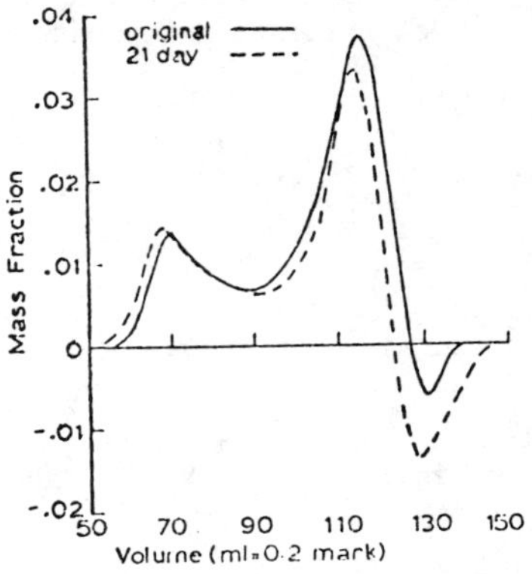

Fig. 4: Resin Composite EXP 105A

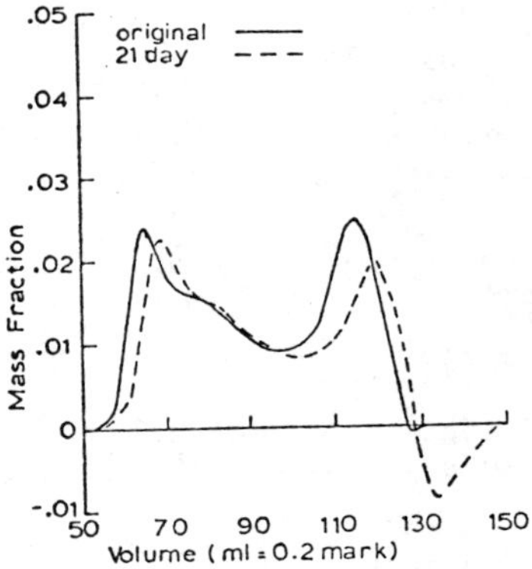

Fig. 5: Resin Composite EXP 105B

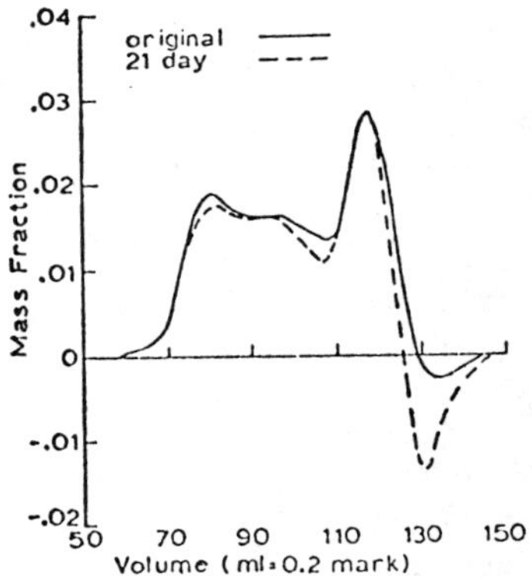

Fig. 6: Resin Composite EXP 110A

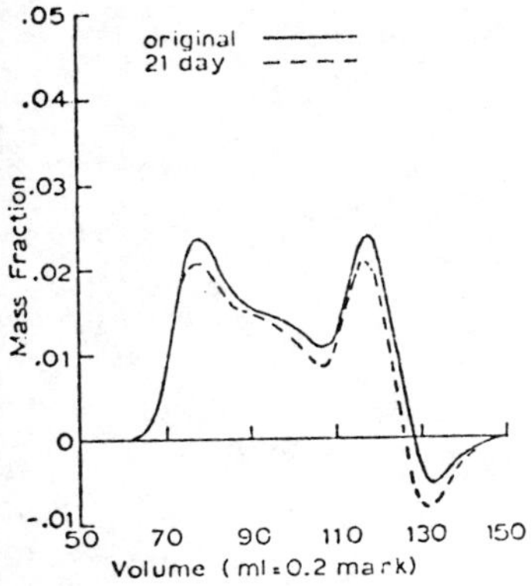

Fig. 7: Resin Composite EXP 110B

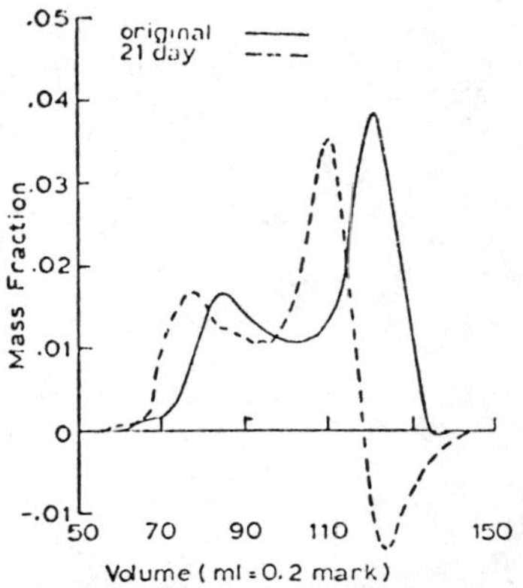

Fig. 10: Resin Composite EXP 114

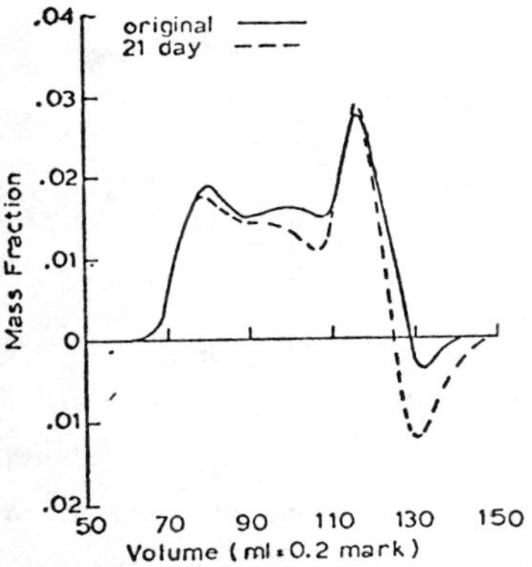

Fig. 8: Resin Composite EXP 110C

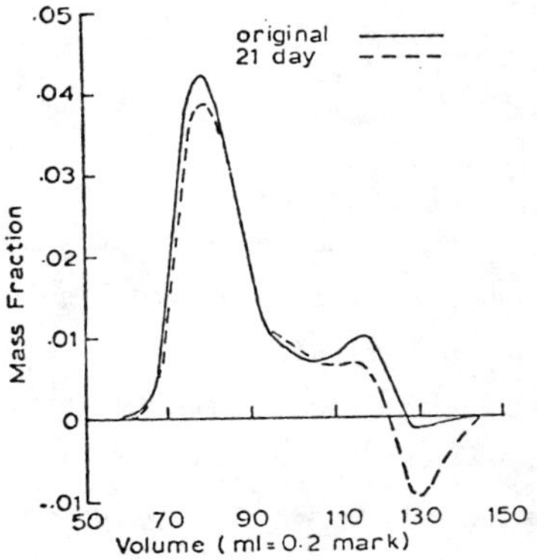

Fig. 11: Resin Composite EXP 119

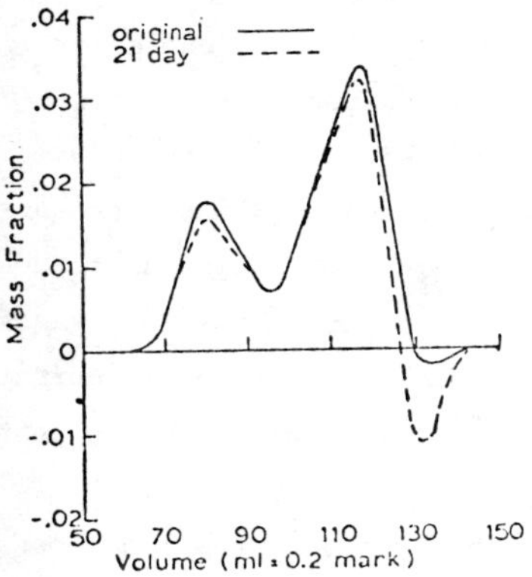

Fig. 9: Resin Composite EXP 112

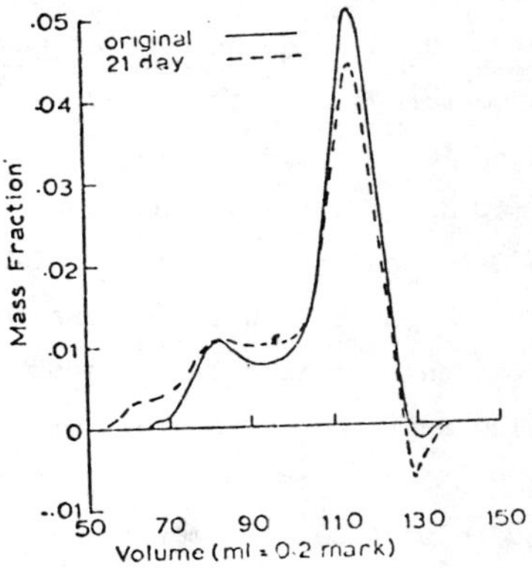

Fig. 12: Resin Composite EXP 120

13th National SAMPE Technical Conference
October 13-15, 1981

THERMOPLASTIC POLYESTER MATRIX COMPOSITES

John Timothy Hartness
University of Dayton Research Institute
Dayton, Ohio 45469
and
Joel A. Dyksterhouse
Air Force Wright Aeronautical Laboratories (AFSC)
Wright-Patterson Air Force Base, Ohio 45433

Abstract

Two thermoplastic polyester resins have been characterized and developed as a matrix for graphite reinforced composites. These resins offer exceptional chemical resistance as well as ease of processability using film. Characterization of neat resin properties including mechanical properties, chemical and environmental resistance, and physical properties have been performed and documented. Bidirectional graphite cloth/ thermoplastic polyester prepreg has been developed and utilized to fabricate composite panels for characterization. Details of fabrication studies and procedures are presented. Composite characterization includes physical and mechanical properties such as shear, flexure, and tension, as well as chemical resistance and environmental durability. A thermoforming study was completed on the composites. All test procedures and results are documented in the paper.

Keywords: Thermoplastic polyesters, composites, graphite, cloth.

1. INTRODUCTION

Continuous fiber graphite reinforced thermoplastic composites have been under development and evaluation for several years, and the potential advantages of thermoplastic matrix composites for aircraft structures are well documented.[1,2,3] These advantages usually include major reductions in fabrication costs, improved repairability, reduced moisture degradation, and less variability in matrix resin constituents. In the past these advantages have been somewhat offset by certain problems. The problems in the past have included lack of chemical resistance as far as the amorphous thermoplastics are concerned and lack of toughness in the semicrystalline materials such as polyphenylene sulfide. The processing of thermoplastic composites has been successfully demonstrated by interlaying film and reinforcement with high temperature (600-700°F) compression molding (100-200 psi).[4,5] The use of this technique has allowed the continued evaluation of the potential of these materials as composite matrix resins. This paper presents the results of two polyester thermoplastic resins.

2. EXPERIMENTAL DISCUSSION AND RESULTS

2.1 Resin Properties

The resins evaluated in this study were poly(ethylene terephthalate), trade name Tenite PET Polyester 7970 from Eastman Chemical Products, and poly(butylene terephthalate), trade name Tenite PBT Polyester 6P4D. These materials represent current thermoplastic polyesters in production. Both materials were supplied in the amorphous state in 10 mil. film. Basic mechanical properties have been generated on both materials. Although both resins revert to their crystalline form upon melt and cool down, there are some differences between the two. Physical properties are shown in Table I for both materials in their crystalline form as supplied by Eastman.

Polyethylene terephthalate (PET) crystallizes somewhat slower than PBT, but since the material is taken above its melt point and slowly cooled, there seems to be no problem with the material crystallizing. Both materials must be used in their crystalline form in order to obtain good chemical resistance. A good deal of chemical resistance data is reported by Eastman. In our study additional chemicals of interest were evaluated. The test procedure involved dead weight loading 10 mil. thick dogbone specimens and exposing them continuously to the agent for a prescribed amount of time or until failure. All specimens were crystallized first by dipping in methylene chloride prior to loading. No failures occurred under the conditions and agents used. Table II shows these conditions. No etching or stress cracking of any of the specimens was observed. Table III shows the results on polyphenylsulfone (Radel), an engineering thermoplastic that has superior chemical resistance to polysulfone (UDEL) and polycarbonate. Polyethylene terephthalate (PET) possesses a low Tg = 110°C (230°F) and a high melt temperature of 260°C (500°F). Polybutylene terephthalate also possesses a low Tg = 40°C (104°F) and a high melt point of 225°C (442°F). As with other semicrystalline polymers, the modulus-temperature behavior is believed to be significantly different from that of amorphous polymers. In the glass transition region, the shear modulus drops to a much less extent. This drop in modulus usually occurs over a broader temperature range. The modulus first exhibits a plateau, then drops slowly up to the melting temperature and drops rapidly thereafter. The composite flexural tests support this conclusion.

2.2 Composite Processing

As with earlier thermoplastic matrix composites, amorphous film was interleaved with bidirectional Thornel 300-8 satin harness graphite cloth with a UC307 finish (thermoplastic finish). The same technique of spot welding the film to the cloth, thus holding the cloth together, was again successful and allowed easily the cutting of any geometric ply orientation. Both resin systems were processed the same. The stack was placed in a matched die mold and compression molded in a press at 273°C (525°F) and 100 psi for 30 minutes. The composite was allowed to cool slowly under pressure to approximately 200°F before removal. Sufficient bleedout of resin resulted in composites with 50-55% fiber volume. Photomicrographs indicate good fiber wetting (Figures 1 and 2). Composite processing parameters were not optimized, although it is expected that no significant changes can be made with respect to temperature. It is possible that shorter times at temperature and pressure may be successful. It is expected that thermoforming should be successful, as the resin is capable of repeated cycles of melting and cooling. Table IV shows the processing conditions.

Consideration should be given to the rate and degree of crystallization for each material. The PBT crystallizes very rapidly and to a greater extent than does the PET.

2.3 Composite Properties

A series of mechanical property tests were performed on composite laminates to determine the resin performance as a matrix material. All composite properties were generated on bidirectional Thornel 300-8 satin harness graphite cloth with a UC307 finish. It was hoped that a unidirectional cloth with a polyester fill could also be evaluated, but cloth data could not be obtained in time to include in this paper. Flexural and tensile properties were evaluated at room and elevated temperature. Tables V through VIII show these results. As can be seen the material maintains its properties with an approximate 50% dropoff at 300°F. The flexural modulus holds up well at 300°F and is not significantly affected even at 350°F. The differences between the two materials in the ±45° tensile strengths may be explained by the rate and degree of crystallization between PBT and PET. Although it is somewhat unfair to compare directly the mechanical properties of a thermoset, such as an epoxy, with those of a thermoplastic, the properties are somewhat similar. There is a greater dropoff at elevated temperature due to the nature of the polyester, but useful properties are observed due to the semicrystalline nature of the materials.

2.4 Hot-Moist Aging

It has been well documented that a hot humid environment brings about a decrease in the mechanical properties of thermoplastic polyesters.(6) Evaluating three grades of PBT, Gardner and Martin have shown that hydrolysis occurs rapidly at high temperature and relative humidities and progressively slows as the temperature and/or humidity are decreased. According to the evaluation equations for making life-cycle predictions at any combination of temperature and humidity were derived from Arrhenius plots. "If a 50% loss in tensile strength constitutes failure, then the PBT examined should be expected to last only three to four years at 50°C (120°F) and 100% relative humidity. Reducing the humidity level to 50% extends the useful life at this temperature to 10-20 years. Predictions based on the tensile strength half-life should not be used where toughness or impact properties are important because PBT embrittles long before the tensile strength half-life is reached." Exactly how and to what degree this study applies to graphite reinforced thermoplastic polyesters remains to be evaluated. There is every reason to expect that there would be an effect. Serious consideration should be given to any application where possible hot humid environments may be encountered for any length of time.

3. CONCLUSIONS

In general, the polyesters appear to have good properties as matrix resins for composites. The ability to form high quality laminates by interleaving polymer film and cloth indicates potential for a variety of product forms and fabrication procedures for producing low-cost structures. Polyesters are both cheap and plentiful. The resin offers excellent chemical resistance, thermal stability, and mechanical properties up to at least 300°F. One major problem with polyesters is their sensitivity to a hot/moist environment.

4. REFERENCES

1. J. T. Hoggatt and A. D. Von Volki, "Evaluation of Reinforced Thermoplastic Composites and Adhesives," Naval Air Systems Command Final Report - Contract N00017-7A-C-0026, March 1975.

2. L. C. May and R. C. Goad, "Manufacturing Methods for Fabrication and Assembly of Advanced Composite Primary Aircraft Structure," U.S. Air Force Materials Laboratory Report No. AFML-TR-75-111, July 1975.

3. J. H. Laakso and J. T. Hoggatt, "Development of a Low-Cost Graphite Reinforced Composite Primary Structural Component," Naval Air Development Center Final Report - Contract N62269-74-C-0368, December 1976.

4. G. Husman and J. Hartness, "Polyphenyl Sulfone Matrix Composites," 24th National SAMPE Symposium and Exhibition, Volume 24, Book 2, pg. 21 (1979).

5. J. Hartness, "Polyphenylene Sulfide Matrix Composites," 25th National SAMPE Symposium and Exhibition, Volume 25, pg. 376 (1980).

6. R. J. Gardner and J. R. Martin, "Effect of Relative Humidity on the Mechanical Properties of Poly(1,4-Butylene Terephthalate)," Journal of Applied Polymer Science, Vol. 25, 2353-2361 (1980).

5. BIOGRAPHY

John T. Hartness is an Assistant Research Chemist and group supervisor of the Composites and Adhesives Group at the University of Dayton Research Institute. Mr. Hartness has been with the Research Institute since 1968. He received his Bachelor of Technology from the University of Dayton in 1974.

Joel A. Dyksterhouse is a Composite Engineering Technician with the U.S. Air Force in the Composites and Adhesives Branch of the Air Force Materials Laboratory.

TABLE I

PHYSICAL PROPERTIES

Crystalline Form

Property, Units	Value	
	PET	PBT
Crystallinity, %	35-40	50
Density, gm/cm^3	1.45	1.31
Temperature Characteristics, °C		
Glass Transition	110±10	40
Softening Point by TMA	245-252	
Crystalline Peak M.P.	<255	--
Melting Point	<257	225

TABLE II

PET AND PBT NEAT RESIN
CHEMICAL RESISTANCE

Chemical	Stress (psi)	Time at Stress
MIL-H-5606C[1]	1000	1 week
Skydrol 500B[2]	1000	1 week
JP-4 AF-Fuel	1000	1 week
Dichloromethane	1000	1 week
MEK	1000	1 week
Acetone	1000	1 week
T-5351 Al[3]	1000	1 week

1 Air Force Hydraulic Fluid

2 Commercial Aircraft Hydraulic Fluid

3 Air Force Paint Stripper

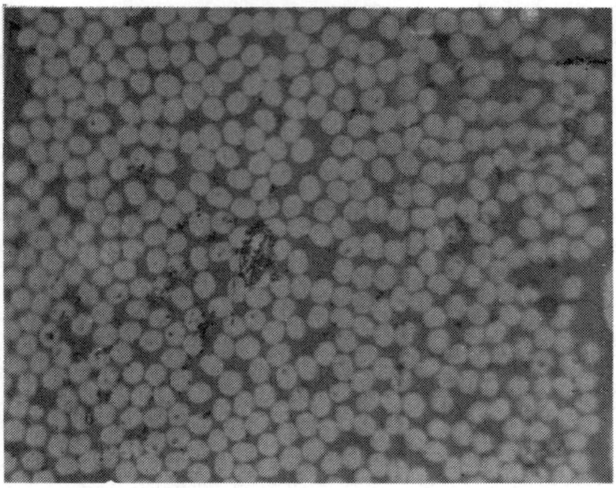

Figure 1. Photomicrograph PET/T-300 Bidirectional Cloth.

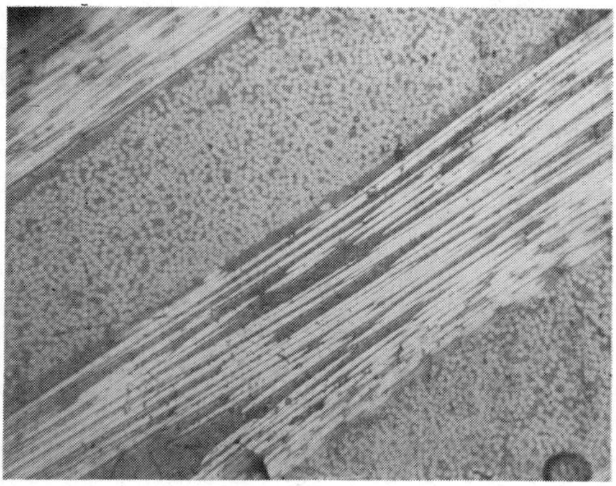

Figure 2. Photomicrograph PBT/T-300 Bidirectional Cloth.

TABLE III

RADEL NEAT RESIN CHEMICAL RESISTANCE

Chemical	Stress (psi)	Rupture Time	Immersion
+MIL-H-5606C	500	No Failure	No Swelling
++Skydrol 500B	500	30 min.	No Swelling
++Skydrol 500B-4	500	15 min.	No Swelling
++Skydrol LD	500	20 min.	No Swelling
++Skydrol LD-4	500	10 min.	No Swelling
Dichloromethane	500	3 min.	--
	1000	10 sec.	--
MEK	500	1 sec.	--
Toluene	500	15 sec.	--
Trichloroethylene	500	1 sec.	--
Acetone	500	20 min.	--
+++T-5351 Al(PS)	500	4 min.	--

\+ Air Force Hydraulic Fluid

++ Commercial Aircraft Hydraulic Fluid

+++ Air Force Paint Stripper

TABLE IV

COMPOSITE PROCESSING

- Heat to 525°F (275°C) as fast as possible using contact pressure.
- Apply 100 psi and hold for 30 min.
- Cool slowly under pressure to 200°F or less.
- Remove from mold.

TABLE V

PET/T-300 BIDIRECTIONAL CLOTH

(UC 307 Finish)(8H Satin)

Property	Condition	Strength $x10^3$ psi	Modulus $x10^6$ psi	Strain to Failure %	Poisson Ratio
Tensile 0° warp direction	RT	61	10.38	.60	0.046
"	200°F	57	11.67	.57	--
"	250°F	43	10.16	.44	--
"	300°F	46	10.47	.56	--
"	350°F	50	8.69	.78	--
Tensile ±45° warp direction	RT	16	2.93	2.37	0.716

TABLE VI

PBT/T-300 BIDIRECTIONAL CLOTH

(UC 307 Finish)(8H Satin)

Property	Condition	Strength $x10^3$ psi	Modulus $x10^6$ psi	Strain to Failure %	Poisson Ratio
Tensile 0° warp direction	RT	59	11.39	.67	.051
"	250°F	34	9.02	.57	.084
"	350°F	33	8.82	.61	.363
Tensile ±45° warp direction	RT	7.61	2.47	.58	.766

TABLE VII

PET/T-300 BIDIRECTIONAL CLOTH

(UC 307 Finish)(8H Satin)

Property	Temp.	Strength x 10^3 psi	Modulus x 10^6 psi
3 pt. Flexure	RT	84.59	7.12
0° warp dir.	200°F	60.09	6.81
	250°F	49.40	5.84
	300°F	46.17	6.06
	350°F	35.31	5.61

55% Fiber Volume

TABLE VIII

PBT/T-300 BIDIRECTIONAL CLOTH

(UC 307 Finish)(8H Satin)

Property	Temp.	Strength x 10^3 psi	Modulus x 10^6 psi
3 pt. Flexure	RT	87.94	6.62
0° warp dir.	200°F	63.36	6.38
	250°F	50.06	5.81
	300°F	43.55	5.58
	350°F	33.83	5.63

60% Fiber Volume

13th National SAMPE Technical Conference
October 13-15, 1981

PHONONOGRAPHY: IT'S ROLE IN NDE

J. Robert Fort
Program Manager, Shurtronics Division of Shur-Lok Corp.

Abstract

To locate and extract "signatures" in the subsurface of layered materials it is necessary to "look" beneath the surface. Fortunately, surfaces are partially transparent to acoustic waves with wavelengths of a few millimeters. Phononography is the art of registering or writing the undulatory movements of mechanical waves.

The approach then is to "ensonify" the subsurface with acoustic wave radiation, to place sensing instruments on the surface, and to process and present the wave data in final visual displays which simulate and pictorially suggest the actual subsurface structures and configurations. Complete processing of signal information involves four operations: search, find, map (or migrate), and focus. The fundamentals of these operations and their sequential order of application depend on the nature of the original wave "illumination". Since vibratory sources can produce either pulses, swept frequency signals, or constant frequency signals, different modes of processing are required for each signal type. It is the purpose of this paper to illustrate, with the minimum of mathematics, how the microcomputer is employed in each case to accomplish the writing of the waves in the desired final forms.

1. THE WRITING OF THE WAVES

Much of signal processing is devoted to the separation of signal waveforms into two or more subclasses. The separation of waveforms according to specific characteristics (signatures) is known as filtering. The signatures are imbedded in a collection of reference waveforms against which the input waveforms are compared(standards). Classically, the comparison process is performed as a set of inner products known variously as correlation, convolution, or filtering. An example of a filtering task is mapping an anomaly onto a focal plane by ensonifying a scattering object (with radiation. The scattered wave radiaation is mapped or filtered onto slot apertures; the phase information is in the form of a pure acoustic envelope. Other tasks include mapping the envelope "back" into the focal plane -- an operation somewhat akin to inverse phase transformation.

In this paper we will review the techniques for performing the operation of correlation, convolution, and narrowband filtering. We will examine and compare the methods of direct implementation of frequency domain versus the phase domain. In particular we will examine several approaches to look for the more efficient filtering task.(1)

In most reflection acoustic signal processing and interpretation we seldom use the wave equation. We do use Huygens' Principle, Fermat's Principle, Snell's Law, ray tracing, and other methodologies. However, these methods do not provide an adequate theory of the interference of waves, a phenomenon of real importance to an understanding of the origin of reflections and guided waves. (2)

Eikonal equations (3) describe the behavior of geometrical optics. The most utilized of these is Fermat's principle of least time. Fermat's principle states that the index of refraction $n(\vec{r})$ in a medium depends on the vector position $(\vec{r})$ of any point in the medium. The ray of acoustic energy chooses its path "l" between travel points in such a way that

$$\int n dl \text{ is a minimum; or} \qquad (1)$$

$$\delta \int_a^b n dl = 0 \qquad (2)$$

$$\text{where } n = \frac{\sin\theta_i}{\sin\theta_r} = \frac{1}{c} \qquad (3)$$

The index of refraction "n" is interpreted as the reciprocal of velocity.

Time of travel along a ray path from "a to b" in a medium is either minimum or maximum or has a stationary value when compared with all adjacent paths.

More general forms of wave equations(4) can be described by the superposition of all their sinusoidal components. These can be reduced to the time-independent Helmholt's equation

$$\nabla^2\psi = k^2\psi = 0 \quad \text{where} \qquad (4)$$

$k = \frac{2\pi}{\lambda}$, $k = \frac{\omega}{c}$ (wave number represents the spatial frequency expressed as 2π times the number of wave cycles per unit length)

ψ is expressed as

$$\psi = \psi_0(x,y,z)e^{-jk_0S(x,y,z)} \qquad (5)$$

where both ψ_0 and S (surface) are functions that are varying slowly with respect to the wave length (λ) of radiation.

Therefore and approximation of the Eikonal equation is:

$$(\nabla S)^2 = n^2 \qquad \nabla = \text{diff. operator}$$

$$\nabla^2 = \frac{\partial^2}{\partial x^2} + \frac{\partial^2}{\partial y^2} + \frac{\partial^2}{\partial z^2}$$

where the index of refraction becomes

$$n = \frac{k^2}{k_0{}^2} .$$

This determines the function S, which allows us to define the surfaces of constant phase via

$$S(x,y,z) = \text{constant}$$

The surfaces of constant phase define the shape of the radiation field, so that the Eikonal equation determines the wave propagation in geometrical acoustics: $\psi_0(x,y,z)$ defines the envelope of the wave.[5]

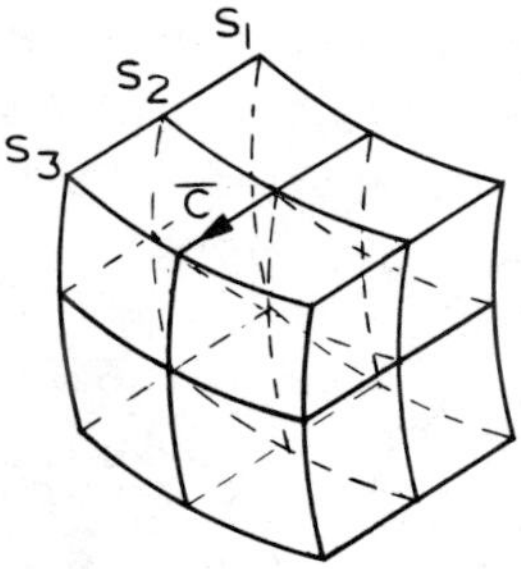

Fig. 1 Three Wavefront Surfaces or Surfaces of Constant Phase

The connecting lines orthogonal to the two surfaces are rays. The lines on the surface are geodesic lines and represent paths of shortest distances on the surfaces. Then the surfaces are a wavelength apart:

$$S_3 - S_2 = \frac{2}{k_0} . \tag{8}$$

One can express the Eikonal equation in terms of "travel times" that correspond to successive wavefront positions:[6]

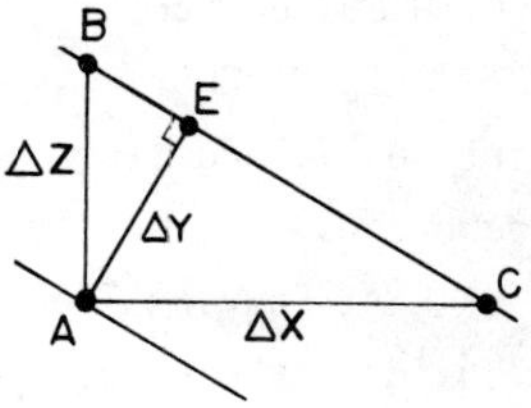

$$\frac{1}{\Delta y^2} = \frac{1}{\Delta x^2} + \frac{1}{\Delta z^2}$$

$$\left(\frac{\Delta T}{\Delta x}\right)^2 + \left(\frac{\Delta T}{\Delta z}\right)^2 = \left(\frac{\Delta T}{\Delta y}\right)^2 = \frac{1}{c^2}$$

Fig. 2 Derivation of the Eikonal Equation in Terms of Travel Times

Now we come to an important application to waveform interpretation -- the backward propagation of waves from the surface to build a subsurface pattern or signature. Subsurface patterns are formed by making acoustic waves recorded at the surface return to their source.

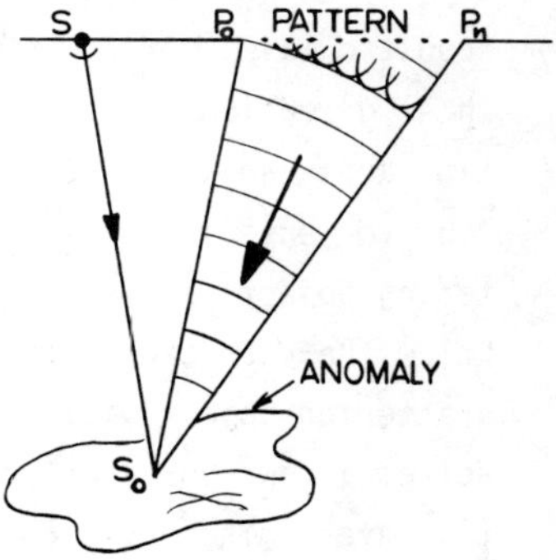

P = position of receiver(s)
S = source radiator (ensonifier)

Fig. 3 Backward Wave Propagation

2. BUILDING SUBSURFACE SIGNATURES IN REVERSE

As a start we will consider only mono-frequencies. Later, by the process of Fast Fourier Transforms, we can "build" any kind of acoustic signal we wish -- pulse, chirp, or whatever is desired. Accordingly, we will use a vibratory source of acoustic waves. Mono-frequency waves will radiate downward toward the subsurface object spaces and ensonify the area by looking through boundaries into subsurface space, "illuminating" the field. In these discussions the following points are necessarily required:

1) The subsurface medium is assumed homogeneous and isotropic with constant wave velocity.
2) Our measurements are scalar; i.e. sound pressure. The scalar wave equation describes the wave propagation.
3) Acoustic waves spread in three dimensions. The acoustic waves radiated from the vibrator ensonify a "scattering object" located below the vibrator. The scattering object will then act as a new source of acoustic waves. The waves radiated from the new source will travel upward to be recorded at various points on an areal map. The measurements at each point will be of two parameters:
   1. The frequency amplitude of the received wave;
   2. The phase of the received wave relative to a common reference, such as the source vibrator.

This set of measurements constitutes a sampled complex linear signature. We use this term because the measurements are at discrete points in the aperture, rather than continuously spread over the receiving points. A signature may be thought of in several ways, e.g.:

1) As a set of frequency amplitudes and phase measurements stored in matrix form in the memory of a microprocessor;
2) As a single parameter scalar plot, "mounted" on a "grey level pedestal".

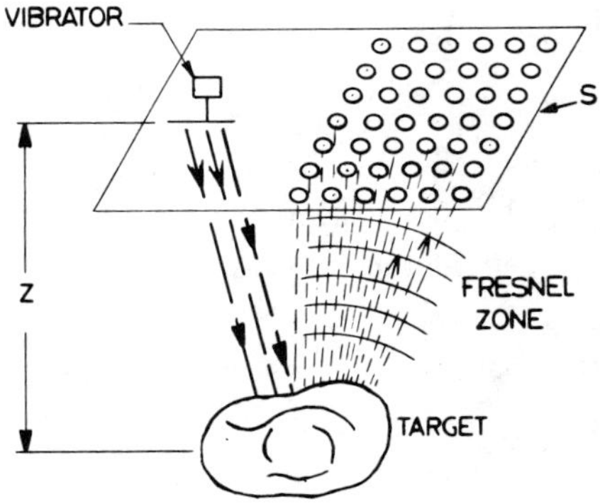

Fig. 4 A 2-D Areal Receiving Aperture

Implementing the back scatter procedures from the target to the surfaces shows the following characteristics:(7)

1) Phase lags are changed to phase leads; this is a simple

function of changing signs.

2) The pattern is directed downward rather than upward.
3) We backward propagate a wave toward the target.
4) We consider a horizontal plane through the target as the focal plane.
5) The reconstructed wave pattern is calculated by Green's theorem which is the expression for the pressure at points along the focal plane of the target. We note that the backward propagated waves from each secondary wave source in the aperture have "added up" in phase at the origin (target). They have come into focus and formed the signature.

How does the concept of backward wave propagation fit into this signature building process and how can we separate the various layers and achieve depth resolution? We will address these questions by examining the interacting considerations of information content and resolution of sonic data.

3. SONIC DATA AS THE PROPAGATION OF WAVES

Spectral magnitude and phase play different roles in the Fourier description of signals. Many of the important features of a signal are contained in the phase domain. This is not true in the frequency/amplitude domain. Contours of crystallographic structures have been diagrammed using phase-only synthesis. (8)

Questions which have not been satisfactorily answered are: what precisely can sonic data tell us about the subsurface; how small a feature in the subsurface can the reflection sonic method resolve, and how does sonic data relate to acoustic data obtained or acquired from "standards". Addressing these questions requires that we first adopt a wave theory viewpoint toward the propagation of a sonic perterbation.

First we resolve a paradox which appears in Figure 5.

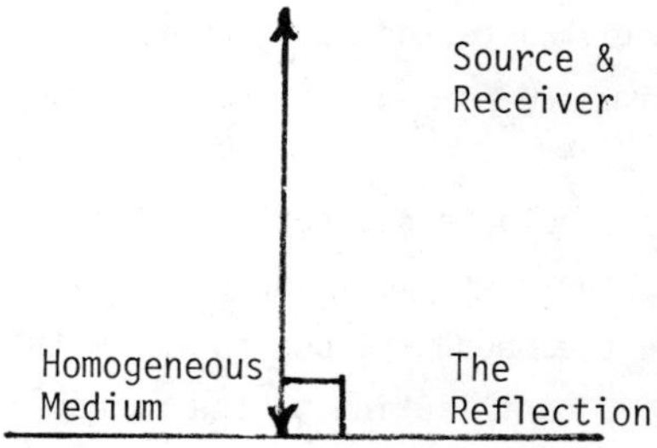

Fig. 5 Ray Path Reflector

If, for example, a source/receiver is located in some place, and according to ray theory we draw a raypath down to the reflecting boundary and back, it would be known as "The Reflection" and the point of contact "The Reflection Point".

At this point we shall redraw this same figure with Huygens' principle of wave theory in mind. Our task will be one of reconciliation.

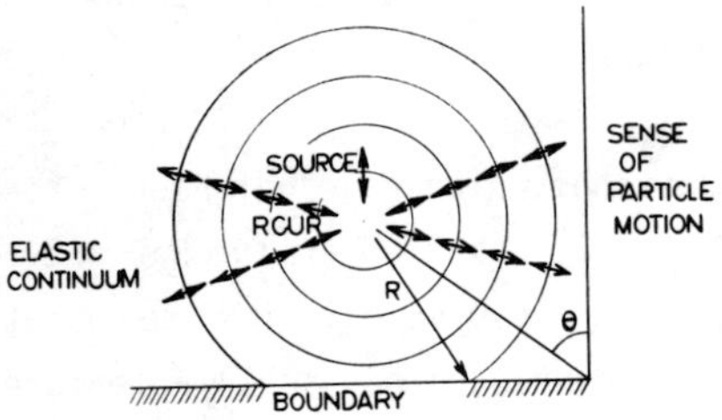

Fig. 6 Wave Front Reflector

Figure 6 shows that in time the propagation wave front will strike the entire boundary and not just a single point. According to Huygens' principle every point of the boundary will act as a source and reradiate energy to the receiver. Hence in wave theory, unlike ray theory, every part of every boundary contributes to the sonogram received anywhere for any source position.

It should be obvious that for a given source location each point on the boundary does not make the same contribution to the receiver. Specifically, then, the magnitude of the contribution for each part of the boundary depends on two factors:

- R - the distance or range from the source to the part of the boundary (to account for divergence effects);
- θ - the inclination of the particle motion (measured normal to the boundary).

When θ is zero for any fixed value of R, the returned energy is a maximum, whereas when $\theta = \pm 90^0$, we will have a minimum contribution. This is intuitively reasonable, since for $\theta = \pm 90^0$ the particle motion will be in a sense parallel to the boundary and not directing any radiation toward the receiver.

Should a surface have a geometry which is similar to the shape of the wave front, then all parts will tend to contribute strongly, giving a strong or focused reflection. This effect is well known, because it is also responsible for the fading of reflection amplitudes which follow from antinodal reflections.

With these concepts in mind the effective zone can be readily defined: this zone is termed the Fresnel zone (see Figure 7).

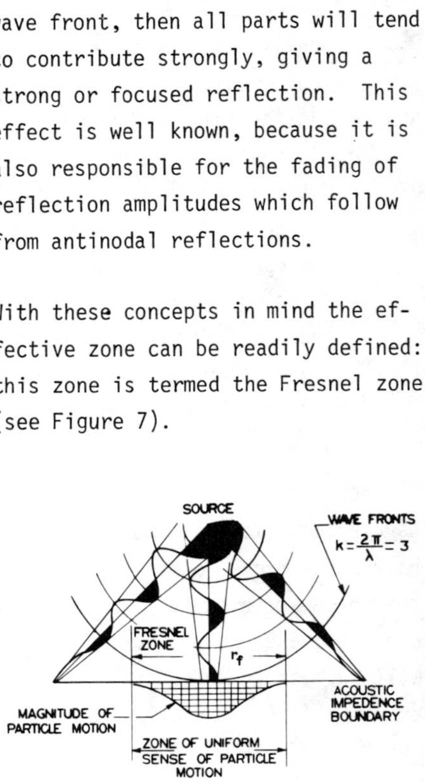

Fig. 7 Fresnel Zone

The radiating wave motion encounters a flat acoustic impedance boundary. The maximum-size region of this boundary, over which the particle motion is all in the same direction, is indicated. A Fresnel zone, although directly related to the size of this region, has a smaller value, because allowance must be made for the respective magnitudes of particle motion within the region.

If the source is at a range R from the reflection and a velocity c over a two-way travel time of t seconds, and the dominant frequency content of the sonic data is f_c, then the Fresnel zone radius r_f is

$$r_f \simeq \frac{c}{4}\sqrt{\frac{t}{f_c}}\,. \qquad (9)$$

This relation is derived using basic geometry and an empirical weighting.

MATERIAL	THICKNESS PER LAYER 10^{-3} INCH	r_f 10^{-3} IN.	VELOCITY 10^4 IN/SEC
KEVLAR	47	135.39	9.38
GRAPHITE - EPOXY	80	329.52	22.83
BORON - EPOXY	51	187.35	12.98
GRAPHITE- RESIN	46	158.77	11.00

Fig. 8 Fresnel Zone Radius

Acoustic ensonification or illumination of the subsurface may then be considered analogous to a "searchlight beam", and the subsurface area illuminated by this beam is considered to be the Fresnel zone. We now can consider the fundamental question, "how small a feature can be compared to a Fresnel zone in order to be detectable, and what are the characteristics of its signal expression in order to facilitate detection?"

4. RESOLUTION CAPABILITIES OF SONIC DATA

As an anomaly approaches the lateral dimensions of a Fresnel zone and becomes smaller, the acoustic signature loses all reflection character and appears in the diffraction form corresponding to a reflecting point. This tells us that thin cracks or voids of small spatial extent have a signature expression but not dominantly of reflective nature and the form of the signatures are alike for bodies smaller than the Fresnel zone. Also their size can be measured by noting the difference in the magnitudes of these events and the smaller extent bodies produce smaller events by occupying only portions of the full Fresnel zone.

As a practical matter the size of the Fresnel zone is initially related to the $\lambda/4$ of the dominant frequency component of the sonic waveform and the near-surface velocity. For the expression r_f the size of Fresnel zone grows in proportion to the ensembled velocities.

5. PHASE-ONLY SYNTHESIS

What if true-magnitude information is eliminated? Have the important characteristics of the waveform been removed? In the discussion outlined above the true-magnitude information is simply replaced by unity-magnitude. With so much information contained in the phase it is natural to consider the possibility of recovering all of the magnitude information from the phase.

A diffraction pattern is a 2-D Fourier transform of an object. The Fresnel diffraction pattern is the Fourier transform multiplied by the PHASE-FACTOR[9], $e^{j(k/2z)(x^2+y^2)}$.

If the input to the medium is a sinusoid, sin ωt, and the signature is sin $(\omega t-\phi)$, then the phase delay t_p is $t_p = \frac{\phi}{\omega}$. Since the continuum acts like a filter, it is analogous to wave propagation theory where each frequency has a different delay. The result will be that the energy envelope's velocity will have a delay which differs from the phase delay of each frequency.

An important aspect of wave propagation theory is the distinction of phase velocity from group velocity; they are analogous to phase delay and group delay. For example, if waves propagate along a 2-D surface, the phase function is

$$\phi(x,y) = k_x(x-x_0) + k_y(y-y_0). \quad (10)$$

(x_0,y_0) is the source input, (x_1,y_1) is the receiver output (where phase is observed), and the symbols k_x and k_y denote the "spatial frequencies", i.e., k_x is $\frac{2\pi}{\lambda}$ measured along the x axis. Theoretical physics provides a relationship between ω and k_x and k_y known as wave numbers.

Since velocity is distance divided by time, we can define the phase velocity along the x axis or direction as

$$(c\ \text{phase})_x = \frac{x - x_0}{\text{phase delay}} = \frac{x - x_0}{\phi/\omega} = \frac{\omega}{k_x} \quad (11)$$

For the x component of group velocity

$$(c\ \text{group})_x = \frac{x - x_0}{\text{group delay}} = \frac{x - x_0}{d\phi/d\omega} = (x - x_0)\frac{d\omega}{d\phi}. \quad (12)$$

In observational physics the velocity one deals with is always the group velocity, and in acoustics impedance is the ratio of pressure to velocity.

In general a wave form is not uniquely defined by its phase. A signal convolved with a zero-phase signal will produce another signal with the same phase. Thus, the phase may uniquely specify a signal to an arbitrary zero-phase factor, but, if a "standard" is available, then a signal may be uniquely defined by its phase.[10]

6. CONVOLUTION & CORRELATION

The equation for computing the linear convolution of a sequence of samples (n) describing a waveform is

$$Y(n) = \sum_{m=0}^{M-1} x(n-m)h(m), \text{ where} \quad (13)$$

$h(m)$ = sequence of length M,
$x(n)$ = sequence > length M.

The corresponding equation for computing the linear correlation of the same two sequences is

$$y(n) = \sum_{m=0}^{M-1} x(n+m)h(m). \qquad (14)$$

The summations indicated represent running weighted sums.

In Fourier transform language a product in the frequency domain corresponds to a convolution in the time domain.

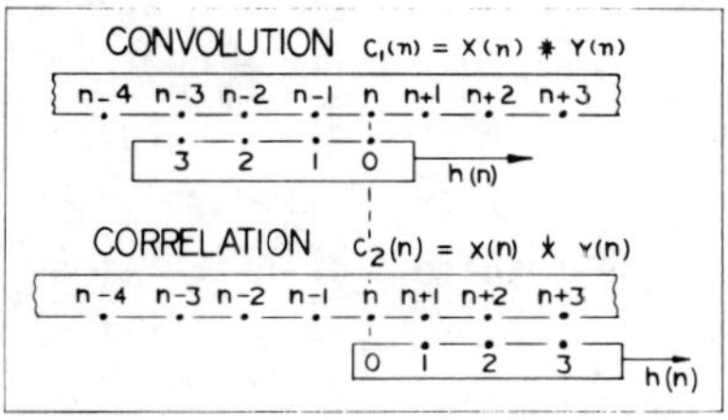

Figure 9

Note the difference of the zero index for the convolution and for the correlation.

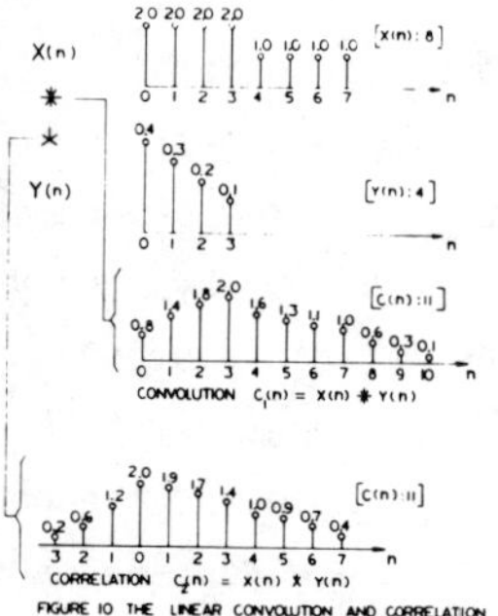

FIGURE 10 THE LINEAR CONVOLUTION AND CORRELATION THE SEQUENCES [x(n):8] [Y(n):4]

Figure 10

As an exercise the calculation of serial products is an entirely feasible procedure. A desk calculator allows successive products to accumulate in the product register and prints the inputs and refults in vertical columns. The answers are written opposite an arrow marked in a place on a movable strip.

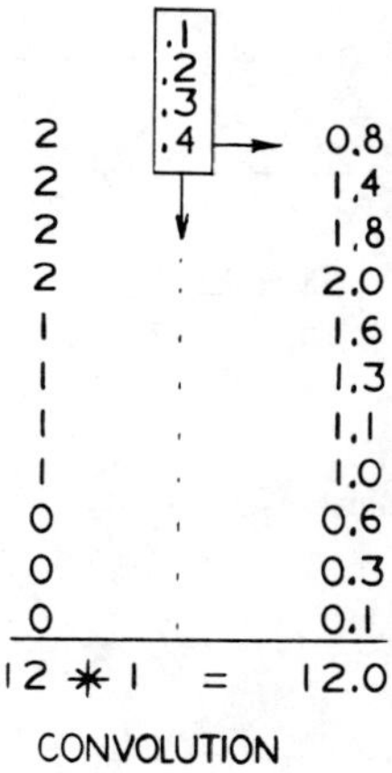

Figure 11 The Convolution Sequence 22221111 * .1.2.3.4

Cross correlation is simpler than convolution, but its properties are less simple. Thus when

$$x(n) * Y(n) = Y(n) * x(n)$$

The cross-correlation operation of x on Y is <u>not</u> the same as that of Y on x; i.e.,

$$x \star Y \neq Y \star x$$

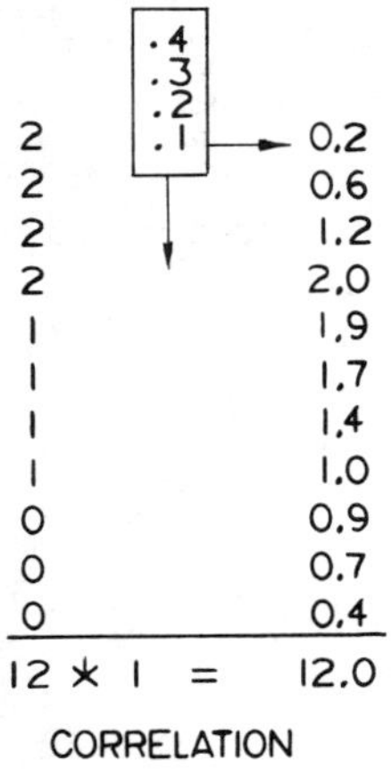

Fig. 12 The Linear Correlation Sequence 22221111 * .1.2.3.4

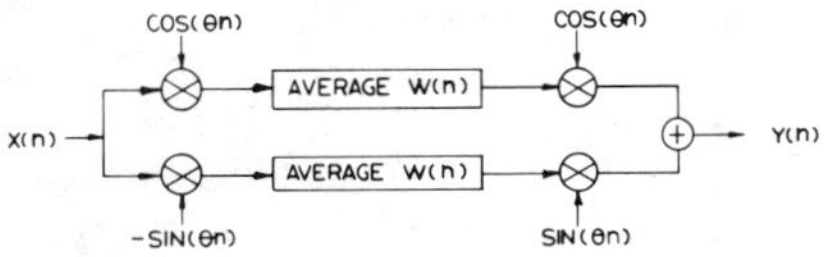

Figure 13

7. THE ROLE OF ACOUSTIC WAVELETS AND INTERPRETIVE PROCESSING

Phase is a rather subtle concept which relates to the alignment in time of the frequency components. The shape of the wavelet and its corresponding phase structure are of greater interpretive significance than the amplitude content of a sonic record as a function of frequency.

Microprocessor based instruments tend to naturally compare "standards" of signatures stored in memory. These patterns are compressed in the phase-transform domain. A deconvolution operator is hetrodyned against the incoming waveforms (Figure 13) and forward transformed into the frequency domain. This operation reduces the size of the signal by decomposition of the symmetry and does a second forward transform. This reduces the signal to the phase domain which is cross-correlated against the phase of the "standard". An inverse transform into the time domain reveals the characteristics of the substructure.

SUMMARY

We have taken you on a fast tour from phononography to waves, from Huygens' Principle to Eikonal equations, from Fermat's Principle to Helmholtz's equation, from wave functions to Fourier transforms, waves in forward gear to waves in reverse gear, from object space to signature space, ray theory to wave theory, convolution to correlation, phase-only synthesis to the Fresnel zone and the point of no return. Now you are on your own. We wish you a pleasant journey!

REFERENCES

(1) F.J. Harris; Convolution, Correlation, and Narrow-band Filtering with the Fast Fourier Transform: Dept of EE & CE, San Diego, CA 92182

(2) Grant & West; Interpretation Theory in Applied Physics: McGraw-Hill, 1965

(3) M. Mizushima; Theoretical Physics, from Classical Mechanics to Group Theory of Microparticles, John Wiley & Sons, 1959, pp 195-196

(4) D. Marcuse; Light Transmission Optics, Bell Labs Series, Van Nostrand & Reinhold Co., 1972 pp 83-87

(5) O.N. Stavroudis; The Optics of Rays, Wavefronts, and Caustics, Academic Press 1972

(6) C.H.Dix; Seismic Prospecting for Oil, Harper Bros. 1952

(7) H.M.A.El-Sum; The Scope of the Symposium in Acoustical Holography, Vol 1, Plenum Press, 1969

(8) M.O. McFarlane; Digital Pictures Fifty Years Ago, Proc. IEEE, Vol 60, pp 768-770, July 1972

(9) T.S. Huang; Digital Holography, Proc. IEEE, Vol 59, pp 1335-1346, Sept. 1971

(10) M.H. Hayes, J.S. Lim, A.V. Oppenheim; Signal Reconstruction from Phase or Magnitude, IEEE Trans. Acoustics, Speech Signal Processing, Vol. ASSP-28, Dec. 1980.

VITAE

J. Robert Fort is Project Manager of Shurtronics, a division of the Shur-Lok Corporation. This division is a specialized engineering-physics firm aiding clients with new product development and other elements of technology advancement.

Personal background includes diversification from nuclear physics, computer development, geophysical exploration programs and signal processing systems for industrial markets.

Fort holds an Engineering-Physics degree from the University of California at Berkeley, followed by data analysis programs at M.I.T. and Cal-Tech, as well as memberships in the I.E.E.E., S.E.G. and A.I.P.G.

13th National SAMPE Technical Conference
October 13-15, 1981

EFFECT OF EXPOSURE TO VARIOUS NATURAL ENVIRONMENTS ON ORGANIC MATRIX COMPOSITES

Eleanor T. Vadala
and
Ronald E. Trabocco
Naval Air Development Center
Warminster, Pennsylvania 18974

Abstract

A study was made of selected graphite organic matrix composite materials exposed to natural weathering at Panama and Warminster.

Commercially available composite prepregs were fabricated into panels using principally a 8 ply layup ($\pm 45^{o}$) and a 16 ply layup (0° $\pm 45^{o}$ 90°). Panels were retrieved periodically and mechanical property determinations were made at R.T. (room temperature) $82.2^{o}C$ ($180^{o}F$) and $121.1^{o}C$ ($250^{o}F$). Tension, flexure, compression, short beam shear and in-plane shear tests were primarily used to assess the damage.

At R.T. the in-plane shear strength property of the $\pm 45^{o}$ panels was unaffected by exposure. At elevated temperature, however, this property of these matrix dominated panels generally decreased with increasing exposure time.

Data indicated that the combination of moisture and elevated temperature caused a decrease in the matrix dominated mechanical properties of the 16 ply panels. Some graphite organic matrix composites degraded at a faster rate than others. Panama exposed panels show a greater degradation in static strength at elevated temperatures than panels exposed at Warminster.

The use of a paint coating seemed to decrease the irreversible degradation due to moisture and ultraviolet irradiation.

A one time exposure aboard different aircraft carriers was made of one of the graphite/epoxy composite systems. The static mechanical properties of the exposed panels were also degraded at elevated temperature.

1. INTRODUCTION

Composite materials are being increasingly utilized in naval aircraft structures. These stiff, strong, lightweight materials have been successfully applied to various types of secondary structure and are currently being applied to primary structure. Composites, however are inhomogenous anisotropic materials which have different defect and damage characteristics from those of the traditional metallic structural materials. A number of different reinforcements have been successfully utilized in composite systems. Early on, fiberglass was used extensively in polyester and epoxy resin composites. Presently, graphite and Kevlar are being utilized to an increasing extent in aircraft composite airframe construction. Those designs employing graphite have ranged from thick skinned strain limited construction to thin skins over honeycomb in a stiffness limited application.

Present day Gr/Ep (graphite/epoxy) systems were originally touted as 176.6°C (350°F) service materials. Increasing evidence disclosed various factors which precluded consideration of these resins for airframe applications in excess of 121.1°C.[1,2,3] The knockdown factor which had received the most attention has been moisture. It is well established that moisture plasticizes the resin matrix at elevated temperatures. The properties affected are those that are matrix or resin dependent such as shear or compression. This effect is reversible if the composite is heated to allow moisture effusion. Typically, in a sound laminate the moisture diffusion and distribution would be expected to follow Fickian behavior. Very thick laminates, greater than 40 plies, take a relatively long time to reach equilibrium saturation levels of absorbed moisture. The SOA 176.7°C curing epoxies such as Hercules AS/3501 and Celanese T300/5208 absorb approximately 1.6% moisture at 95% R.H. and 60°C (140°F). These levels are for unpainted panels. Coating the laminates with epoxy polyamide primer and polyurethane topcoat results in higher moisture saturation levels.

There are other environmental factors than moisture which can have a deleterious effect on organic matrix composites. Ultraviolet radiation can lead to deterioration in organics.[4] The amount of degradation is dependent on the intensity and duration of the exposure. Graphite/epoxies in the uncoated condition usually incur damage due to U.V. radiation, only a few thousandths of an inch deep. This degree of damage has been experienced, in an environment like the Canal Zone, for an exposure of three years. Graphite acts as a U.V. absorber thereby mitigating to some degree the extent of damage. Coating completely precludes U.V. damage of Gr/Ep composites. Thin laminates would be expected to be degraded to a greater extent than thick ones because of the greater percentage of the total thickness being affected.

This study was undertaken to determine the real time effects of natural exposure at various sites on specific graphite reinforced organic matrix composites. The sites of exposure were: Warminster, Pennsylvania and Panama, the Atlantic Ocean site. Residual mechanical properties were determined as a function of exposure at a particular site.

Results are disclosed to show the effects of natural environments on commercially available organic composites. This effort is an engineering study directed at establishing a basis for comparison for composites under specific environmental exposure conditions.

2. EXPERIMENTAL PROCEDURE

Commercially available composite prepregs were fabricated, cured in an autoclave, cut into panels, exposed to natural weathering in Panama and Warminster, Pennsylvania, periodically retrieved and mechanical property determinations made at R.T. 82.2^{o}C and 121.1^{o}C. Tension, flexure, compression, short beam shear and in-plane shear tests were primarily used to assess the effects of exposure.

2.1 Fabrication

The commercial prepreg materials included in this study are presented in Table 1. Eight ply panels ($\pm$ 45) 6 inches by 9 inches, were manufactured for the 300, 800, 900 and 1000 series. The 400 series, 10 ply laminates, consisted of 0° $\pm$ 45 orientation with the 0° ply fabricated on the outside. Sixteen ply panels, (0 $\pm$ 45 90) 8 inches by 9½ inches, were made for the 800, 900 and 1000 series.

Within each series, half of the panels were coated with an epoxy polyamide primer (MIL-C-23377C) followed by a polyurethane (MIL-C-81773B, AS) topcoat for a total thickness of approximately 5 mils. The remaining half of the panels were uncoated.

2.2. Natural Weathering Sites

The average environmental parameters of the Panama[(4)] and Warminster sites are presented in Table 2.

Retrievals were made periodically, approximately at 6 month intervals.

The 300 and 400 series were initiated in January 1977 and the 800, 900 and 1000 series one year later in 1978.

A one time exposure aboard two different aircraft carriers was made of the AS/3501-6 graphite epoxy system. The panels were exposed for 10 months aboard the NIMITZ, a nuclear carrier and for 8 months aboard the US CONSTELLATION, operating in the Pacific theater.

2.3 Physical and Mechanical Tests

The tensile specimens were straight sided 1 inch by 9 inch and were tested to failure.

Flexure specimens, 1/2" by 2-7/8"

were tested to failure in 3 point bending. Compression specimens, 3/4" by 6-1/2", were supported by horizontal tabs to prevent buckling and were tested using a modified ITTRI compression test fixture. Short beam shear specimens, 0.25" by 0.5", were tested to failure in 3 point bending.

The 6" x 9" panels, were cut into five tensile specimens. The 0° tensile strength (in-plane shear) of these $\pm 45^{o}$ panels were made at R.T., $82.2^{o}C$, and $121.1^{o}C$. Two of the 5 tensile specimens tested at R.T. had biaxial strain gages bonded to them in order to determine Poisson's ratio, modulus and strain to failure.

The 8" x 9½" (0 ±45 90) panels were cut into three tensile (one was strain gaged) and compression specimens, and five flexure and short beam shear specimens.

In addition to the mechanical property determinations, weight measurements were made on each panel before and after exposure to determine the percent increase in moisture. Panels retrieved from Panama were sealed in vapor barrier proof bags prior to returning to Warminster.

Visual examinations were performed on all exposed panels. Microexamination and Scanning Electron Microscopy (SEM) were performed on specific specimens.

All data were routinely examined for indications of significant trends. Standard deviations and coefficient of variations were determined for specimen replicates.

3. RESULTS AND DISCUSSION

At R.T., the in-plane shear strength property of the $\pm 45^{o}$ panels was unaffected by exposure. At elevated temperature, however, this matrix dominated property generally decreased with increasing time.

The fiber controlled property of the 16 ply (0 ±45 90) laminates, 0° tension, was basically unaffected by the combination of moisture and elevated temperature. The matrix controlled properties such as shear and flexure, however, were significantly affected. The strength decreases and exposure time increases and the rate of decrease varies with each graphite/matrix system.

In general, the Panama exposed panels showed a slightly greater degradation in static strength at elevated temperature than Warminster exposed panels.

Each graphite/matrix system will be addressed separately and representative data will be presented.

3.1 AS/3501-6 (300, 400, 900 Series)

3.1.1 300 Series, 8 Ply (±45)

The trend of data for increasing exposure time is shown in Figure 1. The in-plane shear strength decreases as exposure time increases for specimens tested at elevated temperature. The rate of increase levels off between 18 to 24 months exposure. R.T. and 82.2°C temperature tests show no adverse affects of outdoor weathering. The increased strength retained at R.T. and 82.2°C evidence post aging behavior that is less marked at the latter temperature.

Warminster exposed panels show approximately the same behavior but to a somewhat lesser degree.

3.1.2 400 Series 10 Ply (0 ±45)

The effect of moisture and temperature on resin dependent properties such as shear is shown in Figure 2. The unpainted panels, tested at R.T., are unaffected by Warminster exposure. At elevated temperatures, however, the coated panels show a degradation effect that is particularly marked at 121.1°C. The 1 to 12 months test data decreases at a faster rate than those tested after one year of exposure. Test results at elevated temperature show some overlapping regards percentage of retained strength.

Flexure test data shows an increase in residual strength at R.T. Elevated temperature tests show a very gradual decrease at 82.2°C to a 85% retention of residual strength after three years; at 121.1°C the loss of strength is closer to 20%.

3.1.3 900 Series, 16 Ply (0 ±45 90) and 8 Ply (±45)

The trend of the 900 series data for the in-plane shear strength of the ±45° panels is similar to the data presented for the 300 series. The R.T. in-plane shear strength is unaffected by exposure. At elevated temperature, the in-plane shear degradation increases with increasing exposure time, especially during the first year. When tested at 121.1°C the Panama exposed panels decrease at a faster rate that the Warminster exposed panels, Figures 3 and 4.

The strength of the fiber controlled property, 0° tension, of the 16 ply painted (0 ±45 90) panels are slightly affected by both exposure (at Panama or Warminster) and test temperature, Figures 5 and 6.

The matrix controlled properties, however, are significantly affected. In particular the flexure strength of the Panama exposed panels, tested at 121.1°C, lose as much as 30% of the residual strength, Figure 7. Warminster exposed panels, tested at 121.1°C do not show any degradation until after about 18 months of exposure and retain approximately 80% of the residual strength after three years of exposure.

The moisture content for the Panama exposed panels was approximately 1% after three years of exposure. The Warminster exposed panels were somewhat less for the same exposure time, about 0.7%. Figure 8 shows the moisture content as a function of time. These moisture contents are representative of the average trend of moisture data for the graphite/epoxy systems included in this study.

With long term exposures there might be a tendency to lose a slight amount of the coating or resin depending on whether the panel was coated or uncoated. This occurrence would tend to decrease weight with increasing exposure time.

Panels of the painted 16 ply (0 $\pm$45 90) AS/3501-6 were subjected to actual shipboard exposure. Aboard the NIMITZ, a nuclear carrier, the environmental exposure was primarily sea spray. The panels on the US CONSTELLATION, operating in the Pacific theater, were exposed to sea spray and stack gases.

As shown in Table 3, the combination of moisture and elevated temperature affected the matrix controlled shear and flexural strength. Percent reduction of the shear strength tested at 121.1°F and after 10 months exposure aboard the NIMITZ is close to 30%.

The greater degree of degradation in the NIMITZ exposure might be attributed in part to the longer time of exposure. A much more significant effect, however, would be the actual environmental conditions during the duration of the exposure. It is known(5) that seasonal variations in humidity can promote significant moisture content differences in Gr/Ep composites, particularly if the surface to edge area ratio is not great.

It is interesting to note that conventional carriers like the CONSTELLATION have SO_2 in the shipboard environment. This apparently did not have any significant effect on resin properties of the composite. This would correlate with previous work conducted by this laboratory showing a salt laden SO_2 environment to have negligible effect on Gr/Ep mechanical properties, excepting the influence of moisture.

3.2 AS/3004 (800 Series)

The 800 series, AS fiber with 3004, a polysulfone matrix, was the only graphite system included in this study that was prepregged with a resin other than an epoxy.

Because of a limited supply of specimens, some elevated temperature tests were eliminated. Panama exposed panels, 8 and 16 ply, were tested at R.T. and 121.1°C. The 8 ply Warminster exposed panels were tested at R.T. and 121.1°C; 16 ply panels were tested at 121.1°C only.

The flexure, shear and tensile strength data of the Panama and Warminster exposed panels did not show much effect of exposure.

The elevated temperature tests of the Warminster exposed panels are interesting in that the matrix shear property does not seem to be significantly affected. Figure 9 shows only a very slight affect on interlaminar shear strength after three years of exposure.

Figure 10 shows the elevated temperature tensile data as a function of exposure time in Warminster. Little change is evident in retained strength of this fiber dominated property.

Panama exposures of this material exhibited similar behavior, reflecting little effect of exposure.

It is interesting to note that both matrix and fiber dominant properties seem to be relatively unchanged after three years of exposure. This material was found to exhibit much lower saturation values of moisture than the epoxies, about 0.3 wt %. The small effect on interlaminar shear strength at elevated temperature is undoubtedly reflected in the moisture not greatly affecting the polymer. This apparent lack of any moisture effect is in direct contrast to the behavior of the epoxy systems exposed.

3.3 T300/5208 (1000 Series)

The mechanical properties of the T300/5208, 1000 series, showed less change with increasing exposure time than the AS/3501-6 system.

The in-plane shear strength of the matrix controlled $\pm 45^{\circ}$ Panama exposed panels was not greatly affected by exposure time or elevated temperature. As shown in Figure 11, better than 90% of the residual shear strength is retained after 18 months of Panama exposure when tested at 121.1°C.

The elevated temperature in plane ($\pm$45) shear strength of the T300/5208 material does not evidence the same degree of degradation as the AS/3501-6 (300, 400 and 900 series) epoxy system.

As shown in Figure 12, there is an apparent deterioration in the in-plane shear strength with an increasing amount of Warminster exposure time, although the amount of reduction is not as much as evidenced with the other epoxy system.

Since these reduced properties are those of coated materials this deterioration can be attributed to moisture.

Data for the 16 ply mechanical tests showed similar trends in that neither the fiber or resin controlled properties were affected by exposure to the same degree as the AS/3501-6 epoxy system.

4. SUMMARY

Three different organic systems were exposed to natural weathering at two different exposure sites reflecting differing environmental conditions. Generally, the 176.6°C curing epoxy systems exhibited sensitivity to moisture particularly in elevated temperature matrix dominant properties. The T300/5208 material exhibited less degradation under these particular conditions than the AS/3501-6 material. Both these systems absorbed approximately 1% moisture after exposure in Panama and approximately .7% moisture after exposure in Warminster at saturation in the coated condition. Room temperature uncoated tensile and flexure strength were affected to some degree by a combination of U.V. radiation and moisture which is more of an erosive degradation. Coated carrier exposure correlates with Panama and Warminster site deterioration data evidencing the plasticizing effects of moisture on Gr/Ep. Carrier stack gas exposure does not appear to have much affect of mechanical properties of Gr/Ep.

Polysulfone reinforced with graphite appears to undergo very little property reductions due to moisture induced degradation. Its elevated temperature matrix dominant properties were affected less by exposure than the epoxy systems. The amount of sorbed moisture found in the AS/polysulfone at equilibrium approached .3 wt % in coated panels exposed in Panama.

5. REFERENCES

(1) J. Hertz, "Investigation into the High Temperature Strength Degradation of Fiber-Reinforced Resin Composite During Ambient Aging," General Dynamics Report No. GDCA-DBG73-005, Contract NAS 8-27435, Final Report, June 1973

(2) J. M. Augl, "The Effect of Moisture on Carbon Fiber Reinforced Epoxy Composites - II Mechanical Property Changes," NSWC/WOL/TR 76-149

(3) G. S. Springer and C. Shen, "Effects of Moisture and Temperature on the Elastic Moduli of Composite Materials," AFML-TR-78-86

(4) R. E. Trabocco, "Natural Exposure of Selected Graphite/Epoxy Composite Material Systems," NADC Report No. 80021-60

(5) P. M. Stifel, "Carrier Testing of Graphite Composites in Contact with Metals on F-4," NADC Report No. 76010-30

6. BIOGRAPHIES

Ronald E. Trabocco, Composite's Group Leader, Aero Materials Division, is responsible for in-house and contractural efforts utilizing composite materials for aircraft applications. Mr. Trabocco holds a Metallurgical Engineering degree and a M.S. (Materials Engineering) from Drexel University. Eleanor T. Vadala, has a B.S. from Juniata College and has been working for the past 3 years in the Composites Group. She has prime responsibility for Composite's environmental and impact degradation efforts.

Table 1. Graphite/Matrix Systems

Series	Fiber	Matrix - type	No. of Plies:	Ply Orientation
300	AS	3501-6 - epoxy	8 ply;	$\pm 45\ (+45\ -45\ +45\ -45)_s$
400	AS	3501-6 - epoxy	10 ply;	$(0\ +45\ -45\ +45\ -45)_s$
800	AS	3004 polysulfone	8 ply: 16 ply:	+ 45 $(+45\ 0\ -45\ 0\ +45\ 0\ -45\ 90)_s$
900	AS	3501-6 - epoxy	8 ply; 16 ply	+45 $(+45\ 0\ -45\ 0\ +45\ 0\ -45\ 90)_s$
1000	T300	5208 - epoxy	8 ply 16 ply;	+45 $(+45\ 0\ -45\ 0\ +45\ 0\ -45\ 90)_s$

Table 2. Environmental Parameters of Exposure Sites

	Fort Sherman (Canal Zone) (Coco Solo)	Naval Air Facility (Warminster, PA)
Precipitation, in.	130	40
Temperature, °F	86 to 87	61 to 43
Relative Humidity, %	82	63 to 90

Table 3. Results of Exposure Aboard US CONSTELLATION and NIMITZ Carriers

	Unpainted Unexposed	US CONSTELLATION Exposure	NIMITZ* Exposure
Tensile Strength			
P.S.I.			
R.T.	55831	62922	58821
82.2°C	57432	61674	63292
121.1°C	61016	60111	57767
Shear Strength			
P.S.I.			
R.T.	6257	6149	6015
82.2°C	5964	6157	6807
121.1°C	6108	5920	4276
Flexural Strength			
P.S.I.			
R.T.	53232	52178	56489
82.2°C	52389	47253	59933
121.1°C	53544	46164	49191

* Nuclear Carrier

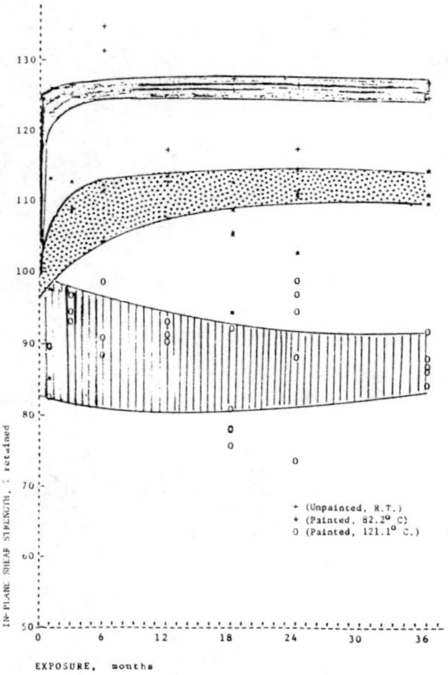

Fig. 1 300 Series, 8 Ply (± 45) Panama, % Retained In-plane Shear Strength

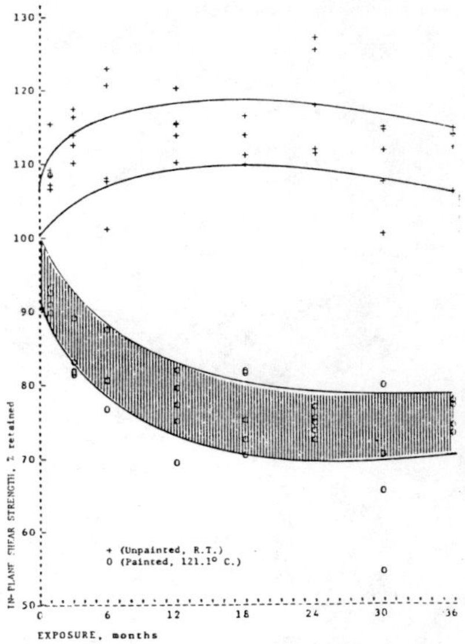

Fig. 3 900 Series, 8 ply (± 45) Panama, % Retained In-plane Shear Strength

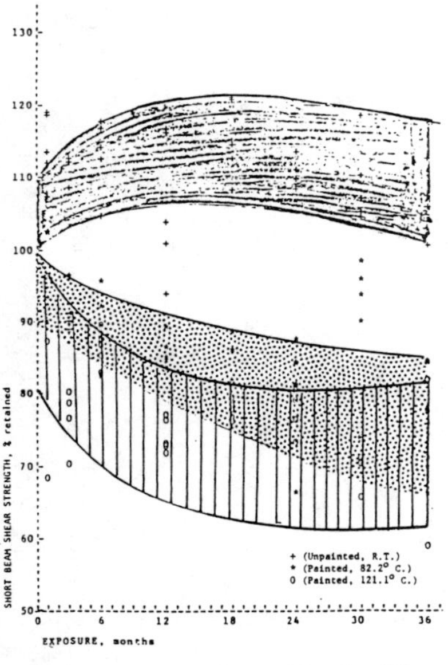

Fig. 2 400 Series, 10 Ply (0 ± 45 90) Warminster, % Retained Short Beam Shear Strength

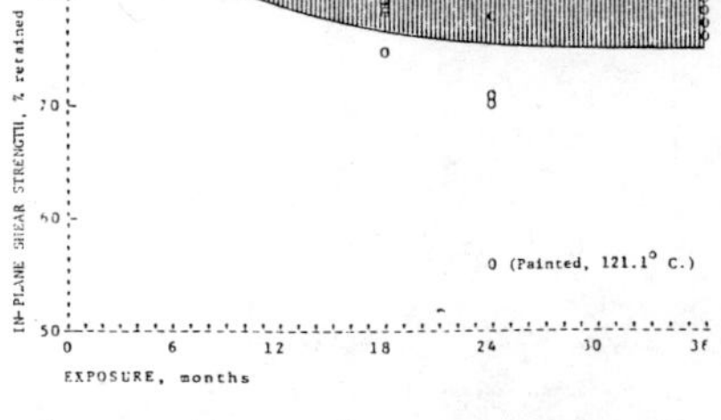

Fig. 4 900 Series, 8 Ply (± 45) Warminster, % Retained In-plane Shear Strength

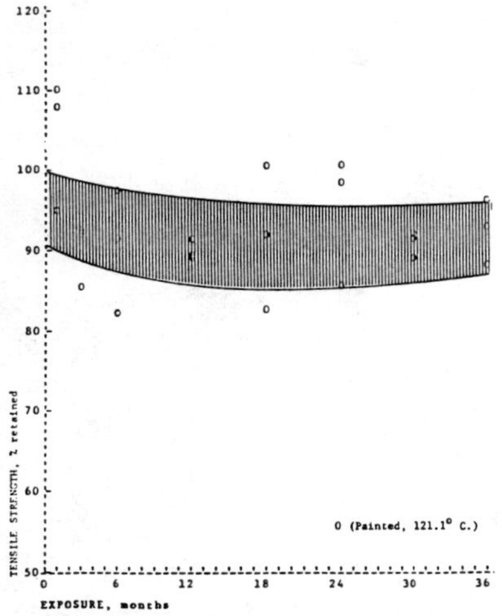

Fig. 5 900 Series, 16 Ply (0 ±45 90) Panama, % Retained Tensile Strength

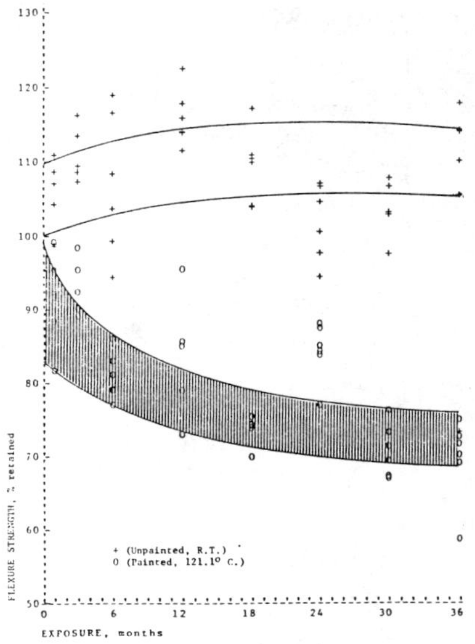

Fig. 7 900 Series, 16 Ply (0 ±45 90) Panama, % Retained Flexure Strength

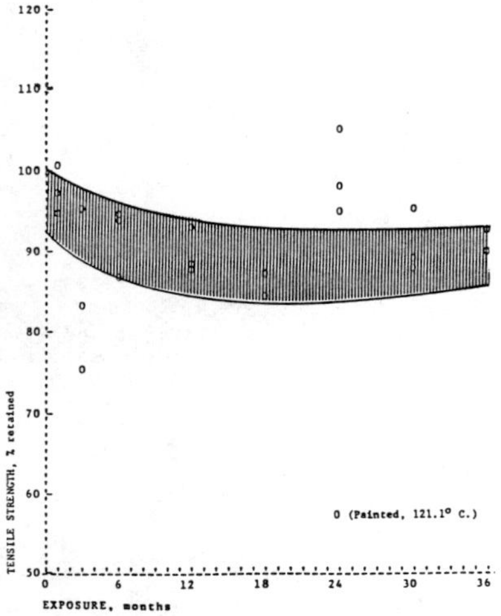

Fig. 6 900 Series, 16 Ply (0 ±45 90) Warminster, % Retained Tensile Strength

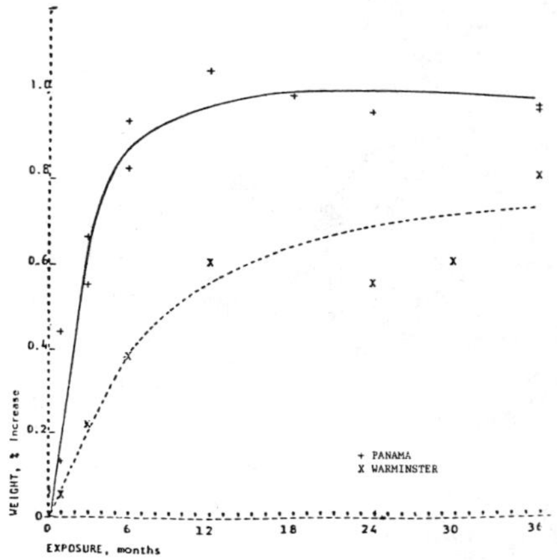

Fig. 8 900 Series, 16 Ply (0 ±45 90) Panama (+) and Warminster (x̄) % Weight Increase, Painted Panels

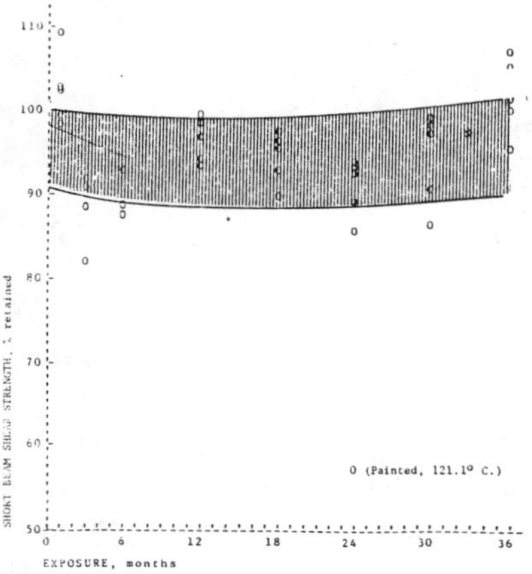

Fig. 9 800 Series, 16 Ply (0 ±45 90)
Warminster, % Retained
Shear Strength

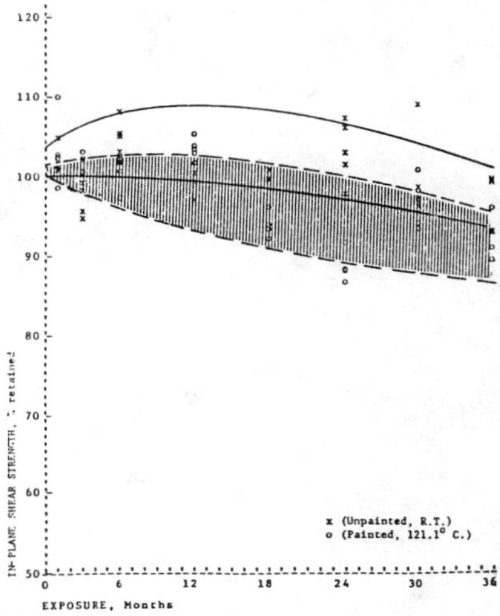

Fig. 11 1000 Series, 8 Ply (+45)
Panama, % Retained
In-plane Shear Strength

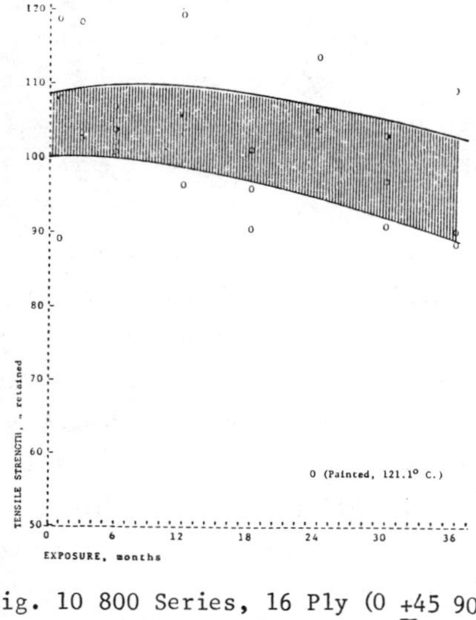

Fig. 10 800 Series, 16 Ply (0 ±45 90)
Warminster, % Retained
Tensile Strength

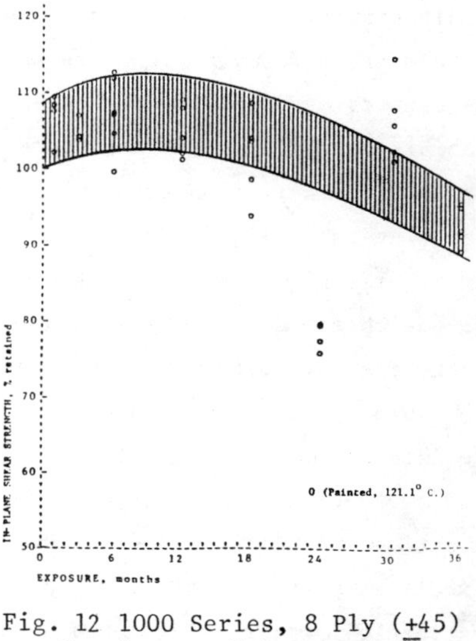

Fig. 12 1000 Series, 8 Ply (±45)
Warminster, % Retained
In-plane Shear Strength

13th National SAMPE Technical Conference
October 13-15, 1981

GLASSY ALLOYS-NEW CLASS OF CORROSION RESISTANT COATINGS

Ronald B. Diegle
BATTELLE
Columbus Laboratories
505 King Avenue
Columbus, Ohio 43201

ABSTRACT

This paper reviews the results of research into the corrosion behavior of glassy alloys prepared by techniques other than liquid quenching. Techniques that are considered are sputtering, ion plating, electrodeposition, laser glazing, and ion implantation. In some cases, glassy alloys prepared by a procedure such as sputtering or electrodeposition exhibit corrosion resistance comparable to their counterparts prepared by rapid liquid quenching. In others, glassy alloys that cannot be obtained by rapid liquid quenching have been made and found to exhibit even better corrosion resistance than that of glassy liquid quenched alloys. Certain compositions of glassy alloys offer excellent pitting resistance and extremely low corrosion rates (less than 1 μm/yr) with as little as 2 or 3 percent of chromium. Therefore, they are ideal candidates for use as protective coatings, and they offer the additional benefit of conservation of chromium and other critical- resource alloying elements.

Keywords: sputtering, ion plating, electrodeposition, laser glazing, ion implantation.

1. INTRODUCTION

The advent of an unconventional class of alloys, commonly referred to as "glassy" or "amorphous", that is extremely corrosion resistant offers the possibility for development of new types of protective coatings. Unlike conventional engineering alloys commonly used for structural and other purposes, these alloys are not crystalline. Their atomic structure has been described as a stop-action photograph of the liquid melt.(1) Although they are sometimes referred to as "amorphous", the existence of some short range order over several atom spacings has prompted the use of the term "glassy" in this paper to distinguish them from truly amorphous materials. Glassy alloys have been prepared and studied since 1960 and the pioneering work of Duwez, et al.[2], although most research directed toward their corrosion behavior and methods for applying them as corrosion resistant coatings has been performed since the mid-1970's.

Certain compositional classes of glassy alloys exhibit truly remarkable resistance to general and localized forms of corrosion, as reviewed by Masumoto and Hashimoto[3] and more recently by Diegle, et al.[4] The majority of research investigations with glassy alloys, whether related to corrosion behavior or other properties, has been performed with filaments or strips prepared by rapid quenching of liquid melts. Alloys of the corrosion resistant variety are typically prepared by quenching a stream of molten melt at rates of the order 10^5 to 10^6 °K and higher. Many experimental techniques have been devised for rapidly quenching liquid metals, such as those described in References 5 through 8. However, there is a growing body of literature describing the properties of glassy alloys prepared by techniques other than rapid liquid quenching, such as sputtering, ion plating, laser glazing, electrodeposition, and ion implantation. Because these latter techniques are particularly suitable for preparing glassy metallic coatings, they are considered below in some detail.

2. COATING PROCEDURES

The bulk of corrosion resistant glassy alloys is currently prepared by rapid quenching from the melt. The geometries are typically filament or strip with thicknesses ranging to about 75 μm and widths to 2 cm and more; in these forms they are not amenable for application as coatings. For example, they cannot be clad by hot rolling procedures because they crystallize and/or undergo embrittlement. They are often too brittle to be clad by cold rolling, and their ability to form protective passive films hinders the cold bonding process. Explosive cladding is a possibility, but this is a highly specialized procedure warranted by only a relatively few applications. The adiabatic heating associated with this process, and the limited ductility of many glassy alloys, place additional limitations on the use of this procedure. However, five techniques have been demonstrated to be suitable for applying glassy alloy coatings. These techniques utilize vapor deposition, surface fusion, deposition from liquid media, and ion implantation as discussed below.

2.1 Sputtering and Ion Plating

Sputtering has been used successfully for preparing glassy alloy coatings with good corrosion resistance. Comparison of anodic polarization behavior of sputtered glassy Fe-base alloys with that of T-304 stainless steel in acidified

0.1 M NaCl is shown in Figure 1.[9]

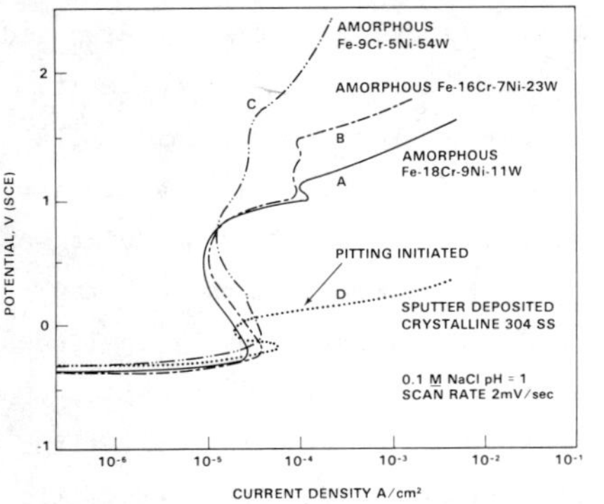

Figure 1. Anodic polarization curves comparing the corrosion behavior of sputter deposited crystalline T304 stainless steel with that of glassy Fe-Cr-Ni-W alloys (Reference 9).

These alloys were prepared by high rate sputter deposition to thicknesses ranging to several hundred μm. The critical pitting potential, E_{cp}, of T304 steel was below 0 V(SCE), whereas E_{cp} of sputtered Fe-18Cr-9Ni-11W, Fe-16Cr-7Ni-23W, and Fe-9Cr-5Ni-54W alloys exceeded 1 V(SCE). The elevation of E_{cp} by over 1000 mV for the sputtered alloys is the result of the homogeneous glassy structure combined with the film-forming properties of tungsten. The alloys crystallized in the temperature range from 600 to 800° C, depending on the concentration of tungsten used for stabilizing the glassy structure. Annealing at 400 C and 500 C for 24 hours slightly increased the corrosion rate in the passive potential range, Figure 2,[10] but it did not decrease E_{cp}.

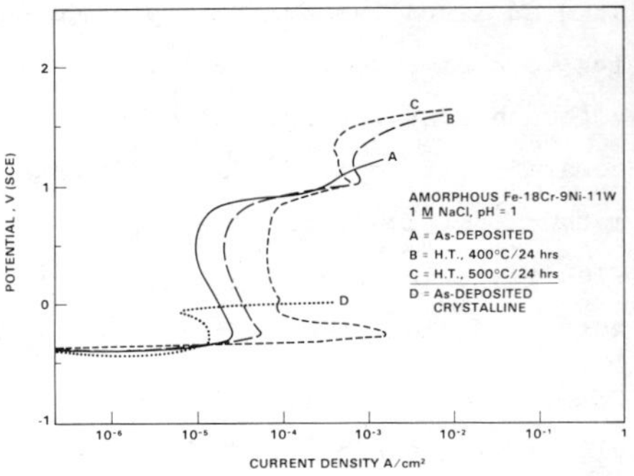

Figure 2. Anodic polarization curves showing the effect of heat treatment on the corrosion behavior of glassy Fe-18Cr-9Ni-11W (Reference 10).

Comparison of a glassy Fe-50W tungsten sputtered alloy with a series of Fe-XW-13P-7C alloys prepared by melt spinning is shown in Figure 3.[11] The addition of W to the

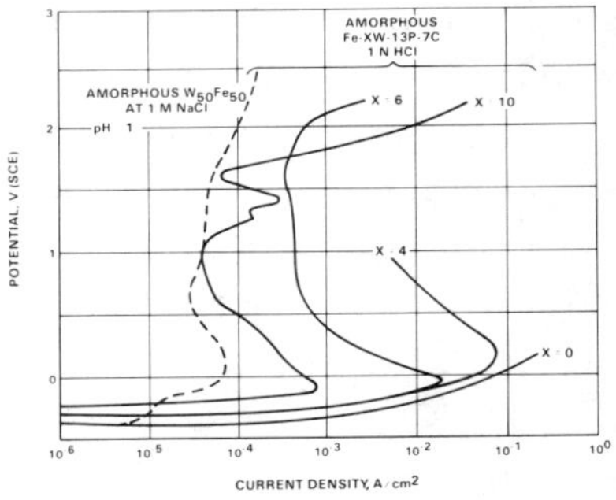

Figure 3. A comparison of the corrosion behavior of glassy Fe-50W with liquid-quenched glassy Fe-XW-13P-7C alloys (Reference 11).

latter alloys decreased the critical current density required for passivation, and it also lowered the corrosion rate (current density) in the passive potential region. However, the sputtered Fe-50W alloy was spontaneously passive throughout the entire potential range of nearly 3 V, and it exhibited the smallest corrosion rate in the passive region. A value of E_{cp} of over 2.5 V(SCE) in pH 1, 1 N NaCl solution indicates excellent resistance to pitting corrosion.

Another investigation utilized glassy Fe-30Ni-2Cr-2V-15P-6B and Fe-30Ni-16Cr-2V-15P-6B alloys.[12] Its purpose was to compare the corrosion behavior of glassy alloys of identical composition, but prepared by both sputtering and liquid quenching techniques. Representative anodic polarization curves are shown in Figures 4 and 5. The 2 Cr

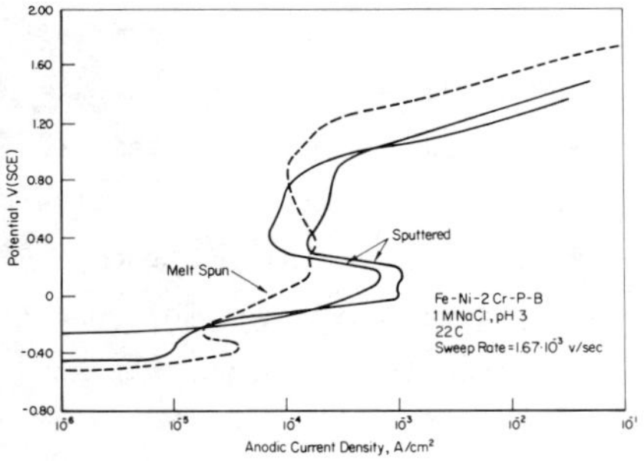

Figure 4. Anodic potentiodynamic polarization curves of sputtered and melt spun 45Fe-30Ni-2Cr-2V-15P-6B alloys in 1 M NaCl, pH 3 (Reference 12).

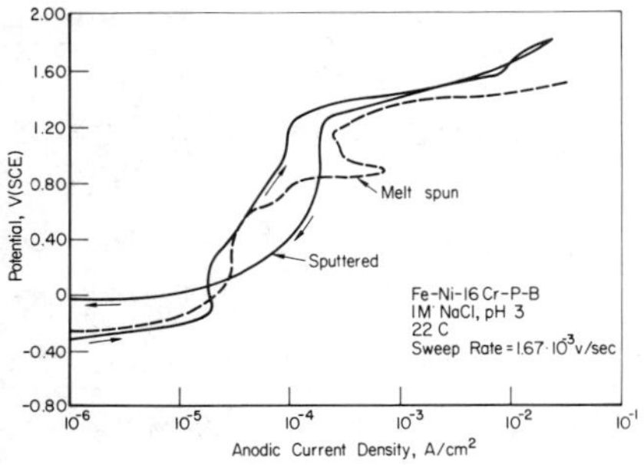

Figure 5. Anodic potentiodynamic polarization curves of sputtered and melt spun 31Fe-30Ni-16Cr-2V-15P-6B alloys in 1 M NaCl, pH 3 (Reference 12).

alloy was spontaneously passive as prepared by melt spinning and spontaneously active as sputtered, but the corrosion rates at passive potentials were virtually identical. More significantly, sputtering resulted in a value of E_{cp} of about 0.9 to 1.0 V(SCE), which compares faborably with 1.2 V(SCE) for the melt spun alloy. The 16 Cr alloy was spontaneously passive when prepared by either technique, and there was initially no difference in corrosion behavior in the passive potential region. E_{cp} in both cases was about 1.2 V(SCE). The corrosion rates in 10 percent $FeCl_3$ solution are shown in Figure 6.[13] For both

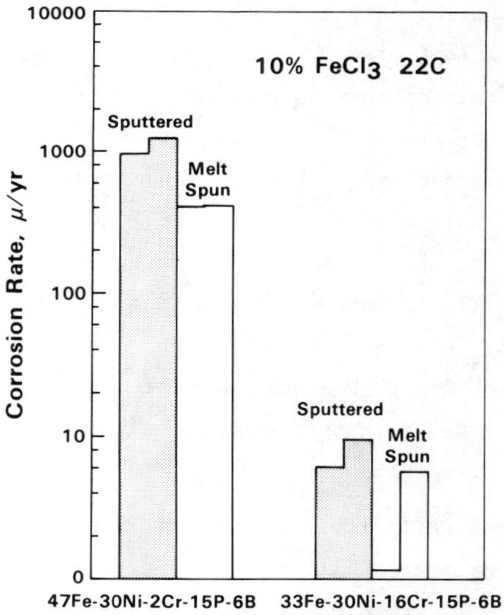

Figure 6. Corrosion rates in 10 percent $FeCl_3$ of glassy alloys prepared by sputtering and melt spinning. (Offset bars indicate results from duplicate experiments; Reference 13).

compositions the corrosion rates of sputtered alloys were a factor of about three qreater than those of melt spun alloys. Increasing the Cr concentration from 2 to 16 atomic percent decreased the corrosion rates by a factor of 100.

Binary glassy alloys of the general composition T_1-T_2 have been prepared, where T_1 is an early transition metal such as Ti, Ta, Mo, or W and T_2 is a late transition metal such as Re, Fe, Co, Ni, or Cu.[14] Corrosion was studied in 1 N NaCl solutions with pH of 1 or 7. Good corrosion resistance was observed with alloys of WFe and TiCu, which possessed E_{cp} values in excess of 2.5 V(SCE) and passive current densities of about 10^{-5} A/cm^2. The corrosion resistance was poor for alloys of MoCo, MoNi, and TaRe containing more than about 50 atomic percent of the T_2 element. The morphology of attack was that of rapid general dissolution, indicating the absence of protective film formation. Concentrations of T_2 elements above critical levels of about 50 atomic percent interfered with the formation of passive films.

An early investigation of rf getter-sputtered alloys was performed with T304 stainless steel with and without additions of 1 to 2 atomic percent of Al and/or Si.[15] Glassy microstructures resulted containing small crystallites 15-30 μm in diameter. Anodic polarization in NaCl solutions was used to show that sputtering effectively increased E_{cp} above that of conventional T304 stainless steel. Si increased E_{cp} to above 1.0 V(SCE), compared with -0.07 V(SCE) for bulk T304. Annealing at 450° C for one hour lowered E_{cp} to about 0.48 V(SCE), but this value was still substantially higher than that measured with conventionally prepared T304 steel.

Recent research[16,17] has shown that ion plating also can be used to prepare glassy Fe-Cr alloys with good pitting resistance in NaCl solution. Anodic current decay behavior for ion plated and liquid quenched glassy alloys was similar in certain potential ranges, indicating similar kinetics of passive film formation. Ion plating, like sputtering, is therefore a technologically important tool for depositing protective, glassy, metallic films.

2.2 Electrodeposition

The ability to electrodeposit glassy corrosion resistant alloys easily would represent a breakthrough in the ability to utilize these materials as protective coatings. Electrodeposition is easier, faster, and more economical than sputtering and ion plating, which are important advantages when commercial utilization is being considered.

Although electroless deposition of glassy Ni-P alloys has been practiced for years, the corrosion resistance of this material is inferior to that of alloy compositions described previously. This inferiority is caused by the absence in the alloy of a good film-forming element, such as Cr, Ti, or W. At present, the author is aware of only one study in which glassy inert alloys have been prepared by electrodeposition.[18] Two types of solutions have been used, namely, sulfate-hypophosphite and chloride-hypophosphite baths. By varying the cathodic current density and bath temperature, alloys have been prepared with the sulfate-hypophosphite bath with general compositions in the range (3-53)Ni-(4-57)Fe-(1-15)Cr-(31-47)P (atomic percent). Similar variations with the chloride-hypophosphite baths resulted in alloy compositions in the range (2-4)Ni-(66-71)Fe-(1-9)Cr-(21-28)P. In both baths a chemical was added to form complex compounds with Fe^{++} and Ni^{++} ions; this was done to shift the deposition potentials in the cathodic direction to facilitate co-deposition with chromium, which has a considerably more electronegative reversible potential. Phosphorus was obtained by reduction of the hypophosphite ion.

Deposits up to several μm thick adhered well to the substrate. Thicker deposits tended to flake off due to high internal stresses resulting from the plating process. It was postulated that the internal stresses may have arisen from absorption of hydrogen or occlusion of basic chromium salt; in either case, the stress is believed to be related to the low cathodic current efficiency of less than 5 percent.

Corrosion tests were performed in 1N HCl and aqua regia at 80° C. Fe-Ni-Cr-P deposits containing less than 7 atomic percent of Cr dissolved rapidly, whereas those containing 7 to 8 atomic percent of Cr dissolved at very slow rates. Those containing more than 7 to 8 atomic percent of Cr did not react at perceptible rates in these acids. This behavior compares favorably with that of glassy Fe-Ni-Cr-P-B and Fe-Ni-Cr-P-C alloys prepared by rapid liquid quenching.

2.3 Laser Glazing

Laser glazing has been used to make glassy surface layers of the general composition Fe-Ni-Cr-P-B.[19] In one series of experiments a laser beam was passed across a polycrystalline ingot of composition 31Fe-30Ni-10Cr-2V-15P-6B to determine whether it could be made glassy. The polycrystalline ingot had been made by conventional melting and homogenizing of components under argon, and then casting. A rectangular piece was removed from the ingot, mounted onto a plate that rotated the piece at a circumferential speed of 320 cm/sec, and then heated by a laser with a 5 kW power input. The ingot sample then was mounted and polished to present a cross sectional view of the laser-fused region and the substrate. The cross section was then swabbed with a 10 percent $FeCl_3$ solution, which rapidly etches the crystalline form of this alloy but barely attacks the glassy form.[20] The cross section showed a surface layer about 50 μm thick that was not perceptibility attacked and which comprised most of the laser-glased region. The layer was heavily cracked, the result of the internal stresses created by the laser glazing procedure. This layer is inferred to have been predominantly glassy in structure because of its resistance to etching by the $FeCl_3$ solution, although X-ray diffraction or electron diffraction measurements were not performed.

In a subsequent series of experiments a poweder of the above alloy composition was sprayed onto a foreign (nickel) substrate, and then laser glazed. The powder was first fused onto the nickel by using a helium shield gas and translating the specimen beneath a laser beam of 8 kW at a rate of 1 cm/sec. Next, laser glazing was performed at a translation speed of 600 cm/sec and power inputs of from 5 to 10 kW. This composite alloy also was evaluated by exposure of a cross section to 10 percent $FeCl_3$ solution. As shown in Figure 7, a thin (about 70 μm) laser glazed surface layer was relatively unetched; severe pitting occurred at the layer-substrate interface, as well as heavy etching of the substrate. Fine microcracks penetrated the layer along its length; no attempt was made to optimize the laser glazing conditions to determine whether microcrack formation could easily be avoided.

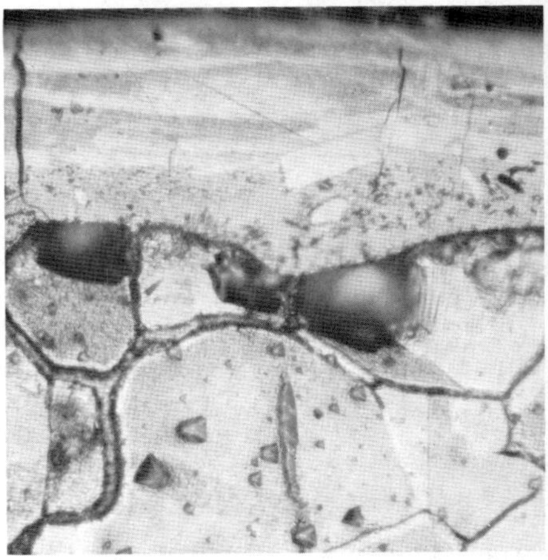

Figure 7. Layer of 31Fe-30Ni-16Cr-2V-15P-6B fused onto nickel substrate by laser glazing, then cross sectioned, polished, and exposed to 10 percent $FeCl_3$ solution. Severe pitting and etching is evident on substrate and at interface and only light etching on glazed layer (Reference 19).

Laser skin melting (LSM) has been used to modify surface properties of alloys, but usually the resulting alloy is microcrystalline rather than glassy. LSM has been applied to austenitic stainless steel,[21] titanium alloys,[22] and aluminum bronze.[23] It has been found to eliminate susceptibility to intergranular corrosion while increasing stress corrosion cracking resistance of sensitized stainless steel, to affect stress corrosion cracking resistance of titanium alloys, and to reduce the selective dissolution of Al from aluminum bronze, respectively. Glassy surface layers were not formed on these alloys as a result of LSM.

2.4 Ion Implantation

It has been demonstrated that ion implantation can produce glassy surface alloys.[24] With increasing ion fluence the target changes from the crystalline to the glassy state as a result of accumulation of lattice defects, and it can be envisioned as a gradual accumulation of disorder. Ni implanted with about 10^{17} ions cm^{-2} of P^+ exhibited a selected area electron diffraction pattern of diffuse rings characteristic of a glassy phase. If implantation is made into targets of thickness about twice the mean projected ion penetration range, the specimen can be rendered glassy through the entire thickness. Transition from the crystalline to glassy state is dose dependent. For the case of P^+ implantation into Ni, Co, and Fe alloys, concentrations of P less than about 10 atomic percent resulted in crystalline films, a concentration of about 15 atomic percent favored a mixed structure, and concentrations above 20-25 atomic percent resulted in a completely glassy structure throughout the film thickness of about 60 μm.

There is ongoing research characterizing the corrosion properties of ion implanted surface layers, which in most cases are crystalline rather than glassy. Agarwal, et al.,[25] surface alloyed T430 stainless steel with Ni^+ ions and showed that the critical and passive current densities in 1 N H_2SO_4 were reduced. Studies in 1N NaCl indicated improved pitting resistance. TEM analysis showed high levels of radiation damage, although selected area electron diffraction indicated a nickel-rich bcc structure. The pitting resistance of an M50 bearing steel (approximate composition: Fe-4Cr-4Mo-09V-0.8C-0.15Si-0.2Mn, minor amounts of P, S, Ni, Co, W, and Cu) was improved by implanting the surface with 20-25 percent of Cr.[26] Singular additions of Cr, Mo, and Ti substantially increased E_{cp} of M50 in chloride-containing solutions, and the highest pitting resistance occurred with a binary implantation of Cr and Mo. The degree of disorder produced by implantation was not reported. Ashworth, et al.,[27] studied the effect of Cr ion implanted into pure Fe. They found that Cr implantation improved the corrosion resistance of iron, as expected; furthermore, the electrochemical behavior of Fe-Cr surface layers produced by ion implantation resembled that of conventionally prepared Fe-Cr alloys. No statements were made concerning the structure of the implanted layers, which are presumed to have been crystalline. In a subsequent study of Ta-implanted Fe,[28] it was shown that Ta-rich surface layers improved the corrosion resistance of iron, which resembled that of an Fe-4.9Cr alloy in an anodic solution. Beneficial effects have also been observed for titanium implanted with Pt^+[29,30] and Pd^+[31].

A particularly relevant study consisted of implanting P^+ into T304 stainless steel. The presence of P with Cr in glassy alloys is known to improve the corrosion resistance substantially relative to Cr-containing glassy alloys without P.[32] P^+ implantation resulted in a microcrystalline surface layer that passivated with a mixed Fe, Cr hydroxide film, which was thinner and contained less Cr than the mixed Fe, Cr oxyhydroxide layer formed on unimplanted steel. There was little variation between the anodic polarization behavior of implanted and unimplanted specimens in 1N H_2SO_4. In 0.1 N NaCl the passive current density was less for the P^+-implanted steel as shown in Figure 8, but the critical pitting poten-

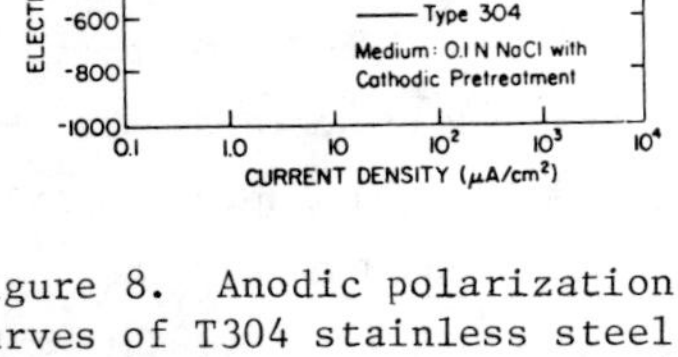

Figure 8. Anodic polarization curves of T304 stainless steel and P^+ implanted T304 stainless steel in 0.1 M NaCl with cathodic pretreatment (Reference 32).

tial was reduced from 0.40 to 0.17 V(SCE). The inferior pitting resistance of the P^+-implanted alloy may have resulted from the absence of a glassy structure, because crystallization of glassy alloys by annealing substantially decreases corrosion resistance.[20,33]

3. CONCLUSIONS

Glassy alloys of certain compositions containing film-forming elements exhibit excellent resistance to localized and general forms of corrosion. Because their corrosion rates can be less than 1 μm/yr, they offer promise as corrosion-resistant coatings for use on base metal substrates. Several techniques have been utilized for applying glassy alloys; these include sputtering, ion plating, electrodeposition, laser glazing, and ion implantation. In many cases glassy alloys prepared by these techniques have corrosion resistance that is comparable to glassy alloys of similar compositions but prepared by rapid liquid quenching. Disadvantages of glassy alloy coatings include a tendency for crystallization at elevated temperatures and susceptibility to hydrogen embrittlement. Nevertheless, they offer promise both as a practical means for corrosion control and conservation of critical-resource metals in a wide variety of applications.

4. REFERENCES

1. Chaudhari, P., Giessen, B. C., and Turnbull, D., Sci. Am. 242, 98 (1980).
2. Duwez, P., Willens, R. H., and Klement, W., J. Appl. Phys. 31, 1136 (1960).

3. Masumoto, T., and Hashimoto, K., Ann. Rev. Mater. Sci. 8, 215 (1978).

4. Diegle, R. B., Sorensen, N. R., Tsuru, T. and Latanision, R. M., submitted for publication in Treatise in Materials Science and Technology, ed. J. C. Scully.

5. Cahn, R. W., Krishnanand, K. D., Laridjani, M., Greenholz, M., and Hill, R., in Rapidly Quenched Metals, Section II, ed. N. J. Grant and B. C. Giessen, Elsevier, Lausanne (1976), pp. 83-86.

6. Davies, H. A., in Rapidly Quenched Metals III, ed. B. Cantor, The Metals Society, London (1978), Vol. 1, pp 1-22.

7. Kavesh, S., in Metallic Glasses, eds. J. J. Gilman and H. J. Leamy, American Society for Metals, Metals Park (1976), pp. 36-73.

8. Maringer, R. E., Mobley, C. E. and Collings, E. W., in Rapdily Quenched Metals, eds. N. J. Grant and B. C. Giessen, The Massachusetts Institute of Technology, Cambridge (1976), pp. 29-36.

9. Wang, R., unpublished research.

10. Wang, R., presented at fall meeting of TMS-AIME, Pittsburgh, October 5-9, 1980.

11. Wang, R., Merz, M. D. and Thomas, M. T., Abstract No. 264, presented at the fall meeting of The Electrochemical Society, Los Angeles, October 14-19, 1979; ECS Extended Abstracts 79-2, p. 680.

12. Diegle, R. B. and Merz, M. D., J. Electrochem. Soc. 127, 2030 (1980).

13. Diegle, R. B. and Lineman, D. M., Interim Technical Report No. ONR-00014-77-C-0488-2 to Office of Naval Research, p 55 (May 1979).

14. Wang, R. Abstract No. 236, presented at the fall meeting of The Electrochemical Society, Hollywood, Florida, October 5-5-10, 1980; ECS Extended Abstracts 80-2, p. 620.

15. Nowak, W. B., Mater. Sci. Eng. 23, 301 (1976).

16. Nowak, W. B. and Okorie, B. A., Abstract No. 255, presented at the fall meeting of The Electrochemical Society, Los Angeles, October 14-19, 1979; ECS Extended Abstracts 79-2, p. 662.

17. Nowak, W. B. and Okorie, B. A., submitted for publication to Corrosion Journal (1981).

18. Li-Qun, R. and Mou-Xian, S., to be published.

19. Ayers, J. and Diegle, R. B., unpublished research (1980).

20. Diegle, R. B., to be presented at Fourth International Conference on Rapidly Quenched Metals, Sendai, Japan, August 24-28, 1981.

21. Anthony, T. R. and Cline, H. E., J. Appl. Phys. 49, 1248 (1978).

22. Moore, P., Kim, C., and Weinman, L., in Applications of Lasers in Materials Processing, American Society for Metals, Metals Park (1979).

23. Draper, C. W., Woods, R. E., and Meyer, L. S., Corrosion 36, 405 (1980).

24. Grant, W. A., Ali, A., Chadderton, L. T., Grundy, P. J., and Johnson, E., in Rapidly Quenched Metals III, ed. B. Cantor, The Metals Society, London (1978), Vol. 1, pp. 63-72.

25. Agarwal, S. B., Wang, Y. F., Clayton, C. R., and Herman, H., Thin Solid Films, Elsevier, Lausanne (1979), Vol. 63, pp. 19-25.

26. Wang, Y. F., Clayton, C. R., Hubler, G. K., Lucke, W. H., and Hirvonen, J. K., ibid. pp. 11-18.

27. Ashworth, V., Baxter, D., Grant, W. A., and Proctor, R. P. M., Corr. Sci. 16, 775 (1976).

28. Ashworth, V., Baxter, D., Grant, W. A., and Proctor, R. P. M., Corr. Sci. 17, 947 (1977).

29. Gleaves, G. A., Dearnaley, G., and Collins, R. A., J. Phys. F: Metal Phys. 7, 2031 (1977).

30. Thompson, N. G., Lichter, B. D., Appleton, B. R., Kelly, E. J., and White, C. W., Ion Implantation Metallurgy, ed. C. M. Preece and T. K. Hirvonen, TMS-AIME (1979), p. 181.

31. Hubler, G. K., and McCafferty, E., Corr. Sci. 20, 103 (1980).

32. Clayton, C. R., Doss, K. G. K., Herman, H., Prasad, S., Wang, Y-F, Hirvonen, J. K., and Hubler, G. K., Ion Implantation Metallurgy, ed. C. M. Preece and J. K. Hirvonen, TMS-AIME (1979), p. 65.

33. Naka, M., Hashimoto, K., and Masumoto, T., Corr. 36, 679 (1980).

BIOGRAPHY

Dr. Ronald B. Diegle is Senior Research Metallurgist in the Corrosion Section of Battelle's Columbus Laboratories, where he directs a wide range of corrosion-related research projects. Between 1977 and 1980 he investigated the localized corrosion behavior of glassy Fe-Ni-Cr-P-B alloys for the Office of Naval Research. Currently he is studying the repassivation kinetics of these alloys, as influenced by phosphorus and chromium. Other interests include development of corrosion monitoring devices and investigating the basic mechanisms of corrosion of steel in soils.

13th National SAMPE Technical Conference
October 13-15, 1981

ELECTROCHEMICAL AND MICROSTRUCTURAL CHARACTERISTICS OF 7000 AND 3000 SERIES POWDER METALLURGY ALUMINUM ALLOYS+

M. J. Topolski, V. Agarwala++ and M. J. Koczak
Drexel University
Department of Materials Engineering
Philadelphia, Pennsylvania 19104

Abstract

The study considers the electrochemical compatibility and microstructural characteristics of powder metallurgy, fully dense alloys, i.e. a high strength 7000 series Al-Zn-Mg-Cu alloy, CT91, and a high modulus Al-Mn alloy. The electrochemical behavior of the test specimens was evaluated utilizing two techniques: (a) a three electrode, potentiostatic, controlled-potential corrosion method and (b) a corrosion couple method. The controlled-potential corrosion method measured the rate of corrosion of the samples at a constant potential of -500mV, -700mV and -900mV. In contrast, the corrosion couple experiment evaluated the effect of the corrosive environment between the Al-Zn-Mg-Cu and Al-Mn alloys. Microstructural characterization illustrates the corrosion effects with varying alloy chemistries. The mode of corrosion for the Al-Zn-Mg-Cu alloy is intergranular and pitting corrosion while the Al-Mn alloy undergoes general pitting corrosion, with the pits initiating in the matrix phase. The results of the experimental study suggest that the Al-Zn-Mg-Cu alloy is slightly more noble than the Al-Mn alloy. Both materials exhibit similar potentials, versus the saturated calomel electrode, in the -800 to -900mV range. The corrosion rates, for the CT91 and Al-Mn alloys appear to be comparable over the range of constant potentials examined and suggests corrosion compatibility. In addition, the corrosion-couple evaluations indicate that the corrosion behavior can be minimized by appropriate design or selection of matrix and dispersoid volume fractions and chemistries.

1. INTRODUCTION

The formulation and development of a high modulus corrosion resistant

---

+ Program support provided by NADC, Warminster, Pa.

++ Located at the NADC, Corrosion Section, Warminster, Pa.

powder processed aluminum duplex alloy very much depends on the electrochemical properties of its constituents. If the matrix alloy and the dispersoid are of a differing nature, it is very likely that an electrochemically noble constituent could lead to corrosion of the other and vice versa on a localized microstructural scale. These corrosion effects could lead to damages like exfoliation, pitting and even enhanced corrosion assisted stress cracking failures. The 7000 series alloys are widely used in high strength structural applications in the aerospace industry and are known to be highly susceptible to stress corrosion cracking, exfoliation and pitting[1-4]. Although powder processed alloys such as CT-91 (or MA87) have shown higher corrosion resistance[4], due to the absence of macrostructural segregation directly attributed to the powder metallurgy techniques. In other words, the alloy homogeneity in powder processed alloys is indicative of insignificant separation of active and noble phases in the matrix which do not contribute to formation of any corrosion cells, i.e. galvanic couples. In contrast to ingot metallurgy process, as in 7075 alloy, the intergranular microstructural corrosion leads to high exfoliation corrosion.

In the present work, where a duplex alloy has been formulated, the high modulus Al-Mn alloy phase is the high modulus dispersoid in the 7000 series CT-91 matrix. It will not hav the same corrosion effects as the intermetallics of Al-Zn-Mg-Cu alloy. Thus, it is important that the corrosion behavior of such constituents be determined and compared in terms of their compatibility. Electrochemical techniques have been most useful in this respect and have been successful in determining the microstructural corrosion susceptibility in significant detail. In particular, the controlled potential measurement has been most applicable in locating the preferential corrosion susceptibility of a microstructure, grain boundary, etc. in most homogeneous alloys[5,6]. The duplex alloys contain a microstructure of a dissimilar chemical composition and crystallography, thus a difference in chemical potential or corrosion potential exists and must be minimized. Electrochemically, these alloys can be made to corrode at these potentials to determine their corrosion kinetics. Hence, such methods have beem employed here to characterize two phase aluminum alloys. As a result, it is the objective of the study to determine the corrosion behavior of a duplex powder processed aluminum alloy in order to evaluate and minimize the corrosion response.

2. EXPERIMENTAL PROCEDURE

2.1 Materials

Processing and recent alloy developments have produced high strength aluminum alloy systems based on the

Al-Zn-Mg-Cu and Al-Mn alloys. By careful control of the constituent particles, size and volume fraction of the dispersoids, aging and grain size, improvement of strength and modulus is possible. Gas atomized powders, of composition shown in Table I, were consolidated into fully dense billets through the process of cold isostatic compaction followed by hot working by Alcoa.

2.2 Corrosion Evaluation

The experimental program consisted of examining the electrochemical behavior of a high strength Al-Zn-Mg-Cu alloy, a 7000 series P/M alloy, and a high modulus Al-Mn alloy. The corrosion environment was a 3.5% solution of NaCl in distilled, de-ionized water open to the laboratory atmosphere at room temperature. The techniques for evaluating the electrochemical behavior of the samples were a three electrode, potentiostatic, controlled-potential corrosion method and a corrosion couple method.

The samples for evaluation of electrochemical behavior were cold mounted in a quick setting epoxy resin. Samples for the three electrode, potentiostatic, controlled-potential corrosion method had one face 1.27cm x 1.27cm (0.5 x 0.5 inch) exposed. Similarly, the corrosion couple method also had one face exposed. One couple consisted of a sample of Al-Zn-Mg-Cu with an exposed area fraction of 84% vs. Al-7.5% Mn with an area fraction of 16%. A second couple consisted of Al-Zn-Mg-Cu (70%) and Al-5% Mn, 70% and 30% area fractions, respectively.

The exposed surface of each test sample was polished using standard metallographic techniques. A number of samples following the corrosion evaluation were etched using Keller's reagent to reveal the microstructure and the mode and mechanism associated with corrosion.

The techniques for evaluating the electrochemical behavior of the samples were a three electrode, potentiostatic, controlled-potential corrosion method, Figure 1, and a corrosion couple method, Figure 2. The three electrode controlled-potential corrosion method incorporates a working electrode, a counter electrode, and a reference electrode. The working electrodes were the test samples mounted on a conventional electrode holder through a teflon gasket. A saturated calomel electrode (S.C.E.) was used as a reference and a platinum electrode was used as a counter. A predetermined constant potential was maintained for the test sample with respect to the S.C.E. reference. The constant potentials and corresponding polarizing currents were maintained potentiostatically using a platinum counter electrode. The corrosion medium was a 3.5% solution of NaCl in distilled, de-ionized water open to the laboratory atmosphere at room temperature. The polarizing current, flowing through the counter electrode, at any instant, is proportional to the rate of corrosion of the sample.

The corrosion couple method employs a reference electrode and a working electrode. The working electrode and the reference electrode consisted of a test sample mounted on a standard electrode holder through a teflon gasket. The currents flowing between the two electrodes were measured using a zero-resistance ammeter. The corrosion medium for both test methods was a 3.5% solution of NaCl in distilled, de-ionized water open to the laboratory atmosphere at room temperature. The corroded test samples were examined using optical microscopy to determine the regions of preferential attack. The test samples were ultrasonically cleaned in distilled water, forced air dried and microscopically examined for various potentials for the controlled potential tests or various exposed areas for the corrosion couple tests.

3. RESULTS AND DISCUSSION

3.1 Controlled Potential Evaluation

The results from the three electrode, controlled-potential corrosion test are shown in Figure 3, to give the corresponding polarizing current density, i.e. anodic polarization curve, for the Al-Zn-Mg-Cu alloy and the Al-7.5% Mn alloy. This indicates that the Al-7.5% Mn generally has a higher corrosion rate than the Al-Zn-Mg-Cu alloy at identical electrode potentials. It is useful now to consider the microstructural alteration at various corrosion potentials for the Al-Zn-Mg-Cu alloy as well as the Al-Mn system. The microstructures of the Al-Zn-Mg-Cu alloy and the Al-7.5% Mn alloy in the as-received condition are shown in Figure 4. The $MnAl_6$ particles, in the Al-7.5% Mn, vary in size from 1μm to approximately 5μm. The grains, in the Al-Zn-Mg-Cu alloy, vary in size from approximately 12μm to 20μm. Figure 5 shows that at a constant potential of -500mV, the Al-7.5% Mn alloy undergoes a general pitting corrosion while the Al-Zn-Mg-Cu alloy shows signs of a faint amount of intergranular corrosion plus some general pitting corrosion. At a potential of -610mV, Figure 5, the intergranular corrosion of the Al-Zn-Mg-Cu alloy is more pronounced, and at a potential of -700mV the Al-7.5% Mn still undergoes general pitting corrosion. Figure 5 also shows that at a potential of -900mV, the Al-Zn-Mg-Cu alloy is no longer on the anodic polarization curve, but the sample still undergoes intergranular corrosion. The intergranular corrosion of the Al-Zn-Mg-Cu alloy at -900mV does not seem to be as deep and severe as the intergranular corrosion at more positive potentials (-610mV). Furthermore, pitting corrosion seemed to decrease, and general surface corrosion has become more predominant.

Figure 5 also indicates that Al-7.5% Mn still undergoes general pitting corrosion, but at a much lower rate. In summary, the results of the controlled-potential corrosion test, Al-Zn-Mg-Cu generally corrodes at a slower rate than the Al-7.5% Mn alloy. Al-Zn-Mg-Cu undergoes intergranular corrosion at more negative potential levels. In contrast, the Al-7.5% Mn generally undergoes pitting corrosion.

3.2 Corrosion Couple Evaluation

In order to simulate the corrosion response of a duplex alloy system, a corrosion couple between the matrix phase, e.g. Al-Zn-Mg-Cu, and the dispersoid phase, e.g. Al-Mn, was evaluated. These results from the corrosion couple test are illustrated in Figures 6 and 7 and depict microstructural results from the first couple where one electrode of Al-Zn-Mg-Cu has a surface area of 3.16 cm^2 (84% area fraction) and the Al-7.5% Mn electrode had an area of 0.58 cm^2 (16% area fraction). The microstructural results from a second corrosion couple where one electrode of Al-Zn-Mg-Cu had a larger area of 2.32 cm^2 (70% area fraction) and the Al-5% Mn electrode has a smaller area of 1.03 cm^2 (30% area fraction).

Since the Al-7.5% Mn electrode contains ∿50% $MnAl_6$ and ∿50% α-aluminum-manganese solid solution, the couple consists of three phases: 84% Al-Zn-Mg-Cu, 8% $MnAl_6$, and 8% α-aluminum-manganese. This couple, after being immersed in a 3.5% solution of NaCl, exhibited a small positive current (≃0.2μA) flow from the Al-Zn-Mg-Cu electrode to the Al-7.5% Mn electrode, where the Al-Mn phase was anodic while the Al-Zn-Mg-Cu was cathodic. After the corrosion couple test, the polished surface of the Al-7.5% Mn sample exhibited corrosion pits aligned in the directionality of the aligned α-Al-Mn matrix. The corrosion couple sample was subsequently etched with Keller's reagent and indicated that the corrosion pits actually initiated in the α-Al-Mn matrix. It is important to note that the corrosion couple photomicrograph, Figure 6, for the Al-7.5% Mn electrode surface is of a corroded sample which has been subsequently etched to reveal what constituent had been responsible for initiation of the corrosion pits. The surface of the Al-7.5% sample after the corrosion couple test, but prior to etching, was featureless except for the directionality aligned pits. The photomicrographs of the Al-Zn-Mg-Cu electrode surface, Figure 6, indicates intergranular corrosion and was not metallographically etched. Considering a second corrosion couple, with a higher area fraction of Al-Mn phase, a similar test was performed and Figure 7 contains the photomicrographs for the couple where the Al-Zn-Mg-Cu comprised 70% of the couple surface and an

Al-5% Mn phase comprised 30% of the couple surface. Since the Al-5% Mn electrode was 34% $MnAl_6$ and 66% α-aluminum-manganese solid solution, the couple actually consisted of three phases: 70% Al-Zn-Mg-Cu, 10% $MnAl_6$, and 20% α-aluminum-manganese. In contrast, this couple showed no current flow between the two electrodes. The Al-Zn-Mg-Cu electrode was neither anodic or cathodic with respect to the Al-5% Mn. The Al-5% Mn electrode surface, after the corrosion couple test, was featureless except for corrosion pits aligned in the matrix direction. Subsequent etching of the electrode surface confirmed that the pits had initiated in the α-aluminum-manganese matrix. As before, the Al-5% Mn electrode surface was etched with Keller's reagent to reveal the phase responsible for initiating corrosion pits. The Al-Zn-Mg-Cu electrode had undergone intergranular corrosion and no subsequent etching was required.

In summary of the results of the corrosion couple test; Al-7.5% Mn is slightly anodic with respect to Al-Zn-Mg-Cu when the couple surface area consisted of 84% Al-Zn-Mg-Cu, 8% $MnAl_6$, and 8% α-aluminum-manganese; Al-5% Mn is neither anodic or cathodic with respect to Al-Zn-Mg-Cu when the couple surface area consisted of 70% Al-Zn-Mg-Cu, 10% $MnAl_6$, and 20% α-aluminum-manganese. Furthermore, all the Al-Mn alloy electrodes underwent pitting corrosion with the pits initiating in the α-aluminum-manganese matrix.

These preliminary corrosion evaluations indicate that the corrosion behavior of a duplex alloy can be altered by varying the volume fraction and composition of the phases present. For instance, the first corrosion couple, Table II, indicated intergranular corrosion of the Al-Zn-Mg-Cu alloy. By comparison, the second corrosion couple apparently suppressed the corrosion between the Al-Zn-Mg-Cu matrix and the high modulus Al-Mn phases. In this manner, these initial results indicate it may be possible to design a duplex alloy system to optimize modulus and strength, as well as corrosion response.

4. CONCLUSIONS AND SUMMARY

This study was directed towards a fuller understanding of the electrochemical behavior of the 7000 or 3000 series powder metallurgy aluminum alloys and considers the electrochemical compatibility. Corrosion evaluations have shown that the Al-Zn-Mg-Cu and the Al-Mn alloys can be made more compatible in a duplex composite alloy. The corrosion couple tests have shown that the volume fractions, or surface area fractions and chemistries of the Al-Zn-Mg-Cu alloy and the Al-Mn alloy can be modified to suppress corrosion. A couple surface of 70% Al-Zn-Mg-Cu and 30% Al-5% Mn had exhibited no current flow between the two constituent alloys and suggest that one constituent alloy did not corrode anodically with respect to the other constituent alloy and vice versa. Also, even though the Al-Zn-Mg-Cu and the Al-5% Mn alloy did not adversely affect each other, each electrode surface had individually undergone corrosion. The Al-Zn-Mg-Cu had undergone intergranular corrosion and the Al-Mn alloys had undergone pitting corrosion in the α-aluminum-manganese matrix. The controlled-potential corrosion tests also have shown that these are the modes of preferential attack typical for each constituent alloy.

ACKNOWLEDGEMENTS

The authors wish to acknowledge the helpful discussions with Drs. John DeLuccia, Gilbert London and Sally Ketcham, Naval Air Development Center, Warminster, Pa. In addition, they appreciate the alloys provided by Alcoa.

TABLE I

Nominal Composition of Powder Metallurgy (P/M) Alloys, Atomic %.

Alloy	Cu	Mn	Mg	Zn	Co	Al
Al-Zn-Mg-Cu	0.6	---	2.3	2.4	0.2	Balance
Al-Mn P/M	---	7.5	---	---	---	Balance
Al-Mn P/M	---	5.0	---	---	---	Balance

TABLE II

Percent Area Fractions of Phases Present in Corrosion Couple

Couple	Al-Zn-Mg-Cu	$MnAl_6$	α-Al-Mn
1	84	8	8
2	70	10	20

REFERENCES

1. Speidel, M.O. and Hyatt, M.V., "Advances in Corrosion Science and Technology", vol. 2, M.G. Fontana and R.W. Staehle, Eds., Plenum Press, New Yor, 1972, p. 115.

2. Staehle, R.W., Forth, A.J. and Van Rooyen, D., Editors, Proceedings of Conference on Fundamental Aspects of Stress Corrosion Cracking, National Association of Corrosion Engineers, Houston, Texas, 1967.

3. Brown, B.F., "Stress Corrosion Cracking in High Strength Steels and in Titanium and Aluminum Alloys", Naval Research Laboratory, Washington, D.C., p. 147, 1972.

4. Brodie, R.W., and Bakow, L., "Investigation of the Mechanical and Corrosion Properties of MA87 Aluminum Powder Alloy", Naval Air Development Center Report No. NADC-78185-60, 1979.

5. Agarwala, V.S. and Murty, Y.V., "A Controlled-Potential Corrosion Study of Al-4.5 Cu Alloy in 3.5% NaCl Solution", Metallography, 10, (1977), p. 451.

6. Galvele, J.R. and Micheli, S.M., "Mechanisms of Intergranular Corrosion of Al-Cu Alloys", Corrosion Science, 10, (1970), p. 795.

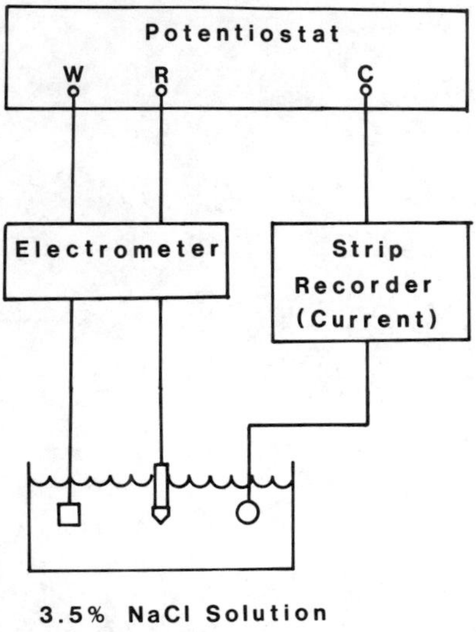

Figure 1. Schematic of the three electrode, potentiostatic, controlled-potential corrosion cell.

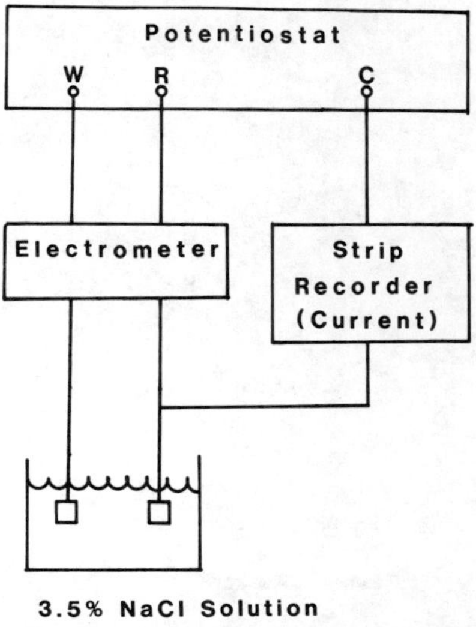

Figure 2. Schematic of the corrosion couple cell.

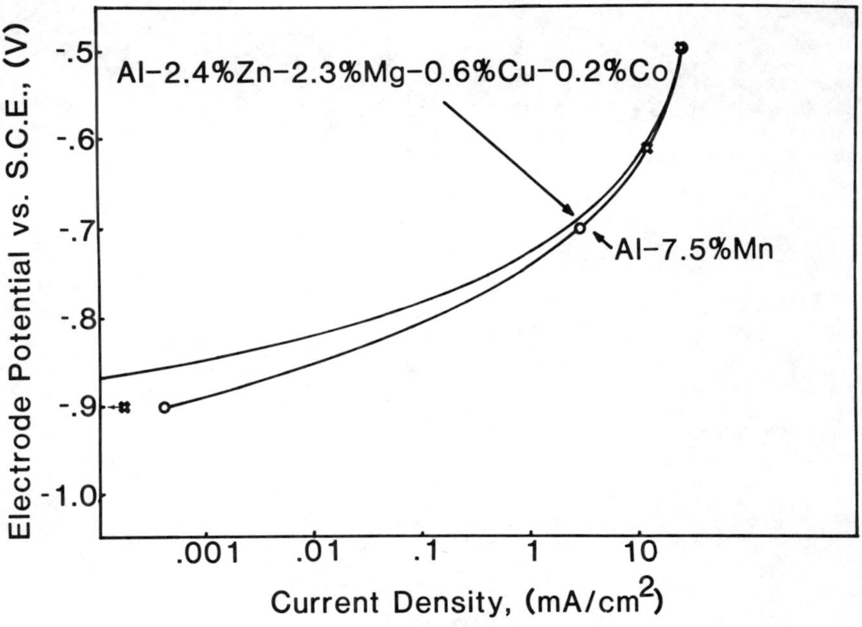

Figure 3. Anodic polarization curve for the Al-Zn-Mg-Cu alloy and the Al-Mn alloy illustrates that the Al-Zn-Mg-Cu alloy is more cathodic and corrodes at a slower rate than the Al-Mn alloy.

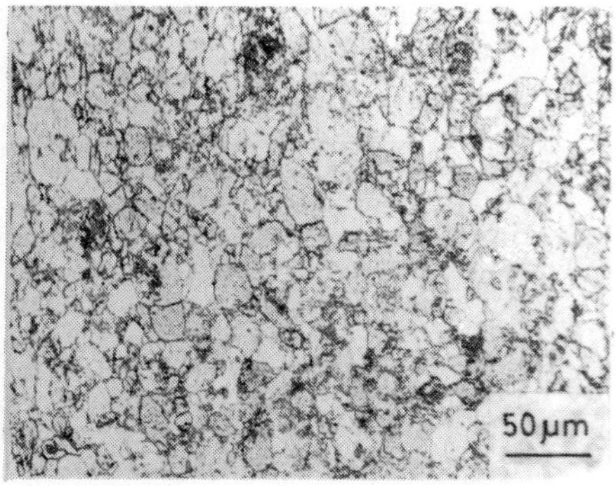

Al-Zn-Mg-Cu alloy.

Al-7.5 atomic % Mn alloy.

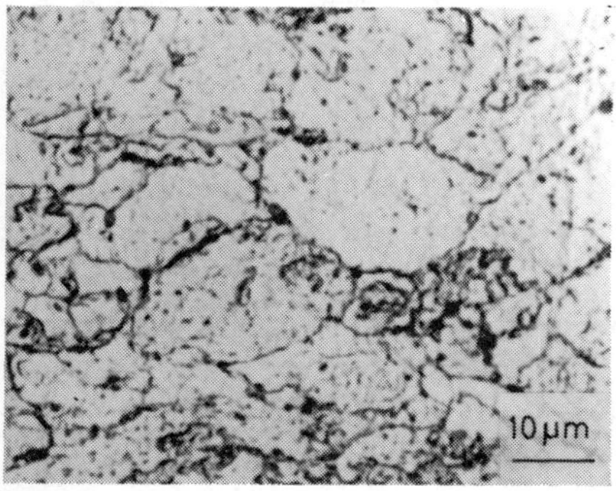

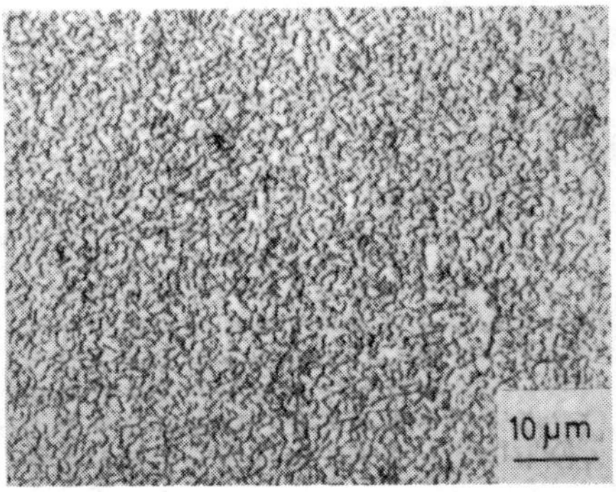

Figure 4. Microstructure of the as-received alloys.

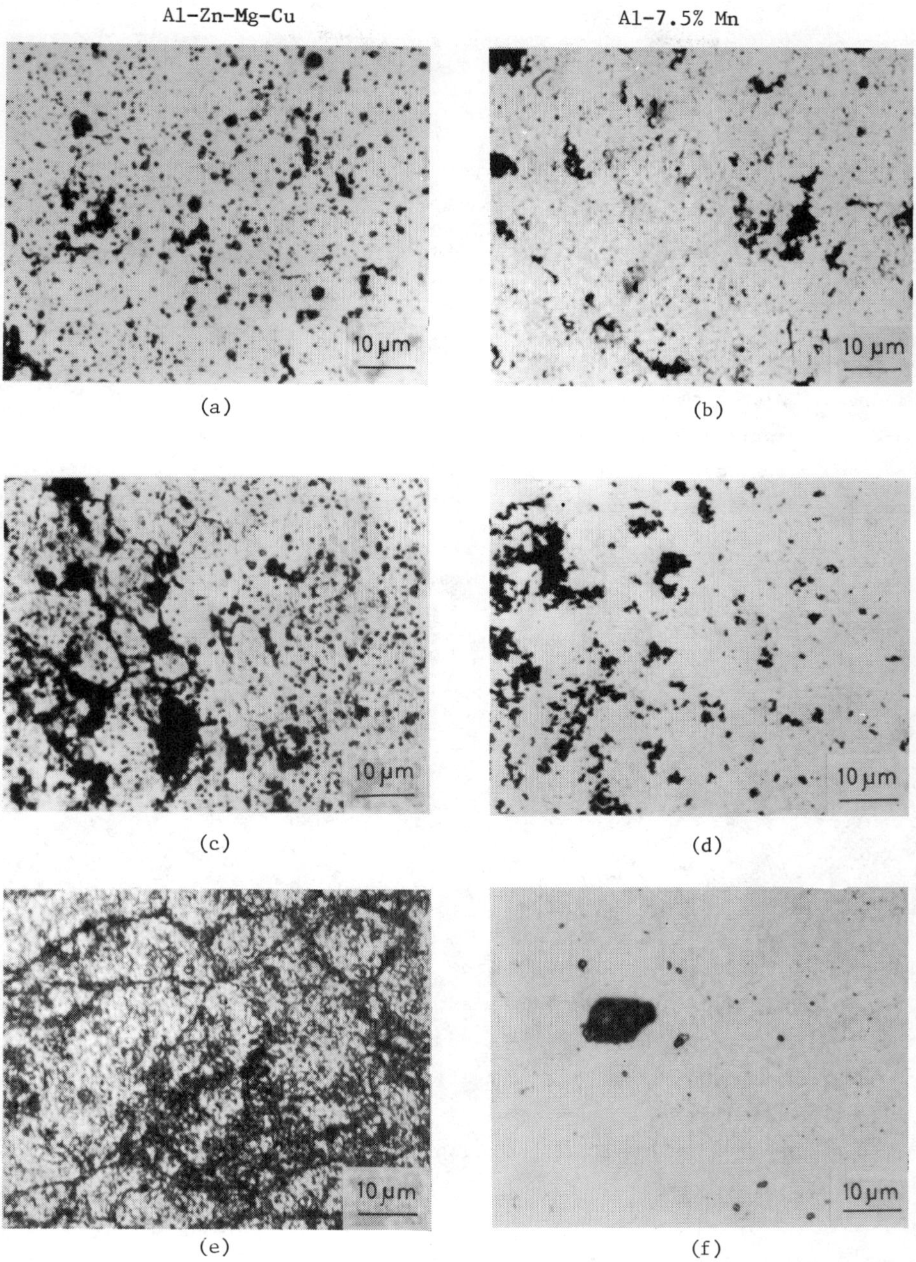

Figure 5. Microstructural results of the Al-Zn-Mg-Cu alloy and the Al-7.5% Mn alloy from the three electrode, potentiostatic, controlled-potential corrosion test. Each sample represents a different potential: (a) E= -500mV, (b) E= -500mV, (c) E= -610mV, (d) E= -700mV, (e) E= -900mV, (f) E= -900mV.

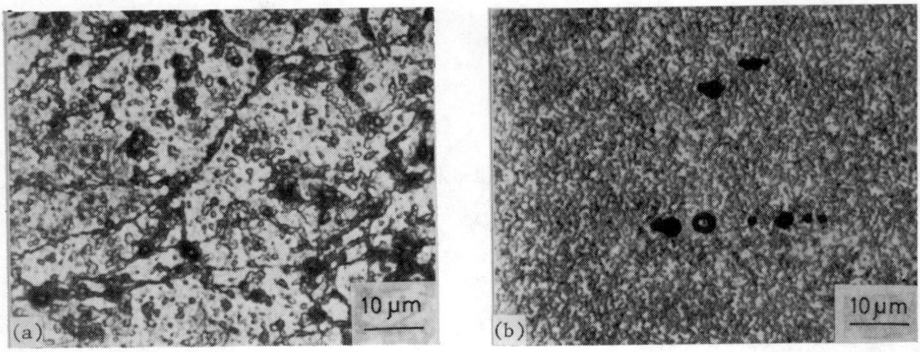

Figure 6. Microstructural results from the corrosion couple test where Al-Zn-Mg-Cu comprised 84% of the couple surface area and Al-7.5% Mn was 16% of the couple surface. The Al-Zn-Mg-Cu sample (a) had a surface area of 84% and is presented in the unetched condition; the Al-7.5% Mn sample (b) had a surface area percent of 16% and was metallographically etched to reveal the phases associated for the corrosion pits.

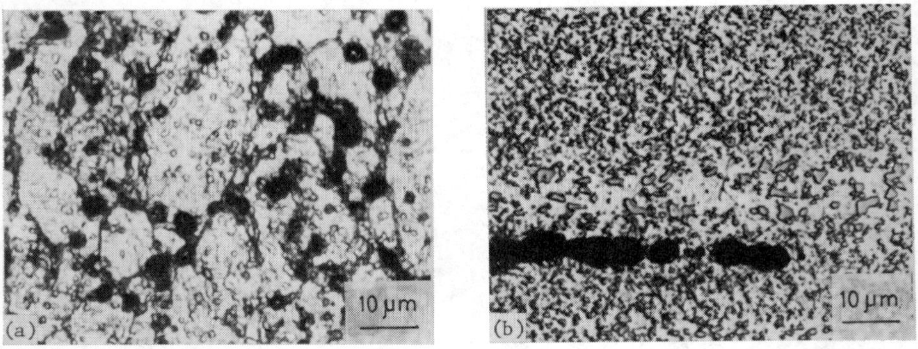

Figure 7. Microstructural results from the corrosion couple test where Al-Zn-Mg-Cu comprised 70% of the couple surface area and Al-5% Mn was 30% of the couple surface. The Al-Zn-Mg-Cu sample (a) had a surface percent of 70% and is presented in the unetched condition; the Al-5% Mn sample (b) had a surface percent of 30% and was metallographically etched to reveal the phase responsible for initiating the corrosion pits.

13th National SAMPE Technical Conference
October 13-15, 1981

RECENT DEVELOPMENTS IN MATERIALS AND PROCESSES FOR AIRCRAFT CORROSION CONTROL

Sara J. Ketcham and John J. De Luccia
Naval Air Development Center
Warminster, Pennsylvania 18974

Abstract

This paper describes advances in materials and processes for aircraft corrosion control currently in use and those under development in the laboratory for future use. Areas covered are corrosion preventive compounds, organic and inorganic coatings, alloy selection and heat treatments.

Available materials highlighted are water displacing compounds, sealant primers, aluminum alloys 7050 and 7010 including a state-of-the-art report on the use of exfoliation and stress corrosion resistant tempers.

Materials under development that offer promise of contributing to future advances include water displacing paints, flexible primer, crack arrestment compounds, powder metallurgy, aluminum alloys and substitutes for cadmium plating.

1. INTRODUCTION

This paper will address material and processes developed by the research and development community that have provided advances in corrosion control of naval aircraft. The subjects to be covered will primarily apply to airframes, although they may also be useful for electronic equipment. First, materials currently available will be discussed, then materials still under development which look promising for the future.

Classical corrosion control involves attacking the problem from three standpoints: interfacial (coatings), internal (metal selection), and external (inhibited environment). All three aspects will be covered here.

2. CORROSION PREVENTIVE COMPOUNDS

The Naval Air Development Center has pioneered in the development of water displacing compounds.(1) These compounds are:

AML-350 - an ultra thin water displacing corrosion preventive compound which dries to a soft film;

AMLGUARD - a water displacing, corrosion preventive which dries to a hard, clear finish.

The need for such materials is occasioned by the difficulty of performing corrosion control procedures in a marine environment. Areas where paint has cracked or chipped leaves bare metal exposed. The mechanism by which these compounds displace water from the surface is shown in Figure 1.[2] The essential elements involved are: (1) the compound spreads over the substrate completely wetting it, (2) it is immiscible with water, therefore does not retain water, and (3) it is preferentially adsorbed, penetrating under the water droplets.

AML-350 is applied to a film thickness of 2 to 5 μm. It is composed of a petroleum sulfonate and a mineral spirits type solvent. The combination of these materials allows the coating to spread over metal and creep under water droplets on a metal surface, thus displacing the water. Eventually the solvent evaporates and the petroleum sulfonate remains, forming a soft, thin, oil-like film. This film isolates the metal from the environment, thus acting as a temporary passive corrosion preventive. The petroleum sulfonate also assists in protection against corrosion by performing as a corrosion inhibitor.

AML-350 is intended for use on internal metallic parts and electrical connectors. It has widespread use on airborne electronic equipment. It is generally not intended for use on external airframe areas because of its soft condition, but under adverse weather conditions or when aircraft are liberally water soaked, it may be applied as a temporary protective measure.

AML-350 is covered by military specification MIL-C-0081309C.[3]

AMLGUARD contains polymeric resins which upon application and cure form a dry, hard film with a thickness of 25.4 to 50.8 μm (1 to 2 mils). It is a combination of organic solvents, silicone and silicone alkyd resins, barium petroleum sulfonate, and several other additives. It also displaces water by spreading over the metal and creeping under the water droplets. Drying occurs via solvent evaporation, leaving a solid film. Although AMLGUARD dries to the touch in 18 hours, it continues to cure for 1 to 3 months, forming a hard, flexible finish. Corrosion protection is provided by the physical barrier of the coating and also by barium petroleum sulfonate and alkyl ammonium organic phosphate performing as corrosion inhibitors. Figure 2 illustrates the protection afforded by AMLGUARD.

This material is intended for temporary use on external aircraft parts where it offers excellent corrosion protection. It has also been recommended for use on leading edges of aircraft wings and helicopter blades and on exhaust and gun blast areas, as well as many other aircraft components such as wheels, wheel wells, cables, landing gear, etc. where erosion resistance and corrosion protection is necessary. AMLGUARD has been successfully used on mild steel, stainless steel, aluminum, magnesium, zinc, cadmium, copper and brass. It should not be used on lubricated surfaces, electrical contact areas, form in place gaskets, removable fasteners or moving parts.

AMLGUARD is covered by military specification MIL-C-85054.[4]

3. SEALANTS

Elastomeric sealants are widely used on naval aircraft to seal out the environment. The most popular at the present time are the polysulfide sealants which contain soluble corrosion inhibitors. These are covered by military specification MIL-S-81733.[5] Ordinary sealants can minimize corrosion of metals in high humidity environments, but cannot prevent it completely because all polymers are permeable to moisture. The inhibitive sealant is very effective when used in faying surfaces and butt joints, for wet installation of fasteners and over fastener patterns and to insulate dissimilar metals (Figure 3[6]).

It is anticipated that the use of elastomeric sealants (polyurethanes and silicones as well as polysulfides) will increase in the future due to their ability to accomodate the dynamic loads imposed on aerospace equipment without cracking.

4. SURFACE TREATMENTS FOR ALUMINUM ALLOYS

With regard to surface treatments for aluminum alloys, there has been a return to anodizing for new weapons systems rather than chromate conversion coatings. Both sulfuric and chromic acid anodizing are being used. Anodized surfaces provide more corrosion protection, abrasion resistance and long-term durability than chromated surfaces. Sealing of anodized coatings has always been recommended and recent work at NADC has re-emphasized the importance of the type of water used to seal non-clad aluminum surfaces. Sealing in tap water resulted in inferior corrosion resistance and paint adhesion, as compared to that of sealing in distilled water.

The paint system used on naval aircraft, the MIL-P-23377 epoxy primer and the MIL-C-82386 polyurethane topcoat, has performed better than any previous system. It is durable, abrasion resistant, retains its gloss well, does not chalk or craze, and is easy to clean. Its main drawbacks are its lack of low temperature flexibility which causes it to crack around fasteners and the fact that both primer and topcoat are two-part systems. Laboratory work to improve the paint system will be described in a later section.

5. ALLOY SELECTION AND HEAT TREATMENT

The advent of the heat treatable 7000 series aluminum alloys provided a boon to aircraft designers. The high modulus, high strength, and low weight of these alloys made them ideal for the high performance of advanced aircraft. Their susceptibilities to intergranular corrosion, exfoliation, and stress corrosion cracking made them less than desirable in marine environments. It was discovered that if these alloys were systematically overaged, their susceptibilities to these forms of attack would be materially lessened if not totally eliminated. Thus the T73 temper was born. As is usually the case with most "fixes," a price had to be paid. There is an approximate 10% strength loss accompanying the T73 overaging treatment. The T73 temper replaced the standard T6 temper on many military aircraft where designs could be altered or the strength loss tolerated.

More recently newer 7000 series alloys have been developed with specific resistance to intergranular attack and environmental embrittlement. Alloys such as 7050 and 7010 contain zirconium replacing chromium. Professor DiRusso in Italy first identified the beneficial effects of this element in Al-Zn-Mg alloys.(7) These newer alloys also pay particular attention to the "cleanliness" of the microstructure. They contain lower concentrations of the tramp elements, iron and silicon, and therefore have a greater fracture toughness. Aluminum alloy 7050 was developed under the guidance of the Naval Air Systems Command by Alcoa.(8) In the T73X temper, it provides the best stress corrosion resistance with the highest strength of any commercial aerospace aluminum alloy.

Alloy 7010, developed by Alcan Plate Ltd., has, in the T73X temper, recently become competitive with 7050.

Some recent work at an Israeli aerospace company indicates that a pre-aging treatment of 7075 alloy renders immunity to stress corrosion cracking ($K_{I_{SCC}}$ is raised by a factor of three) without any sacrifice in strength. This treatment, which is called retrogression and re-aging, bears watching.(9)

The preceding paragraphs have described materials currently in use, or available, to minimize environmental deterioration of airframes. Materials and processes under development in the laboratory will now be addressed.

6. MATERIALS UNDER DEVELOPMENT

6.1 Water Displacing Paint

This material is a pigmented coating which will displace water, dry and subsequently afford corrosion protection. It is composed of a petroleum sulfonate, silicone-alkyd resin, organic solvents, pigments and other organic additives. The mechanism by which this material displaces water is the same as that discussed earlier. This pigmented coating dries to a hard, flexible finish which protects the substrate from corrosion by:

1. The physical barrier of the coating.

2. Corrosion inhibiting pigments; i.e., molybdates and chromates.

This water displacing paint is designed as a touch-up paint for exterior surfaces of aircraft where original paint has cracked or chipped and total repainting is not feasible. Such a situation is confronted on operational aircraft deployed on board aircraft carriers where paint touch-up is necessary but must be completed quickly and efficiently. This paint was designed to be applied during deployment and to last indefinitely until total repainting of the aircraft is necessary.

Reports from an initial evaluation of the paint by the Fleet indicated it performed well. Some settling problems are being resolved.

6.2 Flexible Primer

The current Navy paint scheme for high-performance aircraft includes the application of MIL-P-23377 epoxy primer, MIL-S-8802 or MIL-S-81733 polysulfide sealant, and MIL-C-83286 polyurethane topcoat. This system has several limitations. The primer possesses poor low-temperature flexibility, while the sealant is difficult to apply due to its high viscosity and short pot life. The ideal solution would be a single application of an elastomeric sealant-primer with the adhesion of a primer and the flexibility of a sealant. This would also eliminate the need to handle two separate materials, resulting in a significant cost savings to the government.

A comparative evaluation of a number of inhibited elastomeric coatings including a polysulfide, a polyurethane and an epoxy-polyurethane system is being made. Such properties as the hardness, adhesion, strength, flexibility, erosion resistance, corrosion resistance and ease of application of these materials will be determined and an optimum material selected.

6.3 Development of an Aluminum Plating Process

There is increasing pressure to eliminate the use of cadmium by DoD activities because of its toxicity. While no single coating has been found to replace cadmium in all aircraft applications, aluminum has been found to be a very good alternate coating material in many applications requiring good corrosion resistance and minimal effect on fatigue properties. Only two aluminum coating processes are currently commercially feasible, vacuum deposition and ion vapor deposition. Vacuum deposition has relatively poor covering power and adhesion is often only fair. Ion vapor deposition is proprietary and facilities for its application are complex and cannot meet the demand for the coating. Other methods exist, but they have not been developed sufficiently to be of real commercial value.

NADC is involved in an effort to develop a method to electroplate an aluminum coating from a molten salt bath. To date the process has been scaled up from a small bath to a forty liter bath. A schematic of the bath is shown in Figure 4. It has been demonstrated that a coating of aluminum-manganese can be deposited on aluminum, titanium and steel substrates with conventional pretreatments and with excellent adhesion. The coating can be chromate conversion coated but not anodized.

The molten salt plating process has an advantage over the ion vapor deposition in that the throwing power is better and recesses and holes can be coated. Work is in progress to optimize the plating parameters. Following completion of that phase, a pilot plant production set-up will be attempted.

6.4 Phase Transfer Inhibitors (PTI)

A method was developed in-house by which ions could be solubilized in organic media using phase transfer catalysis (PTC). This method has been used to develop an entirely new vehicle for corrosion inhibitors. These have resulted in inorganic inhibitors incorporated in organic phases; i.e., an oxidizing anion dissolved into an organic non-polar solvent, in this case, mineral spirits. The technique by which this is accomplished has been published elsewhere and will not be repeated here.[10] The inorganic inhibitors being used include sodium dichromate ($Na_2Cr_2O_7 \cdot 2H_2O$), sodium borate ($Na_2B_4O_7 \cdot 10H_2O$), sodium nitrite ($NaNO_2$), sodium molybdate ($Na_2MoO_4 \cdot 2H_2O$), potassium hexachloropalladate and ammonium hexanitrato cerate $(NH_4)_2Ce(NO_3)_6$.

Evaluation of these new phase transferred inorganic inhibitors (PTI) was first carried out for the general corrosion inhibition of steel and aluminum alloys in highly aggressive environments. A salt fog chamber was used and both AISI 1020 steel and 7075 aluminum alloy panels treated with PTI dichromate were tested. A four-day test exposure of such panels showed no observable corrosion of either steel or aluminum alloy specimens. In another test PTI dichromate inhibitor was added into alkyd enamel coating and then tested on 7075 discs in a 5% CaCl salt spray chamber for about four months. The specimen coated with PTI dichromate showed much less attack than the control with only alkyd enamel coating protecting it. A marked improvement in corrosion protection was also found when PTI dichromate was added to epoxy paint primer and temporary preservatives and a corrosion preventive compound (AMLGUARD).

6.5 Crack Arrestment Compounds

The effect of the inhibitors was most interesting in the study of corrosion fatigue, particularly when they were used in combinations. Thus, a number of low cycle corrosion fatigue tests were carried out using notched bend bar specimens (10" x 2" x 1/2" size, fracture mechanics type) fabricated from high strength (1250-1350 MPa) vacuum arc melted 4340 steel. The specimens were fatigue pre-cracked before testing and then mounted on an Instron machine enclosed in a controlled humidity chamber. The specimens were cycled between loads of 1500 to 150 lbs at a frequency of about 0.16 Hz. The load value and the pre-crack length were chosen to yield a minimum fracture mechanics stress intensity factor of approximately 23 MPa mm in dry (i.e., less than 15%) R.H. The test inhibitor or inhibitors were applied by injecting the solution into the pre-cracked notch area of the specimen before and after loading. The specimen was loaded in tension-tension. The crack growth of the specimen was monitored as a function of time and cycles. The details of the mechanical testing and fracture mechanics calculations are described elsewhere. The corrosion fatigue tests were carried out in both the high humidity (moist air) and in chloride containing moist air environments. The PTI inhibitors were then evaluated for their crack arresting properties. The results of these tests are reported in Table I which shows the total fatigue life before overload fracture, the crack growth rate in the region where it is independent of stress intensity factor and the stress intensity factor where it is independent of crack growth rate.

The data in Table I show that PTI dichromate, nitrite and borate are very effective in retarding crack growth rate of 4340 steel by more than order of magnitude when compared to moist air test with no inhibitor present. In terms of fatigue life the inhibition was almost five times more effective. Almost every PTI compound tested; i.e., even cerate and hexachloropalladate, showed significant crack growth inhibition. However, in the presence of chloride (NaCl) the results were entirely different. The dichromate + nitrite + borate system did not show much improvement. But when PTI molybdate was added, the crack growth inhibition properties of the compounds were restored and the fatigue life of specimen was extended by a factor of 4. Figure 5 presents these results graphically.

The use of phase transfer inhibitors in pre-cracked stress corrosion cracking in aluminum alloys is presently being studied. Preliminary tests using dichromate, nitrite, borate, molybdate as well as dichromate, phosphate, silicate inhibitor combinations show the cracking rate to be perceptibly decreased in aluminum alloy 7075-T6 exposed to salt laden moisture (Figure 6).

The field of phase transfer catalysis and organometallic chemistry appears to provide a new class of corrosion inhibitors which will probably have wide applications in materials such as coatings, corrosion preventive compounds and crack arrestment compounds.

6.6 Powder Metallurgy Aluminum Alloys

With regard to alloys and heat treatment, considerable work is in progress on aluminum alloys (and others) made by high density powder metallurgy (P/M) processing rather than conventional ingot metallurgy (I/M) practices of casting and working. P/M processed materials have finer and more homogeneous microstructures. An important advantage of P/M high strength aluminum alloys compared to I/M is that they can be aged to considerably higher yield strength values without losing their resistance to stress corrosion cracking.

Considerable interest is being shown in the Alcoa P/M alloy CT91, a 7XXX composition type for aircraft use. An aircraft company has investigated extrusions made from this material and compared their properties with those of 7075-T6 extrusions. At comparable strength levels, the CT91 extrusions had higher corrosion resistance and improved fatigue properties. A weight payoff study for a V/STOL aircraft showed a 10% weight saving by replacing existing aluminum alloys.

Considerable development work is still going on with P/M alloys, but it is anticipated that their use in aircraft is not long in coming. Surface treatment of these alloys will probably require some investigation since there is evidence they do not anodize like conventional wrought alloys.

This has been a brief discussion of the materials and processes currently being used for corrosion control on naval aircraft. Some of the more important in-house efforts and ongoing research in the Aero Materials Division at the Naval Air Development Center are highlighted. Fighting corrosion is a never ending battle, but even small successes can provide a payoff in reduced maintenance manhours and improved reliability.

7. REFERENCES

1. C. R. Hegedus, "Water Displacing Corrosion Preventive Compounds," Proceedings of Corrosion Control by Organic Coatings Conference, Vol. I, Lehigh University, August 1980, p. 51.
2. R. N. Bolster, "Removal of Fluid Contaminants by Surface Chemical Displacement," Surface Contamination, Vol. 1, Ed. K. L. Mittal, 1979.
3. Military Specification MIL-C-0081309C, Corrosion Preventive Compound, Water Displacing Ultra Thin Film of 1 Mar 77.
4. Military Specification MIL-C-85054, Corrosion Preventive Compound, Water Displacing, Clear (AMLGUARD) of 15 Sep 77.
5. Military Specification MIL-C-81733C, Sealing and Coating, Corrosion Inhibitive of 13 Mar 80.
6. R. Miller, Sealants, Chap. 6, A Handbook of Protective Coatings for Military and Aerospace Equipment, Ed. S. J. Ketcham, First Edition, Houston, TX, to be published by National Association of Corrosion Engineers.
7. E. DiRusso, "Mechanical Properties and Stress Corrosion Resistance of Al-Zn-Mg-Cu Alloys," Instituto Sperimentale dei Metalli Leggeri, Milano, Italy, report of 14 March (1968)
8. J. Staley, H. Hunsicker, and R. Schmidt, "New Aluminum Alloy X7050," Metal Congress, TMS, AIME, October (1971).
9. R. S. Kaneko, Solution for Stress Corrosion Problems with T6 Temper Aluminum, Metal Progress, Vol. 117, No. 5, April 1980, p. 41.
10. V. S. Agarwala and J. J. De Luccia, Inorganic Inhibitors in Organic Media, A New Class of Corrosion Inhibitors, Proceedings 5th European Symposium on Corrosion Inhibitors, Ferrara, Italy, September 1980.

TABLE 1. EFFECT OF FUNCTIONAL PROPERTIES OF VARIOUS CRACK ARRESTMENT INHIBITORS ON LOW-CYCLE FATIGUE OF HIGH STRENGTH 4340 STEEL

	INHIBITORS APPLIED TO NOTCH AREA	CRACK GROWTH RATE MICRO-IN./CYCLE	STRESS INTENSITY FACTOR ΔK KSI $\sqrt{IN.}$	FATIGUE LIFE, CYCLES
	DRY AIR ONLY	12	70	17,000
MOIST AIR ONLY	NO INHIBITOR USED	110	33	1,800
MOIST AIR ONLY	DICHROMATE	42	52	6,800
MOIST AIR ONLY	NITRITE + BORATE	65	35	3,300
MOIST AIR ONLY	HEXAPALLADATE	45	>40	4,000
MOIST AIR ONLY	LANTHANUM NITRATE	50	>40	4,600
MOIST AIR ONLY	DICHROMATE + NITRITE + BORATE	27	55	9,000
MOIST AIR ONLY	CERATE + NITRITE + BORATE	38	45	6,400
MOIST AIR + CHLORIDE	NO INHIBITOR USED	150	32	1,200
MOIST AIR + CHLORIDE	DICHROMATE + NITRITE + BORATE	200	34	1,200
MOIST AIR + CHLORIDE	MOLYBDATE	61	46	4,500
MOIST AIR + CHLORIDE	DICHROMATE + NITRITE + BORATE + MOLYBDATE	28	48	6,500

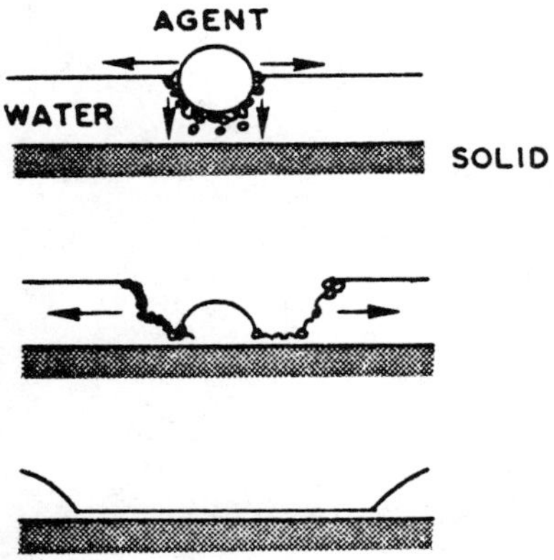

Figure 1. Water Displacement Mechanism (2)

a. Displacing agent applied to surface, mixture with water begins.

b. Agent reaches the surface, pushing water aside.

c. Preferential adsorption of agent displaces water from surface.

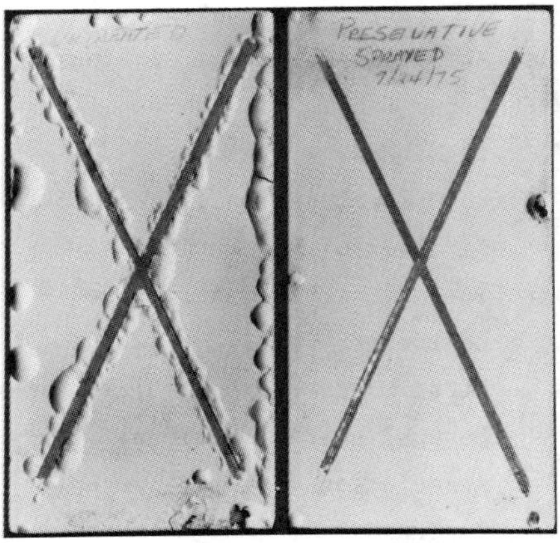

Figure 2. Protection Afforded by AMLGUARD to a Scribed, Painted Aluminum Alloy Panel During 90 Days Exposure to 5% $NACl/SO_2$ Salt Spray

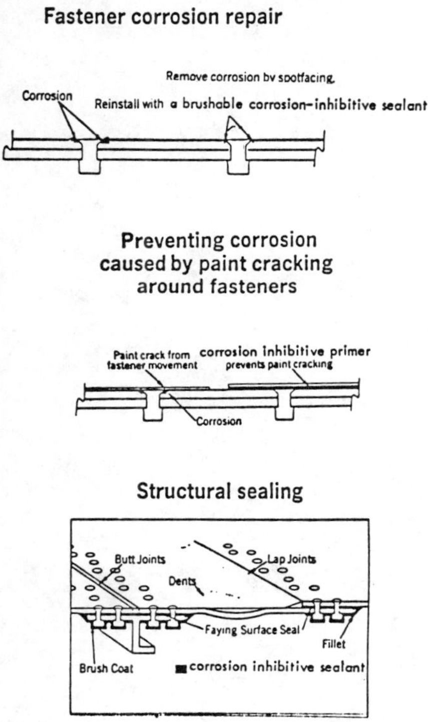

Figure 3. Typical Aircraft Applications of Inhibitive Sealants (6)

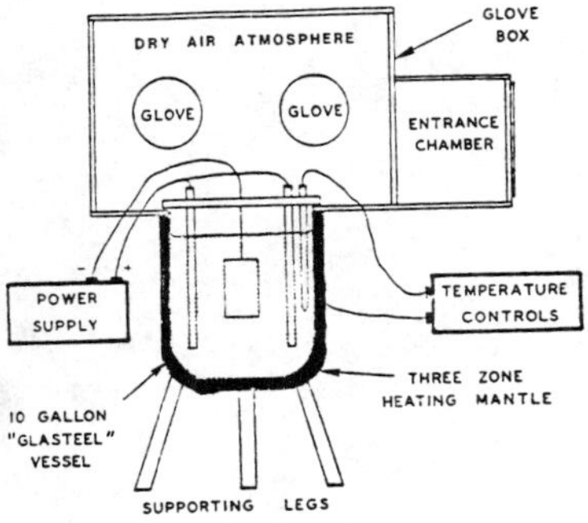

Figure 4. Schematic of Al-Mn Molten Salt Plating Bath

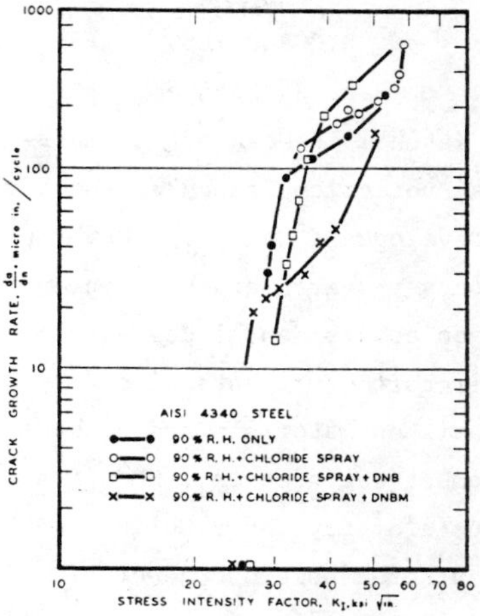

Figure 5. Crack Growth vs. Stress Intensity Factor for Notched Bend Bar Corrosion Fatigue Tests of 4340 Steel in Various Environments.

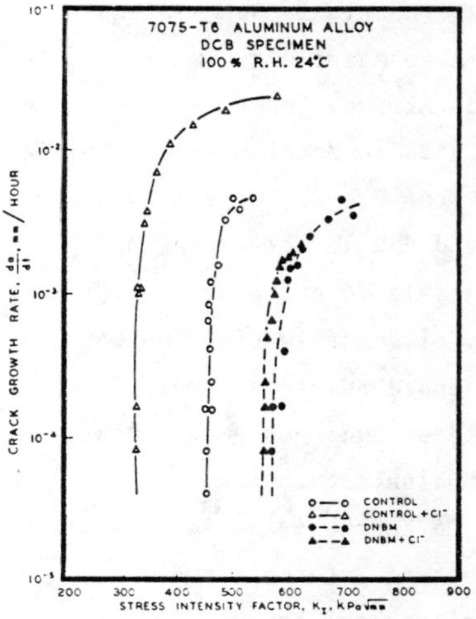

Figure 6. Crack Growth Rate vs. Stress Intensity Factor for DCB Stress Corrosion Specimens of 7075-T6

BIOGRAPHIES

Sara J. Ketcham

Mrs. Ketcham is Head of the Materials Protection Branch at the Naval Air Development Center. The branch conducts research and development work on environmental degradation of aircraft components with emphasis on materials and methods to combat degradation. She is a graduate of Bryn Mawr College and has authored numerous papers in the field of corrosion. She has been certified a corrosion specialist by the National Association of Corrosion Engineers (NACE) and is an active member of the ASTM Committee G-1 on Corrosion of Metals and the NACE Committee T-9 on Corrosion of Military Equipment.

John J. De Luccia

Dr. De Luccia is Superintendent of the Aero Materials Division at the Naval Air Development Center. He has a BS in Metallurgical Engineering from Drexel University and an MS and PhD in Metallurgy and Materials Science from the University of Pennsylvania. He has published numerous papers in the field of hydrogen embrittlement and corrosion prevention.

Dr. De Luccia is an adjunct professor of Materials and Metallurgical Engineering at Drexel University, a Fellow of the American Society for Metals and has been certified a Corrosion Specialist by the National Association of Corrosion Engineers.

13th National SAMPE Technical Conference
October 13-15, 1981

INFLUENCE OF SOME MANUFACTURING VARIABLES ON GRAPHITE CLOTH/EPOXY BOLTED JOINT STRENGTH

J. A. Bailie, S. A. Howard, L. M. Fisher, R. L. Cothern
Lockheed Missiles & Space Co., Inc., Sunnyvale, CA

Abstract

The influence of manufacturing imperfections on the static strength of composite bolted joints was evaluated in a program involving processing and fabrication variations likely to occur in a production environment. Three processing perturbations, namely low and high-fiber volume and high porosity, were induced in 8-ply graphite cloth/epoxy laminates which were subsequently fabricated into single lap bolted joints. Joint strength was fairly insensitive to high-fiber volume and up to 6 percent porosity at both room and elevated temperatures. Fiber volumes as low as 52 percent seriously degraded joint strength. Deliberately machined damage to the holes to simulate chip out also was shown to have little influence on joint strength. In another series of tests, to simulate hole location errors, edge distance was reduced to below current practice. Very little strength reduction resulted at edge distances as low as twice the bolt diameter.

1. INTRODUCTION

A considerable amount of data is available on the strength of nominally fabricated graphite epoxy bolted joints.[1-7] Very little has been published, however, concerning the influence of fiber volume, porosity, and hole quality on composite joint strength. One exception is the study by Pengra and Wood of hole damage in certain tape laminates.[8]

To investigate this problem, a series of 8-ply quasi-isotropic $[45/0]_{2s}$ laminates were fabricated using Fiberite HMF330C graphite cloth/epoxy prepreg. Modifications to layup and fabrication techniques were utilized to produce high and low-fiber volumes, high porosity, and damaged holes that might be considered typical processing imperfections. Single lap mechanical joints fabricated from these laminates were tested in tension shear at room and elevated temperatures and the results compared with tests on nominally processed joints.

It has been demonstrated [3] that there is a difference in tension strength between a narrow single bolt joint and a joint with a row of bolts, whose spacing is equal to the single bolt specimen width. The reason for this is the edges of the narrow specimens are free to move while the material between two bolts in a row is contrained. This results in a different stress concentration factor at the hole. To assess the magnitude of this effect, some high porosity laminates were fabricated into both single bolt joints and into 2 row, 4 bolts per row joints with bolt spacing equal to the single bolt joint width within a row and one inch between rows.

The effect of hole center to edge distance was also examined. The typical edge distance of three to four bolt diameters, used to preclude shear out failures, was reduced to two diameters to determine the importance of this parameter.

2. SPECIMEN FABRICATION

Layup for nominal autoclave cure of 8-ply laminates is as follows:

- Release agent
- Nylon peel ply
- Prepreg
- 3 plies of Burflow 4819 bleeder
- FEP film
- 2 plies of Burflow 4819 breather
- Nylon bag

To obtain the three variations, the following perturbations were introduced.

- High Fiber Volume: Five plies of bleeder cloth were installed instead of three, to increase the amount of resin absorbed in the bleeder.
- Low Fiber Volume: No bleeder plies were used to force the processed laminate to retain all the prepreg resin.
- High Porosity: The layup was nominal but cure took place without vacuum pressure, between heated platens of a press. Absence of vacuum pressure minimizes air and volatiles escape and thereby increases laminate porosity.

Two machining errors were also studied.

- Edge distance variations were studied by drilling the bolt holes at 2d, 3d and 4d from the laminate edge, where d is the bolt diameter.
- Chip out defects in the holes were simulated using the same diamond-coated Woodruff style cutter described by Pengra and Wood. [8] It produces a groove 0.015 in. deep and 0.013 in. wide. In these tests, the groove extended the full 360 degrees around the hole diameter at the laminate midplane. Both high and low fiber volume laminates were utilized in this test series.

Pertinent laminate properties are summarized in Table 1.

It is evident that the processing variations to cause low fiber volume and high porosity were successful. Attempting to produce

Table 1. LAMINATE FIBER VOLUME, POROSITY, AND THICKNESS

Laminate Description	Fiber (Vol. %)	Porosity (%)	Thickness (in.)
Nominal	61 ± 1	<1.1	0.108
High Fiber Vol.	64 ± 2	<1.6	0.106 to 0.108
Low Fiber Vol.	53 ± 1.5	<1.7	0.125
High Porosity	64 ± 2	2.8 to 6.2	0.101 to 0.108

high fiber volume by increasing the number of bleeder plies was much less successful. This was confirmed by supplemental tests that demonstrated that increasing bleeder plies above three is ineffective in both reducing resin content and thereby raising fiber volume of 8-ply laminates.

3. STRUCTURAL TESTING

All tests were single lap joints with one member composite and the other 0.1 in. thick titanium alloy. Eccentricities inherent in single lap joints introduce bending stresses not present in symmetric double lap joints. Hence strength values reported here may tend to be lower than those of similar double lap configurations. Three replicates were tested for each condition. Elevated temperatures were attained by circulating hot air around specimens enclosed by an insulated box, as described elsewhere.[6,7] For all laminates, two widths were tested to obtain bearing and tension failures respectively. The width for the former was 5 bolt diameters, or 0.938 in., while that for the latter was 0.5 in. Data scatter was encouragingly small with the coefficient of variation rarely exceeding 10 percent and frequently being in the neighborhood of 5 percent. Thus only mean values and general trends will be discussed. A sample size of three is inadequate for convincing remarks to be made about design strengths.

To assess the importance of moisture conditioning, two sets of tests were run. In both cases conditioning was done prior to joint assembly to ensure maximum absorption at the bolt hole. In the first series, referred to as "ambient", strength tests were conducted roughly three months after fabrication. Weight measurements suggest they had absorbed roughly 0.3 percent by weight of water. The remaining specimens were conditioned at 60 percent relative humidity until equilibrium was reached. This corresponds to about 0.8 percent weight of water absorbed.

4. TEST RESULTS

4.1 High Fiber Volume Test Results

A summary of the tests is provided in Table 2. As is to be expected from the minor change in processing, the induced strength variation is slight. A general trend is that, at room temperature, both bearing and tension strength reductions occur because of the rise in fiber volume. No explanation is available for the highest temperature bearing strength

Table 2. SINGLE-BOLT-JOINT STRENGTH FOR HIGH- AND NOMINAL-FIBER-VOLUME LAMINATES

Width (in.)	Temp (°F)	Failure Mode	Avg Failure Load (kips)				Ratio (High FV)/(Nominal FV)	
			High FV		Nominal FV			
			Amb	60% rh	Amb	60% rh	Amb	60% rh
0.5	75	Tension	1.58	1.72	1.68	1.84	0.94	0.94
↓	250	↓	1.47	1.62	1.58	1.72	0.93	0.94
↓	310	↓	1.35	1.34	1.38	1.47	0.98	0.91
0.938	75	Bearing	2.55	2.60	2.78	2.81	0.92	0.92
↓	250	↓	2.16	2.14	2.39	2.12	0.90	1.01
↓	310	↓	2.07	1.85	1.90	1.57	1.09	1.17

being raised apparently by increased fiber volume. These minor apparent strength variations are well within the scatter to be expected for such small sample size. This, coupled with the small variations shown in Table 1, suggests that the accidental addition of bleeder plies has a small effect on joint strength.

4.2 Low-Fiber-Volume Test Results

Strength data are provided in Table 3. A clear trend of a temperature dependent strength variation is demonstrated for both tension and bearing failures as a consequence of increased resin content. Tension strength is more strongly influenced than bearing response. This is most evident for the moisture-conditioned specimens. Since tension failures are the most likely in optimally designed multi-bolt row joints, these suggest that low fiber volume in this material is deleterious to joint strength. It should also be noted from Table 1 that the low-fiber-volume laminates are roughly 16 percent thicker than nominal laminates. Hence, comparing joint strength on the basis of stress yields an even clearer degradation than a comparison of loads. In addition, since thickness perturbation causes potential assembly problems, deletion of bleeder plies produces degradations in joint performance.

4.3 High-Porosity Laminates

Data addressing both single-bolt and

Table 3. SINGLE-BOLT-JOINT STRENGTH FOR LOW- AND NOMINAL-FIBER-VOLUME LAMINATES

Width (in.)	Temp (°F)	Avg Failure Load (kips)				Load Ratio (Low FV)/(Nominal FV)		Stress Ratio (Low FV)/(Nominal FV)	
		Low FV		Regular FV					
		Amb	Cond	Amb	Cond	Amb	Cond	Amb	Cond
0.50	75	1.34	1.34	1.68	1.84	0.80	0.73	0.69	0.63
0.50	250	1.33	1.22	1.58	1.72	0.84	0.71	0.72	0.61
0.50	310	1.30	1.02	1.38	1.47	0.94	0.70	0.81	0.60
0.938	75	2.47	2.47	2.78	2.81	0.89	0.88	0.77	0.76
0.938	250	2.04	1.95	2.39	2.12	0.85	0.92	0.73	0.79
0.938	310	1.77	1.63	1.90	1.57	0.93	1.04	0.80	0.90

two-row joints are included in Table 4. The 0.938-in. wide specimens provide bearing strength data. The other three widths involve coupled mode failures. Those joints with eight bolts provide interaction effects between bolts. The eight-bolt joint's failure load was eight times the load per bolt. This parameter was used to allow immediate assessment of bolt interaction effects in multibolt joints. Because of these interactions, the load per bolt will be less than the failure load in a single bolt joint having the same basic parameters.

The last two columns in Table 4 show that, except for the 3.75-in. wide eight-bolt joints at high temperature, porosity has little impact on joint strength. In numerous cases, these experiments suggest porosity actually increases joint strength. This suggests a relief of the stress concentration induced by the bolt hole, a concept described by Hart Smith [3]. However, it is possible that the porosity may tend to act as a "crack stopper" in the resin. Further work is needed. This must include an assessment of porosity on unnotched material samples and on those containing unloaded holes.

4.4 Edge Distance

As in the high-porosity laminates, load per bolt is used as the measure of failure. Results are presented in Fig. 1. It is seen that edge distances as low as twice the bolt diameter do not drastically lower strength for these laminates with the relatively wide bolt spacing used. This shows that they are tolerant of this type of machining error.

4.5 Hole Damage

Results for the high fiber volume laminates were consistent and are summarized in Table 5. As expected,

Table 4. JOINT STRENGTH FOR HIGH- AND NOMINAL-POROSITY LAMINATES

Specimen Width (in.)	No. of Bolts	Temp (°F)	Load Per Bolt (kips) High Porosity Amb	High Porosity Cond	Nominal Amb	Nominal Cond	Ratio (High Porosity)/(Nominal) Amb	Cond
0.656	1	RT	2.09	2.16	2.18	2.22	0.96	0.98
↓	↓	250	1.85	1.81	1.96	1.80	0.94	1.00
		310	1.55	1.46	1.41	1.34	1.10	1.09
2.625	8	RT	1.34	1.38	1.34	1.36	1.00	1.01
↓	↓	250	1.34	1.32	1.30	1.22	1.03	1.08
		310	1.22	1.12	1.15	1.05	1.06	1.07
0.938	1	RT	2.59	2.73	2.78	2.81	0.93	0.97
↓	↓	250	2.27	2.10	2.39	2.12	0.95	0.99
		310	1.95	1.55	1.90	1.57	1.03	0.99
3.75	8	RT	2.03	2.04	2.13	2.00	0.96	1.02
↓	↓	250	1.63	1.60	2.02	1.96	0.81	0.82
		310	1.47	1.22	1.80	1.54	0.82	0.80

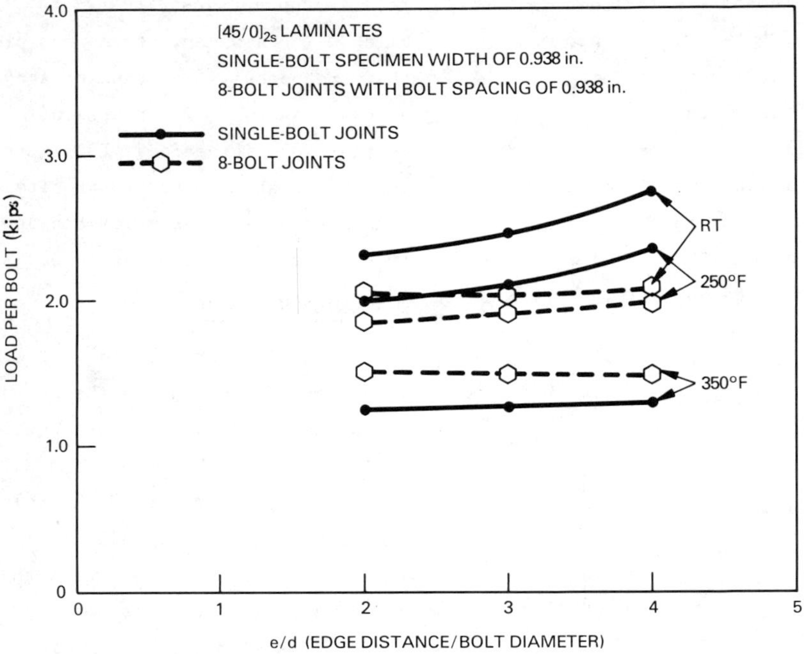

Fig. 1 Influence of Bolt-Edge Distance on Strength of Single- and Two-Row, Eight-Bolt Joints

TABLE 5 INFLUENCE OF DAMAGED HOLES IN HIGH FIBER VOLUME LAMINATES AT 60% RH

Width (in.)	Temp (°F)	Failure Mode	Avg Failure Load (kips) Undamaged	Damaged	Stress Ratio (Damaged)/(Undamaged)
0.5	75	Tension	1.72	1.55	0.90
0.5	250	Tension	1.62	1.46	0.90
0.5	310	Tension Bearing	1.34	1.33	1.00
0.938	75	Bearing	2.60	2.68	1.07
0.938	250	Bearing	2.14	2.16	1.00
0.938	310	Bearing	1.85	–	–

the specimens failing in tension show a roughly 10 percent decrease in strength due to the added stress concentration of the simulated chip out. This is not judged to be a very severe degradation. The specimens involving bearing demonstrate virtually no static strength decrease due to this form of damage. This is different from the data for tape laminates tested at room temperature which suffered degradations up to 14 percent.[8] No quantitative explanation is available for this insensitivity of the holed cloth laminates relative to their tape counterparts. However, cloth laminates have been demonstrated to be less susceptible than similar tape configurations to high interlaminar stresses and potential edge delaminations.[9] Since the cutter is producing a rather gross delamination, it is more likely to degrade tape rather than cloth joint strength.

Data from the low fiber volume laminates was not conclusive. The scatter was markedly greater than in all the other test series, with coefficients of variation reaching as high as 10 percent. Comparing sample averages suggested joint strength was raised by the simulated chip out for some configurations. This is not considered credible, indicating the need for further testing and evaluation.

5. CONCLUSIONS

The test data derived suggests that bolted joints in the graphite cloth epoxy material utilized are not particularly sensitive to four of the five perturbations considered.

It was demonstrated that using the layup procedures and cure cycles chosen, it was not possible to raise the fiber volume significantly above the nominal value. Thus there was no very significant change in joint strength due to this processing variation.

Raising porosity quite markedly was accomplished by not applying any vacuum pressure. This had limited impact on joint strength with some tests indicating porosity may raise joint strength slightly. The reason is believed to be the porosity producing stress concentration relief, as explained by Hart Smith.[4]

Removing all bleeder plies was demonstrated to be an effective method of reducing fiber volume from the nominal of about 61 percent to around 52 percent. This produced a degradation in bolted joint strength reaching as high as 30 percent. This defect also caused a clear thickness increase which meant failure stress level reductions as high as 40 percent.

For joints designed to fail in bearing, the strength was shown to be relatively insensitive to bolt hole edge distance down to twice the bolt diameter.

Using a special cutter to simulate chip out defects in high fiber volume laminates produced a roughly 10 percent reduction in strength for joints failing in tension. No decrease in bearing strength was evident. Attempts to assess the effect of simulated chip out defects in low fiber volume joints produced anomalous results.

Three replicates per test is too small a size to allow definitive remarks to be made about design implications of the types of manufacturing perturbations considered. Clearly, low fiber volume induced by omitting the bleeder plies has serious implications but the material is remarkably tolerant of the other imperfections.

6. REFERENCES

1. Garbo, S. P. and Ogonowski, J. M., "Effect of Variance and Manufacturing Tolerances on the Design Strength and Life of Mechanically Fastened Composite Joints". AFFDL-TR-78-179. December 1978.

2. Hart-Smith, L. J., "Bolted Joints in Graphite Epoxy Composites", NASA CR144899. June 1976.

3. Hart-Smith, L. J., "Mechanically Fastened Joints for Advanced Composites - Phenomelogical Considerations and Sample Analyses".

4. Ramkumar, R. L., "Bolted Joint Design". Proc. of Symposium on Test Allowables for Fibrous Composites. ASTM Conference held at Dearborn, MI on October 3rd and 4th, 1979. To be published.

5. Agarwal, B. L., "Behavior of Multifastener Bolted Joints in Composite Materials", Paper No. 80-0307. Proc. AIAA 18th Aerospace Sciences Meeting. Pasadena, CA. January 14-16, 1980.

6. Bailie, J. A., Duggan, M. F., Bradshaw, N. C. and McKenzie, T. G., "Design Data for Graphite Cloth Epoxy Bolted Joints at Temperatures Up To 450°K". Proc. of ASTM Symposium "Joining of Composite Structures". Minneapolis, Minnesota. April 1980. To be published.

7. Bailie, J. A., Fisher, L. M., Howard, S. A. and Perry, K. G., "Some Environmental and Geometric Effects on the Static Strength of Graphite Cloth Epoxy Bolted Joints".

8. Pengra, J. J. and Wood, R. E., "The Influence of Hole Quality on Graphite Epoxy Composite Laminates". Proc. 21st Structures, Structural Dynamics and Materials Conference, Seattle, WA. May 12-14, 1980. Paper 80-0777, pp. 687-693.

9. Crossman, F. W. and Wang, A.S.D. "Analysis of Free Edge Induced Fracture of Fiber Composite Laminates". Proc. Symposium on "Application of Computer Methods in Engineering". Univ. of Southern California. August 23-26, 1977.

7. BIOGRAPHIES

J. A. Bailie. Mr. Bailie is senior staff engineer in the Loads, Structures, and Dynamics Department of LMSC's Missile Systems Division. He has been project leader for an Independent Development Composite Structures and Manufacturing Program since May 1976. For the past 19 years his assignments have been involved with development to the Polaris, Poseidon, and Trident Fleet Ballistic Missiles. His training includes an apprenticeship at the de Havilland Company in England and a Stanford University Ph D. in Aeronautics.

S. A. Howard. Ms. Howard is a strength engineer with the Structures Department of Lockheed Missile Systems Division. She received her BME in 1975 and MMAE in 1978 both from the University of Delaware. She currently is working in the areas of moisture effects on polymer composites, graphite/epoxy and graphite/polyimide mechanical property characterization, and injection molding materials.

L. M. Fisher. Mr. Fisher is a laboratory analyst in the Experimental Mechanics Organization of the Lockheed Palo Alto Research Laboratory. He has been involved in research and development testing of thin walled and advanced composite structures throughout his seventeen year career. For eight years he was employed in the Department of Aeronautics and Astronautics at Stanford University. The nine most recent years have been at Lockheed. He received his A.A. degree from Foothill Junior College in 1971.

R. L. Cothern. Mr. Cothern is responsible for the composites development group of the Lockheed Missile Systems Division's Advanced Manufacturing Technology organization. His technical specialty is the fabrication of prototype composite components and their tooling. He is an associate manufacturing planner and has been on the staff of Lockheed for 13 years.

13th National SAMPE Technical Conference
October 13-15, 1981

DEVELOPMENT OF MANUFACTURING METHODS FOR JOINING REINFORCED THERMOPLASTICS (TOMAHAWK THERMOPLASTIC COMPOSITE WING)*

G.G. Liskay
Engineering Chemist Senior
General Dynamics Convair Division

Abstract

The development of a reinforced thermoplastic wing for the Tomahawk cruise missile is described. The existing or baseline wing consists of an integral pivot fitting, a frame machined from a 5-inch thick 300-pound aluminum-alloy plate, and adhesively bonded skins. Aluminum alloy plates are chem-milled to required thicknesses and then bonded to the frame. Design and analysis studies indicated that a wing structure using thermoplastic materials would lower production costs, reduce weight, and provide a reduced radar cross section when compared to the baseline metal wing. Studies and analysis indicated that utilization of an injection-molded filled polysulfone frame and unidirectional Kevlar and glass-polysulfone skins was a feasible design concept. A wing was designed using these materials to meet the requirements of the Tomahawk. Adhesive and adhesive-fastener evaluations were conducted and subcomponent specimens were tested. Three full-size cruise missile thermoplastic wings were fabricated. One wing was statically tested to failure; a second was subjected to fatigue testing, then tested to failure; and the third was used for radar cross-section signature measurements. Test results and cost analyses indicate the thermoplastic wing to be a feasible manufacturing concept that also reflects a significant cost savings.

*NASC Contract N00019-76-C-0227,
Program Monitor, Max Stander.

1. INTRODUCTION

Numerous manufacturing technology programs have been conducted in the aerospace industry to incorporate weight savings and reduce costs. Thermosetting composite materials have been used in the majority of these programs for fabricating lightweight components, both semi-structural and structural. Thermoplastics, which previously had been utilized primarily for commercial items only, have recently made inroads into aerospace. Thermoplastics offer a way to achieve cost savings in fabricating reinforced graphite, glass, and Kevlar-reinforced composites. This has been made possible through the development of polysulfone, polyethersulfone, and other thermoplastic resins that compare favorably to high-temperature-resistant epoxies as potential matrices for composites.

Reinforced thermoplastic prepregs can be stored indefinitely while most thermosets have to be refrigerated and possess limited storage life. The thermoplastics can be compacted in relatively short periods of time using hot-press consolidation instead of the lengthy conventional vacuum bag/autoclave curing technique required for standard reactive thermosetting resins. More importantly, where the consolidation is not completed due to equipment failures, and consolidation pressures or temperatures have not been attained, the material is salvageable and can be reprocessed. Furthermore, preconsolidated flat sheet stock can be thermoformed into shapes in the same manner as metals. Another advantage is that the material can be obtained with a net resin content, whereas the majority of thermoset resin systems utilize a "bleeder process" during cure.

In order to show the feasibility of reinforced thermoplastics in the aerospace industry, a Tomahawk cruise missile wing was selected for demonstration.

2. BASELINE DESIGN

The selected baseline cruise missile wing design consists of a glass and graphite fiber-reinforced polysulfone injection-molded frame, Kevlar/polysulfone skins reinforced with S-glass/polysulfone doublers, and an aluminum-alloy wing pivot fitting. Graphite was also considered for use as the wing skin reinforcements. However, with the small quantities involved, the weight savings amounted to 0.6 pounds and was eliminated on the basis of radar cross-section penalty at the long wavelengths. S-glass doublers were selected over Kevlar because its superior compression strength properties resulted in a weight savings of 1.1 pounds. Also, the greatly increased thickness required in Kevlar doublers at the inboard end of the wing would reduce the inside contour dimension at the pivot fitting splice, consequently reducing the efficiency of the fitting's bending section properties.

The pivot fitting is machined from 7050-T73651 aluminum-alloy plate. The wing pivot hub is identical to the existing wing to assure complete interchangeability of the thermoplastic wing with no modifications required to the cruise missile body or wing folding mechanism. The critical splice joint between the fitting and the wing skins and frame is accomplished by a stepped lap joint extending approximately seven inches outboard. The frame, skins leading edge, and pivot fitting are secondary bonded.

3. MATERIALS

Wing Skins

The thermoplastic resin selected for the composite matrix in fabricating the wing skins was P-1700 polysulfone. Selection was based on experience gained at Convair with the P-1700 resin system in prior programs. The P-1700 resin exhibits properties very similar to the 350F epoxy resin systems and, in addition, provides superior toughness.

The consolidation parameters used to fabricate the wing skin materials were developed in a program conducted for NADC, Contract N62269-75-C-0386, titled "Development of Thermoplastic Composite Aircraft Structural Elements." Figure 1 shows the optimum consolidation pressure temperature zones for the P-1700 polysulfone resin system coated on AS graphite fiber.

Laminates, 0.070 to 0.080-inch thick, were fabricated from 5 mil/ply Kevlar-P-1700 and 8-mil S-2 glass-P1700 (supplied by U.S. Polymeric). Mechanical strength properties are listed in Table 1.

Wing Frame

A survey of various thermoplastic materials indicated that a glass-filled polysulfone material exhibited properties and low-cost features that made it a feasible and attractive candidate for the wing frame. Based on radar cross-section (RCS) studies, it was determined that the addition of graphite fibers could enhance the RCS properties. Further analysis of desired strength and RCS properties confirmed the selection of a 28% glass – 2% graphite parts by weight fiber-filled polysulfone molding compound.

Adhesives

Initially, elevated temperature adhesives were evaluated for bonding of the wing component details. However, warpage problems were encountered in the composite after bonding of the Kevlar-polysulfone wing skins to the glass-

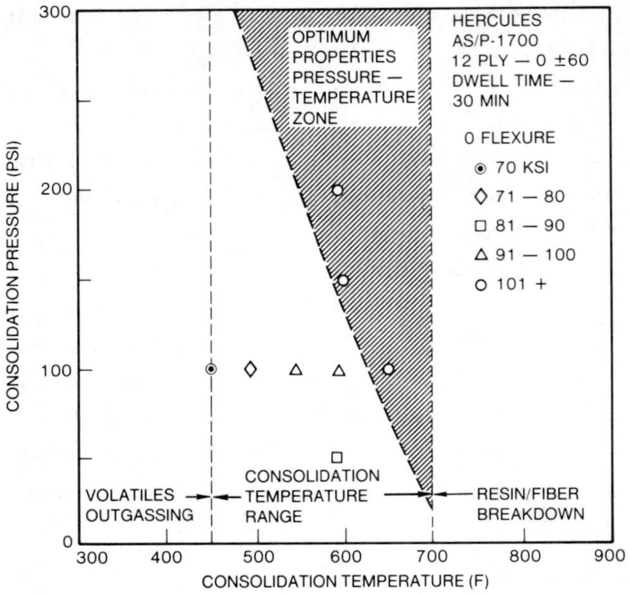

Figure 1. Optimum properties consolidation pressure-temperature zone for P-1700 polysulfone resin.

Table 1. Mechanical properties of unidirectional Kevlar-polysulfone and glass-polysulfone laminates.

MATERIAL	TEMPERATURE, (F)	TENSILE ULT (KSI)	TENSILE MODULUS (PSI × 10^6)	COMPRESSION ULT (KSI)	COMPRESSION MODULUS (PSI × 10^6)	FLEXURAL ULT (KSI)	FLEXURAL MODULUS (PSI × 10^6)	SHORT BEAM SHEAR (KSI)
KEVLAR/1700	−67	190	9.3	42	13.1	78	8.0	4.6
	R.T.	190	9.6	32	8.9	73	7.2	4.8
	+160	172	9.0	30	8.8	60	6.0	4.8
S-GLASS/P1700	−67	170	8.1	128	—	192	8.2	8.5
	R.T.	168	7.4	130	—	190	7.5	9.0
	+160	175	7.2	125	—	184	7.0	9.5

polysulfone doublers due to the differences in the thermal coefficients of expansion. This resulted in selecting a room-temperature curing adhesive, Hysol's EA 9320 epoxy adhesive following an evaluation of various adhesive systems.

4. JOINING EVALUATION

A primary consideration in the thermoplastic wing concept was the development of a joining technique for assembling the various components; i.e., wing skins to doublers, wing skin and doublers to frame, and wing skins to the aluminum pivot fitting. Among those considered were ultrasonic welding, adhesive bonding, and adhesive-bonding plus fasteners.

Following a cursory evaluation, ultrasonic welding was considered impractical for joining of the thermoplastic wing components. After analyzing the wing configuration and the high stress loads imposed on the wing in flight, it was decided to concentrate on adhesive bonding and fastener assembly techniques as the only feasible method for wing component assembly.

Composite joint specimens were designed, simulating the critical wing-to-pivot fitting joint. The design condition is 5,875 lb/in. compression on the upper skin and tension on the lower skin. A tapered or step lap-splice bonded-joint should be capable of at least 1,500 lb/in.^{2} shear strength. A six-inch overlap should therefore approach a splice capable of 9,000 lb/in. This should provide a margin of safety of 0.50 in., which is necessary to account for possible degradation due to environmental effects.

A summary of step-joint analysis and the developmental tests are presented in Table 2. Although these tests indicated that the bonded-joint design was marginal as regards to load requirements, it was decided to pursue the fabrication and testing of subelement verification test specimens.

5 SUBCOMPONENT TESTS

Prior to initiating full-scale wing fabrication, four subcomponent specimens simulating the wing skin-pivot fitting joint were fabricated and tested to verify the design. The specimens were designed so that the wing critical skin loading of 5,435 lb/in. would be reached when the critical shear for the joint area, 3,750 lb/in. is applied to the load wing skin-pivot fitting. The specimen configuration, simulating the critical skin-pivot fitting joint area, is shown in Figure 2.

The test results indicated that fasteners would be necessary to assure the integrity of the critical wing skin - pivot fitting joint to the required ultimate strength of 125% for the Tomahawk cruise missile wing. Table 3 summarizes the results of the subcomponent tests.

6. FULL-SCALE CM THERMOPLASTIC WING FABRICATION

The final phase of this program involved the fabrication of three full-sized Tomahawk CM thermoplastic composite wings, one to be statically tested to failure in the most critical load condition, the second in fatigue, and a third to be used for radar cross-section measurements. The processing parameters and the optimum wing skin-joint fitting concept used in the subcomponent verification specimens were used in the fabrication of the wings.

Wing Tips and Leading Edge Details

The wing tip and leading edge skins were fabricated from Style 143 and Style 281 Kevlar/polysulfone prepregs. Aluminum-matched die tooling was used in the consolidation of these details. These are shown in Figures 3 and 4.

Wing Skins and Doublers

The Kevlar/polysulfone was heat tacked with a

Table 2. Summary of skin to pivot fitting joint development tests.

	SPECIMEN TYPE	LAMINATE	N_x* AT RT (LB/IN.)
1.	SIMPLE OVERLAP BOLTED (4 BOLTS)	143 FABRIC, t = 0.104	3,173
2.	SIMPLE OVERLAP BONDED	143 FABRIC, t = 0.104	6,677
3.	STEP LAP BONDED	** $(45/\text{-}45/\text{-}45/45/0_2)_5$	3,995
4.	STEP LAP BONDED	** $(45/0/\text{-}45/\text{-}45/0/45)_5$	3,155
5.	STEP LAP BONDED PLUS BOLTS	** $(45/\text{-}45/\text{-}45/45/0_2)_5$	5,850
6.	STEP LAP BONDED PLUS BOLTS	** $(45/0/\text{-}45/\text{-}45/0/45)_5$	6,470
7.	STEP LAP BONDED PLUS TAPERED ALUMINUM	** $(45/\text{-}45/\text{-}45/45/0_2)_5$	4,820
8.	STEP LAP BONDED, BOLTED AND TAPERED ALUMINUM	** $(45/\text{-}45/\text{-}45/45/0_2)_5$	6,590
9.	STEP LAP BONDED PLUS TAPERED ALUMINUM	** $(45/\text{-}45/\text{-}45/45)_5 0_{10}$	4,153
10.	STEP LAP BONDED PLUS TAPERED AND PINKED EDGE	** $(45/\text{-}45/\text{-}45/45)_5 0_{10}$	4,350
11.	STEP LAP BONDED PLUS TAPERED AND BOLTED	** $(45/\text{-}45/\text{-}45/45)_5 0_{10}$	6,115
12.	WING TO FITTING JOINT SUB-COMP. BONDED	** $(45/\text{-}45/\text{-}45/45)_5 0_{10}$	6,110
13.	WING TO FITTING JOINT SUB-COMP. BONDED AND BOLTED	** $(45/\text{-}45/\text{-}45/45)_5 0_{10}$	7,402

* DESIGN N_x =5,874 LB/IN. ** ±45-KEVLAR/P1700; 0-GLASS/P1700

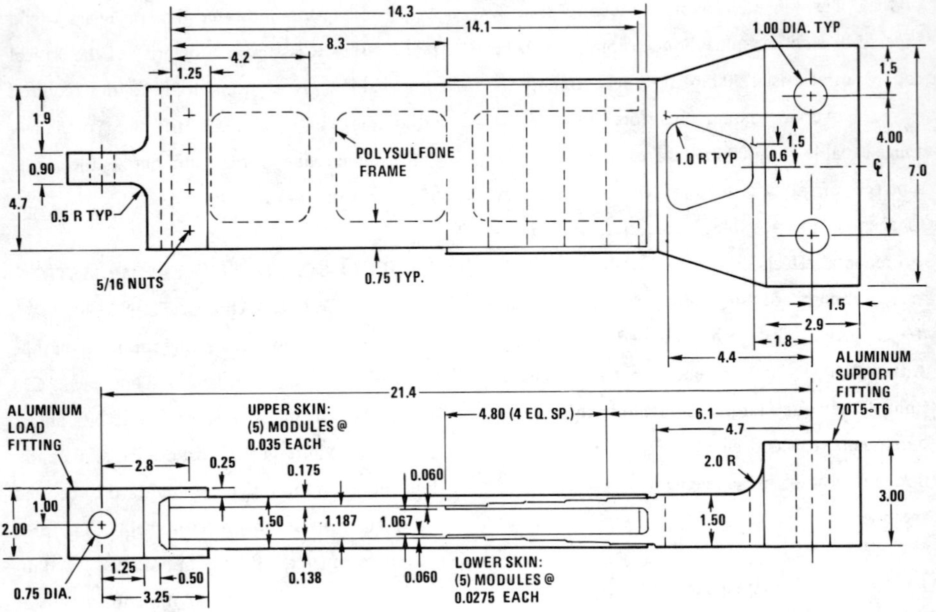

Figure 2. Subcomponent test specimen configuration.

Table 3. Subcomponent test summary.

SPECIMEN I.D.	FASTENER			LOAD SEQUENCE	FAILURE LOAD (LB)	PER-CENT ULT	FAILURE MODE
	TYPE	INSERT	LOCATION				
PJ-1*	8-32	SOLID THREAD	A, B, C, D	UP-DN-FAILUP	3,900	1.01	TEN. IN AL. AT C
PJ-2	SCREW	NONE	A, B, C, D	UP-DN-FAILUP	4,860	1.26	TEN. IN AL. AT C
PJ-3	10-32	COIL THREAD	A, B	UP-DN-FAILUP	4,720	1.22	SHEAR/TEN. AT A TO C
PJ-4	NONE	NONE	—	UP-DN-FAILUP	4,015	1.04	SHEAR/TEN. AT A TO C

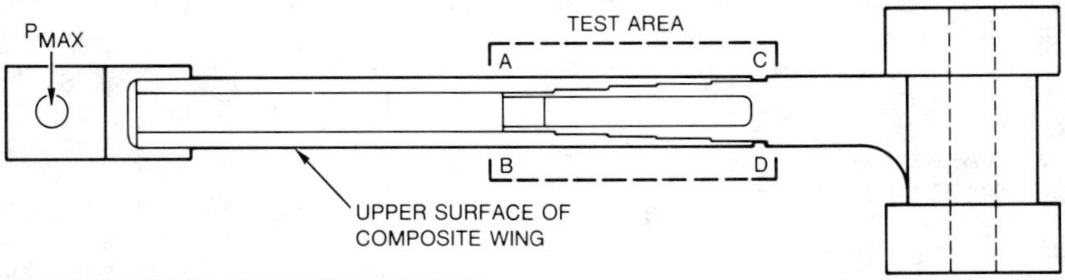

*ALUMINUM PART IS 22% UNDER DRAWING THICKNESS

hot iron into four-ply modules of a ±45° orientation. The Kevlar/polysulfone modules and the glass/polysulfone 2-ply modules (0° orientation) were preconsolidated at 475F and 100 psi. This partial consolidation procedure reduces the thickness to within a few mils from the final 600F consolidated thickness and enables easier stacking of the doubler modules on the die. The glass/polysulfone doublers were then trimmed to fit the steps in the dies.

The die surfaces were coated with two applications of Frekote 33 release agent. After each ap-

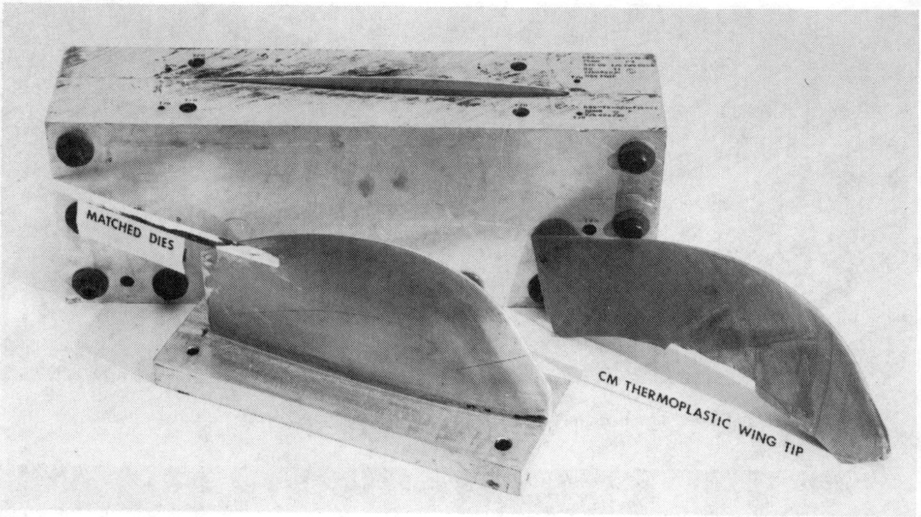

Figure 3. Matched dies and consolidated wing tip detail.

Figure 4. Matched dies and consolidated leading edge detail.

plication, the Frekote was baked on at approximately 400F for about 30 minutes. The upper and lower portions of the electrically heated matched dies are shown mounted in a press in Figures 5 and 6. The electrical heating plate can be seen in position beneath the dies. The glass/PS doublers were placed in the tool and covered with two-mil aluminum foil, previously coated with Frekote on both surfaces. The Kevlar/polysulfone modules were layed over the foil and the press closed to barely contact the Kevlar surface. Figure 7 shows the time-temperature profile used in the press consolidation process.

Wing Frame

The wing frame was injection molded by Envirotech from a 25% glass fiber — 2% graphite fiber molding compound.

Bonding Assembly of Details

Figure 8 shows thermoplastic wing details prior to bonding assembly.

The bonding assembly of the wing was accomplished as follows.

1. **Bonding of Doublers to the Wing Skins —** The faying surfaces of the doublers and wing skins were abraded with 80 grit abrasive and wiped with Freon 12. The EA 9320 adhesive was applied to the faying surfaces and five-mil- thick glass strands were used to produce a minimum five-mil bondline during pressurization. The details were placed in the steel die

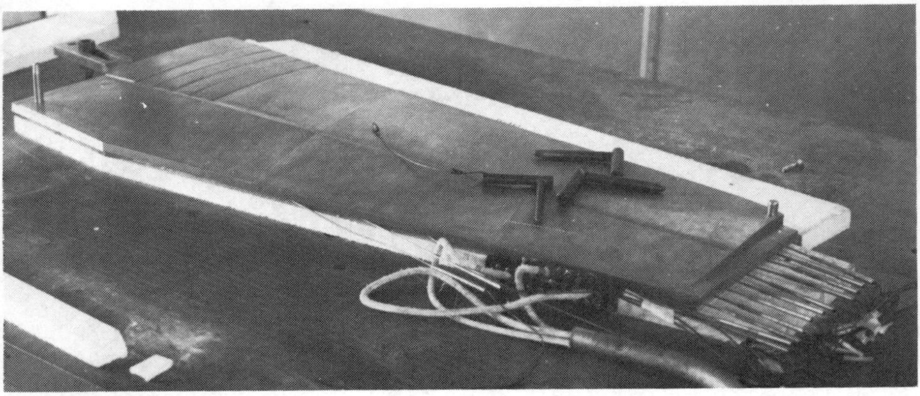

Figure 5. Lower matched steel die — for consolidating upper wing skin and doubler.

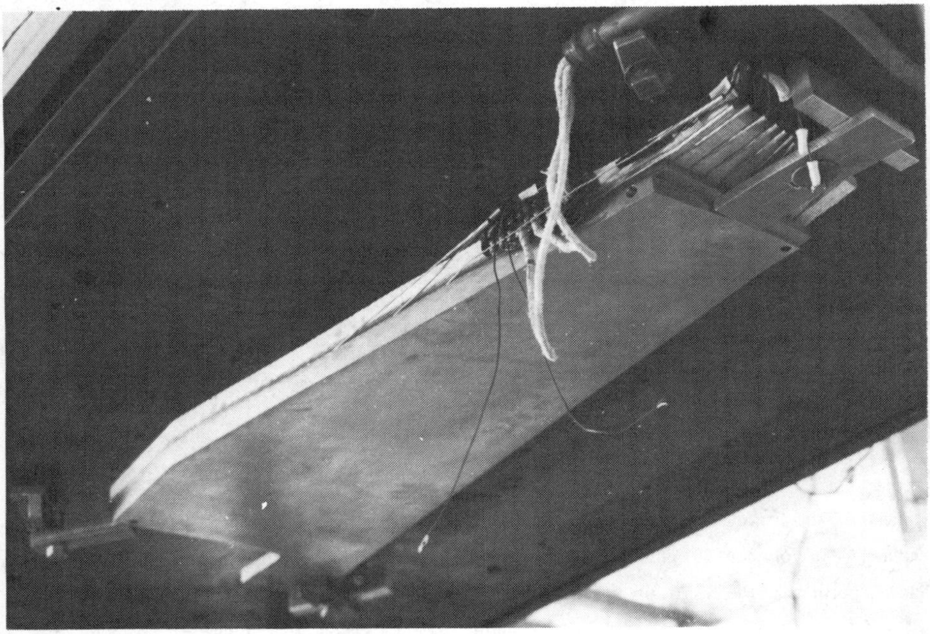

Figure 6. Upper matched steel die — aerodynamic surface.

and vacuum bagged. The skin-doubler assembly was then cured at room temperature for approximately 24 hours under 27-29-inch Hg vacuum. The cured skin-doubler details were trimmed to fit the slotted outline in the frame.

2. **Wing Pivot Fitting-Preparation for Bonding** — The aluminum pivot fitting was cleaned in a caustic solution, rinsed with de-ionized water, etched with the dichromic-sulfuric acid (FPL) etch, rinsed in de-ionized water, and dried at 150F for 45 minutes. The fitting was primed with BR-127 (American Cyanamid), an epoxy-chromate primer, and air dried for 30 minutes and at 250F for 60 minutes.

3. **Bonding of Pivot Fitting to Thermoplastic Frame** — The faying surface of the frame was sanded with a 120-grit paper and wiped with

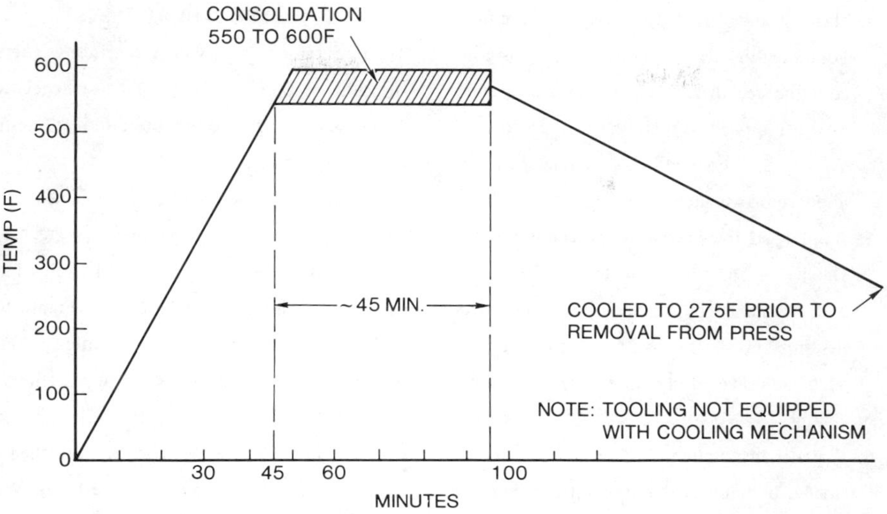

Figure 7. Time-temperature profile for consolidating wing skins and doublers.

Figure 8. Thermoplastic wing details.

Freon 12. EA 9320 adhesive was applied to both the faying surfaces of the pivot fitting and frame, and the frame inserted into the pivot fitting slot. Contact pressure was applied by the use of clamps, and the adhesive was allowed to cure at room temperature for 24 hours.

4. **Bonding of the Doubler-Skin Details to Frame** — The faying surfaces of the doubler-skin details were sanded with 80 grit abrasive, the frame with 120 grit abrasive, and then wiped with Freon 12. EA 9320 adhesive was applied to the faying surfaces of the pivot fitting, frame, and skin-doublers and five-mil

glass strands used to effect a minimum bondline thickness. The details were positioned in the steel dies, then placed in a press, and 8-10 psi applied to mate the faying surfaces. The details cured at room temperature for about approximately 24 hours.

5. **Bonding of the Leading Edge and Wing Tip Details** — The faying surfaces of the details and frame were sanded per procedures described previously, and bonded with the EA 9320 adhesive using discretely positioned clamps.
6. **Fastener Installation** — Pilot holes, 8-10 mils smaller in diameter than the outside diameter of the self-threading fasteners, were drilled in the joint area. The fasteners were coated with EA 9320 adhesive and inserted, using a 30 ft-lb torque.

A completed wing assembly is shown in Figure 9.

7. STRUCTURAL TESTS

The first thermoplastic wing test article was tested statically to failure. The second test article was subjected to fatigue testing prior to static testing to failure.

The test specimens were instrumented with strain gages and acoustic couplers (Wing No. 2 only). Tension pads were installed to apply loads uniformly as shown in the test setup, Figure 10.

Wing No. 1 was subjected to a static structural test to failure by increasing the applied load to 100% of limit load in 20% increments. The steps were repeated. The applied loads were then increased to 125% of limit (ultimate load). Wing No. 2 was subjected to a fatigue spectrum developed for the Air Launched Cruise Missile (ALCM). It was then tested to failure. Wing No. 1 failed at approximately 107% of limit load. Attempts to pinpoint the exact cause of this premature failure have been hampered by the wing

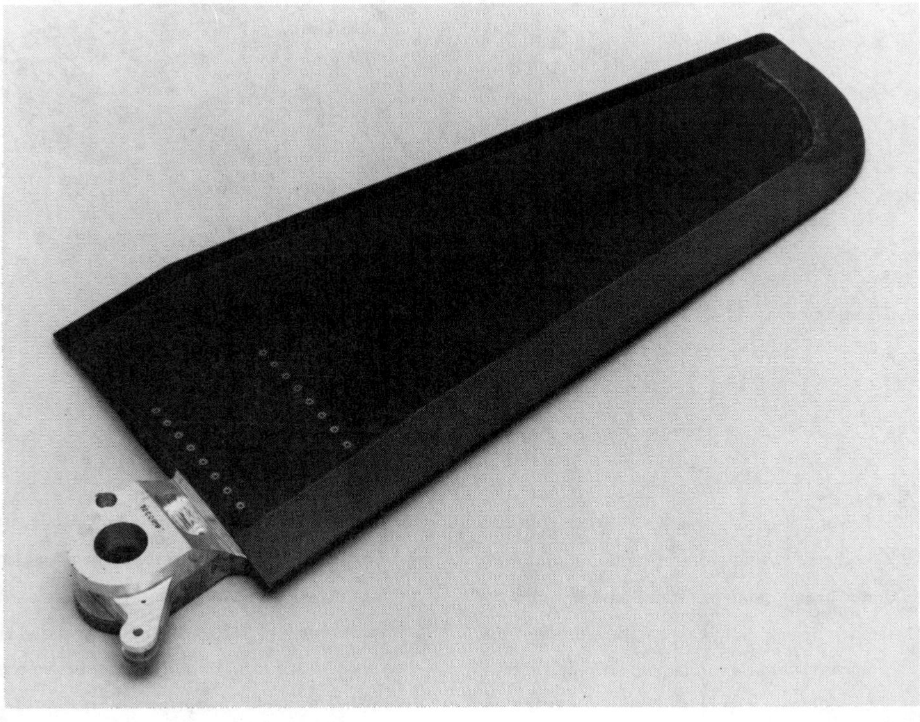

Figure 9. Cruise missile Tomahawk thermoplastic wing.

Figure 10. Test setup for static testing of the thermoplastic wing.

being subjected to a large post-failure loading. This inadvertant loading was the result of a control lever slipping out of its detent and thus causing the loading ram to stroke through its entire motion.

Wing No. 2, following subjection to the fatigue spectrum, failed at 145% of limit load, exceeding the required ultimate of 125%.

8. CONCLUSIONS

1. A thermoplastic wing design and a viable manufacturing process have been generated for meeting the requirements of the Tomahawk Cruise Missile.
2. A cost savings of approximately 50% could be effected by substituting the thermoplastic for the all-metal wing.

3. Although a weight savings was not demonstrated in this program, it is probable that modifications in the pivot and frame could result in a weight savings of 10 to 15%.
4. The radar signature tests on the thermoplastic wing were inconclusive.

9. BIOGRAPHY

Mr. G.G. Liskay is currently a member of the Organics and Composites group in Materials and Process Engineering engaged primarily in the development of advanced composite structures. He was previously employed for almost 20 years at Rockwell International where he worked on the Hound Dog missile, Apollo and Saturn programs, the B-1 Bomber, and the Space Shuttle. He has been a program engineer and manager on cryogenic insulation, radar-absorber materials, and various reinforced composites to effect weight savings on aerospace vehicles.

13th National SAMPE Technical Conference
October 13-15, 1981

FILLER EFFECTS IN COMPOSITES

Thomas H. Ferrigno
Improde
29 Clover Hill Circle
Trenton, NJ 08638

Abstract

Composites of fillers and polymers where the filler is discrete particles of less than 10 to 1 aspect ratio provide many physical properties unobtainable with polymers alone. Filler effects on tensile, flexural and impact strengths and moduli are characterized. Specific examples of composite properties are given.

Keywords: composites, fillers

1. INTRODUCTION

This paper will not discuss certain fundamental aspects of filler utilization such as choice of filler, dispersion, composite rheology and surfation (surface modification/alteration of fillers), which are treated elsewhere.[1] Fillers consist of discrete particles, the most common of which is limestone suitably pulverized and classified for particle size distribution.

A principal property of fillers is packing fraction, P_f, which is the maximum volume of filler in a tightly packed mass, wherein the particles are uniformly coated and dispersed by the matrix or continuous medium which also fills the spaces between particles. Practical aspects and determination are described in reference [2]. At this point, the P_f describes a rigid system as depicted in Fig. 1. At any filler volume, V_f, less than the P_f, the mixture becomes mobile.

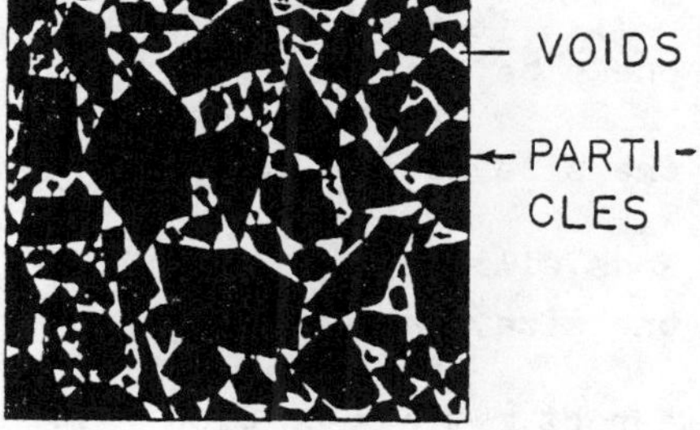

$$P_F = \frac{\text{FILLER VOLUME}}{\text{TOTAL VOLUME}}$$

Fig. 1

2. PHYSICAL PROPERTIES

2.1 Modulus

Modulus is described by the Lewis and Nielsen equation:[3]

$$\frac{M}{M_1} = \frac{1 + AB\,V_f}{1 - B\psi\,V_f} \quad (1)$$

where: $A = k_E - 1$,

$$B = \frac{M_2 / M_1 - 1}{M_2 / M_1 + A}, \text{ and}$$

$$\psi = 1 + \frac{1 - P_f}{P_f^{\,2}}\,V_f$$

M, M_1 and M_2 are moduli of the composite, matrix and filler respectively and may be Young's, shear or bulk moduli. V_f is the fractional filler volume and the Einstein coefficient, k_E, relates to particle shape only as indicated in Table 1. P_f also is included as a guide but for most fillers or mixed fillers is best approximated by oil absorption.

Table 1 [3,4]

	k_E	P_f
Spheres;		
one size, max. packing	2.50	0.74
one size, random packing	2.50	0.64
one size, loose packing	2.50	0.60
Rods, ellipsoids,		
one size, random packing,		
aspect ratio = 2	2.58	0.68
4	3.08	0.60
6	3.80	0.53
10	5.93	0.42
Mixed sizes, irreg. shapes, low surface area	4.0	-
Mixed sizes, plates	5+	-

Fig. 2 describes the relative modulus, M/M_1, effects of a fine limestone filler in two polymers. Note that nylon has a modulus about 6 times that of polyethylene, therefore all actual modulus values of the nylon composite are six times those of the polyethylene composite.

Fig. 2

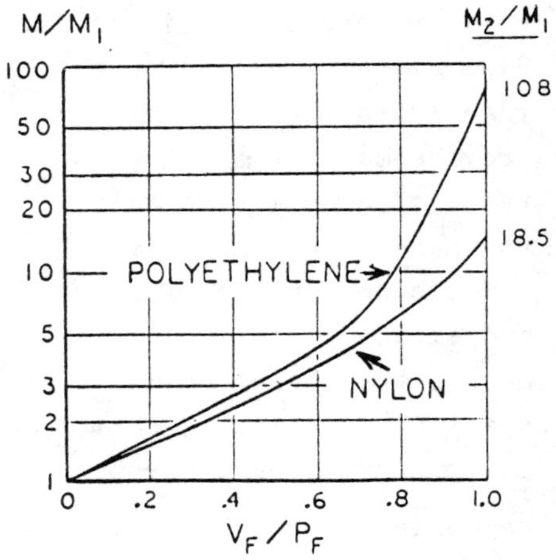

2.2 Tensile and flexural strength

Fig. 3 delineates the effect of relative filler volume, V_f/P_f on relative tensile yield strength; the tensile yield strength of the composite divided by that of the matrix. The temperature of testing determines the effect obtained according to the T_g or second order transition temperature of the matrix.

Significant differences can occur depending on the nature of the filler-matrix bond according to the approximation:

$$T_{yc} = V_{mr} T_{ym} A \qquad (2)$$

where $V_{mr} = 1 - V_f / P_f$

T_{yc} is the yield strength of the composite, V_{mr} is the relative matrix volume and A is an adhesion factor which varies between 0.8 and 1.1 for different types of fillers but may be greater with interfacial bonding or coupling agents.

Flexural strengths follow a similar trend although values are generally higher.

2.3 Impact strength

Fig. 4 is a representation of the cross-section of a filler particle in an extensible matrix at a strain sufficient to cause dilation. Voids which occur at the debonded surfaces cause the familiar stress whitening effect. All Filler particles act as stress concentrators. When adhesion to the matrix is modified or improved and the particles are below a limiting maximum size for the system, considerable improvement in impact resistance can be obtained. This will be ilustrated later with specific data. A filler having a higher P_f will provide greater impact strength than one having a lower P_f when both are compared at the same V_f. Finer fillers, especially those having narrow particle size distributions, will provide more uniform stress concentration and will improve the impact resistance of tough polymers such as PVC.

Fig. 3

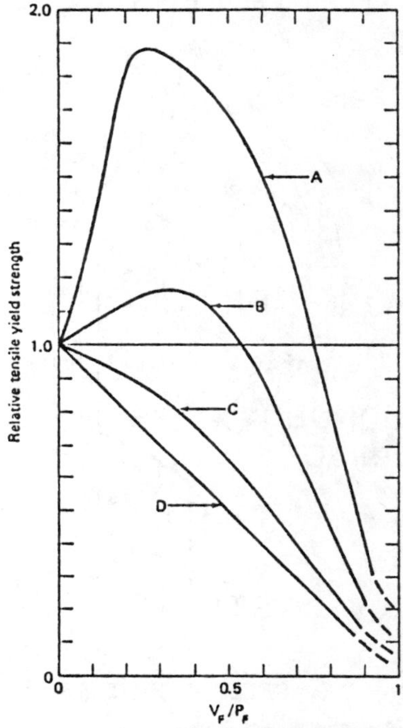

Relative tensile yield strength = tensile yield of composite/tensile yield of matrix. V_f/P_f = relative packing volume; volume of filler in composite/maximum packing volume of the filler.

A = Above T_g of non-polar, crystallizing polymers with high aspect ratio, high surface area fillers.

B = Above T_g of non-polar, crystallizing polymers with low surface area fillers.
Below T_g of polar, rigid polymers with with reinforcing fillers; fibers, flakes, and those treated with coupling agents.

C = Below T_g of polar, rigid polymers with low surface area fillers.

D = Above T_g of polar polymers with low surface area fillers.

Fig. 4

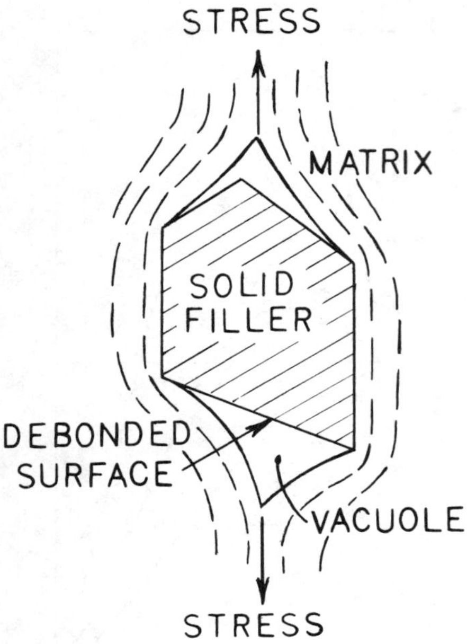

3. GENERAL EFFECTS

Table 2 is a general guide to filler use. Note that the matrix may have to be modified to obtain the indicated property and that the filler effect is a function of its relative filler volume; V_f/P_f.

4. EXAMPLES

The data of Table 3 indicate the improvements in impact resistance obtainable with the Gardner falling tup tester using untreated and surfied commercial limestone fillers. Atomite and Kotamite are 3 µm and Supercoat is 1 µm in average particle size. With a brittle polymer such as polypropylene, sufficiently fine fillers improve impact resistance but the stearate coating provides exceptional improvement, especially at low temperatures. All of the fillers increase toughness, moduli and deflection temperatures, resulting in significant property improvements for the composites.

Table 4 illustrates the effect of particles larger than 15 µm on impact resistance. Rigid PVC is a tough material, but is readily degraded by very low concentrations of large particles. Low concentrations of much finer top-size fillers improve the impact resistance of rigid vinyl house siding.

5. REFERENCES

(1) Handbook of fillers and Reinforcements for Plastics, Katz and Milewski, editors, pp 11-58, Van Nostrand Reinhold Co., NY (1978)
(2) Ferrigno, T. H., Poly. Eng. & Sci., Jan. 1978, Vol. 18, No. 1, p.33.
(3) Nielsen, L. E., Mechanical Properties of Polymers and Composites, Vol. 2, p. 388 Marcel Dekker, Inc., NY(1974)
(4) Nielsen, L. E., Ind. Eng. Chem., Fund, 13, No. 1 1974
(5) Thompson, Weinman & Co., Tech. Bulletin T-5-14a
(6) Thompson, Weinman & Co., Tech. Bulletin T-5-9a

Table 2

Physical Property of Composite	MAXIMUM IMPROVEMENT BY FILLER at	when V_f/P_f is	FILLER Bond/ Matrix	Type	Size Range	RELATIVE FILLER EFFECT ON MATRIX TYPE Rigid	Flexible
Modulus, M	P_f	high	max	solid, max M	broad	least	most
Tensile strength	V_f/P_f	low	max	fiber	narrow	least	most
Flexural strength	V_f/P_f	medium	max	fiber	narrow	least	most
Elongation	V_f/P_f	low	min	sphere	narrow, fine	least	most
Tear strength	V_f/P_f	low	max	fiber	narrow		
Impact strength	V_f/P_f	low	min	fiber	narrow, fine	most	least
Compressive strength	P_f	high	max	solid, max M	broad	least	most
Creep	P_f	high	max	fiber	broad	least	most
Hardness	P_f	high	max	solid, hi Mohs	broad	least	most
Coefficient/ friction	P_f	high	max	solid	broad	least	most
Abrasion resistance	P_f	high	max	solid, hi Mohs	broad	most	least
Density	P_f	high		sphere, low density	broad		

TABLE 3 (5)

40% BY WT LIMESTONE FILLERS IN HOMOPOLYMER POLYPROPYLENE

FILLER	% STEARIC ACID	GARDNER FT-LB / INCH 22 °C	IMPACT, 0 °C	FLEX. MOD., k PSI	DEF. TEMP. °C *
NONE	-	6.7	0	275	100
ATOMITE	0	37.	5.3	391	111
KOTAMITE	1	114.	9.6	347	116
SUPERCOAT	2	174.	7.6	359	106

* At 66 psi

Table 4 (6)

Effect of Filler Coarseness on Impact Resistance

	Coarseness	phr	Tensile yield*	% Elong.	Impact**
Supermite	100% - 15 µm	10	7486	149	80
Supermite + 2% Duramite	100% - 44 µm, 99.5% -15 µm	10	7212	168	56

* k psi

** % passing 1.87 in. lb/mil, Gardner IG-1120. No filler = 80

6. BIOGRAPHY

Tom Ferrigno has a B. S. in chemistry and, since 1971, has operated a consulting firm; Improde, an acronym for imaginative product development. He specializes in the development and use of fillers in plastics, paints, rubber and related applications. He has 30 U. S. patents, has authored two books and chapters in several, and numerous technical society papers. His lectures for Plastics Fair are devoted to the utilization of fillers.

13th National SAMPE Technical Conference
October 13-15, 1981

CHARACTERIZATION OF JUTE FIBER REINFORCED COMPOSITES

I. W. Sorensen and H. W. Chang
Bendix Advanced Technology Center
Columbia, Maryland

Abstract

Jute fibers were characterized using thermal analysis and microscopy. Static and dynamic mechanical as well as thermal properties of jute and glass cloth composites with a polyester matrix were compared with those of the resin itself. Study of the fracture surface morphology was used to suggest a mechanism by which jute composites exhibit more ductility than those reinforced with glass. The cost effectiveness of jute in applications requiring minimum reinforcement cost per unit modulus or strength attained was demonstrated.

1. INTRODUCTION

Natural as opposed to man-made fibers offer the potential of lower "cost" composites especially from energy and resource depletion standpoints. At the same time, certain properties of natural fibers such as jute appear to be roughly equivalent to glass fiber. (See Table 1). The current price of jute fluctuates rather widely but is still typically lower than glass fiber on a per pound basis. In light of these considerations, it would be reasonable to expect that composites using such fibers could be made cost-effective for many applications.

A number of attempts at fabrication of jute fiber reinforced plastics have been made with mixed results. Winfield[2] concluded that hybrid jute glass reinforcement combined with calcium carbonate filler in a polyester resin provided acceptable cost-effectiveness in low cost housing construction. However, it was felt that jute reinforcement alone was insufficient.

The current effort was directed toward obtaining a broader understanding of the physical properties of jute fiber composites especially in comparison with glass fiber composites fabricated under similar conditions.

2. EXPERIMENTAL PROCEDURE

The matrix resin used was Dow Corning's Durakane 790 containing calcium carbonate filler, zinc stearate and an Mg0-vinyl ester slurry. Cloths of jute fiber (Bengalon) and glass fiber were used to compression-mold composites with the polyester matrix resin. Fiber loadings were determined to be 48 V% and 35 V% for the glass and jute respectively.

The jute fibers themselves were studied using microscopy for shape, size and irregularity. TGA (Thermogravimetric Analysis) in air and nitrogen was used as an indicator of thermal stability of the fiber.

The molded composites (jute/polyester; glass/polyester; and neat resin) were subjected to a variety of tests as follows:

(i) Densities of the composites and cured neat resin were determined.

(ii) Convention mechanical properties (tensile strength and modulus) were obtained on an Instron testing machine.

(iii) TGA using the Dupont 1090 System (Heating rate of 10°C/min in air) and nitrogen.

(iv) TMA (for thermal expansion coefficient) was performed on the composites.

(v) Dynamic Mechanical Properties were obtained over a broad temperature range using a Rheometrics Dynamic Spectrometer in the 3-point bending mode. Parameters obtained included G' (storage modulus), G" (loss modulus) and tan (dissipation factor).

(vi) Fracture surface morphologies for the jute composites were obtained by SEM.

3. RESULTS AND DISCUSSION

3.1 Jute Fiber Characterization

3.1.1 Microscopy. Jute fibers are comprised of elongated cellulose cells with aspect ratios of about 75-200. A typical fiber cross-section is shown schematically in Figure 1 and consists of half a dozen such cells arranged to form a somewhat flattened ellipse with major and minor axes of about 60 and 30 respectively. The size and shape of these fibers is significant in determining packing geometries of the fibers incorporated in a composite. The fiber surface can be fairly rough as indicated in Figure 2. This may be expected to provide improved physical interlocking of the fiber and matrix. Some evidence of a weak intercellular bond is provided in Figure 2 where one of the cells is observed to be separating from the rest of the fiber. Further evidence of this weak bond is provided in the fracture surface morphology studies.

3.1.2 Thermal Analysis (TGA). Since it is a biological organic material, the temperature capability of jute is expected to be a limiting factor in its application. In this respect, it should behave much like wood, itself a cellulosic material. In addition to temperature, water absorption is expected to be significant but the same limitation for wood has not prevented its use as an important structural material.

The TGA thermogram shown in Figure 3 illustrates the degradation (as evidenced by weight loss) that occurs in jute fiber subjected to heating in air. Initially, up to about 100°C there is a weight loss that varies from 3 to 7% from sample to sample. This may be attributed to loss of absorbed water and the differences observed appear to be due to the sample's exposure to humidity prior to testing. After this point, there is some small additional weight loss (rate of approximately 0.01%/°C) below 250°C which is likely due to loss of more tightly-bonded water. Samples tested in nitrogen display similar behavior to this temperature. Above 250°C, decomposition of the jute occurs at an accelerating rate with an associated weight loss greater in air than nitrogen because of reduced oxidation of the molecular units in the inert atmosphere. For practical purposes, we may define an upper limit for temperature of exposure to be around 250°C. The actual maximum use temperature would probably be somewhat less. However, the degradation of the mechanical properties of the resin with increasing temperature would be the limiting factor in most low cost composite applications, and thus jute's heat resistance would be acceptable.

Jute fiber costs are quite variable. However, much of the cost is related to yarn and fabric manufacture. The cost may be significantly reduced if the jute is used in "sliver" form, the stage prior to twisiting and yarn manufacture. Indeed, the untwisted form of jute is desirable for maximizing the tensile modulus of the composite. However, this must be balanced against the processability of the resulting weakened fibrous tow. A solution to this problem may be hybridization using a small amount of continuous high strength filament (e.g. Kevlar).

3.2 Characterization of Molded Composites

3.2.1 Densities, Tensile Strengths and Moduli.

The results for densities and tensile properties are summarized in Table 2. The properties in two orthogonal directions in the plane were averaged.

It is clear that the glass composite is superior to the jute in room temperature modulus and strength by a factor of between 3 and 4 in both cases. Still the jute composite represented a significant improvement over the neat resin (75% stronger and twice the modulus).

When the specific gravities of the materials are considered (specific strength and stiffness) as well as the volume fraction loading, the gap between the glass and jute composites narrows significantly. More importantly, the costs of the fiber required to achieve a particular volume fraction loading (and hence properties) may be accounted for as indicated in the last two rows of Table 2. Here, because of the low fiber density for jute compared to glass, the jute composites offer a significant advantage.

3.2.2 Thermal Analysis (TGA, TMA). TGA curves for the jute and glass composites as well as the base resin are shown in Figure 4. As expected, the jute composite is degraded at lower temperatures than the glass composite and appears less thermally stable than the base polyester resin itself. This is due to the higher concentration of inert filter in the base resin. The principal concern at temperatures below 200°C is the loss of moisture in the jute. This can have a beneficial effect on the modulus (by improving the bonding between adjacent crystalline regions in the cellulose) but may pose a problem related to swelling. In practice, thermal degradation should not occur since the low cost resins of interest have usable mechanical properties only at temperatures significantly below 200°C.

TMA results gave thermal expansion coefficients (below 120°C) of 4.2 x 10^{-6} $(°C)^{-1}$ and 5.3 x 10^{-6} $(°C)^{-1}$ for the glass and jute composites respectively parallel to the fiber direction in the ply. Above 120°C, there is shrinkage that is especially severe in the jute composite as more tightly bound water is evolved.

3.2.3 Dynamic Mechanical Properties. The results of the dynamic mechanical testing performance in the three point bending mode are shown in Figures 5 to 7 for the neat resin, jute and glass composites respectively. Plotted are G', the storage modulus; G", the loss modulus and tan δ, the loss factor. G' represents the elastic component of the dynamic modulus while G" is a measure of the viscous element. The maximum in the loss factor curve can generally be related to the onset of large scale molecular motion that accompanies the glass transition in polymers.

The results indicated that both glass and jute reinforcement improve the heat resistance of the neat resin itself. The ratio of G' for the glass composite to G' for the jute composite at room temperature is about 2.7 which is comparable to the tensile moduli ratios obtained even though the dynamic test mode was three-point bending. The increase in T_g, the glass transition temperature (as indicated by the peak in G") is dramatic in the composites: to 130°C and 141°C for the jute and glass composites respectively.

The neat resin itself had a relatively low T_g of about 55°C. It is clear that the principal limiting factor in this case is the temperature capability of the resin. For most envisaged applications the difference in temperature stability between the jute and glass composites would not be significant.

3.2.4 Fracture Surface Morphology. A representative SEM photomicrograph of a fractured jute composite surface (Figure 8) illustrates that brittle failure in the jute occurs only across an individual cell. The apparently weak bond between cells has some crack stopping or diverting capability as evidenced by the differing sites of failure from cell to cell within a single fiber. The bond between the jute and the resin appears to be somewhat better than the inter-cellular bond. The increased fracture energy absorption that results may be borne out by some preliminary impact test results that indicate better ductility in the jute as compared to the glass-reinforced composites.

4. CONCLUSIONS

(i) Jute as a reinforcement for a low cost composite is not as effective as glass on a simple modulus or strength basis. However, if cost and fiber density are considered, jute appears to be a viable alternative to glass for many applications where cost per unit modulus or strength is paramount.

(ii) The thermal resistance of jute-reinforced composites is matrix-limited.

(iii) Fracture of a jute-reinforced composite involves an additional energy absorption mechanism whereby the weak inter-cellular bond within the fiber causes crack-stopping and deflection.

(iv) The moisture absorption and desorption in the jute fibers is significant and special precautions may be necessary to minimize the effects.

REFERENCES

(1) S. A. Wainwright et al, "Mechanical Design in Organisms", John Wiley and Sons, New York 1976, pp. 101.

(2) Winfield, Armand G and Winfield Barbara L., Advances in Chemistry Series Number 135, pp. 207-218.

BIOGRAPHIES

H. W. Chang and I. W. Sorensen are both Members of the Technical Staff at the Bendix Advanced Technology Center which recently evolved from the Bendix Research Laboratories.

After receiving his doctorate in physical chemistry, Dr. Chang joined the Bendix Research Laboratories where he worked on various aspects of both carbon/carbon and polymer matrix composites.

Dr. Sorensen obtained his degree in Materials Engineering and has also worked at Bendix in the areas of carbon technology and fiber reinforced polymers.

ACKNOWLEDGEMENT

The authors are grateful for the assistance provided in the procurement of fiber samples by Mr. Whitney Bird of R. L. Prichard and Co.

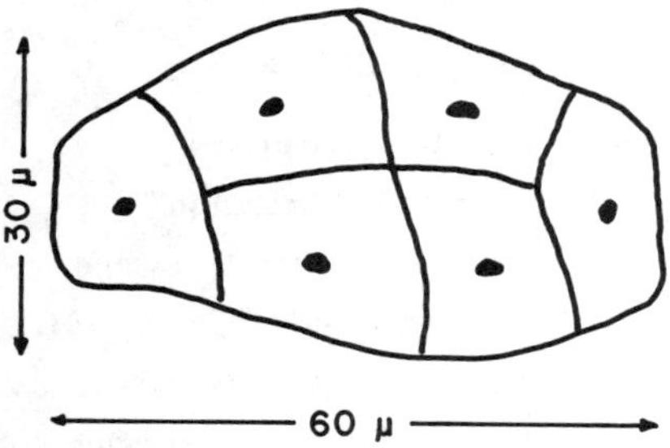

Figure 1. Schematic Cross-Section of Jute Fiber Showing Cellular Arrangement

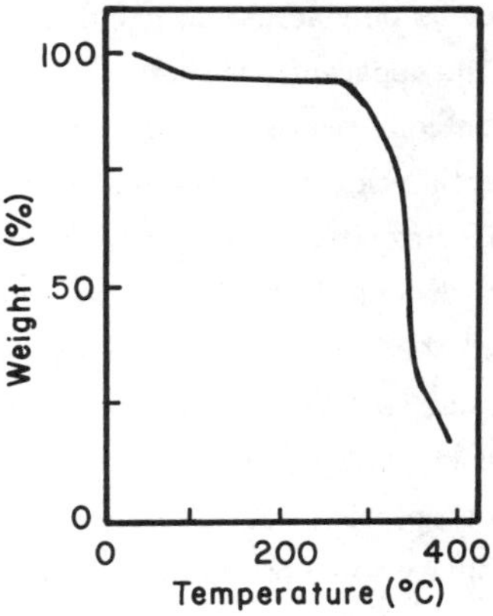

Figure 3. Jute Fiber Weight As A Function of Temperature (TGA)

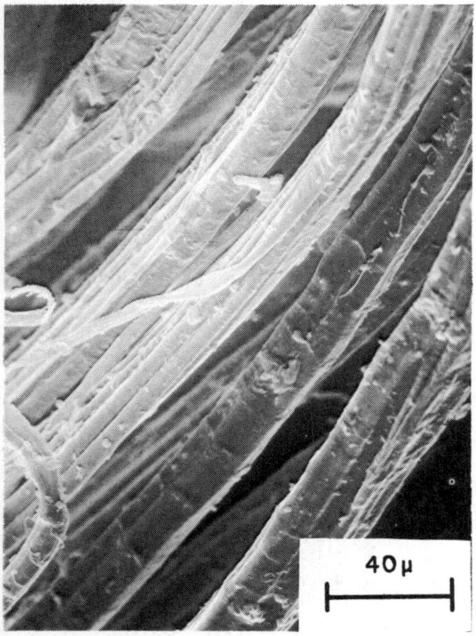

Figure 2. Representative Jute Fibers (SEM)

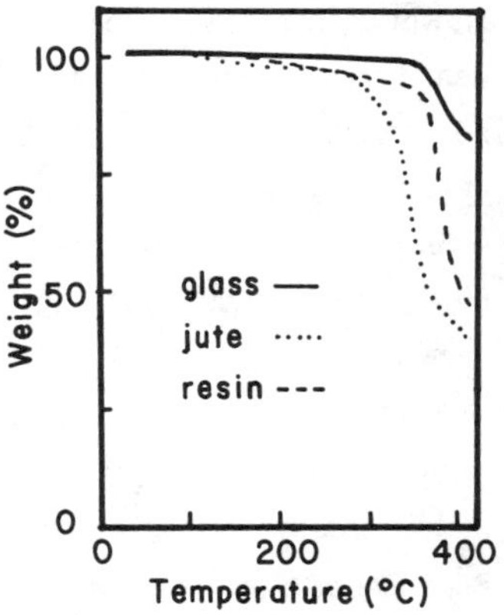

Figure 4. TGA Curves For Resin Alone and Reinforced with Glass and Jute

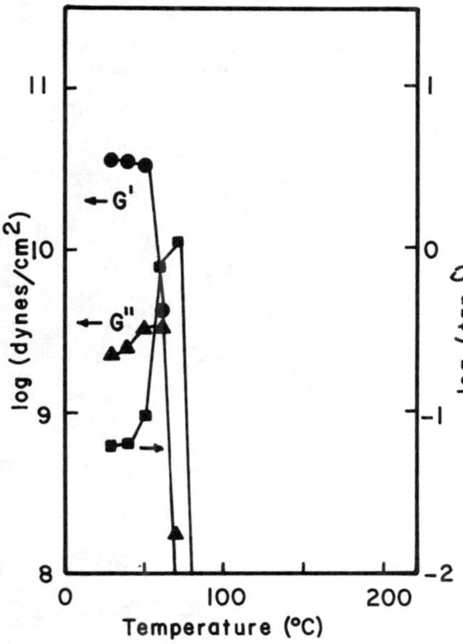

Figure 5. Dynamic Mechanical Properties of Base Resin As A Function of Temperature

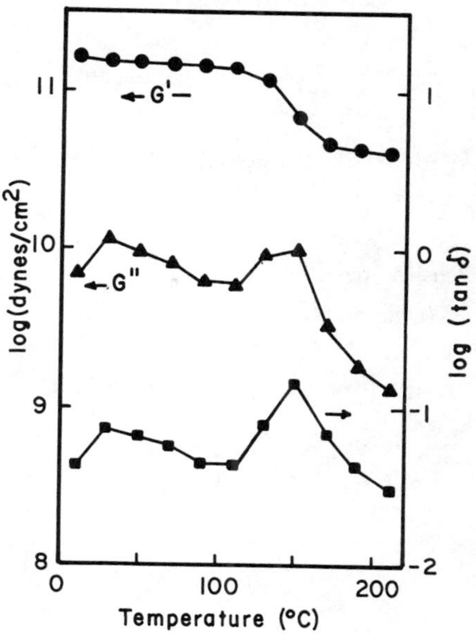

Figure 7. Dynamic Mechanical Properties of Glass-Reinforced Resin

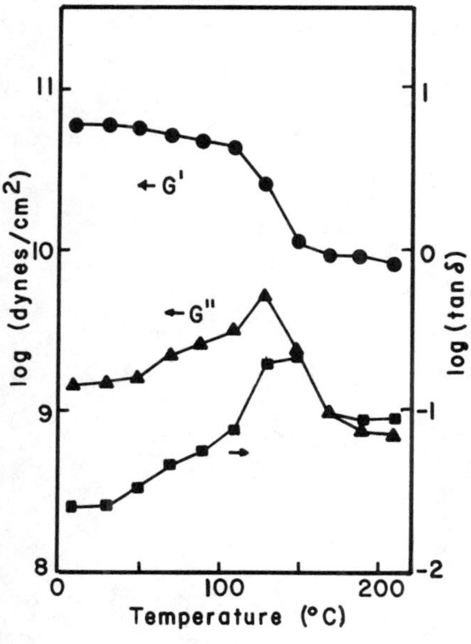

Figure 6. Dynamic Mechanical Properties of Jute-Reinforced Resin

Figure 8. Representative Jute Composite Fracture Surface (SEM)

Table 1. Comparative Mechanical Properties of Jute and Glass Fiber

	Jute (1)	E Glass
Tensile Modulus psi (dyne cm^{-2})	8.7×10^6 (6×10^{11})	10.5×10^6 (7.2×10^{11})
Tensile Strength psi (dyne cm^{-2})	1.25×10^5 (8.6×10^9)	5×10^5 (3.4×10^{10})
Extensibility %	3.8% (wet)	4.8

Table 2

	Cured Resin	Jute/Resin	Glass/Resin
Specific Gravity (SG)	1.17	1.14	1.72
Ultimate Tensile Strength (UTS) (dyne cm^{-2})	2.8×10^8	4.8×10^8	1.8×10^9
Tensile Modulus (E); (dyne cm^{-2})	3.5×10^{10}	7.6×10^{10}	2.4×10^{11}
Specific Strength (UTS/SG) (dyne cm^{-2})	2.4×10^8	4.2×10^8	1.0×10^9
Specific Modulus (E/SG) (dyne cm^{-2})	2.8×10^{10}	6.9×10^{10}	1.4×10^{11}
Specific Strength Normalized to 48V% loading (dyne cm^{-2})		5.8×10^8	1.0×10^9
Specific Modulus Normalized to 48% Loading (dyne cm^{-2})		9.5×10^{10}	1.4×10^{11}
Specific Strength per $ Reinforcement Used (dyne cm^{-2} $\$^{-1}$)		1.0×10^{12}	4.7×10^{11}
Specific Modulus Per $ Reinforcement Used (dyne cm^{-2} $\$^{-1}$)		1.7×10^{14}	6.6×10^{13}

13th National SAMPE Technical Conference
October 13-15, 1981

EXTENDING RESINS WITH CALCIUM CARBONATE
- 1981 UPDATE -

Paul E. Kummer
Technical Director
Cyprus Thompson-Weinman Co.

Abstract

This paper focuses on plastic compounding developments which have benefited from the use of calcium carbonate filler in the last several years. The influence of $CaCO_3$ filler variables on compound performance is highlighted, where applicable.

Calcium carbonate is the most widely used extender pigment or filler by the plastics compounding and fabricating industry.[1] Polyvinyl chloride, polyolefins, phenolics, polyesters, and epoxies are all resins which are compounded with calcium carbonate. Loading levels vary from just a few parts to over 50% filler.

This broad usage can be attributed to both economic and performance considerations. Calcium carbonate is readily available at favorable pound-volume costs. Its properties include ease of wetting and dispersion by many resins as well as whiteness, natural purity, softness, resistance to heat, and controllable particle size distribution.[2]

Current commercial grade qualities include finenesses not available a decade ago and organic treatments of the filler surface to achieve ease of wetting and dispersion, and consequently higher loadings. The most common of these surface treatments are stearic acid and calcium stearate, a technology which was pioneered in Europe.[2]

1. COMPOSITION AND PROPERTIES

Calcium Carbonate, $CaCO_3$, is one of our earth's abundant minerals. Well known forms include limestone, marble, calcite, chalk, and aragonite. Dolomitic limestone formula is $CaCO_3 \cdot MgCO_3$ Most deposits contain some impurities, such as silicates or pyrite, which may be largely removed by beneficiations. $CaCO_3$ for filler use is also manufactured by precipitation, utilizing recarbonating of lime water process. Properties applicable to plastics compounding include density of 2.7, hardness 3 Mohs, refractive indices 1.48+ and 1.66–. Solubility in water is only 0.0013 gm/100 gm water.

$CaCO_3$ can be compounded in plastics at all common processing temperatures, since CO_2 evolution from molecular breakdown does not normally occur below 800°C.

2. AVAILABLE GRADES

Classification and/or grinding methods are used to produce numerous degrees of fineness and, with few exceptions, resultant products range from one-tenth to 10 times the weight median particle size. Graph I shows typical grades in common use by plastics industry. Precipitated grades for plastics use have relatively quite narrow particle size distribution and weight median particle size ranging from 0.2 to 2.0 micrometers (microns).

3. APPLICATIONS

Glass-reinforced polyester compounds and flexible PVC materials use the highest volumes of $CaCO_3$. Rigid (unplasticized) PVC and polyolefin applications are growing.

3.1 Flexible PVC

20 to 65 phr of 3 micrometer $CaCO_3$ is common in imitation leather and other film applications. 1 micrometer grades, while more expensive, produce superior physical properties, especially important in thin guage films. Graph II illustrates this for tear strength.

3.2 Rigid PVC

Extruded pipe and construction profiles and injection molded fitting are examples of this application. The important property of impact resistance is affected by $CaCO_3$ fineness. Reference Tables I and II showing 10 and 30 parts loading respectively, the Impact designation represents percentage of specimens surviving Gardner Constant Energy Drop of 1.87 inch pounds per mil thickness.

3.3 Sheet Molding Compound (SMC)

$CaCO_3$ loading levels up to 200 phr are used in the paste which is combined with chopped glass on a 2 to 3 parts paste to 1 part glass ratio. The glass contributes the strength and other important physical properties. The filler contributes economy and reduced shrinkage after compression molding. Median weight particle size of 3 to 6 micrometers makes up greatest volume.

3.4 Bulk Molding Compound (BMC)

System chemistry is not unlike SMC, but handling techniques permit higher loading levels, up to 400 phr filler. Preferred $CaCO_3$ grades similar to SMC.

3.5 Polyethylene

Low- and high density polyethylene accept $CaCO_3$ loadings readily. Our work in HDPE demonstrates definite advantage for the stearate coated grades. Graph III, Gardner Drop Weight Impact Test, shows superior performance for coated grades, especially above 30% filler.

3.6 Polypropylene

$CaCO_3$ is easily compounded into this versatile resin. Very significant improvement in impact resistance is attributed to the presence of $CaCO_3$, especially the stearate coated grades. This advantage carries over to freezing temperatures. See Table III.

4. ENERGY CONSIDERATIONS

Using polypropylene resin system, an investigation of the influence of $CaCO_3$ filler variables on plastic compound processibility and energy consumption was made. As might be expected, the finer grade with its concomitant higher surface area, requires more energy for equal performance. However, the (lubricant) coated grades more than offset this and deliver performance which can represent meaningful greater production per unit energy cost or cost reduction per unit output. See Table IV.

5. CURRENT ACTIVITY

High energy and resin feedstock costs continue to encourage investigations which incorporate fillers in plastic compounding applications. Filler technology includes examination of coupling agents, in situ, or as interface modifications produced by the filler manufacturer. Calcium carbonate plays an active part in these endeavors.

6. REFERENCES

(1) Chas. H. Kline & Co., Fairfield, N.J. "Reinforcements and Fillers for Plastics", 1980 (Marketing Research).

(2) H.S. Katz and J.V. Milewski,eds. "Handbook of Fillers and Reinforcements for Plastics" (New York: Van Nostrand Reinhold, 1978).

BIOGRAPHY

Paul E. Kummer received a B.S. in Chemistry from Union College (N.Y.) in 1943 and an M.S. from Stevens Institute of Technology in 1950. He joined Thompson-Weinman in 1946, has held assignments in technical service and marketing, and is currently Technical Director. He is a member of the Technical Association of the Pulp & Paper Industry, Fillers and Pigments Committee; Society of the Plastics Industry, Fillers and Additives Committee; and ASTM, Pigment Specification Committee D01-3100 ; Pulverized Limestone Association, Technical Committee.

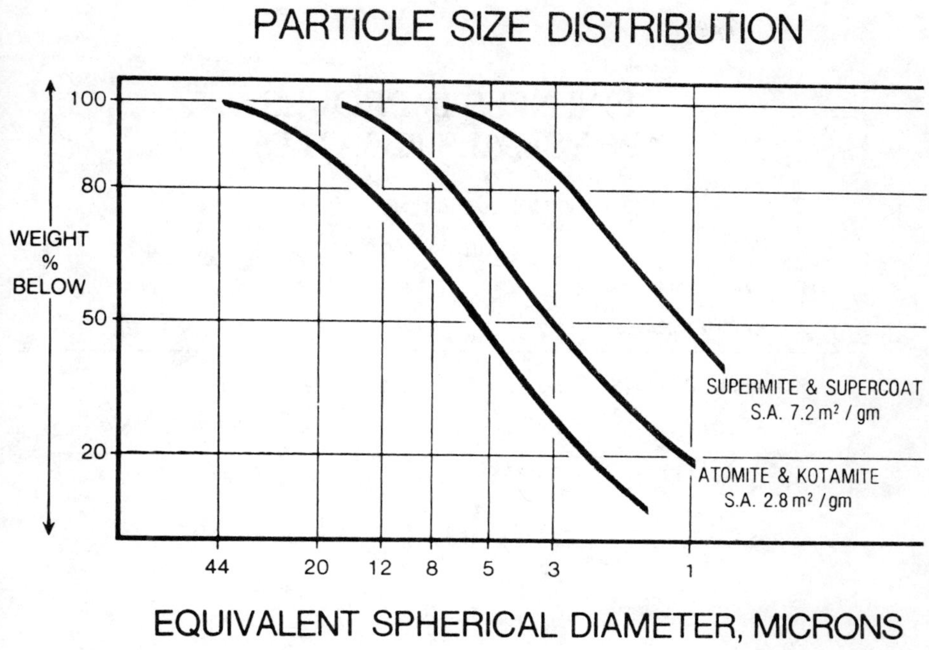

GRAPH I

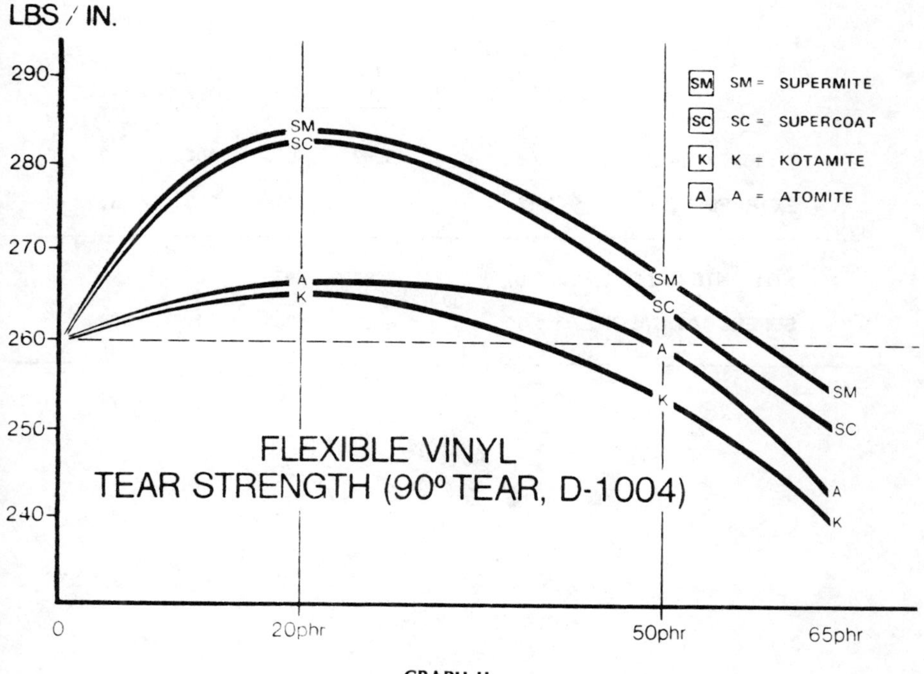

GRAPH II

FINENESS IMPROVES PHYSICALS (COATED)

	MICRON P.S. 50%	PHR	TENS STR YIELD	TENS STR BREAK	% ELONG BREAK	IMPACT
P-1 FINE CHALK	2.5	10.0	7240	7456	171	44
EXTRAFINE CHALK	1.2		7254	7136	142	70
KOTAMITE CALCITE	3.0	10.0	7120	7298	159	62
SUPERCOAT CALCITE	1.0		7194	7288	156	86
n			7	3	3	25

TABLE I

FINENESS IMPROVES PHYSICALS (COATED)

	MICRON P.S. 50%±	PHR	TENS STR YIELD	TENS STR BREAK	% ELONG BREAK	IMPACT
P-1 FINE CHALK	2.5	30	5989	5866	109	0
EXTRAFINE CHALK	1.2		6047	6006	120	52
KOTAMITE CALCITE	3.0	30	6083	5959	105	0
SUPERCOAT CALCITE	1.0		6063	5858	110	68
n			7	3	3	25

TABLE II

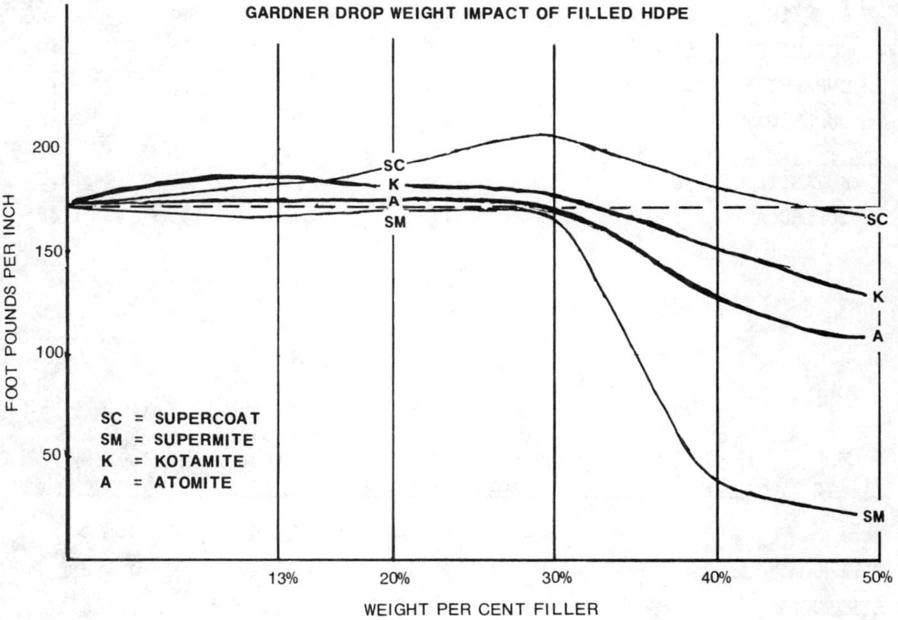
GARDNER DROP WEIGHT IMPACT OF FILLED HDPE
FOOT POUNDS PER INCH
200
150
100
50
SC
K
A
SM
SC = SUPERCOAT
SM = SUPERMITE
K = KOTAMITE
A = ATOMITE
13%
20%
30%
40%
50%
WEIGHT PER CENT FILLER

GRAPH III

Polypropylene Resin	Tensile(Y) PSI	% Elong.	Izod (ft.lbs/in) Notched	Unnotched	SPI Gardner (ft.lbs/in) Rm. Temp.	0°C
Unfilled	4850	28	.40	16.0	82	2.0
40% filled w.						
ATOMITE	4220	27.5	.53	9.6	106	11
KOTAMITE	4230	35	.70	12.9	175	22.8
SUPERMITE	4330	26	.70	10.3	71	5.8
SUPERCOAT	4030	50	.78	12.1	193	23
50% filled w.						
KOTAMITE	3810	33	.58	8	108	13.2
SUPERCOAT	3630	38	.74	8.4	159	11.8

TABLE III

Filler	u Mean P.S.	% Coating Level	% Weight Loading	% Volume Loading	*Pressure Required for C & E Output **kgf/cm^2	**Power at C & E Joules/ Second
None	-	-	-	-	15.8	0.169
SUPERCOAT	1	2	40	18	17.1	0.182
SUPERCOAT	1	2	50	25	17.3	0.184
KOTAMITE	3	1	40	18	17.8	0.190
KOTAMITE	3	1	50	25	19.4	0.206
ATOMITE	3	0	40	18	21.8	0.232
SUPERMITE	1	0	40	18	22.5	0.240
ATOMITE	3	0	50	25	25.5	0.271

*C & E Output = Constant and Equal output of 6.5 cc/min or 120 sec^{-1}.

**The power consumed by the thermoplastic compound is derived from the potential energy of the piston load.

TABLE IV

13th National SAMPE Technical Conference
October 13-15, 1981

INTEGRALLY WOVEN COMPLEX SHAPES FOR MULTIDIMENSIONALLY REINFORCED COMPOSITES

Robert Adolph Florentine
The CUMAGNA Corporation
26 S. Wakefield Road, Norristown, Penna.

Abstract

MAGNAWEAVE process produces, on a continuous loom, multidimensional preforms, suitable for reinforcing composites. The weave eliminates delamination as a failure mode; it permits the selective orientation of reinforcing fibers to maximize the efficiency of the reinforcing fibers to maximize the efficiency of the reinforcing fibers. Applications in the aerospace naval systems, and automotive industries require the use of stiffened sections, whose integrity suffers from unsatisfactory bonding or attachment resulting in failure at the critical load points. The MAGNAWEAVE process solves that problem by integrally weaving the stiffeners into the original shape. Examples include stiffened panels for aircraft fuselages, stiffeners in missile bodies, and turbine blades.

Details of the weaving process are presented, with emphasis on weaving complex and irregular shapes. The process reflects the varying requirements along a part, with orientations programmed to meet that variation.

The application of MAGNAWEAVE to parts requiring varying thickness is reviewed, and an analysis for the case of decreasing thickness is presented, accomplished integral with the weave, internal to the part, isolation damage initiation points within the weave. The weave heals the point of thickness reduction and produces a continuous and changing thickness. The process is reversed for thicknesses to grow.

1. INTRODUCTION

The successful application of reinforced composites to complex shapes is restricted by the following limitations inherent in existing processes

(a) Reinforcing fibers cannot be oriented to produce the shapes, and retain that orientation during impregnation and cure;

(b) Reinforcing systems bear weak delamination planes that limit the loads;

(c) Integrally woven preforms cannot be produced;

(d) Attaching components of the complex shape reduce or eliminate the advantage of using composites;

(e) Delamination, shear failure, limited third direction strength and modulus, all are compounded as failure mechanisms in complex shapes. The MAGNAWEAVE process has the inherent capacity to meet all these requirements, and produce complex shapes that are integrally reinforced, continuously fabricated, and have the capacity for low manufacturing costs.

Figure I describes the basic MAGNAWEAVE loom, in the Cartesian system. It is, in effect, a movable creel, with supplies of yarn elements being moved in regular fashion to produce a multidimensional skeleton. The fiber orientation in the final part is controlled by the arrangement in

2. THE MAGNAWEAVE PROCESS

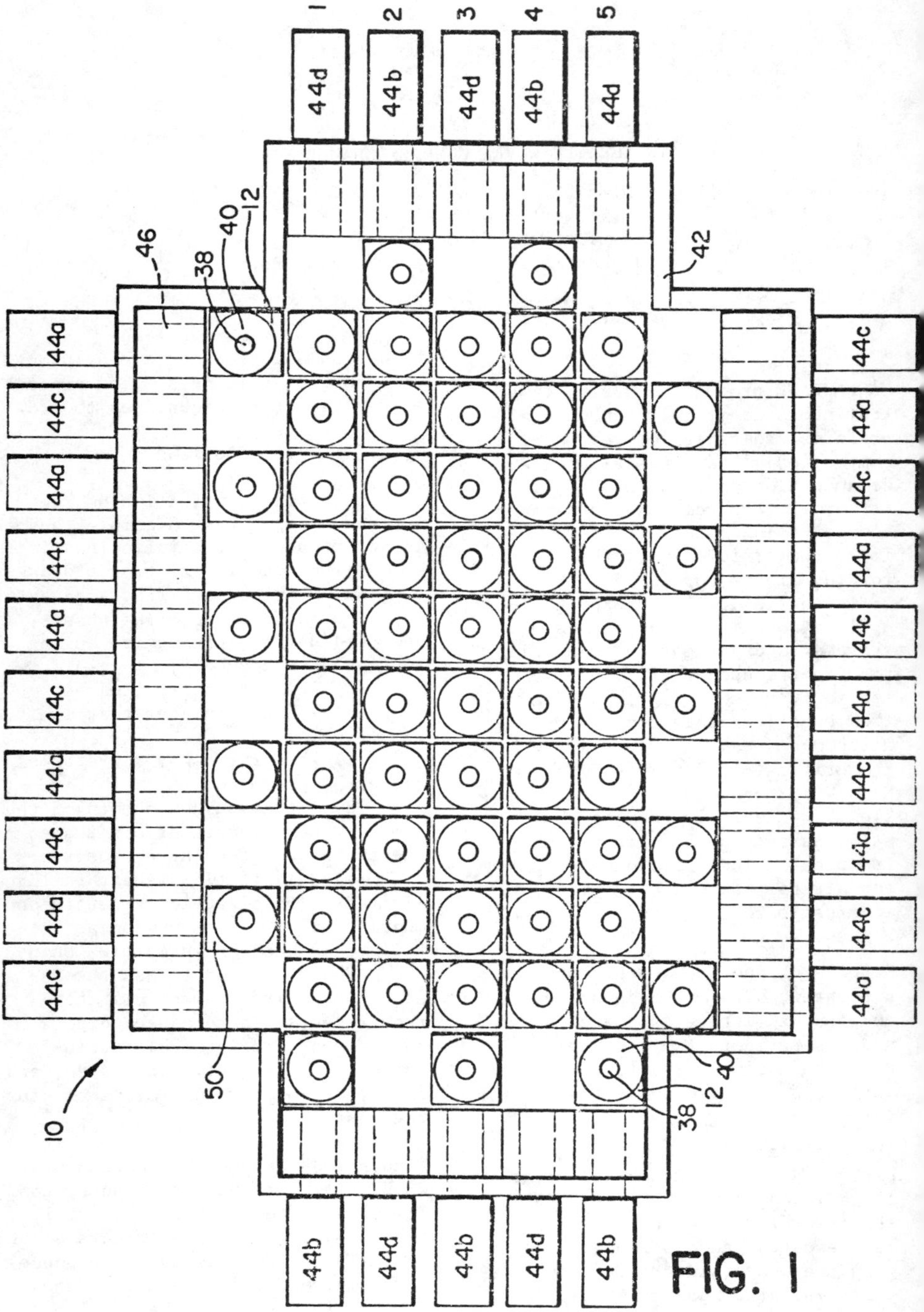

FIG. 1

the loom that determines the relative motion of the yarns in two directions and the degree of combing in the resultant weave. The ultimate fiber orientation is the resultant of these three orthogona 1 movements; the yarn direction, because of the three components, is never orthogonal, although it can be as close to orthogonal as 5°.

Manipulation of the loom mechanics produces a skeleton whose fiber orientation can be controlled for maximum efficiency in the use of the reinforcement. Depending on the structural (and thermal) requirements, a weave orientation is selected that meets all the property requirements in all directions of interest. To achieve this goal, some of the directions will exceed the requirement for properties.

The construction so produced is, in effect, a three dimensional braid. Braids have no warp yarns; all yarns pass completely from one end of the construction to the other. The preform so produced is a closed system, in that the edges of the part formed in the weaving operation do not have free ends.

Because of the motion, the resultant part contains no plane where delamination can occur. Delamination failure is eliminated. As a corrollary, the interlaminar shear properties are increased dramatically, since such shear failure must be from spallation, or yarn breakage. Propagation of such breakage must also be by a continued breaking of yarns, rather than slicing through the relatively weak matrix. Any incipient damage is trapped in the yarns surrounding it.

2. STIFFENED PANELS

Perhaps the most elementary application of the MAGNAWEAVE process for complex shapes is the fabrication of panels stiffened by vertical "hat sections". Currently, processes are attempting to produce such shapes by forming the panels by normal laminate construction, the hat sections are formed separately over mandrels, and partly cured. The stiffeners are placed in position on the panels, and the panel-hats are "cocured", in the hope that the matrix will join the system. In practice, the shear loads for which the systems are designed fail the parts, at the bond line, as predicted, at the low shear level of the resin.

MAGNAWEAVE produces a stiffened panel in integrally weaving the panel with the stiffeners, to produce a single shape for impregnation and cure. In this instance, the MAGNAWEAVE loom, which is really a set of modules that can be arranged to produce the cross section of interest, is rearranged to include in the loom motion the paths that will produce the stiffeners, where they are required. In weaving the parts, appropriately shaped mandrels are placed to form the surface against which the weave is combed. The resultant skeleton is a panel, with stiffeners along its length, integrally a part of the skeleton. The weave density across the "interface" (not an interface with MAGNAWEAVE) is identical with that in any other part of the piece

3. MAGNAWEAVE HYBRIDS

Hybrid fiber systems can profitably be employed with MAGNAWEAVE systems. The process has the inherent capacity to separate the weave patterns of two separate yarns, appropriately loaded into the loom. The result is a synergistic hybrid MAGNAWEAVE, where the weave itself will contribute to the property of interest, and the presence of the binary fiber system adds its properties to the system as well. As a single example, longitudinal stiffness can be enhanced by loom modification alone. This results in a percentage of the yarns remaining unidirectional, while the rest of the yarns weave around them in a normal MAGNAWEAVE. The result is a part with enhanced longitudinal stiffness. If those unidirectionally controlled fiber positions are filled with a stiff yarn, the result is a stiffened part whose longitudinal stiffness depends on both the hybrid yarn and the weave it is used to provide the hybrid weave.

4. CIRCULAR WEAVING-MAGNA SWIRL

The MAGNAWEAVE concept is adaptable to parts of circular symmetry; the MAGNASWIRL looms address that requirement. A polar coordinate system is employed to direct the yarn motion; combing determines the third component for yarn direction, as before. The yarn travel is circumferential, in rings, and radial along spokes; the rings and spokes correspond to the rows and columns of the Cartesian system.

5. FLANGED CYLINDERS

In circular systems, flanges on cylinders or cones are similar to stiffeners in panels. The MAGNASWIRL process produces flanged parts with the flanges integral parts of the cylinders, or cones. In this example, the circular weaving is combed against a conical mandrel of the desired geometry (inner contour). The loom is loaded with rings exceeding those to weave the wall of the cone. Until the flange formation is required, these extra rings do not participate in the weaving. When the flange is to be initiated, these rings are activated, and the weaving occurs over the enlarged number of rings now in action. The result is a thickening of the preform, in the exact shape of the flange required. When the desired flange length is woven, the extra rings are deactivated and the loom returns to weaving the original shape. It is evident that several flanges may be woven on the same circular part.

6. DIMENSIONAL CHANGES

Time and space prohibit a detailed discussion of the loom mechanics whereby the thickness of a piece can be gradually and continuously changed, either to decrease or to increase its dimension. Suffice it to say, that the weave operation acts with the number of rows or columns acting at any time; by decreasing the number of rows, the thickness of the part is decreased. So, too, with a change in width. The principal works equally well with Cartesian or Cylindrical coordinate systems.

The advantage of the process is that the dimensional change does not produce a discontinuity at the surface, as with laminate processes, or post lamination machining. As a consequence, although the thickness changes, the outer contour of the part remains as a closed weave, with no delamination edges. The yarn discontinuity is buried in the part, and cannot contribute to crack initiation because of the discontinuity. The dimensional change is a gradual one, as little as one yarn diameter at a time.

7. CONCLUSION

MAGNAWEAVE and MAGNASWIRL complement each other; together they are introducing a revolutionary approach to the potential for composites in structural applications. The elimination of delamination, the ability to maximize yarn efficiency, and the ability to weave complex shapes simply, cheaply, and integrally, is providing the designer with a capability of exceptional range.

The examples cited in this paper serve to illustrate the potential of the processes to meet extreme requirements simply and effectively. The ultimate potential for their application depends on the imagination of the designers, in any field, who can recognize their needs, and see the way in which a MAGNAWEAVE or a MAGNASWIRL part gives them the satisfaction of that need.

13th National SAMPE Technical Conference
October 13-15, 1981

REPLACEMENT OF SULFURIC-CHROMIC ACID (FPL) ETCH BY 3-DIMENSIONAL ABRASIVE SURFACE CONDITIONING FOR PRETREATMENT OF AEROSPACE ALLOYS BEFORE H_3PO_4 ANODIZATION

Alphonsus V. Pocius
Adhesives, Coatings & Sealers Division, 3M
St. Paul, MN. 55144
and

Joseph J. Claus
Building Services and Cleaning Products Division, 3M
St. Paul, MN. 55144

Abstract

An investigation was made into the possibility of replacing FPL Etch before H_3PO_4 anodization by Scotch-Brite(R) 3-dimensional abrasive surface conditioning as a pretreatment for aerospace alloys before adhesive bonding. Initial mechanical properties of adhesive bonds prepared with AF-163-2K/EC-3960 showed the two surface treatments to be essentially identical. Environmental exposure of these adhesive bonds as well as wedge tests were run and these indicated that durable adhesive bonds would result from the new surface treatment. Surface analysis was preformed on Scotch-Brite(R) treated surfaces in comparison to FPL etched surfaces and the former was found to give a more elementally clean surface as well as a thinner oxide. This study seems to indicate that Scotch-Brite(R) 3-dimensional abrasive surface conditioning is essentially equivalent to the FPL Etch before H_3PO_4 anodization in providing a surface for making durable adhesive bonds.

1. INTRODUCTION

The most widely used surface treatment of aluminum for adhesive bonding is the FPL (Forest Products Laboratory) etch. It is used as a final treatment as well as a pretreatment before anodization. However, the FPL etch contains significant quantities of $Na_2Cr_2O_7$ and there is a great deal of concern regarding the use of Cr(VI) containing solutions both from the viewpoint of health of the materials handler and also from the viewpoint of having to dispose of large amounts of hazardous waste. Further, todays metal treatment lines are quite energy intensive because they require maintenance of large baths at high temperatures. This report describes an initial evaluation of an alternative pretreatment for aluminum prior to adhesive bonding which addresses both the problem of Cr(VI) and high energy usage. The pretreatment procedure we propose for consideration is the use of Scotch-Brite(R) 3-dimensional abrasive surface conditioning (hereinafter call "brushing") followed by anodization.

2. EXPERIMENTAL

2.1 Materials. The aluminum used in this work was commercially obtained 2024T-3 and 7075T-6 bare alloy. Two Scotch-Brite(R) brushes were used. These were the Flat Sheet Deburring Brush 1A Medium (hereinafter called Brush 1) and the Heavy Duty Printed Circuit Board Brush (hereinafter called Brush 2). The adhesive used in this work was AF-163-2K (0.06 wgt.). This adhesive has been previously described in the literature (1) with regard to its resistance to prebond high humidity exposure. The primer used was EC-3960, a chromated 250°F curing epoxy based primer. The Scotch-Brite(R) brushes, adhesive and primer were all taken from typical lots manufactured by 3M.

2.2 Procedure. Although it is common practice to alkaline degrease or solvent degrease aluminum parts before further treatment, in our study no alkaline or solvent degrease was used. The aluminum panels were processed as received on a SOMACA brand roller fed finishing machine. The machine was equipped with one of the Scotch-Brite(R) brushes mentioned above in the working head position. The panels were run through at a feed speed of either 10 or 20 feet/min. and at 2 different operating pressures of 0.05 or 0.1 horsepower inch. This "pressure" unit is derived from the amperage drawn by the finishing machine as the sample is being processed. The sample and brush were flooded with tap water during processing at a rate of 8 gallons/min. The samples were dried as they emerged from the machine by a bank of air knives. Each side of the sample was passed through the machine once. A number of samples were retained for surface analysis.

One to four hours after brushing the samples were H_3PO_4 anodized. In the interveneing period the samples were wrapped in Kraft paper. The anodization procedure is that described by McMillan (2). Within 24 hours after anodization, the panels were spray primed with EC-3960. The primer was allowed to air dry for 1/2 hour and was then cured for 1 hour at 250°F. Within a week after the primer was cured, the samples were bonded together using AF-163-2K. The cure conditions were 4-5°F/min. rise rate to 235°F, 90 min. at 235°F, 35 psi applied pressure. A vacuum was drawn on the autoclave tray until the temperature reached 150°F.

Samples were also prepared using an alkaline degrease (Oakite 164) and the FPL Etch followed by H_3PO_4 anodization.

2.3 Environmental Exposure. The wedge test samples were made up according to Marceau (5b). They were exposed either to 1 day at 140°F/100% RH or 7 days salt spray. The 1/2 square inch overlap shear specimens and the 1/2 inch wide floating roller peel specimens were subjected to either 30 days salt spray or 30 days 140°F/100% RH. These samples were exposed to these environments in an unstressed condition. The salt fog cabinet was manufactured by Harshaw and it was maintained according to ASTM B-117 standards (5% salt solution).

2.4 Surface Analysis. Representative samples of brushed and FPL etched aluminum were surface analyzed by a Physical Electronics Model 545 Scanning Auger Microprobe equipped with a 3M Secondary Ion Mass Spectrometer (SIMS).

3. RESULTS AND DISCUSSION

Scotch-Brite(R) 3-dimensional abrasives are a unique family of low density abrasive products made from three basic materials: fiber, resin and abrasive mineral. The basic form of these products is an open, non-woven, synthetic fiber web that is uniformly impregnated with abrasive particles. The abrasive particles are held in place by a high strength resin bonding system. These components are oriented in a manner that allows the mineral to follow along a surface in a spring-like manner (see Figure 1).

One major advantage of this type of construction is called "controlled cut". Controlled cut is the ability of this abrasive media to follow over contours of the surface removing only surface contaminants and a small amount of substrate material. In fact these 3-dimensional abrasive products are known to work to a depth of about 5000Å. This number is on the order of the surface roughness generated by these products.

The material in web form can be

converted into wheels and brushes which are used for various surface conditioning applications. Scotch-Brite(R) 3-dimensional abrasives can deburr, clean and finish surfaces without changing the geometry of the part. An example of this is shown in Figure 2. Scotch-Brite(R) surface conditioning has been extensively used on steel and the aerospace industry has in recent years accepted its use in a system for deburring flat sheet aluminum. The work in this report indicates that Scotch-Brite(R) surface conditioning also has further important applications in the aerospace industry.

3.1 Surface Analysis. Certain of the samples prepared as described above (without anodization) were subjected to Auger Electron Spectroscopy (AES) and SIMS surface analysis. The data comparing brushed aluminum and FPL etched aluminum reveal interesting contrasts and similarities. Figure 3 shows typical AES data obtained in this work. Figure 3 compares the AES from FPL Etched 2024T-3 bare aluminum to the same alloy surface conditioned by Brush 1. It is easy to note that all of the same elements are present on both surfaces with the exception of copper.

Detailed analysis of the AES data allows us to list the elemental surface compositions of the various alloy surfaces as ratios with respect to the aluminum concentration. This is shown in Table I. This table also includes a determination of the thickness of the oxide layer. This determination was made by sputtering the surface with an ion beam while monitoring the variation of the AES oxygen signal as a function of depth. The contrast between the oxide thicknesses indicates the most striking difference between the types of aluminum and between the types of treatment. In the as received state, the 7075T-6 aluminum has an oxide layer about 50% deeper than the 2024T-3 alloy. However, after treating the two types of aluminum with either the FPL etch or by brushing, the oxide forms a significantly thinner layer, that is less than 200Å in depth. Of these treatments, brushing limits the oxide layer growth to a thickness consistently below 100Å.

Table III shows the elemental composition of the first 20Å of surface as determined by SIMS. These, once again, are shown as ratios with respect to aluminum. SIMS information is inherently more sensitive than Auger. Combining the information in Tables I and II. we see that the FPL etched and brushed surface elemental compositions are similar with the following exceptions. The FPL etched surface seems to have a larger amount of Cu, Na and F while the brushed surface has a slightly larger amount of C. Especially notable is the fact that the AES shows a significant amount of Cu on the FPL etched surface, but none was detectable on the brushed surface. The signicicance of the presence of these elements on the surface has not been investigated but may be of concern as additional work is carried out.

The conclusion we must draw is that processing the aluminum alloys used in aircraft structure by Scotch-Brite(R) 3-dimensional abrasive surface conditioning produces a surface whose chemical composition is similar to that produced by the FPL etch. However, it is very likely that the surface morphology produced by the FPL etch (3) does not occur in the 3-dimensional abrasive conditioned surfaces. This is perhaps discernable from the oxide thickness data. It is apparent from the data in the following section, though, that the FPL etch surface is not a prerequisite for obtaining an anodized aluminum surface that is adequate for adhesive bonding.

3.2 Initial Mechanical Properties. Tables III and IV show the initial floating roller peel and overlap shear mechanical properties of the bonds created in this work. The floating roller peels were done on 1/2" wide specimens (0.025" to

0.063" for 2024T-3, 0.020" to 0.063" for 7075T-6) while the overlap shears were 1/2 square inch in overlap (0.063" to 0.063" for both alloys). Also, a sample of an FPL Etched/H_3PO_4 anodized alloy is included for comparison. The initial results are all excellent. The -67°F floating roller peel results seemed to be a little low but this was found to be due to excessive primer thickness. The data at the bottom of Table III is a re-run of the experiment with thinner primer and as can be seen the results are excellent. It has been shown in the literature (4) that surface pretreatment can have a significant effect on peel strength. The data in Table III shows that the initial peel strength of the brushed/anodized surfaces is equivalant to the FPL Etched/anodized surfaces. Overlap shear properties are not very greatly affected by surface preparation. The results in Table IV show consistently excellent results. In addition, there seems to be no apparent effect of feed speed or pressure. Failure was always either in the primer or the adhesive in both types of specimens.

3.3 Environmental Exposure and Durability. There has been substantial work in the literature on the relationship between surface preparation and durability of adhesively bonded structure (5). In order to investigate this with the surfaces we have prepared, we exposed our overlap shear and floating roller peel specimens, unstressed to two aggressive environments. These were high temperature along with high humidity and also salt spray. The mechanical properties of these specimens were tested at 75°F and the results are shown in Tables V and VI. Once again the overlap shear results are uniformly good. The floating roller peel results on 2024T-3 seem to show some degradation of properties. However, these tests were rerun (data at the bottom of Table V) and excellent results were obtained. Once again it seems as though the primer thickness had an effect since the data at the bottom of Table V is taken from samples at a thinner primer coating than that at the top of Table V. Failure in all cases was either in the primer or the adhesive.

In order to attempt to predict durability of adhesively bonded structure having brushed/anodized metal surfaces, we also made wedge test specimens. These were made up according to Marceau (5b). McMillan (2) has discussed the relationship of wedge test results to adhesive bond durability. The basic relationship is that small crack extension ($<$ 0.75") after 4 hrs. environmental exposure is indicative of a durable bond. We have performed these measurements after 24 hours of exposure to 140°F/100% RH and 7 days of salt spray. The results are shown in Table VII. The crack extensions are uniformly small and the crack appears to propogate through the adhesive. Based upon literature reference (2) we would predict that all of the surface preparations would result in durable bonded structure. In addition, there once again seems to be no apparent effect of the surface conditioning process parameters.

4. CONCLUSIONS

4.1 Scotch-Brite(R) brushing yields a surface which is apparently more elementally clean than does the FPL etch.

4.2 Brushing leads to an oxide layer thickness less than 1/2 the thickness of the FPL etch oxide layer.

4.3 Brushed aluminum alloys anodize well.

4.4 Brushed/anodized aluminum alloys can be used to make adhesively bonded structure that is apparently equivalent in performance to an FPL Etched/anodized alloy.

5. REFERENCES

1. C. J. Almer and A. V. Pocius, Nat'l. SAMPE Tech. Conf. Proc., 12, 924 (1980).

2. J. C. McMillan, in "Bonded Joints and Preparation for Bonding", AGARD Lecture Series No. 102, Harford House, London, 1979, pp 7-1 to 7-30.

3. J. D. Venables, D. K. McNamara, J. M. Chen, T. S. Sun, and R. L. Hopping, Appl. Surf. Sci., 3, 88 (1979).

4. P.F.A. Bijlmer, Metal Finishing, December, 1971, pp 34-39.

5. a. A. W. Bethune, SAMPE J., July, 1975, pg. 4.
b. J. A. Marceau, Y. Moji, J. C. McMillan, Adhesives Age, Oct., 1977, p. 28.
c. J. D. Minford, Adhesives Age, March, 1978, p. 17.

6. BIOGRAPHIES

Alphonsus V. Pocius received his B.A. in chemistry from Knox College, Galesburg, IL in 1970 and a Ph.D. in physical chemistry from the University of Illinois in Champaign-Urbana in 1974. He joined 3M in 1974 and has worked and published in the areas of polymer and adhesive characterization, surface chemistry and corrosion research.

Joseph J. Claus received a B.S. from St. Louis University in 1964. For the past nine years he has worked as a research and development scientist at the Building Services and Cleaning Products Division of 3M. He has co-authored papers dealing with deburring of aerospace metal parts and the surface characterization of auto body steel and its relationship to corrosion.

Table I

Auger Electron Spectroscopy Data on Variously Treated Aluminum

Surface Preparation	Oxide Depth (A)	C/Al	O/Al	Mg/Al	Cu/Al
2024T-3 Bare Aluminum Alloy					
None	380	2.6	6.9	5.5	-
FPL etch	165	0.2	3.0	0.03	0.2
Brush 1	80	0.6	2.6	0.02	-
Brush 2	65	0.3	2.5	0.04	-
7075T-6 Bare Aluminum Alloy					
None	570	4.2	7.1	5.7	-
FPL etch	110	0.2	2.8	0.02	0.07
Brush 1	65	0.4	3.2	0.07	0.05
Brush 2	55	0.5	3.2	0.08	0.05

Table II

Secondary Ion Mass Spectroscopy Data on Variously Treated Aluminum

(Data was collected in the first 20A of surface)

2024T-3 Bare Aluminum Alloy

Surface Preparation	Elemental Ratios					
	C/Al	O/Al	Mg/Al	Cu/Al	Na/Al	F/Al
None	1.1	14.1	1.4	0.02	0.57	0.57
FPL Etch	0.04	0.67	0.04	0.002	0.61	0.24
Brush 1	0.02	0.22	0.07	0.002	0.17	0.03
Brush 2	0.03	0.37	0.06	0.001	0.15	0.04

7075T-6 Bare Aluminum Alloy

Surface Preparation	C/Al	O/Al	Mg/Al	Cu/Al	Na/Al	F/Al
None	0.16	5.6	3.4	0.006	0.47	0.88
FPL Etch	0.003	0.4	0.07	0.003	0.59	0.17
Brush 1	0.006	0.35	0.19	0.002	0.16	0.07
Brush 2	0.004	0.29	0.15	0.002	0.17	0.05

Table III

Initial Floating Roller Peel Results (pounds/in. width)

Alloy	Brush	Feed Speed (ft./min.)	Applied Pressure (HP/in.)	Test Temperature (°F) -67	75
2024T-3	1	10	0.05	54	70
		10	0.1	43	66
		20	0.05	52	67
		20	0.1	40	73
	2	10	0.05	43	65
		10	0.1	49	70
		20	0.05	47	64
		20	0.1	54	66
------------FPL Etch----				50	77
7075T-6	1	10	0.05	51	68
		10	0.1	48	68
		20	0.05	51	67
		20	0.1	53	70
	2	10	0.05	49	71
		10	0.1	46	67
		20	0.05	47	66
		20	0.1	53	69
------------FPL Etch----				51	71
2024T-3	1	20	0.1	67	79
	2	20	0.1	61	74
------------FPL Etch----				58	75

Table IV

Initial Overlap Shear Results (pounds/sq. in.)

Alloy	Brush	Feed Speed (ft./min.)	Applied Pressure (HP/in.)	Test Temperature (°F) -67	75	160
2024T-3	1	10	0.05	7100	6650	3970
		10	0.1	6930	6570	4165
		20	0.05	7010	6675	4180
		20	0.1	7110	6280	3800
	2	10	0.05	6860	6560	4140
		10	0.1	7100	6700	4200
		20	0.05	7060	6455	4130
		20	0.1	7100	6690	4100
------------FPL Etch----				7050	6540	4040
7075T-6	1	10	0.05	8800	7400	4900
		10	0.1	8700	6600	4500
		20	0.05	8250	6900	4500
		20	0.1	9000	6700	4700
	2	10	0.05	8600	6700	4420
		10	0.1	8950	6650	4480
		20	0.05	7000	5800	4100
		20	0.1	8450	6200	4300
------------FPL Etch----				9120	6850	4740

Table V

Floating Roller Peel After Environmental Exposure (pounds/in. width) Measured at 75°F

Alloy	Brush	Feed Speed (ft./min.)	Applied Pressure (HP/in.)	After 30 Days at 140°F/100% RH	After 30 Days Salt Spray
2024T-3	1	10	0.05	61	61
		10	0.1	59	50
		20	0.05	64	59
		20	0.1	47	49
	2	10	0.05	55	55
		10	0.1	60	60
		20	0.05	56	54
		20	0.1	63	59
		FPL Etch		65	72
7075T-6	1	10	0.05	67	67
		10	0.1	66	68
		20	0.05	65	64
		20	0.1	70	69
	2	10	0.05	72	76
		10	0.1	64	66
		20	0.05	69	64
		20	0.1	69	71
		FPL Etch		72	71
2024T-3	1	20	0.1	80	78
	2	20	0.1	75	72
		FPL Etch		76	73

Table VI

Overlap Shear Strength After Environmental Exposure (pounds/sq. in.) Measured at 75°F

Alloy	Brush	Feed Speed (ft./min.)	Applied Pressure (HP/in.)	After 30 Days at 140°F/100% RH	After 30 Days Salt Spray
2024T-3	1	10	0.05	6110	6425
		10	0.1	6235	6600
		20	0.05	6165	6225
		20	0.1	6125	6375
	2	10	0.05	6110	6425
		10	0.1	6235	6600
		20	0.05	6165	6225
		20	0.1	6125	6375
	FPL Etch			5800	6150
7075T-6	1	10	0.05	7250	7000
		10	0.1	6400	6500
		20	0.05	6600	7050
		20	0.1	6700	7150
	2	10	0.05	6240	7000
		10	0.1	6600	7000
		20	0.05	6100	6280
		20	0.1	6360	6100
	FPL Etch			7150	7000

Table VII

Wedge Test Results (1/4" thick alloy adherends)

Alloy	Brush	Feed Speed (ft./min.)	Applied Pressure (HP/in.)	Crack Growth after 24 Hrs. at 140°F/100% RH (inches)	Crack Growth after 7 days Salt Spray (inches)
2024T-3	1	10	0.05	0.044	0.063
		10	0.1	0.047	0.022
		20	0.05	0.016	0.055
		20	0.1	0.044	0.047
	2	10	0.05	0.079	0.030
		10	0.1	0.04	0.028
		20	0.05	0.045	0.035
		20	0.1	0.047	0.063
	FPL Etch			0.044	0.031
7075T-6	1	10	0.05	0.056	0.024
		10	0.1	0.055	0.033
		20	0.05	0.063	0.048
		20	0.1	0.056	0.024
	2	10	0.05	0.063	0.033
		10	0.1	0.054	0.040
		20	0.05	0.079	0.033
		20	0.1	0.069	0.055
	FPL Etch			0.049	0.030

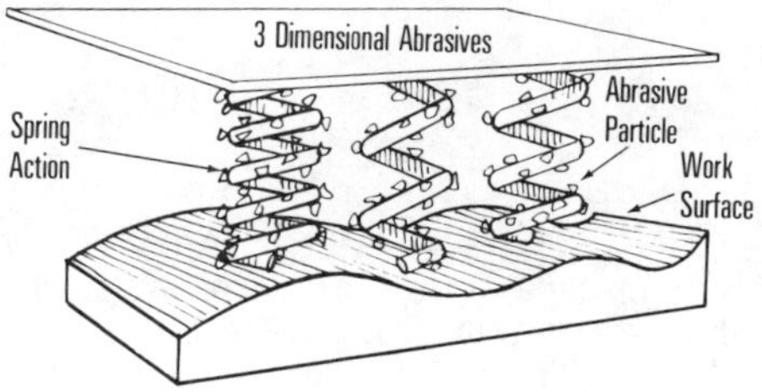

Figure 1. Schematic of 3-Dimensional Abrasive Action

Figure 2. Deburring Capability of 3-Dimensional Abrasives

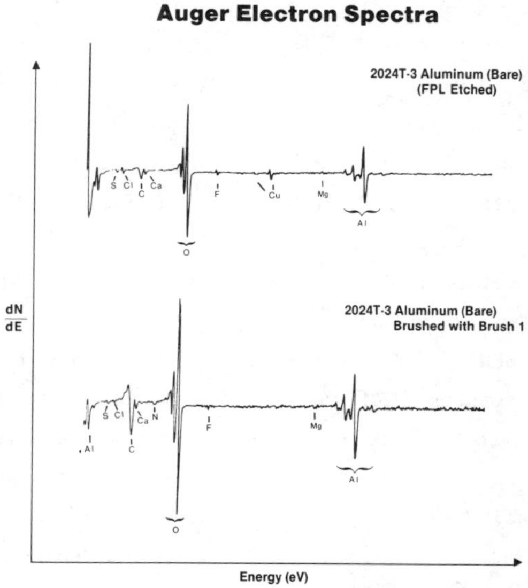

Figure 3.

13th National SAMPE Technical Conference
October 13-15, 1981

PRE-PRODUCTION EVALUATION OF A NONCHROMATED ETCHANT FOR PREPARING ALUMINUM ALLOYS FOR ADHESIVE BONDING

Narvel L. Rogers
Chemical and Process Laboratory
Bell Helicopter Textron, Texas

Abstract

Aluminum alloy surfaces treated with a nonchromated etchant developed at U.S. Army Armament Research and Development Command, Dover, N. J. are compared directly to surfaces treated with a standard chromated (FPL) etchant. Surfaces were evaluated for physical and chemical characteristics as well as bondability. The etchant was also used as a pre-treatment prior to both chromic acid anodize (CAA) and phosphoric acid anodize (PAA). Solution control parameters and usable life were investigated. Bondability was established with four different adhesive systems and MMM-A-132 qualification testing was accomplished. The Scanning Electron Microscope (SEM) and Transmission Electron Microscope (TEM) were used to investigate surface topography.

1. INTRODUCTION

Historically, chemical solutions for surface treatment of aluminum alloys have incorporated the use of hexavalent chromium compounds. This is especially true in the area of surface preparation prior to adhesive bonding. Processes such as sulfuric acid-sodium dichromate (FPL) etch, and chromic acid anodize are typical examples. Although these processes have proven to be highly satisfactory and are used extensively throughout industry, they have a major shortcoming--their effect on the environment. The considerations concerning the use of hexavalent chromium are much more important today than ever before. EPA regulations governing the control and disposal of toxic waste (chromium compounds) are more restrictive than ever, and chromium is one of those metals in critical supply in the United States. In an effort to provide suitable solutions without chromates, personnel at the Army Armament Research and Development Command (ARRADCOM), Dover, N. J., began studying the possibility of a substitute for the FPL etchant. In 1976, Bill Russell, of Picatinny Arsenal, reported the results of a preliminary study of a chromate-free process[1]. A second report was made in 1978[2] that indicated good, durable adhesive bonds had been produced in the laboratory without the use of chromates. In late 1978, Bell Helicopter Textron started a program, under sponsorship of ARRADCOM, to further evaluate the nonchromated etchant. The etchant solution evaluated is composed of sulfuric acid and ferric sulfate. It has been designated the P2 etch. The evaluation was accomplished under Department of the Army Contract DAAK10-78-C-0398 and was reported in ARLCD-CR-80008 dated April 1980[3].

2. DISCUSSION

2.1 Experimental Approach

Any surface preparation process worthy of consideration must be capable of producing an acceptable adhesive bonded joint. For this evaluation, an acceptable joint was defined as one capable of passing the tests required for structural adhesives. Naturally, an acceptable surface preparation process must not degrade the properties of the basic metal and the joint produced must be durable. With these criteria in mind, a program was initiated that was intended to study solutions prepared at different concentration levels. The characteristics of these solutions were evaluated along with their effect on the mechanical properties of three aluminum alloys. Prepared surfaces were studied and immediate bond strengths were determined. Results were compared to those obtained from joints exposed to various environments prior to testing. Storage life of the prepared surfaces prior to bonding was evaluated. In each instance, where feasible, surfaces prepared by the FPL etch process were used as controls.

2.2 Solution Studies

Solutions were prepared at the concentration levels shown in Table 1. The 10 percent above nominal solution contains some undissolved ferric sulfate. This is primarily because the solubility of ferric sulfate in sulfuric acid solutions decreases as the acid concentration increases. At 25°C a 2.2 N sulfuric acid solution will hold 250 g/l of ferric sulfate. Whereas, if the acid concentration is increased to 6.6 N, the solution will become saturated with ferric sulfate when only 146 g/l are added[4].

Solutions were analyzed for both sulfuric acid and ferric iron content. Then, weighed amounts of 2024 aluminum alloy turnings were added and allowed to dissolve.

The concentrations were again determined. The results of analysis are shown in Tables 2 and 3. It should be noted that the ferric iron content drops rapidly after about 3 g/l of 2024 aluminum has been added. These and other data obtained later were used to establish working control limits for

TABLE 1. CONCENTRATION OF P2-ETCH SOLUTION

Solution No.	Constitutent Sulfuric Acid g/l	Ferric Sulfate g/l	Concentration Level
1	407.5	165.0	+10%
2	388.5	157.5	+ 5%
3	370.0	150.0	Nominal
4	351.5	142.5	- 5%
5	333.0	155.0	-10%

TABLE 2. SULFURIC ACID CONTENT

Solution Identification	Normality as H_2SO_4
Low Limit Solution (-10%)	
No Al Added	8.5
0.1 g/l Al	7.8
0.5 g/l Al	7.5
1.0 g/l Al	7.5
5.0 g/l Al	7.0
10.0 g/l Al	6.5
High Limit Solution(+10%)	
No Al Added	9.7
0.1 g/l Al	9.7
0.5 g/l Al	9.5
1.0 g/l Al	9.5
5.0 g/l Al	9.1
10.0 g/l Al	7.0

TABLE 3. FERRIC IRON CONTENT

Solution Identification	Grams/Liter of Ferric Iron (Fe+++)
Low Limit Solution (-10%)	
No Al Added	28.48
0.1 g/l Al	28.38
0.5 g/l Al	26.52
1.0 g/l Al	22.62
5.0 g/l AL	0.42
10.0 g/l Al	0.28
High Limit Solution (+10%)	
No Al Added	35.73
0.1 g/l Al	33.41
0.5 g/l Al	30.99
1.0 g/l Al	28.62
5.0 g/l Al	5.44
10.0 g/l Al	0.42

the P2 etch solution. These are:

Sulfuric Acid	6.5 to 9.5 N
Ferric Iron	22 to 35 g/l

2.3 Effect on Basic Metal

The high (+10%) and low (-10%) concentration solutions were used to determine the amount of metal removed at different immersion times and a constant temperature. Three different alloys were used and an FPL etch solution was used for comparision. The P2 etch solution is more aggressive than the FPL etchant in all tests. Typical results are shown in Table 4. However, additional testing indicated that by lowering the temperature to 62.8°C (145°F) and reducing the immersion time to 8 minutes the weight loss can be controlled within limits which are comparable to those expected from the FPL etch process. The effect of the solutions on mechanical properties and topography were evaluated next. Standard tensile test coupons were prepared by treatment in the P2 etchant and the FPL etchant. There was no significant reduction in the properties of the three alloys tested (2024, 6061 and 7075 clad). The

TABLE 4. WEIGHT LOSS VS ETCH TIME

Alloy and Time	Solution: FPL mg/dm^2(mg/ft^2)	Solution: P2 +10% mg/dm^2(mg/ft^2)
12 min 68.9°C		
2024	41.9 (389)	47.9 (538)
6061	31.0 (288)	50.8 (472)
7075	28.5 (265)	87.4 (812)
16 min 68.9°C		
2024	65.7 (610)	121.2 (1126)
6061	49.8 (463)	73.4 (682)
7075	33.6 (312)	87.7 (814)
20 min 68.9°C		
2024	90.2 (838)	172.4 (1602)
6061	72.7 (675)	96.7 (898)
7075	35.8 (333)	110.2 (1024)

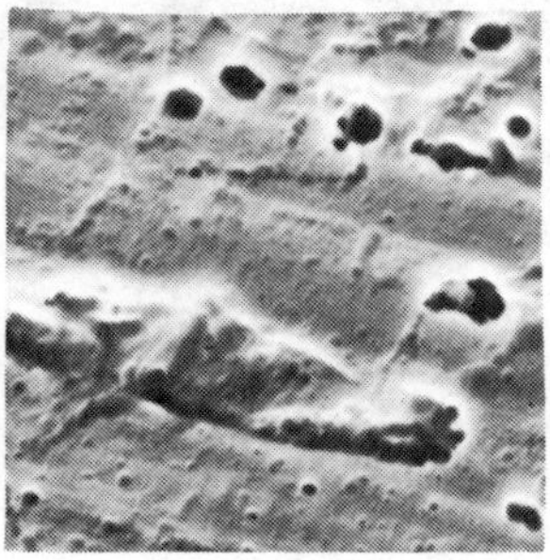

Figure 1. 2024 etched in FPL solution for 12 minutes at 64°C (147°F).

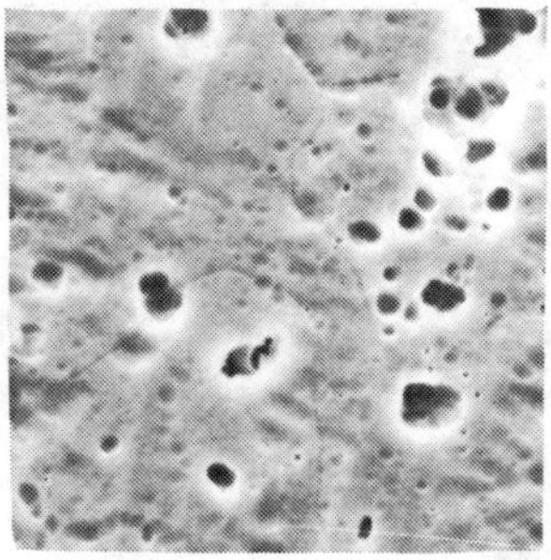

Figure 2. 2024 etched in P-etch solution for 12 minutes at 66°C (151°F).

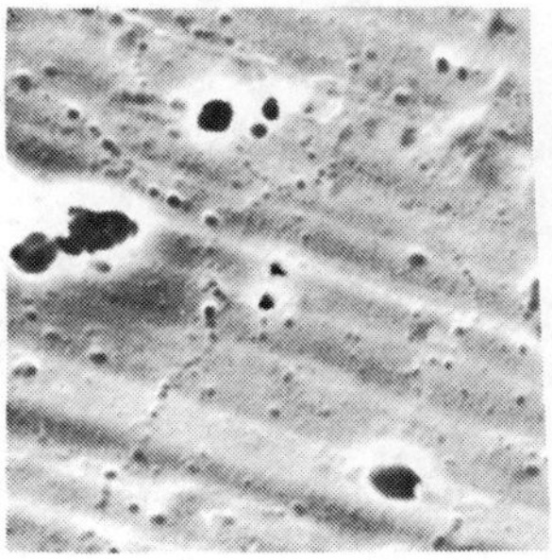

Figure 3. 6061 etched in FPL solution for 12 minutes at 63°C (145°F).

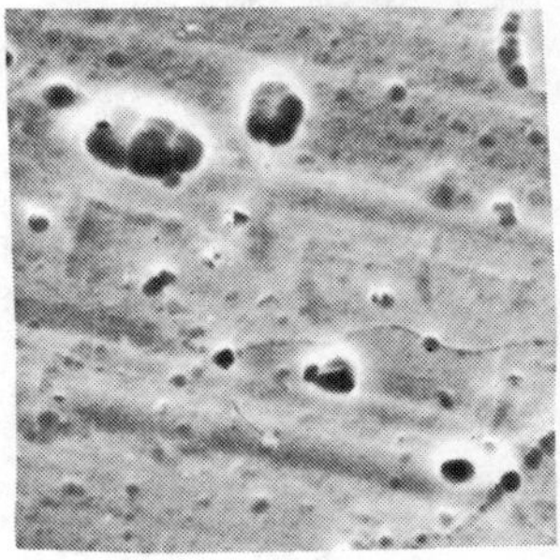

Figure 4. 6061 etched in P2 etched solution (-10%) for 12 minutes at 68°C (155°F).

Figure 5. 7075 Etched in FPL solution for 12 minutes at 63°F (145°F).

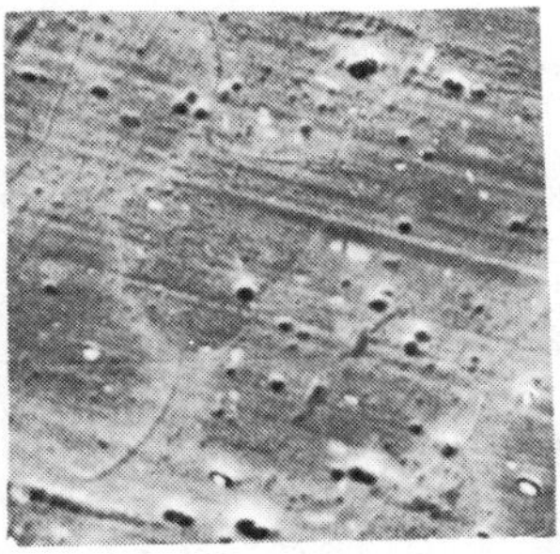

Figure 6. 7075 etched in P2 etch solution (-10%) for 12 minutes at 68°C (155°F)

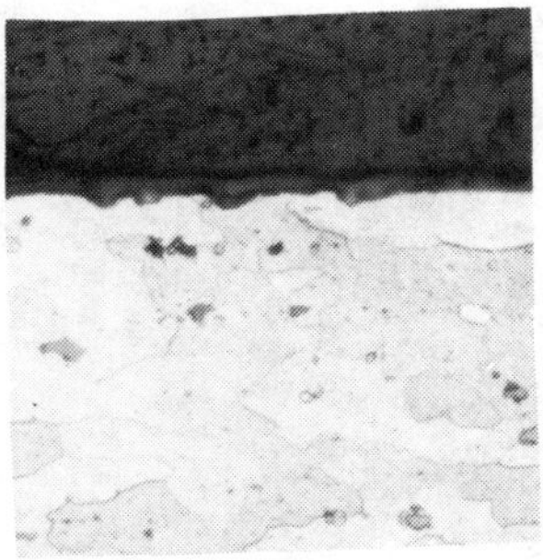

Figure 7. 2024 etched in FPL solution.

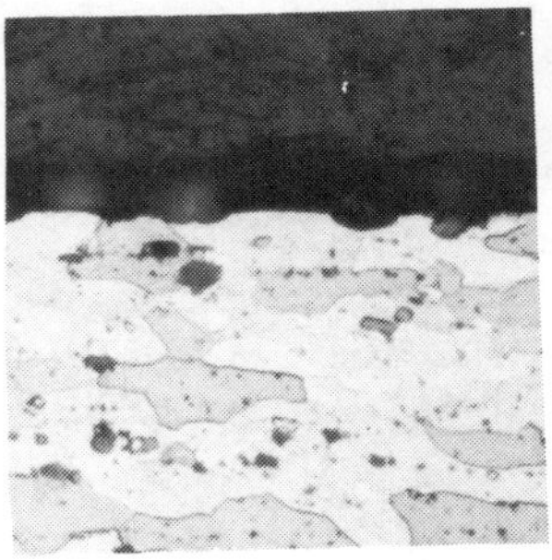

Figure 8. 2024 etched in P2 etch (-10%) solution.

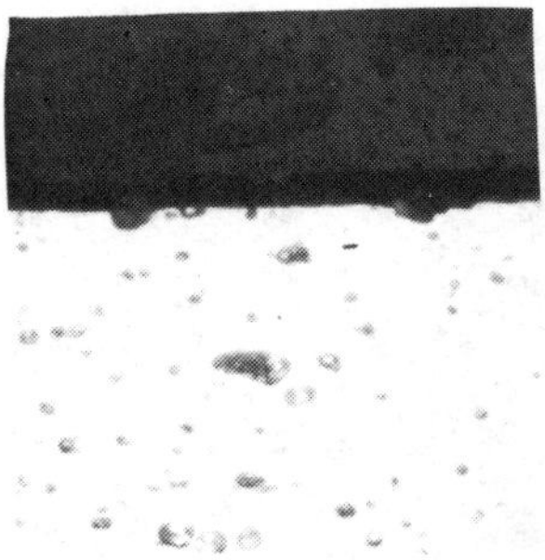

Figure 9. 6061 etched in P2 etch (-10%) solution.

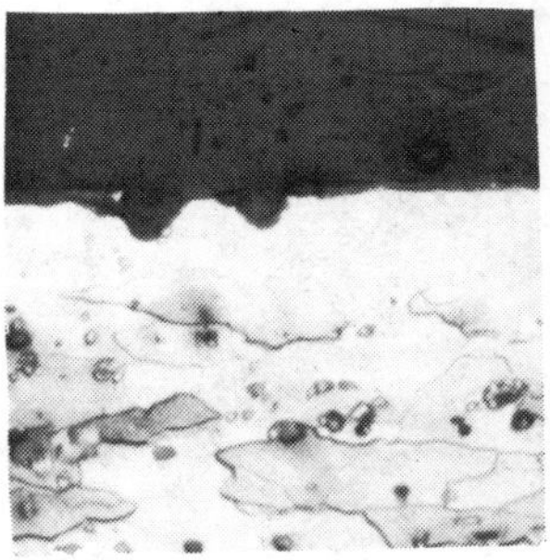

Figure 10. 7075 clad etched in P2 etch. (-10%) solution.

topography of specimens was examined using the SEM. Figures 1 through 10 are typical surface conditions (at 1000X) which can be expected from the FPL and P2 etchants.

2.4 Bondability

The initial bonding tests were conducted using three different adhesive systems and 2024-T3 (bare) aluminum alloy test panels. The panels were prepared by using the FPL etch process and three different concentrations of the P2 etch solution. The results of these tests indicated that there is no significant difference in the adhesive bonding properties of the surfaces tested. Therefore, testing of the surfaces was continued. Tanks of about 20-gallon capacity were charged with the P2 etch solution. This solution was used alone to prepare specimens, and it was also used as a preliminary deoxidizer prior to chromic acid anodize (CAA) and phosphoric acid anodize (PPA). Test panels of 2024-T3 aluminum were prepared and then bonded with FM 1000, N 1113, FM 53, and EA 9628 adhesive systems. Both lap shear and metal-to-metal peel tests were performed at room temperature, -67°F and +180°F. Tables 5 through 8 show typical results of these tests. It should be noted that there are significant variations in the bond strengths produced with a particular adhesive system on the different surfaces. The next series of tests were conducted to determine the effects of various environmental exposure on the durability of adhesive-bonded joints produced. For these tests, the P-2 etch solution was scaled up to a 50-gallon tank. The solution was used to prepare both lap shear and metal-to-metal peel specimens for bonding with N1113 and EA 9628 adhesive systems. After bonding, the test coupons were exposed to environments as follows:

Salt Spray	30 days
Water Soak	30 days
Humidity 95±5% 120°F	30 days
JP-4 Fuel Immersion	7 days
MIL-F-5606 Hydraulic Fluid	7 days
MIL-L-23699 Turbine Oil	7 days

Control specimens were tested prior to exposure. Again, there was no significant difference in the surfaces produced by FPL etch, P2 etch, P2 etch plus CAA and P2 etch plus PAA. Typical results are shown in Tables 9 and 10.

2.5 Storage Life

Storage life prior to adhesive primer application was also evaluated. This was accomplished by preparing a group of test panels using both the P2 and FPL etchants. Both 2024 (bare) and 7075 (clad) were used. After surface preparation, one panel from each group was primed and bonded. The remaining test coupons were wrapped in paper and stored in the laboratory. Panels were removed from storage at intervals of 2, 4, 8, 12, 24, 48, 72,

96, 168 and 240 hours for priming and bonding. Each test panel was cut into five test coupons. Two coupons were tested for peel strength, and the remaining three coupons were immersed in water at 140°F for seven days prior to testing. There was no significant difference between the surfaces produced by the FPL etch and the P2 etch. Tables 11 and 12 show typical results.

TABLE 5. BONDABILITY TESTING

Adhesive System: FM53/BR53
Lap Shear Strength
(Units: Psi)

Surface & Test Condition	Average
FPL Etch (Control)	
RT	4926
180°F	3482
-67°F	5844
P2-Etch (High Conc.)	
RT	4653
180°F	3584
-67°F	5878
P2-Etch (Low Conc.)	
RT	4838
180°F	3560
-67°F	5909
FPL Etch/Chromic Anodize	
RT	4476
180°F	3552
-67°	5006
P2-Etch Chromic Anodize	
RT	4294
180°F	3573
-67°F	4765
P2-Etch/Phosphoric	
RT	4910
180°F	3458
-67°F	5530

TABLE 6. BONDABILITY TESTING

Adhesive System: FM53/BR53
Peel Strength (Units: Pli)

Surface & Test Condition	Average
FPL Etch (Control)	
RT	88
180°F	87
-67°F	92
P2-Etch (High Conc.)	
RT	87
180°F	81
-67°F	41
P2-Etch (Low Conc.)	
RT	84
180°F	89
-67°F	85
FPL Etch/Chromic Anodize	
RT	76
180°F	71
-67°F	61
P2-Etch/Chromic Anodize	
RT	83
180°F	71
-67°F	37
P2-Etch/Phosphoric Anodize	
RT	81
180°F	71
-67°F	81

TABLE 7. BONDABILITY TESTING

Adhesive System: EA9628/EA9202
Lap Shear Strength (Units: Psi)

Surface & Test Condition	Average
FPL Etch (Control)	
R	4492
180°F	3553
-67°F	5023
P2-Etch (High Conc.)	
RT	4321
180°F	3107
-67°F	4980
Surface & Test Condition	Average
P2-Etch (Low Conc.)	
RT	4437
180°F	3619
-67°F	5007
FPL Etch/Chromic Anodize	
RT	4225
180°F	2986
-67°F	4257
P2-Etch/Chromic Anodize	
RT	4041
180°F	3239
-67°F	4192
P2-Etch/Phosphoric Anodize	
RT	4391
180°F	3526
-67°F	4721

TABLE 8. BONDABILITY TESTING

Adhesive System: EA9628/EA9202
Peel Strength (Units: Pli)

Surface & Test Condition	Average
FPL Etch (Control)	
RT	56
180°F	68
-67°F	49
P2-Etch (High Conc.)	
RT	59
180°F	63
-67°F	54
P2-Etch (Low Conc.)	
RT	68
180°F	43
-67°F	67
FPL Etch/Chromic Anodize	
RT	73
180°F	67
-67°F	57
Surface & Test Condition	Average
P2-Etch/Chromic Anodize	
Rt	63
180°F	65
-67°F	57
P2-Etch/Phosphoric Anodize	
RT	64
180°F	68
-67°F	54

TABLE 9. MM-A-132 QUALIFICATION TESTS

Adhesive System: N1113/N6726
U. S. Units - PSI

Surface & Exposure	Lap Shear Control	Strength After Exposure
FPL Etch		
Salt Spray	4343	4995
Water	4314	4221
Humidity	4550	4407
Fuel	4684	5122
Hydraulic	4824	5069
Lubricant	4768	5296
P2 Etch		
Salt Spray	4608	4716
Water	4603	4707
Humidity	4778	4106
Fuel	4634	4921
Hydraulic	4491	5085
Lubricant	4945	4962
CR Anodize		
Salt Spray	4368	4923
Water	4204	4348
Humidity	4410	4339
Fuel	4560	5129
Hydraulic	4970	4873
Lubricant	4682	4949
PA Anodize		
Salt Spray	4711	5177
Water	4205	5186
Humidity	4784	5087
Fuel	4514	5268
Hydraulic	4871	5605
Lubricant	5055	5167

TABLE 10. MM-A-132 QUALIFICATION TESTS

Adhesive System: N1113/N6726
U. S. Units: PLI

Surface & Exposure	Peel Control	Strength After Exposure
FPL Etch		
Salt Spray	75	73
Water	75	72
Humidity	85	75
Fuel	75	77
Hydraulic	77	78
Lubricant	85	78
P2 Etch		
Salt Spray	85	72
Water	78	76
Humidity	75	73
Fuel	75	75
Hydraulic	75	75
Lubricant	75	71
CR Anodize		
Salt Spray	80	76
Water	83	79
Humidity	78	72
Fuel	80	79
Hydraulic	77	79
Lubricant	80	75
PA Anodize		
Salt Spray	87	86
Water	84	86
Humidity	85	87
Fuel	80	87
Hydraulic	88	84
Lubricant	82	84

TABLE 11. STORAGE LIFE PRIOR TO BONDING

Alloy: 2024 Bare
Surface Treatment FPL Etch
Adhesive System: N1113/N6726
Peel Strength (Units: Pli)

Out-Time Before-Priming	No-Water Soak Average	7-Day Soak 140°F Average
Immediately	79	77
2 hours	79	75
4 hours	75	71
8 hours	82	69
12 hours	80	78
24 hours	84	85
48 hours	86	65
72 hours	72	72
96 hours	77	75
168 hours	75	75
240 hours	74	73

TABLE 12. STORAGE LIFE PRIOR TO BONDING

Alloy: 2024 Bare
Surface Treatment: P2 Etch
Adhesive System: N1113/N6726
Peel Strength (Units: Pli)

Out-Time Before-Priming	No-Water Soak Average	7-Day Soak 140°F Average
Immediately	76	75
2 hours	79	73
4 hours	78	64
8 hours	78	75
12 hours	83	69
24 hours	84	83
48 hours	74	62
72 hours	75	70
96 hours	73	75
168 hours	76	76
240 hours	83	81

2.6 Surface Studies

Each of the three alloys (2024, 6061 and 7075) were examined using the SEM at magnifications of 100X, 300X and 1000X and the TEM at magnifications of 7000X, 29,000X and 87, 000X. Surface in the as-received condition, as well as those after FPL etch, P2 etch, CAA, and PAA were studied. Figures 11 through 16 show typical examples of these surfaces. In addition to the physical appearance of the surfaces, the chemical characteristics were also studied.

Energy Dispersive Analysis (EDS) was used in conjunction with the SEM to determine if elements such as chomium, iron, and sulfur were present on the prepared surfaces. There were no significant amounts of these elements found on any of the surfaces.

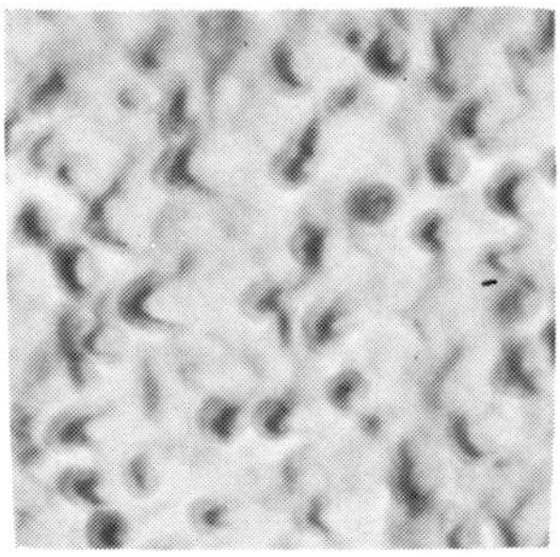

Figure 11. 87,000x
7075 FPL Etch

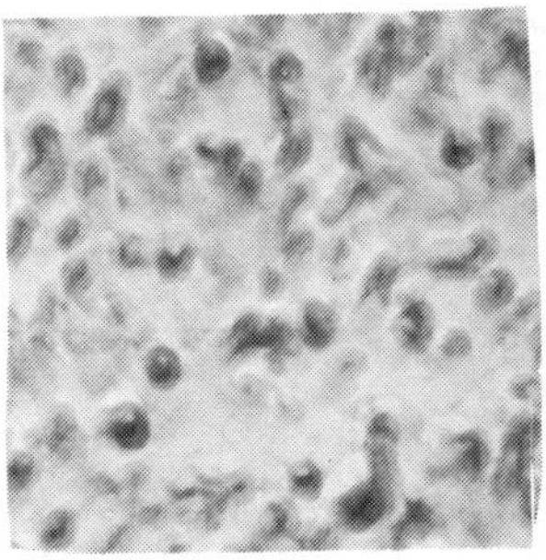

Figure 12. 87,000x
7075 P2 Etch

Figure 13. 87,000X
7075 FPL Etch + PAA

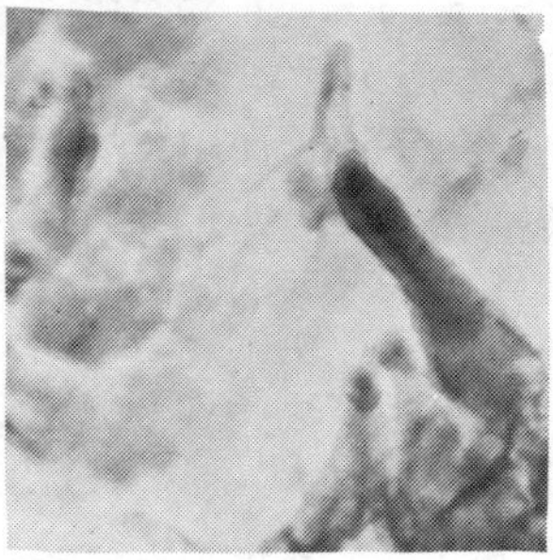

Figure 14. 87,000X
7075 P2 Etch + PAA

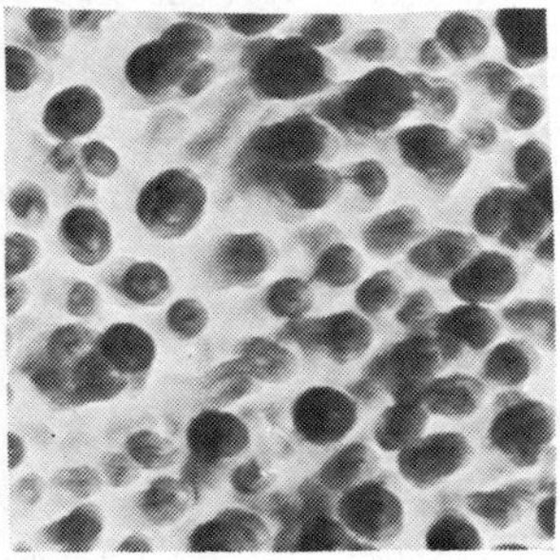

Figure 15. 87,000X
7075 FPL Etch + CAA

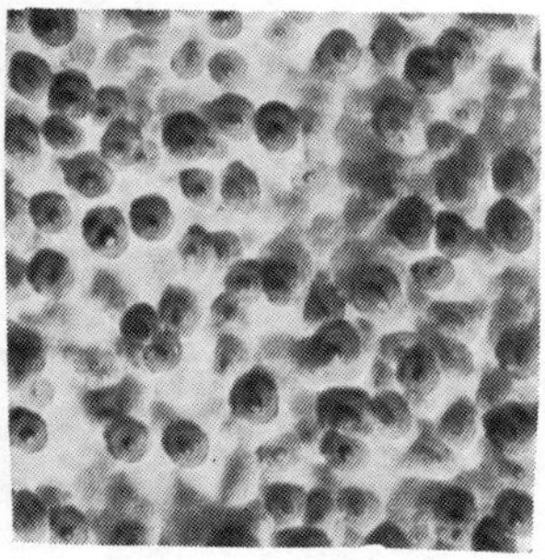

Figure 16. 87,000X
7075 P2 Etch + CAA

Ion Scattering Spectrometry (ISS) was also used to examine the surfaces of prepared specimens. Although a slight amount of iron was found on the surface of the 7075 clad alloy specimens, it was not considered to be contamination from the P2-etch solution.

3. CONCLUSIONS

The P2-etch treatment process:

- Does not degrade the properties of the aluminum alloys tested beyond acceptable limits.
- Produces surfaces that are receptive to adhesives.
- Provides surfaces that produce adhesive bond strengths (90° peel and lap shear) equal to or better than those produced by the FPL etch.
- Provides an acceptable pretreatment (deoxidizer) prior to either phosphoric acid anodize (PAA) or chromic acid andodize (CAA).
- When used alone or coupled with PAA will provide a chromate-free surface treatment process for preparing aluminum alloys for adhesive bonding.
- Provides a process that does not require elaborate waste disposal procedures as is the case with solutions containing hexavalent chromium compounds.

4. REFERENCES

1. Russell, William J., CHROMATE FREE PROCESS FOR PREPARING ALUMINUM FOR ADHESIVE BONDING, presented at the Symposium on Durability of Adhesive Bonded Structures, Picatinny, Arsenal, U. S. Army Research and Development Command, Dover, New Jersey, October 27-29, 1976.
2. Russel, W. and Garnis, E., CHROMATE-FREE METHOD OF PREPARING ALUMINUM SURFACES FOR

ADHESIVE BONDING: AN ETCHANT COMPOSITION OF LOW TOXICITY, ARLCD Technical Report 78001 U. S. Army Armament Research and Development Command, Dover, New Jersey, May 1978.

3. Rogers, Narvel., and Russell William, EVALUATION OF NON-CHROMATED ETCH FOR ALUMINUM ALLOYS (P2-ETCH). ARLCD Technical Report CR-80008, U.S. Army Armament Research and Development Command, Dover, New Jersey, April, 1980.

4. Siedell, Atherton, Ph.D., SOLUBILITIES OF INORGANIC AND METAL ORGANIC COMPOUNDS, Volume 1, D. Van Nostrand Company, Incorporated, New York, 1940.

5. BIOGRAPHY

Narvel Rogers has been involved in aerospace for over 25 years. He has been active in the development and control of surface preparation process at Bell Helicopter Textron since 1956. He has prepared a number of papers for presentation and holds a patent for a surface treatment process. He is an active member of SAMPE and is the Director for the Dallas-Ft. Worth Chapter. He is also a member of the National Association of Corrosion Engineers and is active in Technical Practices Committee T-9, Corrosion of Aerospace Equipment.

13th National SAMPE Technical Conference
October 13-15, 1981

ANALYTICAL TECHNIQUES FOR MATERIALS SURFACE CHARACTERIZATION

Stuart M. Lee
Ford Aerospace & Communications Corp.
Palo Alto, California

Abstract

Analytical problems routinely solved today would have been major research programs or unattainable just a relatively short time ago when the state of available analytical techniques was limited to the optical microscope, wet chemical and spectrographic analysis. Contrast this with today's profusion of instrumental acronyms, including ESCA, SAM, SIMS, ISS and UPS which are available to the materials scientist. Surface analysis of materials is the ultimate approach. Photons, electrons, atoms, molecules or ions can be used as probes in a variety of combinations and in an imposing array of analytical instrumentions. Each surface analytical technique is unique with its own capabilities and corresponding specificity. Analysis of the surface layer is usually the first step which may then lead to the application of multiple instrumentation techniques performed. These remarkable instrumental techniques are capable of detecting and identifying abnormalities in the femtogram (10^{-12} to 10^{-15}) range. Surface analysis not only identifies the chemical elements present, but also determines their concentration, distribution, surface depth and type of bonding involved.

1. INTRODUCTION

Surfaces represent a mere portion of the total volume of most solids. However, the local surface composition is very often the critical factor in determining a material's performance. Top surfaces consisting of a single atomic layer can dictate the observed physio-chemical characteristics. In addition, since physical and chemical interactions of solids can take place through interfaces of solid state related phenomena such as diffusion, corrosion, nucleation, conduction, adhesion and catalysis can be studied with surface analytical equipment. Through the utilization of instrumental analytical techniques, the surface structure, composition and electronic energy states has emerged as well as the detection and identification of trace materials. Surface analytical techniques, besides identifying the chemical elements present, show their concentration, distribution, surface depth and the structural bonding involved. Many surface analytical techniques are available. However, with this wide array of sophisticated analytical surface techniques, questions have arisen regarding usage, selectivity and limitations. Scanning Auger Microprobe (SAM) analysis is becoming largely the domain of the materials scientist, whereas ESCA is nurtured primarily by the chemist. It is the purpose of the following discussion to present the Auger and ESCA surface analytical approaches, provide background, operational information, define guide lines for their utilization and illustrate each technique with actual case histories.

2. SURFACE CHARACTERIZATION

Surfaces free of chemisorbed species such as water, nitrogen, oxygen, and solvents are rare. Gold is often thought of as a clean surface, but when analyzing a clean gold surface it is found to contain three monolayers of carbon, containing 60% of gold. Good quality silicon wafers, when analyzed, indicate 20-25% surface carbon.[1] Therefore, for a surface analytical technique to be meaningful, it should contain or provide for a method for the removal by etching of absorbed surface contaminants. The importance of cleaning is demonstrated by a comparative cleaning of bare silicon surfaces by liquid reagents, ion sputtering and plasma etching.[2] Surface analysis revealed that liquid reagent cleaning yielded silicon surfaces heavily contaminated with carbon, which could be minimized with oxidizing agents, Sputtering with argon gave an atomically clean surface with active sites which, when removed from the vacuum chamber, was capable of absorbing surface impurities while plasma etching proved to be the most effective.

2.1 Scanning Electron Microscope (SEM)

Mention surface analysis and some people think of the SEM as the ultimate instrument for surface analysis. The SEM enables the analyst to obtain unsurpassed, three-dimensional, high magnification photographs showing microtopography and indicating possible anomolies. However, when utilized for surface analysis, it represents only a first step because the SEM cannot analyze films thinner than 10,000A° (40 micro-inches). Elemental compositional analysis is obtained by subjecting the x-rays generated by the SEM electron beam (from the cubic micron probed area) to analysis (EDX). EDX examines fluorescent x-rays generated from the sample by electron or x-ray beam excitation. The escape depths of these x-rays are much deeper (μm) as compared to the Auger electrons and photo-electrons (A^o). For this reason, the EDX data are more representative of bulk constituents present at high concentrations. The analytical features of this instrument are limited to the lack of clean vacuum because of their operation at 10^{-7} to 10^{-9} torr pressure range, the inability to utilize "bakeout" because of organic slip rings seals which preclude reliable surface analysis because of the residual pressure in unbaked vacuum system can contain many active gases and poorer analytical sensitivity because of design constraints, i.e., positioning of electron beam axis in which only a portion of the available analyzer solid angle is utilized leads to shadowing, markedly reducing sensitively. In addition, lens design imposes limits of specimen dimensions and analysis area. The "bulk mode" trace analytical technique (3) has increased the detection limits for the transition elements.

2.2 Scanning Auger Microprobe (SAM)

The SAM is a non-destructive analytical technique for surface analysis which can give unbelievable results as far as analyzing thin films and surfaces are concerned. This highly sophisticated instrument is ideal for the determination of the composition of very thin surface films down to 0.1% of a monolayer. It can handle films as thin as 10 A^o, and in addition, by successive sputtering, determine the total thickness and profile of any film. The analysis is based on the ejection of secondary, or what is known as Auger electrons, a phenomenon discovered by Pierre Auger in the 1920's. The technique involves surface irradiation with a primary beam of high energy electrons (10 KeV), resulting in the formation of ions via loss of electrons from deep-lying atomic energy levels. The process of transformation can be represented as shown in Figure 1.

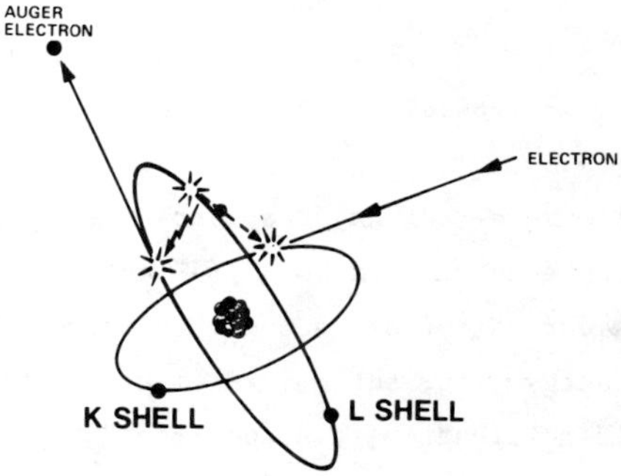

Figure 1
Auger Process Transformation

First the energetic electron ionizes an atom by removing the inner K shell electron. The electron in the adjacent L shell moves to fill the K shell vacancy and simultaneously, it gives up the energy of that transition (from L to K) to another L electron which becomes the ejected Auger electron. The Auger electrons are well defined and appear as sharp peaks in the electron energy distribution providing a fingerprint which can be used for identification of elements from lithium to thorium (3 to 90) as shown in Figure 2.

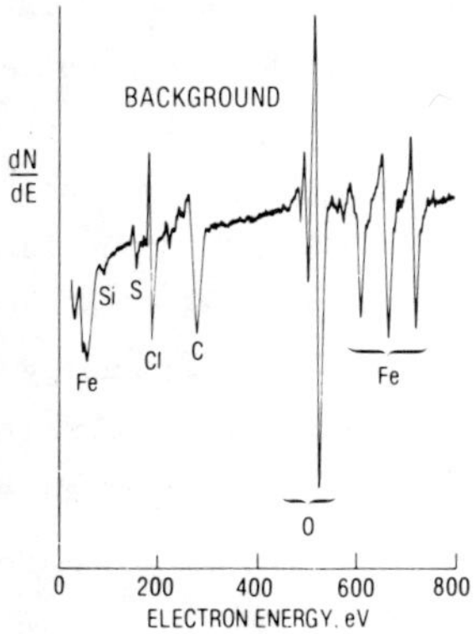

Figure 2
Typical Auger Spectrum

The resultant Auger spectra are represented as a dN (E)/dE curve, a derivative of secondary electron energy distribution. Elemental identification is accomplished by measuring the energy of these Auger or secondary electrons. The primary Auger electrons lie in the 0-2500 eV range and originate from KLL, LMM and MNN transitions. Vacuum requirements for the SAM require an ultra-high capability vacuum (10^{-10} torr) because SAM typically probes the top few atomic layers and therefore an environment is required that precludes disposition of residual gases. Insulating materials may cause charging problems which can be minimized by either of two techniques: lowering the energy of the incident electron beam or using a bombardment beam at grazing angle of incidence.

<u>2.2.1 The Case of the Aluminum Spike</u> A technique which has added dramatically to SAM, allows for sputter-etching a specimen surface with a simultaneous analysis. Data is acquired relative to specimen composition as a function of depth into the specimen. This technique involves bombardment of the specimen surface with a beam of argon or xenon ions which sputter-etch the specimen surface while simultaneously bombarding the specimen with an electron beam yield Auger analytical data. It is important that the sputter-etched area be much larger than the area of the electron beam so that the region of analysis is in the center of a relatively flat-bottomed crater. This precludes analysis of the sides of the crater where the depth resolution would be seriously impaired. In typical present-day equipment, the ion beam diameter is on the order of

a few millimeters while the electron beam can be focused down to 25 microns (cf., Figure 3).

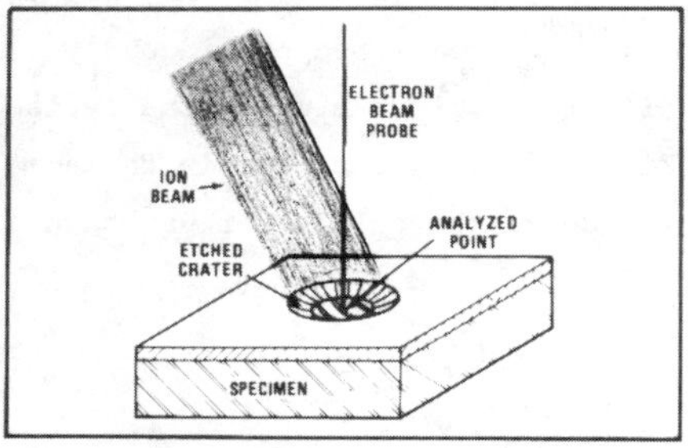

Figure 3
Ion Sputter Etching

"Aluminum spiking" is a phenomenon which occurs after the aluminum contact layer has been deposited and the device is heated to promote aluminum diffusion into the silicon layer to obtain improved electrical contact. Defects in the silicon permit the diffusion of the aluminum through the p layer into n type silicon causing a direct short in high leakage currents in the device as shown in Figure 4.

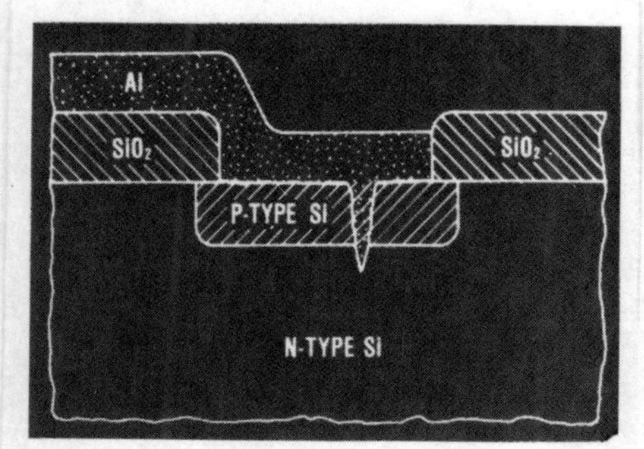

Figure 4
Aluminum Spiking

Ion sputter-etching was used to expose the Al-Si interface, Auger analysis identified the "Al spike" as an Al-Si alloy with a background of silicon after ion etching was continued to a depth greater than 0.5 μm below the "planar contact level".

<u>2.2.2 The Case of Contact Irregularities</u> Auger surface images or elemental maps are also produced which are invaluable for metallurgical grain boundary segregation, surface diffusion, and then film studies. For example, contact irregularities were observed in a microwave transistor as shown in Figure 5.

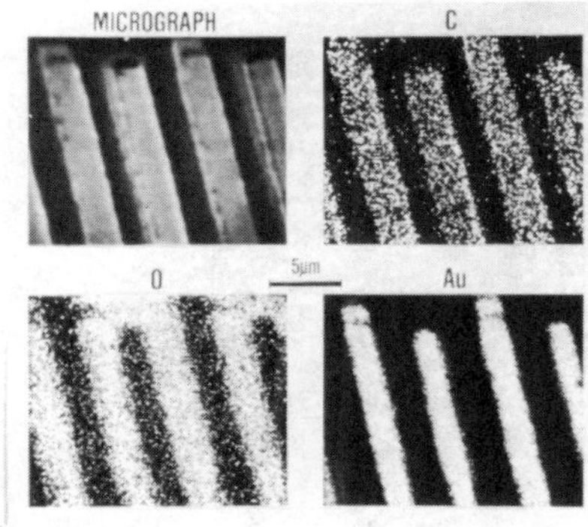

Figure 5
Microwave Transistor -
Au-Mo Metallization

A secondary electron micrograph shows the topography of the emitter and base contact windows consisting of 2.5 μm wide emitter and base strips with 0.6 μm thick gold layer over 0.1 μm molybdenum. The gold

Auger image delineates the gold metallization layer and its carbon contaminant while the oxygen Auger image maps the silicon dioxide passivation layer between the emitter and base contacts.

2.2.3 The Case of the Ineffective Molybdenum Metal Barrier Layer.

Sputter-etching was used to expose the interfaces below the surface of the device for Auger microanalysis. The amplitudes of the Au, Mo, and Si peaks were monitored continuously while ion sputter-etching the device (Figure 6A). A decrease in the amplitude of the gold Auger peak and a

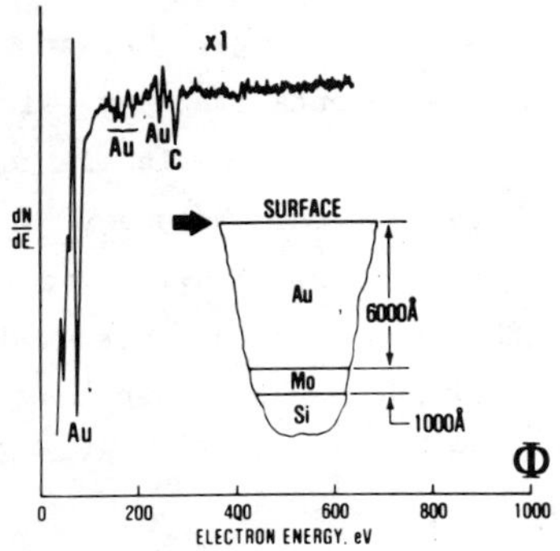

Figure 6A Surface

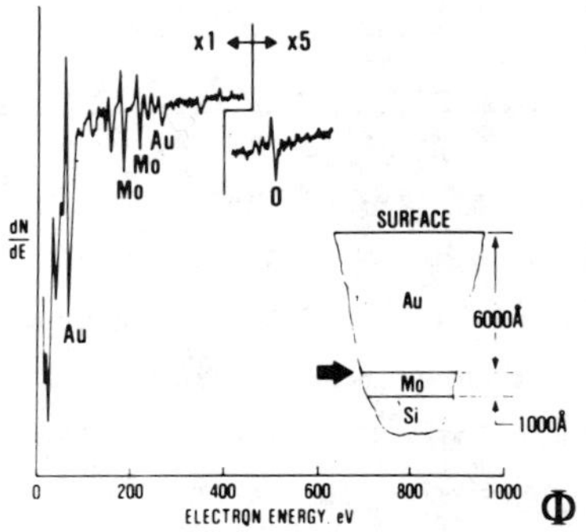

Figure 6B Au-Mo Interface

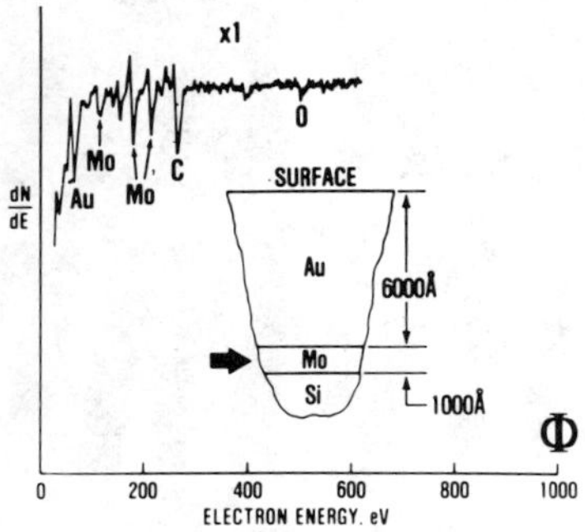

Figure 6C Mo Layer

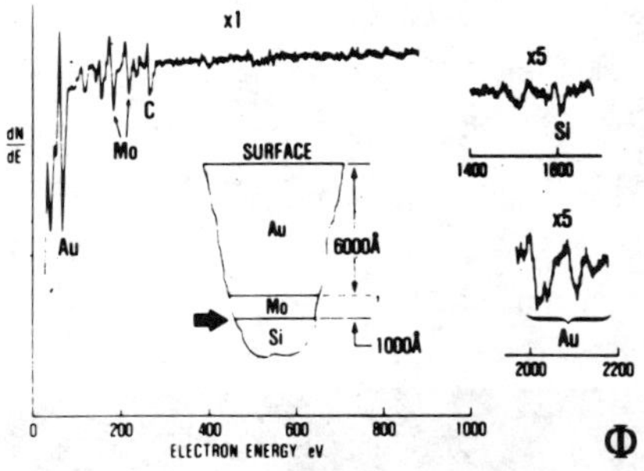

Figure 6D Mo-Si Interface

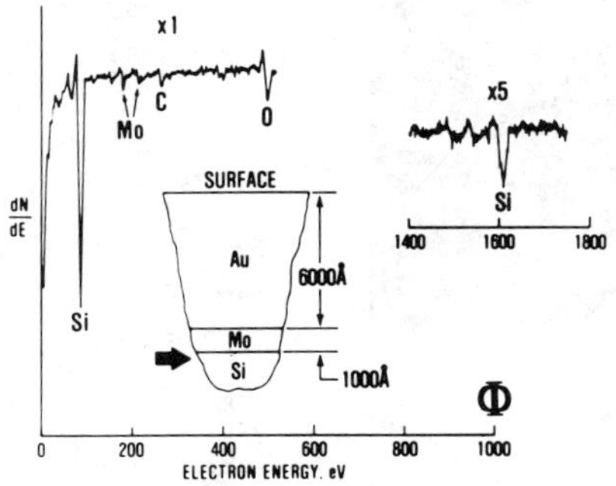

Figure 6E Si Substrate

Figure 6 Microwave Transistor-Auger Spectrum Au-Mo Metallization simultaneous increase in the molybdenum Auger peak signals the presence of the Au-Mo interface(Figure 6B); ion sputter-etching was stopped and an Auger spectrum recorded to determine the elemental composition of the Au-Mo interface. This procedure was repeated until the selected interfaces are analyzed. Auger spectra obtained on the Au surface, at the Au-Mo interface, in the "bulk" of the Mo layer(Figure 6C) and at the Mo-Si interface (Figure 6D) and into the Si substrate (Figure 6E). The Au surface is clean, with only a small amount of carbon present. The Auger spectrum obtained at the Au-Mo interface (Figure 6B) indicated the presence of a thin oxide layer which can adversely affect the contact resistance of the metallization. The spectrum recorded in the "bulk" molybdenum film (Figure 6C) suprisingly showed the presence of a significant amount of gold and carbon, because the molybdenum layer is supposed to act as a barrier to the diffusion of gold into the silicon. The presence of carbon in the molybdenum film probably indicates a contaminated metal-deposition system. The Auger spectrum obtained near the Mo-Si interface showed a dramatic increase in the concentration of Au over that present in the "bulk" Mo thin film, some carbon contamination is also noted at the Mo-Si interface. The results of the Auger microanalysis of the microwave transistor contact indicated the molybdenum metal layer was not acting as an effective barrier to gold diffusion for the process conditions used to make the device.

<u>2.2.4 Identification of Solid Film Lubricant on Spherical Bearing (Case History)</u>. Spherical bearings had been coated with an unknown dry film lubricant. It was imperative to ascertain the identity of the dry film lubricant before the port could be accepted for a specific space application. The manufacturer could have used two candidate dry film lubricants. An Auger "fingerprint" spectra was made of both known candidate materials and the unknown bearing surface. The results show the lubricant to be Lubeco K350 as shown in comparative Figures 7A and 7B.

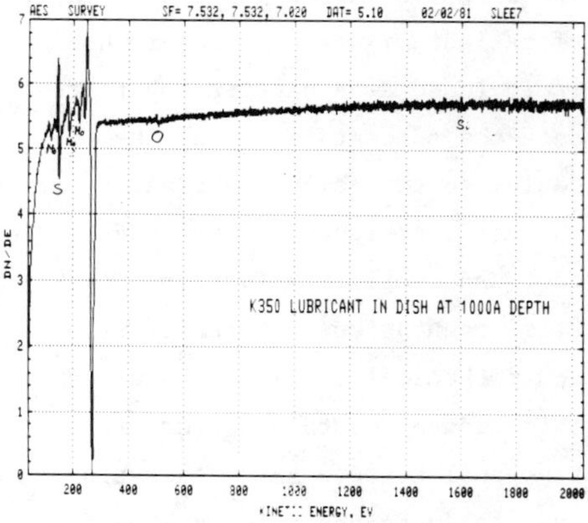

Figure 7A Unknown Lubricant

Figure 7 Auger Comparative Spectra
(Contd. on next page)

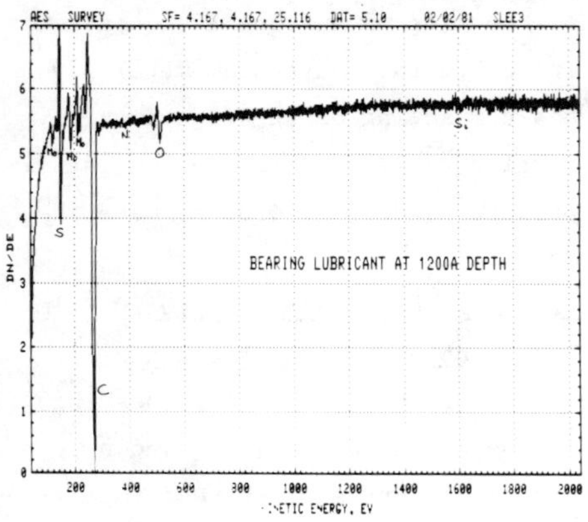

Figure 7B Lubeco K350

Figure 7 Auger Comparative Spectra

2.2.5 The Case of Solar Cell Interlayer Corrosion. Trimetal metallizations are routinely used to form electrical contacts in solar cells. Multilayer metallization on silicon consists of a titanium layer which is applied first to act as a bond for a several hundred Angstrom layer of palladium, the next layer which is included as a barrier and to reduce the susceptability of the titanium to corrosion and finally, a third silver layer to seal the contact and facilitate further electrical connections. Deterioration and mechanical stability resulted as evidenced by contact failure in tape-peel tests. Auger analysis was undertaken to characterize the distribution of oxygen in the Ti layer of a Ag/Pd/Ti/Si (111) metallization.[4] Most of the silver was removed by chemical dissolution. An Auger depth profile utilizing sputtering was then made of the remaining metallization while monitoring, by multiplexing, the Ag, C, O, Pd, Ti, and Si Auger signals (peak-to-peak amplitudes in the derivative mode) as shown in Figure eight.

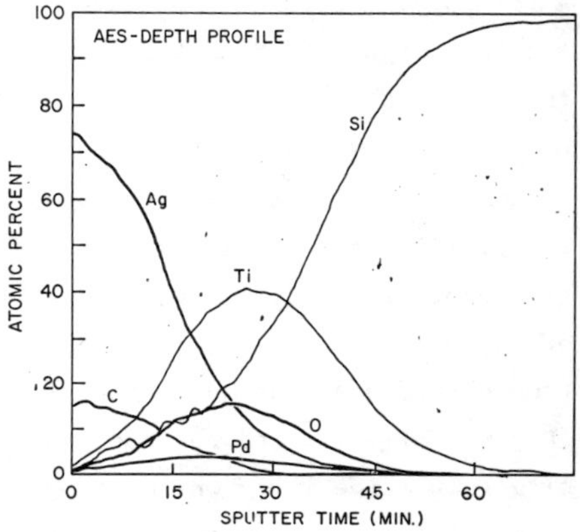

Figure 8

Solar Cell Auger Depth Profile

Note the oxygen relative to the titanium profile in which the O-line appears to track the Ti-line. No similar correlations are found with Ag, Pd, C or Si and it appears that the oxygen is combined with the Ti. This observation has been corroborated by other instrumental analytical techniques. Pores in the Ag layer provide access for oxygen bearing molecules to reach Ti layer as shown in Figure 9 on the next page.

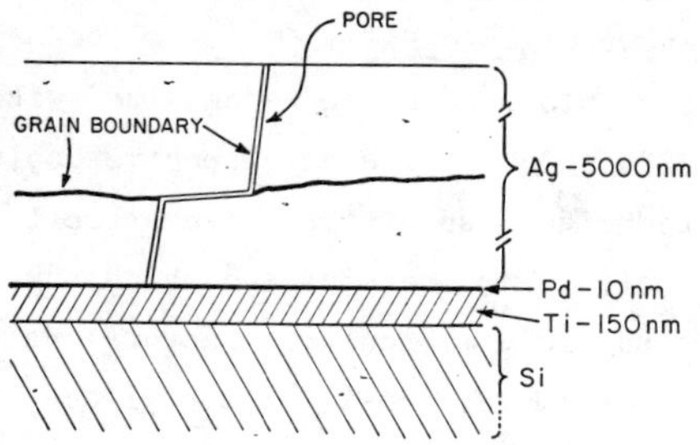

Figure 9
Diagrammatic Representation
of Pores in Ag Layer

There the oxidation will progress radially in the Ti boundary and longitudinally into the Ti layer with the highest degree of oxidation taking place at the pore site.

2.3 Electron Spectroscopy for Chemical Analysis (ESCA) or X-Ray Photoelectron Spectroscopy (XPS).

The previous analytical methods have utilized electrons or noble gas ions for bombarding surfaces. A unique technique, which produces information about composition and bonding within the first few atomic layers of surface, known as ESCA, employs x-rays as shown in Figure 10.

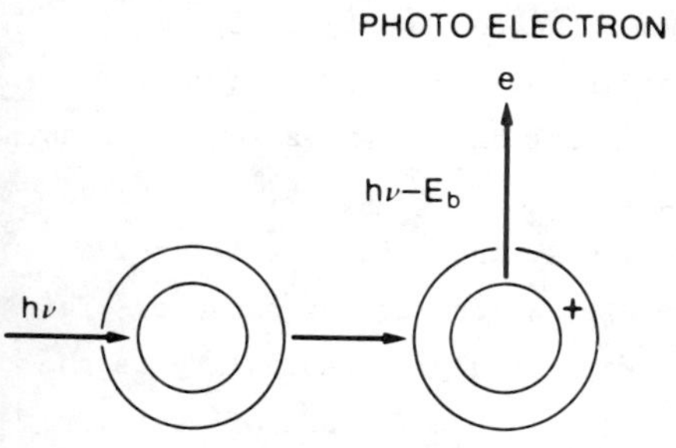

Figure 10
Photoelectron Emission Process

The hardware and data collecting for the ESCA is similar to the Auger except than an x-ray tube replaces the electron gun of the SAM. In ESCA, a specimen under vacuum is irradiated with monochromatic x-rays, emitting electrons from inner or core orbitals, with kinetic energies characteristic of the parent atoms as well as incident x-ray energy. The total absorbed x-ray photon energy ($h\nu$) is equal to kinetic energy (E_K) and electron binding energy (E_B) which is the work expended or amount of energy needed to remove an electron from a core level or inner orbital to the Fermi level in its removal from the atom.

$$h\nu = E_K + E_B$$

Knowing the total absorbed x-ray photom energy and measuring the kinetic energy with electron spectrometer allows the direct determination of the binding energy which is characteristic of each element. A spectrum is thus obtained with distinguishing peaks for each constituent element by plotting total number of ejected electrons versus kinetic energy. ESCA can thus detect elements above helium and analyze films down to a surface depth of 20A^{o}. ESCA is a most useful technique for the determination of chemical modification of surfaces. It has provided new insights into photo-oxidation and weathering of polymeric systems, such as paints and the effects on modification of adhesive properties.

2.3.1 Case of Differentiation Between the Metal and the Metal Oxide. In many instances, the corrosion scientist needs to differentiate chemical species in a thin (1-2 mm) surface film to identify the corrosion initiator and the nature of the first films formed in the corrosion process. It becomes a comparatively straight forward matter to distinguish the metallic form of a corrosion element from its lowest valency state with ESCA. Oxidation of most metals shifts spectral peaks to a higher binding energy. The metallic states of Cr, Mn, Fe, Co and Ni have photoelectron binding energies shifted 2-3 eV below the lowest oxide binding energies.(5) A rough rule of thumb in many cases is that a shift of 1 eV occurs per unit change in oxidation state. Aluminum shows this expected chemical shift of about 3 eV which is shown when comparing the ESCA spectra, in Figure 11, of FPL etched and unetched 2024 T3 aluminum.

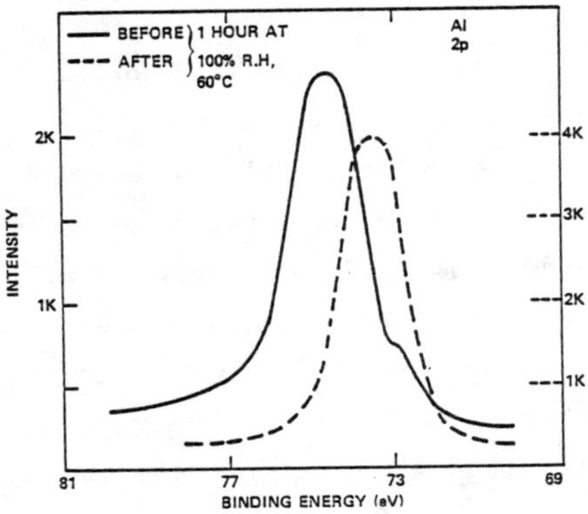

Figure 11

ESCA Spectrum of FPL Etched 2024 T3 Al

2.3.2 Case of Identifying an Unknown Organic Film. Chemical bonding information can be obtained with ESCA which is the only spectroscopic technique useful for non-destructive analysis of organics and which can also yield molecular structural data. For example, a film of methyl acetate shows three characteristic carbon peaks due to the different bonding as shown in Figure 12.

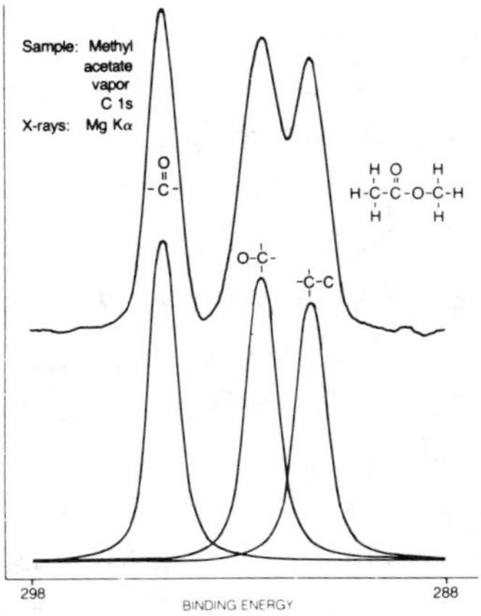

Figure 12

ESCA Spectrum of Methyl Acetate Vapor

In addition, since ESCA analyzes only the top twenty to fifty Angstroms of the sample, valuable chemical data is revealed about thin film organic coatings, surface additives, and molecular structure modifications at the surface. This surface analysis capability is the most valuable feature offered for polymer studies. The technique of identifying a thin organic film begins with a survey scan of a polymeric film such as Mylar in Figure thirteen.

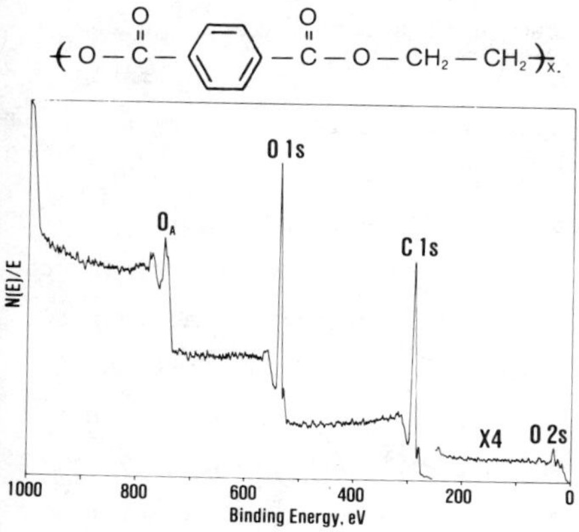

Figure 13
Survey Scan of Mylar

Note the O 1s photoelectron line occurs at about 532 eV and the C 1s line at about 285 eV. The high resolution data for both the C 1s and O 1s lines appear in Figure 14. and respectively with the various chemical state assignments designated above each peak.[7]

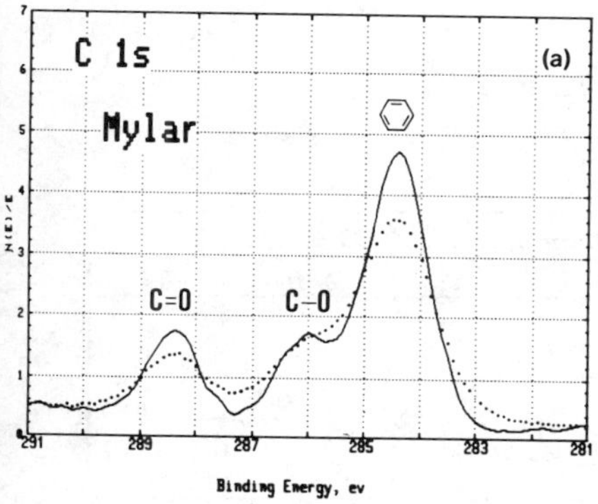

Figure 14
High Resolution Scan
of Mylar

ESCA has been shown to be capable of polymer identification using a collection of standard spectra on known polymeric materials.

<u>2.3.3 Case of Metal-Polymer Delamination</u>. A tin plated brass substrate coated with a polyester insulation suffered delamination. Investigation revealed the fact that bond failure occured in substrates that were exposed to air for 30 minutes or longer after tin plating without polyester coating. Samples of both good and bad material were delaminated and both sides subjected to ESCA analysis. The results from the good sample showed the polymer side of the interface was mostly C and O without detectable amounts of Sn, Cu or Zn from the plated brass. However, the ESCA spectra from the sample exhibiting poor bonding revealed not only C and O but, also detectable amounts of oxidized Zn on the polymer side of the interface. Clearly, the presence of Zn indicated a cohesive failure occurring in the metal surface and not an adhesive failure at the metal-to-polymer surface. To determine if the Zn was a surface contaminant or had diffused throughout the tin layer, depth-composition profiles were conducted. The bad sample contained ZnO in the tin. Additional studies indicated that Zn, accelerated by moisture and heat, will diffuse rapidly through the Sn plate and form an oxide. A weak

adhesion bond arises from the brittle fracture of the ZnO layer and lastly the polyester coating, if immediately applied after the Sn plating, retards the diffusion of Zn. For more thorough discussion of this failure mechanism see Reference (8).

2.3.4 Case of the Kinetic Contaminent. Conventional analytical techniques were unable to solve this problem concerning polybutadiene, a tacky polymer employed as as film adhesive and protected with a release material. Previously aluminum had been employed as the backing, but recently Mylar was being substituted because of its low cost, pliability and handling. However, it was observed that after about an hour had elapsed between the removal of the film from the Mylar backing and application of the polymer film to glass or even itself, bond failures occured. ESCA analysis of a pulled apart sample of polybutadiene and the Mylar revealed a silicone contaminent on the Mylar. Analysis showed that the amount of silicon on the polybutadiene increased from 0.5 atomic % when immediately separated to 1.5 atomic % after one hour as given in Table 1. The explanation for this kinetic situation can be explained by noting the polybutadiene prior to use was in contact with the Mylar. The contaminant then diffused from the Mylar surface into the polybutadiene. However, when the contaminated Mylar was removed from the polymer most of the diffused silicone migrated back to the polymer surface within an hour.

TABLE I

Polybutadiene Adhesion (9)

(Migration of Silicone to Surface)

Element (Atomic %)	Exposure Time (Hours) 0	0.5	1.0	17.0
Oxygen	1.5	2.9	3.5	3.4
Silicon	0.5	1.0	1.5	1.6
Carbon	97.8	96.1	95.0	95.0

3. MULTI-INSTRUMENTATION

As already shown, each technique of surface analysis has its own complex of capabilities, definitive application areas and limitations. Frequently, the information required cannot be provided in an optimum way by a single technique. As a result a "marriage" of convenience applying two or more techniques and utilizing designed instrumentation has taken place. A significant increase of synergism to real-world problems has resulted, making several types of data from the same surface. Equipment for these multiple data instrumentation techniques allow the investigator to perform several types of analyses in a shorter period on a specimen without "breaking" vacuum and precludes contamination. The SAM/ESCA combinations are invaluable for providing a comprehensive analysis. A primary reason for combining Auger spectroscopy with ESCA is to employ the chemical bonding analysis capability of ESCA with the SAM which provides compositional information on a point, a line or an area and as a function of depth. With the Auger a well-defined electron beam excites the sample with a small analytical area, usually less than a micron. In contrast, ESCA employs a broad diffure x-ray beam for excitation so that the analysis area is usually larger than 2 mm in diameter. Now the larger the surface area, the more limited the sputtering rate and as a result, Auger depth profiling is faster and is preferred for thick coatings (hundreds of A^{o} or micrometers). Utilizing the Auger for depth profiling of highly insulating materials may result in severe charging problems, making it difficult to obtain special data. In such cases, the use of ESCA profiles may be preferred. Unfortunately, combination design does not involve a mere joining of components. Auger and ESCA do both require an electron analyzer. However, an analyzer design for ESCA requires signal collection from a large uniformly illuminated sample area which may prove marginal in Auger where the signal originates from a point source. On the other hand, ESCA instrumentation utilizes a slow scan of the electron spectrum, while the Auger requires high speed. SAM/ESCA combinations for performing dual analyses are becoming popular[(10)] as shown in the PHI Model 550 in Figure 15.

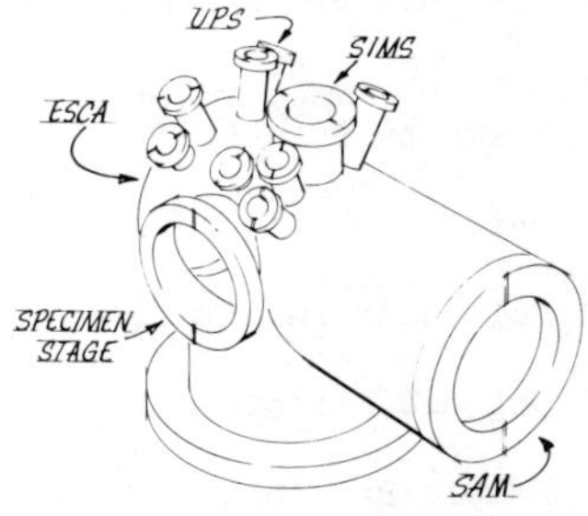

Figure 15

PHI Model 590 Test Chamber Drawing With Accessories & Other Components

4. SUMMARY

The failure analyst need not be perplexed in selecting a technique for surface failure analysis. It is true that the present available methods employ a variety of surface excitation means, viz., electrons, x-rays, and noble gas ions. The SEM's capabilities are more widely known having become a welcome extension of the optical microscope which allows for super magnification, great depth of field and surface element identification. However, if compositional information on a point a line or an area is desired as a function of depth, the SAM is the answer. On the other hand, if detailed information on chemical bonding be required, the usage of ESCA is unquestioned.

TABLE 2

COMPARATIVE SUMMARY

INFORMATION	AUGER	ESCA
Probing Particle	Electron	X-Ray, Photons
Particle Detected	Auger Electrons	Photoelectrons
Diameter (A^o)	5000 - 250,000	Complete
Depth (A^o)	10	10-30
Vacuum (Torr)	10^{-10}	10^{-8}
Sputter Rate (A^o/Min)	≥ 0.005 m	≥ 2 mm
Analysis vs. Sputter	Simultaneous or Sequential	Usually Sequential
Elements Detected	Except H & He	Except H
Detection Limits (g)	10^{-10}	10^{-10}
Molecular Information	No	Extensive
Organic Detection	No	Yes
Sample Problems	Charging	Size

5. REFERENCES

1. S.M. Lee, Electronic Pack. & Prod., 17, 72 (1977).
2. M.G. Yang, et. al., J. Electrochem. Soc., 122, 675 (1975).
3. S.M. Lee, IEEE Advanced Techniques of Failure Analysis, Nov. 6-9, 1978, Los Angeles, CA, pp. 1-8.
4. E.N. Sickafus, et. al., J. Thin Solid Films 78, 49 (1981).
5. C.O. Wagner, et. al., "Handbook of X-Ray Photoelectron Spectroscopy", Perkin-Elmer Corp., Eden Prairie, MN, 1979.
6. J.S. Noland, et. al., 1st SAMPE Intercontinental Conference, Jan. 12-14, 1981, Cannes, France.
7. Perkin-Elmer, "Introduction to Polymer Studies with ESCA", No. 7905, 8-30-79.
8. Perkin-Elmer, "Metal-Polymer Adhesion", No. 8001, 8-5-80.
9. M.V. Zeller & B.F. Phillips, Ind. Res. & Dev., March 1981, pp. 139-140.
10. S.M. Lee, "Film Deposition Techniques", "Encyclopedia of Chemical Technology", Vol. 10, p. 279, John Wiley & Sons, NY, 1980.

6. BIOGRAPHY

Dr. Stuart M. Lee, Senior Staff Scientist, Ford Aerospace & Communications Corporation, is concerned with the development and improvement of materials for advanced electronics used in satellite systems. Formerly Senior Technical Specialist at the Autonetics Division of Rockwell International Corporation and previously Manager of the Chemical Research and Development Department at Electro-Optical Systems, Inc. (Xerox Corporation), Senior Research Chemist and Project Leader at Allied Chemical Corporation and Aerojet-General Corporation. Dr. Lee received his Ph.D degree in Physical-Organic Chemistry from Florida State University as a Research Corporation Predoctoral Fellow. He is the Editor of the SAMPE Journal and a member of the Editorial Board of "The International Journal of Polymer Engineering" and authored fifty-five publications, including 10 U.S. and foreign patents, "Film Deposition Techniques" (Vol. 10) "Encyclopedia of Chemical Technology", "Epoxy Resins - Science and Technology" Chapter 9, Electrical and Electronic Applications.

13th National SAMPE Technical Conference
October 13-15, 1981

FPL ETCH SOLUTION LIFETIME

D.K. McNamara and J.D. Venables
Martin Marietta Laboratories
1450 South Rolling Road
Baltimore, Maryland 21227
and
R.L. Hopping and L. Cottrell
Martin Marietta Aerospace
Baltimore Division
Baltimore, Maryland 21220

Abstract

We examined the influence of etch solution usage on the surface morphology of aluminum prepared for adhesive bonding by the Forest Products Laboratory process to determine the effective lifetime of the solution. Solution chemistry was monitored for both the original and the optimized FPL etch solution formulations as a function of the level of dissolved aluminum. We used the amount of chemical component replenishment versus aluminum content to derive a model of the chemical reactions occurring during processing. With high-resolution scanning electron microscopy, we determined the effectiveness of the etch solution by monitoring the oxide morphology produced by it. Accelerated testing based on the reaction model produced good oxide morphologies to very high levels of solution "use." Comparison of controlled laboratory tests with in-plant production runs supports the solution modeling and establishes practical limits for solution usage.

1. INTRODUCTION

Although the Forest Products Laboratory (FPL) treatment[1] for preparing aluminum surfaces for adhesive bonding has been in use for many years, very little is known about the role played by the components of the "etching" solution and the influence of reaction by-products on the process. An "optimized" FPL process that improved bonding performance[2] was developed empirically without

Keywords: adhesive bonding, FPL etch, aluminum oxide morphology changing the "black-magic" understanding of the process. A major question that confronts users of either the original or the optimized solution is when they should replace the solution to avoid degradation in bonding performance. Present literature regarding solution monitoring(3) is very vague on this point.

Our previous work(4) has shown that an aluminum surface treated by the FPL process develops a dense, microscopic, aluminum-oxide "whisker" structure, as shown in Figure 1. The absence of this structure causes considerably reduced bond strength and durability. Monitoring the oxide morphology gives a good indication of the effectiveness of the surface preparation, allowing study of a wide range of solution variables. Using this diagnostic tool, we investigated the effect that increasing the solution age has on process effectiveness for both the original and optimized FPL solutions, on two aluminum substrates -- 2024 bare and 7075 clad with 7072 alloy.

2. PROCEDURE

2.1 Morphology Examination

We examined the surface morphology of test coupons with a JEOL 100 CX scanning transmission electron microscope operating in the high resolution scanning mode. After applying a ~ 5-nm Pt coating to prevent charging due to the electron beam, we took stereoscopic micrographs at 50,000x magnification to clearly reveal the morphology present.

2.2 Test Solutions

We prepared the original FPL solutions with 16.3% by volume of concentrated sulfuric acid and 30 g/l sodium dichromate dihydrate, and prepared the optimized solutions with 17.3% by volume of concentrated sulfuric acid and 60 g/l sodium dichromate dihydrate. We aged the solutions by dissolving 2024 aluminum sheets in them, then replenishing the acid and dichromate levels according to Methods A and B of ASTM standard D2574. We degassed the test coupons in MEK, alkaline-cleaned them with Turco 4215-S, and then etched them in the test solutions for 15 minutes at 65°C. Specific gravity and the amount of chemical replenishment were recorded for each solution; and elemental aluminum, chromium, and copper levels were measured with an atomic absorption spectrometer.

The amount of replenishment necessary for a given quantity of dissolved aluminum was determined for solutions "aged" with aluminum additions up to 0.9 wt% Al. We simulated levels of up to 2.0 wt% Al by adding the calculated reaction by-products directly to the starting solution. This avoided the necessity for multiple replenishments and long waiting times for complete dissolution of the aluminum.

2.3 Production Monitoring

To provide a data base for comparison, we collected replenishments and specific gravity records from an original and an optimized production FPL etch tank and took periodic samples for atomic absorption analysis. Values from T-peel, climbing-drum-peel, and wedge tests were compiled to correlate solution age and bonding effectiveness.

3. RESULTS AND DISCUSSION

Figure 2 presents data for sulfuric acid replenishment as a function of dissolved aluminum for four solutions. N-4 represents a production run with an optimized solution. N-3 corresponds to an original FPL formulation, while the beaker-test curve represents both optimized and original solution tests done in the laboratory. Figure 3 is a similar plot of chromium additions versus aluminum content. Since the data fall on a straight line, activity of the solution does not appear to change with age. Furthermore, since the lab test values from both solutions fall on one line, reactions undergone in the optimized solution are not fundamentally different from those in the original solution. The reasons for higher levels of replenishment in the production tanks will be discussed shortly.

The micrographs in Figure 1 offer a comparison of oxide morphologies prepared on 2024 aluminum in optimized solutions, one without dissolved aluminum and related by-products, and one with 0.9 wt% aluminum in solution. The two micrographs viewed in stereo have nearly indentical features. Study of all the micrographs, from both the original and optimized solutions and on both of the aluminum alloy substrates, showed no significant effect on morphology due to solution aging.

The slopes of the replenishment curves can be used, with appropriate factors for the number of molecules per unit weight or volume, to determine the molecular balance of the reactions occurring. For each seven H^+ that are oxidized, one Al atom is dissolved, and for every Cr^{+6} to Cr^{+3} conversion, one Al atom is involved. These ratios are identical to those found by Biljmer[5] who investigated a similar solution.

On the basis of these ratios, we propose that the following reactions represent the chemical steps occurring during FPL processing:

$$2Al + 4H_2SO_4 + Na_2Cr_2O_7 \rightarrow Al_2O_3 + Na_2SO_4 + Cr_2(SO_4)_3 + 4H_2O \quad (1)$$

$$Al_2O_3 + 3H_2SO_4 \rightarrow Al_2(SO_4)_3 + 3H_2O \quad (2)$$

Of course, most of the listed compounds are actually ions in solution.

This model, which involves the formation of Al_2O_3 in the first reaction and dissolution of the oxide in the second, can explain many puzzling features of the FPL process. For example, we observe a somewhat thicker and more dense oxide on specimens treated in optimized solutions, which we believe accounts for its improved bonding performance. With the higher sodium dichromate level used in the optimized solution, the first reaction can be expected to proceed more quickly, resulting in the greater amount of oxide left on the surface. Since the total thickness of the remaining oxide is < 60 nm and almost 2,000 nm of aluminum is removed during normal processing, small variations in oxide thickness would have a negligible effect on the chemical balance.

Furthermore, the distinctive whisker morphology can be explained as the result of the competition between the oxide formation and dissolution reactions. Pores formed in the initial oxide surface film, perhaps by local variations in chemical potential, extend into the aluminum substrate through the thin oxide at the pore bottom. Simultaneously, the oxide is dissolved away at the top, removing the thin walls connecting the pores and leaving the thicker regions at junctures of multiple walls, which become the whiskers.

Using the reaction model, test solutions could be quickly prepared to very high equivalent levels of dissolved aluminum by adding the appropriate reaction by-products to a fresh solution. This procedure does not change the level of active solution components, eliminating the need for their measurement and replenishment. This avoids a significant problem for achieving similarly high solution levels by dissolving aluminum sheet, since the active components would be completely exhausted long before the aluminum had completely dissolved.

Test coupons were prepared in optimized solutions "aged" up to 2.0 wt% aluminum using the model reaction compounds and ratios. The oxide surfaces continued to exhibit a normal FPL morphology, with no sign of degradation caused by the high level of reaction by-products. However, these highly aged solutions did develop considerable amounts of precipitate as they were cooled from operating to ambient temperatures. Figure 4 shows the large volume fraction of these solutions occupied by precipitate at room temperature.

This precipitation begins to occur in solutions with a specific gravity > 1.37, and can pose a number of problems. Removing the corrosive and toxic sludge from a large tank is much more difficult than simply pumping out the liquid. It can also explain why bonding problems might occur long before the oxide morphology appears to be affected. We believe the onset in precipitation is related to the "difficulty in rinsing off" the etch solution noted by production-line operators. As parts are removed from the hot etch solution, precipitation from the cooling remnants of solution dragout could occur before rinsing is completed, leaving behind a contaminated surface. The volume of contaminant required to obscure the microscopic structure and prevent the beneficial mechanical interlocking between the aluminum oxide and adhesive would be very small. Tests are presently underway using a small mock-up of the production line to simulate normal rinsing procedures and to investigate this hypothesis.

The differences found in solution replenishment rates for the production tanks compared to those of the laboratory tests must be explained to have confidence in the solution reaction model and in the dumping criteria established by the onset of precipitation. We noted that graphite block heaters used in the production tank were dissolving in the etch solution. Various amounts of carbon black were mixed with test solutions to determine whether this was the cause of the increased replenishment levels. Such a cause would explain different degrees of divergence from the laboratory tests, since the total replenishment would be a function of the length of time the material was exposed to carbon in the tank and the amount of aluminum processed, which would vary considerably with production schedules. From these tests, we determined a ratio between amount of acid and the dichromate replenishment necessary due to the carbon, which matched the ratio of the components added in excess of those expected for the amount of aluminum. This finding strongly supports the hypothesis.

Strength of bond values from the three production tests -- T-peel, climbing-drum-peel, and wedge test -- also showed no variation with solution age. Production etch solutions are generally replaced when their specific gravity reaches ~ 1.36, usually due to complaints by operators about poor rinsing. The good correspondence between laboratory tests and production experience further supports the solution reaction model.

4. CONCLUSIONS

Aging studies on standard and optimized FPL solutions indicate that no significant change in oxide morphology occurs up to very high levels of dissolved aluminum (~ 2 wt% Al). Long before this level is reached, however, we observed that reaction by-products precipitate from the solution as it is cooled to room temperature. Since this effect may cause rinsing problems, we conclude that the solution should not be used beyond the point at which its specific gravity reaches a value of 1.37, which is equivalent to 1.0 wt% dissolved Al. We also recommend that carbon not be used in contact with the solution (e.g., carbon block heaters) because the two react, causing accelerated aging of the solution.

5. REFERENCES

1. H.W. Eickner and N.E. Schowalter, "A Study of Methods for Preparing Clad 24S-T3 Aluminum Alloy Sheet Surface for Adhesive Bonding," Forest Products Laboratory Report No. 1813, May 1950.
2. R.H. Olsen, "Implementation of Phosphoric Acid Anodizing at Air Force Air Logistic Centers," 11th National SAMPE Technical Conference Proceedings, Boston, Massachusetts, November 1979, p. 770.
3. ASTM Standard D-2674-72 (1979) "Sulfochromate Etch Solution Used in Surface Preparation of Aluminum, Analysis of," Part 22, Annual Book of Standards, 1980, p. 791.
4. J.D. Venables, D.K. McNamara, J.M. Chen, T.S. Sun, and R.L. Hopping, "Oxide Morphologies on Aluminum Prepared for Adhesive Bonding," Appl. of Surf. Sci., 1979, Vol. 3, No. 1, pp. 88-98.
5. P.F.A. Biljmer, "Pickling Aluminum Alloys for Adhesive Bonding," Metal Finishing, 1971, Vol. 69, No. 12, p. 34.

6. BIOGRAPHIES

David K. McNamara is a research associate in the Materials and Surface Science Department at Martin Marietta Laboratories. He obtained a B.A. degree in physics from Ohio Wesleyan University in 1969. After working on instrumentation and systems design at Mound Laboratories and Brookhaven National Laboratories, he joined Martin Marietta in 1973 where he has studied fabrication techniques and properties of silicon nitride ceramics as well as the role of surface preparations in adhesive bonding.

John D. Venables is the Manager of the Materials and Surface Science Department at Martin Marietta Laboratories. He has been with the Laboratories since 1964, and is responsible for studying properties of the transition metal carbides, nitrogen ceramics, and adhesive bonded structures. Dr. Venables was a Research Fellow at the University of Warwick in 1968-1969 and obtained his Ph.D. degree there in 1971. He has been a consultant on the Viking Mars Lander program and the Space Shuttle program.

Russell L. Hopping obtained the degree of Mechanical Engineer at Cornell University in 1940. He is a registered Professional Engineer in Maryland and Colorado. Mr. Hopping worked at the Baltimore and Denver divisions of Martin Marietta Aerospace from 1946 until his retirement in May 1981. He held various supervisory positions in the Matador and other early missile programs, in solid propellants and warheads, in nuclear power systems and space systems, in lunar and earth orbital space programs, and in Titan launch vehicle and Skylab programs. Most recently, he was in charge of the bonding and surface-preparation studies at the Baltimore division.

Lee Cottrell is an Associate Manufacturing Engineer in the Advanced Manufacturing Research and Technology department at the Baltimore division of Martin Marietta Aerospace. He is presently responsible for surface preparation aspects of adhesive bonding research, and investigates metal finishing and organic coating. He received his B.S. degree in Chemistry from the University of Maryland at College Park in 1980.

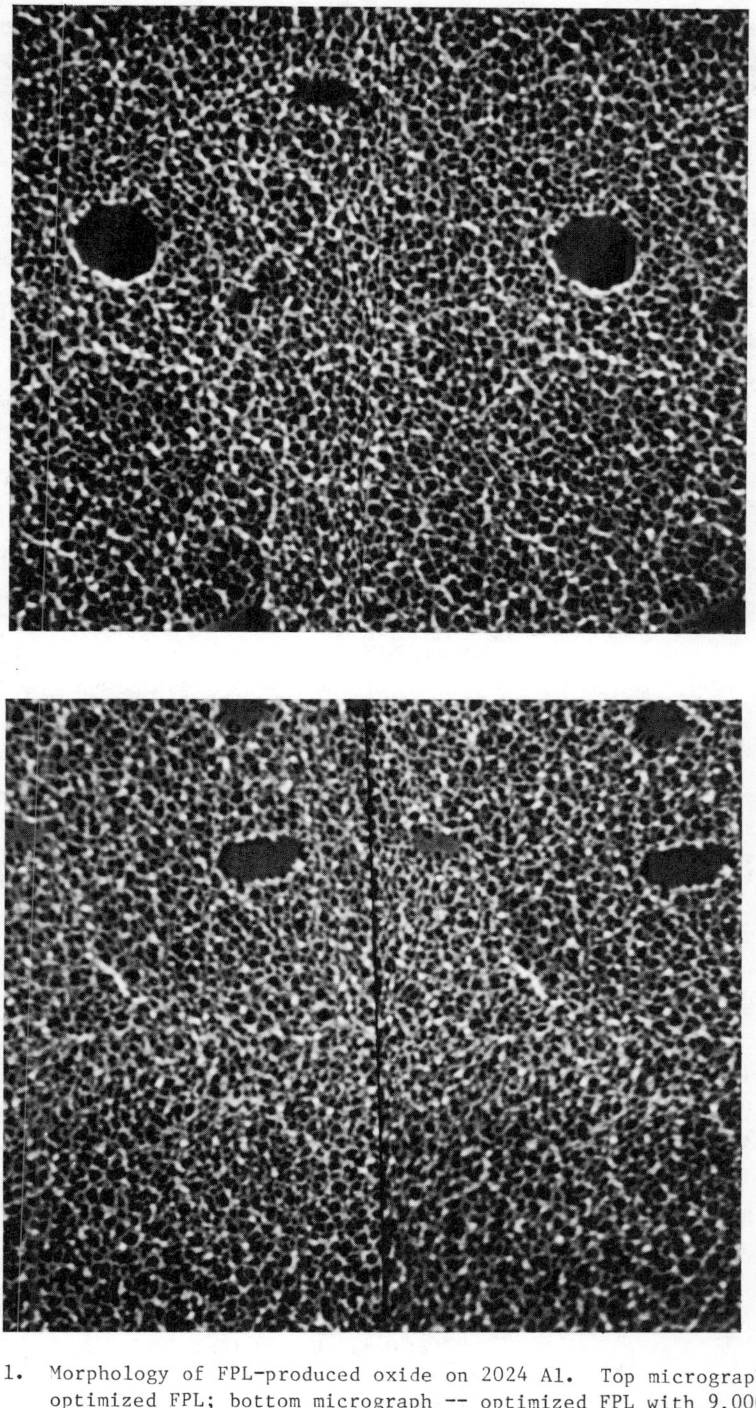

Figure 1. Morphology of FPL-produced oxide on 2024 Al. Top micrograph -- optimized FPL; bottom micrograph -- optimized FPL with 9,000 ppm of Al added. No significant change occurs from using an aged test solution.

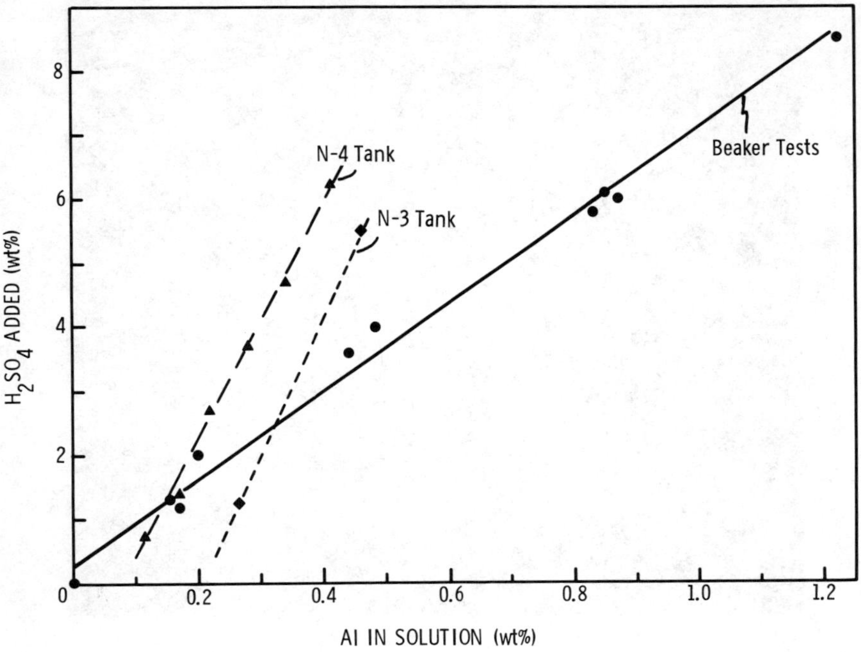

Figure 2. Sulfuric acid additions required in FPL solutions with increasing Al content.

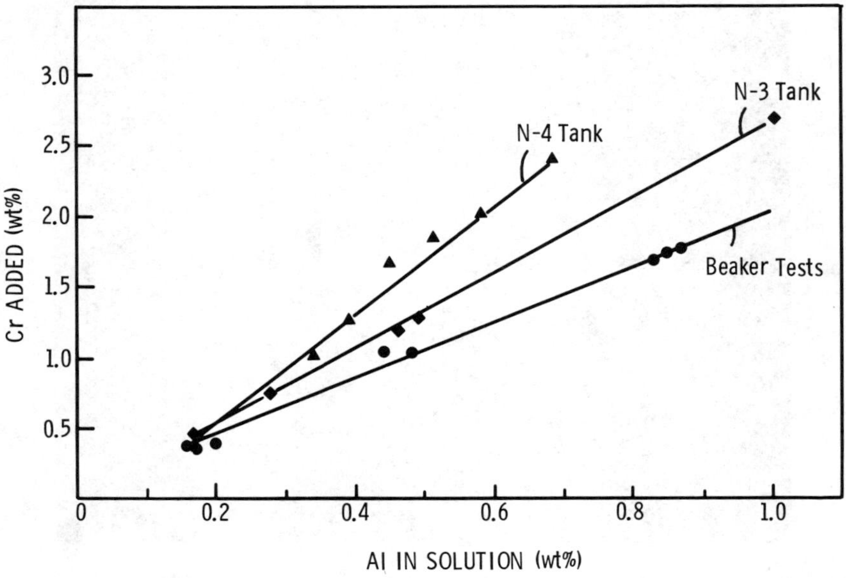

Figure 3. Cr^{+6} additions (as sodium dichromate) required in FPL solutions with increasing Al content.

Figure 4. Room-temperature precipitation from FPL solutions of different aluminum contents. Bottom row -- precipitate; top row -- supernatant liquid.

13th National SAMPE Technical Conference
October 13-15, 1981

EFFECT OF RINSE WATER PH ON ADHESIVE BOND DURABILITY

S. P. Kodali, R. C. Curley. L. Cottrell
Advanced Manufacturing Research and Technology
Martin Marietta Aerospace
Baltimore Division
103 Chesapeake Park Plaza
Baltimore, Maryland 21220
and
B. M. Ditchek, D. K. McNamara, J. D. Venables
Martin Marietta Laboratories
1450 South Rolling Road
Baltimore, Maryland 21227

Abstract

During processing of Al alloys for adhesive bonding, etch solution drag over from the etch tank may accumulate in the immersion rinse tank and lower the pH of the rinse water. The effect of variations of rinse tank parameters on surface oxide morphology and bond durability was determined. The variables considered were pH of rinse water, length of time immersed in the bath, spray rinse after immersion, and closed or open electrical circuit between the test sample and the stainless steel rinse tank. The rinse water pH was varied by adding appropriate amounts of etch solution. It was found that for the range of variables encountered in typical production practice there is no effect of rinse water pH on the surface morphology of aluminum adherends or bond durability. A marked effect on surface morphology was found with long immersion times in low pH solutions. It was shown that long rinse times in low pH water dissolved part of the oxide, leaving a smooth surface that may have poor bonding properties.

1. INTRODUCTION

Prior to adhesive bonding, aluminum alloys, in general, are processed through a series of cleaning steps such as solvent wiping, vapor degreasing and alkaline cleaning before they are Forest Product Laboratory (FPL) etched (dipped in a mixture of sodium dichromate, sulfuric acid, and water).

During the FPL etching, it is recognized [1] that previously formed oxide on aluminum surfaces dissolves and a controlled oxide forms. The etching reactions are postulated elsewhere.[2] In a manufacturing environment, complex formed parts and sheet metal details are cleaned and etched by dipping racked parts in the appropriate solutions. After the parts are etched, they are rinsed in deionized water by immersing the racks in a tank. Afterwards, they are spray rinsed and dried in an oven.

During the FPL etch process, etch solution drag over from the etch tank may accumulate in the immersion rinse and lower pH of the water. The purpose of the present study was to determine the effect of rinse water pH on adhesive bond durability.

2. Experimental Procedure

To determine the effect of rinse water pH on wedge test durability and surface morphology, the following test matrices were followed. Table I was followed for preparing wedge tests and Table II was followed for preparing morphology specimens.

2024-T3 clad panels (6" x 6" x 0.125") and strips for surface analysis were processed through an FPL cleaning line. Specimens were processed using the following sequence of steps; solvent (MEK) wipe, alkaline clean (Turco 4215S), cold water immersion rinse, etching in an "optimized" FPL etch [3] solution for 15 minutes at 155 $\pm$ 2°F and then rinsing. The etch solution concentration was 22.1 fl.oz/gal of H_2SO_4 and 8.0 oz/gal of $Na_2Cr_2O_7$. The immersion rinse water pH was varied from 3.0 to 2.0 by adding appropriate amounts of etch solution to the immersion rinse tank. The immersion rinse time was varied from 3 to 10 minutes and the spray rinse time was varied from 10 to 3 minutes to yield a constant rinse time of 13 minutes for the wedge test panels. However, for the strips from which surface morphology and chemical composition were to be determined, the immersion time varied from 3 to 60 minutes with and without a constant spray rinse time of 3 minutes. After the final rinse all panels and strips were dried in an air circulating oven at 140° $\pm$ 10°F for 15 minutes. For each pH and immersion time, three separate runs of wedge test panels were made.

All the wedge test panels were primed with BR-127 adhesive primer, manufactured by American Cyanamid Company, to yield a dry film thickness of 0.0003 $\pm$ 0.0001". Then the wedge test panels were bonded with FM 123-2 adhesive film, product of American Cyanamid Company, at 80-100 psi and 250° $\pm$ 10°F. for approximately 1 hour. The panels were cut into 1" x 6" wedge test specimens, wedged and exposed to 140°F. and 100% relative humidity for 1 hour. After noting the crack extensions, the wedge test specimens were split open to observe the failure mode in the crack extension area.

Electron micrographs were taken of the strips using the high resolution mode of a JEOL 100 CX Scanning Transmission Electron Microscope (STEM). All samples were coated with a thin (~100 Å) layer of platinum to prevent sample charging during examination. The chemical composition of surfaces was determined by Electron Spectroscopy for Chemical Analysis (ESCA) using a PHI Model 548 electron spectrometer.

3. Results and Discussion

Surface preparation for adhesive bonding is often evaluated by means of the wedge test,[4] a test method where the propagation under controlled conditions of stress, humidity and temperature of a crack introduced into a bond line is measured. A well prepared surface will exhibit crack extensions in the range of 0.1 to 0.2 inches. Crack extension in excess of 1.0 inch have been correlated with in service failure of aircraft bondments[4]. An arbitrary criterion of less than 0.75 inch crack extension is often used to distinguish between satisfactory and unsatisfactory FPL etched surfaces.

The average crack extension values after 1 hour at 140°F and 100% relative humidity from fifteen specimens at various pH values and immersion times are given in Table III. It can be seen that all the crack extension values were lower than 0.2 inch. After splitting the wedge test samples open, the failure mode was found to be always cohesive in the crack extension area.

An assessment of etched surface bonding quality based on surface morphology for various rinse water pH values and immersion times are given in Table IV.

Figure 1, shows the surface morphology of a sample immersed for 5 minutes in rinse water at a pH of 2.25. Figure 2 shows the surface morphology of a sample immersed for 15 minutes in rinse water at a pH of 2.75. The surface morphologies shown in Figure 1 and 2 are quite similar to that which is obtained with just a pure deionized water spray rinse. In a production environment the normal variation in the rinse water pH has been found to be from 5.5 to about 2.5. Of course, this depends on the number of racks processed and the volume of the overflow in the immersion tank. For the 5 - 10 minute rinse time normally used in production, surface morphology analyses and wedge test bond durability measurements indicate that there is no effect on quality of surface preparation from rinse water pH in the range 5.5 to 2.5.

However, for long immersion times in low pH rinse water, the surface morphology was degraded by the dissolution of the oxide as evidenced by Figure 3. This dissolution of oxide leaves the surface smooth which is thought to have poor bonding properties.

High resolution ESCA analyses of the surfaces were conducted to determine the degree of hydration of the surface by monitoring the chemical shifts and peak heights of oxygen and aluminum. Pure Al_2O_3 should have an O/Al concentration ratio of 1.5, while hydrated surfaces should have ratios greater than 2. In this study, values for this ratio ranged from 1.6 to 1.9, probably reflecting water or CO_2 absorbed on the surface. Since no evidence of the distinctive "cornflake" structure found on hydrated surfaces (5) was evidenced in the STEM and since ESCA values do not indicate uniform hydration on the surface, the altered morphology does appear to be due to dissolution rather than transformation.

When similar tests were performed omitting the final spray rinse, surface contamination was observed by ESCA. Above pH 2.75 very little contamination was seen when the rinse solution was allowed to dry on the surface, but the majority of the surface was covered with a residue of etch solution components at pH 2.25. In situations where the effectiveness of the final spray rinse may be minimal, low pH of the immersion rinse tank may affect the bond durability.

Since the rinse tank walls are generally made out of stainless steel, it was thought the electrical contact between the sample and the tank walls might allow a flow of current that would encourage chemical reactions. While electrical potential between the sample and the tank walls did change with the pH of rinse water, no effect on morphology was seen by closing the circuit.

4. CONCLUSIONS

- For the actual immersion times involved in production practice, there seems to be no detrimental effect of rinse water pH on the durability of adhesive bonds.
- Morphology and chemical composition of surfaces correlate well with durability results.
- Long immersion times in low pH rinse solutions degrades the morphology of FPL etched surfaces.
- Effective final spray rinse seems to be important to obtain surfaces without contamination for good adhesive bond durability.

5. REFERENCES

1. J. D. Venables, D. K. McNamara, J. M. Chen, T. S. Sun and R. L. Hopping, "Oxide Morphologies on Aluminum Prepared for Adhesive Bonding", Applications of Surface Science, 1979, Vol. 3, No. 1, pp.88-98.

2. D. K. McNamara, J. D. Venables, R. L. Hopping and L. Cottrell, "FPL Etch Solution Lifetime", to be presented at the 13th National

SAMPE Technical Coference, October 13-15,1981, Mt.Ponoco, PA.

3. R. H. Olsen, "Implementation of Phosphoric Acid Anodizing at Air Force Logistic Centers", Proc. 11th National SAMPE Technical Conference, November 1979, p.770.

4. J. A. Marceau, Y. Moji, J. C. McMillan, "A Wedge Test for Evaluating Adhesive Bonded Surface Durability", Proc. 21st National SAMPE Symposium, April 1976, pp. 332-254.

5. J. D. Venables, D. K. McNamara, J. M. Chen, B. M. Ditchek, T. I. Morgenthaler, T. S. Sun and R. L. Hopping, "Effect of Moisture on Adhesively Bonded Aluminum Structures", Proc. 12th National SAMPE Technical Conference, 1980, pp. 909-923.

6. BIOGRAPHIES

Surya P. Kodali is the Manager of Research and Development at the Baltimore Division. He has been with Baltimore Division since 1978 and is responsible for the planning and implementation of research and development programs. Dr. Kodali obtained his Ph.D. degree in Engineering Science (Physical Metallurgy) from Washington State University in 1977. Since joining Baltimore Division, he has been actively involved in bonding technology research along with advanced composites and welding technologies.

Robert C. Curley is the Director of the Advanced Manufacturing Research and Technology Department at the Baltimore Division of Martin Marietta Aerospace. He obtained a B. S. in Applied Chemistry from the California Institute of Technology in 1957. He has extensive experience in the application of composites and adhesive bonding to aircraft and missile structure. Since joining Martin Marietta in 1980, he has been in charge of the development of materials, processing and manufacturing to support current and future programs.

Lee Cottrell is an Associate Manufacturing Engineer in the Advanced Manufacturing Research and Technology Department at the Baltimore Division of Martin Marietta Aerospace. He is presently responsible for surface preparation aspects of adhesive bonding research, and investigates metal finishing and organic coatings. He received his B. S. degree in Chemistry from the University of Maryland at College Park in 1980.

Brian M. Ditchek obtained his Ph.D. degree at Northwestern University in 1977. He was a National Research Council Postdoctoral Associate at the National Bureau of Standards from 1977 to 1978 where he studied fatigue hardening of copper single crystals. Dr. Ditchek joined Martin Marietta Laboratories in 1978 and has since engaged in research on new high performance silicon nitride ceramics and the bondability of Al and Ti alloys.

David K. McNamara is a research associate in the Materials and Surface Science Department at Martin Marietta Laboratories. He obtained a B. A. degree in Physics from Ohio Wesleyan University in 1969. After working on instrumentation and systems design at Mound Laboratories and Brookhaven National Laboratories, he joined Martin Marietta in 1973 where he has studied fabrication techniques and properties of silicon nitride ceramics as well as the role of surface preparations in adhesive bonding.

John D. Venables is the Manager of the Materials and Surface Science Department at Martin Marietta Laboratories. He has been with the Laboratories since 1964, and is responsible for studying properties of the transition metal carbides, nitrogen ceramics, and adhesive bonded structures. Dr. Venables was a Research Fellow at the University of Warwick in 1968-1969 and obtained his Ph.D. degree there in 1971. He has been a consultant on the Viking Mars Lander program and the Space shuttle program.

Table I. Test Matrix for Wedge Test

Immersion Time	Rinse Time	pH 2.00	pH 2.25	pH 2.50	pH 2.75	pH 3.00
3 min	10 min	x	x	x	x	x
5 min	8 min	x	x	x	x	x
10 min	3 min	x	x	x	x	x

Table II. Matrix for Morphology Samples

Immersion Time	Rinse w/D Water	Circuit Closed	ESCA	pH 2.00	pH 2.25	pH 2.50	pH 2.75	pH 3.00
3 min	●	▲	▲	x	x	x	x	x
	▲	▲	▲	x	x	x	x	x
5 min	●	▲	▲	x	x	x	x	x
	▲	▲	▲	x	x	x	x	x
15 min	●	●	▲	x	x	x	x	x
	●	▲	▲	x	x	x	x	x
	▲	▲	●	x	x	x	x	x
30 min	●	▲	▲	x	x	x	x	x
	▲	▲	▲	x	x	x	x	x
60 min	●	▲	▲	x	x	x	x	x
	▲	▲	▲	x	x	x	x	x

Legend: ● Yes
▲ No

Table III. Average wedge test crack extensions in 1 hour at 140° F and 100% RH for various pH values and immersion times

Number	pH of Rinse Water	Immersion Rinse Time (min)	Spray Rinse Time (min)	Average 1 Hour Crack Extension (in.)
1	3.00	3	10	.10
2	3.00	5	8	.06
3	3.00	10	3	.06
4	2.75	3	10	.08
5	2.75	5	8	.08
6	2.75	10	3	.09
7	2.50	3	10	.05
8	2.50	5	8	.04
9	2.50	10	3	.06
10	2.25	3	10	.07
11	2.25	5	8	.11
12	2.25	10	3	.09
13	2.00	3	10	.09
14	2.00	5	8	.07
15	2.00	10	3	.09

Table IV. Assessments of Bonding Quality of Surface Morphology

		Immersion Rinse Solution pH				
		2.0	2.25	2.5	2.75	3.0
	3	Good				
Immersion	5	Good	Good			
Time	15	Fair	Good	Good	Good	Good
(min)	30	Poor	Fair	Fair	Fair	Good
	60	Poor	Poor	Poor	Poor	Fair

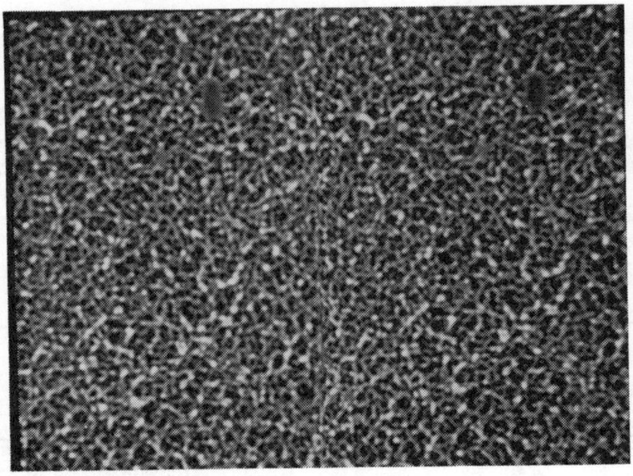

Figure 1. Stereomicrograph of 2024-T3 clad surface that was etched and immersed for 5 minutes in a rinse water pH of 2.25. 50,000x.

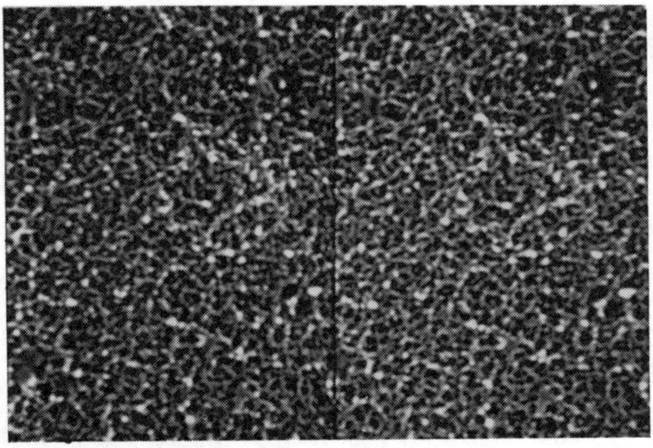

Figure 2. Stereomicrograph of 2024-T3 clad surface that was etched and immersed for 15 minutes in a rinse water pH of 2.75. 50,000x.

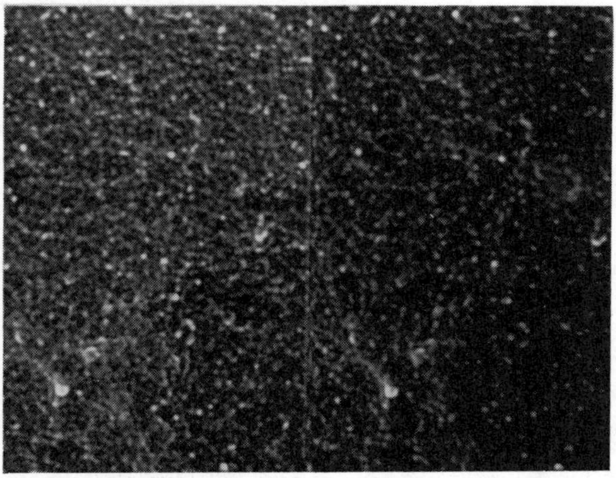

Figure 3. Stereomigraph of 2024-T3 clad surface that was etched and immersed for 60 minutes in a rinse water pH of 2.00. 50,000x.

13th National SAMPE Technical Conference
October 13-15, 1981

EFFECT OF WEATHERING ON THE MECHANICAL PROPERTIES OF SHEET MOLDING COMPOUNDS

Paul McElroy, Ronald Allred and David Roylance
Massachusetts Institute of Technology
Cambridge, Massachusetts 02139

Abstract

Sheet molding compounds are highly filled, thickenable, short glass fiber reinforced, polyester matrix composites. These systems offer economic and property advantages in a wide variety of structural applications, including support structures for solar energy collection systems. In this application, the long-term weatherability of the material is of great concern. This paper will discuss the effect of outdoor weathering on engineering mechanical properties, notably the flexural stiffness and strength and the dynamic mechanical spectrum. Results are presented for a graded series of sheet molding compounds of differing weatherability, and chemical mechanisms responsible for the observed weathering are proposed.

1. INTRODUCTION

Sheet molding compounds (SMC) form a class of relatively low cost, versatile composite materials. These considerations have led to their use in many diverse applications including automotive, electrical and sporting goods. Similar requirements have also shown SMC to be potentially attractive in energy applications such as focusing solar collectors(1,2). Temperature-humidity cycling of sub-scale SMC parabolic collectors has shown that SMC has the necessary dimensional stability for focusing applications(3); however, before these materials may be considered for large scale use, their resistance to outdoor weathering needs to be investigated.

Some SMC formulations have been shown to exhibit surface degradation and eventual reduction of mechanical properties upon exposure to chemical(4) and outdoor weathering(5,6) environments. It is expected that the environmental stability of SMC can be altered by variations of the base resin chemistry and choice of inorganic filler materials. This paper presents the results of an investigation to determine the effects of outdoor weathering and laboratory ultraviolet (UV) exposures on the mechanical response of five different formulations of SMC. Tensile, flexural, and dynamic mechanical properties are presented for exposures up to 18 months. Moisture absorption and swelling results are also presented.

2. MATERIALS

Sheet molding compound is a combination of glass fiber reinforcements dispersed in a thermosetting resin paste. The resin paste is usually an unsaturated polyester or vinyl ester resin modified with fillers, chemical thickening agents and other materials. These are mixed in a formulation that provides desirable processing and molding characteristics with a good balance of physical and mechanical properties.

Laboratory-fabricated SMC plaques have been formulated for differing degrees of weatherability. Two polyester resins and one vinyl ester resin were used as the basis to make five SMC materials for exposure tests. A general purpose polyester resin containing orthophthalic and maleic anhydrides with diethylene and propylene glycols in a molar ratio of 1.5:1.0:0.32:2.18 was used for one set of unstabilized SMC plaques. SMC based upon a special purpose polyester resin containing a 1.0:1.0:2.0 molar ratio of isophthalic and maleic anhydrides and neopentyl glycol was formulated in an unstabilized resin mixture and two ultraviolet stabilized resin mixtures.

The UV stabilized SMC composites contain 0.1% and 0.2%, respectively, of 2(2-hydroxy-5-T-octylphenyl) benzotriazole. The general purpose polyester resin contains styrene monomer and the special purpose resin contains styrene monomer plus 10% methyl methacrylate monomer. The vinyl ester resin is made of bisphenol-A epoxy, terminated with methacrylic acid and crosslinked with styrene. All of the SMC composites are low profile materials containing polyvinyl acetate resin in the formulations. Each SMC plaque contains proportioned calcium carbonate filler, zinc stearate internal mold release agent, and magnesium oxide thickener. Chopped random glass fibers are added at a level of 25% by weight of the total composite.

Moisture absorption measurements were made on two commercially available SMC systems designated 9220-30 and 9640-50 by Haveg Industries, Sate Fe Springs, Ca. The exact formulations of these materials are propriatary to Haveg, but are believed to be orthophthalic anhydride base resins with polyvinyl acetate low shrink additive, $CaCO_3$ filler and similar thickener, release agent, catalyst, etc. to the model compounds discussed above. The 9220-30 may have a flexibilized polyester system. The 9640-50 is partially filled with glass microballoons to reduce density.

3. WEATHERING

SMC plaques were exposed to outdoor

weathering in the Boston area on a 45° south facing rack, and to laboratory UV exposure using a S-1 mercury sunlamp. All of the SMC plaques under investigation show varying degrees of discoloration, microcracking, fiber bloom and erosion from outdoor exposure.

UV-induced or hydrolytic bond cleavage, crosslinking and plasticization are the potential weathering mechanisms which can be monitored using mechanical property analyses[5,7]. These deterioration processes are sometimes competing mechanisms and can be surface or bulk phenomena. SMC has a resin rich surface when molded, yet the highly filled nature of this material results in a filler-rich bulk matrix.

Ultraviolet radiation primarily affects the resin in SMC by causing either chain scission or increased crosslinking. Moisture absorption also primarily affects the resin matrix where it interrupts interchain hydrogen bonds and acts as a plasticizer. Moisture plasticization effects are generally reversible[8]; however, in SMC, a synergistic interaction can be envisioned with combined UV and moisture. Such a mechanism would entail enhanced moisture absorption through a UV degraded surface layer, followed by hydrolysis of calcium carbonate to calcium hydroxide and subsequent separation of filler particles from the resin matrix which would lead to increased rates of surface erosion. Because weathering effects are primarily matrix dominated, the glass transition temperature (Tg) of the SMC base resin could possibly be used as an indicator of which deterioration process is operative during exposure.

4. MECHANICAL AND THERMOMECHANICAL PROPERTIES

Physical and mechanical property evaluation of SMC provides a basis for examination of material of different formulations. Weathering would be expected to affect some property measurements more than others. Flexural properties of a free standing composite SMC structure are important when wind loads are expected[1]. Surface deterioration is most detrimental in SMC when the structure is subjected to bending or torsional loads during exposure. Cracks that develop in a weakened surface under tensile stress enhance water absorbtion into the bulk of the SMC composite, possibly resulting in delamination of fiber reinforcements and filler erosion from the bulk.

Figure 1 shows how the tensile and flexural moduli of the general purpose polyester resin based SMC deteriorated over an 18 month outdoor weathering period. The flexural modulus is seen to be markedly more sensitive to outdoor degradation than the tensile modulus. This is a general trend among all of the SMC formulations under investigation, and similar results, obtained for the vinyl ester resin based SMC, appear in Figure 2. The exposed surface of the SMC flexural specimens was placed in tension during testing and the effects of photo-initiated chemical change and physical deterioration of the surface are more pronounced than in the tensile tests.

The data in Figures 1 and 2 represent the mean values obtained for the mechanical properties using 10 to 25 test replications. The decrease in tensile modulus from initial values after 18 months of outdoor exposure was 16.4% for the special purpose polyester resin SMC with 0.2% UV stabilizer and 31.7% for the unstabilized vinyl ester SMC. The other SMC formulations show decreases in tensile modulus intermediate between these two values.

A statistical evaluation of the mechanical properties of the SMC formulations reveals a high degree of variability about the mean values. Coefficients of variation of 15-20 percent in strength are common for as-molded SMC composites. Figure 3 presents a probability plot of the changes in flexural strength of vinyl ester resin SMC from outdoor exposure. It can be seen that the mean strength value decreases progressively at each 6 month testing interval. The coefficient of variation for the specimens tested after 12 months of exposure is considerably less than the other sets of data and shows up on the plot as a steeper slope. This trend is consistent among all of the SMC materials. It was noticed visually that degradation of the plaques removed at 12 months was more uniform than those tested after 6 months outdoors. The mean flexural strength in Figure 3 has decreased 63.3% from its initial value after 18 months outdoors. The general purpose polyester SMC showed a decrease of 54.4% and the special purpose polyester SMC containing 0.2% UV stabilizer decreased 40.9% in flexural strength over the same exposure period.

Thermomechanical tests of dynamic properties are a valuable means of monitoring surface degradation. Torsional pendulum analysis (TPA) measures specimens in low frequency torsional oscillation (approximately 1 Hz) and is sensitive to surface changes in weathered SMC because the shear stress is maximum there. Weathering of polymers often changes the character of the log decrement curve, which is a measure of energy dissipation of the material. The glass transition temperature occurs near the maximum energy loss and will shift to a lower temperature with a sufficient amount of chain scission or plasticization and increase to a higher temperature if photoinduced crosslinking dominates. Figure 4 is a plot of the dynamic mechanical response of vinyl ester

SMC after different outdoor exposure periods. The glass transition temperature is seen to decrease approximately 6°C, a weak indication of chain scission, and the log decrement curve is seen to become increasingly broader with increasing exposure. This could be due to the combined dynamic mechanical response of low molecular weight fractions formed in the resin during exposure or could represent an interfacial dissipation between fibers, fillers and resin matrix[9,10]. This effect is not seen in the dynamic mechanical behavior of the neat resins used to formulate the different SMC's.

The dynamic mechanical behavior of the general purpose polyester resin SMC appears in Figure 5. After 18 months of outdoor exposure, Tg broadens to a slightly higher temperature, indicating that cross-linking dominates during outdoor exposure of general purpose SMC. The peak at 45°C is the Tg of polyvinyl acetate low profile resin.

5. MOISTURE ABSORPTION

Moisture absorption and moisture-induced swelling behavior were investigated to aid in interpretation of the weathering data. Three specimens of the 9220-30 and 9640-50 SMC compositions obtained from Haveg Industries were conditioned in each of six constant relative humidity environments. Moisture absorption was monitored by weight gain measurements. Swelling was determined from micrometer measurements at a given moisture weight gain. Constant relative humidities of 27, 54, 67, 76, 96 and 100 percent were maintained over salt solutions or deionized water held at 24°C. Specimens 50 mm square were ground from 300 mm square molded flat panels. The as-molded thickness of 3 mm was maintained.

Moisture absorption occurred rapidly in all specimens. At relative humidities less than 96 percent, 85 percent of equilibrium moisture content was reached in 300 hours. Moisture content slowly increased to an apparent equilibrium value at 4000 hours and remained virtually constant through 24,500 hours of exposure. These results indicate that the SMC formulations tested have an interpenetrating void network which allows most of the matrix access to moisture without diffusion over long distances. An interpenetrating network of voids is expected in these materials due to the high thermal expansion of the thermoplastic low-shrink additives[11].

Specimens exposed at 96 and 100 percent humidities did not level off after 300 hours, but rather continued to pick up moisture throughout 24,500 hour exposure period. A plot of equilibrium moisture content (or 24,500 hour weight gain) versus relative humidity is given in Figure 6. The increased absorption at the higher humidities is readily apparent. A similar result may be seen in the data reported in Ref. 12.

Swelling behavior did not follow the trend shown in Figure 6 for absorbed moisture. Measured values of moisture swelling coefficients β(% strain/% moisture absorbed), were relatively constant over the humidity range examined up to 76 percent. Beta values of 2 and 0.05 were determined for the thickness and in-plane directions of the 9220-30 specimens up to 76 percent humidity. Values for the 9640-50 were 1 and 0.05. The higher transverse moisture expansion of the 9220-30 may indicate that it has a more flexible resin system.

Specimens exposed to 96 and 100 percent humidity showed much lower values of through-the-thickness β-0.3 for the 9220-30 and 0.4 for the 9640-50. The observed drop in moisture expansion for the specimens showing high moisture is not entering the matrix phase which would appear as increased swelling. Examination of the specimen surfaces with an optical microscope revealed an extensive network of microcracks on the as-molded surfaces of the specimens conditioned at high humidities. No change was apparent on the surfaces of specimens conditioned at 76 percent relative humidity and below.

A plausible mechanism for the enhanced moisture absorption at high relative humidities (Figure 6) may be proposed in which swelling strains act to microcrack the resin-rich SMC surfaces and to debond the filler and/or glass reinforcement. Moisture would collect in the cracked and debonded regions and appear as anamolous absorbed water. The reason for the increased absorption exhibited by the 9640-50 compared to the 9220-30 (Figure 6) at high humidity is unknown; however, one may speculate on one of a combination of two mechanisms. The 9640-50 may contain a more rigid matrix system that would sustain a higher crack density and subsequently lead to increased moisture absorption, or, a reaction might be occuring between moisture and the calcium carbonate filler in the 9220-30.

Calcium carbonate is known to react with atmospheric moisture to form calcium hydroxide and CO_2 gas[13]. Such a reaction in $CaCO_3$ filled SMC would result in a weight loss from the evolution of CO_2. The lower weight gain values for the 9220-30 system seen in Figure 6 could also be due to an actual weight loss in the specimens in addition to the moisture weight gain. Weight losses have been observed with similar SMC formulations immersed in water at elevated temperatures[12,14]. Loss of unreacted resin monomer could also lead to weight loss under those conditions[15]. Determination of the exact mechanism of moisture interaction in these systems requires further study.

The observations of moisture interaction with SMC indicate that moisture absorption occurs rapidly and that it alone can lead to degradation in the form of microcracking of the matrix phase. In combination with UV in a weathering environment, it is likely that surface cracking will be enhanced. Moisture induced swelling strains would increase the probability of chain scission upon the absorption of a UV photon[16] or, swelling would lead to increased cracking in an embrittled surface. Both mechanisms have been observed in the model SMC systems investigated in this study. Moisture may also act to enhance spallation of degraded surface material and thereby expose additional matrix material to ultraviolet radiation.

6. CONCLUSIONS

Outdoor weathering has been shown to substantially reduce the tensile and flexural properties of five model sheet molding compounds (SMC). Flexural properties, which are surface sensitive, are particularly affected by outdoor exposure. Losses of up to 50 percent in flexural strength were observed after 12 months outdoor exposure. Little subsequent degradation was observed after 18 months of weathering. Dynamic mechanical results indicate that the vinyl ester based SMC degrades by a chain scission mechanism, whereas the general purpose polyester systems undergo further crosslinking upon UV exposure. Moisture absorption alone is shown to induce surface microcracking, and may act synergystically with UV to accelerate weathering of SMC.

References

1. R.C. Reuter and R.E. Allred, "Structural Optimization of Sheet Molding Compound Line Focusing Collectors," Proc. 24th National SAMPE Symposium and Exhibition, San Francisco, CA (May 8-10, 1979), pp. 866-877.
2. R.L. Champion and R.E. Allred, "Development of Sheet Molding Compound Solar Collectors with Molded-In Silvered Glass Reflective Surfaces," Sandia Laboratories Report SAND80-0702, December 1980.
3. R.E. Allred, D.W. Miller, and B.L. Butler, "Environmental Testing of Solar Reflector Structures," Proc. 1979 International Solar Energy Society Congress, Atlanta, GA (May 28-June 1, 1979), paper F-II-C-2.
4. W. Grisch and J. Maxel, "Performance of SMC in Severe Environments," Proc. 32nd Reinforced Plastics/Composites Institute Tech. Conf., Society of the Plastics Industry, Washington, D.C. (February 8-11, 1977), paper 5-A.
5. R. Gruenwald and O. Walter, "Fifeteen Years Experience with Sheet Molding Compound," Proc. 30th Reinforced Plastics/Composites Institute Tech. Conf., Society of the Plastics Industry, Washington, D.C. (February 4-7, 1975), paper 1-C.
6. W. Grisch, "SMC Use in Corrosion and Weathering Applications," Proc. 33rd Reinforced Plastics/Composites Institute Tech.Conf., Society of the Plastics Industry Washington, D.C. (February 7-10, 1978), paper 6-C.

7. D.K. Roylance and M.E. Roylance "Weathering of Fiber-Reinforced Epoxy Composites," Polymer Engr. and Sci., 18, 4 (March 1978), pp. 249-254.

8. N. Fried, in Mechanics of Composite Materials, F.W. Wendt, et al. , eds., Pergammon, NY, 1970, pp. 813-861.

9. R.F. Gibson and A. Yau, "Complex Moduli of Chopped Fiber and Continuous Fiber Composites: Comparison of Measurements with Estimated Bounds," J. Compt. Mat'ls., 14 (April 1980), pp. 155-167.

10. J. Collister and M. Gruskiewicz "Dynamic Mechanical Properties of Fiber Filled Unsaturated Polyester Resins," presented at ASTM Symposium on Short Fiber Reinforced Composite Materials.

11. R. Burns, "Polyester Molding Compounds - Review of Progress," Polymer-Plastics Tech. Engr., 10 (1978), pp. 165-198.

12. A.C. Loos, G.S. Springer, B.A. Sanders and R.W. Tung, "Moisture Absorption of Polyester-E Glass Composites," J. Comp. Mat'ls., 14 (April 1980), pp. 142-154.

13. Huheey, J.E., Inorganic Chemistry: Principles of Structure and Reactivity, Harper and Row New York, pp. 654-55, 1972.

14. R.E. Allred, Sandia Laboratories, unpublished data.

15. Roskott, L. and Croenendall, A.A.M., "Residual Styrene Content-What Does This Mean To The Polyester Processor," Proc. 31st Reinforced Plastics/Composites Institute Tech. Conf., Society of the Plastics Industry, Washington, D.C., (February 1978), paper 5-B.

16. S.N. Zhurkov, "Kinetic Concept of the Strength of Solids," Int'l. J. Fracture Mech., 1 (1965), pp. 311-323.

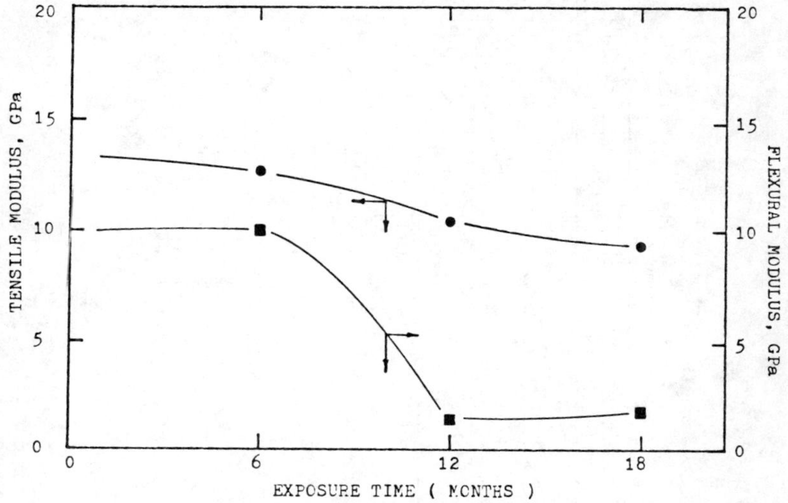

FIGURE 1. EFFECT OF WEATHERING ON THE MEAN VALUES OF TENSILE AND FLEXURAL MODULI FOR GENERAL PURPOSE POLYESTER RESIN BASED SMC, 25% GLASS CONTENT.

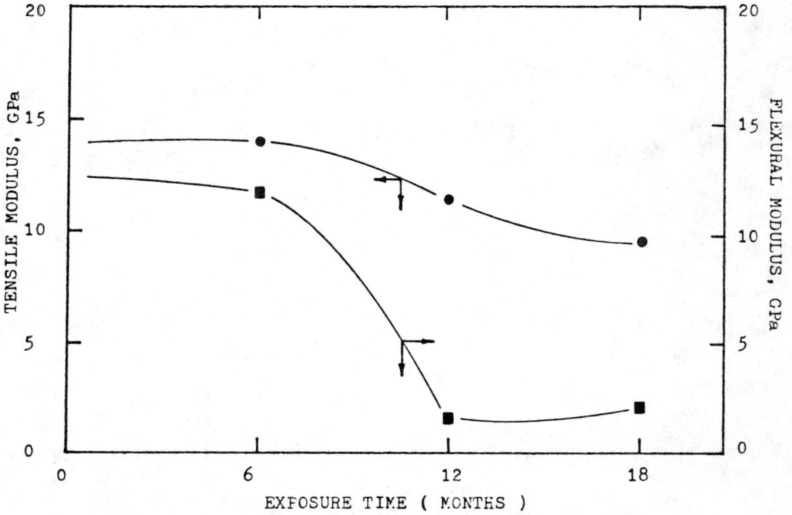

FIGURE 2. EFFECT OF WEATHERING ON THE MEAN VALUES OF TENSILE AN FLEXURAL MODULI FOR VINYL ESTER RESIN SMC, 25% GLASS. THE FLEXURAL MODULUS HAS DECREASED 86.6% OF ITS INITIAL VALUE AFTER 18 MONTHS OUTDOOR EXPOSURE.

FIGURE 3

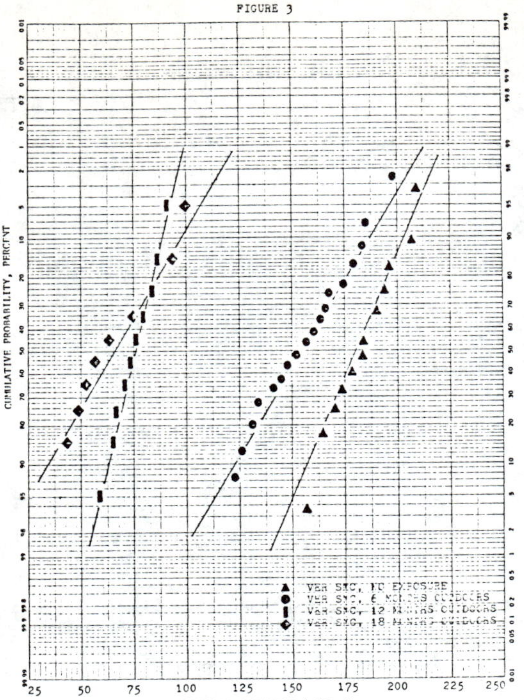

PROBABILITY PLOT OF THE PERCENT CUMULATIVE PROBABILITY VERSUS THE FLEXURAL STRENGTH FOR THE VINYL ESTER SMC, 25% GLASS, OVER AN 18 MONTH OUTDOOR EXPOSURE PERIOD. THE MEAN VALUE FOR EACH DATA SET IS AT 50% CUMULATIVE PROBABILITY AND THE COEFFICIENT OF VARIATION IS REPRESENTED BY THE SLOPE OF THE LINE THROUGH THE DATA POINTS.

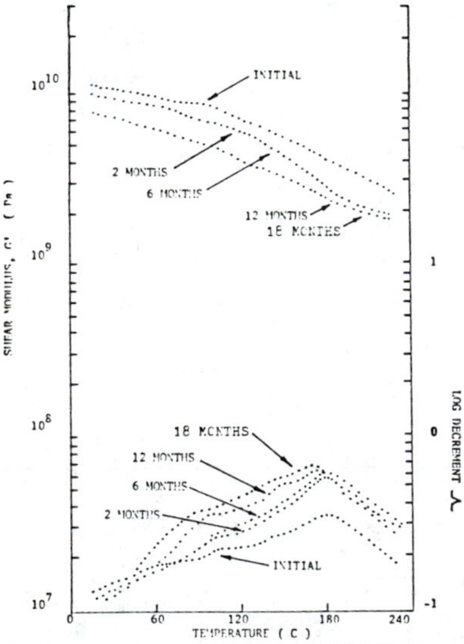

FIGURE 4. DYNAMIC MECHANICAL SPECTRUM OF VINYL ESTER RESIN SMC, 25% GLASS FOR AN OUTDOOR EXPOSURE TIME OF 18 MONTHS.

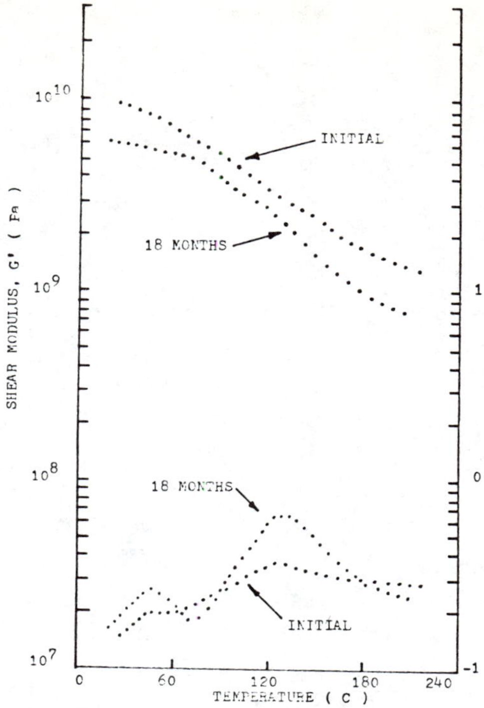

FIGURE 5. DYNAMIC MECHANICAL SPECTRUM OF GENERAL PURPOSE POLYESTER RESIN SMC, 25% GLASS CONTENT, AFTER 18 MONTHS WEATHERING.

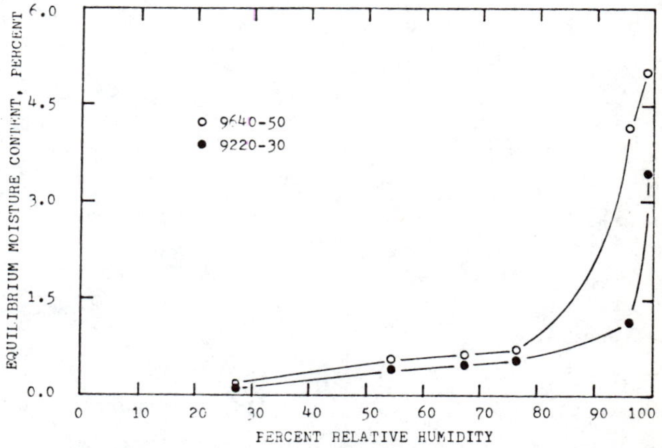

FIGURE 6. EQUILIBRIUM MOISTURE ABSORPTION OF GLASS FILLED (9640-50) AND CALCIUM CARBONATE FILLED (9220-30) SHEET MOLDING COMPOUND SYSTEMS VERSUS RELATIVE HUMIDITY EXPOSURE.

NAME INDEX

NAME INDEX